Published for
OXFORD INTERNATIONAL AQA EXAMINATIONS

International GCSE
MATHEMATICS
Extended

Steve Fearnley
June Haighton
Steve Lomax
Peter Mullarkey
James Nicholson
Matt Nixon

OXFORD
UNIVERSITY PRESS

OXFORD
UNIVERSITY PRESS

Great Clarendon Street, Oxford, OX2 6DP, United Kingdom

Oxford University Press is a department of the University of Oxford. It furthers the University's objective of excellence in research, scholarship, and education by publishing worldwide. Oxford is a registered trade mark of Oxford University Press in the UK and in certain other countries

© Oxford University Press 2016

The moral rights of the authors have been asserted

First published in 2016

All rights reserved. No part of this publication may be reproduced, stored in a retrieval system, or transmitted, in any form or by any means, without the prior permission in writing of Oxford University Press, or as expressly permitted by law, by licence or under terms agreed with the appropriate reprographics rights organization. Enquiries concerning reproduction outside the scope of the above should be sent to the Rights Department, Oxford University Press, at the address above.

You must not circulate this work in any other form and you must impose this same condition on any acquirer

British Library Cataloguing in Publication Data
Data available

978-0-19-837587-6

14

Paper used in the production of this book is a natural, recyclable product made from wood grown in sustainable forests. The manufacturing process conforms to the environmental regulations of the country of origin.

Printed and bound by CPI Group (UK) Ltd, Croydon, CR0 4YY

Acknowledgements
The publishers would like to thank the following for permissions to use their photographs:

Cover: blickwinkel/Alamy

All photos by Shutterstock

Although we have made every effort to trace and contact all copyright holders before publication this has not been possible in all cases. If notified, the publisher will rectify any errors or omissions at the earliest opportunity.

Links to third party websites are provided by Oxford in good faith and for information only. Oxford disclaims any responsibility for the materials contained in any third party website referenced in this work.

The manufacturer's authorised representative in the EU for product safety is Oxford University Press España S.A. of El Parque Empresarial San Fernando de Henares, Avenida de Castilla, 2 – 28830 Madrid (www.oup.es/en or product.safety@oup.com).
OUP España S.A. also acts as importer into Spain of products made by the manufacturer.

Contents

1 Calculations 1
1.1 Place value and rounding 4
1.2 Adding and subtracting 8
1.3 Multiplying and dividing 12
Summary 16
Review 17
Assessment 1 18

2 Expressions
2.1 Simplifying expressions 22
2.2 Indices 26
2.3 Expanding and factorising 1 30
2.4 Algebraic fractions 34
Summary 38
Review 39
Assessment 2 40

3 Angles and polygons
3.1 Angles and lines 44
3.2 Triangles and quadrilaterals 48
3.3 Symmetry 52
3.4 Congruence and similarity 56
3.5 Polygon angles 60
Summary 64
Review 65
Assessment 3 66

4 Handling data 1
4.1 Representing data 1 70
4.2 Averages and spread 1 74
4.3 Frequency diagrams 78
Summary 82
Review 83
Assessment 4 84

5 Fractions, decimals and percentages
5.1 Fractions and percentages 88
5.2 Calculations with fractions 92
5.3 Fractions, decimals and percentages 96
Summary 100
Review 101
Assessment 5 102
Life skills 1: The business plan 104

6 Formulae and functions
6.1 Formulae 108
6.2 Functions 112
6.3 Equivalences in algebra 116
6.4 Expanding and factorising 2 120
Summary 124
Review 125
Assessment 6 126
Revision 1 128

7 Working in 2D
7.1 Measuring lengths and angles 132
7.2 Area of a 2D shape 136
7.3 Transformations 1 140
7.4 Transformations 2 144
Summary 148
Review 149
Assessment 7 150

8 Probability
8.1 Probability experiments 154
8.2 Theoretical probability 158
8.3 Mutually exclusive events 162
Summary 166
Review 167
Assessment 8 168

9 Measures and accuracy
9.1 Estimation and approximation 172
9.2 Calculator methods 176
9.3 Measures and accuracy 180
Summary 184
Review 185
Assessment 9 186

10 Equations and inequalities
10.1 Solving linear equations 190
10.2 Quadratic equations 194
10.3 Simultaneous equations 198
10.4 Approximate solutions 202
10.5 Inequalities 206
Summary 210
Review 211
Assessment 10 212

11 Circles and constructions
11.1 Circles 1 216
11.2 Circles 2 220
11.3 Circle theorems 224
11.4 Constructions and loci 228
Summary 232
Review 233
Assessment 11 234
Life skills 2: Starting the business 236

12 Ratio and proportion
- 12.1 Proportion 240
- 12.2 Ratio and scales 244
- 12.3 Percentage change 248
- Summary 252
- Review 253
- Assessment 12 254
- **Revision 2** 256

13 Factors, powers and roots
- 13.1 Factors and multiples 260
- 13.2 Powers and roots 264
- 13.3 Surds 268
- Summary 272
- Review 273
- Assessment 13 274

14 Graphs 1
- 14.1 Equation of a straight line 278
- 14.2 Properties of quadratic functions 282
- 14.3 Kinematic graphs 286
- 14.4 Gradients and areas under graphs 290
- 14.5 Exponential and trigonometric functions 294
- Summary 298
- Review 299
- Assessment 14 300

15 Working in 3D
- 15.1 3D shapes 304
- 15.2 Volume of a prism 308
- 15.3 Volume and surface area 312
- Summary 316
- Review 317
- Assessment 15 318
- **Life skills 3: Getting ready** 320

16 Handling data 2
- 16.1 Averages and spread 2 324
- 16.2 Box plots and cumulative frequency graphs 328
- 16.3 Scatter graphs and correlation 332
- 16.4 Time series 336
- Summary 340
- Review 341
- Assessment 16 342

17 Calculations 2
- 17.1 Calculating with roots and indices 346
- 17.2 Exact calculations 350
- 17.3 Standard form 354
- Summary 358
- Review 359
- Assessment 17 360
- **Revision 3** 362

18 Pythagoras, trigonometry, vectors and matrices
- 18.1 Pythagoras' theorem 366
- 18.2 Trigonometry 1 370
- 18.3 Trigonometry 2 374
- 18.4 Pythagoras and trigonometry problems 378
- 18.5 Vectors 382
- 18.6 Matrices 386
- 18.7 Combining transformation matrices 390
- Summary 394
- Review 395
- Assessment 18 396

19 The probability of combined events
- 19.1 Venn diagrams 400
- 19.2 Sample spaces 404
- 19.3 Tree diagrams 408
- 19.4 Conditional probability 412
- Summary 416
- Review 417
- Assessment 19 418
- **Life skills 4: The launch party** 420

20 Sequences
- 20.1 Linear Sequences 424
- 20.2 Quadratic sequences 428
- 20.3 Special sequences 432
- Summary 436
- Review 437
- Assessment 20 438

21 Units and proportionality
- 21.1 Compound units 442
- 21.2 Converting between units 446
- 21.3 Direct and inverse proportion 450
- 21.4 Growth and decay 454
- Summary 458
- Review 459
- Assessment 21 460

22 Differentiation
- 22.1 Rates of change 464
- 22.2 Differentiating polynomials 468
- 22.3 Tangents to curves 472
- 22.4 Turning points 476
- Summary 480
- Review 481
- **Revision 4** 482

Formulae 484
Key phrases and terms 485
Answers 487
Index 558

About this book

This book has been specially created for the Oxford AQA International GCSE Mathematics examination (9260). It has been written by a team of teachers and consultants, including those with examining experience.

In each chapter the lesson are organised in pairs. The first lesson is focussed on helping you to master the basic skills required (AO1) whilst the second lesson applies these skills in questions that develop your reasoning and problem solving abilities.

Throughout the book, four-digit MyiMaths codes are provided allowing you to link directly, using the search bar, to related lessons on the MyiMaths website where you can see the topic from a different perspective, work independently and revise.

At the end of a chapter you will find a summary of what you should have learnt, together with a review section that allows you to test your fluency in the basic skills. Depending on how well you do, a *What next?* box provides suggestions on how you could improve further. Finally there is an Assessment section where you can practise exam-style questions.

At the end of the book you will find a guide to key phrases and terms and a full set of answers to all the exercises.

We wish you well with your studies and hope that you enjoy this course and achieve exam success.

1 Calculations 1

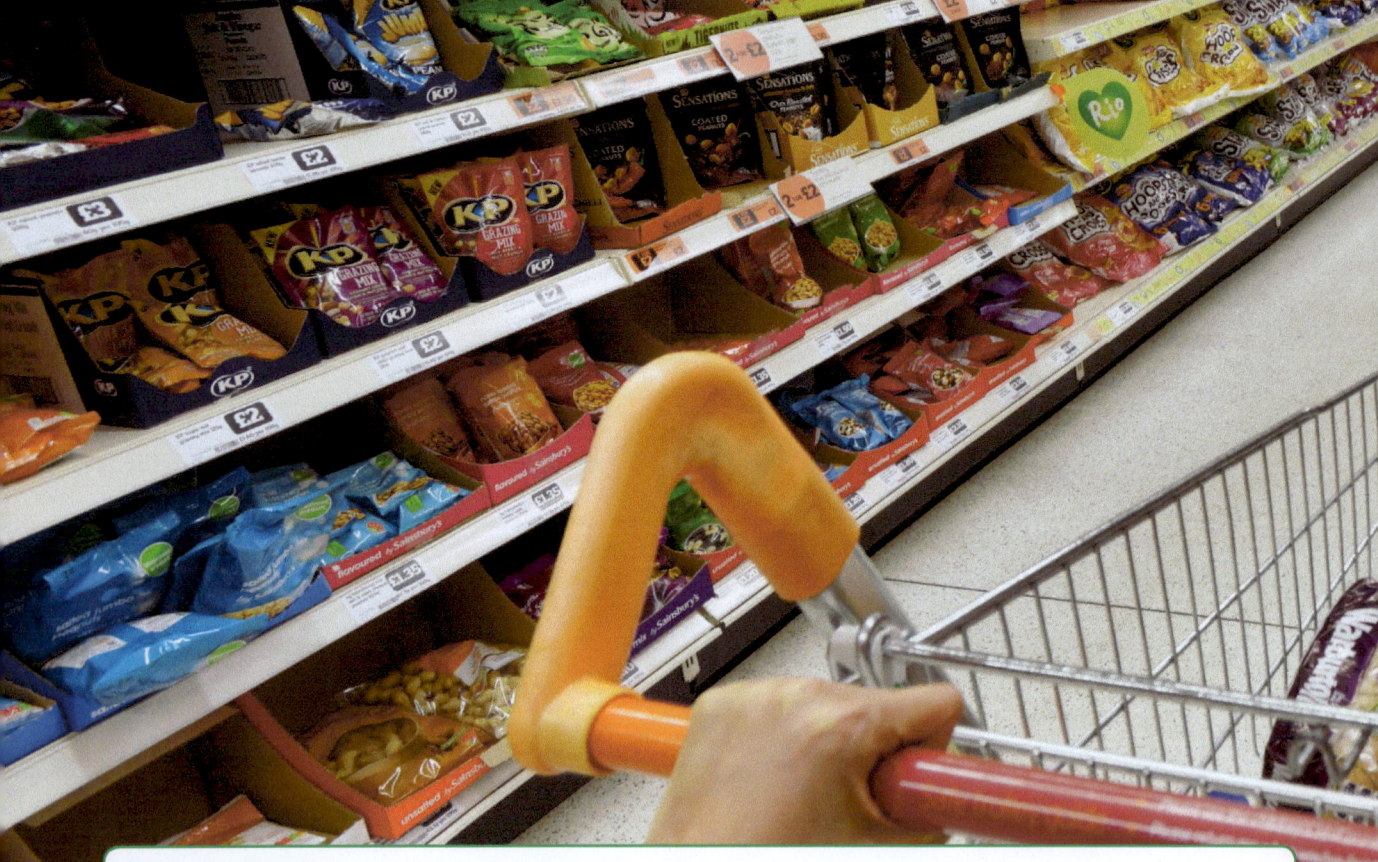

Introduction
When you go shopping in a supermarket, you are presented with hundreds of products, often looking similar, as well as lots of different offers. You need to be able to do arithmetic in your head to ensure that you are keeping within your budget and also that you choose the best value offer.

What's the point?
Being able to add, subtract, multiply and divide doesn't just mean that you're good at maths at school – it means that you can confidently look after your own finances in the real world.

Objectives
By the end of this chapter you will have learned how to…
- Order positive and negative integers and decimals.
- Round numbers to a given number of decimal places or significant figures.
- Use mental and written methods to add, subtract, multiply and divide with positive and negative integers and decimals.
- Use BIDMAS to complete calculations in the correct order.

Check in

1. Write in words the value of the digit 4 in each of these numbers.
   - **a** 4506
   - **b** 23 409
   - **c** 200.45
   - **d** 13.054

2. **a** Sketch a number line showing values from -5 to $+5$ and mark these numbers on it.
   - **i** -3
   - **ii** $+4$
   - **iii** -2.5
   - **iv** 0

   b Write this set of directed numbers in ascending order.
   $+5, -2.4, -3, +6, 0, +1.5, -1.8$

Chapter investigation

Ternary uses a base-three system and the digits 0, 1 and 2 to write numbers.

In ternary, you use units, threes, nines, twenty-sevens, eighty-ones ... to record place value.

100 = 1 eighty-one + 0 twenty-sevens + 2 nines + 0 threes + 1 unit
100 is 10201 in ternary

Investigate these ternary calculations.

$1 + 1 = 2$ $1 + 2 = 10$ $2 + 2 = 11$
$1 \times 1 = 1$ $1 \times 2 = 2$ $2 \times 2 = 11$
$10 \times 12 = 120$ $11 \times 11 = 121$ $12 \times 12 = 221$
$101 \times 12 = 1212$ $202 \times 21 = 12012$ $121 \times 21 = 11011$

Show that the rules for long multiplication and division apply to ternary numbers.

SKILLS

1.1 Place value and rounding

- < means less than
- > means greater than
- ≤ means less than or equal to
- ≥ means greater than or equal to
- ≠ means *not* equal to

EXAMPLE

Place the correct symbol <, > or = between the numbers in each pair.
- **a** 5.07 5.7
- **b** 397 379
- **c** −10 5
- **d** −19 −24
- **e** $\frac{3}{2}$ 1.5

a 5.07 < 5.7   **b** 397 > 379   **c** −10 < 5   **d** −19 > −24   **e** $\frac{3}{2}$ = 1.5

- To round a number look at the next, smaller, digit
  - next digit = 0, 1, 2, 3 or 4 round down
  - next digit = 5, 6, 7, 8 or 9 round up.

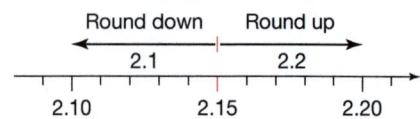

EXAMPLE

Round 72 456.0374 to the nearest
- **a** ten
- **b** hundred
- **c** thousand
- **d** tenth
- **e** hundredth
- **f** thousandth.

a 72 460   **b** 72 500   **c** 72 000   **d** 72 456.0 (1 dp)   **e** 72 456.04 (2 dp)   **f** 72 456.037 (3 dp)

- When rounding to **significant figures**, count from the first non-zero digit.

EXAMPLE

a Round these numbers to 2 dp.
  - **i** 34.567
  - **ii** 3.887 126
  - **iii** 215.587 54

dp means 'decimal places' and **sf** means 'significant figures'.

b Round these numbers to 2 sf.
  - **i** 39.54
  - **ii** 217
  - **iii** 0.000 455
  - **iv** 12 019
  - **v** 25.505

a **i** 34.57   **ii** 3.89   **iii** 215.59
b **i** 40   **ii** 220   **iii** 0.000 46   **iv** 12 000   **v** 26

- Multiplying or dividing a number by a power of 10 changes the place value of each digit.

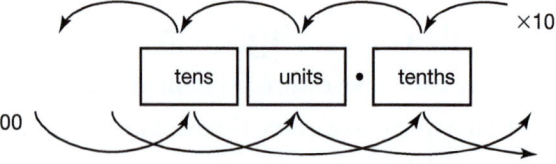

Multiplying by 10 moves the digits one place to the left. Dividing by 100 moves the digits two places to the right.

EXAMPLE

Work out   **a** 3.72 ÷ 100   **b** 0.0349 × 10 000   **c** 17.3 ÷ 1000

a Move the digits 2 places right.
0.0372

b Move the digits 4 places left.
349

c Move the digits 3 places right.
0.0173

Number   Calculations 1

Fluency

Exercise 1.1S

1. Write these numbers in words.
   a. 1307
   b. 29 006
   c. 300 000
   d. 605 030

2. Write these numbers in figures.
   a. Eight thousand and forty-three
   b. Seventy million
   c. Two hundred thousand and fifty-one
   d. Two thousand and ten

3. Write these sets of numbers in ascending (increasing) order.
   a. 0.3, 3.1, 1.3, 2, 1, 0.1
   b. 607, 77.2, 27.6, 7.06, 6.07
   c. 7.83, 7.3, 7.8, 7.08, 7.03, 7.38
   d. 4.2, 8.24, 8.4, 4.18, 2.18, 2.4

4. Write these sets of numbers in descending (decreasing) order.
   a. 6008, 682.8, 862.6, 6000.8, 8000.6
   b. 47.9, 94.7, 49.7, 79.4, 74.9, 97.4
   c. 16.7, 18.16, 16.18, 17.16, 18.7, 17.6
   d. 1.06, 13.145, 1.1, 2.38, 13.2, 2.5

5. Use one of the symbols $<$, $>$ or $=$ to complete these statements.
   a. 250 □ 205
   b. 1.377 □ 1.73
   c. $\frac{3}{8}$ □ 0.4
   d. -17 □ -71
   e. -0.09 □ -0.089
   f. $\frac{5}{8}$ □ 0.625

6. Explain which number in each pair is bigger.
   a. 4.52 and 4.05
   b. 5.5 and 5.05
   c. 16.8 and 16.75
   d. 16.8 and 16.15

7. Say whether each statement is true or false.
   a. $4.1 < 4$
   b. $6.33 < 6.333$
   c. $0.23 \leq 0.24$
   d. $-2.3 \geq -2.4$
   e. $5.31 < 5.31$
   f. $5.31 \leq 5.31$

8. Round these numbers to the nearest
   i. 10   ii. 100   iii. 1000
   a. 3048
   b. 1763
   c. 294
   d. 51
   e. 43
   f. 743
   g. 2964
   h. 1453
   i. 17
   j. 24 598
   k. 16 344
   l. 167 733

9. Round these numbers to
   i. 1 decimal place   ii. 2 decimal places.
   a. 39.114
   b. 7.068
   c. 5.915
   d. 512.715
   e. 4.259
   f. 12.007
   g. 0.833
   h. 26.8813
   i. 0.082 93

10. Round these numbers to the nearest
    i. tenth   ii. hundredth   iii. thousandth.
    a. 0.07
    b. 15.9184
    c. 127.9984
    d. 887.172
    e. 55.144 55
    f. 0.007 49

11. Round these numbers to one significant figure.
    a. 157
    b. 2488
    c. 4.66
    d. 13.77
    e. 0.000 453
    f. 121 450

12. Round these numbers to two significant figures.
    a. 483
    b. 1206
    c. 488
    d. 13 562
    e. 533
    f. 14 511
    g. 0.355
    h. 0.421
    i. 0.0566
    j. 0.004 673
    k. 1.357
    l. 0.000 004 152

13. Round each number to the accuracy given in brackets.
    a. 9.732 (3 sf)
    b. 0.362 18 (2 dp)
    c. 147.49 (1 dp)
    d. 28.613 (2 sf)
    e. 0.5252 (2 sf)
    f. 4.1983 (2 dp)
    g. 1245.4 (3 dp)
    h. 0.004 25 (3 dp)
    i. 273.6 (2 sf)
    j. 459.973 14 (1 dp)

14. Multiply these numbers by 10.
    a. 16.7
    b. 24.8
    c. 0.716
    d. 1.095
    e. 243
    f. 281.3

15. Divide these numbers by 10.
    a. 214
    b. 67.3
    c. 4106
    d. 200.7
    e. 6.025
    f. 86

16. Calculate
    a. 13.06×100
    b. $208.5 \div 100$
    c. 1.085×1000
    d. $2487 \div 1000$
    e. $0.008 \div 10$
    f. $0.006\,19 \times 1000$
    g. $45.13 \div 1000$
    h. $0.000\,045 \times 100$

17. Calculate
    a. $1.76 \times 10 \times 100$
    b. $9.3 \times 100 \div 10$

1001, 1005, 1013, 1072

APPLICATIONS

1.1 Place value and rounding

RECAP
- When rounding: five or more rounds up, four or less rounds down.
- For decimal places count from the decimal point.
- For significant figures count from the first *non-zero* digit.
- To multiply or divide by 10 use place value and move the digits one place to the left or right.

HOW TO

To solve a problem involving place value or rounding
1. Read the question and think what to do.
2. Apply your knowledge of place value and rounding.
3. Answer the question.

EXAMPLE

Ajani charges a customer £352.46, but has mixed up two digits. They should be swapped around. His mistake costs him £3.96. What should he have charged?

① Since £1 < mistake < £10, the units column (2) must be wrong.
② Swapping the 2 and 4 would leave the 6, so the mistake would end in zero. So it is the 6 and 2.
The correct amount is £356.42 ③

EXAMPLE

An engineer measures the thickness of four sheets of metal.
2.05 mm   2.033 mm   2.4 mm   2.303 mm
a   If she piled up 100 of the thinnest sheets, how high would the pile be?
b   If her measurement was inaccurate by 0.001 mm, between what limits would the pile be?

① Put the measurements in ascending order.
   2.033   2.050   2.303   2.400
a   100 × 2.033 = 203.3 mm   ②
b   2.033 − 0.001 = 2.032,   100 × 2.032 = 203.2 mm
   2.033 + 0.001 = 2.034,   100 × 2.034 = 203.4 mm
The pile is between 203.2 mm and 203.4 mm high. ③

EXAMPLE

The following distances were recorded in a long jump competition.
MacLane 5.89 m   Neyman 5.98 m   Ockham 6.12 m
Pell   6.03 m   Quillen 5.09 m   Ricci   5.8 m
a   Minh-Ha says 'the gap between first and last is over ten times the gap between first and second'. Is she correct?
b   Sze-Kie says 'if the results were given to 1 dp then there would be a joint second place'. Is she correct?

a ① Put the results in descending order.
   6.12 6.03 5.98 5.89 5.80 5.09
   First − second = 6.12 − 6.03 = 0.09 ②
   First − last = 6.12 − 5.09 = 1.03
   10 × 0.09 = 0.90 < 1.03
   Minh-Ha is correct. ③

b ① Round the results to 1 dp
   6.1 6.0 6.0 5.9 5.8 5.1 ②
   Sze-Kie is correct. ③

Number   Calculations 1

Reasoning and problem solving

Exercise 1.1A

> Don't use a calculator for this exercise. Practise your arithmetic!

1. Soren has given a customer a bill for $356.28. He realises he has mixed up the 6 and the 2. How much does he have to pay back to the customer?

2. Veneer is a thin sheet of attractive wood. A joiner has a pile of 10 sheets of oak veneer. Each sheet is 0.5 mm thick.
   a. How thick is the pile?
   b. The joiner glues a sheet of veneer on to the top of different blocks of wood. What is the new overall thickness of
      i. a 5 cm block
      ii. a 3.5 cm block
      iii. a 12.25 cm block?
   c. What would be the new thicknesses of the blocks above if he glues veneer on all four sides of the blocks?

3. Votes for four politicians were declared.
   A 25 958   B 2705
   C 26 057   D 5651
   The local newspaper decides to round these off to the nearest 1000 in its report.
   a. What would each result be reported as?
   b. What would each result be if they were rounded to the nearest 100?

4. These are the times recorded in a 100 m sprint race.
   Adams 12.37 s   Bolyai 12.35 s
   Carroll 13.72 s   d'Arcy 11.09 s
   Eckert 11.33 s   Fisher 11.9 s
   a. Mikael says 'the winner is quarter of a second faster than their nearest rival'. Is this true?
   b. Carroll's personal best time is 12.17 s. If he had run this time, what would have been his position?
   c. Which two runners had the closest times?

5. Garvan is resizing photographs to make thumbnail pictures. He decides to divide lengths by 50 and then round to the nearest whole number.
   What width and height would these pictures become as thumbnails?
   a. 4288 × 2848   b. 2197 × 1463
   c. 3648 × 2746   d. 6032 × 4502
   e. His resize screen allows him to put in a "percentage of original width and height". What percentage would he put in the box to divide by 50?

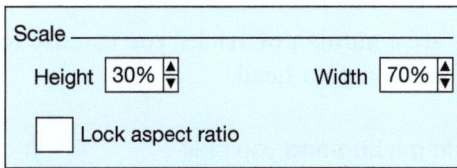

*6. Ian gets his homework back.

> Find $\sqrt{3} \times 25^3$, rounding your answer to 3 significant figures.
> $\sqrt{3} \times 25^3 = 1.73 \times 15600 = 26988$
> $= 27000$ (3 sf)

p.260

What mistake has Ian made?

7. How many correct statements can you make using one of these symbols
   $<$ $\leq$ $=$ $>$ $\geq$
   and one of these pairs of numbers?
   3.118   and   3.112
   4.5   and $\frac{9}{2}$
   3.004   and   2.9961

8. A number x satisfies
   $x = 1.5$ (2 sf) and $x \neq 1.50$ (3 sf).
   What possible values can x take?

9. A lift has a safe maximum load of 350 kg. Four people give their weights to these accuracies.
   80 kg (10 kg)   95 kg (2 sf)
   96.5 kg (1 dp)   72 kg (1 kg)
   Is it safe for them all to get into the lift together? Show your working.

SKILLS

1.2 Adding and subtracting

A number with a plus or minus sign is a **directed number**. You can extend the basic rules of addition and subtraction to include negative numbers.

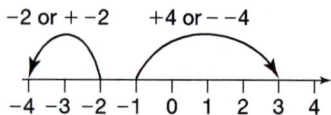

- Adding a **negative** number counts as subtraction.
- Subtracting a negative number counts as addition.

EXAMPLE

Calculate **a** $-5 + -6$ **b** $4 - -2$

a $-5 + -6 = -5 - 6 = -11$ **b** $4 - -2 = 4 + 2 = 6$

Think of a negative number as a debt. Adding a debt makes you worse off. Taking away a debt makes you better off.

There are a number of 'tricks' you can use to help you do a calculation in your head.

- In **partitioning** you split the number into smaller parts.
- In **compensation** you 'round' the number and include a correction.

EXAMPLE

a Use partitioning to calculate
   i $61.9 + 7.2$ **ii** $61.9 - 7.2$

b Use compensation to calculate
   i $21.6 + 3.8$ **ii** $21.6 - 3.8$

a $7.2 = 7.0 + 0.2$
   i $61.9 + 7 + 0.2$
      $= 68.9 + 0.2$
      $= 69.1$
   ii $61.9 - 7 - 0.2$
      $= 54.9 - 0.2$
      $= 54.7$

b $3.8 = 4.0 - 0.2$
   i $21.6 + 4 - 0.2$
      $= 25.6 - 0.2$
      $= 25.4$
   ii $21.6 - 4 + 0.2$
      $= 17.6 + 0.2$
      $= 17.8$

You might know 'tricks' of your own. Can you say how they work?

- Always use an estimate to check the result of a written calculation.

EXAMPLE

Calculate these using a written method.

a $102.773 + 28.47$ **b** $26.44 - 1.105$

a Estimate = $100 + 30 = 130$

```
  1 0 2 . 7 7 3
+   2 8 . 4 7 0
---------------
  1 3 1 . 2 4 3   ≈ 130 ✓
      1 1 1
```

b Estimate = $26 - 1 = 25$

```
   2 6 . 4 ³4̸ ¹0
 -    1 . 1 0 5
---------------
   2 5 . 3 3 5   ≈ 25 ✓
```

Add zeros to help you line the digits up in the correct columns.

Number Calculations 1

Fluency

Exercise 1.2S

1 Calculate
   a $+8 - -14$
   b $-1 + -11$
   c $-9 - -7$
   d $+3 + -17$
   e $+8 - -4$
   f $+13 + -1$
   g $+48 - +29$
   h $-19 + +4$
   i $+34 + -23$
   j $-104 + +43$
   k $+208 - -136$
   l $+347 + -298$

2 Calculate
   a $-4.5 + -6.3$
   b $-2.8 - -3.5$
   c $+5.6 - -7.9$
   d $-9.4 + +8.7$
   e $-26.5 + -11.7$
   f $+45.9 - -66.8$

For questions **3** to **6**, work out the answer in your head; use jottings to indicate your method.

3 a $4.7 + 5.3$
   b $4.7 + 5.4$
   c $3.6 + 6.7$
   d $6.8 + 4.3$
   e $7.5 + 8.9$
   f $2.7 + 4.8$

4 a $3.55 + 4.22$
   b $2.13 + 3.12$
   c $3.18 + 0.42$
   d $3.72 + 0.18$
   e $1.42 + 0.71$
   f $8.39 + 4.65$

5 a $6.6 - 4.1$
   b $7.5 - 5.4$
   c $8.1 - 5.9$
   d $7.2 - 3.3$
   e $3.1 - 1.7$
   f $7.3 - 6.6$

6 a $2.16 - 1.42$
   b $1.51 - 0.46$
   c $6.39 - 4.88$
   d $15.46 - 8.32$
   e $5.17 - 4.09$
   f $4.29 - 3.65$

For questions **7** to **10** use a written method.

7 a $23.45 + 12.51$
   b $13.44 + 21.17$
   c $77.33 + 19.02$
   d $65.47 + 38.95$
   e $8.152 + 6.779$
   f $5.426 + 4.975$
   g $11.625 + 14.586$
   h $25.341 + 38.495$

8 a $24.72 - 14.04$
   b $1.52 - 1.09$
   c $6.149 - 2.052$
   d $16.64 - 15.88$
   e $5.23 - 3.11$
   f $17.45 - 13.26$
   g $6.41 - 4.37$
   h $23.6 - 17.9$

9 a $1.09 + 1.54$
   b $0.09 + 0.36$
   c $14.52 + 9.8$
   d $13.92 + 0.8$

10 a $4.5 - 0.53$
   b $3.085 - 2.99$
   c $16.3 - 3.86$
   d $112.14 - 53.8$

11 Work out these calculations using an appropriate written or mental method. Show your method clearly.
   a $5.8 - 3.2$
   b $16.73 - 8.87$
   c $9.6 - 3.7$
   d $109.54 - 17$
   e $2.37 - 1.4$
   f $26.25 - 1.98$

12 Work out these calculations using a standard written method. Check your answers using estimates.
   a $21.864 - 7.968$
   b $104.87 - 85.42$
   c $417.48 - 57.69$
   d $24.503 - 16.82$
   e $19.21 - 18.884$
   f $102.01 - 90.59$

13 Use a calculator to check your answers to question **12**.

14 Work out these calculations using an appropriate method; show your workings.
   a $8.6 - 4.5$
   b $26.4 + 13.8$
   c $18 - 6.712$
   d $15.808 - 9.84$
   e $4.008 - 3.116$
   f $4.109 - 3.64$

15 Work out these calculations using an appropriate method; show your workings.
   a $23.11 + 31.24 + 45.53$
   b $14.58 + 13.35 + 1.056$
   c $26.87 + 13.11 - 25.44$
   d $4.162 + 5.516 - 1.0023$
   e $231.72 + 45.07 + 180.96$
   f $32.63 + 128.54 - 67.38$
   g $94 - 62.103 - 12.095$
   h $0.0127 + 11.9874 - 54.0706$

16 Omar's grandad is helping him with his homework but he uses a 'funny' method.
   a Explain why the method works.
   b Use the method to calculate.
       i $67.362 - 23.454$
       ii $60.009 - 12.347$

$$\begin{array}{r} 5\,{}^1 2.\,{}^7 8\,{}^1 0 \\ -\ 1\ 9.\,{}^6 5\,4 \\ \hline 4\ 3.\,2\,6 \end{array}$$

APPLICATIONS

1.2 Adding and subtracting

RECAP
- You can use partitioning and compensation to help you do calculations in your head.
- Always use an estimate to check an exact calculation.
- Use written methods to do exact calculations: use the decimal point to line up digits with the same place value.

Line up digits in the correct columns and work from right to left.

```
  63.70          ⁵¹²¹⁶¹
+  9.85        ⁶³.⁷0
―――――          -  9.85
  73.55        ―――――
  ¹ ¹           53.85
```

HOW TO

To solve a problem involving addition or subtraction
① Read the question and decide which calculations need doing.
② Use an appropriate written or mental method.
③ Use an estimate to check each exact calculation.

EXAMPLE

Katie obtained these marks on Paper 1 of her maths exam: 3, 5, 2, 7, 5, 8, 11, 4, 5, 0, 7

a Paper 1 is out of 100.
How many marks did she lose?

b To pass you need 140 marks on Papers 1 and 2.
What score does Katie need on Paper 2?

① Calculate 100 − total marks

a Paper 1 total = (3 + 5 + 2) + (7 + 5 + 8) + (11 + 4 + 5) + 7 ②
= 10 + 20 + 20 + 7 = 57
100 − 57 = 43.
She lost 43 marks.

① Calculate 140 − total marks

b 140 − 57 = 83 ②
She needs 83 marks.

It sometimes helps to group the numbers into multiples of 10.

EXAMPLE

Duman receives two payments for work done of €568.42 and €76. He also has to pay a bill for €265.16.

a How much is he paid?
b What does he have left after paying his bill?

① Add the payments.

a
```
  5 6 8 . 4 2
+   7 6 . 0 0
―――――――――――
  6 4 4 . 4 2
```
€644.42

② Use zeros to line up columns.

③ 570 + 80 = 650 ✓

① Subtract the bill.

b
```
    ⁵ ¹³ ¹   ³ ¹
    6 ₄4 . ₄2
  − 2 6 5 . 1 6
  ―――――――――――
    3 7 9 . 2 6
```
€379.26

② Take care when carrying.

③ 650 − 270 = 380 ✓

Number Calculations 1

Reasoning and problem solving

Exercise 1.2A

1. Vinay wants to score 200 runs in the cricket season. With one game left to play his scores are

   7, 15, 23, 5, 8, 1, 12, 9, 0, 42 and 5.

   What must he score in the final game?

2. Stefan wants to buy a new mountain bike.

   FOREST BIKE Co
   Viking   £129.99
   Bullet   £159.99
   Diamond   £225
   Grifter   £255.50

   He has £280 to spend.
   How much would he have left if he bought
   a a Viking
   b a Bullet
   c a Diamond
   d a Grifter?
   e He really wants a FastTrax which costs £359. How much more does he need to save?

3. Three friends go into a shop to buy some books. The prices are

   'Ants' $6.78   'Bees' $5.42
   'Cows' $2.99   'Ducks' $1.28

   Each friend pays with $20 and gets the following change.
   a Rebekah $4.81
   b Suzie $10.23
   c Tara $8.60

   What books did each of the three friends buy?

4. Copy and complete this addition pyramid.

   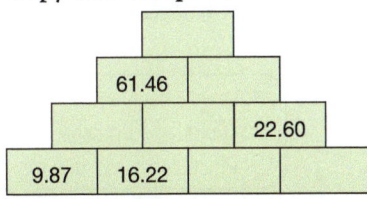

   Each number is the sum of the two numbers below it.

5. Here are the populations of four villages

   Thornton 4675   Norton   12250
   Brawton   562   Lockton   853

   a What is the range of population (the difference between biggest and smallest)?
   b What is the total population?
   c New houses are built in Brawton. It is no longer the smallest of the four villages. At least how many people have moved in?

6. Five friends run a league table of points scored on a game they all play.

Name	Score	Ahead of next person by
Asha	6523	
Bryony	12 653	
Callum	7350	
Dora	964	
Edward	7000	
TOTAL		

   a Copy the table, but put the friends in order, with the leader at the top.
   b Fill in the differences in the third column.
   c Fill in the total.
   d Dora is determined to be number one. How many more points must she score?

7.
1026	724	1448
432	342	522

   a Find a pair of numbers in this table where their total is twice their difference.
   b Can you find a pair where the total is three times the difference?
   c Can you find four numbers where one pair adds up to twice the total of the other pair?

8. A printer has printed some fuzzy characters, making it difficult to read some numbers in a calculation. Work out what the numbers are.

   □2□□ + 3□73 = 6322

   3□.58 − □7.7□ = 8.□9

SKILLS

1.3 Multiplying and dividing

- Multiplying or dividing a positive number by a negative number gives a negative number.
- Multiplying or dividing a negative number by a negative number gives a positive number.

EXAMPLE

Use a mental method to work out these calculations.

a 21×-73 **b** $-392 \div -9$

a $21 \times 73 = 20 \times 73 + 73$
$= 1460 + 73$
$= 1533$
$21 \times -73 = -1533$

b $392 \div 9 = 360 \div 9 + 32 \div 9$
$= 40 + 3r5$
$= 43r5$
$-392 \div -9 = 43r5$

- To multiply or divide decimals
  - **estimate** the answer
  - perform the arithmetic without decimal points
  - use the estimate to place the decimal point.

EXAMPLE

Calculate

a 18.5×7.9 **b** $47.52 \div 1.8$

a Estimate $20 \times 8 = 160$

```
    1 8 5
  ×   7 9
  1 2₅9₃5 0
+ 1 6₇6₄5
  1 4₁6 1 5
```

b Estimate $50 \div 2 = 25$

```
        2 6 4
   18)4 7 5 2
     - 3 6       2 × 18 = 36
       1 1 5
     - 1 0 8     6 × 18 = 108
           7 2
         - 7 2   4 × 18 = 72
             0
```

Use the estimates to check your answers and place the decimal points.

$18.5 \times 7.9 = 146.15$ $47.52 \div 1.8 = 26.4$

You need to know your tables. 2, 5 and 10 are easy. That just leaves 36 other answers to learn in your tables up to 12 × 12.

- The order of precedence for **operations** is given by **BIDMAS**:
  Brackets, **I**ndices, **D**ivision, **M**ultiplication, **A**ddition, **S**ubtraction

When you have square roots or fractions add brackets to show which terms go together.

EXAMPLE

Work out the value of each expression.

a $4 + 3 \times 2$ **b** $5 + 3^2$ **c** $\sqrt{5 + 4 \times 11}$

a $4 + 3 \times 2$ M
$= 4 + 6$ A
$= 10$

b $5 + 3^2$ I
$= 5 + 9$ A
$= 14$

c $\sqrt{(5 + 4 \times 11)}$ B, M
$= \sqrt{(5 + 44)}$ B, A
$= \sqrt{49} = 7$ I

Number Calculations 1

Fluency

Exercise 1.3S

1 Calculate these multiplications.
- a $+5 \times -5$
- b $+4 \times -8$
- c $-8 \times +9$
- d $-4 \times +5$
- e -3×-10
- f -7×-7
- g $+8 \times +2$
- h $+5 \times -4$
- i $-2 \times +9$
- j -13×-2
- k $-7 \times +6$
- l $+12 \times -4$

2 Calculate these divisions.
- a $-18 \div +9$
- b $-20 \div +4$
- c $-30 \div -6$
- d $-12 \div -3$
- e $-66 \div +3$
- f $+47 \div -47$
- g $-80 \div -2$
- h $+24 \div +6$
- i $-45 \div -9$
- j $-51 \div +3$
- k $+57 \div -19$
- l $-81 \div -3$

For questions **3** to **5**, work out the answer in your head; use jottings to indicate your method.

3 Calculate these multiplications.
- a 14×7
- b 19×8
- c 21×13
- d 17×19
- e 11×28
- f 21×29
- g 51×49
- h 37×8
- i 71×16
- j 59×61

4 Calculate these divisions.
- a $1176 \div 21$
- b $4662 \div 18$
- c $5728 \div 8$
- d $583 \div 11$
- e $374 \div 17$
- f $28\,380 \div 66$
- g $66\,600 \div 222$
- h $2520 \div 45$
- i $6375 \div 125$
- j $9963 \div 81$

5 Calculate these multiplications and divisions.
- a 31×0.3
- b $49 \div 0.07$
- c $3.66 \div 0.3$
- d $4.24 \div 0.4$
- e $13.9 \div 0.03$
- f $3.9 \div 0.03$
- g $171 \div 0.3$
- h 5.2×0.08
- i $0.006\,25 \div 50$
- j 0.011×0.0011

6 Use a written method to work out the answers to questions **3** to **5**. Check that you get the same answers with both methods.

For questions **7** to **9**, use a written method.

7 Calculate these multiplications.
- a 4.7×5.3
- b 1.53×2.8
- c 21.6×4.9
- d 33.65×3.89
- e 21.58×1.99
- f 42.77×8.64
- g 0.666×33.3
- h 0.456×0.0789

8 Calculate these divisions.
- a $34.83 \div 9$
- b $5.425 \div 7$
- c $7.328 \div 8$
- d $451.8 \div 60$
- e $54.39 \div 3$
- f $58.65 \div 17$
- g $66.4 \div 16$
- h $185.76 \div 24$
- i $7.752 \div 1.9$
- j $3.055 \div 1.3$

9 Calculate these divisions giving your answers to two decimal places.
- a $14.73 \div 2.8$
- b $51.99 \div 1.8$
- c $193.8 \div 0.14$
- d $1013 \div 5.77$
- e $23.78 \div 0.83$
- f $65.79 \div 0.59$
- g $10.73 \div 250$
- h $10\,594 \div 12.3$

10 Use a calculator to check your answers to questions **7–9**.

11 Given that $43 \times 67 = 2881$, find
- a 4.3×6.7
- b 430×0.067
- c $2881 \div 670$
- d $28.81 \div 430$
- e $2.881 \div (0.43 \times 0.67)$
- f $(4.3 \times 0.67) \div 0.028\,81$

12 Evaluate these expressions.
- a $(5^2 + 3) \times 7$
- b $(9 - 7)^2$
- c $(5 - 3) \times (4^2 - 7)$
- d $(5^2 - 8)^2$

13 Evaluate these expressions.
- a $(4 + 7)^2$
- b $(6 + 7) \times 9 \div 3$
- c $\dfrac{6 \times (5^2 - 13)}{4}$
- d $\sqrt{100 - 2 \times 6^2}$
- e $\dfrac{28}{4} + \sqrt{100 - (9^2 + 5 \times 2)}$
- f $\sqrt{28 + 4^2 - (10 - 2)} + 4 \times 3$
- g $\dfrac{10\sqrt{4^2 + 20} + 2^2 \times 3}{(8 + 6) \div 7 + 7}$
- h $1 + 2(1 - 3(1 + 4(1 - 5(6 + 7))))$

1011, 1068, 1916, 1917

APPLICATIONS

1.3 Multiplying and dividing

RECAP
- You can use partitioning and factorisation to help you do calculations in your head.
- Use integer arithmetic to carry out exact calculations.
- Use an estimate to check exact calculations and to place the decimal point.

×	+ve	−ve
+ve	+ve	−ve
−ve	−ve	+ve

B	Brackets
I	Indices
D	Division
M	Multiplication
A	Addition
S	Subtraction

HOW TO

To solve a problem involving multiplication or division
1. Read the question and decide which calculations need doing.
2. Use an appropriate written or mental method.
3. Use an estimate to check each exact calculation.

EXAMPLE

Miss Georgiou is organising a school trip for year 10. There are 176 students and 12 staff. Coaches can carry 56 people. If Miss Georgiou hires enough coaches for everybody then how many spare places will there be?

① Find total number of passengers ÷ 56 and *round up*.

$176 + 12 = 188$

② Use mental methods.

$188 ÷ 56 = 3 + $ remainder   An exact result is not needed.

$4 × 56 = 4 × 50 + 4 × 6 = 200 + 24 = 224$

$224 - 188 = 36$

There will be 36 spare places. Remember to answer the question!

EXAMPLE

Add brackets to make this calculation correct.
$4 × 5 + 5 × 6 = 150$

① There are three possibilities, try each in turn.

$4 × (5 + 5 × 6) = 4 × (5 + 30) = 4 × 35 = 140$ ✗

$4 × (5 + 5) × 6 = 4 × 10 × 6 = 240$ ✗

$(4 × 5 + 5) × 6 = (20 + 5) × 6 = 25 × 6 = 150$ ✓

Without brackets
$4 × 5 + 5 × 6 = 20 + 30 = 50$

EXAMPLE

Maíra obtains $4.46 per hour for her part time job. One weekend she earns $57.98. She works two hours longer on Saturday than she does on Sunday. How long does she work for each day?

① Calculate the total hours worked and set up an equation.

$57.98 ÷ 4.46 = 13$ ② Use a calculator.   ③ Check $60 ÷ 4 = 15$ ✓

Let x = number of hours worked on Sunday

$x + (x + 2) = 13$

$2x = 11$

$x = 5.5$

She works for seven and a half hours on Saturday and five and a half hours on Sunday.

Number Calculations 1

Reasoning and problem solving

Exercise 1.3A

1. Rugby league teams have 13 players in the starting line-up and four substitutes. Eight teams are in a tournament.
   a. How many players are there altogether?
   b. Each team has six forwards, and one substitute forward. How many forwards are in the tournament?

2. Nancy receives £6 pocket money per week.
   a. How much does she receive in February?
   b. How much does she receive in a year? Assume 1 year = 52 weeks.
   c. She multiplies the February amount by 12, but gets the wrong answer for a year. Why?

3. Duha buys some decorative gravel for her garden. She decides on Scottish pebbles.

   Scottish pebbles
   20 kg bag $5.75
   5 bags $22.50
   3 bags cover 1 m²

   She measures her garden. It is 8 m².
   a. How many bags should she order?
   b. How much will this cost?
   c. Her friend suggests she orders an extra bag. Why is this a good idea?

4. The maths department orders some exercise books for the new term. There are 30 graph books in a pack and 25 lined books in a pack. There are 1228 students in the school. Graph packs cost €12.80, lined packs €11.25.

   The department decides to buy enough to give each student 1 graph book and 1 exercise book. What is the cost of the department's order?

5. Add brackets to make these calculations correct.
   a. $8 - 4 \times 3 - 1 = 8$
   b. $16 \div 2 \times 3 - 2 = 4$
   c. $5 \times 3 - 3 \times 9 \div 3 - 6 = 4$
   d. $4 - 2 \times 2^2 \div 2^2 = 4$
   e. $2 + 3^2 \times 4 + 3 = 65$

6. Ivan is tiling his floor with a mixed tile pattern.

Packs of 12	
Small square	£12.95
Packs of 6	
Medium	£11.65
Large	£20.95
Rectangle	£18.95

   The tile company says that for 100 square feet he will need 24 small squares, 25 medium squares, 25 large squares and 26 rectangles. His room is 14 feet wide by 25 feet long.
   a. How many packs of each size will Ivan need?
   b. How much will the tiles cost?
   c. The company suggest adding 20% to each quantity to allow for wastage. What is the cost if he does this?

*7. Jal's car manual says his car does 42 mpg (miles per gallon). A gallon is 4.55 litres. A litre of diesel costs $1.38. Jal drives from Jalandhar to New Delhi, a distance of 215 miles. How much should it cost for diesel?

8. **Try this**

   Think of a number between 100 and 999. Repeat it to make it a 6-digit number. Enter this number on your calculator. Now divide it by 13.
   And then by 11.
   And then by 7.
   Do you end up with what you first thought of?
   Does it work for any 3-digit number?

   Why? Hint: work out $7 \times 11 \times 13$

Summary

Checkout
You should now be able to...

		Test it Questions
✓	Order positive and negative integers and decimals.	1, 2
✓	Round numbers to a given number of decimal places or significant figures.	3
✓	Use mental and written methods to add, subtract, multiply and divide with positive and negative integers and decimals.	4 – 7
✓	Use BIDMAS to complete calculations in the correct order.	8, 9

Language	Meaning	Example
Place value	The value of a digit according to its position in a number.	123.4 2 means 2 tens = 20 4 means 4 tenths = $\frac{4}{10}$

Rounding	Expressing a number to a given degree of accuracy.	103.67 = 103.7 (1 dp)
		      = 100 (1 sf)
		0.0055 = 0.0 (1 dp)
		      = 0.006 (1 sf)

Decimal places	The number of digits after the decimal point.	
Significant figures	The number of digits after the first non-zero digit.	
Directed number	A number with a plus or minus sign in front of the number	... −4 −3 −2 −1 0 +1 +2 +3 ....
Negative	A number that is less than zero	3 − 6 = −3
Estimate	An approximate calculation or a judgement of a quantity.	Estimate 68.89 × 21.1 ≈ 70 × 20 = 1400 Exact = 1453.579

Partitioning	Splitting a larger number into smaller numbers which add up to the original number.	85 + 25.6 = 85 + (15 + 10.6)
		         = 100 + 10.6
		         = 110.6

Compensation	Replacing a number by a simpler approximate value and a correction.	158 − 18.9 = 158 − (20 − 1.1)
		          = (158 − 20) + 1.1
		          = 139.1

Operations	Rules for processing numbers.	Addition, subtraction, multiplication and division.
Order of operations	The order in which operations have to be carried out to give the correct answer to a calculation.	2 + 4 × 3 − 1 = 2 + 12 − 1 = 13 (2 + 4) × 3 − 1 = 6 × 3 − 1 = 18 − 1 = 17 (2 + 4) × (3 − 1) = 6 × 2 = 12 2 + 4 × (3 − 1) = 2 + 4 × 2 = 2 + 8 = 10
BIDMAS	An acronym for the correct order of operations: **B**rackets, **I**ndices (or powers), **D**ivision or **M**ultiplication, **A**ddition or **S**ubtraction.	

Review

1. Copy the numbers and write > or < between them to show which is larger.
   a. 24.3 ☐ 24.5
   b. −0.5 ☐ −0.9
   c. 0.5 ☐ 0.06
   d. 1.456 ☐ 1.46

2. Are the following statements true or false? In cases where the statement is false write a true statement relating the two expressions.
   a. $0.85 \div 10 > 85 \div 1000$
   b. $0.85 \div 10 \geqslant 85 \div 1000$
   c. $1 \leqslant \frac{7}{5} \leqslant 1\frac{1}{2}$
   d. $3^2 = 3 \times 2$
   e. $9^2 \neq 9 + 9$

3. Round
   a. 45.892 to 1 decimal place
   b. 0.0752 to 2 decimal places
   c. 0.0854 to 2 significant figures
   d. 1521 to 1 significant figure
   e. 78025 to 3 significant figures
   f. 3.529 to 2 significant figures.

4. Work out the value of these expressions.
   a. 83×100
   b. 2.59×10
   c. 0.31×1000
   d. 0.05×100
   e. $764 \div 10$
   f. $5490 \div 1000$
   g. $8.5 \div 100$
   h. $0.08 \div 10$

5. Calculate these using written or mental methods.
   a. $627 + 3215$
   b. $27.3 + 164.7$
   c. $302.8 + 6.52$
   d. $0.34 + 52.713$
   e. $8124 - 398$
   f. $104.1 - 69.5$
   g. $1589.4 - 672$
   h. $18.31 - 8.4$

6. Calculate these using written or mental methods.
   a. 18×30
   b. 3.5×18
   c. $3.36 \div 6$
   d. $650 \div 1.3$
   e. 8.31×6.2
   f. 1.083×2.45
   g. $15.2 \div 8$
   h. $406 \div 1.4$

7. Work out these calculations involving negative numbers.
   a. $17 + -28$
   b. $-189 + -52$
   c. $9.3 - -3.3$
   d. $-0.62 + 0.19$
   e. 14×-3
   f. -1.1×-5
   g. $170 \div -10$
   h. $-6.5 \div -1.3$

8. Evaluate these without using a calculator.
   a. $19 - 3 \times 6$
   b. $35 + 25 \div 5$
   c. $8 \times (17 - 3)$
   d. 7×4^2
   e. $3.2 + \sqrt{49}$
   f. $\sqrt{6 + 5 \times 6} \div 0.2$

9. Use your calculator to work these out.
   a. $12.78 + 6.41 \times 8.32 - 9.3$
   b. $\dfrac{3 - \sqrt{(-3)^2 - 4 \times (4) \times (-1)}}{2}$

What next?

Score		
0 – 4		Your knowledge of this topic is still developing. To improve, look at MyiMaths: 1001, 1005, 1007, 1011, 1013, 1068, 1072, 1916, 1917
5 – 8		You are gaining a secure knowledge of this topic. To improve your fluency look at InvisiPens: 01Sa – p
9		You have mastered these skills. Well done you are ready to progress! To develop your problem solving skills look at InvisiPens: 01Aa – f

Assessment 1

1. Carli tries to order each set of numbers from smallest to largest.
   One pair of numbers in each list is in the wrong order.
   Say where Carli has made a mistake and put each list of numbers in order, starting with the smallest.

   a 0.8 1.9 3.3 44 303 57.6 [2]

   b −0.07 −2.19 30 43.56 188.0 194.7 [2]

2. Carli then tries to put these numbers in ascending order
   $42 \div 100, 0.3 \times 10, 4236 \div 1000, 516 \div 10, 42 \times 100, 216 \times 1000$
   Has she ordered the numbers correctly? Give your reasons. [4]

3. The world's tallest man, Robert Wadlow, was 271.78 cm (2 dp) tall.
   The world's tallest woman, Yao Defen, is 233.34 cm (2 dp) tall.

   a A challenger to the world's tallest man record measured his height as 271.8 cm to one decimal place.
      Has the challenger definitely beaten the world record? Give your reasons. [1]

   b A challenger to the world's tallest woman record measured her height as 233.341 cm.
      Has the challenger definitely beaten the world record? Give your reasons. [1]

4. Dave is 36 and Nadya is 44. Nadya says that she and Dave are the same age to one significant figure.

   a Is Nadya correct? [1]

   b Will Dave and Nadya be the same age to one significant figure in one year's time? Give your reasons. [2]

   c How old will Dave and Nadya be the next time their ages are the same to one significant figure? [2]

5. As the Earth spins on its axis, everything on the Earth's surface moves with it.
   The distance travelled in one day due to the Earth's rotation is $3.142x$, where x is the diameter of the circular path.
   Abena lives on the equator and Edward lives in the UK.
   $x = 12756$ for Abena and $x = 8134$ for Edward.

   a Write both values of x to two significant figures. [2]

   b Use your answers to part **a** to estimate how much further Abena travels than Edward in one day. [3]

   c Explain how using values of x correct to one significant figure would affect the estimate in part **b**. [3]

6. Jasmine's bike has wheels of circumference (i.e. perimeter) 2.5 m.
   When Jasmine cycles to school, the wheels go round 850 times.
   How far does Jasmine cycle to school? [2]

7. Work out these problems and include units in your answer.

   a A bag of sweets weighing 113 g includes wrappings totalling 0.5 g.
      Each sweet weighs 4.5 g.
      How many sweets are in the bag? [2]

   b A football stadium has 135 400 m² of seating for its fans. Each fan is allowed 5.4 m² of space.
      How many fans, to the nearest 1000, is the stadium capable of holding? [2]

Number Calculations 1

8 There are 10 questions in a quiz.
   A *correct* answer scores 3 points. A *wrong* answer *loses* 2 points.
   Any question not answered loses 1 point. A negative total is possible.

   a Write down the maximum and minimum points any player can score. [2]
   b Rabiyah answers eight of the ten questions. Five are correct. How many points does Rabiyah score? [3]
   c Describe three different ways of scoring −10 points. [3]

9 To find your BMI (Body Mass Index) you use this process:

Name	Mass (kg)	Height (m)	BMI
Sue	57.5	1.7	
Clive	105.8	1.95	
Kivi		1.77	19.7
Henry	71.3		21.3

   Copy and complete the table. Give your answers to 1 decimal place. [6]

10 A supermarket stocks packets of the new breakfast cereal Maltibix.
   Each packet of Maltibix holds 650 g inside a cardboard box weighing 68 g.
   Fifty boxes, each holding 36 of these packets are delivered to the supermarket.
   Does the mass of the delivery exceed 1000 kg? [4]

11 In the number grids shown, the number in each cell is the sum of the two cells above it.
   Copy and complete the grids shown.

   a 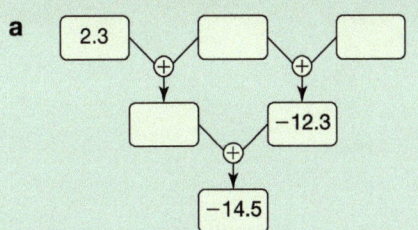   b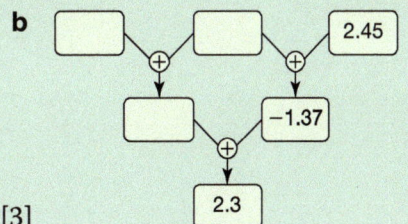
   [3]   [3]

12 A magic square is a square grid of numbers where each number
   is *different*. The sum of the numbers in each row, each column
   and each diagonal is the same.
   Fill in the missing values in the magic square.

   [4]

13 Lana's garden is rectangular and measures 12.5 m by 9.2 m.
   The garden is to be sown with grass seed.
   The gardener needs 25 g of grass seed for each square metre of ground.
   Grass seed costs $5.35 per kg. The gardener estimates that he will
   need $20 to buy the grass seed.
   Lana disagrees with the gardener's estimate.
   She gives the gardener $15 to buy the grass seeds.
   Which estimate do you agree with? Explain your answer. [4]

14 The *reciprocal* of a number n is $1 \div n$.
   a What is the only number which is the same as its reciprocal? [1]
   b What is the only number which has no reciprocal? Explain your answer. [1]
   c Nisreen says that every positive number is greater than its reciprocal.
      Find an example that disproves Nisreen's claim. [1]

15 Write one pair of brackets in each calculation to make the answer correct.
   a $3 + 4 \times 5 = 35$ [1]   b $4 \div 3 + 5 = 6\frac{1}{3}$ [1]   c $5(2^3 + 0.4) \div 4 - 3 \times -1 = 6$ [2]

2 Expressions

Introduction
The real world is messy, complicated and always in motion. Algebra is a vital part of maths because it attempts to describe aspects of the world, such as fluid flow or the forces acting on a suspension bridge. Through equations and formulae, algebra provides mathematical models to describe real-world situations, from which understanding can be gleaned and predictions made. Algebra can only be applied if assumptions are made to simplify the situation. Although simplifying the situation means that the model is only an approximate reflection of the real world, it helps us to understand the forces that lie behind it.

What's the point?
Without algebra, you would not be able to work with large mechanical forces. There would be no skyscrapers or suspension bridges. You would also not be able to understand electronics, so there would be no tablets or mobile phones.

Objectives
By the end of this chapter, you will have learned how to …
- Use algebraic notation and simplify expressions by collecting like terms.
- Substitute numbers into formulae and expressions.
- Use the laws of indices.
- Multiply a single term over a bracket.
- Take out common factors in an expression.
- Simplify algebraic fractions and carry out arithmetic operations with algebraic fractions.

Check in

1. Work out these multiplications and divisions mentally.

   a 15×3
   b 4×13
   c $(-2) \times 13$
   d 14×14
   e $56 \div 8$
   f $91 \div 7$
   g $150 \div (-3)$
   h $1200 \div 40$

2. Explain why the answer to each of these questions is 15.

   a $9 + 3 \times 2$
   b $24 \div 3 + 7$
   c $(3 + 2) \times 3$
   d $3^3 - 4 \times 3$

3. Find the highest common factor of these number pairs.

   a 6 and 9
   b 8 and 12
   c 20 and 30
   d 12 and 18
   e 24 and 52
   f 50 and 75
   g 99 and 132
   h 7 and 14

Chapter investigation

Think of a number between 1 and 10.
- Double it.
- Add 4.
- Halve your answer.
- Take away the number you first thought of.

What do you notice? Investigate why this is the case.

Can you invent different instructions that give similar results?

SKILLS

2.1 Simplifying expressions

Collecting like terms and using the rules of algebraic notation lets you simplify **expressions**.

- Like terms have the same letter or symbol.

The expression $4a + 3b - 2a + b$ is simplified by collecting the 'like terms' '$4a$' and '$-2a$' and, '$+3b$' and '$+b$'.

$4a + 3b - 2a + b = 2a + 4b$

These expressions are simplified using the rules of algebraic notation.

$b \times 4 \times a = 4ab \qquad x \times x \times x = x^3 \qquad p \div 2 = \dfrac{p}{2} \text{ or } \dfrac{1}{2}p$

> By convention you write the **coefficient** first and then the **variables** in alphabetical order.

EXAMPLE (p.8)

Simplify these expressions.

a $4p + p + 2p$
b $4a - 3b + 2a + 5b$
c $5x - 2y - 3 + 3x$
d $3a + 4 + 2y$

a $7p$
b $(4a) - 3b + 2a + 5b = 6a + 2b$ Collect 'a' and 'b' terms separately
   $4a + 2a = 6a \qquad -3b + 5b = 2b$
c $(5x) - 2y - 3 + 3x = 8x - 2y - 3$ Collect 'x', 'y' and numbers separately
   $5x + 3x = 8x$
d This cannot be simplified.

> Be careful with the negative signs.

> It can be useful to circle the individual terms to ensure the sign stays with term.

EXAMPLE

Simplify these expressions.

a $y^2 + 3y^2$
b $2ab + 5a - 3b$
c $ab^2 + 6ab^2$
d $5ab^2 + 2a^2b - 3ab^2$

a $4y^2$
b This cannot be simplified.
c $7ab^2$
d $(5ab^2) + 2a^2b - 3ab^2 = 2ab^2 + 2a^2b$ Collect the 'ab^2' and 'a^2b' terms separately.

EXAMPLE

Simplify these expressions.

a $2 \times y \times w \times 3$
b $\sqrt{3} \times a \times 4 \times a$
c $4 \div p$
d $\dfrac{\sqrt{3} \times c \times 10}{5 \times d}$

a $6wy$
b $4\sqrt{3}a^2 \quad a \times a = a^2$
c $\dfrac{4}{p}$
d $\dfrac{2\sqrt{3}c}{d} \quad 10 \div 5 = 2$

EXAMPLE (p.104)

Evaluate these expressions.

a $7a$ when $a = -2$
b $\dfrac{p}{6}$ when $p = 42$
c $\dfrac{6}{p}$ when $p = -3$
d $-2w^2$ when $w = -2$

a $7a = 7 \times -2 = -14$
b $\dfrac{p}{6} = \dfrac{42}{6} = 7$
c $\dfrac{6}{p} = \dfrac{6}{-3} = -2$
d $-2w^2 = -2 \times (-2)^2 = -8$

Algebra   Expressions

Fluency

Exercise 2.1S

1. Simplify $x + 3y + 4x + 5y$.
2. Simplify $a + 3b - 4a - b$.
3. Oday thinks that $3d + 2e + 4d + 5e = 14de$. Write the correct answer.
4. Simplify these expressions.
   a. $a + 2b + a$
   b. $4t + 3s + 2t$
   c. $d + 2e + e + 3e$
   d. $4y + 2y + w + 7w$
   e. $2p + f + 3p + 3f$
   f. $e + 2f + 3g + 4h$
   g. $5k + 4 + 2d + 2k$
   h. $8y + 5 + 2y + 11$
   i. $3p + 4 + 2p + 2q$
   j. $8b + 5 - 2a + 6ab$
5. Simplify these expressions.
   a. $d + 5d - 2d$
   b. $4q - 2q + q$
   c. $4a - 2a - a + 6a$
   d. $y - 3y - y + 7y$
   e. $4t + t - t - 3t$
   f. $e - 2e - 3e + 4e$
   g. $5k + 4 - 2k$
   h. $8y - 5 - 2y + 11$
   i. $8p - 4 - 12p + 11$
   j. $8b + 4d - 12 + 4bd$
6. Ian thinks that $k + k + k = k^3$.
   Eric thinks that $k + k + k = k3$.
   Write the correct answer.
7. Wisam thinks that $a^2 + a^2 + a^2 = 6a$. Write the correct answer.
8. Simplify these expressions.
   a. $d^2 + d^2$
   b. $b^2 + 3b^2$
   c. $4a^3 + 5a^3$
   d. $2y^2 + 4y + y^2$
   e. $4t^3 + 5t^3 - 6t^3$
   f. $6m^2 - 4m^2 + 5m$
   g. $5k^3 + 4 - 2k^3$
   h. $8y^2 - 5 - 2y + 11$
9. Simplify these expressions.
   a. $ad^2 + ad^2$
   b. $3ab^2 + 10ab^2$
   c. $4xy^3 + 5xy^3$
   d. $2xy^2 + 4xy + xy^2$
   e. $4st^3 + 5s^3t - 6st^3$
   f. $6m^2n - 4mn^2 + 5mn^2$
   g. $5gh^2 + 4g^2h - 2gh$
   h. $8y^2 - 5y - 2x + 11x^2$
10. State whether $2a^2 + 3b^2$ can be simplified.
11. Simplify these expressions.
    a. $3 \times a \times b$
    b. $b \times 3 \times a$
    c. $a \times b \times 3$
    d. $b \times 3 \times 2$
    e. $10 \times a \times 2$
    f. $p \times 4 \times q$
    g. $\sqrt{3} \times p \times 2$
    h. $a \times 4 \times a$
    i. $a \times b \times a$
    j. $a \times b \times a \times b$
    k. $2 \times b \times a \times b$
    l. $a \times 3 \times a \times \sqrt{2} \times b$
12. Simplify these expressions.
    a. $3 \div a$
    b. $a \div 3$
    c. $10 \div 2p$
    d. $2m \div 5$
    e. $y \div \sqrt{3}$
    f. $10b \div 2\sqrt{3}$
13. Simplify these expressions.
    a. $\dfrac{3 \times b \times 8}{6 \times a}$
    b. $\dfrac{\sqrt{3} \times b \times 8}{2}$
14. Evaluate these expressions when $y = 12$.
    a. $2y$
    b. $20 + y$
    c. y^2
    d. $3y^2$
    e. $\dfrac{y}{3}$
    f. $\dfrac{24}{y}$
15. Evaluate these expressions when $p = -2$.
    a. $2p$
    b. $p + 6$
    c. p^2
    d. $3p^2$
    e. $\dfrac{p}{10}$
    f. $\dfrac{10}{p}$
*16. Simplify $2(b - 3) - 5(4 - b)$.
17. Simplify the expressions in the grid and find the 'odd one out' for each row.

$3p + 2q + p + 5q$	$6p + 3q - 2p + 4q$	$5p - 3q - p + 5q$
$2m \times 3n$	$2 \times n \times m \times 5$	$6mn$
$\dfrac{24cd}{12c}$	$\dfrac{2d^2}{d^2}$	$\dfrac{2d^2}{d}$
$2n - 8$	$3m + 2n - m - 2m$	$3n - 2 - 6 - n$

1178, 1179, 1186

APPLICATIONS

2.1 Simplifying expressions

RECAP
- Use algebraic notation to write expressions concisely and consistently.
- Collect like terms to simplify an expression.
- To evaluate an expression, replace the variables by the given numbers, restore any missing × signs and use the rules of arithmetic.

Don't write the multiplication symbol ×.

$x + x + x = 3 \times x = 3x$

Use indices $y \times y \times y = y^3$ but *not* y^1 for y.

Write divisions as fractions.

$x \div y = \dfrac{x}{y}$

Write in the order: numbers then letters alphabetically.

$z \times 3 \times y \times x^2 = 3x^2yz$

HOW TO

To write expressions and equations in algebra
1. Give every '**unknown**' a letter and write it down. Translate the words into letters and symbols.
2. Use algebraic notation and collect like terms to simplify your expression. You may need to substitute values into your expression.
3. Interpret your answer in the context of the question.

EXAMPLE

Find simplified expressions for
a the perimeter
b the area
of this shape.

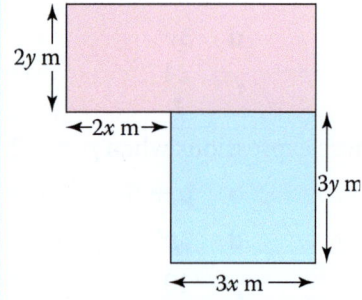

Width pink rectangle = 2x + 3x = 5x

a Perimeter = 5x + 2y + 3y +
       3x + 3y + 2x + 2y
     = (10x + 10y) m

b Area pink rectangle = 2y × 5x = 10xy
   Area blue rectangle = 3y × 3x = 9xy
   Total area = 10xy + 9xy
             = 19xy m²

① Find any missing lengths.
② Collect like terms.
③ Include the units.

① Find the area of each part of the composite shape.
② Collect like terms.
③ Include the units.

EXAMPLE

The formula $v^2 = u^2 + 2as$ connects final speed, v, with initial speed, u, acceleration, a, and distance, s.

Find the final speed of a racing car that is stationary at the start line and then accelerates at 6 m/s² for 300 metres.

① The car is stationary, so initial speed, $u = 0$.
② Substitute the values into the formula.

$u = 0, a = 6$ and $s = 300$.
$v^2 = u^2 + 2as$
$ = 0^2 + 2 \times 6 \times 300$
$ = 3600$

③ Find v and include units.

$v = \sqrt{3600} = 60$ m/s

Algebra Expressions

Reasoning and problem solving

Exercise 2.1A

1 a Write a simplified expression for the
   i perimeter
   ii area of this rectangle.

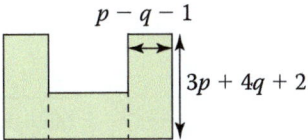

   b What are the measurements of a rectangle with perimeter $6x + 4y$ and area $6xy$?

2 This U-shape is made from three identical rectangles. What is its perimeter?

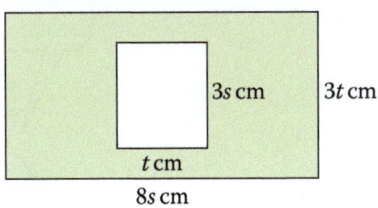

3 Find the area of this shape.

   3s cm   3t cm
   t cm
   8s cm

4 The bill for a mobile phone is calculated using the formula

   Cost in pounds = $0.05 \times$ number of texts + $0.10 \times$ minutes of calls

   a Use the formula to work out the bills for
      i 40 texts and 20 minutes of calls
      ii 5 texts and 70 minutes of calls
      iii 32 texts and 15 minutes of calls.
   b What is the cost, in pence, for a call lasting 1 minute?

5 The cost of hiring a van is given by the formula $C = 25d + 40$ where C is the cost in pounds and d is the number of days.

   How much more does it cost to hire a van for 10 days than for 3 days?

6 Three students tried to simplify $3m + 5$. Which of them did it correctly?

Sara	Paul	Abdul
$3m + 5 = 8m$	$3m + 5 = 15m$	$3m + 5 = 3m + 5$

7 Hanan and Paul evaluated $2x^2$ when $x = 6$. Who was right? Give your reasons.

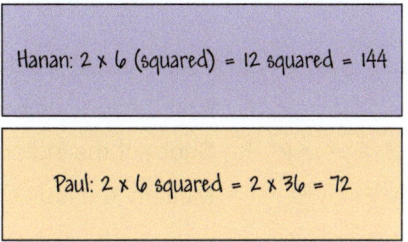

8 Complete these algebraic pyramids. The expression in a cell is a sum of the expressions in the two cells below it.

   a
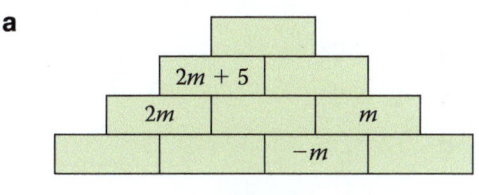

   b
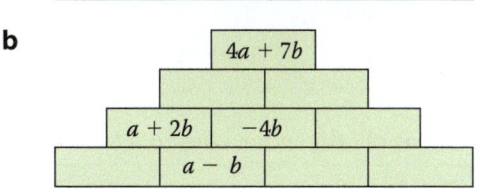

9 Rearrange each set of cards to make a correct statement.

   a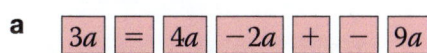

   b $\boxed{(b + 4c)}\ \boxed{-}\ \boxed{(3b + 7c)}\ \boxed{=}$
      $\boxed{(5b + c)}\ \boxed{+}\ \boxed{(3b - 2c)}$

   c $\boxed{e}\ \boxed{-}\ \boxed{+}\ \boxed{8e}\ \boxed{4d}\ \boxed{\div}\ \boxed{2}\ \boxed{3d \times 4}\ \boxed{+}$
      $\boxed{=}\ \boxed{16d + 3e}$

10 Enter the following expressions into this 3×3 magic square so that the sum of the terms in each row and each column is the same expression. Three terms have been entered for you.

$x + 2y - 2$	$2x - y + 8$	$3y - 8$
$2y - 6$	$2x + y + 4$	$x + 2$

	$y - 4$	$2x + 6$
	$x + y$	

SKILLS

2.2 Indices

Repeated multiplications of the same number or letter can be written using **index form**.

2^4 is in index form and represents $2 \times 2 \times 2 \times 2$. 2 is the **base** and 4 is the **index** or **power**. The plural is indices.

> The bases must be the same in order to use the index laws.

- Indices obey a set of rules known as the index laws.
  - $y^a \times y^b = y^{a+b}$ Add the indices when multiplying expressions with the same base.
  - $y^a \div y^b = y^{a-b}$ Subtract the indices when dividing expressions with the same base.
  - $(y^a)^b = y^{ab}$ Multiply the indices when using powers with an expression.

EXAMPLE

Simplify these expressions, leaving your answer in index form.

a $2^6 \times 2^3$
b $2^6 \div 2^3$
c $(2^3)^2$

'Index form' means leave the answer in the form a^n.

a $2^6 \times 2^3 = 2^{6+3} = 2^9$

$2^6 \times 2^3 = (2 \times 2 \times 2 \times 2 \times 2 \times 2) \times (2 \times 2 \times 2)$
$= 2 \times 2 \times 2 \times 2 \times 2 \times 2 \times 2 \times 2 \times 2$
$= 2^{6+3} = 2^9$

b $2^6 \div 2^3 = 2^{6-3} = 2^3$

$2^6 \div 2^3 = \dfrac{2^6}{2^3} = \dfrac{2 \times 2 \times 2 \times 2 \times 2 \times 2}{2 \times 2 \times 2} = 2^{6-3} = 2^3$

c $(2^3)^2 = 2^{3 \times 2} = 2^6$

$(2^3)^2 = (2 \times 2 \times 2)^2 = (2 \times 2 \times 2) \times (2 \times 2 \times 2) = 2^{3 \times 2} = 2^6$

> Multiply and divide the numbers as normal.

You can apply the index laws to letters or combinations of letters and numbers.

EXAMPLE

Simplify these expressions, leaving your answer in index form.

a $a^3 \times a^5$ **b** $b^6 \div b^3$ **c** $4c^7 \times 3c^5$ **d** $12p^6 \div 4p^2$ **e** $(3p^2q^3)^3$

a $= a^{3+5}$
$= a^8$

b $= b^{6-3}$
$= b^3$

c $= 4 \times 3 \times c^7 \times c^5$
$= 12 \times c^{7+5}$
$= 12c^{12}$

d $= \dfrac{12 \times p^6}{4 \times p^2}$
$= 3 \times p^{6-2}$
$= 3p^4$

e $= 3^3 \times (p^2)^3 \times (q^3)^3$
$= 27 \times p^{2 \times 3} \times q^{3 \times 3}$
$= 27p^6q^9$

EXAMPLE

Simplify $\dfrac{8a^6 \times \sqrt{2}a^4}{2a^2}$

$\dfrac{8a^6 \times \sqrt{2}a^4}{2a^2} = \dfrac{8 \times \sqrt{2}}{2} \times \dfrac{a^6 \times a^4}{a^2} = 4\sqrt{2} \times \dfrac{a^{6+4}}{a^2}$ Add indices for multiplication.

$= 4\sqrt{2} \times a^{10-2} = 4\sqrt{2}a^8$ Subtract indices for division.

- Indices can be defined for zero, negative and fractional powers.
  - $y^0 = 1$ Anything to the power zero is 1.
  - $y^{-2} = \dfrac{1}{y^2}$ Negative powers correspond to reciprocals.
  - $y^{\frac{1}{2}} = \sqrt{y}$, $y^{\frac{1}{3}} = \sqrt[3]{y}$, $y^{\frac{1}{4}} = \sqrt[4]{y}$ Fractional powers correspond to roots which may be surds.

EXAMPLE

Simplify

a x^0
b $(-2x)^{-2}$
c $(9x^2)^{\frac{1}{2}}$
d $(x^6)^{\frac{2}{3}}$

a 1

b $\dfrac{1}{(-2x)^2} = \dfrac{1}{(-2)^2 \times x^2} = \dfrac{1}{4x^2}$

c $\sqrt{9x^2} = \sqrt{9} \times \sqrt{x^2} = 3x$

d $x^{6 \times \frac{2}{3}} = x^{\frac{12}{3}} = x^4$

Algebra Expressions

Fluency

Exercise 2.2S

1. Write $2 \times 2 \times 2 \times 2 \times 2 \times 2$ using index notation.

2. Evaluate these expressions.
   a. 2^4
   b. $5^4 \times 5^7$

3. i. Evaluate these expressions.
   ii. Use your calculator to check your answers.
   a. 5^3
   b. 2^{10}
   c. 1^5
   d. 6^2
   e. $(-1)^3$
   f. $\left(\frac{1}{2}\right)^2$

4. i. Simplify these expressions leaving your answers in index form.
   ii. Use your calculator to check your answers.
   a. $3^2 \times 3^4$
   b. $2^2 \times 2^7$
   c. $5^2 \times 5^5$
   d. $6^5 \times 6^2$
   e. $7^4 \times 7^5$
   f. $11^3 \times 11^7$
   g. $5^7 \div 5^3$
   h. $3^{10} \div 3^4$
   i. $8^9 \div 8^4$
   j. $2^{14} \div 2^7$
   k. $6^4 \div 6^8$
   l. $4^3 \div 2^7$

5. i. Simplify these expressions leaving your answers in index form.
   ii. Use your calculator to check your answers.
   a. $(2^3)^2$
   b. $(5^4)^6$
   c. $(3^3)^7$
   d. $(8^2)^8$
   e. $(7^4)^{-2}$
   f. $\left(\left(\frac{1}{2}\right)^2\right)^3$

6. Firas thinks that $a^5 \times a^2 = a^{10}$.
   Do you agree with Firas? Give your reasons.

7. Simplify these expressions, leaving your answers in index form.
   a. $a^2 \times a^4$
   b. $y^2 \times y^8$
   c. $b^6 \times b^{-2}$
   d. $p^{-3} \times p^{-4}$
   e. $h^4 \times h^8$
   f. $s^3 \times t^9$
   g. $x^6 \times y^3 \times x^9 \times y^2$
   h. $x^7 \times y^2 \times x^{-2} \div y^3$
   i. $\dfrac{p^4 \times q^8}{p^0 \times q}$
   j. $\dfrac{p^3 \times q^7 \times r^4 \div q^2}{r^3 \times p \times q^3}$

8. Rula thinks that $4y^5 \times 2y^2 = 6y^7$ because 'the index rules say that you add the powers when two terms are multiplied'.
   Do you agree with Rula? Give your reasons.

9. Simplify these expressions, leaving your answers in index form.
   a. $3x^5 \times x^2$
   b. $5y^2 \times y^5$
   c. $4b^2 \times 3b^6$
   d. $\sqrt{2}p^4 \times \sqrt{2}p^7$
   e. $5h^5 \times 6h^6$
   f. $4s^3 \times \sqrt{3}t^4$

10. Diego thinks that $12p^{12} \div 3p^4 = 9p^8$ because 'the index rules say subtract the powers when two terms are divided.'
    Do you agree with Diego? Give your reasons.

11. Simplify these expressions, leaving your answers in index form.
    a. $10y^6 \div 5y^2$
    b. $6a^9 \div 3a^3$
    c. $20k^7 \div 4k^3$
    d. $18p^8 \div 6p^3$
    e. $4x^{10} \div 8x^4$
    f. $\sqrt{6}y^8 \div \sqrt{2}y^4$

12. Simplify these expressions, leaving your answers in index form.
    a. $(a^3)^2$
    b. $(y^2)^6$
    c. $(k^3)^5$
    d. $(p^7)^8$
    e. $(a^2)^4$
    f. $(q^4)^2$
    g. $(4a^3)^2$
    h. $(3y^2)^6$
    i. $(2k^3)^5$
    j. $(\sqrt{2}p^7)^4$
    k. $(\sqrt{3}a^3)^3$
    l. $(\sqrt{5}a^4)^5$

13. Simplify these expressions.
    a. $(a^{-4})^2$
    b. $(b^{-3})^{-2}$
    c. $(3c^{-3})^3$
    d. $10d^3 \times 6d^{-4}$
    e. $9e^3 \div 3e^{-3}$
    f. $(3f^3 \times 2f^{-7})^2 \div 18f$
    g. $((g^4)^2)^5$
    h. $18(h^{-4})^4 \div (3h^{-8})^2$

14. Simplify these expressions.
    a. $(5a)^0$
    b. $5a^0$
    c. $(4p)^{-2}$
    d. $(2p)^{-4}$
    e. $(25a)^{\frac{1}{2}}$
    f. $(8d)^{\frac{1}{3}}$

*15. Simplify these expressions.
    a. $\left(\dfrac{8x^2 \times \sqrt{2x^{\frac{1}{2}}}}{4x^3}\right)^3$
    b. $\left(\dfrac{y^{-2} \times 2y^4}{32y^{\frac{2}{3}}}\right)^{\frac{4}{5}}$

16. Solve these equations.
    a. $x^4 = 625$
    b. $4^x = 64$
    c. $2^x = 0.25$
    d. $3^x = \dfrac{1}{9}$

APPLICATIONS

2.2 Indices

RECAP

- Index notation is a way to write a number as a power of a base number. For integer powers it allows repeated multiplication to be written concisely.
- Index laws can be used to simplify expressions having the same base.

  $x^m \times x^n = x^{m+n}$ $x^m \div x^n = x^{m-n}$ $(x^m)^n = x^{mn}$

- Indices have meaning when they are not positive integers.

  $y^0 = 1$ $y^{-n} = \dfrac{1}{y^n}$ $y^{\frac{1}{n}} = \sqrt[n]{y}$

base → index
$3^4 = 3 \times 3 \times 3 \times 3 = 81$
$5p^7 \times 8p^{-4} = 5 \times 8 \times p^{7+(-4)}$
$= 40p^3$

HOW TO

To solve problems involving indices
1. Read the question and be prepared to use your knowledge of other topics.
2. Apply the index laws.
3. Answer the question.

EXAMPLE

A rectangle has area $27x^3y^4$ and width $9xy^2$. What is the perimeter of the rectangle?

$A = 27x^3y^4$ $9xy^2$

1. To find the perimeter you need to know the length of the rectangle.
   Area = width × length
2. Length = $\dfrac{\text{area}}{\text{width}} = \dfrac{27x^3y^4}{9xy^2} = \dfrac{27}{9} \times x^{3-1} \times y^{4-2} = 3x^2y^2$ Treat the numbers, x-terms and y-terms separately.
3. Perimeter = $9xy^2 + 3x^2y^2 + 9xy^2 + 3x^2y^2 = 18xy^2 + 6x^2y^2$

EXAMPLE

Find the value of x in these equations.

a $81^x \times (3^x)^2 = 9^3$ b $8 \times 2^{3x} \times 4^x = \dfrac{1}{4}$ c $\dfrac{6^{2x} \times (36^x)^2}{216^x} = 1$

Find a numbers that can be used as a base for each term in the equation.

a $81 = 9^2 = 3^4, 9 = 3^2$
$(3^4)^x \times (3^x)^2 = (3^2)^3$
$3^{4x} \times 3^{2x} = 3^6$
$3^{6x} = 3^6$
$6x = 6$
$x = 1$
Hint: Because the bases are the same, the indices must be equivalent

b $8 = 2^3, 4 = 2^2$
$2^3 \times 2^{3x} \times (2^2)^x = \dfrac{1}{2^2}$
$2^{3+3x} \times 2^{2x} = 2^{-2}$
$2^{3+5x} = 2^{-2}$
$3 + 5x = -2$
$x = -1$
Hint: Because the bases are the same, the indices must be equivalent

c $36 = 6^2, 216 = 6^3$
$\dfrac{6^{2x} \times ((6^2)^x)^2}{(6^3)^x} = 1$
$6^{2x} \times 6^{4x} = 6^{3x}$
$6^{6x} = 6^{3x}$
$6x = 3x$
$x = 0$
Hint: Because the bases are the same, the indices must be equivalent

Algebra   Expressions

Reasoning and problem solving

Exercise 2.2A

1. Find and simplify an expression for the area of each shape.

   a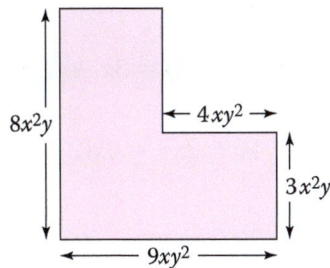

   b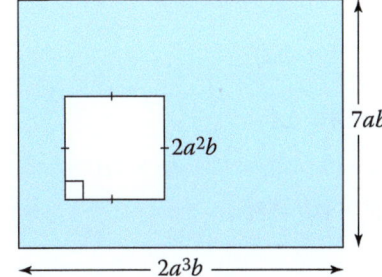

2. Find the perimeter of these rectangles.
   a. Area $= 64p^6q^8$, width $= 4p^3q^4$
   b. Area $= 5s^4t^{-2}$, width $= 10s^2t^{-1}$
   c. Area $= 12uv^4$, width $= 3u^2v^2$

3. A square has area $169a^4b^{12}$. What is the perimeter of the square?

4. A square has perimeter $28a^3b^{-1}$. What is the area of the square?

5. Write a simplified expression for the area of this triangle.

   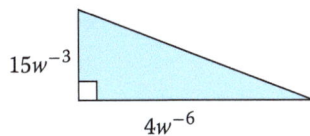

6. Rearrange each set of cards to make a correct statement.

   a. $($ $=$ a^6b^2 a^3 $)^2$ $\times$ b^2
   b. c^4d $\div$ $($ c^7d^3 $=$ $)^{\frac{1}{2}}$ c^3d c^5d^3
   c. $\div$ e^4f^{-6} $\times$ $e^{-1}f^2$ $e^{-2}f^6$ e^3f^{-2} $=$

7. Simplify this expression. $(-xy^2)^{101}$

8. A cube has sides of length $3x^2$. Write an expression for its volume.

9. a. True or false? $3^x \times 3^y$ simplifies to give 9^{x+y}. Give your reasons.
   *b. Is it ever true that $3^x \times 3^y = 9^{x+y}$? Give your reasons.

10. Which expression is the odd one out? Give your reasons.

    $5t^2 \times 10t^{-4} \div (5t^{-3})^2$ | $\dfrac{2}{t^6}$ | $\left(\dfrac{4t^2}{16t}\right) \times 8t^{-7}$

11. Find the value of x in these equations.
    a. $(2^2)^x \times 2^{3x} = 32$
    b. $\left(\dfrac{3^{7x} \times 3^{5x}}{3^{8x}}\right)^2 = 81$
    c. $\dfrac{16^{-x} \times 4^{4x}}{8^3} = 32$
    d. $\dfrac{9^{2x} \times (3^x)^2}{27 \times 3^{4x}} = \dfrac{1}{9}$

12. a. If $u = 3^x$, show that $9^x + 3^{x+1}$ can be written as $u^2 + 3u$.
    b. Write an expression in terms of x for
       i. $u^3 + 9u$.
       ii. $u^2 - \dfrac{1}{u}$.
    c. Write $81^x - 9^{x-1}$ in terms of u.

13. By applying the laws of indices to these expressions

    $x \div x$ | $1 \div x$ | $x^{\frac{1}{2}} \times x^{\frac{1}{2}}$

    justify the interpretation of

    x^0 | x^{-1} | $x^{\frac{1}{2}}$.

14. By the laws of indices
    $x^{\frac{2}{3}} = (x^2)^{\frac{1}{3}} = \sqrt[3]{x^2}$
    $= (x^{\frac{1}{3}})^2 = (\sqrt[3]{x})^2$.

    Use this result to simplify these expressions.
    a. $8^{\frac{2}{3}}$
    b. $64^{\frac{2}{3}}$
    c. $(27x^6)^{\frac{2}{3}}$
    d. $(16x^8)^{\frac{3}{4}}$

15. The expressions on the edges of this triangle are the product of the expressions at the vertices.

    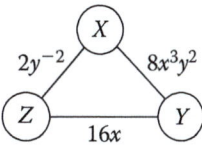

    $XY = 8x^3y^2$ etc.

    Find X, Y and Z.

SKILLS

2.3 Expanding and factorising 1

Expressions can be **expanded** by multiplying a single term over a bracket or **factorised** by finding common factors.

- To expand an expression multiply *every* term inside the brackets by the term outside the brackets.

  $3(x + 2) = 3x + 6$

- To factorise an expression find the *largest* **common factor** which will divide all the terms and take this outside using brackets.

  $4x + 2xy = 2x(2 + y)$

Expanding is the opposite of factorising.

EXAMPLE

Expand and simplify these expressions.

a $4(y + 5)$
b $p(p - 3)$
c $4b(b - 3 + 2d)$

a $4(y + 5)$
   $= 4 \times y + 4 \times 5$
   $= 4y + 20$

b $p(p - 3)$
   $= p \times p - p \times 3$
   $= p^2 - 3p$

c $4b(b - 3 + 2d)$
   $= 4b \times b - 4b \times 3 + 4b \times 2d$
   $= 4b^2 - 12b + 8bd$

Brackets can be expanded using the 'grid method'.

×	y	+5
4	4y	+20

×	p	−3
p	p^2	−3p

×	b	−3	2d
4b	$4b^2$	−12b	8bd

EXAMPLE

Expand and simplify $6(t + 2) - 5(t - 2)$

$6(t + 2) - 5(t - 2) = 6t + 12 - 5t + 10$
$= t + 22$

$-5(t - 2) = -5 \times t - 5 \times -2 = -5t + 10$

Collect like terms $6t - 5t = t$ and $+12 + 10 = +22$

Factorising is the opposite of expanding.

EXAMPLE

Factorise *fully* these expressions.

a $3x + 15$
b $p^2 - 7p$
c $12a^2bc - 30ab^2c$

a $3x + 15$
   $= 3 \times x + 3 \times 5$
   $= 3(x + 5)$
   3 is both a factor of 3x and 15

b $p^2 - 7p$
   $= p \times p - 7 \times p$
   $= p(p - 7)$
   p is both a factor of p^2 and 7p

c $12a^2bc - 30ab^2c$
   $= 2 \times 6 \times a \times a \times b \times c + 5 \times 6 \times a \times b \times b \times c$
   $= 6abc(2a - 5b)$
   6, a, b and c are all factors of both $12a^2bc$ and $30ab^2c$

You can always check your factorisation by multiplying out the brackets.

Always check that you have factorised completely:
$12a^2bc - 30ab^2c = 2abc(6a - 15b)$
but it isn't complete.

Algebra   Expressions

Fluency

Exercise 2.3S

1 Expand these expressions.
  a $4(y + 2)$
  b $6(b + 7)$
  c $7(3 + y)$
  d $12(3d + 5)$
  e $3(8 + 5t)$
  f $\frac{1}{2}(w + 10)$

2 Sandra thinks that $5(x + 4) = 5x + 4$. Work out the correct answer.

3 Expand these expressions.
  a $-4(x + 5)$
  b $-6(b + 3)$
  c $-(t + 2)$
  d $-3(8 + d)$
  e $-10(3t + 8)$
  f $-8(9 + 4w)$

4 Expand these expressions.
  a $-3(x - 5)$
  b $-2(b - 8)$
  c $-(t - 8)$
  d $-7(10 - d)$
  e $-9(2t - 9)$
  f $-8(6 - 5w)$

5 Expand these expressions.
  a $y(y + 2)$
  b $b(b - 7)$
  c $-y(y + 3)$
  d $d(5 + 2d)$
  e $-t(8 - 3t)$
  f $w(2s + 5)$

6 Karl thinks that $x(x + 4) = x^2 + 4$ for all x. Work out the correct answer.

7 Expand these expressions.
  a $y(y^2 - 2)$
  b $b^2(b - 6)$
  c $3y(y + 3)$
  d $2d(d - 5)$
  e $7t(t - 8)$
  f $9w(7 - w)$
  g $ab(a + 5b)$
  h $\frac{5t}{4}(2s + 3t)$

8 Expand and simplify these expressions.
  a $4(x + 3) + 5(x + 6)$
  b $8(y + 3) + 5(y - 3)$
  c $5(t - 2) + 8(t + 2)$
  d $9(p - 3) + 4(p - 4)$
  e $6b(b - 1) - 7b(b - 2)$
  f $2m^2(m - 3) - 7m^2(4 - m)$
  g $12(3x - 4) + 2x(3x - 7) - 5(x^2 - 15)$
  h $5r(t - 2s) - 2s(3t - 5r) + t(6s - 5r)$
  i $r(s + t) + s(t + r) + t(r + s)$
  j $r(s - t) + s(t - r) + t(r - s)$

9 For each pair of terms find their highest common factor. The first part has been done for you.
  a $12x$ and 3 Answer 3
  b $16y$ and 8
  c $35z$ and -28
  d $6p$ and 6
  e q^3 and q^2
  f $24r^2$ and $9r$
  g $18s^3t^2$ and $45s^2t^2$

10 Factorise fully these expressions.
  a $4p + 8$
  b $5y + 10$
  c $3d + 21$
  d $9k + 72$
  e $6b + 24$
  f $6w + 54$

11 Factorise fully these expressions.
  a $6b + 24bc$
  b $6w + 54wy$
  c $16ab - 40b$
  d $15q - 45p$
  e $p^2 + 8p$
  f $y + 6y^2$
  g $w + 4w^3$
  h $ab - 4ab^2$
  i $6b^2 + 24bc$
  j $12x^2yz + 36xy^2z$
  k $15mn - 5m + 10m^3$
  l $mufc + rfc$
  m $mickey + mouse$

12 Ahmed thinks that $12p + 20pq$ factorised completely is $2(6p + 10pq)$. Work out the correct answer.

13 Factorise these expressions.
  a $(p + 1)^2 + 2(p + 1)$
  b $(2q + 3)^3 + (2q + 3)^2$
  c $4(4r + 2)^2 + 8(4r + 2)$
  d $27s^2t(u + v)^4 + 18s^3(u + v)^3$

APPLICATIONS

2.3 Expanding and factorising 1

RECAP

- Expand means multiply each term inside a bracket by the term outside the bracket.
- Factorising is the 'opposite' of expanding brackets. To factorise an expression, look for the highest **common factor** for all the terms.

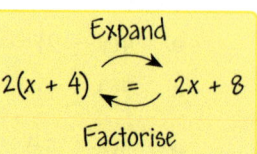

Expand
$2(x + 4) = 2x + 8$
Factorise

HOW TO

To solve a problem involving algebraic manipulation

1. Read the question carefully. Give any unknown values a letter. Decide whether to expand ...or... factorise.
2. Expand the brackets ...or... find the HCF of all of the terms and factorise.
3. Collect like terms ...or... check your answer by expanding.
4. Answer the question.

EXAMPLE

A rectangle of width x cm has length 1 cm more than the width. Its area is 192 cm².

Show that $x^2 + x = 192$.

Explain why x must lie between 13 and 14.

It often helps to sketch a diagram.

1. You are told that the length is 1 cm more than the width.
   Length = $(x + 1)$ cm
   Area = length × width
   $192 = (x + 1) \times x$
   $192 = x(x + 1)$
2. $192 = x^2 + x$
4. Substitute $x = 13$ and $x = 14$ into $x^2 + x$.
   $13^2 + 13 = 169 + 13 = 182$ < 192
   $14^2 + 14 = 196 + 14 = 210$ > 192
   x must be more than 13 and less than 14. $13 < x < 14$

EXAMPLE

A rectangle has area = $(6x^3y + 3xy)$ cm².

Write down two possible pairs of values for the length and width of the rectangle.

1. Write the area as a product of factors: area = length × width.
2. $6x^3y + 3xy = 3xy(2x^2 + 1)$ = $3 \times x \times y \times (2x^2 + 1)$
4. Length = $3xy$, width = $(2x^2 + 1)$ or   Length = 3, width = $xy(2x^2 + 1)$

There are 8 possible solutions; 16 if you distinguish between length and width.

EXAMPLE

Find the missing numbers, p and q, in this question.

$p(3x - 4) = 12x + q$

1. Expand and then compare the coefficients of x and the constant terms.
2. $3px - 4p = 12x + q$
   $3 \times p = 12$ and $-4 \times p = q$
4. $p = 4$ $\Rightarrow$ $q = -4 \times 4$
                              $= -16$

Algebra   Expressions

Reasoning and problem solving

Exercise 2.3A

1. All three students have completed their factorisations incorrectly.
   Say what they have done wrong.

   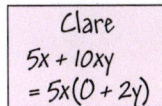
   Clare
   $5x + 10xy$
   $= 5x(0 + 2y)$

   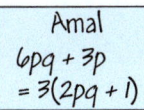
   Amal
   $6pq + 3p$
   $= 3(2pq + 1)$

   Vicky
   $21p + 14pq$
   $= 7p(14 + 7q)$

2. Factorise fully these expressions.
   a. $ax + bx + ay + by$
   b. $cd + bd + cm + bm$
   c. $a^2 + ab + 2a + 2b$
   d. $cd + ce - me - md$
   e. $a^2b + c^3d^4 + a^2d^4 + bc^3$
   f. $xy + 2x^2 + x^3 + 2y$

3. Fill in the missing numbers in these equations
   a. $7(3x - p) = qx + 28$
   b. $p(5x - 2) + q = 15x$
   c. $3(2x - p) + q(1 - 2x) = 2(x + 4)$
   d. $p(4x + 7) + 5(3x - 2) = 27x + q$
   e. $6(2x + py - 1) + qx = r(x + 3y - 3)$
   f. $2x(px + 3) + q(2x - 3) = 8x^2 + r$

4. Write a factorised expression for
   a. the perimeter of this rectangle
      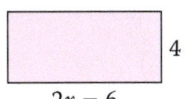
      (4, $2x - 6$)
   b. the perimeter of a regular hexagon with sides $5b + 10$.

5. The area of the rectangle is $15\,\text{cm}^2$. Show that $6(x - 3) = 0$.
   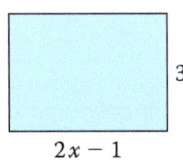
   (3, $2x - 1$)

6. The answer is $20x + 35$.
   a. Write down a possible question of the form $a(bx + c)$ where a, b and c are integers.
   b. How many possible questions can you have using one pair of backets?
   c. Write a possible question that uses the sum of two pairs of brackets.

7. a. Write an expression involving brackets for the area of this trapezium.
   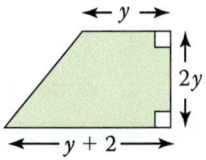
   b. Expand and simplify your expression.

8. Use factorisation to help you to evaluate these without a calculator.
   a. $2 \times 1.86 + 2 \times 1.14$
   b. $3 \times 5.87 - 3 \times 0.37$
   c. $5.86^2 + 5.86 \times 4.14$
   d. $3.32 \times 6.68 + 3.32^2$

9. Show that the shaded area of this rectangle is $2(4x + 5)$.
   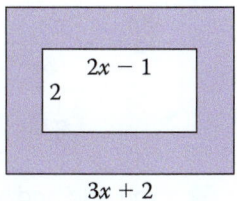

10. a. Pick a number and follow these instructions.

i	ii
Double the number. Add 4. Halve the total. Take away your original number.	Double the number. Add 1. Multiply the total by 5. Take 5 from the total. Divide the total by 10.

    b. What do you notice? Does the same thing happen if you pick a negative number, fraction or decimal?
    c. Use algebra to explain what is happening.

11. a. Expand $z(x + y)$.
    b. Substitute $z = x + y$ into your answer for part **a** and expand and simplify the resulting expression.
    c. Hence or otherwise expand $(x + y)^2$.
    d. Expand
       i. $(2x + 3)^2$
       ii. $(x + 4)(2x + 5)$.

*12. Factorise $x^2 + 7x + 12$.

13. Show that the perimeter of rectangle A is three times the perimeter of rectangle B.

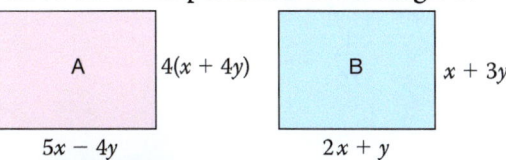

SKILLS

2.4 Algebraic fractions

- An algebraic fraction is a fraction whose **numerator** and/or **denominator** are **expressions**.
- The rules for algebraic fractions are the same as those for numerical fractions.

> Here are two algebraic fractions.
> $$\frac{x^2 + 2x}{x}$$
> $$\frac{x^2 + 3x}{x - 3}$$

Algebraic fractions can be simplified using factorisation and cancelling the terms in the denominator and numerator by common factors.

EXAMPLE

Simplify these fractions.

a $\dfrac{4x^6y^2}{8x^2y^2}$ b $\dfrac{x^2 + 2x}{x}$ c $\dfrac{x - 3}{x^2 - 3x}$ d $\dfrac{(x + 2)^2}{x + 2}$

a $= \dfrac{4x^6y^2}{8x^2y^2} = \dfrac{1}{2}x^4$

b $= \dfrac{x(x + 2)}{x} = x + 2$

c $= \dfrac{x - 3}{x(x - 3)} = \dfrac{1}{x}$

d $= \dfrac{(x + 2)(x + 2)}{x + 2} = x + 2$

- You add and subtract algebraic fractions by finding a **common** denominator.

EXAMPLE

Simplify these expressions.

a $\dfrac{7x + 3}{3} - \dfrac{4x - 5}{2}$ b $\dfrac{4}{5} + \dfrac{3}{2x + 3}$ c $\dfrac{2}{x + 5} + \dfrac{3}{x + 10}$

a $= \dfrac{2(7x + 3) - 3(4x - 5)}{3 \times 2}$
$= \dfrac{14x + 6 - 12x + 15}{6}$
$= \dfrac{2x + 21}{6}$

b $= \dfrac{4(2x + 3) + 5 \times 3}{5(2x + 3)}$
$= \dfrac{8x + 12 + 15}{5(2x + 3)}$
$= \dfrac{8x + 27}{5(2x + 3)}$

c $= \dfrac{2(x + 10) + 3(x + 5)}{(x + 5)(x + 10)}$
$= \dfrac{2x + 20 + 3x + 15}{(x + 5)(x + 10)}$
$= \dfrac{5x + 35}{(x + 5)(x + 10)}$
$= \dfrac{5(x + 7)}{(x + 5)(x + 10)}$

- When multiplying or dividing factorise and then cancel common factors.

EXAMPLE

Simplify these expressions

a $\dfrac{x}{3x + 6} \times \dfrac{x + 2}{x^2}$ b $\dfrac{3}{x} \div \dfrac{3x - 6}{x^3}$

a $= \dfrac{x}{3(x + 2)} \times \dfrac{x + 2}{x^2} = \dfrac{1}{3x}$

b $= \dfrac{3}{x} \times \dfrac{x^3}{3x - 6}$
$= \dfrac{3}{x} \times \dfrac{x^3}{3(x - 2)}$
$= \dfrac{x^2}{x - 2}$

> To do a division, turn the second fraction upside down and multiply.

Algebra   Expressions

Fluency

Exercise 2.4S

1 Simplify these fractions.
- a $\dfrac{3}{9}$
- b $\dfrac{33}{57}$
- c $\dfrac{125}{750}$
- d $\dfrac{51}{170}$
- e $\dfrac{168}{672}$
- f $\dfrac{91}{1001}$

2 Simplify these expressions.
- a $\dfrac{4s^2t^3}{12s^2t}$
- b $\dfrac{p^7q^7}{q^4r^3}$
- c $\dfrac{(10x^2y)^2}{15x^3y}$

3 Simplify these expressions.
- a $\dfrac{4x+6}{2}$
- b $\dfrac{x^2+5x}{x}$
- c $\dfrac{x+6}{x^2+6x}$
- d $\dfrac{3x+12}{x+4}$

4 David is simplifying $\dfrac{5x+35}{x+4}$

$$\dfrac{5x+35}{x+4} = \dfrac{5+35}{4} = \dfrac{40}{4} = 10$$

Do you agree with David's solution?

5 Simplify these fractions.
- a $\dfrac{x^2+2x}{x+2}$
- b $\dfrac{p^2-3p}{p-3}$
- c $\dfrac{y-5}{y^2-5y}$
- d $\dfrac{6y^2}{y^3-y^2}$
- e $\dfrac{3x^2+6x}{x^2+2x}$
- f $\dfrac{3x^3+4x}{x^2+2}$

6 Simplify these fractions.
- a $\dfrac{(y+2)^2}{y+2}$
- b $\dfrac{(x-4)^2}{x-4}$
- c $\dfrac{x+3}{(x+3)^2}$
- d $\dfrac{(p-1)^3}{p-1}$
- e $\dfrac{(y+4)^3}{(y+4)^2}$
- f $\dfrac{b-2}{(b-2)^4}$

7 Add or subtract these fractions.
- a $\dfrac{3}{8}+\dfrac{2}{8}$
- b $\dfrac{4}{7}-\dfrac{1}{7}$
- c $\dfrac{2}{3}+\dfrac{1}{2}$
- d $\dfrac{7}{9}-\dfrac{5}{6}$
- e $\dfrac{7}{18}+\dfrac{4}{27}$
- f $\dfrac{19}{24}-\dfrac{3}{56}$

8 Add or subtract these expressions.
- a $\dfrac{3p}{7}+\dfrac{2p}{7}$
- b $\dfrac{4q}{9}+\dfrac{2q}{3}$
- c $\dfrac{4}{3r}-\dfrac{2}{3r}$
- d $\dfrac{6}{s}-\dfrac{5}{s^2}$
- e $\dfrac{12}{5t}+\dfrac{7}{4t^2}$
- f $\dfrac{6}{5u}-\dfrac{5}{v}$

9 Paloma writes

$$\dfrac{2}{x}+\dfrac{4}{2x+3} = \dfrac{6}{3x+3} = \dfrac{6}{3(x+1)} = \dfrac{2}{x+1}$$

Do you agree with Paloma?

10 Simplify these expressions.
- a $\dfrac{5}{x+2}+\dfrac{2}{x}$
- b $\dfrac{3}{x}+\dfrac{5}{x-1}$
- c $\dfrac{3}{x+2}+\dfrac{2}{2x+3}$
- d $\dfrac{x}{x+2}+\dfrac{2}{x-3}$
- e $\dfrac{4}{x-2}-\dfrac{2}{x+3}$
- f $\dfrac{4}{2x-2}-\dfrac{x}{x-3}$

11 Multiply or divide these fractions.
- a $\dfrac{2}{3}\times\dfrac{4}{3}$
- b $\dfrac{5}{8}\div\dfrac{3}{8}$
- c $\dfrac{6}{11}\times\dfrac{2}{3}$
- d $\dfrac{9}{13}\div\dfrac{9}{11}$
- e $\dfrac{21}{36}\times\dfrac{27}{14}$
- f $\dfrac{19}{24}\div\dfrac{3}{56}$

12 Multiply or divide these expressions.
- a $\dfrac{7x}{2}\times\dfrac{3x}{5}$
- b $\dfrac{9x}{11}\div\dfrac{3x}{22}$
- c $\dfrac{6}{x}\times\dfrac{5}{2x}$
- d $\dfrac{4x^2y}{9z}\div\dfrac{2xy^2}{27}$
- e $\dfrac{16yz^3}{5x^2}\times\dfrac{15yx^2}{8z}$
- f $\dfrac{9y}{x^2z^2}\div\dfrac{4xz}{21y^2}$

13 Simplify these expressions.
- a $\dfrac{5}{x+2}\times\dfrac{2}{x}$
- b $\dfrac{3}{x}\times\dfrac{x}{x-1}$
- c $\dfrac{3}{x+2}\times\dfrac{3x+6}{2x+3}$
- d $\dfrac{x}{x+2}\times\dfrac{x^2+2x}{x^2}$
- e $\dfrac{x}{x+2}\div\dfrac{x}{2}$
- f $\dfrac{5y-10}{15}\div\dfrac{y-2}{3y}$
- g $\dfrac{4x+12}{7x}\times\dfrac{5x^2-10x}{6-9x}\times\dfrac{21}{2x-4}$

APPLICATIONS

2.4 Algebraic fractions

RECAP

- Algebraic fractions follow the same rules as numerical fractions.
  - To simplify a fraction cancel common factors from the numerator and denominator.
  - To add or subtract fractions use a common denominator.
  - When multiplying or dividing fractions start by factorising and then cancelling common factors.

This lesson contains material covered later in the course, so could be left until after Chapter 21 as a useful recap in algebra.

$$\frac{4x^2 + 2xy}{6x^3} = \frac{2^1 x^1 (2x + y)}{6^3 x^{3\,2}} = \frac{2x + y}{3x^2}$$

$$\frac{2}{x} - \frac{3x + 2}{4x} = \frac{2 \times 4 - 3x - 2}{4x} = \frac{6 - 3x}{4x} = \frac{3(2 - x)}{4x}$$

$$\frac{2}{x} \div \frac{3x + 2}{4x} = \frac{2}{x^1} \times \frac{4x^1}{3x + 2} = \frac{8}{3x + 2}$$

HOW TO

To simplify expressions that involve algebraic fractions

① Apply to algebraic fractions the same techniques you use when calculating with numerical fractions.

② Use your knowledge of collecting like terms and factorising to simplify the fractions.

EXAMPLE (p.434)

The first two terms of a **linear sequence** are $\frac{2}{x + 1}$ and $\frac{3}{x + 2}$. Find the third term in this sequence.

A linear sequence is one like 1, 4, 7, 10, 13, ... The difference is always +3.

Find the difference between the terms.

$$\frac{3}{x + 2} - \frac{2}{x + 1} = \frac{3(x + 1) - 2(x + 2)}{(x + 1)(x + 2)} = \frac{3x + 3 - 2x - 4}{(x + 1)(x + 2)} = \frac{x - 1}{(x + 1)(x + 2)}$$

Third term = second term + difference.

$$\frac{3}{x + 2} + \frac{x - 1}{(x + 1)(x + 2)} = \frac{3(x + 1) + x - 1}{(x + 1)(x + 2)} = \frac{3x + 3 + x - 1}{(x + 1)(x + 2)} = \frac{4x + 2}{(x + 1)(x + 2)}$$

$$= \frac{2(2x + 1)}{(x + 1)(x + 2)}$$

EXAMPLE (p.136)

The big rectangle is an **enlargement** of the small rectangle. x is a positive number.

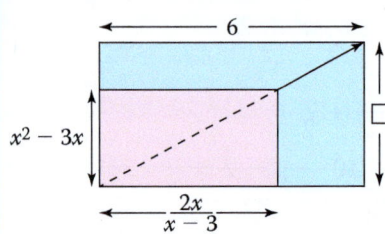

a Find the height of the big rectangle.

b Explain why x cannot be less than or equal to 3.

a Find the 'scale factor' for the enlargement.

Scale factor $= 6 \div \frac{2x}{x - 3} = \frac{6^3 \times (x - 3)}{2^1 x} = \frac{3(x - 3)}{x}$

Height $= (x^2 - 3x) \times \frac{3(x - 3)}{x} = x(x - 3) \times \frac{3(x - 3)}{x}$

$= 3(x - 3)^2$

b Width of small rectangle > 0

Therefore $x - 3 > 0$

$x > 3$

Algebra Expressions

Reasoning and problem solving

Exercise 2.4A

This exercise could be left until after Chapter 21.

1 Abbi and Bez have both got the wrong answers. Find their mistakes and give the correct answers.

a
$$\frac{6}{x-5} - \frac{3}{2x+1} = \frac{6(2x+1) - 3(x-5)}{(x-5)(2x+1)}$$
$$= \frac{12x + 6 - 3x - 15}{(x-5)(2x+1)}$$
$$= \frac{9x - 9}{(x-5)(2x+1)}$$
$$= \frac{9(x-1)}{(x-5)(2x+1)}$$

b
$$\frac{3x+6}{2x-5} \div \frac{4x-10}{x^2+2x} = \frac{3(x+2)^1 \times 2(x-5)^1}{(2x-5)^1 \times x(x+2)^1}$$
$$= \frac{6}{x}$$

2 Sort these expressions into equivalent pairs. Which is the odd one out? Create its pair.

A $\frac{5x}{12} - \frac{3x}{12}$ B $\frac{x}{6} + \frac{x}{4}$

C $\frac{2}{3}x - \frac{1}{3}x$ D $\frac{x}{3} - \frac{x}{4}$ E $\frac{x}{6}$

F $\frac{x}{12}$ G $\frac{4x^2}{12x}$

3 The first two terms of a linear sequence are $\frac{x+4}{3}$ and $\frac{x+5}{4}$.
Find the third term in the sequence.

4 Find the missing term in this linear sequence.
$\frac{2}{x+3}$, ———, $\frac{2}{2x-3}$

5 Show that these three expressions form a linear sequence.
$\frac{2}{(x+1)}$, $\frac{x+3}{(x+1)^2}$, $\frac{4}{(x+1)^2}$

6 Find the sum of these expressions.
a $\frac{x-2}{12}$ $\frac{x+1}{6}$ $\frac{x}{4}$
b $\frac{3}{x}$ $\frac{4x+5y}{xy}$ $\frac{4x}{y^2}$

7 a Find an expression for the perimeter of this rectangle.

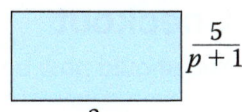

b Give reasons why p must be greater than 3.

8 The big rectangle is an enlargement of the small rectangle.

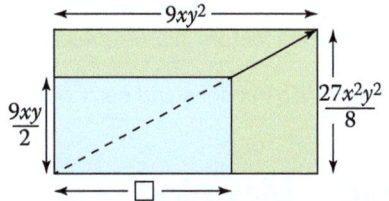

a Find the width of the small rectangle.
b Find the perimeter of the small rectangle.

9 The big rectangle is an enlargement of the small rectangle.

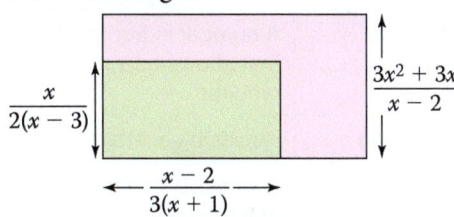

a What is the area of the big rectangle?
b Why must x not equal 2?

10 Complete this grid so that the product of the entries in each row and column is 1.

$\frac{(x+1)}{6}$		
	$\frac{3}{5}$	
$\frac{3x+3}{2}$		$\frac{10}{x+1}$

11 Simplify this expression.
$$\frac{3x+6}{x-1} \times \frac{x+4}{6x-18} \times \frac{2x-2}{5x+20} \div \frac{x+2}{5x-15}$$

12 This shape is made from three identical rectangles.

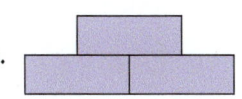

The rectangles have width $\frac{2}{(2x-3)}$ and height $\frac{3}{(5-2x)}$. What is the perimeter of the shape?

Summary

Checkout
You should now be able to...

Test it
Questions

✓ Use algebraic notation and simplify expressions by collecting like terms.	1 – 3
✓ Substitute numbers into formulae and expressions.	4, 5
✓ Use the laws of indices.	6, 7
✓ Multiply a single term over a bracket.	8
✓ Take out common factors in an expression.	9
✓ Simplify algebraic fractions and carry out arithmetic operations with algebraic fractions.	10, 11

Language	Meaning	Example
Expression	A collection of letters, numbers and operations.	$5x - 2$
Term	One of the component parts in an expression.	Terms: $5x$ and -2
Variable	An unknown quantity represented by a letter.	Variable: x
Coefficient	A number in front of a variable or unknown letter that shows how many of that variable or letter are required.	Coefficient of x: 6
Substituting	Replacing a letter with a numerical value.	If $x = 2$, $5x - 2 = 5 \times 2 - 2 = 8$
Like terms	Terms that contain exactly the same combination of variables. It is usual to collect like terms.	$5x^2 + 3x - 7x = 5x^2 - 4x$
Index/Indices **Base** **Power**	In index notation, the index or power shows how many times the base has to be multiplied by itself. The plural of index is indices.	$5^3 = 5 \times 5 \times 5$ (Power or index; Base)
Index laws	A set of rules for calculating with numbers written in index notation.	$a^m \times a^n = a^{m+n}$ $3^2 \times 3^5 = 3^7$ $a^m \div a^n = a^{m-n}$ $5^6 \div 5^2 = 5^4$ $(a^m)^n = a^{mn}$ $(2^3)^4 = 2^{12}$
Expand **Simplify**	Multiply out brackets. Collect like terms.	$5x(5 + x) + 10(x - 3)$ $= 25x + 5x^2 + 10x - 30$ $= 5x^2 + 35x - 30$
Factorise	Rewrite an expression using brackets by taking out the highest common factor.	$10xy^2 + 5x^2y = 5xy \times 2y + 5xy \times x$ $= 5xy(2y + x)$

Algebra Expressions

Review

1. In the expression
   $$7x^2 + 3x - 10$$
   a. how many terms are there
   b. what is the largest coefficient
   c. how many variables are there?

2. Simplify these expressions.
   a. $y \times 13$
   b. $y \times 7 \times x$
   c. $x + x + x$
   d. $y \times y \times y$
   e. $2 \times x \div 4$
   f. $4yx + yx$

3. Simplify these expressions.
   a. $4a - 3b + 2a$
   b. $7 - 5a - 5 + 7b$
   c. $4a + 4a^2$
   d. $3a^2b + 2a^2b - 5ab^2$
   e. $6ab + 3b - 9$
   f. $8a^2 \div 4 + 2a - 3a^2$

4. Evaluate these expressions when $x = 2$ and $y = -7$.
   a. $5x$
   b. $3y$
   c. $-2x$
   d. $-4y$
   e. $3x^2$
   f. xy
   g. $\dfrac{4x}{6}$
   h. $\dfrac{-21}{y}$

5. $v^2 = u^2 + 2as$
   a. Find v when $u = 8$, $a = 3$ and $s = 6$.
   b. Find s when $v = 12$, $u = 9$ and $a = 9$.

6. Simplify these expressions.
   a. $a^3 \times a^4$
   b. $b^8 \div b^3$
   c. $(c^3)^2$
   d. $4d^3 \times 5d^6$
   e. $8e^9 \div 2e^3$
   f. $18f^6 \times 2f^2 \div 6f^8$

7. Simplify these expressions.
   a. $x^6 \times x^4 \times x^2 \div x^7$
   b. $\dfrac{y^6 \times y^3 \times y^3}{y^5 \times y^5}$
   c. $(2z^2)^3$
   d. $(27x^2y)^0$
   e. $14u^7 \times 2u^{-3}$
   f. $10p^3 \div 2p^{-5}$
   g. $(3r^{-2})^3$
   h. $(2s^2t^{-3})^{-2}$

8. Expand the brackets in these expressions.
   a. $5(2a + 3)$
   b. $3(6b - 3c)$
   c. $-4d(8d - 2c)$
   d. $y(y + 3) - 2y(y + 1)$

9. Factorise fully these expressions.
   a. $5x^2 + 10x$
   b. $21ab^2 - 14a$
   c. $30p^2 + 15pq^2 - 45pq^3$

10. Simplify these algebraic fractions.
    a. $\dfrac{2x^2 + 4x}{8x}$
    b. $\dfrac{x^2 + x}{(x + 1)^2}$

11. Simplify these expressions involving algebraic fractions.
    a. $\dfrac{1}{a} + \dfrac{2}{a}$
    b. $\dfrac{2b}{5} - \dfrac{3b}{8}$
    c. $\dfrac{5}{6c} + \dfrac{2}{3c}$
    d. $\dfrac{3}{4d} - \dfrac{1}{2d^2}$
    e. $\dfrac{3}{d} + \dfrac{5}{f}$
    f. $\dfrac{3a}{b} \times \dfrac{5b}{3a^2}$
    g. $\dfrac{3a^2}{b} \div \dfrac{9a}{b^2}$
    h. $\dfrac{a}{a+1} - \dfrac{a}{a+2}$

What next?

Score		
	0 – 4	Your knowledge of this topic is still developing. To improve, look at MyiMaths: 1033, 1045, 1064, 1149, 1151, 1155, 1164, 1178, 1179, 1186, 1247, 1301
	5 – 9	You are gaining a secure knowledge of this topic. To improve your fluency look at InvisiPens: 02Sa – g
	10 – 11	You have mastered these skills. Well done you are ready to progress! To develop your problem solving skills look at InvisiPens: 02Aa – f

Assessment 2

1. The formula for the curved surface area of a cone is $A = \pi r l$, where r is the radius of the base and l is the slant height.

   a. Find A for a cone with base radius 5 cm and slant height 10 cm. [2]

   b. Find r when $A = 45\,\text{m}^2$ and $l = 4\,\text{m}$. [2]

   c. Find l when $A = 126.4\,\text{inches}^2$ and $r = 12.3\,\text{inches}$. [2]

2. The area of a trapezium is given by the formula $A = \dfrac{(a+b) \times h}{2}$ where a and b are the parallel sides and h is the height.

   Find the area of the trapezium where $a = 2z^2$, $b = 3z^2$ and $h = 4z$. [3]

3. Sasha divides the square STUV into four rectangles SADC, ATED, DEUB and CDBV.

   a. Sasha says that in simplest form

   Area of SADC $= 4y$
   Area of ATED $= 28$
   Area of DEUB $= 7y$
   Area of CDBV $= y^2$

   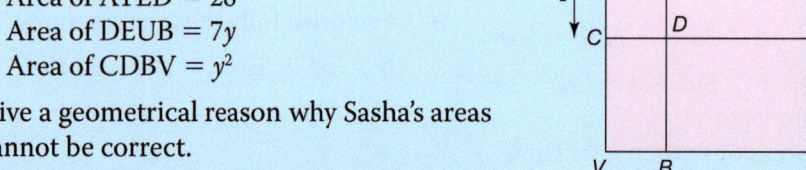

   Give a geometrical reason why Sasha's areas cannot be correct. [3]

   b. Give the correct area of the rectangle CDBV, in its simplest form. [3]

4. You can measure the height of a cliff, H m, by dropping a stone from the top and timing it until it hits the sea. The formula is $H = 5t^2$ where t is the time in seconds.

   a. Use this formula to find the height of the cliff if you hear the splash after

      i   4 seconds [2]   ii   7 seconds. [2]

   b. A cliff is 320 m high.
   How long will it take to hear the splash if you drop a stone from the top? [2]

5. Bengt tried to factorise the expressions below.

   For each expression, decide if Bengt

   - factorised the expression completely
   - factorised the expression, but not completely
   - factorised the expression incorrectly.

   In the last two cases, give the correct answer.

   a. $vwx + xyz - vxz = x(vw + yz - vz)$ [1]

   b. $a^2bc^3 - ab^4c^2 + a^5b^3c^4 = ab(ac^3 - b^3c^2 + a^4b^2c^4)$ [2]

   c. $12p^3q^2r^9 - 18p^3q^5r^5 - 30p^5q^7r^4 = p^3q^2r^4(12r^5 - 18q^3r - 30p^2q^3r)$ [3]

   d. $14pq + 21q^2 - 56pq^2 = 7q(2p + 3q - 8pq)$ [2]

   e. $g^3 + g^2 - g = g(g^2 + g + 1)$ [2]

   f. $g^3 + g^2 - g + 1 = g(g^2 + g - 1)$ [2]

   g. $4(p - q) - 6(p - q)^2 = (p - q)(4 - 6(p - q))$ [2]

   h. $3(y + 2z)^2 + 9(y + 2z)^3 = 3(y + 2z)^2(3y + 6z)$ [3]

   i. $4(x^2 - 3x + 2) - 6(x^2 - 3x + 2) = -2(x^2 - 3x + 2)$ [3]

Algebra   Expressions

6 Romeo buys Dara a gift. It is a cuboid with a square base of side y cm and height 20 cm.
  a Calculate, in terms of y, the total surface area of the cuboid in its simplest form. [3]
  b The cuboid has volume 200 cm³. Romeo says that the *exact* value of y is 10.
  Is he correct? Give your reasons. [3]

7 Wanda is W years old.
  a Wanda has twin brothers who are 5 years less than twice Wanda's age.
    i Find an expression for the total of the ages of the three children, giving your answer in its simplest form. [3]
    ii Factorise your answer to part **i**. [1]
  b Wanda's dad's age is three years more than four times Wanda's age.
  Her mum is 2 years younger than her dad.
    i Find an expression for the sum of her parents' ages, giving your answer in its simplest form. [4]
    ii Factorise your answer to part **i**. [1]

8 An L-shaped room has width w, and a floor area of $w(3w - 1)$.
Greta draws a plan of the room.
Find the values of x and w. [7]

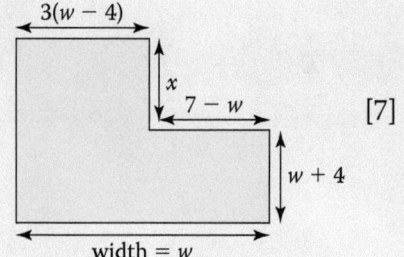

9 A pyramid has a rectangular base with width $2p$ and length $5p$. Two of the triangular sides have area $2p(3p + 2)$ each and the other two have area $5p(p - 3)$ each.
  a Find a simplified expression for the total surface area of the pyramid. [4]
  b Factorise your answer to part **a**. [1]

10 Simplify these fractions.
  a $\dfrac{55}{99}$ [1]   **b** $\dfrac{55}{100}$ [1]   **c** $\dfrac{26}{36}$ [1]   **d** $\dfrac{4q}{10}$ [1]
  e $\dfrac{15x^2}{3x}$ [1]   **f** $\dfrac{8pq}{16q^2}$ [1]   **g** $\dfrac{(3ab)^4}{6a^2b^4}$ [2]   **h** $\dfrac{2z-8}{12}$ [2]

11 Write each of the following expressions as a single fraction in its simplest form.
  a $\dfrac{z-3}{6} + \dfrac{2z+1}{8}$ [4]   **b** $\dfrac{7-2y}{5} - \dfrac{6y-2}{7}$ [4]
  c $\dfrac{4}{x+1} + \dfrac{3}{x+2}$ [4]   **d** $\dfrac{7}{2p+3} - \dfrac{4}{3p-2}$ [4]
  e $\dfrac{2a}{a-2} + \dfrac{3a}{2a+9}$ [4]   **f** $\dfrac{3}{(2x+1)^2} - \dfrac{4}{(2x+1)^3}$ [3]

12 **a** Find an expression for the perimeter of this triangle.
  Write your answer in its simplest form. [5]
  b The base is 5 metres long. Find the value of w. [2]
  c Use this value of w to find the lengths of the other two sides. [3]

3 Angles and polygons

Introduction
Tiling is a fascinating topic that is highly mathematical, involving angles and shapes. There are some wonderful tiling patterns to be found in architecture, particularly in Islamic palaces and mosques. The tiling shown in this picture is from the Alhambra Palace in Granada, Spain.

What's the point?
An understanding of angles and shapes allows us to create beautiful things.

Objectives
By the end of this chapter, you will have learned how to …
- Use angle facts including at a point, on a line, at an intersection and for parallel lines.
- Use bearings to specify directions.
- Identify types of triangle and quadrilateral and use their properties.
- Identify and use symmetry.
- Identify congruent shapes and use congruence to prove geometric results.
- Identify similar shapes and use similarity to find lengths and areas.
- Use the properties of polygons including interior and exterior angles for regular polygons.

Check in

1. Draw a grid with axes from −3 to +5.
   Plot these points on the grid.

   a (1, 3)   **b** (3, −2)   **c** (4, 5)   **d** (5, 2)

   e (−3, 3)   **f** (0, 4)   **g** (−1, 0)   **h** (−3, −2)

2. The diagrams show angles on a straight line or at a point.
   Work out the missing angles.

   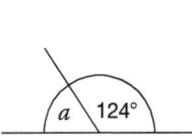 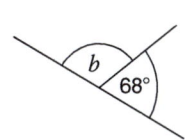 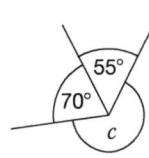 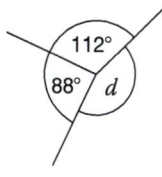

3. Work out the missing angle in these shapes.

   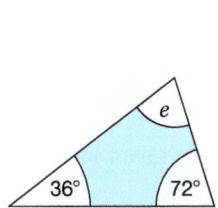 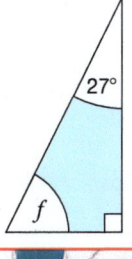 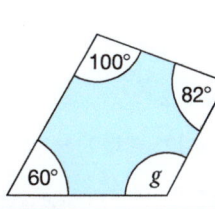 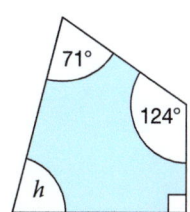

Chapter investigation

A quadrilateral has been drawn on a 3 × 3 square dotty grid.

How many different quadrilaterals can you find?

What if the grid was extended to 4 × 4?

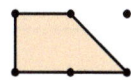

43

SKILLS

3.1 Angles and lines

A **right angle** is 90°. An **obtuse** angle is more than 90° but less than 180°.

An **acute** angle is less than 90°. A **reflex** angle is more than 180° but less than 360°.

- Angles at a point add up to 360°.
- Angles at a point on a straight line add up to 180°.
- Vertically opposite angles are equal.

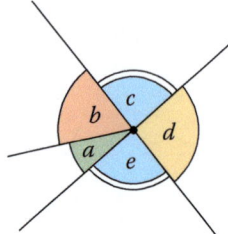

◀ Angles at a point
$a + b + c + d + e = 360°$

◀ Angles on a straight line
$a + b + c = 180°$, $d + e = 180°$
$e + a + b = 180°$ and $c + d = 180°$

◀ Vertically opposite angles
$c = e$ and $a + b = d$

EXAMPLE

Work out the values of x, y and z.
Give reasons for your answers.

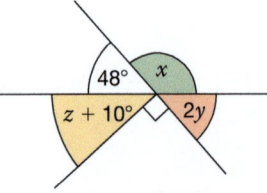

$x = 180° - 48°$ Angles on a straight line add to 180°.
$ = 132°$

$2y = 48°$ Vertically opposite angles are equal.
$y = 24°$

$z + 10° = 180° - 90° - 48°$ Angles on a straight line add to 180°.
$z + 10° = 42°$
$z = 32°$

There is often more than one way to find an angle.
For y, $2y + 132° = 180°$
For z, $z + 10° + 90° = 132°$

Perpendicular lines meet at a right angle.

Parallel lines are the same distance apart everywhere along their length.

Arrows show that lines are parallel.

A **transversal** is a line that crosses parallel lines.

- **Alternate** angles are equal.
- **Corresponding** angles are equal.
- **Interior** angles add up to 180°.

Angles that add up to 180° are said to be **supplementary**.

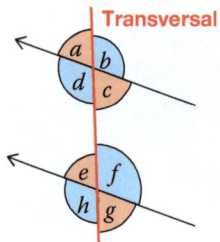

Transversal

◀ Alternate angles
$d = f \qquad c = e$
Look for a Z or Σ shape.

◀ Corresponding angles
$a = e \quad b = f \quad c = g \quad d = h$
Look for a F or ⊣ shape.

◀ Interior angles
$d + e = 180° \qquad c + f = 180°$
Look for a ⊏ or ⊐ shape.

EXAMPLE

Find angle *EYB* and angle *BYF*, giving reasons for your answers.

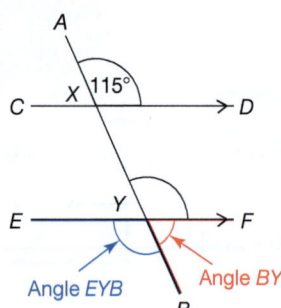

Angle $XYF = 115°$ Corresponding angles are equal.

Angle $EYB = 115°$ Vertically opposite angles are equal.

Angle $BYF = 180° - 115°$ Angles on a straight
$ = 65°$ line add to 180°.

Geometry Angles and polygons

Fluency

Exercise 3.1S

1 Work out the angles marked by letters. Give a reason for each answer. State whether each answer is acute, obtuse or reflex.

   a 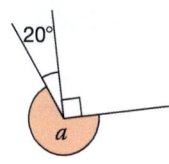   **b**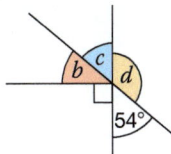

2 Calculate the value of each letter, giving reasons for your answers.

   a 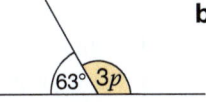   **b**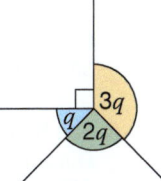

   c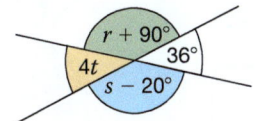

3 a Find the unknown angles. Give a reason for each answer.

   i   **ii**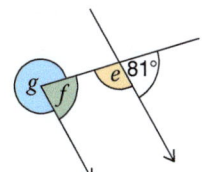

  b Write down the letters that represent
   i acute angles   **ii** obtuse angles
   iii reflex angles
   iv pairs of supplementary angles.

4 Find the value of each letter. Give reasons for your answers.

   a   **b**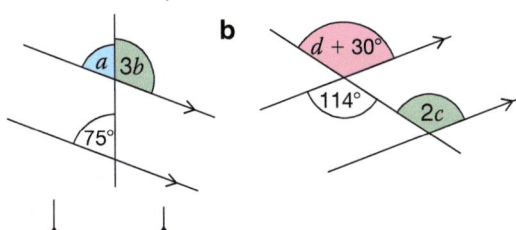

   c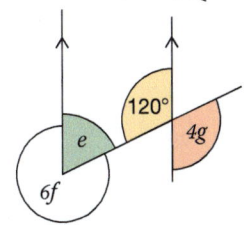

5 Find the size of each angle marked with a letter. Give a reason for each answer.

   a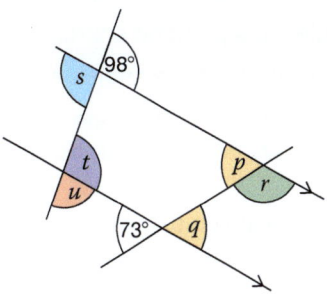

   b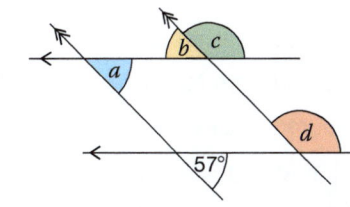

6 In the diagram, BC is perpendicular to CD and angle $ABC = 24°$.

Find angle CDE.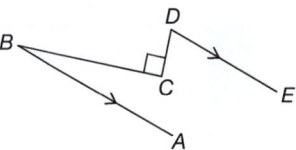

> Consider adding another parallel line through C.

7 Angle $PQR = 138°$ and angle $RST = 152°$.

Find angle QRS.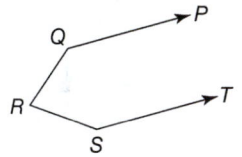

*8 The transversal XY cuts the parallel lines PQ and RS at M and N respectively.

Angle $QMN = 74°$.

Find angle MNR, angle MNS and angle YNR, giving reasons for your answers.

*9 AB is parallel to CD.

EH cuts AB at F and CD at G.

Angle $FGD = 96°$.

Find the following angles, giving reasons for your answers.

  a angle EFB **b** angle GFB
  c angle AFG **d** angle CGH
  e angle DGH

1082, 1086, 1109

APPLICATIONS

3.1 Angles and lines

RECAP
- Angles at a point add up to 360°.
- Angles at a point on a straight line add up to 180°.
- Vertically opposite angles are equal.
- For parallel lines, alternate angles are equal, corresponding angles are equal and interior angles add to 180°.

Compass rose: North, N; North-west, NW; North-east, NE; West, W; East, E; South-west, SW; South-east, SE; South, S.

Directions can be given using compass points or using angles.

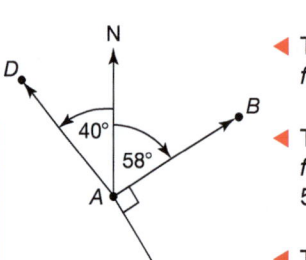

◀ The bearing of B from A is 058°.

◀ The bearing of C from A is 58° + 90° = 148°.

◀ The bearing of D from A is 360° − 40° = 320°.

- A three-figure **bearing** is
  - measured from North
  - measured clockwise
  - a 3-figure angle.

"Put the north line at the point where the bearing is from."

HOW TO

To solve problems involving angles and parallel lines
1. Draw a sketch. Include any angles that you know or are given.
2. Look for parallel lines or places where angles meet at a point or on a line.
3. Use rules from the Skills section to find the angle(s) you need.

EXAMPLE

Find the angle between the hands of a clock at 17:15.

2. The 12 equal angles at the centre add up to 360°.
   $x = 360° ÷ 12 = 30°$ Angles at a point add to 360°.
3. Angle $= 2x + \dfrac{x}{4}$ 15 minutes is quarter of an hour.
   $= 60° + 7.5°$ 17:15 is quarter past five.
   $= 67.5°$

EXAMPLE

The bearing of a ship from a lighthouse is 245°.

a   Find the bearing of the lighthouse from the ship.

b   The ship moves so that its bearing from the lighthouse increases. What happens to the bearing of the lighthouse from the ship?

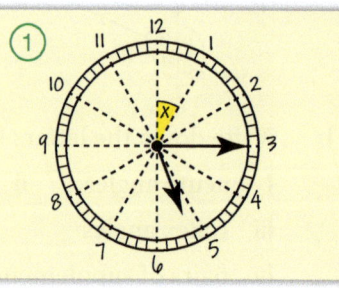

1. For a bearing *from* the ship, draw a north line at S.
   $x = 245° − 180° = 65°$ 2. Angle on a straight line is 180°.
   $y = x = 65°$ 3. Alternate angles are equal.
   Bearing is 065°.

b   The bearing of the lighthouse from the ship increases by the same amount.   Increase in x = increase in y.

Geometry   Angles and polygons

Reasoning and problem solving

Exercise 3.1A

1   Four swimmers set out from a boat as shown.

    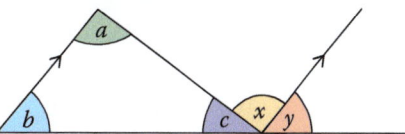

    a   Write each direction as a three-figure bearing.

    b   Neil then turns anti-clockwise to swim east. Work out the angle he turns through.

2   a   Find the angle between the hands of a clock at each of these times.

        i   7 o'clock   ii   half past one
        iii 10:45   iv   20:25

    b   Find the angle turned through between 03:10 and 11:24 by

        i   the minute hand
        *ii the hour hand.

3   a   The bearing of Aton from Barnum is 054°. Find the bearing of Barnum from Aton.

    b   The bearing of Padsey from Mulfield is 296°. Find the bearing of Mulfield from Padsey.

    c   The bearing of a plane from an airfield is 167°. On what bearing should the plane fly to go to the airfield?

    d   A ship leaves a port on a bearing of 195°. The captain is told to return to the port. On what bearing must the ship travel?

4   The diagram shows three ports P, L and Y and the capital C of an island.

    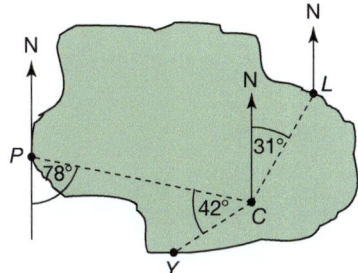

    a   Find the bearing of each port from the capital.

    b   Find the bearing of the capital from each port.

5   A radar station, R, is due west of another radar station, S.

    The bearing of a plane, P, from R is 162°.

    The bearing of P from S is 205°.

    On which bearings should the plane travel to go to S followed by R?

6   A boat sails north-east from a port P to a buoy B. Then the boat sails on a bearing of 298° to a lighthouse L due north of P.

    Find the bearings on which the boat needs to travel

    a   to return directly from L to P

    b   to retrace its journey from L to B, then from B to P.

7   The diagram shows a triangle with angles a, b and c. One side of the triangle is extended and a parallel line added.

    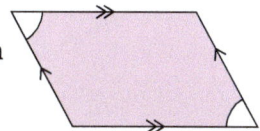

    Use this diagram to prove that the angles of a triangle add up to 180°.

8   In parallelogram $PQRS$, angle $P = 64°$.

    a   Use the rules for parallel lines to find the other angles.

    b   Write down two general rules that apply to the angles of a parallelogram.

*9 Jan says that adding 180° to the bearing of A from B gives the bearing of B from A.

    a   Find the range of bearings for which this is true.

    b   Find another rule that works for the other bearings.

    c   Explain how this shows that when the bearing of A from B increases, the bearing of B from A also increases by the same amount.

10 Prove that the opposite angles in a parallelogram are equal.

SKILLS

3.2 Triangles and quadrilaterals

- The sum of the angles of a triangle = 180°.
- The **exterior angle** of a triangle = the sum of the **interior opposite angles**.
- The sum of the angles of a quadrilateral = 360°

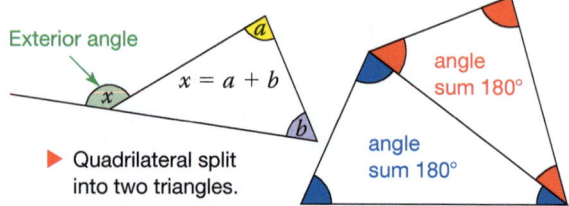

▶ Quadrilateral split into two triangles.

Triangles and quadrilaterals are **two-dimensional** (2D) shapes – they lie in a **plane**.

Triangles
In a **scalene** triangle, the sides and angles are all different.
An **isosceles** triangle has 2 equal sides and 2 equal 'base' angles.
An **equilateral** triangle has 3 equal sides. Each angle is 60°.

Quadrilaterals
In a **parallelogram** both pairs of opposite sides are parallel.
A **trapezium** has only 1 pair of parallel sides.
A **rhombus** is a parallelogram with 4 equal sides.
A **kite** has 2 pairs of equal adjacent sides.
A **rectangle** is a parallelogram whose angles are all right angles.
A **square** is a rectangle with 4 equal sides.

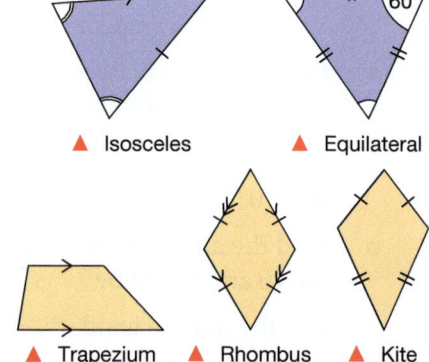

▲ Isosceles   ▲ Equilateral

▲ Trapezium   ▲ Rhombus   ▲ Kite

You should always give the most specific name that you can for a shape.

EXAMPLE

a Find **i** ∠ACB   **ii** ∠ADC

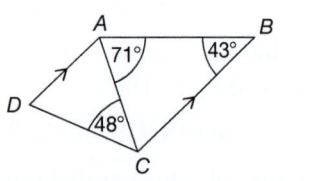

b What special type of triangle is ADC?

a **i** ∠ACB = 180° − 71° − 43° = 66°
Angle sum of triangle ABC = 180°.

ii ∠DAC = 66° Alternate angles are equal.
∠ADC = 180° − 48° − 66° = 66°
Angle sum of triangle ADC = 180°.

b Triangle ADC is isosceles as it has 2 equal angles.

EXAMPLE

a Find **i** angle DEF
          ii angle EFC
          iii angle CBD
          iv angle ABD.

b What special type of quadrilateral is ABCD?

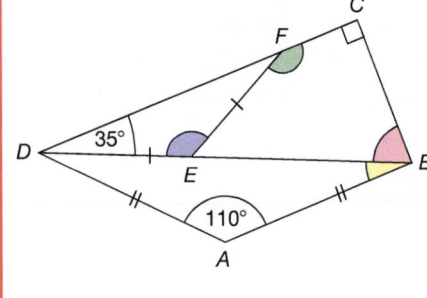

a **i** Triangle DEF is isosceles, so angle EFD = 35°.
∠DEF = 180° − 35° − 35° = 110°
Angle sum of isosceles triangle with equal base angles.

ii ∠EFC = 35° + 110° = 145°
Exterior angle of triangle EFD = sum of interior opposite angles.

iii ∠CBD = 180° − 90° − 35° = 55°
Angle sum of triangle CBD = 180°

iv Triangle ABD is isosceles, so the base angles are equal.
∠ABD = ½(180° − 110°) = 35°
Angle sum of isosceles triangle with equal base angles.

b ∠ABC = ∠ABD + ∠CBD
     = 35° + 55°
     = 90°
DC is parallel to AB so ABCD is a trapezium.

Geometry   Angles and polygons

Fluency

Exercise 3.2S

1 a Find the lettered angles.

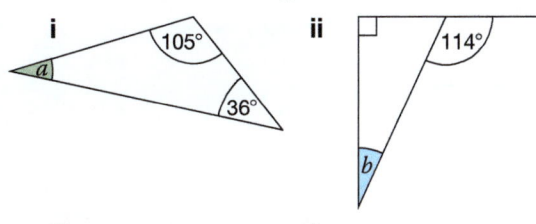

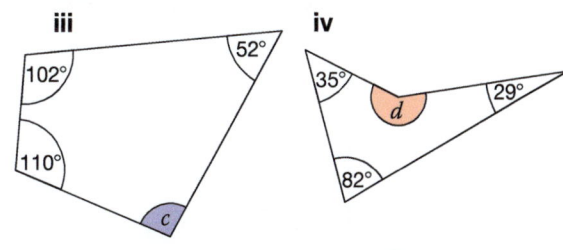

b Write down the letter that marks
  i an obtuse angle
  ii a reflex angle.

2 Find the angles marked by letters.

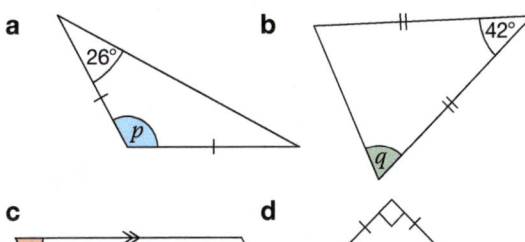

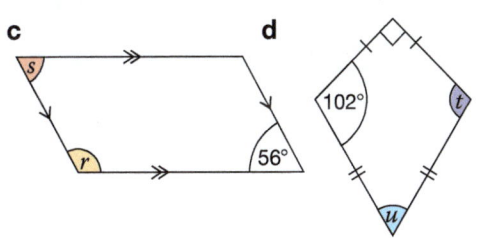

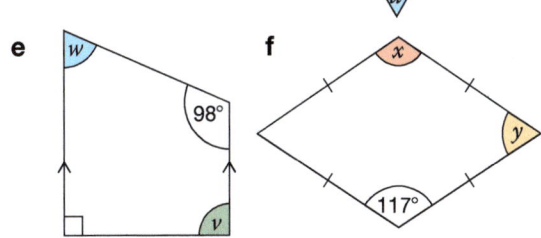

3 Work out the angles marked by letters.

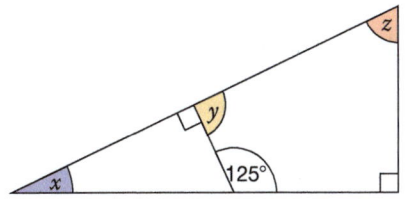

4 Find the value of x, y and z.

a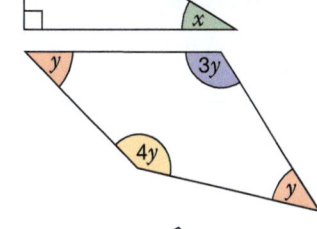

b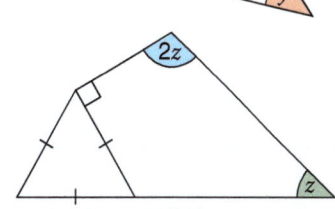

c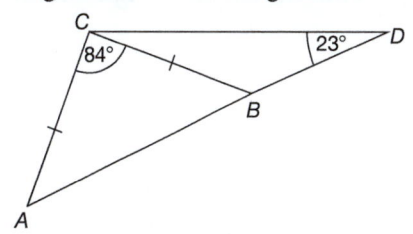

5 a Find the following angles. Give reasons.
  i angle ABC **ii** angle BCD

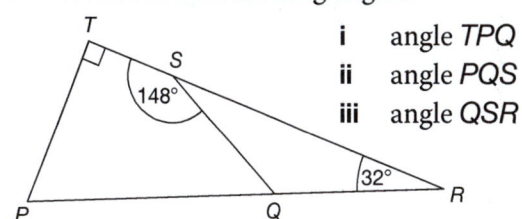

b Is triangle BCD isosceles? Give a reason.

6 a Work out the following angles.
  i angle TPQ
  ii angle PQS
  iii angle QSR

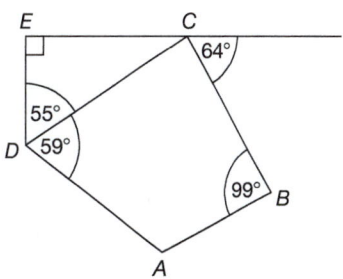

b What special type of triangle is QRS?

7 a Find the missing angles of quadrilateral $ABCD$.

b What special type of quadrilateral is it?

1080, 1102, 1130, 1141

APPLICATIONS

3.2 Triangles and quadrilaterals

RECAP
- The sum of the angles of a triangle = 180°.
- The exterior angle of a triangle equals the sum of the interior opposite angles.
- The sum of the angles of a quadrilateral = 360°

▲ Engineers often work with triangles and quadrilaterals in their designs.

HOW TO

To solve a problem involving triangles, quadrilaterals and parallel lines
① When appropriate, sketch a diagram or use the one given. Mark (or look for) known angles, equal or parallel sides.
② Identify parallel lines, angles on a line, special shapes, etc.
③ Use geometrical properties to answer the questions.

EXAMPLE

In quadrilateral ABCD, angle A is 16° more than angle B, angle B is 16° more than angle C and angle C is 16° more than angle D.

Find the smallest angle and the largest angle in this quadrilateral.

Not all problems need a diagram. It would be difficult to draw a good sketch of this quadrilateral.

Let $\angle D = x$.   D is the smallest angle.
Then $\angle C = x + 16°$, $\angle B = x + 32°$ and $\angle A = x + 48°$
② $x + 48° + x + 32° + x + 16° + x = 360°$ Angle sum of a quadrilateral is 360°.
$4x + 96° = 360°$
$4x = 264°$
$x = 264° \div 4 = 66°$

To check the answer, see if all the angles add to 360°.

③ The smallest angle $D = 66°$.
The largest angle $A = 66° + 48° = 114°$.

EXAMPLE

Tao writes '$x + y = 98°$ Corresponding angles'
a Find the correct value of $x + y$. Give reasons.
b What mistake has Tao has made?

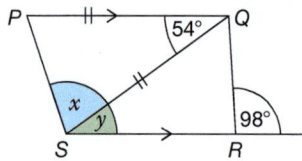

a ② PQ is parallel to SR.   PQS is an isosceles triangle.
③ $y = 54°$ Alternate angles.
$x = \dfrac{180° - 54°}{2} = 63°$ Equal base angles in isosceles triangle.
$x + y = 63° + 54° = 117°$

b ② Tao thought $x + y$ and 98° were corresponding angles.
③ Tao thought that PS was parallel to QR.

This is the **converse** rule: If corresponding angles are equal, then the lines are parallel.

Geometry   Angles and polygons

Reasoning and problem solving

Exercise 3.2A

1 Find the angles marked by letters.

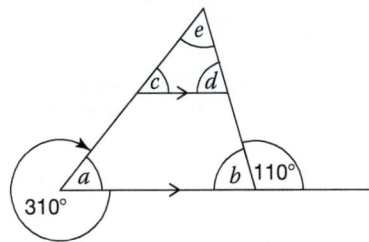

2 Find the angles marked by letters.

a

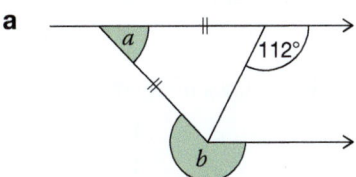

b

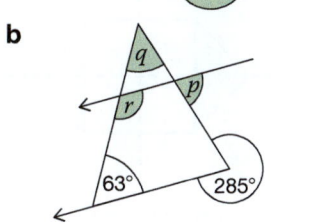

3 Find angle *RST*.

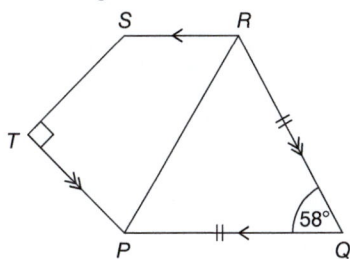

4 In triangle *ABC*, angle *A* is 25° more than angle *B* and angle *B* is 25° more than angle *C*.

Find the angles of this triangle.

5 In quadrilateral *PQRS*, angle *P* is 10° less than angle *Q*, angle *Q* is 10° less than angle *R* and angle *R* is 10° less than angle *S*.

Find all the angles in this quadrilateral.

6 Explain why a triangle cannot have more than one obtuse angle.

7 a Anna says a square is a type of rhombus.
   b Felicity says a square is a type of rectangle.
   c Malha says a square is actually a type of parallelogram.
      Discuss which of these statements are correct.

8 a A parallelogram has vertices at (0, 1), (2, 1) and (3, 2).
      Find the coordinates of the other vertex.
   b A kite has vertices at (2, 7), (4, 6) and (5, 7).
      Find the coordinates of the other vertex.
   c A rhombus has vertices at (4, 5), (6, 6) and (8, 5).
      Find the coordinates of the other vertex.

9 One angle of a kite is 130° and another angle is 50°.

Show that the other angles must both be 90°.

*10 *ABC* is an equilateral triangle.
    BC is extended to a point *D* and *CDEA* is a parallelogram.

Find the angles of this parallelogram.

*11 *PQRS* is a square. *PTS* is an equilateral triangle.

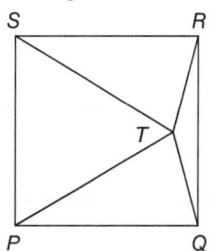

Calculate angle *QTR*.

12 Describe the triangles you can make by cutting these quadrilaterals along one or more diagonals.
   a kite   b rhombus
   c rectangle   d square

13 Describe the shapes that are made by joining the mid-points of the sides of each of the following quadrilaterals.
   a a square   b a rectangle
   c a rhombus   d a kite
   e a parallelogram
   f a quadrilateral in which all the angles are different.

SKILLS

3.3 Symmetry

p.140
- A 2D shape has **reflection symmetry** about a line through its centre if reflecting it in that line gives an identical-looking shape.
- A shape has **rotation symmetry** if rotating it through an angle greater than 0° and less than 360° about its centre gives an identical-looking shape. The **order of rotation symmetry** is the number of ways a tracing of the shape would fit on top of it as the tracing is rotated through 360°.

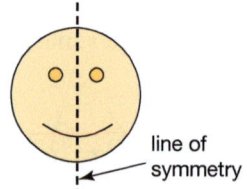
line of symmetry

Shapes that are not symmetrical have rotation symmetry of order 1 because a rotation of 360° always produces an identical-looking shape.

EXAMPLE

Describe the symmetry of these patterns.

a 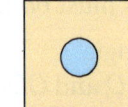 b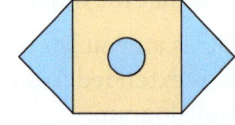

a 4 lines of symmetry 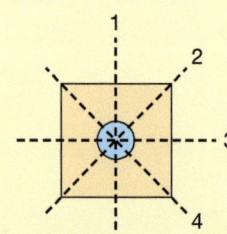 and rotation symmetry, order 4

b 2 lines of symmetry 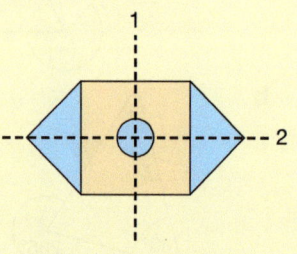 and rotation symmetry, order 2

p.218

The diagrams show the names of parts of a circle.

Every diameter of a circle is a line of symmetry.

The order of rotation symmetry is infinite.

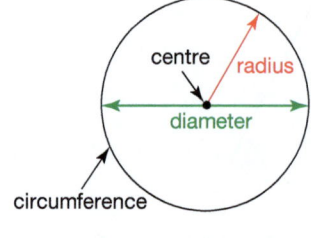
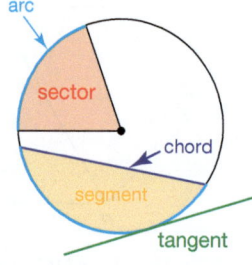

EXAMPLE

The diagram shows a tangent AB touching a circle at point P.

Draw and describe the line of symmetry.

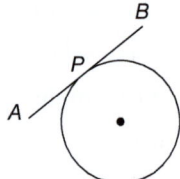

The line of symmetry is the diameter through P.

Or say the line of symmetry is the line through P at 90° to the tangent.

Geometry   Angles and polygons

Exercise 3.3S

1. For each shape write down
   i   the order of rotation symmetry
   ii the number of lines of symmetry.

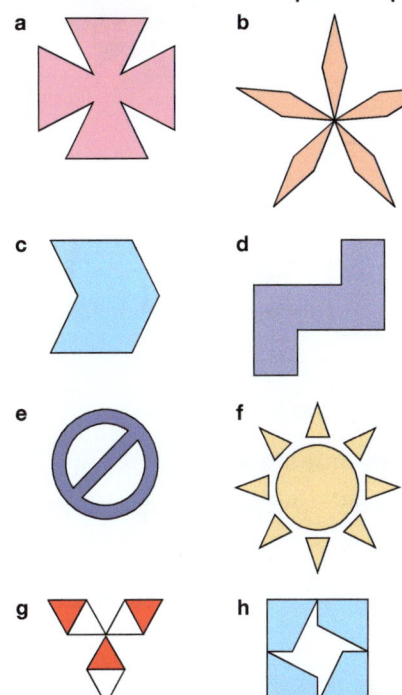

2. In each part, describe the symmetry.

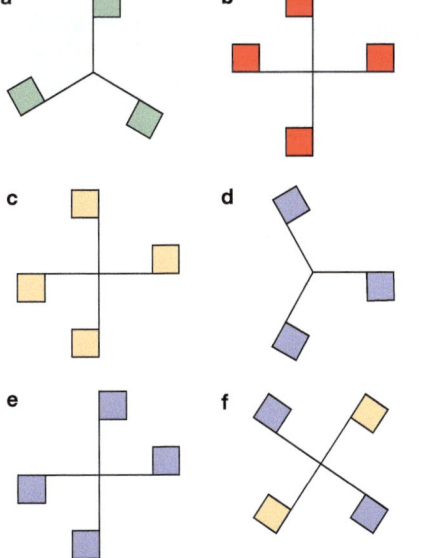

3. a Copy and complete this table.

Shape	Number of lines of symmetry	Order of rotation symmetry
Equilateral triangle		
Isosceles triangle		
Scalene triangle		
Rectangle		
Square		
Parallelogram		
Rhombus		
Kite		
Trapezium		

   b List the quadrilaterals in the table for which each statement below is always true.

   i   Both pairs of opposite sides are equal.
   ii   Both pairs of opposite angles are equal.
   iii Both diagonals are lines of symmetry.
   iv   The diagonals are equal.
   v   The diagonals cross at right angles.
   vi   The diagonals cut each other in half.

4. The diagrams show parts of a circle. In each case describe the symmetry.

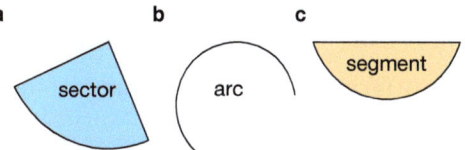

5. Describe the symmetry of each company logo.

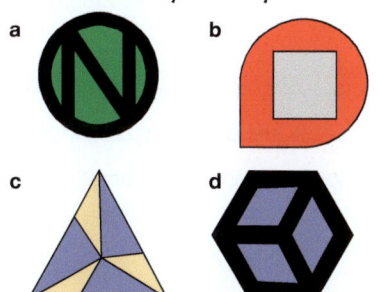

APPLICATIONS

3.3 Symmetry

RECAP (p.140)
- 2D shapes can have reflection symmetry and/or rotation symmetry.
- The order of rotation symmetry is the number of ways a tracing of the shape would fit on top of it as the tracing is rotated through 360°.

Union Flag

Tile

Road sign

HOW TO To solve a symmetry problem
1. Draw a diagram (unless one is given).
2. Consider reflection and/or rotation symmetry.
3. Answer any questions.

EXAMPLE

Two angles of this **isosceles trapezium** are equal to 48°.

Find x.

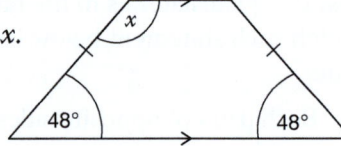

② The trapezium has reflection symmetry.
The other unknown angle is also x.
$2x = 360° - 48° - 48°$ Angle sum of quadrilateral = 360°.
$= 264°$
$x = 264° \div 2$
$= 132°$
The other angles are both 132°.

> An **isosceles** trapezium has 2 equal sides and 2 pairs of equal angles.

EXAMPLE

On a copy of this diagram, shade 4 squares so the result has rotation symmetry of order 2, but no lines of symmetry.

There are many possible answers. You can use tracing paper to check.

EXAMPLE (p.278)

$(3, 2)$ and $(-4, -1)$ are corners of a 2D shape. The line $y = -1$ is a line of symmetry.

a Write down the coordinates of the other corners.   **b** What special type of shape is this?

a
1. Draw axes and plot the given points.
2. Draw the line $y = -1$ through points that have a y coordinate of -1.
   Use symmetry in the line to find the other corner.
3. There is just one other corner at $(3, -4)$.

b The shape is an isosceles triangle.

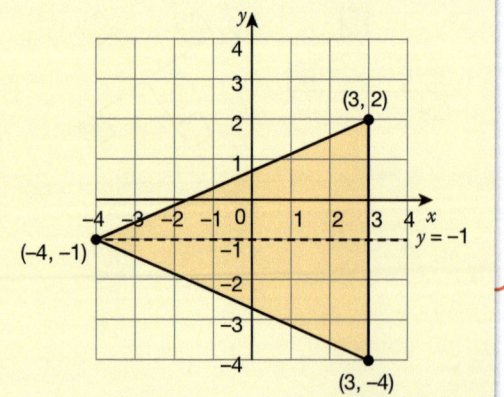

Geometry   Angles and polygons

Reasoning and problem solving

Exercise 3.3A

1. Describe the symmetry in these road signs.

   a 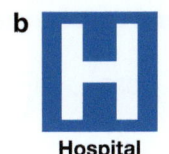 No stopping b H Hospital c  Ice or snow

2. Describe the symmetry in the following flags.

   Suriname   UK

3. Copy and complete each shape so the dotted line is a line of symmetry.

   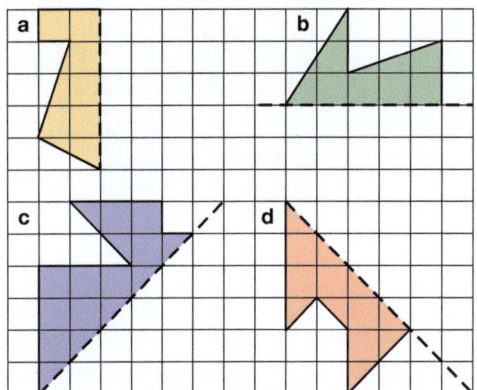

4. a Which type of triangle has 1 line of symmetry, but no rotation symmetry?

   b Name a special quadrilateral that has rotation symmetry of order 2, but no lines of symmetry.

5. Copy and complete these crossword patterns so that they have rotation symmetry of order 2.

   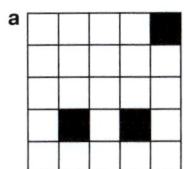   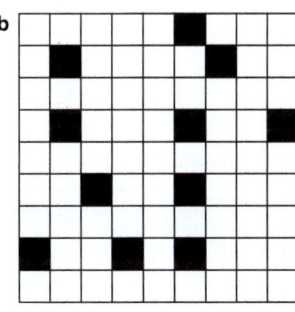

6. a Copy the diagram and shade squares so the result has rotation symmetry order 2, but no lines of symmetry.

   b Shade squares on another copy of the diagram so the result has one line of symmetry, but no rotation symmetry.

7. a On a copy of this diagram, shade triangles so the result has one line of symmetry, but no rotation symmetry.

   b Alter another copy so the result has rotation symmetry order 2, but no lines of symmetry.

8. a On a copy of the diagram, shade 4 trapeziums so the result has rotation symmetry of order 2 and 2 lines of symmetry.

   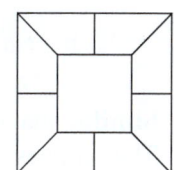

   b On another copy of the diagram, shade 4 trapeziums so the result has rotation symmetry of order 4, but no lines of symmetry.

9. Points $(-2, 3)$ and $(2, -1)$ are corners of a 2D shape. The line $x = 1$ is a line of symmetry.

   a Find the coordinates of the other corners.

   b What special type of shape is this?

10. Points $(-2, 4)$ and $(-4, -2)$ are corners of a shape. The x and y axes are both lines of symmetry.

    a Draw the shape and write down the coordinates of the other corners.

    b Write down the order of rotation symmetry of the shape.

    c Write down the equations of the other lines of symmetry.

11. Points $(-3, 2)$ and $(3, 0)$ are the diagonally opposite corners of a square. Draw the square and write down the coordinates of the other two corners.

*12 Draw a shape for which the lines $y = x$ and $x + y = 3$ are both lines of symmetry.

SKILLS

3.4 Congruence and similarity

Congruent shapes are exactly equal in size and shape: equal sides and equal angles.

You can prove that two triangles are congruent by showing that they have

- Three equal sides (SSS)
- Two sides and the angle *between them* equal (SAS)
- Two angles and a **corresponding** side equal (ASA)
- A right angle, the hypotenuse and another side equal (RHS)

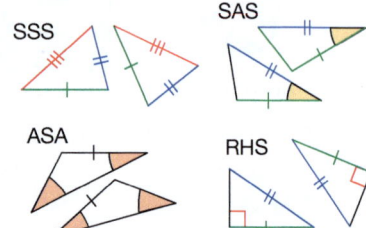

EXAMPLE

a Is triangle *A* congruent to triangle *B*?
b Is triangle *X* congruent to triangle *Y*?

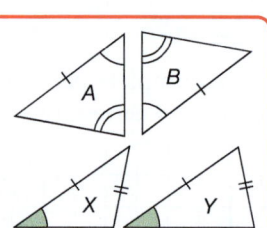

a Yes   ASA   The sides marked equal are corresponding sides – opposite the double-marked angle.
b No   The equal angles are not *between* the two sides.

Similar shapes are the same shape but different in size.
Their angles are equal.
The sides are multiplied or divided by the same **linear scale factor**.

- The linear scale factor $= \dfrac{PQ}{AB} = \dfrac{QR}{BC} = \dfrac{RP}{CA}$

  For the area and volume of similar 2D and 3D shapes:

- **area scale factor** = (linear scale factor)^2
- **volume scale factor** = (linear scale factor)^3

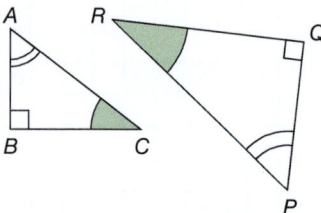

▲ Triangles *PQR* and *ABC* are similar because they have equal angles. The corresponding sides are *PQ* and *AB*, *QR* and *BC*, and *RP* and *CA*.

EXAMPLE

a Show that triangle *ADE* is similar to triangle *ABC*.
b Calculate the length of *CE*.
c The area of triangle *ADE* is 48 cm^2
    Find the area of triangle *ABC*.

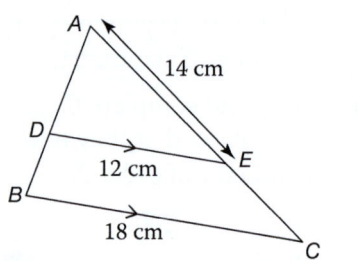

a $\angle ADE = \angle ABC$ Corresponding angles.
   $\angle AED = \angle ACB$ Corresponding angles.
   $\angle DAE = \angle BAC$ The same angle.
   Triangles *ADE* and *ABC* are similar.   Equal angles.

b The scale factor $= \dfrac{BC}{DE} = \dfrac{18}{12} = \dfrac{3}{2} = 1.5$
   $AC = 1.5 \times 14$ $CE = AC - AE = 21\,cm - 14\,cm$
   $\quad\;\; = 21\,cm$ $\quad\;\;\, = 7\,cm$

BC and *DE* are corresponding sides.
AC and *AE* are corresponding sides.

c The area of triangle $ABC = 2.25 \times 48 = 108\,cm^2$ The area scale factor $= 1.5^2 = 2.25$

Geometry   Angles and polygons

Fluency

Exercise 3.4S

1. From the grid, write down which shapes are
   a. congruent   b. similar

   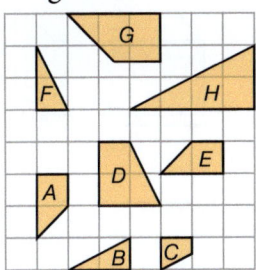

2. In each part state whether the triangles are congruent. Give reasons.

   a 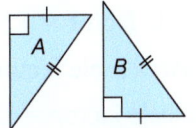   b

   c 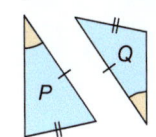   d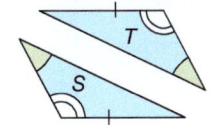

3. State which two triangles are congruent. Explain your answer.

   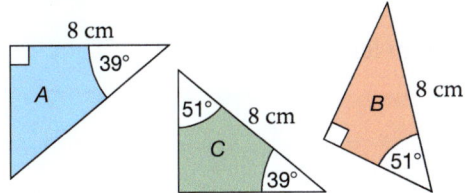

4. Three squares have sides of length
   A 2 cm   B 4 cm   C 16 cm.

   How many copies of
   a. square A will fit inside square B?
   b. square B will fit inside square C?
   c. square A will fit inside square C?

5. Two solids are similar in shape. The linear scale factor is 2.5

   Find the scale factor for the
   a. surface area   b. volume.

6. a. Show that triangle ABC is similar to triangle DEC.
   b. Calculate
      i. DE
      ii. AC.

   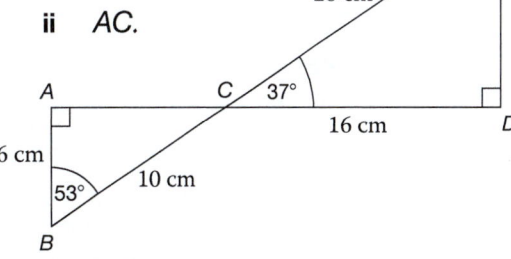

7. a. Calculate the length of
      i. CD
      ii. BD.

   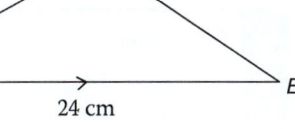

   b. The area of triangle ABC is 24 cm². Calculate the area of triangle CDE.

8. The diagram shows the dimensions of a cylinder. A larger cylinder is similar in shape, with diameter 10 cm.

   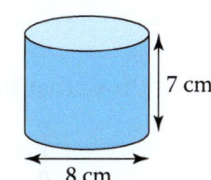

   a. Find the height of the larger cylinder.
   b. The area of the top of the smaller cylinder is 50 cm². Find the area of the top of the larger cylinder.
   c. The volume of the smaller cylinder is 352 cm³. Find the volume of the larger cylinder.

*9. a. Calculate the length of PS.
    b. The area of triangle PQR is 48 cm². Calculate the area of trapezium PQTS.

    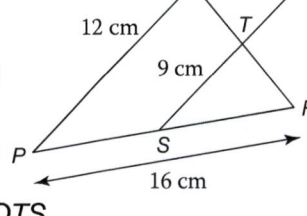

1119, 1148

APPLICATIONS

3.4 Congruence and similarity

RECAP

- Congruent shapes are identical. All angles and lengths are the same.
- Triangles are congruent if these measurements are equal
  - SSS   3 sides
  - SAS   2 sides and the angle between them
  - ASA   2 angles and a corresponding side
  - RHS   a right angle, the hypotenuse and 1 other side.
- Similar figures are the same shape but different in size. All angles are the same but lengths are all multiplied or divided by the same linear scale factor.
- The area scale factor = (linear scale factor)^2
- For solids, the volume scale factor = (linear scale factor)^3

Similar cuboids

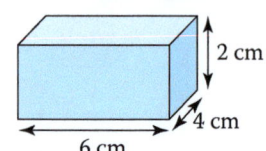

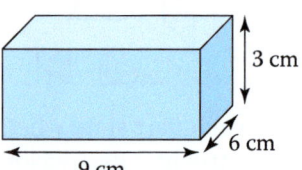

Linear scale factor = 1.5
Surface Area scale factor = 1.5^2
Volume scale factor = 1.5^3

HOW TO

1. Sketch a diagram (unless one is given) – include known values.
2. Look for congruent or similar shapes: you may need to **prove** congruence or similarity.
3. Use congruence to prove facts about shapes.
4. Use similarity to work out lengths, areas or volumes of objects that are enlarged or reduced.

EXAMPLE

Prove that the opposite sides of a parallelogram are equal in length.

③ Prove triangles *PQR* and *RSP* are congruent.
∠SPR = ∠QRP   Alternate angles
∠SRP = ∠QPR   Alternate angles
Side PR is in both triangles
Triangles PQR and RSP are congruent   ASA
Therefore PQ = RS and QR = SP by congruence.

①
② Adding a diagonal gives a pair of triangles.

EXAMPLE

A food manufacturer makes 1 kg packs of icing sugar as shown.

The manufacturer wants a similar smaller pack to hold just 125 g.

Work out the dimensions of the smaller pack.

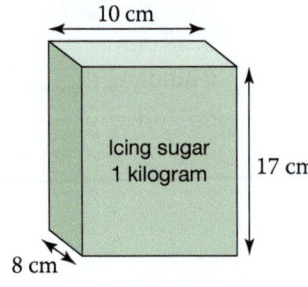

② The volume scale factor will be same as that for the weights.

④ Volume scale factor
= 1000 ÷ 125 = 8
1 kg = 1000 g
Linear scale factor = $\sqrt[3]{8}$ = 2
Lengths in small pack
= lengths in large pack ÷ 2

The smaller pack will be 5 cm long, 4 cm wide and 8.5 cm tall.

You can check.
5 × 4 × 8.5 = 170
10 × 8 × 17 = 1360
and 170 = 1360 ÷ 8

Geometry   Angles and polygons

Reasoning and problem solving

Exercise 3.4A

1. PQR is an isosceles triangle. S is the mid-point of PR.

   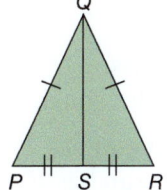

   a. Prove that triangle PQS is congruent to triangle RQS.

   b. Prove that angle PSQ is a right angle.

2. In rectangle $ABCD$, AB is parallel to DC, AD is parallel to BC and all the angles are 90°.

   a. Prove that triangle ABD is congruent to triangle CDB.

   b. Write down the angle that is equal to

      i. angle ADB ii. angle BDC.

3. In the kite $ABCD$, $AB = AD$ and $BC = CD$. Prove that angle B = angle D.

4. A diagram is 10 cm wide and 12 cm high. A photocopier is used to reduce the size of the diagram.

   a. Find the new height of the diagram when the new width is 8 cm.

   b. Find the new width of the diagram when the new height is 8 cm.

5. A miller sells cornflour in small packs as shown. The miller wants a similar catering pack to hold 64 times as much.

   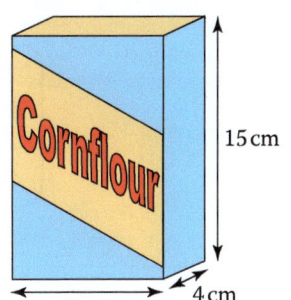

   Work out the dimensions of the catering pack.

6. A shop sells picture frames in the following sizes.

Size	Width	Height
Mini	24 cm	30 cm
Small	30 cm	40 cm
Medium	40 cm	50 cm
Large	50 cm	70 cm
Extra Large	60 cm	80 cm

   Which of these sizes are similar in shape? Explain your answer.

7. a. Prove that the diagonals of a rhombus are perpendicular.

   b. i. Name two other quadrilaterals that have perpendicular diagonals.

      ii. Use congruency to prove this.

8. This planter holds 10 litres of compost.

   Find the dimensions of a similar planter that holds 80 litres of compost.

9. Two containers are similar. When full, the smaller container holds 2560 ml and the larger container holds 5 litres.

   a. The height of the smaller container is 40 cm.

      Find the height of the larger container.

   b. The area of the label on the larger container is 1100 cm².

      Find the area of the label on the smaller container.

*10. a. Find

   i. EF

   ii. BC

   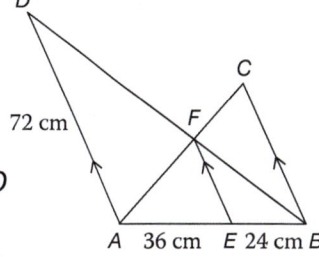

   b. The area of triangle ABD is 1950 cm². Find the area of triangle ABC.

*11. Triangles PQR and PUS are both equilateral.

   Prove that triangle PUQ is congruent to triangle PSR.

   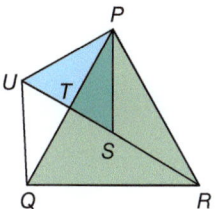

SKILLS

3.5 Polygon angles

A **polygon** is a two-dimensional shape with straight sides.
A **regular** polygon has all its sides equal and all its angles equal.

A **pentagon** has 5 sides, a **hexagon** has 6 sides, a **heptagon** has 7 sides, an **octagon** has 8 sides, a **nonagon** has 9 sides and a **decagon** has 10 sides.

- The sum of the interior angles of *any* polygon
  = (number of sides − 2) × 180°
- The sum of the exterior angles of *any* polygon = 360°
- Exterior angle of a *regular* polygon = 360° ÷ number of sides.
- At each vertex: interior angle + exterior angle = 180°

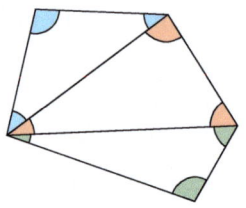

▲ Pentagon (5 sides)
Angle sum = 3 × 180°
         = 540°

EXAMPLE

Find the largest angle in this heptagon.

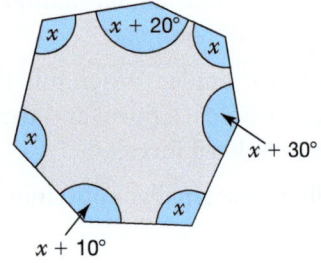

A heptagon has 7 sides. It can be split into 5 triangles.

Angle sum of a heptagon = 5 × 180° = 900°

$7x + 60° = 900°$
$7x = 900° − 60° = 840°$
$x = 840° ÷ 7 = 120°$

The largest angle = $x + 30°$ = 150°

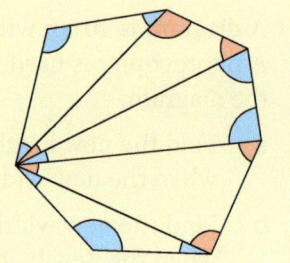

EXAMPLE

a Find the size of each interior angle in a regular hexagon.
b Describe the symmetry of a regular hexagon.

a Sum of 6 equal exterior angles = 360°
Exterior angle = 360° ÷ 6 = 60°
Interior angle = 180° − 60° = 120°

Or

Sum of interior angles = (6 − 2) × 180° = 720°
Interior angle = 720° ÷ 6 = 120°

b A regular hexagon has 6 lines of symmetry and rotational symmetry of order 6.

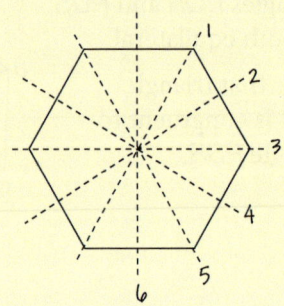

Going round a polygon, the total angle turned through = 360°

You can use tracing paper to check line and rotational symmetry.

Geometry   Angles and polygons

Fluency

Exercise 3.5S

1. A pentagon has angles of 116°, 89°, 101° and 153°. Find the other angle.

2. Two of the angles in an octagon are right angles. The other angles are all equal.

   Find the size of these angles.

3. Show that *x* and *y* are equal

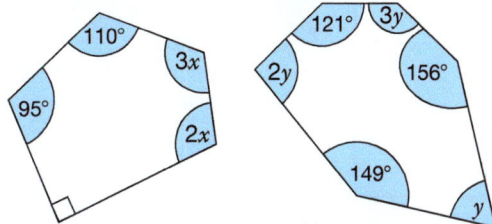

4. Copy and extend this table for regular polygons with up to 10 sides.

   Complete the angle columns.

Regular polygon	No. of sides	Exterior angle	Interior angle
Triangle	3		
Quadrilateral	4		
Pentagon	5		

5. A dodecagon is a polygon with 12 sides.

   Eleven of the angles of a dodecagon are equal to 154°. Calculate the other angle.

6. A regular polygon has 15 sides.

   a. Use the angle sum of this polygon to work out the size of an interior angle.

   b. Use the sum of the exterior angles to check your answer to part **a**.

7. Describe the symmetry of

   a. a regular pentagon

   b. a regular octagon

   c. a regular decagon

8. Find the size of the smallest angle of this pentagon.

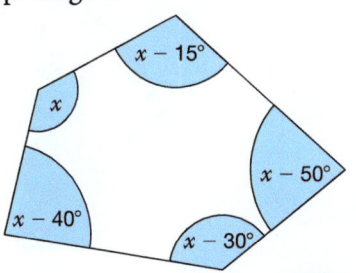

9. The diagram shows a regular nonagon, divided into congruent triangles.

   a. Find the angles marked *x* and *y*.

   b. Use the value of *y* to check your answers in question **4** for a regular nonagon.

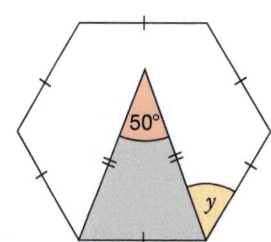

10. Calculate angle *y*.

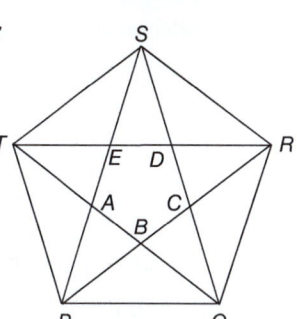

11. *PQRST* is a regular pentagon.

    Diagonals are drawn as shown.

    Calculate angle *APB*.

*12. *ABCDEFGHI* is a regular nonagon.

    The sides *AB* and *DC* are extended until they meet at point *X*.

    Calculate angle *BXC*.

*13. *PQRSTUVW* is a regular octagon.

    Calculate angle *RWS*.

*14. *ABCDEF* is a regular hexagon.

    Show that angle *ACD* is a right angle

15. Find the number of sides of a regular polygon that has exterior angles of 12°.

APPLICATIONS

3.5 Polygon angles

RECAP
- The sum of the exterior angles of any polygon is 360°.
- The sum of the interior angles of an n-sided polygon is $(n - 2) \times 180°$
- Interior angle + exterior angle = 180°

▲ Designers can create elaborate patterns using simple polygons.

HOW TO To solve problems involving polygons
1. Draw a sketch (if needed).
2. Decide whether to use the interior or exterior angle sum.
3. Find the answer.

EXAMPLE

A regular polygon has interior angles of 156°.
How many sides does the polygon have?

② Find the exterior angle.
Exterior angle = 180° − 156° = 24°   Angles on a straight line.
③ For a regular polygon, exterior angles = 360° ÷ n.
Number of exterior angles = 360° ÷ 24° = 15
The polygon has 15 sides.   The number of sides = the number of angles.

① Sides of polygon
156°
Exterior angle

EXAMPLE

Explain why regular 12-sided polygons
a   will not tessellate on their own
b   will tessellate with equilateral triangles.

A tessellation is a tiling of the plane in which there are no gaps or overlaps.

a ② Shapes tessellate if the sum of the angles where they meet is 360°.
Find the interior angle of the polygon.

① Interior angle / Exterior angle

③ Exterior angle = 360° ÷ 12 = 30°   Sum of exterior angles = 360°.
Interior angle = 180° − 30° = 150°   Angles on a straight line.

Sum of 2 interior angles = 300°   <360°
Sum of 3 interior angles = 450°   >360°

Sum of interior angles = 10 × 180° = 1800°
Each interior angle = 1800° ÷ 12 = 150°

The polygons do not fit together at a point − they do not tessellate.

b ③ Each angle in an equilateral triangle = 60°
300° + 60° = 360°
Two regular 12-sided polygons and an equilateral triangle will fit together at a point.
If the sides are the same length then they will tessellate.

Geometry   Angles and polygons

Reasoning and problem solving

Exercise 3.5A

1. A regular polygon has acute interior angles. Name the shape.

2. The diagram shows part of a regular polygon.

   Calculate the number of sides of the polygon.

3. A regular polygon has interior angles of 177°. How many sides does this polygon have?

4. A regular polygon can be made by fitting together squares and regular hexagons. How many sides does it have?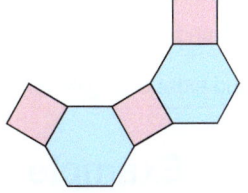

5. A regular polygon can be made by fitting together squares and regular pentagons in an alternating pattern. How many sides does it have?

6. The diagram shows a regular hexagon attached to a square.

   a Calculate angle z.
   b Name one or more regular polygons that would fit exactly into this space.

   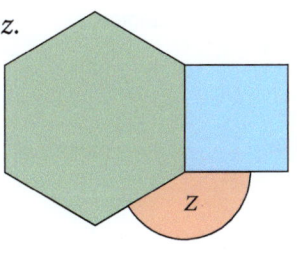

7. The interior angle of a regular polygon is four times the size of the exterior angle.
   a Find the number of sides.
   b Name the polygon.

8. a Is it possible to have a regular polygon with the given angle as its exterior angle? If so, find the number of sides.

   i 10°   ii 9°
   iii 8° iv 7°

   b Write down a general rule to describe when a regular polygon with a given exterior angle is possible.

*9 $PQRST$ is a regular pentagon. Prove that PS is parallel to QR.

*10 $ABCDEFGH$ is a regular octagon. Prove that $ACEG$ is a square.

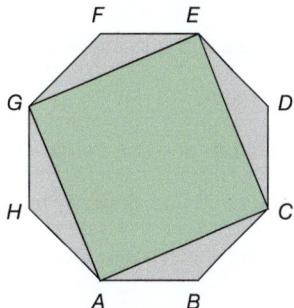

11. a Find combinations of two or more regular shapes that will fit together to tile the plane without gaps or overlaps.
    b Research Schlafli Tessellations.

12. The diagram shows a **concave** hexagon.

    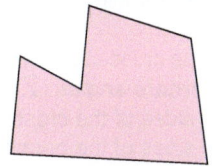

    a Is it true that the sum of the interior angles of an n-sided concave polygon is $(n-2) \times 180°$?
    b Investigate the 'exterior' angles of concave polygons. Can you find a general rule?

*13 $ABCDE...$ is a regular polygon with n sides. Write an expression for the size of angle ACD in terms of n.

*14 In the diagram $ABCDEF$ is a regular hexagon and $ABGHI$ is a regular pentagon. Find angle HGC.

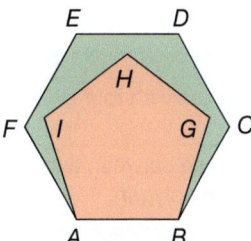

Summary

Checkout
You should now be able to...

	Test it Questions
✓ Use angle facts including at a point, on a line, at an intersection and for parallel lines.	1, 2
✓ Use bearings to specify directions.	3
✓ Identify types of triangle and quadrilateral and use their properties.	4 – 7
✓ Identify and use symmetry.	5a
✓ Identify congruent shapes and use congruence to prove geometric results.	8
✓ Identify similar shapes and use similarity to find lengths and areas.	9
✓ Calculate the properties of polygons including interior and exterior angles for regular polygons.	10

Language	Meaning	Example
Acute angle	0 < acute angle < 90°	
Right angle	right angle = 90°	
Obtuse angle	90° < obtuse angle < 180°	
Reflex angle	180° < reflex angle < 360°	
Alternate, corresponding and interior angles	When a line crosses a pair of parallel lines, **alternate angles** (Z) lie on opposite sides of the crossing line and opposite sides of the parallel lines. **Corresponding angles** (F) lie on the same side of the crossing line and the same side of the parallel lines. **Interior angles** (⊏) lie on the same side of the crossing line and opposite sides of the parallel lines as shown	
Three-figure bearing	A direction defined by a three-figure angle measured clockwise from north.	North-east is 045° North-west is 315°
Congruent	Exactly the same shape and size.	
Similar	The same shape but different in size.	
Scale factor	The ratio of corresponding lengths in two similar shapes.	A and B are similar; the scale factor is 2. A and C are congruent.
Polygon	A 2D shape with straight edges.	Triangle, square, hexagon.
Quadrilateral	A polygon with four sides.	Square, rectangle, rhombus, parallelogram, trapezium, kite.
Interior angle	The angle between two adjacent sides inside a polygon.	
Exterior angle	The angle between one side of a polygon and the next side extended.	

Geometry Angles and polygons

Review

1 Calculate the size of angles *a*, *b* and *c*.

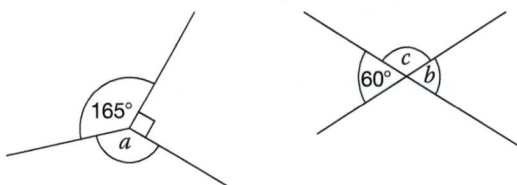

2 Calculate the size of angles *a*, *b* and *c*. Give a reason for each answer.

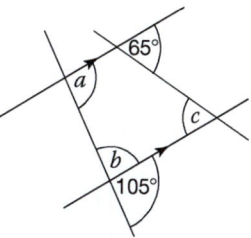

3 Find these three-figure bearings.

  a The bearing of A from B is 063°. Work out the bearing of B from A.

  b The bearing of C from D is 141°. Work out the bearing of D from C.

  c The bearing of E from F is 205°. Work out the bearing of F from E.

4 Calculate the size of angle *a* in this triangle.

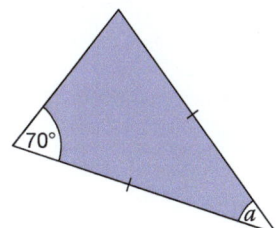

5 a Describe the symmetry of this kite.

  b Calculate the size of angle *a*.

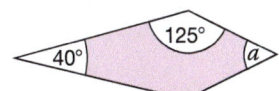

6 Draw coordinate axes with *x* and *y* from 0 to 6. Now plot these points: A (1, 1), B (3, 3), C (5, 1).

Join up the dots to form a triangle. What type of triangle is this?

7 Which quadrilateral is being described below?

  a One pair of parallel sides and no equal angles.

  b Two pairs of equal angles, two pairs of parallel sides and all sides equal.

8 Are these two triangles congruent? Give a reason for your answer.

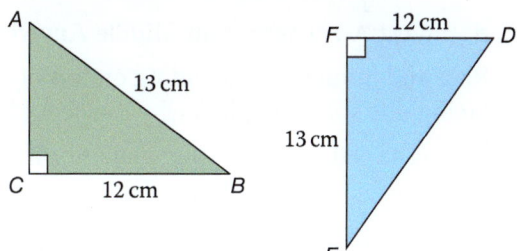

9 These hexagons are similar. Find *x* and *y*.

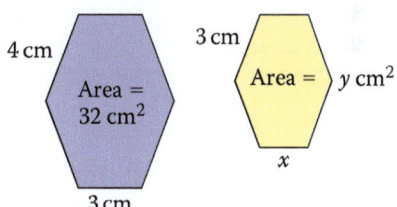

10 a A heptagon has seven sides. What do the interior angles of a heptagon add up to?

  b Prove that the sum of the exterior angles of a regular pentagon is 360°.

What next?

Score		
	0 – 4	Your knowledge of this topic is still developing. To improve, look at MyiMaths: 1080, 1082, 1086, 1100, 1102, 1109, 1119, 1130, 1141, 1148, 1320
	5 – 9	You are gaining a secure knowledge of this topic. To improve your fluency look at InvisiPens: 03Sa – k
	10	You have mastered these skills. Well done you are ready to progress! To develop your problem solving skills look at InvisiPens: 03Aa – h

Assessment 3

1. **a** What is the angle between the hour hand and the minute hand of a clock at five minutes past seven? [5]

   b Yon-Hee claims that between 1:05 pm and 1:06 pm the hour and minute hands of a clock are pointing in the same direction. Is she correct? Explain your answer. [6]

2. The cruise ship 'Black Watch' sails on a bearing of 135°. To avoid a storm it changes course to a bearing of 032°. What angle has it turned through? [2]

3. Three villages, East Anywhere, Middle Anywhere and West Anywhere, are the vertices of the triangle shown. The bearing of East Anywhere from West Anywhere is 075°.

   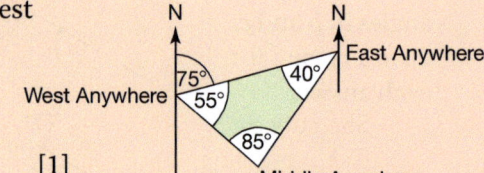

   Write down the bearing of:

   a Middle Anywhere from West Anywhere [1]

   b West Anywhere from East Anywhere [2]

   c Middle Anywhere from East Anywhere [2]

   d West Anywhere from Middle Anywhere. [3]

4. Rafa and Sunita were standing on top of a hill. Rafa was facing due north and Sunita was facing due south. Explain why they could see each other. [1]

5. Sienna says that the missing angles have these values:

   $a = 46°$ $b = 46°$ $c = 139°$ $d = 41°$

   $e = 92°$ $f = 88°$ $g = 128°$ $h = 101°$

   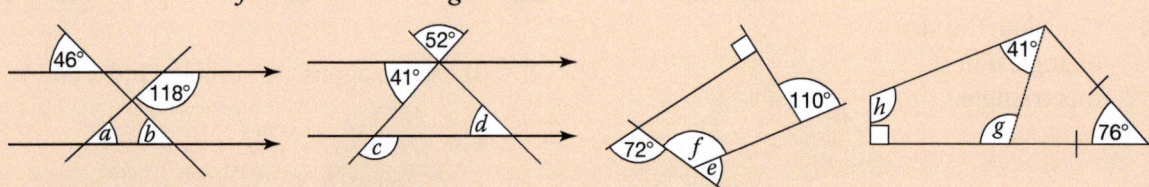

   Decide if her value for each angle is correct or incorrect. For the incorrect angles, give the correct value. Give reasons for your answers. [12]

6. Manuel made the following statements about triangles. Write down if each of his statements is true or false. If the statement is true draw an example. If the statement is false explain why it is false.

   a Some triangles have two obtuse angles. [2]

   b Some triangles have one obtuse angle and two acute angles. [2]

   c Some triangles have one right angle, one acute angle and one obtuse angle. [2]

   d Some triangles have two acute angles and one right angle. [2]

7. **a** Natasha draws a pentagon with angles of 125°, 155° and 74°. She wants to draw the other two angles so that they are are equal. What size should she draw these two angles? [4]

   b Shivani draws an octagon that has 5 angles each of 114°. She wants to draw the other three angles so that they are equal. What size should she draw these three angles? [4]

   c Tyler draws a hexagon with three angles of 137° each. He wants to draw the remaining three angles so that they have the ratio $w : w + 120 : w - 30$. What value should he use for w? [5]

   d Dean draws a regular polygon containing 60 sides. What size should he use for the interior angle? [2]

Geometry Angles and polygons

8   Rose has designed a patio for her garden in the shape of a regular pentagon of side 8 m.

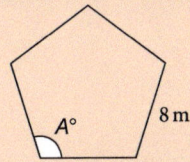

   a   Calculate angle *A*. [2]
   b   Rose borders the patio with square paving slabs of side length 2 m, as shown.

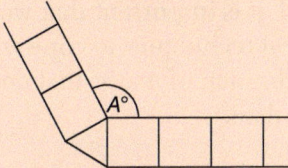

   She cuts each triangular corner slab from one square slab.
   Calculate the number of slabs needed. [2]
   c   Using the value of *A* that you found in part **a**, calculate
       the three angles in the triangular corner slab. [4]

9   In a photo a statue's height is 19.2 cm and its legs measure 8 cm.
    The statue is actually 288 cm tall.
    How long are its legs? [4]

10 a   Show that triangles *PQR* and *PTS* are similar. [2]
    b   Find length *RQ*. [2]

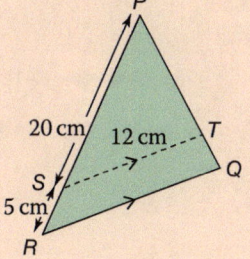 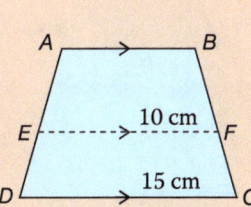

    c   Show that trapeziums *ABCD* and *ABFE* are not similar. [4]

11 Triangle *BCG* has area 36 cm²

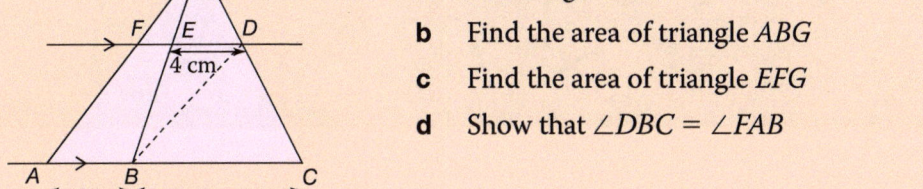

    a   Find length *FE* [2]
    b   Find the area of triangle *ABG* [5]
    c   Find the area of triangle *EFG* [4]
    d   Show that ∠*DBC* = ∠*FAB* [2]

12 Draw a shape that has
    a   rotation symmetry of order 3, but no lines of symmetry
    b   3 lines of symmetry, but no rotation symmetry. [4]

4 Handling data 1

Introduction

In the modern world, there is an ever-increasing volume of data being continually collected, analysed, interpreted and stored. It was estimated that in 2007, there were 295 exabytes (or 295 billion gigabytes) of data being stored around the world. It has increased significantly since then. To put this into perspective, if all that data were recorded in books, it would cover the area of China in 13 layers of books. With all this information being generated, it is important that we have the mathematical techniques to cope with it. Statistics is the branch of maths that deals with the handling of data.

What's the point?

Availability of data helps us to understand the world around us. Scientists look for trends in global warming and consumers looking at cost comparison data to provide information on purchasing decisions.

Objectives

By the end of this chapter, you will have learned how to …
- Construct and interpret two-way tables, bar charts and pie charts.
- Calculate the mean, median and mode of a data set.
- Calculate the range and inter-quartile range of a data set.
- Use averages and measures of spread to compare data sets.
- Use frequency tables to represent grouped data.
- Construct histograms with equal or unequal class widths.

Check in

1 Find

 a $\frac{1}{2}$ of 45   **b** $\frac{1}{2}$ of (41 + 1)   **c** $\frac{1}{4}$ of 24   **d** $\frac{3}{4}$ of (27 + 1)

2 List these two sets of numbers in ascending order.

 a 45, 66, 89, 101, 98, 55, 112, 90, 29, 65

 b 11.3, 8.9, 7.88, 9.0, 8.78, 8.95, 11.25, 10.9, 8.87, 9.8.

Chapter investigation

What is the average age of students in your class?

SKILLS

4.1 Representing data 1

You can use a **bar chart** to display data.

Bar charts give a visual picture of the size of each category.

The bars can be horizontal or vertical.

- **Bar-line charts** are a good way to display (discrete) numerical data.

You can use a **pie chart** to display data.

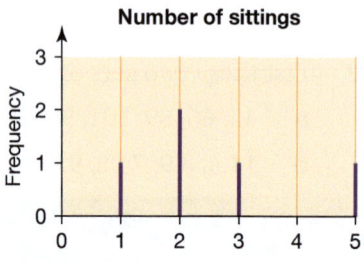

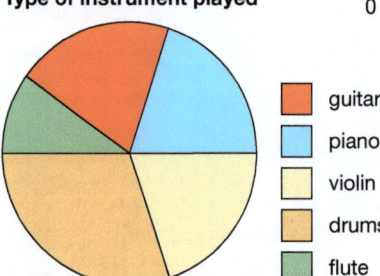

p.238
- A pie chart shows the **proportion** or fraction of each category compared to the whole circle.

EXAMPLE

60 vehicles are shown in the pie chart.

Calculate the numbers of cars, vans, buses and lorries.

Vehicles parked in the High St.

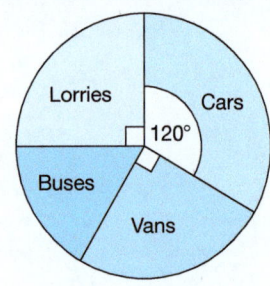

60 vehicles represents 360°.
1 vehicle represents 360° ÷ 60 = 6°
The angle for buses is 360° − (90° + 90° + 120°) = 60°
Number of cars   = 120 ÷ 6 = 20
Number of vans   = 90 ÷ 6 = 15
Number of buses = 60 ÷ 6 = 10
Number of lorries = 90 ÷ 6 = 15
Total number of vehicles =   60

Use a protractor to measure the angles in the pie chart.

EXAMPLE

A two-way table links two types of information.

A class of 30 students study either History or Geography.

There are 13 girls in the class, with 8 girls and 9 boys studying Geography.

How many boys study History?

① Draw a two-way table to show the information that you know.
② Use the totals to fill in the other entries.

	History	Geography	Total
Boys		9	30 − 13 = 17
Girls	13 − 8 = 5	8	13
Total		9 + 8 = 17	30

↓

	History	Geography	Total
Boys	17 − 9 = 8	9	17
Girls	5	8	13
Total	30 − 17 = 13	17	30

③ Read the answer from the table.

8 boys study History.

Statistics   Handling data 1

Fluency

Exercise 4.1S

1. A group of 40 students study either French or Spanish.
   There are 17 boys in the group, with 15 girls and 10 boys studying French.
   How many girls study Spanish?

2. In a traffic survey, the colour and speed of 100 cars are recorded. The results are summarised in the two-way table.

	Not speeding	Over the speed limit
Red	5	55
Not red	10	30

   Maria claims that drivers of red cars tend to break speed limits. Use the two-way table to decide whether you agree with Maria.

3. The number of Bank Holidays in different countries is shown in the table.
   Draw a bar chart to show this information.

Country	Number of Bank Holidays
UK	8
Italy	16
Iceland	15
Spain	14

4. The cost to fly to certain resorts is given in the frequency table.
   Draw a bar chart to show this information.

Resort	Cost ($)
Lisbon	90
Crete	100
Malta	80
Menorca	70
Madagascar	110

5. Seven boys and five girls attend an after school homework club.
   a. Calculate the angle one student represents in a pie chart.
   b. Calculate the angles to represent boys and girls.
   c. Draw a pie chart to show the information.

6. The weather record for 60 days is shown in the frequency table. This gives the predominant weather for that particular day.

Weather	Number of days
Sunny	15
Cloudy	18
Rainy	14
Snowy	3
Windy	10

   a. Calculate the angle that one day represents in a pie chart.
   b. Calculate the angle of each category in the pie chart.
   c. Draw a pie chart to show the data.

7. A school open day runs from 10 am to 4 pm. A teacher has offered to help.
   She spends these times in each class.

Class	Time
1	30 mins
2	25 mins
3	35 mins
4	80 mins
5	70 mins
6	60 mins
7	60 mins

   Draw a pie chart to show this information.

8. In one week a car dealer sells 18 cars of three different types: diesel, petrol and electric.

   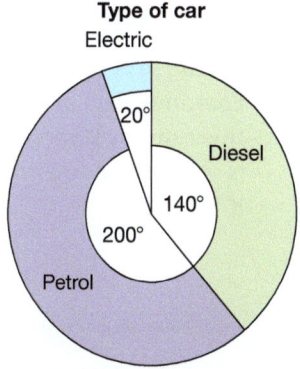

   a. Calculate the angle that represents one car.
   b. Calculate the number sold that are
      i. diesel   ii. petrol   iii. electric.

APPLICATIONS

4.1 Representing data 1

RECAP
- You can use a bar chart or a pie chart to display data.
- A bar chart uses a bar to show the size of each category.
- A pie chart uses a circle to give a visual picture of all the data. The sectors of the circle represent the size of each category. The size of the sector angle is proportional to the frequency.

HOW TO
1. Read information from the diagram. Make sure that you read the scale or key carefully.
2. Use the information from the diagram to answer the question.

EXAMPLE

A class are asked to name one favourite pet. The results are shown in the table.

Pet	Dog	Cat	Guinea pig	Other
Number of students	16	9	12	3

Eve uses her computer to create this graph.
Make two criticisms of the graph.

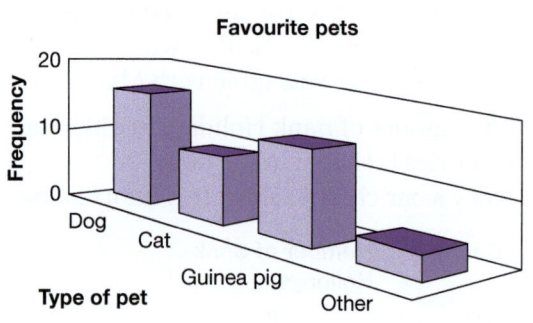

1. Compare the graph with the frequency table and graph in 4.1 Skills. Look at the scale of the graph.
   The scale chosen makes it is difficult to read information from the graph.
2. Bar charts should let you compare the different categories.
   The perspective makes it difficult to compare the heights of the bars.

EXAMPLE

Janelle asks people in Easton and Westville to name their favourite fruit. The results are shown in the pie chart.

Janelle says that there *are more people* in Easton whose favourite fruit is banana than there are in Westville. Explain why Janelle is wrong and rewrite her statement so that it is true.

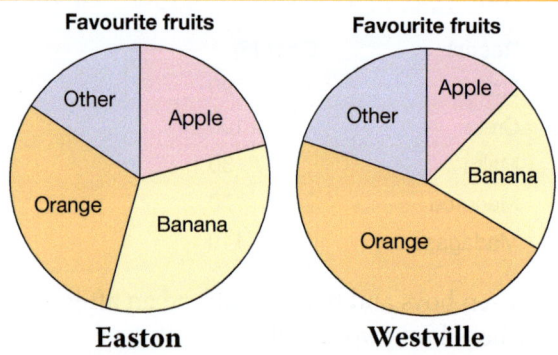

1. Pie charts don't show the frequency of each category.
2. Janelle doesn't know how many people in Westville are represented in the pie chart. Although the sector is larger for Easton, it could represent fewer people.
   Pie charts show the proportion of each category.
   There <u>is a higher proportion of people</u> in Easton whose favourite fruit is banana than there is in Westville.

Statistics   Handling data 1

Exercise 4.1A

1 The number of students in different year groups is given.
   Marnie draws this bar chart for her data.

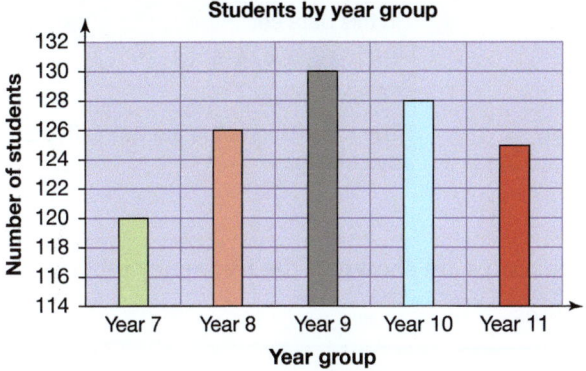

   Give one reason why Marnie's graph is misleading.

2 Feroz records the number of jars of sauce sold in his deli.

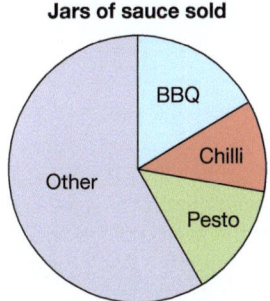

   He draws a pie chart to show the information. Josh wants to compare the sales of each flavour he sells. Give three disadvantages of Feroz's chart.

3 The number of drinks sold during lunchtime in a shop are shown in the table.

Drink	No. sold	Drink	No. sold
Apple	6	Lemonade	3
Blackcurrant	5	Mocha	2
Cappuccino	8	Orange	7
Cola	12	Pineapple	4
Espresso	3	Tea	9
Latte	5	Water	9

   a Give one disadvantage of using a pie chart to display the data.
   b Give one advantage of using a bar chart to display the data.

4 The pie charts show the number of Year 11 students studying languages at two nearby schools.

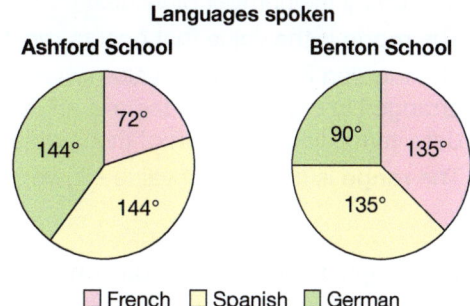

   a Jermaine says twice as many students at Ashford school study German compared to those studying Spanish. Do you agree with Jermaine?
   b Lydia says it must be the case that more students at Ashford schools study French than at Benton school. Do you agree with Lydia?

5 The **dual bar chart** shows the number of letters delivered in one week to No. 10 and No. 12.

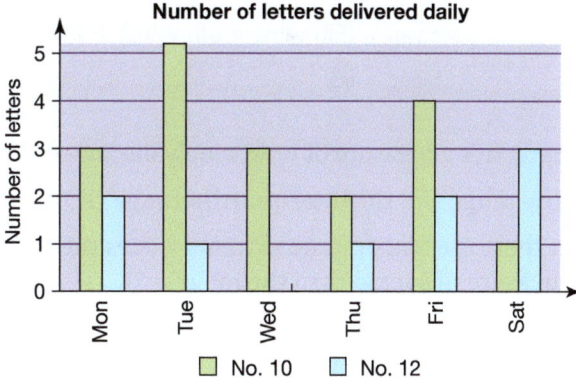

   a Draw an accurate pie chart to show the number of letters delivered to No. 10.
   b Draw an accurate pie chart for No. 12.
   c Rocco wants to compare the proportion of letters received on each day for the two houses. Should he use the bar chart or the pie charts? Give your reasons.
   d Ahmed want to compare the number of letters received each day for the two houses. Should he use the bar chart or the pie charts? Give your reasons.

SKILLS

4.2 Averages and spread 1

- The **mean** of a set of data is the total of all the values divided by the number of values.
- The **mode** is the value that occurs most often.
- The **median** is the middle value when the data is arranged in order. Where there is an even number of data items, find the mean of the middle two values.
- The **range** is the highest value – lowest value.

> Outliers are values that lie outside most of the other values of a set of data.
> In this data set
> 1, 1, 2, 2, 3, 4, 4, 4, 16
> 16 is an outlier.
> The mean and range are both affected by outliers.

EXAMPLE

Ten people took part in a golf competition. Their scores are shown in the frequency table.
Calculate the mean, mode and median of the scores.

You can calculate directly from the frequency table.

Score	Frequency	Score × Frequency
67	1	67
68	4	272
69	3	207
70	1	70
71	1	71
	10	687

68 + 68 + 68 + 68 or 68 × 4.

The total of all the scores of the 10 golfers.

Mean = 687 ÷ 10 = 68.7
Median = (5th value + 6th value) ÷ 2 = (68 + 69) ÷ 2 = 68.5
Mode = 68   The mode is 68, not 4.

> With an even number of values (10) the median is the averge of the two middle values (5th and 6th). Don't forget to put the values in size order first!

Spread is a measure of how widely dispersed the data are.

The range and the **interquartile range** (IQR) are measures of spread.

If there are one or more extreme values the IQR is a better measure of spread than the range.

- IQR = upper quartile − lower quartile

> **Lower quartile**
> $= \frac{1}{4}(n + 1)$th value.

> **Upper quartile**
> $= \frac{3}{4}(n + 1)$th value.

EXAMPLE

Randa collected data on the number of times her friends went swimming in one month.

4 7 22 1 6 2 1 5 6 6 4

Work out the interquartile range.

The outlier (22) does not affect the IQR.

Put the data in order.   1 1 ② 4 4 5 6 6 ⑥ 7 22
There are 11 pieces of data.

Lower quartile = $\left(\frac{11 + 1}{4}\right)$th value   Upper quartile = $\left(\frac{3(11 + 1)}{4}\right)$th value

= 3rd value = 2   = 9th value = 6

IQR = 6 − 2 = 4

Statistics Handling data 1

Fluency

Exercise 4.2S

1. Calculate the mean for each set of numbers.
   - **a** 4, 9, 7, 12
   - **b** 8, 11, 8
   - **c** 3, 2, 2, 3, 0
   - **d** 1, 9, 8, 6
   - **e** 2, 2, 4, 1, 0, 3
   - **f** 23, 22, 25, 26
   - **g** 17, 19, 19, 20, 18, 17, 16
   - **h** 103, 104, 105
   - **i** 14, 10, 24, 12
   - **j** 4, 6, 7, 6, 4, 4, 3, 7, 3, 6

2. **a** Find the median of each set of numbers.
   - **i** 7, 8, 8, 5, 4, 3, 3
   - **ii** 11, 12, 10, 9, 9, 10, 26
   - **iii** 38, 35, 30, 37, 34, 32, 36
   - **iv** 101, 98, 103, 97, 99, 97, 95
   - **v** 3, 2, 0, 0, 1, 2, 1, 2, 3

   b Which data sets contain outliers? Have the outliers affected the median?

3. Calculate the mode and range of each set of numbers.
   - **a** −6, 0, 1, 1, 1, 1, 2, 2, 2, 3, 3, 4, 4
   - **b** 5, 5, 6, 6, 6, 7, 7, 7, 7, 8, 8, 8, 8, 8
   - **c** 10, 11, 11, 11, 12, 12, 13, 14
   - **d** 21, 22, 23, 24, 24, 25, 25, 25, 36
   - **e** 8, 8, 8, 9, 9, 10, 11
   - **f** 4, 3, 5, 5, 6, 6, 4, 3, 4, 5, 6, 5, 3

   Which data sets contain outliers? Have the outliers affected the mode or range?

4. The numbers of flowers on eight rose plants are shown in the frequency table.

Number of flowers	Tally	Frequency
3	||||	4
4	||	2
5	||	2

   a List the eight numbers in order of size, smallest first. Calculate the mean, mode, median and range of the eight numbers.

   b Use the table to calculate the mean, mode and median. Check your answers against your answers for part **a**.

5. The number of days that 25 students were present at school in a week are shown in the frequency table.

Number of days	Tally	Frequency
0		0
1	||||	4
2	卌 |	6
3	||	2
4	卌	5
5	卌 |||	8

   a List the 25 numbers in order of size, smallest first. Calculate the mean, mode, median and range of the 25 numbers.

   b Use the table to calculate the mean, mode and median. Check your answers against your answers for part **a**.

6. For these sets of numbers work out the
   - **i** range
   - **ii** mode
   - **iii** mean
   - **iv** median
   - **v** interquartile range.

   - **a** 5, 9, 7, 8, 2, 3, 6, 6, 7, 6, 5
   - **b** 45, 63, 72, 63, 63, 24, 54, 73, 99, 65, 63, 72, 39, 44, 63
   - **c** 97, 95, 96, 98, 92, 95, 96, 97, 99, 91, 96
   - **d** 13, 76, 22, 54, 37, 22, 21, 19, 59, 37, 84
   - **e** 89, 87, 64, 88, 82, 88, 85, 83, 81, 89, 90
   - **f** 53, 74, 29, 32, 67, 53, 99, 62, 34, 28, 27, 27, 27, 64, 27
   - **g** 101, 106, 108, 102, 108, 105, 106, 109, 103, 105, 107, 104, 104, 105, 105

7. **a** Subtract 100 from each of the numbers in question **6g** and write down the set of numbers you get.

   b Work out the
   - **i** range
   - **ii** mode
   - **iii** mean
   - **iv** median
   - **v** interquartile range.

   c Compare your answers for the measures of spread in part **b i** and **v** and **6g i** and **v**. What do you notice?

APPLICATIONS

4.2 Averages and spread 1

RECAP
- To calculate the mean of a set of data, add all the values and divide by the number of values.
- To find the median, arrange the data in order and choose the middle value.
- To find the mode, choose the value that occurs most often.
- To find the range, subtract the smallest value from the largest value.
- To find the interquartile range, subtract the lower quartile from the upper quartile.

The mean, median and mode are averages. The range and interquartile range are measures of spread.

HOW TO

To compare data sets
1. Calculate the mean, median, mode, range or interquartile range for each set of data.
2. To compare the averages, look at the mean, median and mode.
3. To compare the spread, look at the range or interquartile range.

EXAMPLE

A team of 7 girls and a team of 8 boys did a sponsored run for charity.
The distances the girls and boys ran are shown.

Girls

Distance (km)	Frequency
1	3
2	2
3	1
4	1
5	0

Boys

3, 5, 5, 3, 4, 4, 5, 5
all distances in kilometres

By calculating the mean, median and range, compare each set of data.

1. Calculate the mean, median and range of each set of data.

Girls

Mean = (3 + 4 + 3 + 4 + 0) ÷ 7 = 14 ÷ 7 = 2

Median = 2   (the 4th distance is the middle value)

Range = 4 − 1 = 3   (highest value − lowest value)

Boys

Mean = 34 ÷ 8 = 4.25

Median = (4 + 5) ÷ 2 = 9 ÷ 2 = 4.5   (the mean of the 4th and 5th distances)

Range = 5 − 3 = 2   (highest value − lowest value)

2. Compare the mean and median.

The mean and median show the boys ran further on average than the girls.

3. Compare the range.

The range shows that the girls' distances were more spread out than the boys' distances.

Statistics   Handling data 1

Reasoning and problem solving

Exercise 4.2A

1 The number of bottles of milk delivered to two houses is shown in the table.

	Sat	Sun	Mon	Tue	Wed	Thu	Fri
Number 45	2	0	1	1	1	1	1
Number 47	4	0	2	2	2	2	2

a Calculate the range for Number 45 and Number 47.

b Use your answers for the range to compare the number of bottles of milk delivered to each house.

2 The number of cars at each house on Ullswater Drive are

2 4 1 0 1 2 1 2 3 2

a Copy and complete the frequency table.

Number of cars	Tally	Number of houses
0		
1		
2		
3		
4		

b Calculate the mean, mode and median number of cars for Ullswater Drive. The mean, mode and median number of cars at each house on Ambleside Close are

Mean	Mode	Median
0.7	0	1

c Use the mean, mode and median to compare the number of cars on Ullswater Drive and Ambleside Close.

3 The range of these numbers is 17.
Find two possible values for the unknown number.

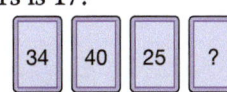

4 Monica has five numbered cards.
One of the cards is numbered −2.6
Monica's cards have
- range = 7.2
- median = 3.5
- mode = 4.1

Write down the five numbers on Monica's cards.

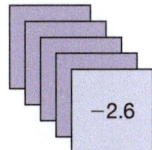

5 Write down three sets of seven numbers that have a median 6, range 14 and interquartile range 5.

6 A set of five numbers satisfies these criteria:

Range = 10   Median = 8   Mode = 6

Explain why the mean of the numbers must be between 8.8 and 10.4.

7 The bar chart shows the test results for a class of 20 students.

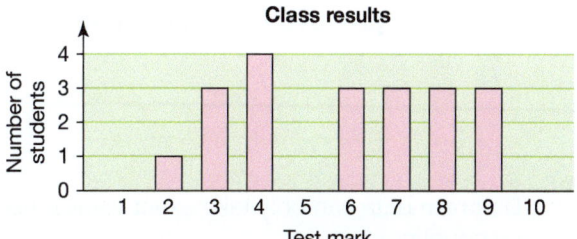

a Calculate the mean, mode, median, range and interquartile range.

b If a new student joined the class and got a mark of 10, how would this affect your answer to part **a**?

8 There are nine passengers on a bus.
The mean age of the passengers is 44 years old.
Sawsan gets on the bus.
Sawsan is 14 years old.
Find the new mean age of the passengers on the bus.

9 Reuben counted the raisins in 21 boxes.
The mean number of raisins per box was 14.1 (1 dp).
Reuben records the information for 20 of the boxes in the table.

Number of raisins	Number of boxes
13	5
14	8
15	7

Find the number of raisins in the last box.

*10 The mean mark in a statistics test for a class was 84%. There are 32 students in the class, 12 of whom are girls.
The mean mark in the test for the girls was 93%. Work out the mean mark in the statistics test for the boys.

77

SKILLS

4.3 Frequency diagrams

Some surveys produce data with many different values.

You can **group** data into **class intervals** to avoid having too many individual values.

EXAMPLE

The exam marks of class 10A are shown:

35	47	63	25	31	8	19	55	47	14
24	36	56	61	15	43	22	50	66	10
36	45	18	20	53	31	40	60	44	47

Complete a **grouped frequency table**.

Mark	Tally	Frequency
1–20	|||| ||	7
21–40	|||| ||||	9
41–60	|||| ||||		11
61–80	|||	3

|||| = 5

Check that the frequencies add to 30.

- **Discrete** data can only take exact values (usually collected by counting).

- **Continuous** data can take any value in a given range (usually collected by measuring). Continuous data cannot be measured exactly.

The number of students in a class is discrete data. The heights of the students is continuous data.

You must be careful if the data is **grouped**.

You can use a **bar chart** to display grouped discrete data.

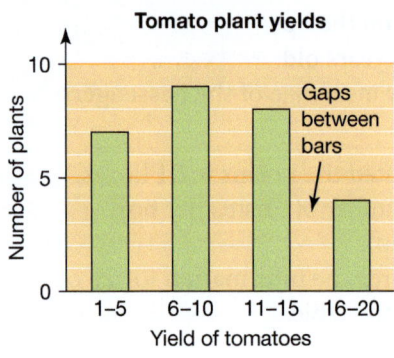

You can use **a histogram** to display grouped continuous data.

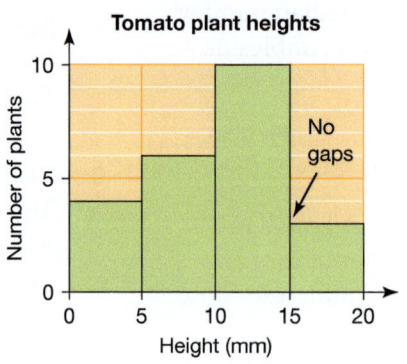

EXAMPLE

The times taken, in seconds, to run 100 m are shown in the table.

Time (seconds)	Number of people
0 < t ≤ 10	0
10 < t ≤ 20	4
20 < t ≤ 30	6
30 < t ≤ 40	3

Draw a frequency diagram to illustrate this information.

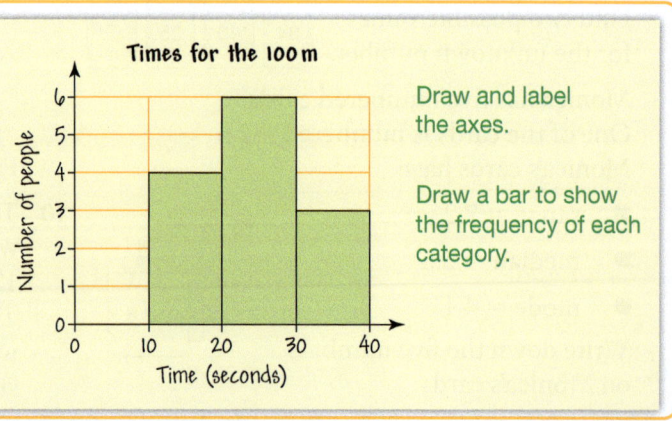

Draw and label the axes.

Draw a bar to show the frequency of each category.

Statistics   Handling data 1

Fluency

Exercise 4.3S

1 a Copy and complete the frequency table using these test marks.

```
 8 14 21  4 15 22 25 24 15 11
10 17 24 20 13 16 12  9  3 14
20 10 16 15  7 23 23 14 15 16
 8  2  9 19 12 10 10 20 13 13
15 17 11 14 19 20 23 23 24  5
```

Test mark	Tally	Frequency
1 – 5		
6 – 10		
11 – 15		
16 – 20		
21 – 25		

b Calculate the number of people who took the test.

2 a Copy and complete the frequency table using these masses of people, given to the nearest kilogram.

```
48 63 73 55 59 61 70 63 58 67
46 45 57 58 63 71 60 47 49 51
53 61 68 65 70 60 52 59 50 49
48 47 63 61 58 71 53 51 60 70
```

Mass, w (kg)	Tally	Number of people
$45 \leq w < 50$		
$50 \leq w < 55$		
$55 \leq w < 60$		
$60 \leq w < 65$		
$65 \leq w < 70$		
$70 \leq w < 75$		

b Calculate the number of people in the sample.

3 a Copy and complete the frequency table using these heights of people.

```
153 134 155 142 140 163 150 135
170 156 171 161 141 153 144 163
140 160 172 157 136 160 134 154
176 154 173 179 160 152 170 148
151 165 138 143 147 144 156 139
```

Height, h (cm)	Tally	Number of people
$130 < h \leq 140$		
$140 < h \leq 150$		
$150 < h \leq 160$		
$160 < h \leq 170$		
$170 < h \leq 180$		

b Draw a histogram to show the heights.

4 The depth, in centimetres, of a reservoir is measured daily throughout April.

Draw a histogram to show the depths.

Depth, d (cm)	Number of days
$0 < d \leq 5$	1
$5 < d \leq 10$	5
$10 < d \leq 15$	14
$15 < d \leq 20$	8
$20 < d \leq 25$	2

5 The exam marks of 36 students are shown in the frequency table.

Draw a bar chart to show the exam marks.

Exam mark (%)	Number of students
1 to 20	4
21 to 40	8
41 to 60	13
61 to 80	6
81 to 100	5

6 The histogram shows the best distances, in metres, that athletes threw a javelin in a competition.

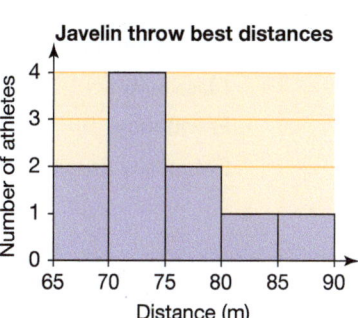

Javelin throw best distances

a State the number of athletes who threw the javelin
  i between 65 and 70 metres
  ii between 80 and 85 metres.

b In which class interval was the winner?

c What is the modal class interval?

d Calculate the total number of athletes who threw a javelin.

APPLICATIONS

4.3 Frequency diagrams

RECAP
- You can group data into class intervals when there are many values.
- Discrete data can only take exact values (usually collected by counting), for example the number of students in a class.
- Continuous data can take any value (collected by measuring), for example the heights of the students in a class.
- You can use a bar chart to display grouped discrete data.
- You can use a histogram to display continuous data.

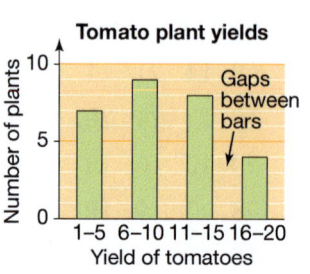

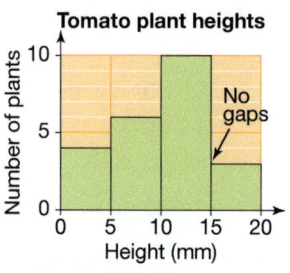

The bars in a histogram can have different widths.

The area of each bar represents the frequency.

The height of the bar represents the **frequency density**.

- Frequency density = $\dfrac{\text{Frequency}}{\text{Class width}}$

HOW TO

To draw a histogram
1. Find the class width of each interval.
2. Calculate the frequency density.
3. Draw a histogram. There are no gaps between the bars.

EXAMPLE

The table shows the time taken to complete a simple jigsaw.

Draw a histogram to represent the data.

Time, t seconds	$40 \leq t < 60$	$60 \leq t < 70$	$70 \leq t < 80$	$80 \leq t < 90$	$90 \leq t < 120$
Frequency	6	6	10	7	6

Time, t seconds	$40 \leq t < 60$	$60 \leq t < 70$	$70 \leq t < 80$	$80 \leq t < 90$	$90 \leq t < 120$
Class width	20	10	10	10	30
Frequency	6	6	10	7	6
Frequency density	0.3	0.6	1	0.7	0.2

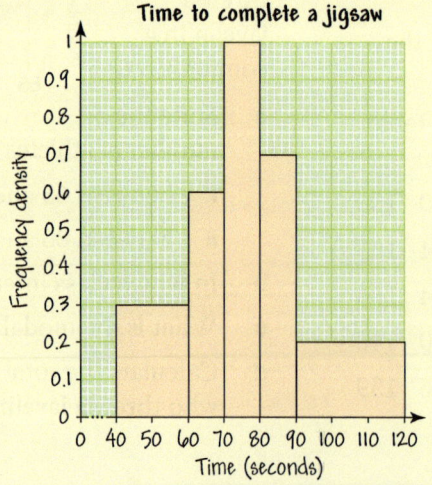

Add rows to the table to calculate class width and frequency density.

Reasoning and problem solving

Exercise 4.3A

For each set of data in questions **1–3**

a Copy and complete the table. **b** Draw a histogram to represent the data.

1 Reaction times of a sample of students.

Time, *t* seconds	$1 \leq t < 3$	$3 \leq t < 4$	$4 \leq t < 5$	$5 \leq t < 6$	$6 \leq t < 9$
Class width					
Frequency	12	17	19	11	18
Frequency density					

2 Amounts spent by the first 100 customers in a shop one Saturday.

Amount spent, £*a*	$0 \leq a < 5$	$5 \leq a < 10$	$10 \leq a < 20$	$20 \leq a < 40$	$40 \leq a < 60$	$60 \leq a < 100$
Class width						
Frequency	6	10	23	29	24	8
Frequency density						

3 Times dog owners spend on daily walks.

Time, *t* minutes	$10 \leq t < 20$	$20 \leq t < 40$	$40 \leq t < 60$	$60 \leq t < 90$	$90 \leq t < ?$
Class width					30
Frequency	8				9
Frequency density		0.8	1.4	1.3	

4 The histogram shows the times a sample of students spent watching TV one evening.

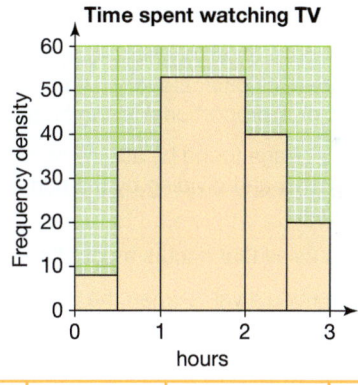

a How many students spent longer than $2\frac{1}{2}$ hours watching TV?

b Copy and complete the frequency table for the data.

Time, *t* hours	$0 \leq t < 0.5$	$0.5 \leq t < 1$	$1 \leq t < 2$	$2 \leq t < 2.5$	$2.5 \leq t < 3$
Frequency					

c How many students were in the sample?

5 The incomplete table and histogram give some information about the masses, in grams, of a sample of apples.

a Use the information in the histogram to work out the missing frequencies in the table.

b Copy and complete the histogram.

Mass, *g* grams	Frequency
$120 \leq g < 140$	8
$140 \leq g < 150$	6
$150 \leq g < 155$	
$155 \leq g < 160$	
$160 \leq g < 165$	
$165 \leq g < 175$	16
$175 \leq g < 185$	12
$185 \leq g < 200$	6

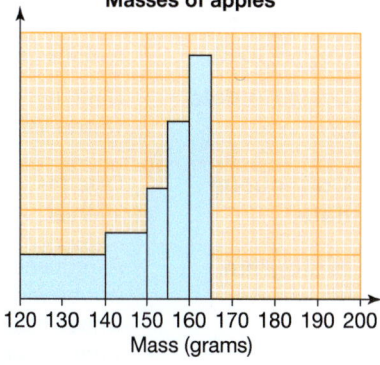

1196, 1197

Summary

Checkout
You should now be able to...

		Test it Questions
✓	Construct and interpret two-way tables, bar charts and pie charts.	1, 2
✓	Calculate the mean, median and mode of a data set.	3, 4
✓	Calculate the range and interquartile range of a data set.	3, 4
✓	Use averages and measures of spread to compare data sets.	4
✓	Use frequency tables to represent grouped data.	5
✓	Construct histograms with equal or unequal class widths.	5

Language	Meaning	Example
Two-way table	A table that links information about two different categories.	<table><tr><td></td><td>Boy</td><td>Girl</td></tr><tr><td>Year 10</td><td>73</td><td>67</td></tr><tr><td>Year 11</td><td>59</td><td>81</td></tr></table>
Bar-chart	A way of displaying data where the height of each bar represents the frequency.	See page 70.
Bar-line chart	A way of displaying data where the length of each line represents the frequency.	See page 70.
Pie chart	A circular chart in which each sector represents one category. The angle of each sector is proportional to the frequency.	See page 70
Mean	An average found by adding all the values together and dividing by the number of values.	Data: 1, 1, 2, 2, 4, 5, 7, 8, 8, 8, 9 Mean = $(1 + 1 + 2 + 2 + 4 + 5 + 7 + 8 + 8 + 8 + 9) \div 11$ $= 55 \div 11 = 5$
Mode	The value that occurs most often.	Mode = 8
Median	The middle value when the data is arranged in order of size. If there is an even number of data, the median is the mean of the middle two values.	Median = 5
Range	The difference between the largest value and the smallest value.	Range = $9 - 1 = 8$
Lower/Upper quartile	The middle value or median of the lower/upper half of the set of data.	Lower Quartile = $\frac{11 + 1}{4}$ = 3rd value = 2 Upper Quartile = $\frac{3(11 + 1)}{4}$ = 9th value = 8
Interquartile range (IQR)	The difference between the upper quartile and the lower quartile.	Interquartile range = $8 - 2 = 6$
Discrete data	Data that can only take values from a set whose values can be listed.	Shoe sizes 4, $4\frac{1}{2}$, 5, $5\frac{1}{2}$, 6, ... Discrete data is usually counted.
Continuous data	Data that can take any value within an interval.	Heights 1.63 m, 1.42 m, 1.565 m ..... Continuous data is usually measured.
Histogram	A frequency diagram using rectangles whose width is the **class interval** and whose height is the **frequency density** and whose *area* is the frequency.	See page 74.

Review

1. A class of 36 students choose to study either French or Spanish, but not both.
   There are 15 boys in the class.
   7 boys study French.
   19 students study Spanish.
   How many girls study Spanish?

2. The table shows the amount of money spent by a business in one year on different services.

Service	Expenditure ($1000s)
Warehouses	120
Corporate/Finance	60
Staff training	210
Transport	45
IT support	105

   a Construct a pie chart to show this information.

   A second business spends the same proportion on each service but has a larger budget.
   The second business spends $252 000 on staff training.

   b What is the total service budget of the second business?

   c Draw a bar chart to show the expenditure of the second council.

3. Anelle counts the number of seconds people can hold their breath for.

   25, 32, 39, 41, 17, 23, 29, 37, 35, 40, 72, 39, 31, 39, 42

   a Calculate the
      i mean   ii range of the data.
   b Calculate the
      i median   ii upper quartile
      iii lower quartile of this data.
   c What is the interquartile range?

4. The ages of members of a choir are given below.

   Women: 31, 39, 27, 56, 58, 60, 28, 34, 37, 31, 43

   Men: 52, 61, 35, 47, 48, 25, 30

   a For the men's ages find
      i the median
      ii the interquartile range.
   b For the women's ages find
      i the median
      ii the interquartile range.
   c Compare the two sets of data.

5. The table shows the lengths of the cars in a car park. Represent this data in a histogram.

Length, L (m)	Frequency
$3.4 < L \leq 3.8$	20
$3.8 < L \leq 4.2$	30
$4.2 < L \leq 4.4$	18
$4.4 < L \leq 4.6$	20
$4.6 < L \leq 5$	10

What next?

Score			
	0 – 2		Your knowledge of this topic is still developing. To improve, look at MyiMaths: 1192, 1193, 1196, 1197, 1202, 1205, 1206, 1207, 1214, 1254
	3 – 4		You are gaining a secure knowledge of this topic. To improve your fluency look at InvisiPens: 04Sa – l
	5		You have mastered these skills. Well done you are ready to progress! To develop your problem solving skills look at InvisiPens: 04Aa – g

Assessment 4

1. A manager recorded the time that staff members spent on personal emails during working hours.

Time (nearest min.)	5	6	7	8	9	10	11	12
Frequency	75	52	47	31	22	18	12	6

   a Draw a pie chart to represent this data [5]

   b Find the

       i mode [1]   ii median [2]

       of the time spent on personal emails.

   c Estimate the mean time spent on personal emails, giving your answer to the nearest second. [4]

2. Display the following information in a two-way table.

   In a group of 35 people, 15 wear glasses and 5 are left-handed. Four are left-handed *but don't* wear glasses. [4]

3. The bar chart shows the different types of coins in Zesiro's till.

   a Zesiro adds 12 £1 coins, 14 ten-pence coins and 7 two-pence coins to the till. Add this data to a copy of the bar chart. [2]

   b Which coin is found most often in the till? [1]

   c What is the value of the median coin in the till? [2]

   d Find the mean value of the coins in the till. [4]

   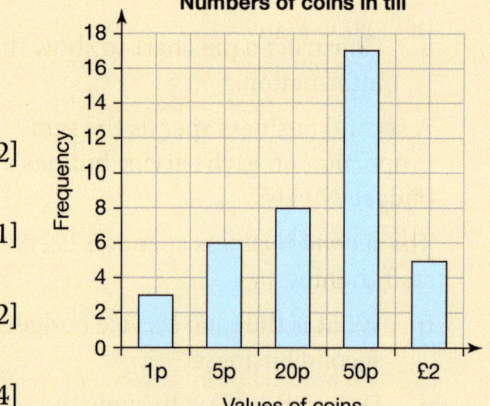

4. This pie chart show how 1200 hospital staff members get to work in summer.

   a Which is the modal form of travel? [1]

   b Which is the least favourite form of travel? [1]

   c What percentage of staff take the bus to work? [2]

   d How many staff drive to work? [2]

   e How many staff cycle to work? [2]

   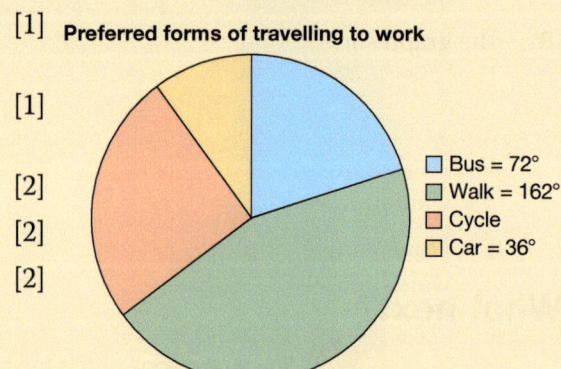

Statistics   Handling data 1

5 Tanya records the distances driven in 10 journeys (in miles).

   56   113   88   67   163   90   88   109   135   121

   a Calculate the
      i mean [2]   **ii** median [1]   **iii** mode. [1]

   b Tanya claims that she averages 100 miles per day.
   Is her claim reasonable? Give reasons for your answer. [2]

   c Calculate
      i the range [1]   **ii** the interquartile range. [3]

   d Tanya realises that one distance recorded as 88 should have been 98.
   Without performing any additional calculations, decide what effect this will have on the mean, median, mode, range and interquartile range of the distances.
   Give a reason for each of your answers. [5]

6 The table shows information about the number of children and pets in different families.

   a How many families have one pet? [1]

   b How many two children families have 1 pet? [1]

   c How many families are there where the number of pets is the same as the number of children in the family? [1]

		Number of pets					
		0	1	2	3	4	5
Number of children in family	1	3	5	4	1	0	1
	2	4	2	2	0	0	0
	3	1	0	2	1	0	0
	4	0	1	0	0	0	0
	5	2	1	0	0	0	0

   d What is the modal number of pets? [1]

   e Calculate the mean number of pets per family. [4]

7 The table shows the age distribution of the members of a club.

   a Draw a histogram to illustrate this data. [6]

   b How many members are younger than 20 years old? [1]

   c How many members are at least 45? [1]

   d How many members are between 20 and not more than 35 years old? [1]

Age, Y (years)	Frequency
$10 \leq Y < 15$	14
$15 \leq Y < 20$	18
$20 \leq Y < 25$	26
$25 \leq Y < 35$	33
$35 \leq Y < 45$	28
$45 \leq Y < 60$	24
$60 \leq Y < 80$	7

8 The graph shows the ages of all the people staying in a hotel.
There were 50 guests present who were 10 years old or younger.

The classes are $0 \leq$ Age < 10, ...

   a How many people aged 10 to 20 were in the hotel? [2]

   b How many people aged 50 to 80 were in the hotel? [1]

   c How many people were in the hotel altogether? [2]

5 Fractions, decimals and percentages

Introduction
Many food and drink products that you buy come with nutrition information clearly displayed on the labelling. This is usually in the form of percentages, for example 'saturated fat 5%, total carbohydrate 12%, calcium 9%, etc.'. Product manufacturers are expected by law to display this information, and the labels often tend to have a similar format.

What's the point?
Awareness of what you are eating and drinking is important in achieving a healthy, balanced diet and percentages allow you to monitor this.

Objectives
By the end of this chapter, you will have learned how to ...
- Find fractions and percentages of amounts.
- Add, subtract, multiply and divide with fractions and mixed numbers.
- Convert between fractions, decimals (including recurring decimals) and percentages.
- Order fractions, decimals and percentages.

Check in

1 Shade five copies of this diagram to represent each of these fractions.

  a $\frac{1}{2}$ **b** $\frac{2}{3}$ **c** $\frac{3}{4}$ **d** $\frac{5}{6}$ **e** $\frac{7}{12}$

2 Cancel each of these fractions down to their simplest terms.

  a $\frac{2}{4}$ **b** $\frac{15}{20}$ **c** $\frac{8}{10}$ **d** $\frac{95}{100}$ **e** $\frac{6}{8}$

3 Write notes to show how you would find a mental estimate for each of these calculations.

  a 19.2×28.9 **b** $355.72 \div 58.91$ **c** $1206 - 816$ **d** $6987 + 6039$

Chapter investigation

$\pi = 3.14159265358979323846$ (20 dp)

Find an approximation to π in the form $3 + \frac{1}{a}$

Find another approximation to π in the form $3 + \dfrac{1}{a + \frac{1}{b}}$

Which approximation is more accurate?

SKILLS

5.1 Fractions and percentages

- You find fractions of a quantity by multiplying.

Two thirds of 5 is $\frac{2}{3} \times 5 = \frac{2 \times 5}{3} = \frac{10}{3} = 3\frac{1}{3}$

Reminder
Types of fraction
Proper $\frac{a}{b} < 1$ $\frac{3}{4}$
Improper $\frac{a}{b} > 1$ $\frac{4}{3}$
Mixed Number $1\frac{1}{3}$

- When the quantity and the denominator of the fraction have a **common factor**, you should **cancel** this factor before multiplying.

Two thirds of 24 is $\frac{2}{\cancel{3}_1} \times \cancel{24}^8 = \frac{2}{1} \times 8 = 16$

EXAMPLE

Calculate **a** $\frac{3}{4}$ of 28 **b** $\frac{5}{8}$ of 6 **c** $\frac{4}{9}$ of 12 **d** $\frac{5}{9}$ of 25.5

a $\frac{3}{\cancel{4}_1} \times \cancel{28}^7 = 3 \times 7 = 21$ **b** $\frac{5}{\cancel{8}_4} \times \cancel{6}^3 = \frac{5 \times 3}{4} = \frac{15}{4} = 3\frac{3}{4}$

c $\frac{4}{\cancel{9}_3} \times \cancel{12}^4 = \frac{16}{3} = 5\frac{1}{3}$ **d** $\frac{5}{9} \times 25.5 = \frac{5}{\cancel{9}_3} \times \frac{\cancel{51}^{17}}{2} = \frac{85}{6} = 14\frac{1}{6}$

$25.5 \times \frac{2}{2} = \frac{51}{2}$

- A percentage (%) is a fraction with a denominator of 100. It tells you how many parts per 100.

To convert a percentage to a decimal divide it by 100 and to a fraction write it over 100.

- To find a percentage of an amount, multiply the quantity by an equivalent decimal or fraction.

$40\% = \frac{40}{100} = \frac{2}{5}$ or 0.4 so 40% of £30 is $\frac{2}{5} \times 30$ or $0.4 \times 30 =$ £12

- You can use mental methods to find percentages of amounts.

To find 50%, divide by 2. $50\% = \frac{50}{100} = \frac{1}{2}$ To find 25%, divide by 4. $25\% = \frac{25}{100} = \frac{1}{4}$

To find 10%, divide by 10. $10\% = \frac{10}{100} = \frac{1}{10}$ To find 1%, divide by 100. $1\% = \frac{1}{100}$

EXAMPLE

Calculate **a** 45% of 60 cm **b** 34% of 85 kg **c** 16% of €25

a 50% of 60 = 30   50% is a half.
5% of 60 = 3   Divide by 10.
45% of 60 cm = 30 cm − 3 cm
= 27 cm   Include units.

b 34% of 85 kg = 0.34 × 85   Use a decimal equivalent.
= 28.9 kg   By calculator.
= 29 kg (2 sf)

c 16% of €25 = $\frac{\cancel{16}^4}{\cancel{100}_1} \times \cancel{25}^1$ Cancel twice.
= €4

Number   Fractions, decimals and percentages

Fluency

Exercise 5.1S

1. Calculate these fractions of a quantity. Give your answers as fractions or mixed numbers.
   - **a** $\frac{1}{2}$ of 7.5
   - **b** $\frac{1}{5}$ of 8
   - **c** $\frac{1}{3}$ of 10
   - **d** $\frac{1}{3}$ of 2.6
   - **e** $\frac{2}{5}$ of 6
   - **f** $\frac{1}{5} \times 8.4$

2. Calculate these fractions of a quantity.
   - **a** $\frac{1}{2}$ of 20
   - **b** $\frac{1}{4}$ of 84
   - **c** $\frac{1}{3}$ of 36
   - **d** $\frac{1}{5}$ of 65
   - **e** $\frac{2}{3}$ of 33
   - **f** $\frac{1}{8} \times 64$

> For questions **3**, **4** and **5** show your working, including cancelling common factors.

3. A reel holds 60 m of wire when new. Two-fifths of the wire has been used.
   - **a** What length has been used?
   - **b** What length is left on the reel?

4. Calculate the amount of liquid in these containers.
   - **a** A 40 litre barrel that is $\frac{3}{8}$ full.
   - **b** A 240 cl jar that is $\frac{3}{4}$ full.
   - **c** A 120 cl glass that is $\frac{2}{5}$ full.
   - **d** A 750 ml bottle that is $\frac{2}{3}$ empty.

5. Calculate these distances.
   - **a** $\frac{5}{8}$ of 48 m
   - **b** $\frac{2}{9}$ of 36 km
   - **c** $\frac{4}{7}$ of 28 mm
   - **d** $\frac{3}{4}$ of 120 m

> For questions **6** and **7** give your answers as fractions or mixed numbers.

6. Calculate these fractions of a quantity.
   - **a** $\frac{1}{6}$ of 10
   - **b** $\frac{1}{4}$ of 22
   - **c** $\frac{3}{10}$ of 15
   - **d** $\frac{1}{12}$ of 8
   - **e** $\frac{4}{9}$ of 21
   - **f** $\frac{2}{5} \times 72$

7. Calculate these distances.
   - **a** $\frac{1}{9}$ of 24 miles
   - **b** $\frac{5}{6}$ of 40 miles
   - **c** $\frac{5}{18}$ of 45 miles
   - **d** $\frac{3}{20}$ of 25 miles

8. Calculate these fractions of a quantity. Give your answers as decimals to an appropriate degree of accuracy.
   - **a** $\frac{5}{12}$ of 16 m
   - **b** $\frac{4}{9}$ of 12 mm
   - **c** $\frac{3}{22}$ of 64 cm
   - **d** $\frac{3}{14}$ of 104 km

9. Work out these percentages in your head.
   - **a** 25% of 42
   - **b** 90% of 140
   - **c** 20% of 1200
   - **d** 11% of 900
   - **e** 60% of 500
   - **f** 30% of 440
   - **g** 30% of $750
   - **h** 55% of 1800 m

10. Calculate these percentages.
    - **a** 9% of 1500
    - **b** 13% of 700
    - **c** 31% of 2400
    - **d** 36% of 50
    - **e** 43% of 900
    - **f** 6% of 3200
    - **g** 23% of 4800 mm
    - **h** 61% of 3200 kg
    - **i** 39% of €3700
    - **j** 17% of £2900

11. Write these percentages as equivalent decimals.
    - **a** 50%
    - **b** 60%
    - **c** 25%
    - **d** 51%
    - **e** 64%
    - **f** 22%
    - **g** 15%
    - **h** 70%
    - **i** 7%
    - **j** 8.5%
    - **k** 0.15%
    - **l** 0.01%

12. Calculate these percentages using an appropriate method.
    - **a** 15% of 38
    - **b** 25% of 800
    - **c** 27% of 59
    - **d** 96% of 104
    - **e** 41% of 41
    - **f** 80% of 25

13. Calculate these percentages.
    - **a** 16% of €24
    - **b** 63% of $85
    - **c** 93% of $15
    - **d** 42% of £405
    - **e** 88% of £32
    - **f** 6% of €265

14. Calculate
    - **a** $\frac{3}{5}$ of $\frac{1}{6}$ of €90
    - **b** 30% of $\frac{1}{6}$ of £54
    - **c** 25% of 35% of $48

1030, 1031, 1046

89

APPLICATIONS

5.1 Fractions and percentages

RECAP
- To find a fraction of a quantity, multiply the quantity and the fraction.
- To find a percentage of a quantity, multiply the quantity written as an equivalent decimal or fraction by 100.

Multiplying by a fraction is the same as dividing by the denominator, then multiplying by the numerator.

HOW TO

To solve problems involving fractions and percentages
1. Read the question and decide what you need to do.
2. Decide which method, written, mental or calculator, to use.
3. Answer the question and don't forget to include any units.

EXAMPLE

VAT (Value Added Tax) is added on to the cost of items in some countries. VAT is 20% of the item's price. Donna bought a camera for £360 and a case for £17.65. She was *not* charged VAT. How much money did Donna save by buying the items VAT-free?

① Calculate 20% of the price of each item. Add the two amounts.

10% is £36   ② 10% is the same as ÷ 10.   10% is £1.765   ② Divide by 10, don't round.
20% is £72   Double it.   20% is £3.53   Double it.

£72 + £3.53 = £75.53.   Donna saved £75.53.   ③

EXAMPLE

Abda earns $840 per week.
She budgets to put 50% aside for general spending and split the rest into four shares.

i $\frac{1}{4}$ for food   ii $\frac{1}{5}$ for travel   iii $\frac{3}{10}$ for electricity   iv $\frac{3}{8}$ for rent

a How much does she put aside for each item.
b Abda has made a mistake. What is the problem?

① Multiply $840 by each fraction.   ② You can do these without a calculator.

a 50% is $420

i $\frac{1}{4}$ of $420 = $105   Divide by 4.   ii $\frac{1}{5}$ of $420 = $84   Divide by 5.
iii $\frac{1}{10}$ of $420 = $42   Divide by 10.   iv $\frac{1}{8}$ of $420 = $52.50   Divide by 8.
    $\frac{3}{10}$ of $420 = $126   Multiply by 3. $\frac{3}{8}$ of $420 = $157.50 Multiply by 3.

b $105 + $84 + $126 + $157.50 = $472.50 > $420   Add up the total.
The fractions she chose add up to more than one whole.   ③

EXAMPLE

Tamer has €42. He spends 30% of it on Monday. On Tuesday, he spends $\frac{3}{4}$ of the remainder. How much does he spend on Tuesday?

① Work out how much is left after Monday.   Then calculate $\frac{3}{4}$ of the remainder.

30% of €42 = 0.3 × 42   29.4 × 0.75 = 22.05   ② $\frac{3}{4}$ = 0.75
0.1 × 42 = 4.2   3 × 4.2 = 12.6   He spends €22.05 on Tuesday.   ③
42 − 12.6 = 29.4   ② Calculate 70%.

Number   Fractions, decimals and percentages

Reasoning and problem solving

Exercise 5.1A

1. Egor works overtime for two weeks. He earns $190 in the first week, and $224.50 in the second. He pays a 20% tax on each amount.

   How much tax has he paid at the end of the two weeks?

2. Luc and Jill fetched 24 pints of water in their bucket. Jill drank a tenth of this, Luc a quarter. They gave three eighths to Hayal, and a fifth to Jill's mother.

   How much water was left over?

3. Phil works out that yesterday he spent a third of the day asleep, $\frac{3}{8}$ of the day watching TV, a tenth eating, and a quarter at school.

   Is he correct? How can you tell?

4. Jayne earns €320 a week. She spends $\frac{3}{5}$ of the money and deposits the remainder into a savings account. How much money will Jayne save in a year?

5. The label on a chocolate bar says it has mass 220 g and contains 19% saturated fat. If one bar is made up of 32 squares of chocolate, how many grams of fat are in 4 squares?

6. Stefan is drawing a pie chart to display information on students' favourite lessons. He uses the fractions of the total number of students who like each subject to work out the angles in degrees for each sector.

   Favourite subject

English	$\frac{1}{4}$	History	$\frac{1}{12}$
French	$\frac{1}{10}$	Maths	$\frac{3}{8}$
Geography	$\frac{1}{8}$	Other	☐

   What angle is the 'Other' sector?

7. Mrs Kim has a conservatory built, which costs £12 000. She pays 15% of this as a deposit. She has to pay the rest in 24 equal monthly payments.
   How much is each monthly payment?

8. Shapes A, B and C are identical.

   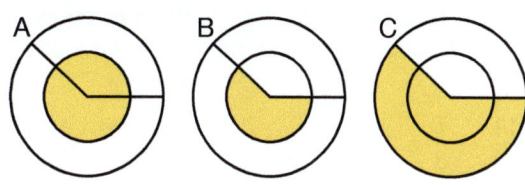

   $\frac{2}{3}$ of A is shaded. 60% of C is shaded. What fraction of B is shaded?

9. Latif is given some money. He spends $\frac{1}{3}$ of it on Monday and 45% of what is left on Tuesday. What fraction of the total did he spend on Tuesday?

10. To get a driving licence, you have to pass *both* the theory and practical test. You cannot take the practical test if you fail the theory test. Two hundred people applied for a driving licence: 80% of them passed the theory test and $\frac{3}{4}$ of the people who took the practical test passed.
    How many people got a licence?

11. At a gym, 70% of the members are women. 50% of the women use the swimming pool. 80% of the men also use the pool. What percentage of the members use the pool?

12. Farmer Macdonald has died leaving his 17 horses to his three sons. The only condition is that the oldest son gets *exactly* half the horses, the middle son *exactly* $\frac{1}{3}$ and the youngest son *exactly* $\frac{1}{9}$. Seeing them struggle to do this their neighbour lends them her horse. Suddenly they can divide up the horses and still be able to give the neighbour's horse back to her. Is this possible? Give your reason.

SKILLS

5.2 Calculations with fractions

The easiest way to add and subtract fractions is to make sure they have a **common denominator**.

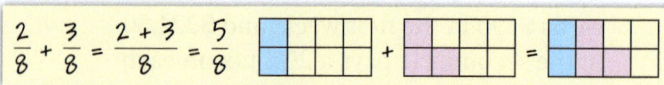

$$\frac{2}{8} + \frac{3}{8} = \frac{2+3}{8} = \frac{5}{8}$$

- To add or subtract fractions with different denominators, change them to equivalent fractions with a common denominator.

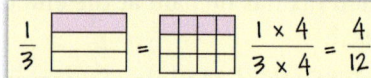

$$\frac{1}{3} = \frac{1 \times 4}{3 \times 4} = \frac{4}{12}$$

p.256 $\dfrac{11}{12} - \dfrac{1}{3} = \dfrac{11-4}{12} = \dfrac{7}{12}$ The **lowest common denominator** is the lowest common multiple (LCM) of 12 and 3, which is 12.

You should write your answer in its simplest form by cancelling common factors, so that the numerator and denominator are as small as possible.

If your answer is an improper fraction, change it to a mixed number.

$$\frac{3}{4} + \frac{2}{5} = \frac{15}{20} + \frac{8}{20} = \frac{23}{20} = 1\frac{3}{20} \qquad 23 = 20 + 3$$

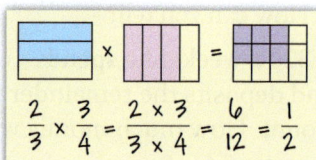

$$\frac{2}{3} \times \frac{3}{4} = \frac{2 \times 3}{3 \times 4} = \frac{6}{12} = \frac{1}{2}$$

- To multiply proper and improper fractions, cancel any common factors, and then multiply the numerators and then the denominators.

To multiply by a mixed number, first convert it to an improper fraction.

- To divide by a fraction, multiply by its **reciprocal**.

The reciprocal of a number is what you multiply the number by to get 1.

The reciprocal of a fraction is the fraction 'turned upside down'.

$$\frac{2}{3} \div \frac{5}{6} = \frac{2}{{}_1\cancel{3}} \times \frac{\cancel{6}^2}{5} = \frac{4}{5}.$$

EXAMPLE

Work out each of these calculations.

a $\dfrac{3}{5} + \dfrac{11}{16}$ b $\dfrac{2}{7} - \dfrac{1}{4}$ c $\dfrac{3}{8} \times \dfrac{5}{9}$ d $\dfrac{3}{5} \div 8$ e $\dfrac{3}{8} \div \dfrac{6}{7}$ f $2\dfrac{4}{5} \div 3\dfrac{1}{2}$

a $\dfrac{3}{5} + \dfrac{11}{16} = \dfrac{48 + 55}{80} = \dfrac{103}{80} = 1\dfrac{23}{80}$

The LCM of 5 and 16 is 80.

c $\dfrac{\cancel{3}}{8} \times \dfrac{5}{\cancel{9}_3} = \dfrac{1}{8} \times \dfrac{5}{3} = \dfrac{5}{24}$

Cancel common factors before multiplying.

e $\dfrac{3}{8} \div \dfrac{6}{7} = \dfrac{\cancel{3}}{8} \times \dfrac{7}{\cancel{6}_2} = \dfrac{7}{16}$

The reciprocal of $\dfrac{6}{7}$ is $\dfrac{7}{6}$.

b $\dfrac{2}{7} - \dfrac{1}{4} = \dfrac{8-7}{28} = \dfrac{1}{28}$

The LCM of 7 and 4 is 28.

d $\dfrac{3}{5} \div 8 = \dfrac{3}{5} \times \dfrac{1}{8} = \dfrac{3 \times 1}{5 \times 8} = \dfrac{3}{40}$

The reciprocal of 8 is $\dfrac{1}{8}$.

f $2\dfrac{4}{5} \div 3\dfrac{1}{2} = \dfrac{14}{5} \div \dfrac{7}{2} = \dfrac{\cancel{14}^2}{5} \times \dfrac{2}{\cancel{7}_1} = \dfrac{4}{5}$

Change mixed numbers to improper fractions first.

Number   Fractions, decimals and percentages

Fluency

Exercise 5.2S

1 Calculate these additions and subtractions.
 a $\frac{1}{5}+\frac{2}{5}$ b $\frac{1}{4}+\frac{3}{4}$ c $\frac{3}{8}+\frac{5}{8}$
 d $\frac{4}{5}-\frac{1}{5}$ e $\frac{5}{6}-\frac{1}{6}$ f $\frac{9}{10}-\frac{3}{10}$

2 Find these equivalent fractions.
 a $\frac{2}{3}=\frac{\square}{30}$ b $\frac{3}{7}=\frac{\square}{42}$ c $\frac{7}{9}=\frac{\square}{45}$
 d $\frac{5}{8}=\frac{\square}{40}$ e $\frac{21}{63}=\frac{\square}{9}$ f $\frac{5}{6}=\frac{35}{\square}$

3 Calculate these additions.
 a $\frac{1}{5}+\frac{1}{10}$ b $\frac{2}{3}+\frac{1}{6}$ c $\frac{2}{5}+\frac{3}{20}$
 d $\frac{1}{8}+\frac{1}{4}$ e $\frac{1}{4}+\frac{3}{5}$ f $\frac{2}{7}+\frac{3}{8}$

4 Convert these improper fractions into mixed numbers.
 a $\frac{5}{4}$ b $\frac{9}{5}$ c $\frac{13}{8}$ d $\frac{17}{4}$

5 Convert these mixed numbers into improper fractions.
 a $1\frac{3}{4}$ b $1\frac{7}{16}$ c $1\frac{5}{9}$ d $2\frac{4}{7}$

6 Calculate these additions and subtractions.
 a $1\frac{1}{5}+2\frac{1}{10}$ b $3\frac{1}{4}-\frac{1}{8}$
 c $4\frac{3}{7}+3\frac{1}{2}$ d $5\frac{4}{9}+2\frac{3}{7}$
 e $3\frac{3}{5}-2\frac{1}{4}$ f $2\frac{1}{2}-1\frac{3}{4}$
 g $3\frac{3}{4}-1\frac{4}{5}$ h $7\frac{3}{7}-2\frac{1}{2}$

7 Calculate these multiplications. Give your answers in their simplest form.
 a $\frac{3}{4}\times\frac{1}{5}$ b $\frac{2}{3}\times\frac{2}{9}$
 c $\frac{2}{7}\times\frac{1}{4}$ d $\frac{5}{16}\times\frac{4}{5}$
 e $\frac{5}{9}\times\frac{4}{7}$ f $\frac{7}{8}\times\frac{2}{21}$
 g $\frac{4}{5}\times\frac{3}{13}$ h $\frac{8}{35}\times\frac{7}{24}$
 i $\frac{2}{3}\times\frac{6}{7}\times\frac{5}{12}$ j $\frac{6}{28}\times\frac{8}{15}\times\frac{35}{48}$

8 Calculate these divisions, giving your answers as fractions in their simplest form.
 a $3\div4$ b $6\div8$ c $8\div10$
 d $9\div5$ e $22\div7$ f $22\div8$

9 Calculate these divisions.
 a $\frac{5}{8}\div4$ b $\frac{3}{4}\div6$ c $\frac{2}{3}\div7$
 d $\frac{3}{16}\div9$ e $4\div\frac{1}{5}$ f $6\div\frac{2}{3}$
 g $5\div\frac{2}{5}$ h $11\div\frac{3}{7}$ i $51\div\frac{17}{3}$

10 Calculate these divisions.
 a $\frac{1}{8}\div\frac{3}{8}$ b $\frac{1}{5}\div\frac{4}{5}$ c $\frac{1}{14}\div\frac{3}{7}$
 d $\frac{1}{10}\div\frac{2}{5}$ e $\frac{2}{3}\div\frac{3}{4}$ f $\frac{5}{8}\div\frac{7}{9}$
 g $\frac{1}{12}\div\frac{3}{8}$ h $\frac{2}{7}\div\frac{3}{4}$ i $\frac{72}{154}\div\frac{16}{33}$

11 Calculate these multiplications and divisions.
 a $1\frac{1}{2}\times\frac{3}{4}$ b $2\frac{3}{4}\times\frac{2}{5}$
 c $1\frac{1}{5}\times2\frac{1}{5}$ d $1\frac{2}{3}\times1\frac{4}{5}$
 e $2\frac{7}{8}\div\frac{3}{5}$ f $4\frac{1}{4}\div3\frac{1}{2}$
 g $5\frac{3}{8}\div2\frac{3}{4}$ h $9\frac{1}{3}\div2\frac{1}{4}$

12 Calculate
 a $3\frac{7}{8}+2\frac{1}{4}$ b $3\frac{7}{8}-3\frac{1}{4}$
 c $5\frac{1}{2}\times1\frac{7}{8}$ d $2\frac{1}{2}\div3\frac{3}{4}$

13 Calculate
 a $\dfrac{\frac{3}{4}+\frac{1}{2}}{\frac{5}{8}-\frac{1}{4}}$ b $\left(\frac{3}{4}-\frac{1}{8}\right)\left(\frac{3}{4}+\frac{1}{8}\right)$
 c $\dfrac{1\frac{1}{2}-\frac{7}{8}}{2\frac{3}{4}-1\frac{1}{6}}$ d $\left(2\frac{3}{5}-1\frac{3}{4}\right)\left(1\frac{7}{8}+3\frac{1}{4}\right)$

14 Check your answers to questions **6–13** using a calculator.

APPLICATIONS

5.2 Calculations with fractions

RECAP
- To add and subtract fractions, convert to equivalent fractions with a common denominator, then add or subtract the numerators.
- To multiply fractions, convert any mixed numbers to improper fractions, cancel any common factors, and then multiply the numerators and the denominators.
- Dividing by a fraction is the same as multiplying by its reciprocal.

$$6\frac{1}{4} - \frac{5}{6} = \frac{25}{4} - \frac{5}{6}$$
$$= \frac{75 - 10}{12}$$
$$= \frac{65}{12} = 5\frac{5}{12}$$

$$6\frac{1}{4} \div \frac{5}{6} = \frac{{}^5\cancel{25}}{{}_2\cancel{4}} \times \frac{\cancel{6}^3}{\cancel{5}_1}$$
$$= \frac{15}{2} = 7\frac{1}{2}$$

HOW TO
To solve problems involving calculations with fractions
1. Read the question and decide what you need to do ($+, -, \times, \div$).
2. Use the appropriate written method(s) or your calculator.
3. Answer the question, give your answer in its simplest form and include any units.

EXAMPLE

Hani works in a cafe. At the end of the night he has three measuring jugs with milk left in. One has half a litre in it, the second $1\frac{1}{4}$ litres, the third has $\frac{1}{3}$ of a litre. He tries to pour all the milk into one bottle.

Will the milk fit in a 2 litre bottle?

1. Add the three fractions together to check if sum ≤ 2.
2. $\frac{1}{2} + 1\frac{1}{4} + \frac{1}{3} = \frac{6}{12} + \frac{15}{12} + \frac{4}{12}$ A common denominator is 12.
   $= \frac{25}{12} = 2\frac{1}{12}$ Convert to a mixed number.
3. It does not fit in the bottle because $2\frac{1}{12} > 2$.

EXAMPLE

Tamara's answer is incorrect. Show how to get the correct answer and explain what she did wrong.

$\left(\frac{3}{4} + \frac{7}{5}\right) \times \left(\frac{4}{3} \div \frac{2}{3}\right) = \frac{10}{9} \times \frac{2}{1} = \frac{20}{9}$ ✗

1. Find the correct answer using BIDMAS and the rules for fractions.
2. Use a common denominator.   Multiply by the reciprocal.

$\frac{3}{4} + \frac{7}{5} = \frac{15 + 28}{20} = \frac{43}{20}$ $\frac{4}{3} \div \frac{2}{3} = \frac{{}^2\cancel{4}}{{}_1\cancel{3}} \times \frac{\cancel{3}^1}{\cancel{2}_1} = 2$

$\frac{43}{{}_{10}\cancel{20}} \times \cancel{2}^1 = \frac{43}{10} = 4\frac{3}{10}$ Simplify.   Your working answers the 'show' part of the question.

3. In the first bracket, she added the numerators and the denominators.

EXAMPLE

Rachael sews together hexagonal patches to make a quilt. Each triangle is equilateral.

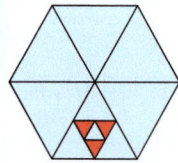

What fraction of the patch do the red triangles represent?

1. The red triangles are a fraction, $\frac{3}{4}$, of a fraction, $\frac{1}{4}$, of a fraction, $\frac{1}{6}$, of the hexagon so multiply.
2. $\frac{\cancel{3}^1}{4} \times \frac{1}{4} \times \frac{1}{\cancel{6}_2} = \frac{1}{32}$
3. The red triangles are $\frac{1}{32}$ of the patch.

Number   Fractions, decimals and percentages

Reasoning and problem solving

Exercise 5.2A

1. Gil cuts three pieces of piping from a pipe six inches long. The pieces measure $1\frac{1}{2}$ inches, $2\frac{1}{4}$ inches and $1\frac{5}{8}$ inches.

   How much pipe is spare?

2. Asif writes

   $$\left(\frac{1}{2} + \frac{1}{3}\right) \div \left(\frac{1}{1\frac{1}{2}} - \frac{1}{2\frac{1}{2}}\right) = \frac{2}{9} \quad \times$$

   a. What is the correct answer?

   b. Suggest what Asif has done wrong.

3. Beatrice works overtime one weekend. She is paid at *time and a quarter*, that is, she is paid $1\frac{1}{4}$ hours for every hour she works. She works for 4 hours on Saturday, and $5\frac{1}{2}$ hours on Sunday. She gets paid $5 per hour.

   How much money does she earn from the overtime?

4. Here are eight fractions.

   | $\frac{1}{2}$ | $\frac{3}{4}$ | $\frac{3}{5}$ | $\frac{1}{8}$ | $\frac{4}{3}$ | $\frac{5}{2}$ | $\frac{3}{8}$ | $2\frac{1}{2}$ |

   a. Which pairs add up to a whole number?

   b. Which pair multiply to make 1?

   c. Which pair multiply to make $1\frac{1}{2}$?

   d. Find a pair where one is twice the other.

   e. What is the largest value minus the smallest value?

   f. How many times does the smallest fraction go into the largest?

   g. Which two are reciprocals of one another?

5. By writing the fractions with a common denominator put these fractions in

   a. ascending order

      i. $\frac{3}{7}, \frac{7}{8}, \frac{5}{14}$ ii. $\frac{2}{3}, \frac{5}{6}, \frac{2}{7}$

   b. descending order.

      i. $\frac{2}{5}, \frac{3}{8}, \frac{3}{4}, \frac{17}{40}$ ii. $\frac{5}{6}, \frac{11}{24}, \frac{7}{12}, \frac{5}{8}$

6. Is $\frac{1}{2}$ closer to $\frac{4}{9}$ or $\frac{6}{11}$?

   Give your reason.

7. Here are three identical shapes.

   $\frac{4}{7}$ of A is shaded. $\frac{3}{5}$ of C is shaded.

   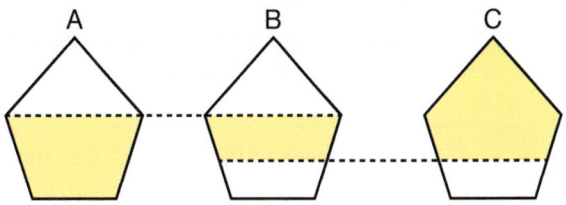

   How much of B is shaded?

8. Choose from this list the two fractions with the lowest product. Explain your choice.

   $\frac{1}{2} \quad \frac{3}{8} \quad \frac{2}{5} \quad \frac{4}{7} \quad \frac{1}{10}$

9. If $a > b$, then $\frac{1}{a} > \frac{1}{b}$.

   Is this statement true? Give your reason.

10. Fill the empty squares to make each statement true horizontally and vertically.

$\frac{1}{2}$	+	$\frac{1}{6}$	=	
×		+		★
	×		=	
=		=		=
$\frac{3}{20}$	×		=	$\frac{1}{10}$

    Which operation must replace '★'?

11. $\frac{5}{8} \times \frac{51}{11} \times \frac{\square}{45} \times \frac{3}{7} \div \frac{34}{44} = 1$

    Find the missing number.

12. Three friends have different ways of dividing fractions.

    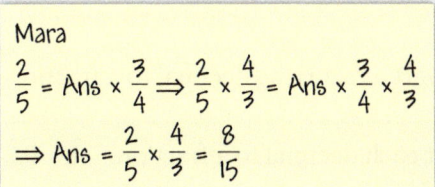

    Show how the three methods work for these divisions.

    a. $\frac{4}{5} \div \frac{3}{7}$ b. $\frac{a}{b} \div \frac{c}{d}$

SKILLS

5.3 Fractions, decimals and percentages

- To convert a fraction to a decimal divide the numerator by the denominator.
- To convert a fraction to a percentage write it as a decimal and multiply by 100%.

$$100\% = \frac{100}{100} = 1$$

You use dot notation to show which digits recur.

$0.12\dot{3} = 0.12333...$

$0.1\dot{2}\dot{3} = 0.1232323...$

$0.\dot{1}2\dot{3} = 0.1231231...$

EXAMPLE

Convert these fractions to **i** decimals **ii** percentages.

a $\frac{3}{8}$ **b** $\frac{1}{12}$

a **i** $\frac{3}{8} = 3 \div 8 = 8\overline{)3.0^60^40} = 0.375$ **b** **i** $\frac{1}{12} = 1 \div 12 = 12\overline{)1.00^40^40...} = 0.08\dot{3}$

A terminating decimal.   A recurring decimal.

ii $0.375 \times 100\% = 37.5\%$ **ii** $0.08\dot{3} \times 100\% = 8.\dot{3}\%$

- A fraction will be a **recurring** decimal if the denominator has **prime factors** other than 2 and/or 5 otherwise it will be a **terminating** decimal.

A terminating decimal stops, e.g. 0.625

EXAMPLE

Say if these fractions are terminating or recurring decimals.   **a** $\frac{4}{30}$ **b** $\frac{7}{16}$

a $30 = 2 \times 3 \times 5 \Rightarrow$ recurring decimal   3 is a prime factor.
b $16 = 2^4 \Rightarrow$ terminating decimal

- To convert a **terminating** decimal to a fraction, write the decimal as a fraction with **denominator** 10, 100, 1000, ... according to the number of decimal places.
- To convert a percentage to a fraction, divide the percentage by 100.

EXAMPLE

Convert these to fractions.   **a** 0.306   **b** 45%   **c** 32.5%   **d** 0.52

a $0.306 = \frac{306}{1000} = \frac{153}{500}$ **b** $45\% = \frac{45}{100} = \frac{9}{20}$

c $32.5\% = \frac{32.5}{100} = \frac{65}{200} = \frac{13}{40}$ **d** $0.52 = \frac{52}{100} = \frac{26}{50} = \frac{13}{25}$

You can convert a recurring decimal to a fraction by using basic algebra.

EXAMPLE

Convert each decimal to a fraction.   **a** $0.\dot{3}$ **b** $0.\dot{1}\dot{2}$

a Let $x = 0.3333...$
$10x = 3.3333...$
$9x = 3$ $10x - x = 3.\dot{3} - 0.\dot{3}$
$x = \frac{3}{9} = \frac{1}{3}$

b Let $x = 0.121212...$
$100x = 12.121212...$
$99x = 12$ $100x - x = 12.\dot{1}\dot{2} - 0.\dot{1}\dot{2}$
$x = \frac{12}{99} = \frac{4}{33}$

Number   Fractions, decimals and percentages

Fluency

Exercise 5.3S

1. Write these fractions as
   i decimals ii percentages.
   a $\frac{1}{2}$ b $\frac{3}{4}$ c $\frac{2}{5}$ d $\frac{1}{10}$
   e $\frac{1}{5}$ f $\frac{1}{4}$ g $\frac{1}{8}$ h $\frac{5}{8}$

2. Convert these fractions to percentages.
   a $\frac{5}{8}$ b $\frac{4}{5}$ c $\frac{7}{8}$ d $\frac{3}{5}$
   e $\frac{3}{8}$ f $\frac{1}{8}$ g $\frac{3}{16}$ h $\frac{7}{32}$

3. Try to do question 2 without a calculator.

4. Convert these fractions to
   i decimals ii percentages
   a $\frac{1}{16}$ b $\frac{7}{25}$ c $\frac{7}{125}$ d $\frac{3}{40}$
   e $\frac{7}{16}$ f $\frac{1}{32}$ g $\frac{41}{80}$ h $\frac{33}{160}$

5. Which of these fractions will make recurring decimals? Explain your answer.
   a $\frac{22}{25}$ b $\frac{17}{20}$ c $\frac{8}{11}$ d $\frac{2}{5}$

6. Write each of these recurring decimals using 'dot' notation.
   a 0.111... b 0.555...
   c 0.755 55... d 0.346 346 346...
   e 0.765 656 5... f 0.16161616...

7. Convert each fraction to
   i a decimal ii a percentage
   a $\frac{1}{3}$ b $\frac{1}{6}$ c $\frac{2}{3}$ d $\frac{1}{7}$
   e $\frac{1}{9}$ f $\frac{5}{6}$ g $\frac{1}{11}$ h $\frac{1}{12}$

8. Convert these fractions to decimals.
   a $\frac{3}{7}$ b $\frac{3}{16}$ c $\frac{17}{80}$ d $\frac{5}{9}$
   e $\frac{4}{25}$ f $\frac{5}{7}$ g $\frac{41}{125}$ h $\frac{11}{12}$

9. Try to do questions 7 and 8 without a calculator.

10. Convert these percentages to decimals.
    a 43% b 86% c 94%
    d 45.5% e 3.75% f 105%

11. Convert these decimals to fractions, using a mental method.
    a 0.5 b 0.25 c 0.2
    d 0.125 e 0.75 f 0.9

12. Convert these decimals to fractions.
    a 0.51 b 0.43 c 0.413
    d 0.719 e 0.91 f 0.871

13. Convert these decimals to fractions. Give your answers in their simplest forms.
    a 0.32 b 0.55 c 0.44
    d 0.155 e 0.64 f 0.265

14. Convert these percentages to fractions.
    a 49% b 53% c 73%
    d 81% e 37% f 19%

15. Convert these percentages to fractions.
    a 55% b 62% c 84%
    d 65% e 72% f 18.5%

16. Convert these recurring decimals to fractions.
    a $0.\dot{7}$ b $0.\dot{6}$ c $0.\dot{1}\dot{5}$
    d $0.5\dot{1}$ e $0.\dot{1}2\dot{3}$ f $0.1\dot{4}\dot{8}$
    g $0.0\dot{3}$ h $0.1\dot{5}$ i $0.02\dot{7}$
    j $0.0\dot{6}\dot{3}$ k $0.2\dot{4}\dot{5}$ l $0.1\dot{4}0\dot{7}$

17. Convert these percentages to fractions.
    a $33.\dot{3}\%$ b $1.\dot{1}\%$ c $0.0\dot{4}\%$
    d $20.\dot{1}\dot{8}\%$ e $0.18\dot{5}\%$ f $10.0\dot{6}\dot{3}\%$

18. Write $0.\dot{0}12\,345\,67\dot{9}$ as a fraction.

19. Which of these values is largest?
    a 45% or $\frac{4}{9}$ b $\frac{1}{6}$ or $1.\dot{6}\%$
    c 0.123 or $\frac{10}{81}$ d $0.1\dot{7}0\dot{4}$ or $\frac{7}{40}$

APPLICATIONS

5.3 Fractions, decimals and percentages

RECAP
- You can convert between fractions, decimals and percentages.
- To convert a recurring decimal to a fraction you can use algebra.

Let $x = 0.\dot{4}\dot{5}$
$100x = 45.4545...$
$x = 0.4545...$
$99x = 45$
$x = \dfrac{45}{99} = \dfrac{5}{11}$

HOW TO
To solve problems involving fractions, decimals and percentages
1. Read the question and decide what you need to do.
2. Apply appropriate conversion techniques.
3. Answer the question and include units if appropriate.

EXAMPLE

Vita is comparing her marks in four exams. Each mark is given in a different way.

A 15 out of 20   **B** 92%   **C** $0.\dot{6}$ **D** $\dfrac{8}{9}$

Vita says 'the difference between my best two marks is over twice as big as the difference between my worst two marks'. Is Vita correct? Give your reason.

① Order the marks by converting them all to percentages.

② **A** 15 out of 20 = $\dfrac{15}{20} = \dfrac{75}{100}$ = 75% **B** 92%   Already a percentage.

C $0.\dot{6}$ = 66.$\dot{6}$%   0.666... × 100%
= 67% (2 sf)

D $\dfrac{8}{9}$ = 8 ÷ 9 = $9\overline{)8.0^80...}$ 0.8 8...
= 88.$\dot{8}$%   0.888... × 100%
= 89% (2 sf)

B > D > A > C
B − D = 92% − 89% = 3%
A − C = 75% − 67% = 8%
3% < 2 × 8% so Vita is wrong. ③

EXAMPLE

A tennis club increased its membership from 36 to 42.

What percentage of the 42 members are new?

$\dfrac{42 - 36}{42} = \dfrac{6}{42} = \dfrac{1}{7}$
$1 ÷ 7 = 0.14285...$
= 14.3% (1 dp)

① Find the increase as a fraction of the original.
② Convert to a decimal then a percentage.
③ ×100%

EXAMPLE

Find $0.\dot{5}\dot{4} + 0.\dot{4}\dot{5}$.

① Change the recurring decimals to fractions and then find the sum.
Let $x = 0.\dot{5}\dot{4}$ and $y = 0.\dot{4}\dot{5}$
② Use your knowledge of place value and algebra to eliminate the recurring digits.

$100x = 54.\dot{5}\dot{4}$ 2 dp so multiply by 100. $100y = 45.\dot{4}\dot{5}$
$99x = 54$ $100x - x$ $99y = 45$
$x = \dfrac{54}{99} = \dfrac{6}{11}$ $y = \dfrac{45}{99} = \dfrac{5}{11}$

③ $0.\dot{5}\dot{4} + 0.\dot{4}\dot{5} = \dfrac{6}{11} + \dfrac{5}{11} = \dfrac{11}{11} = 1$

Number Fractions, decimals and percentages

Reasoning and problem solving

Exercise 5.3A

1. Violet is doing a survey to find the percentage of girls in classes in her year group. She lists the fraction of each class that are girls.

10A	15 out of 30	10B	48%
10C	$\frac{3}{5}$	10D	0.72

   Which class has the highest fraction of
   a girls   b boys?

2. Write these numbers in ascending order.
   a 33.3%   0.33   33 $33\frac{1}{3}$%
   b 0.45   44.5%   0.454 $0.\dot{4}$
   c $0.2\dot{3}$ 0.232   22.3%   23.22%   0.233
   d $\frac{2}{3}$ 0.66 $0.6\dot{5}$ 66.6%   0.6666
   e $\frac{1}{7}$ 14%   0.142 $\frac{51}{350}$ 14.1%
   f 89% $\frac{5}{6}$ $0.8\dot{6}$ 0.866 $\frac{6}{7}$

3. The population of a village in 2001 was 200. By 2011 it was 232. Assuming no one left the village between 2001 and 2011, what fraction of the population in 2011 were not living in the village in 2001?

4. A supermarket surveys its customers about waiting times. 46 out of 800 customers said they were forced to queue more than 5 minutes to be served, the rest said it was less than 5 minutes. Can the supermarket managers fairly claim that '95% of our customers queue for less than 5 minutes'? Explain your answer.

5. Two bookshops are advertising special offers.

   Shop A Book sale! 25% off

   Shop B Buy one get one half price!* *Offer applies to lower price book.

   You want to buy two books. Which shop should you go to if
   a the books are the same price
   b the books are different prices?
   You must give your reasons.

6. A box has these dimensions.
   $a = \frac{3}{4}$ of 1 m
   $b = 0.4\dot{5}$ m
   $c = 0.55\%$ of a

   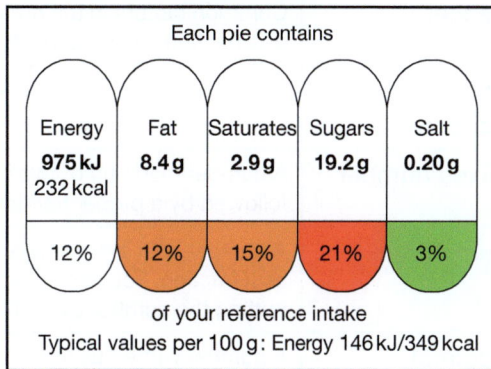

   What is the volume of the cuboid as a fraction of 1 m³?

7. Is $\frac{1}{2}$ closer to $0.\dot{3}\dot{6}$ or $0.\dot{6}\dot{3}$? Show your working.

8. Given that $0.\dot{5}\dot{7} = \frac{19}{33}$ write $0.3\dot{5}\dot{7}$ as a fraction.

9. Write these numbers as fractions.
   a $0.\dot{9}$ b $0.4\dot{9}$
   Comment on your answers.

*10. A food label shows nutritional information for an individual apple pie.

Each pie contains				
Energy 975 kJ 232 kcal	Fat 8.4 g	Saturates 2.9 g	Sugars 19.2 g	Salt 0.20 g
12%	12%	15%	21%	3%

   of your reference intake
   Typical values per 100 g: Energy 146 kJ/349 kcal

   *a The bottom figures show that 19.2 g of sugar is 21% of a recommended daily amount. How many grams is a recommended daily amount?
   *b What are the recommended daily amounts of fat, saturates and salt?
   *c The label says that 100 g of pie would provide 349 kcal. Angela has lost the bit of the label with the total mass of the pie. How many grams does the pie weigh?

11. Solve these equations.
    a $0.\dot{8}\dot{8} + x = 1$
    b $0.3\dot{5} + 0.\dot{7}x = 0.7$

Summary

Checkout
You should now be able to...

Test it
Questions

✓ Find fractions and percentages of amounts.	1, 2
✓ Add, subtract, multiply and divide with fractions and mixed numbers.	3 – 6
✓ Convert between fractions, decimals (including recurring decimals) and percentages.	7 – 11
✓ Order fractions, decimals and percentages.	12

Language	Meaning	Example
Fraction	A number of equal parts of the whole.	$\frac{3}{4}$
Denominator	The number of equal parts in the whole.	Denominator 4
Numerator	The number of equal parts in the fraction.	Numerator 3
Common factor	A factor that is shared by two or more numbers or terms.	$15 = 3 \times 5 \qquad 35 = 5 \times 7$ 5 is a common factor of 15 and 35
Cancel	Common factors in the numerator and denominator of a fraction can be cancelled.	$\frac{15}{35} = \frac{3}{7}$
Improper fraction	A fraction with a larger numerator than denominator.	$\frac{4}{3} \qquad 4 > 3$
Mixed number	A number made up of two parts: a whole number followed by a proper fraction.	$1\frac{1}{3}$
Percentage	A number with a denominator of 100 when written as a fraction. You can always substitute % by putting the number over 100.	$75\% = \frac{75}{100} = \frac{3}{4}$
Decimal	A number written using tenths, hundredths, thousandths, etc.	$0.125 = \frac{1}{10} + \frac{2}{100} + \frac{5}{1000} = \frac{1}{8}$
Terminating	A decimal with a finite number of digits.	$0.125 = \frac{1}{8}$
Recurring	A decimal with a repeating pattern that goes on forever.	$0.\dot{8}\dot{1} = 0.818181... = \frac{9}{11}$
Reciprocal	The reciprocal of a number is what you multiply it by to get 1.	Reciprocal of 5 is $\frac{1}{5}$ Reciprocal of $\frac{2}{3}$ is $\frac{3}{2}$

Number Fractions, decimals and percentages

Review

1 Calculate these amounts.
  a $\frac{1}{7}$ of 28
  b $\frac{3}{8}$ of 40
  c $\frac{4}{9}$ of 27
  d $1\frac{1}{5}$ of 20

2 Calculate these amounts.
  a 35% of 60
  b 85% of 12
  c 2.5% of 40
  d 20% of 25

3 Convert these mixed numbers to improper fractions.
  a $2\frac{6}{7}$
  b $1\frac{5}{11}$

4 Convert these improper fractions to mixed numbers.
  a $\frac{20}{9}$
  b $\frac{50}{7}$

5 Write down the reciprocal of these numbers.
  a 5
  b $\frac{1}{3}$
  c $\frac{2}{7}$
  d $1\frac{2}{5}$

6 Calculate these expressions giving each answer in its simplest form.
  a $\frac{3}{5} \times \frac{1}{6}$
  b $9 \times \frac{3}{8}$
  c $\frac{2}{9} \div \frac{1}{3}$
  d $12 \div \frac{2}{3}$
  e $\frac{1}{12} + \frac{5}{12}$
  f $\frac{5}{6} - \frac{1}{3}$
  g $\frac{2}{11} + \frac{1}{3}$
  h $2\frac{3}{4} - 1\frac{2}{7}$

7 Convert these numbers to decimals.
  a $\frac{4}{5}$
  b $\frac{17}{20}$
  c $\frac{3}{100}$
  d $\frac{15}{40}$
  e 22%
  f 0.4%
  g 200%
  h 5.5%

8 Convert these numbers to fractions in their simplest form.
  a 0.7
  b 0.214
  c 0.36
  d 0.01
  e 12%
  f 150%
  g 44.4%
  h 0.2%

9 Convert these numbers to percentages.
  a $\frac{3}{5}$
  b $\frac{13}{20}$
  c $\frac{7}{1000}$
  d $\frac{18}{40}$
  e 0.35
  f 0.8
  g 0.09
  h 1.8

10 Write these fractions as recurring decimals.
  a $\frac{2}{9}$
  b $\frac{6}{7}$

11 Write these recurring decimals as fractions in their simplest form. Show your method.
  a $0.\dot{8}$
  b $0.2\dot{3}$

12 Write these numbers in ascending order.
  a $\frac{9}{16}$ $\frac{5}{8}$ $\frac{2}{3}$
  b 22.2% $0.\dot{2}$ $\frac{1}{5}$

What next?

Score		
	0 – 4	Your knowledge of this topic is still developing. To improve, look at MyiMaths: 1015, 1016, 1017, 1030, 1031, 1040, 1046, 1047, 1063, 1066, 1074
	5 – 10	You are gaining a secure knowledge of this topic. To improve your fluency look at InvisiPens: 05Sa – i
	11 – 12	You have mastered these skills. Well done you are ready to progress! To develop your problem solving skills look at InvisiPens: 05Aa – f

Assessment 5

1. The cost of buying a DVD in 2012 was $20. This figure was made up as shown in the table.

Labour	$\frac{7}{25}$
Materials	$\frac{1}{10}$
Advertising	$\frac{2}{5}$
Profit	P

   a Find the value of P. [2]

   b Calculate the actual cost of each of the four parts. [2]

   By 2014 the costs had altered. The actual cost of labour had decreased by $\frac{1}{10}$ and the actual cost of materials had decreased by $\frac{3}{40}$. The actual profit had increased by $\frac{2}{11}$ and advertising had not changed.

   c Calculate the new price of the DVD. [3]

2. Cai asked some friends which e-mail provider they use.

Provider	Number of boys	Number of girls
Space	5	0
Cyber	1	2
Chat-chat	3	6
World Telecom	8	8
Blue sky	8	4
	Total 25	Total 20

   a Which provider do 20% of the girls use? [1]

   b Which provider do 20% of the boys use? [1]

   c Which provider do 20% of the total number of pupils use? [2]

   d Cai said: 'In my survey, World Telecom was equally popular with both boys and girls.' Was Cai correct or incorrect? Give your reason. [2]

3. a What percentage of the word MISSISSIPPI is made up by the letter S? [2]

   b Old Macdonald had a farm. He had 21 goats, 8 lambs, 150 sheep, 8 llamas, 27 turkeys, 35 chickens and 1 rooster.
      What percentage of his livestock were chickens? [2]

4. In the 4 × 100 m relay, the first 3 runners took, respectively, 25%, $\frac{7}{25}$ and 0.35 of the total time.

   a What proportion of the time was taken by the 4th runner?
      Give your answer as a decimal. [2]

   b Which runner ran the fastest leg? [1]

   c Which runner ran the slowest leg? [1]

5. G. Russet has an orchard. The orchard contains $3\frac{3}{4}$ hectares of apple trees. Today he needs to treat $\frac{2}{5}$ of the area for disease prevention. What area does he need to treat? [3]

6. In a football match the goalkeeper kicked the ball from the goal line for $\frac{5}{9}$ of the length of the pitch and a player then kicked it a further $\frac{7}{20}$. If the length of the pitch is 90 yards, how many yards further is the opposite goal line? [4]

Number Fractions, decimals and percentages

7 **a** In her garden, Lizzie planted flowers in 27% of the total area. She covered 145.8 m² of the garden with flowers. What is the total area of her garden? [2]

    b 36% of the remainder of her garden was taken up by the patio. Find the area of the lawn, to the nearest m². [4]

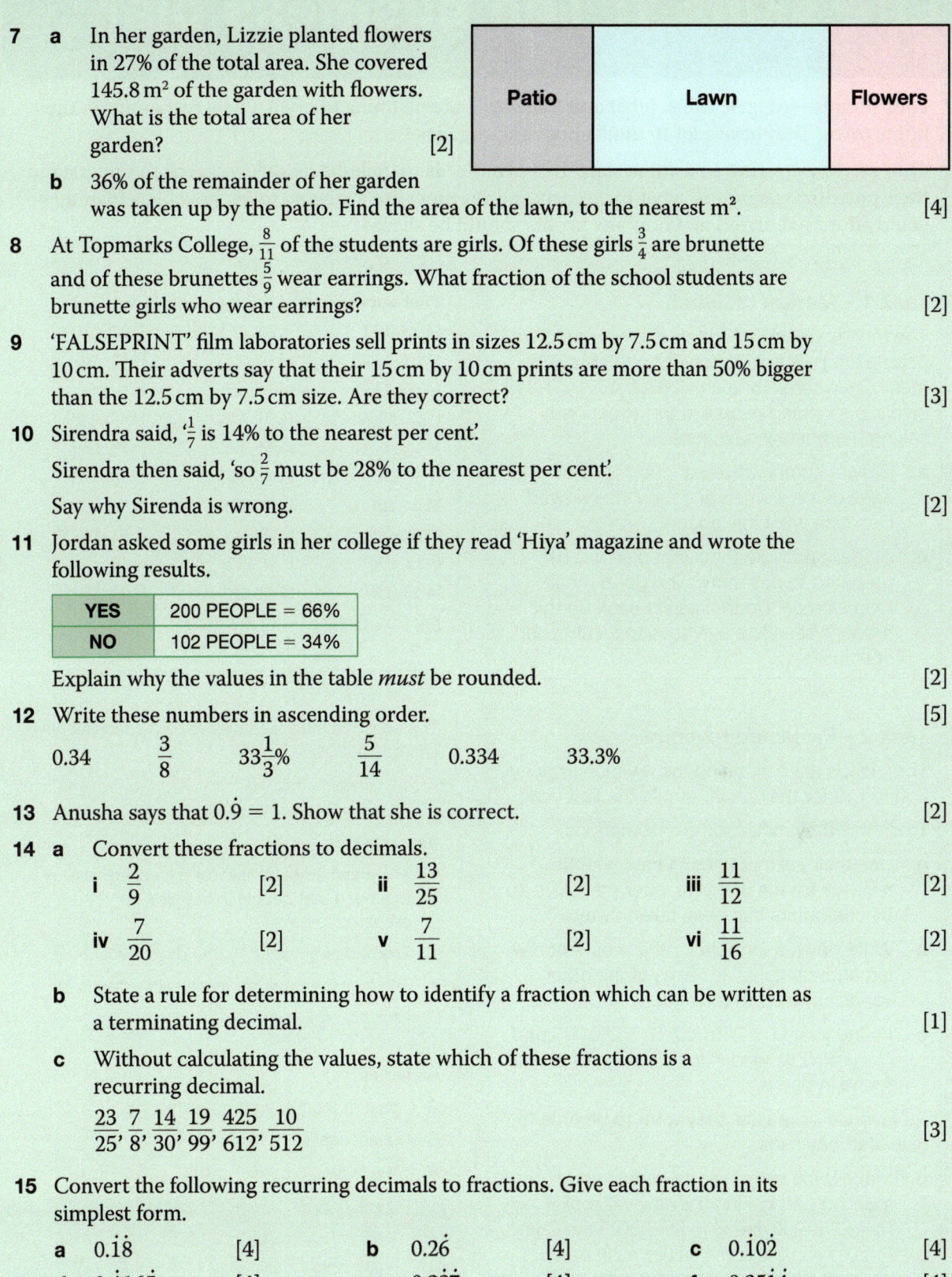

8 At Topmarks College, $\frac{8}{11}$ of the students are girls. Of these girls $\frac{3}{4}$ are brunette and of these brunettes $\frac{5}{9}$ wear earrings. What fraction of the school students are brunette girls who wear earrings? [2]

9 'FALSEPRINT' film laboratories sell prints in sizes 12.5 cm by 7.5 cm and 15 cm by 10 cm. Their adverts say that their 15 cm by 10 cm prints are more than 50% bigger than the 12.5 cm by 7.5 cm size. Are they correct? [3]

10 Sirendra said, '$\frac{1}{7}$ is 14% to the nearest per cent'.

    Sirendra then said, 'so $\frac{2}{7}$ must be 28% to the nearest per cent'.

    Say why Sirenda is wrong. [2]

11 Jordan asked some girls in her college if they read 'Hiya' magazine and wrote the following results.

YES	200 PEOPLE = 66%
NO	102 PEOPLE = 34%

    Explain why the values in the table *must* be rounded. [2]

12 Write these numbers in ascending order. [5]

    0.34 $\frac{3}{8}$ $33\frac{1}{3}\%$ $\frac{5}{14}$ 0.334   33.3%

13 Anusha says that $0.\dot{9} = 1$. Show that she is correct. [2]

14 **a** Convert these fractions to decimals.

    i $\frac{2}{9}$ [2]   **ii** $\frac{13}{25}$ [2]   **iii** $\frac{11}{12}$ [2]

    iv $\frac{7}{20}$ [2]   **v** $\frac{7}{11}$ [2]   **vi** $\frac{11}{16}$ [2]

    b State a rule for determining how to identify a fraction which can be written as a terminating decimal. [1]

    c Without calculating the values, state which of these fractions is a recurring decimal.

    $\frac{23}{25}, \frac{7}{8}, \frac{14}{30}, \frac{19}{99}, \frac{425}{612}, \frac{10}{512}$ [3]

15 Convert the following recurring decimals to fractions. Give each fraction in its simplest form.

    a $0.\dot{1}\dot{8}$ [4]   **b** $0.2\dot{6}$ [4]   **c** $0.\dot{1}0\dot{2}$ [4]

    d $0.\dot{4}16\dot{5}$ [4]   **e** $0.2\dot{2}\dot{7}$ [4]   **f** $0.35\dot{1}\dot{4}$ [4]

Life skills 1: The business plan

Four friends – Abigail, Mike, Juliet and Raheem – are planning to open a new restaurant in their home town. They have a lot to think about and organise!

They start by creating a business plan. This plan needs to include: market research to understand their potential customers, what their costs and revenues are expected to be, how big a loan they could afford to borrow, and how any profits should be shared.

Task 1 – Market research

The friends decide to investigate how much people are prepared to pay for a three-course meal. They carry out a small pilot survey. This involves stopping people in the street and asking them a few questions.

a Draw a comparative bar chart to show the ages of the women and men interviewed. Describe what this shows.

b Abigail and Mike think that men will be prepared to pay more for a good meal than women. Do the results back up this theory? Calculate averages to justify your conclusions.

Pilot survey results (15 men and 15 women)

M 24, £33	F 20, £23	F 22, £25
M 37, £36	M 62, £33	F 47, £36
M 47, £35	M 42, £32	F 19, £16
F 52, £32	M 31, £22	M 66, £25
M 26, £24	M 55, £40	F 38, £35
F 18, £20	M 39, £35	M 40, £30
M 21, £21	F 23, £21	M 20, £30
F 58, £40	F 35, £32	F 32, £28
F 22, £30	F 61, £37	F 28, £20
M 23, £27	M 51, £27	F 44, £34

Key Gender (M/F) Age (years), amount prepared to spend

Task 2 – Projected revenue

The friends are estimating the revenue (money coming in) for their restaurant in the first year.

To do this they make some assumptions.

a Use their assumptions to estimate the revenue for the first year, after the VAT paid by customers has been taken away.

b Write down a formula for the profit (money left after costs), P, in terms of the other variables listed to the right.

c In one year G = £40 000, S = £80 000 and C = £50 000. Find P for the value of R you found in part **a**.

As revenue increases, they want to be able to give staff pay rises.

d Make S the subject of the formula you found in part **b**. Find the new value of S for R = £250 000 and G = £50 000. Take C and P as having the same values as in part **c**.

The cost of a meal includes Value Added Tax (VAT) charged at 20%.

Assumptions
- The mean amount paid for a meal equals the mean that people in the pilot survey are prepared to pay.
- The restaurant is open 364 days a year.
- The mean number of meals sold a day is 25.

Variables
G = cost of food bought by restaurant
S = salary costs C = other costs
R = revenue P = profit

Life skills

Age band	Persons
16–24	67 400
25–49	166 100
50–64	64 100
65 and over	58 000

▲ Age distribution of the population in their home town.

Ownership shares

Abigail $\frac{2}{5}$
Raheem $\frac{1}{4}$

Mike and Juliet both own an equal share of the remainder.

Task 3 – The survey

Following from the pilot survey, the friends decide to do a much larger survey based on a sample of 200 people from their home town. In their sample, they want the numbers of people in the different age groups to be in the same ratio as in the whole population. The table shows the population of the town by age.

How many people from each age group should they include in their sample?

Task 4 – Shares in the business

The friends invested different capital (initial amounts of money) into the business. Based on this, they each own shares that determine the fraction of profit they are entitled to.

a What fraction of the business do Mike and Juliet each have?

b Draw a pie chart to show how much of the business each person owns.

c How much profit would each owner get from a yearly overall profit of £50 000?

Five year repayment formula

C = amount borrowed A = amount repaid each year

i = annual interest rate (AIR), expressed as a decimal

$$C = \frac{A}{1+i} + \frac{A}{(1+i)^2} + \frac{A}{(1+i)^3} + \frac{A}{(1+i)^4} + \frac{A}{(1+i)^5}$$

Task 5 – The business loan

The friends decide to take out a business loan in order to equip the restaurant.

a If the maximum they can repay each year is £5000, how much can they borrow at an AIR of

   i 6%   ii 8%?

b If they borrow £30 000, how much will each yearly repayment be at an AIR of

   i 6%   ii 8%?

c If they borrow £30 000 but cannot afford to repay more than £8000 per annum, what is the maximum interest rate they can borrow at?

d Mike complains that the repayment formula is too long.
   Abigail tells him that it can also be written as

   $$C = \frac{A}{1+i}\left(\frac{1 - (\frac{1}{1+i})^5}{1 - (\frac{1}{1+i})}\right)$$

   Raheem says that the formula can be simplified further.

   $$C = \frac{A}{i}\left(1 - \left(\frac{1}{1+i}\right)^5\right)$$

   Starting from Abigail's formula, show that Raheem is correct.

e Mike then complains he can't use it to find the amount to repay. Show that he can, by re-arranging the formula in part **d** to make A the subject.

p.104

6 Formulae and functions

Introduction
Nurses often use mathematical formulae when they are administering drugs: converting from one unit to another, calculating the amount of a drug based on somebody's mass, or working out concentrations from solutions. Working with formulae is a topic within algebra and is a good example of the practical use of mathematics.

What's the point?
The ability to apply a mathematical formula accurately when calculating a patient's dose of a particular medicine is vitally important.

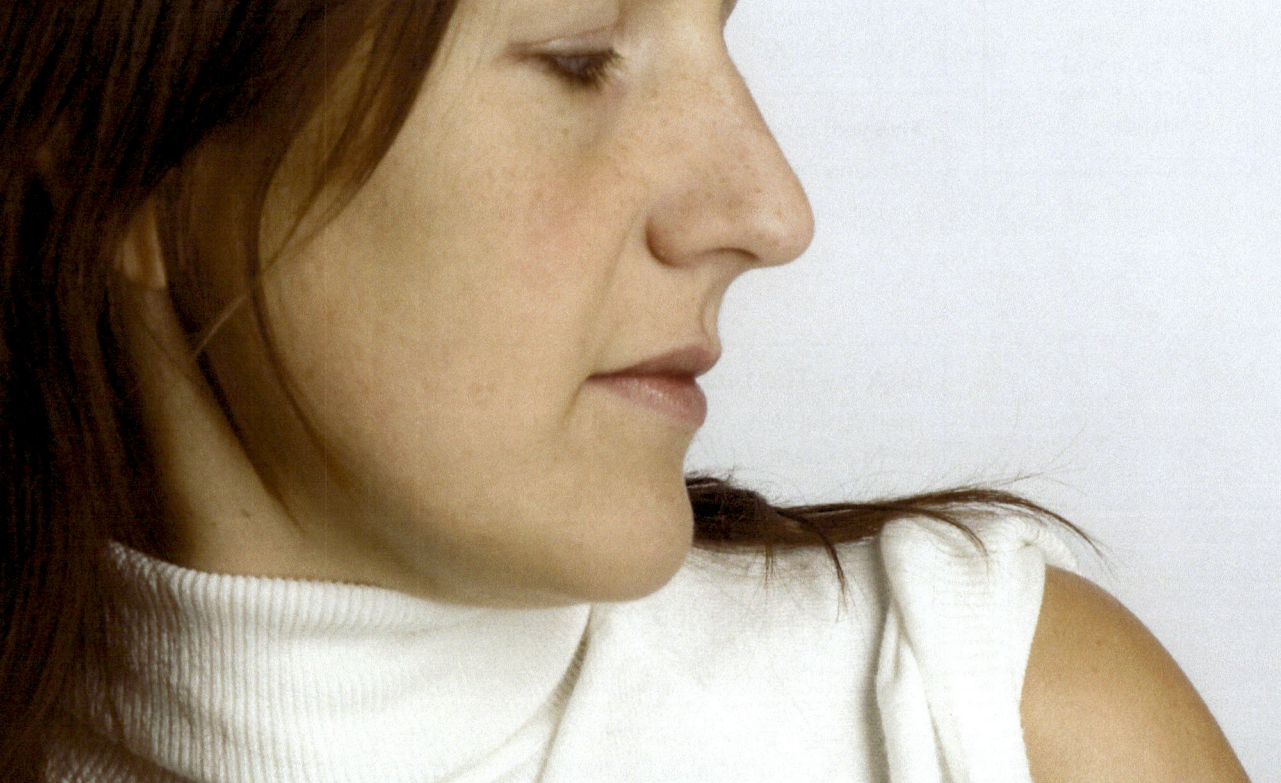

Objectives
By the end of this chapter, you will have learned how to …
- Substitute values into formulae and rearrange formulae to change their subject.
- Write an equation to represent a function and find inputs and outputs. Find the inverse of a function and construct and use composite functions.
- Use the terms expression, equation, formula, identity, inequality, term and factor.
- Construct proofs of simple statements using algebra.
- Expand brackets to get a quadratic expression and factorise quadratics into brackets.

Check in

1 a Find the highest common factor of these sets of numbers.

    i 24 and 36   **ii** 30 and 75   **iii** 56 and 72

    iv 90, 180 and 225   **v** 17, 28 and 93

b What is the highest common factor of $2x$, x^2 and $5x^3$?

2 Write these statements using the rules of algebra.

    a I think of a number, multiply it by 4 and add 7.

    b I think of a number, subtract 6 and then multiply by 3.

    c I think of a number, multiply it by itself then subtract this from 10.

    d I think of a number, treble it, take away 6 then divide this by two.

    e I think of a number and multiply it by itself five times.

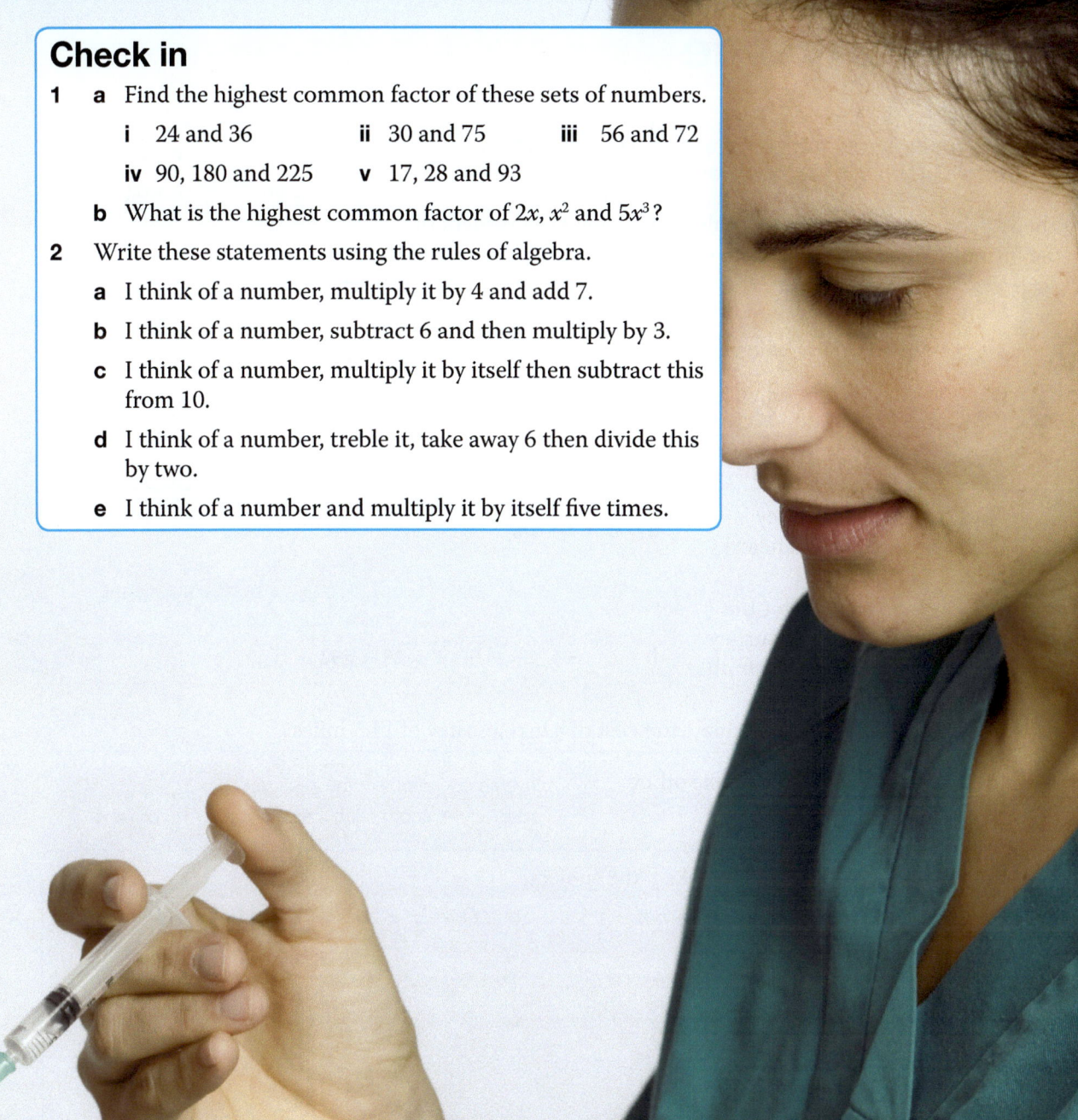

Chapter investigation

This grid of numbers uses each of the numbers from 1 to 9.

4	9	2
3	5	7
8	1	6

Every row, column and diagonal adds up to 15. It is called a magic square.

Create your own magic square.

SKILLS

6.1 Formulae

- A **formula** is an equation representing the relationship between two or more **variables**.

$$F = ma \qquad v = u + at$$
$$A = \tfrac{1}{2}bh \qquad P = 2(a + b)$$

The variable on the left of the equals sign is called the **subject** of the formula. Formulae are used to calculate the value of a variable.

EXAMPLE

Using these formulae, find the value of s.

a $s = 2x + y^2$ when $x = 4$, $y = 5$
b $s = \sqrt{2x} + p^2$ when $x = 4.5$, $y = -3$

Substitute the values and use the BIDMAS rules.
a $s = 2x + y^2 = 2 \times 4 + 5^2 = 8 + 25 = 33$
b $s = \sqrt{2x} + p^2 = \sqrt{9} + (-3)^2 = 3 + 9 = 12$

Formulae can be **derived**.

EXAMPLE

The cost of a taxi journey is €2 per mile plus €3 standard charge.

a Derive a formula for the cost, €C, of a taxi journey as a function of the number of miles, m, travelled.
b Use the formula to calculate the cost of a taxi journey of 14.5 miles.

a $C = 3 + 2m$
A 1 mile journey costs €3 + 1 × €2
A 2 mile journey costs €3 + 2 × €2
A journey costs €3 plus the number of miles multiplied by 2.

b $C = 3 + 2 \times 14.5 = 3 + 29 = €32$

Formulae can be rearranged to make other variables the subject of the formula.

Inverse operations come in pairs:
addition ↔ subtraction   multiplication ↔ division

EXAMPLE

Rearrange these formulae to make p the subject.

a $A = 2p + d$ b $T = mp - c$ c $Q = \dfrac{p}{4} + 2$ d $M = \dfrac{p + b}{a}$

$p \to \times 2 \to + d \to A$ 　　$p \to \times m \to -c \to T$ 　　$p \to \div 4 \to +2 \to Q$ 　　$p \to +b \to \div a \to M$

a $A - d = 2p$ $-d$
$\dfrac{A - d}{2} = p$ $\div 2$
$p = \dfrac{A - d}{2}$

b $T + c = mp$ $+c$
$\dfrac{T + c}{m} = p$ $\div m$
$p = \dfrac{T + c}{m}$

c $Q - 2 = \dfrac{p}{4}$ -2
$4(Q - 2) = p \times 4$
$p = 4(Q - 2)$

d $aM = p + b$ $\times a$
$aM - b = p$ $-b$
$p = aM - b$

EXAMPLE

Make y the subject of these formulae.

a $A = 2 - \sqrt{y}$
b $V = y^2$
c $py + q = ry + s$

a $A + \sqrt{y} = 2$
$\sqrt{y} = 2 - A$
$y = (2 - A)^2$

b $y = \pm\sqrt{V}$

c $py - ry + q = s$
$py - ry = s - q$
$y(p - r) = s - q$
$y = \dfrac{s - q}{p - r}$

If the intended subject is negative, add terms to make the subject positive.

Square root both sides, include +ve and −ve square roots.

Ensure all terms in y are on one side.
Factorise $py - ry$
Divide both sides by $(p - r)$

When rearranging a formula, focus on the new subject and reverse the operations.

Algebra   Formulae and functions

Fluency

Exercise 6.1S

1 Using the given formulae, calculate the value of F when $a = 9$.

    a $F = 2a + 5$ **b** $F = 4a - 6$

    c $F = 2(a + 5)$ **d** $F = a^2$

    e $F = 2a^2$ **f** $F = \sqrt{a}$

2 Using the formula $s = ut + \frac{1}{2}at^2$, complete this table.

	u	a	t	s
a	4	5	2	
b	−3	2	4	
c	5	−2	3	
d		2	−4	8
e	−2		6	78

3 The formula to convert temperatures in °C to temperatures in °F is $F = \frac{9}{5}C + 32$.
Convert a temperature of 25 °C to °F.

4 For each situation derive a formula for A.

    a A is equal to 6 more than p.

    b A is equal to 5 less than m.

    c A is equal to double k, divided by 7.

    d A is equal to 6 add 5 lots of t.

    e A is equal to 3 times d add 7.

    f A is equal to t add 7, multiplied by 3.

    g A is equal to the square root of y, add 8.

5 The time for a pendulum to complete a cycle can be calculated using this formula

$$T = 2\pi\sqrt{\frac{L}{g}}$$

where L is the length of the pendulum and $g = 9.8 \, \text{m/s}^2$ is the acceleration due to gravity.

    a   Calculate T when $L = 25$ cm.

    b   Find the length of a pendulum with a time of 14.5 s.

6 Rearrange each of these formulae to make x the subject.

    a $y = x + 3$ **b** $y = x - c$

    c $y = 5x$ **d** $y = 3x + p$

    e $y = \frac{x}{5}$ **f** $y = \frac{x}{c} + d$

    g $y = x^2$ **h** $y = \sqrt{x}$

    i $y = x^3$ **j** $y = \sqrt[3]{x}$

    k $y = \sqrt{x} + 2$ **l** $T = \frac{x^2 + 2}{5}$

7 Rearrange each of these formula to make p the subject.

    a $A = 2p + 5$ **b** $A = 2p + d$

    c $A = 3p - 6$ **d** $H = 4pd$

    e $H = \frac{p}{4} + t$ **f** $H = \frac{p + t}{4}$

    g $W = \sqrt{\frac{p}{5}}$ **h** $D = w(w - p)$

8 Frederik has rearranged the formula $y = 10 - x$ to make x the subject.
His answer is $x = y - 10$.
Frederik's answer is wrong. Give the correct answer.

9 Rearrange each of these formula to make x the subject.

    a $mx + n = ox + p$

    b $ax - b = cx + d$

    c $f + ex = g - hx$

    d $r - sx = t - ux$

*10 Rearrange this formula to make x the subject.

$$y = \frac{x + 2}{x - 3}$$

11 Make x the subject of these formulae.

    a $p(x - y) = q(z - x)$

    b $\frac{x + t}{x - 5} = k$

    c $\frac{x + p}{q - x} = \frac{3}{4}$

    d $\sqrt{\frac{x - a}{x + b}} = 4$

    e $\frac{4}{x^2 - m} = \frac{3}{n^2 - 2x^2}$

*12 Make R the subject of this formula.

$$\frac{2p + 10}{25} = \frac{R}{15 + R}$$

APPLICATIONS

6.1 Formulae

RECAP
- A formula is an equation representing the relationship between two or more variables.
- To evaluate a formula, replace each variable by its numerical value and use the BIDMAS rules to evaluate the arithmetic expression.
- Formulae can be rearranged by 'reversing the operations'.

$$E = \tfrac{1}{2}mv^2 \quad \leftarrow \text{Subject}$$
$$2E = mv^2$$
$$\tfrac{2E}{m} = v^2$$
$$v = \pm\sqrt{\tfrac{2E}{m}}$$

HOW TO

To rearrange a simple formula
1. Identify the order of operations in the formula.
2. Undo the formula by applying the correct inverse operations one step at a time.
3. Continue until the formula has just the required subject on one side.

If the new subject of a formula appears more than once then first simplify the formula so that the new subject only appears once.

EXAMPLE

The formula for the volume of a sphere is $V = \tfrac{4}{3}\pi r^3$ where r is the radius of the sphere.

a The radius of a power ball is 2.3 cm. Find the volume of rubber in the power ball.
b Rearrange the formula to make r the subject.

a $V = \tfrac{4}{3} \times \pi \times 2.3^3 = 50.965...$ Use BIDMAS.

The volume of rubber is 51.0 cm³ (1 dp)   Include the units in the final statement.

b $V = \dfrac{4\pi r^3}{3}$

$3V = 4\pi r^3$

$\dfrac{3V}{4\pi} = r^3$

$r = \sqrt[3]{\dfrac{3V}{4\pi}}$

① The order of operations is $r \to r^3 \to \times 4\pi \to \div 3$.
② Multiply both sides by 3. Divide both sides by 4π.
③ Cube root both sides.

You can treat r^3 as a single term because powers have priority over multiplication.

EXAMPLE

A car mechanic has a £45 callout charge plus £40 for every hour she works.
a Write a general formula for the number of hours worked, h, as a function of the fee, F.
b How long did a job last if the fee was £185?

a $F = 45 + h \times 40$
$F - 45 = 40h$
$h = \dfrac{F - 45}{40}$

① Create a formula for F
② then change the subject.

b $h = \dfrac{185 - 45}{40} = \dfrac{140}{40}$

③ Substitute $F = 185$ into the formula. Include any units.

$= 3.5$ hours

Algebra Formulae and functions

Reasoning and problem solving

Exercise 6.1A

1. The surface area of a sphere is given by $S = 4\pi r^2$ where r is the radius of the sphere.
   a. The radius of the earth is 6378 km. Find the surface area of the earth.
   b. Rearrange the formula to find an expression for r.
   c. The surface area of the moon is 37.93 million km². What is its radius?

2. The volume of a cylinder is given by the formula $V = \pi r^2 h$.
   a. Rearrange the formula to make r the subject.
   b. If $V = 72\pi$ m³ and $h = 4$ m find r.

3. Rearrange each formula to make h the subject.
   a. The surface area of a cylinder is given by $S = 2\pi r^2 + 2\pi rh$.
   b. The surface area of a cone is given by $S = \pi r^2 + \pi r\sqrt{h^2 + r^2}$.

4. Rearrange the lens formula $\frac{1}{v} + \frac{1}{u} = \frac{1}{f}$
   a. to make f the subject
   b. to make v the subject.

5. Here are some facts about electricity.

Fact	Formula
Power = Voltage × Current	$P = VI$
Voltage = Current × Resistance	$V = IR$
Charge = Current × Time	$Q = It$
Energy = Voltage × Charge	$E = VQ$

   a. Rearrange $P = VI$ to make V the subject.
   b. Write a formula for Resistance in terms of Voltage and Current.
   c. Write a formula for Energy in terms of Voltage, Current and Time.
   d. Write a formula for Energy in terms of Current, Resistance and Charge.
   e. Write a formula for E in terms of I, R and t.

6. A bank sells US dollars for £0.64 plus a commission fee of £6.99
   a. Write a formula for the cost, £c, of buying d dollars.
   b. Write a formula for the number of dollars bought given the cost.
   c. How many dollars were bought if the cost was £48.59?

7. Heron's formula for the area, A, of a triangle with sides a, b and c is
   $$A = \sqrt{s(s-a)(s-b)(s-c)}$$
   where s is the semiperimeter $s = \frac{a+b+c}{2}$
   a. Find the area of these triangles.

   i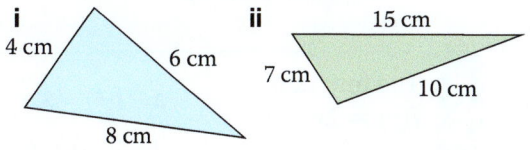

   ii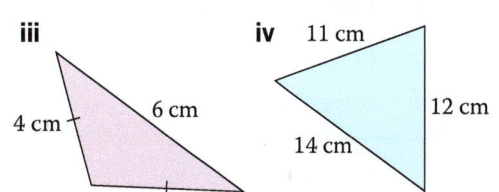

   iii

   iv

   *b. Rearrange the formula for A to make c the subject.

8. The volume of a torus ('doughnut') with inner radius r and outer radius R is given by this formula.
   $$V = \frac{\pi^2(R+r)(R-r)^2}{4}$$
   a. Find V when $R = 10$ cm and $r = 4$ cm.
   *b. If $V = 352\pi^2$ mm³ and the difference between the inner and outer radius is 8 mm, find the inner and outer radius.

SKILLS

6.2 Functions

- A **function** is a relation between a set of **inputs**, the 'domain', and a set of **outputs**, the 'range', such that each input is related to an output.

For a function f, input x gives output $f(x)$.

EXAMPLE

Find the functions shown in these **mapping diagrams**.

a 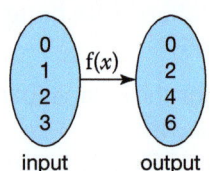   b 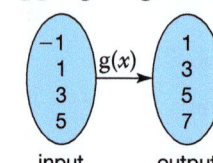   c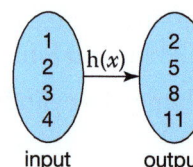

a $f(x) = 2x$
The rule is 'double the input'.

b $g(x) = x + 2$
The rule is 'add 2 to the input'.

c $h(x) = 3x - 1$
The rule is 'multiply the input by 3 and subtract 1.

EXAMPLE

The function $f(x) = 2x + 1$
a find $f(3)$
b find $f(-2)$
c solve $f(x) = 0$.

a $f(3) = 2 \times 3 + 1 = 7$ $f(3)$ means the input to the function $f(x)$ is $x = 3$.
b $f(-2) = 2 \times -2 + 1 = -3$ $f(-2)$ means the input to the function $f(x)$ is $x = -2$.
c $2x + 1 = 0$ $f(x) = 0$ means the output to the function $f(x)$ is 0.
   $2x = -1$
   $x = -\frac{1}{2}$

The inverse of a function $f(x)$ is written $f^{-1}(x)$.

EXAMPLE

The function $g(x) = 4x - 3$.
Find $g^{-1}(x)$.

$y = 4x - 3$
$y + 3 = 4x$
$\frac{y + 3}{4} = x$
$g^{-1}(x) = \frac{x + 3}{4}$

If a function maps the input, x, to an output, y, then the inverse function maps the output, y, to the input, x.

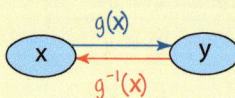

Functions can also be combined to form **composite functions** $fg(x) = f(g(x))$.

EXAMPLE

The function $f(x) = x^2$ and $g(x) = 2x + 3$.
a Find i $ff(3)$ ii $fg(2)$ iii $gf(x)$.   b Solve $fg(x) = 2gf(x)$.

The order is important.
$fg(x) = f(g(x))$
$\neq gf(x) = g(f(x))$.

a i $f(3) = 3^2 = 9$
     $ff(3) = f(9) = 9^2 = 81$
  ii $g(2) = 2 \times 2 + 3 = 7$
     $fg(2) = f(7) = 7^2 = 49$
  iii $f(x) = x^2$
     $gf(x) = g(x^2) = 2 \times x^2 + 3 = 2x^2 + 3$

b $fg(x) = 2gf(x)$
   $(2x + 3)^2 = 2(2x^2 + 3)$
   $4x^2 + 12x + 9 = 4x^2 + 6$
   $12x = -3$
   $x = -\frac{1}{4}$

Algebra   Formulae and functions

Fluency

Exercise 6.2S

1. Find the missing functions.

   a, b, c, d, e, f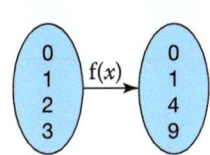

2. The function $f(x) = 2x + 1$. Find

   a $f(2)$ b $f(5)$

   c $f(-3)$ d $f(0)$

3. The function $g(x) = 2x - 4$. Solve these equations for x.

   a $g(x) = 0$ b $g(x) = 10$

   c $g(x) = -6$ d $g(x) = -4$

4. Complete the mapping diagram for the function $f(x) = 5x - 2$.

   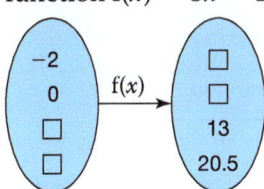

5. The function $h(x) = \dfrac{1}{x+2}$. Evaluate

   a $h(2)$ b $h(0)$

   c $h(0.5)$ d This is invalid

6. Find the inverse of these functions.

   a $f(x) = x + 4$ b $g(x) = x - 3$

   c $h(x) = \dfrac{x}{2}$ d $f(x) = 5x$

   e $g(x) = 2x - 3$ f $h(x) = 7x + 2$

   g $f(x) = \dfrac{x}{3} - 4$ h $f(x) = \dfrac{x+5}{2}$

7. Find the inverse of these functions.

   a $f(x) = \dfrac{1}{x}$ b $g(x) = 2 - x$

   Comment on your answers.

8. The functions $f(x)$ and $g(x)$ are defined as $f(x) = 2x + 3$ and $g(x) = 4x$. Find

   a $ff(2)$ b $gg(5)$

   c $fg(3)$ d $gf(3)$

   e $fgf(1)$ f $gfg(1)$

   g $fg(x)$ h $gf(x)$

9. The functions $f(x)$ and $g(x)$ are defined as $f(x) = x^2$ and $g(x) = 3x + 1$.

   a Find $fg(x)$ b Find $gf(x)$

   c Solve $fg(x) = 3gf(x)$

10. The function $f(x) = 4x + 5$. Show that $ff^{-1}(x) = x$.

*11. Using the functions and inverse functions from question 7, explain why the graph of $f^{-1}(x)$ is a reflection of the graph of $f(x)$ in the line $y = x$.

12. By considering the function $f(x) = x^2$, investigate why it is important to 'restrict the domain' for inverse functions.

13. The functions f and g are defined as $f(x) = 2x + 1$ and $g(x) = 3x - 2$. Find

    a $f^{-1}(x)$ b $g^{-1}(x)$

    c $fg(x)$ d $(fg)^{-1}(x)$

    e $g^{-1}f^{-1}(x)$

    f Comment on your result for part e.

Did you know...

The Swiss mathematician Leonard Euler was one of the most prolific mathematicians ever. He introduced lots of new notation including f(x) for a function.

APPLICATIONS

6.2 Functions

RECAP
- A function, f, is a relation between a set of inputs, the 'domain', and a set of outputs, the 'range', such that each input, x, is related to *one* output, $f(x)$.
- Functions can be combined to form composite functions $fg(x) = f(g(x))$.
- The inverse of a function $f(x)$, written $f^{-1}(x)$, satisfies $ff^{-1}(x) = x$.

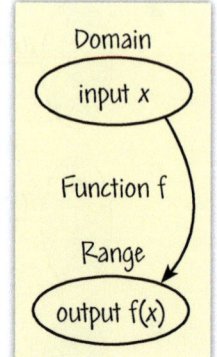

HOW TO

To represent a function given in words
1. Write the function as a mapping diagram.
2. Write the function as a formula.
3. Plot the inputs and outputs on a graph.

EXAMPLE

The **mapping diagram** shows information about the function 'multiply by 2 and then add 1'.

a Use the function to find the missing numbers.
b Create a formula that describes the function.
c Write the function using function notation.
d Plot the inputs and outputs on a graph.

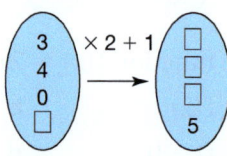

a ①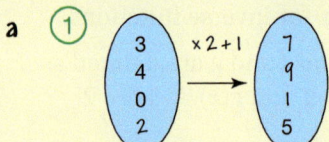

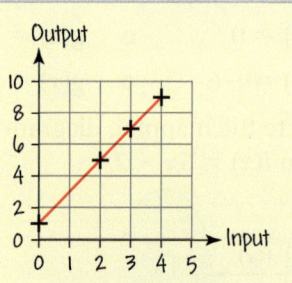

b ② $y = 2x + 1$ where x = input and y = output
c $f(x) = 2x + 1$ It is also true that $y = f(x)$
d ③ Plot points with coordinates (input, output).
   The points could be joined with a straight line.

The line on the graph shows the pattern of all possible inputs and outputs.

EXAMPLE

Use the information given to find g where g is a linear function.
$g(x) = mx + c$ for some numbers m and c.
$fg(x) = 3 - 10x$ and $f(x) = 5x - 2$

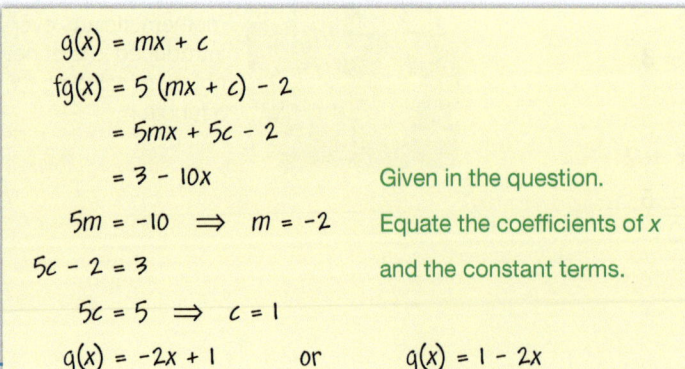

$g(x) = mx + c$	
$fg(x) = 5(mx + c) - 2$	
$\quad = 5mx + 5c - 2$	
$\quad = 3 - 10x$	Given in the question.
$5m = -10 \Rightarrow m = -2$	Equate the coefficients of x and the constant terms.
$5c - 2 = 3$	
$5c = 5 \Rightarrow c = 1$	
$g(x) = -2x + 1 \quad$ or $\quad g(x) = 1 - 2x$	

Algebra Formulae and functions

Reasoning and problem solving

Exercise 6.2A

1.
   i. Copy and complete each mapping diagram.
   ii. Write a formula to describe the function.
   iii. Write the formula using function notation.
   iv. Plot the inputs and outputs on a graph.

   a, b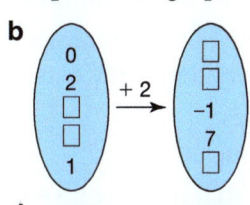

   c, d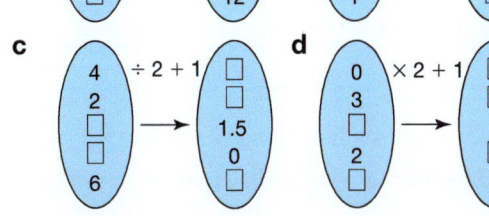

   For questions **2** and **3** you can use graphing software.

2.
   i. Find the inverse of each function in **1**. Write your solution using function notation.
   ii. Plot the graph of each function, and its inverse, on the same grid. Use equal scales on the x- and y-axes.
   iii. Add the line $y = x$ to each of your graphs. What do you notice? Suggest a reason for your observations.

3. Plot the graph of each of these functions and their inverses.

   a. $y = 5 - x$
   b. $y = \dfrac{2}{x}$
   c. $y = -\dfrac{10}{x}$
   d. $y = -x$

   Comment on your answers.

4. Are these the graphs of functions? Give your reasons.

   a 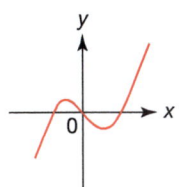   b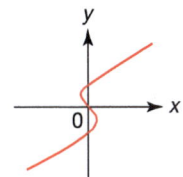

5. The graph shows the inverse of a function.

   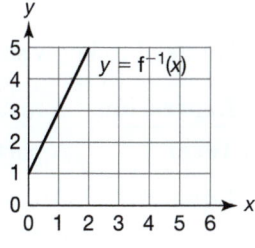

   a. Copy the diagram and add the graph of $y = f(x)$.
   b. Write a formula for $f(x)$.

6. The graph shows the inverse of a function $f^{-1}(x)$.

   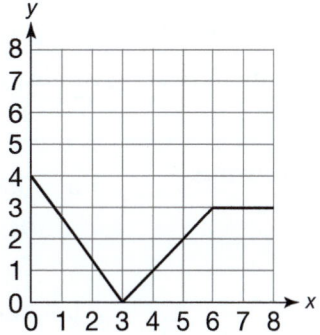

   On a copy of the graph add the function $y = f(x)$.
   Comment on your answer.

7. f is a linear function, $f(x) = mx + c$.

   a. If f is its own inverse, $f^{-1}(x) = f(x)$, find $f(x)$. **There are two solutions.**
   b. If $ff(x) = f(x)$ find $f(x)$.

8. The 'index functions' are defined by $f_n(x) = x^n$, where n is a number.

   a. Show that $f_n f_m(x) = f_{nm}(x)$.
   b. Hence or otherwise prove
      i. $f_n f_m(x) = f_m f_n(x)$
      ii. $f_n^{-1}(x) = f_{\frac{1}{n}}(x)$.

9. Use the information given to find g. In each case g is a linear function, $g(x) = mx + c$, for some numbers m and c.

   a. $f(x) = 2x + 3$ $fg(x) = 6x + 7$
   b. $f(x) = 5 - 8x$ $fg(x) = 21 - 4x$
   c. $f(x) = \dfrac{7 - 2x}{5}$ $fg(x) = \dfrac{15 - 6x}{5}$
   *d. $f(x) = 6x$ $(fg)^{-1}(x) = \dfrac{x + 30}{18}$

SKILLS

6.3 Equivalences in algebra

You need to know some key vocabulary used in algebra.

Expression	A meaningful collections of mathematical symbols.	$x^2 - x$
Term	A part of an expression separated by + or − signs.	x^2 and $-x$
Equation	An expression containing *one* = sign and at least one unknown. Equality is only true for *particular* values of the unknown.	$x^2 - x = 2$ $x = 2$ or -1
Identity	An equation that is true for *every* value of the unknown. The = sign is usually replaced by ≡ for 'identically equal to'.	$x^2 - x \equiv x(x - 1)$
Formula	An equation involving several variables; often associated with a real application.	$E = mc^2$
Inequality	A statement relating two expressions that may *not* be equal.	$x^2 - x < 3$

EXAMPLE

Use one of these words –
 expression,
 equation,
 identity,
 formula,
 inequality
– to describe the following.

a $3x - 2 = 5$
b $4x - 2 \leq 5$
c $x(x + 2) \equiv x^2 + 2x$
d $F = \frac{9}{5}C + 32$
e $3x - 2$

a Contains an = sign. How many values of x is it true for?
  $3x - 2 = 5 \Rightarrow 3x = 7 \Rightarrow x = 2\frac{1}{3}$ Only true for one value of x.
  Equation
b Inequality Contains one of the inequality symbols: <, ≤, >, ≥.
c Contains an ≡ sign. Is it true for all values of x?
  $x(x+2) = x \times x + x \times 2$
      $= x^2 + 2x$
  Identity
d Contains an = sign and two variables F and C.
  Formula It is a formula for converting temperatures in Celsius to temperatures in Fahrenheit.
e Does not contain an equals, identity or inequality sign.
  Expression

EXAMPLE

Prove this identity.
$4(x + 2) \equiv 4x + 8$

$4(x + 2) \equiv 4 \times x + 4 \times 2$
      $\equiv 4x + 8$

×	x	+2
4	4x	+8

EXAMPLE

Find the values of a, b and c such that
$4(2x - 1) - 6(x + 2) \equiv a(bx + c)$.

$4(2x - 1) - 6(x + 2) \equiv 8x - 4 - 6x - 12$
         $\equiv 2x - 16$
         $\equiv 2(x - 8)$
$a = 2, b = 1$ and $c = -8$

EXAMPLE

Write an identity for these expressions.
a $(a + b)^2$
b $(x + 1)(x + 2)(x + 3)$

a $(a + b)^2 \equiv (a + b)(a + b)$
      $\equiv a^2 + ab + ab + b^2$
      $\equiv a^2 + 2ab + b^2$

b $(x + 1)(x + 2) \equiv x^2 + 3x + 2$
  $(x^2 + 3x + 2)(x + 3)$
      $\equiv x^3 + 3x^2 + 2x + 3x^2 + 9x + 6$
      $\equiv x^3 + 6x^2 + 11x + 6$

Algebra Formulae and functions

Fluency

Exercise 6.3S

1 Match the correct cells.

$x + 3 < 10$	Equation
$x(x + 3) \equiv x^2 + 3x$	Formula
$2x + 1 = 6$	Identity
$x^2 + 3x$	Inequality
$v = u + at$	Expression

2 Write an example of
  a an expression   b an equation
  c a formula   d an inequality
  e an identity.

3 Are these identities?
Give reasons for your answers.
  a $4(a + 2) \equiv 4a + 2$
  b $3(x + 2) \equiv 3x + 6$
  c $5(y - 2) \equiv 5y - 10$
  d $y(y + 3) \equiv 2y + 3$
  e $x(x - 4) \equiv x^2 - 4x$

4 Prove that these are all identities.
  a $5a + 10 \equiv 5(a + 2)$
  b $3x + 12 \equiv 3(x + 4)$
  c $5y - 15 \equiv 5(y - 3)$
  d $y^2 + 3y \equiv y(y + 3)$
  e $x^3 - 4x^2 \equiv x^2(x - 4)$

5 Prove that these are identities.
  a $4(a + 2) + 2(a + 1) \equiv 6a + 10$
  b $3(x + 2) + 4(x - 1) \equiv 7x + 2$
  c $5(y - 2) + 3(y - 3) \equiv 8y - 19$
  d $y(y + 3) + 2(y + 3) \equiv y^2 + 5y + 6$
  e $x(x - 4) + x(x + 2) \equiv 2x^2 - 2x$

6 Find the values of a and b such that
  a $2(x + 2) + 5(x + 1) \equiv ax + b$
  b $3(x - 2) + 4(x + 3) \equiv ax + b$
  c $5(y + a) + 3(y - b) \equiv 8y - 19$
  d $y(y + a) + 2(y + b) \equiv y^2 + 3y + 4$
  e $x(x - 4) + 2x(x - 3) \equiv ax^2 - bx$
  f $4(y - 3) - 2(5 - y) \equiv ay + b$

7 Are these identities?
Explain your answers.
  a $(a + 2)(a + 5) \equiv a^2 + 7a + 7$
  b $(x + 3)(x + 4) \equiv x^2 + 7x + 7$
  c $(b + 2)(b + 6) \equiv b^2 + 8b + 12$
  d $(y - 2)(y + 3) \equiv y^2 + y - 6$
  e $(2y + 2)(y - 3) \equiv 2y^2 + y - 6$
  f $(p - 2)(3p - 3) \equiv 3p^2 - 5p - 6$

8 Fill in the blanks to complete these identites.
  a $24p + 8 \equiv 8(\square + \square p)$
  b $18q - 27 \equiv \square(\square q + \square)$
  c $25r + 55 \equiv \square(\square r + \square)$
  d $4s^2 - 12s \equiv \square s(\square s + \square)$
  e $21t + 28t^2 \equiv \square t(\square + \square t)$
  f $54u^2v + 18u^3 \equiv \square u^2(\square u + \square v)$

9 Classify these expressions as equations, identities or formulae.
  a $V = \frac{4}{3}\pi r^3$ b $x + yz = yz + x$
  c $3f = 4(f - 2) + 8$ d $a + bc = d$
  e $x - 1 = y$ f $x^n x^m = x^{n+m}$
  g $20 - p = -(p - 20)$
  h $a^2 + b^2 = c^2$ i $s = ut + \frac{1}{2}at^2$
  j $3(x - 2)^2 = 27$
  k $4a + 2(a - 3) = 6(a - 1)$

10 Create a flow chart that uses a sequence of Yes or No questions to classify a collection of mathematical symbols as an expression, equation, identity, formula or inequality.

Did you know…

$$e^{i\pi} + 1 = 0$$

Euler's identity is often voted the most beautiful expression in mathematics. Can you see why this might be so?

1150, 1247, 1938, 1942   SEARCH

APPLICATIONS

6.3 Equivalences in algebra

RECAP
- Letters can either represent variables or unknowns.
- An expression is a meaningful collection of mathematical symbols.
- An equation is an expression that uses one '=' symbol and at least one unknown.
- A formula is an equation that describes the connection between two or more variables.
- An identity is an equation that is true for every possible value of the unknown.
- An inequality is a statement about two expressions that are not equal.

EXAMPLE

Look at this diagram.

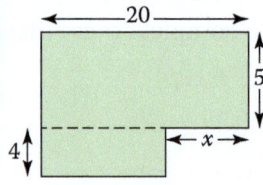

a State an inequality that can be derived from the shape.
b Create an identity by considering the area of the shape.
c The area is 144 square units. State an equation that could be solved to find the value of x.

a $x < 20$ Part < whole.
b Calculate the area in two ways as a sum or as a difference.

$20 \times 5 + 4 \times (20 - x) \equiv 20 \times (5 + 4) - 4 \times x$
$100 + 4(20 - x) \equiv 180 - 4x$

c $180 - 4x = 144 \implies 45 - x = 36$
$\implies x = 45 - 36 = 9$

HOW TO

To prove a statement is true
1. Use algebra to create statements about the information in the problem.
2. Use the rules of algebra to manipulate your expression.
3. Make a final conclusion using correct symbols.

Establishing an identity is a form of proof.

- To prove a statement is true, you need to show that it works for *all* cases.
- To prove a statement is false, you need to find *one* **counter-example**.

EXAMPLE

Prove that the sum of two **consecutive** odd numbers is an even number.

An even number is a multiple of 2: $2n$.
An odd number is one more than an even number: $2n + 1$.
Consecutive odd numbers:
$2n + 1, (2n + 2), 2n + 3$.
$(2n + 1) + (2n + 3) = 4n + 4$ ①
$= 4(n + 1)$ ②
$= 2 \times 2(n + 1)$
The sum is even because it is a multiple of 2. ③

EXAMPLE

For every integer value of n, $n^2 + n + 41$ is prime. Show that this statement is false.

Try different values of n until you find one that does not work.

$n = 1$ $n^2 + n + 41 = 1^2 + 1 + 41 = 43$ prime ✓
$n = 2$ $= 2^2 + 2 + 41 = 47$ prime ✓
$n = 3$ $= 3^2 + 3 + 41 = 53$ prime ✓
Try $n = 41$ 41 will appear in each term.
$n^2 + n + 41 = 41^2 + 41 + 41 = 1763$
$= 41(41+1+1)$
$= 41 \times 43$ – not prime ✗
The statement is false because it does not work for $n = 41$. (or for $n = 40$)

Algebra Formulae and functions

Reasoning and problem solving

Exercise 6.3A

1 For each of the following diagrams
   i   create an identity
   ii write an equation.

   a 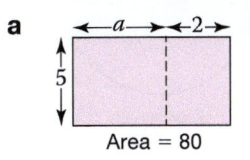   b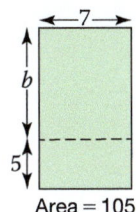

   c 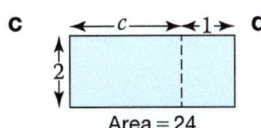   d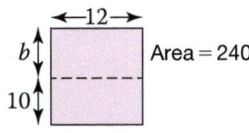

2 For each of these diagrams
   i   state an inequality that can be derived from the shape
   ii create an identity by considering how to find the area
   iii use your answers to parts **i** and **ii** to find a possible value for the area
   iv write an equation for your value. Swap with a partner and solve each other's equations.

   a 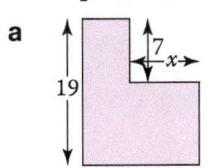   b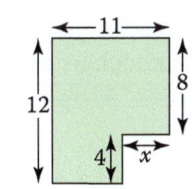

3 Create an identity using each of these diagrams.

   a 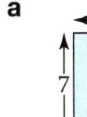   b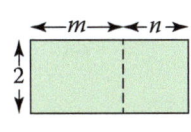

   c 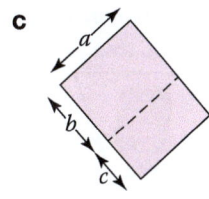 d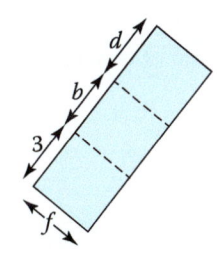

   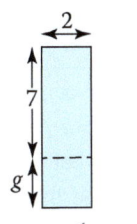

3 e $\begin{array}{cc} a & a \\ \end{array}$ (square with sides a, a)   f $\begin{array}{cc} a & b \\ \end{array}$ (rectangle with sides a, b)

4 Decide whether each of the following statements is always true, sometimes true or never true
   a An equation is a formula.
   b A formula is an equation.
   c A formula is a function.
   d A function is a formula.
   e An inequality is an equation.
   f An inequality is a function.
   g An equation uses the letter 'x'.
   h A formula is written using algebra.

5 The sum of five consecutive integers is exactly divisible by 5.
   a Check this statement is true for three examples.
   b Prove that the statement is always true.

   > Let the numbers be
   > $n, n + 1, n + 2, n + 3$ and $n + 4$.

6 Prove that each of these statements is false.
   a Adding 7 to an integer gives an odd number.
   b The sum of two primes is always odd.
   c $x^2 > x$ for any number x.
   d The product of two consecutive integers is always odd.

7 Prove each of these statements.
   a The square of any even number is a multiple of 4.
   b If you multiply two consecutive integers and subtract the smaller number you always get a square number.
   c The product of an odd number and an even number is an even number.
   d The sum of two odd numbers is even.

SKILLS

6.4 Expanding and factorising 2

Quadratics can be **expanded** by multiplying two **binomials** or **factorised** by finding common factors.

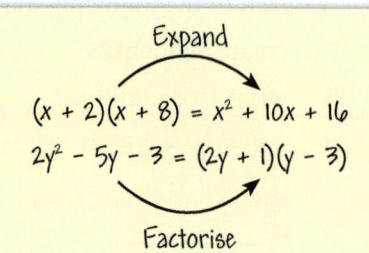

Expand
$(x + 2)(x + 8) = x^2 + 10x + 16$
$2y^2 - 5y - 3 = (2y + 1)(y - 3)$
Factorise

- Expanding quadratics involves removing brackets to create an equivalent expression.
- Factorising quadratics involves finding common factors and using brackets.
- Three binomials can also be multiplied together.

EXAMPLE

Expand and simplify these expressions.

a $(x + 5)(x + 2)$ **b** $(x + 5)(x - 2)$ **c** $(2x - 5)(x - 2)$ **d** $(3x - 2)^2$

Each term in the first bracket is multiplied by each term in the second bracket.

a $(x + 5)(x + 2)$ **b** $(x + 5)(x - 2)$ **c** $(2x - 5)(x - 2)$ **d** $(3x - 2)(3x - 2)$
$= x^2 + 5x + 2x + 10$ $= x^2 + 5x - 2x - 10$ $= 2x^2 - 5x - 4x + 10$ $= 9x^2 + 2 \times (-6x) + 4$
$= x^2 + 7x + 10$ $= x^2 + 3x - 10$ $= 2x^2 - 9x + 10$ $= 9x^2 - 12x + 4$

Be careful with negative numbers and collect like terms to simplify the answer.

There are several ways to help you remember what to do.

Treat one bracket as a single term.
$(x + 5) \times x + (x + 5) \times 2$
$x^2 + 5x + 2x + 10$

Use a grid.

×	x	+5
x	x^2	+5x
-2	-2x	-10

Use FOIL.
First $2x \times x$
Outer $2x \times -2$
Inner $-5 \times x$
Last -5×-2

Use a smiley face.
$(3x - 2)(3x - 2)$

Factorising is the opposite of expanding – so you can always check by expanding!

EXAMPLE

Factorise fully these quadratic expressions.

a $x^2 + 11x + 18$ **b** $y^2 + 3y - 10$ **c** $3a^2 - 11a + 6$ **d** $x^2 - 4$

You will need to work systematically to find factors that work.

a $x^2 + 11x + 18$ **b** $y^2 + 3y - 10$ **c** $3a^2 - 11a + 6$ **d** $x^2 - 4$
$= (x + 9)(x + 2)$ $= (y + 5)(y - 2)$ $= (3a - 2)(a - 3)$ $= (x + 2)(x - 2)$

There are also ways to help you find the factors of a quadratic.
You can use a partially filled grid.
What two numbers have a product of −10 and a sum of 3?

×	y	?
y	y^2	
?		−10

Here is another method that works well when the coefficient of x^2 is not 1.
$3a^2 - 11a + 6$ Multiply $3 \times 6 = 18$.
Find factors such that $18 = -2 \times -9$ and $-2 + -9 = -11$.
$= 3a^2 - 2a - 9a + 6$ Split the linear term in two.
$= a(3a - 2) - 3(3a - 2)$ Factorise the pairs.
$= (a - 3)(3a - 2)$ Factorise again.

This is an example of a special type of factorisation known as the 'difference of two squares'.

Algebra   Formulae and functions

Fluency

Exercise 6.4S

1 Copy and complete these identities.
   a $(x + 6)(x + 2) \equiv x^2 + \square x + \square$
   b $(x + 5)(x + 3) \equiv x^2 + \square x + \square$
   c $(x - 2)(x + 8) \equiv x^2 + \square x - \square$
   d $(x - 6)(x + 2) \equiv x^2 - \square x - \square$
   e $(x - 5)(x + 3) \equiv x^2 - \square x - \square$
   f $(x - 8)(x + 2) \equiv x^2 - \square x - \square$

2 Expand and simplify these expressions.
   a $(x + 5)(x + 2)$ b $(x + 4)(x + 3)$
   c $(y + 2)(y - 7)$ d $(y + 6)(y - 6)$
   e $(b - 3)(b + 8)$ f $(b - 9)(b + 9)$
   g $(a + 3)^2$ h $(a - 7)^2$

3 Expand and simplify these expressions.
   a $(2x + 5)(x + 2)$ b $(3x + 4)(x - 3)$
   c $(2y + 2)(3y + 7)$ d $(3y + 6)(3y - 6)$
   e $(4b - 3)(2b + 8)$ f $(5b - 9)(5b + 9)$
   g $(2a + 3)^2$ h $(3a - 7)^2$

4 Copy and complete these identities.
   a $(x - 6)(x - 2) \equiv x^2 - \square x + \square$
   b $(x - 5)(x - 3) \equiv x^2 - \square x + \square$
   c $(x - 8)(x - 2) \equiv x^2 - \square x + \square$

5 Expand and simplify these expressions.
   a $(x - 5)(x - 2)$ b $(x - 4)(x - 3)$
   c $(y - 2)^2$ d $(y - 6)^2$

6 Expand and simplify these expressions.
   a $(3x - 5)(x - 2)$ b $(2x - 4)(x - 3)$
   c $(3x - 5)(2x - 2)$ d $(2x - 4)(4x - 3)$
   e $(5x - 2)^2$ f $(2x - 5)^2$

7 Using the answers to Question 2, or otherwise, expand and simplify the following
   a $(x + 5)(x + 2)(x + 3)$
   b $(x + 4)(x + 3)(x + 1)$
   c $(y + 2)(y - 7)(y + 3)$
   d $(y + 6)(y - 6)(y + 4)$
   e $(b - 3)(b + 8)(b - 2)$
   f $(b - 9)(b + 9)(b - 2)$
   g $(a + 3)^3$ h $(a - 7)^3$

8 Factorise fully these quadratic expressions.
   a $x^2 + 5x$ b $x^2 + 7x$
   c $2x^2 + 12x$ d $12x^2 + 6x$

9 Factorise fully these quadratic expressions.
   a $x^2 + 5x + 6$ b $x^2 + 7x + 10$
   c $a^2 + 11a + 24$ d $a^2 + 13a + 12$
   e $x^2 + 20x + 91$ f $x^2 + 20x + 19$
   g $x^2 + x - 6$ h $x^2 + 3x - 10$
   i $y^2 + y - 12$ j $y^2 + 2y - 15$
   k $b^2 - 11b - 12$ l $a^2 - 5a - 24$
   m $p^2 - 9p + 18$ n $x^2 - 19x - 20$

10 Factorise fully these quadratic expressions.
   a $2x^2 + 7x + 3$ b $2y^2 + 5y + 3$
   c $3b^2 + 9b + 6$ d $3b^2 + 11b + 6$
   e $2x^2 + x - 6$ f $2x^2 + 4x - 6$
   g $19x + 4x^2 + 12$ h $4x^2 - 6x - 4$
   i $15x^2 - 31x + 14$
   j $18x^2 + 37x - 20$

11 Factorise these quadratic expressions.
   a $x^2 - 9$ b $y^2 - 25$
   c $b^2 - 100$ d $h^2 - 81$
   e $y^2 - 64$ f $a^2 - 225$
   g $4x^2 - 100$ h $9x^2 - 64$
   i $5b^2 - 125$ j $16p^2 + 25$

12 Factorise these expressions.
   a $x^2 - \dfrac{1}{4}$ b $x^2 - \dfrac{49}{81}$
   c $400x^2 - 169$ d $x^2 - 2$
   e $9x^2 - 4y^2$ f $25x^2 - \dfrac{1}{9}y^2$
   g $\dfrac{16}{49}x^2 - \dfrac{64}{81}y^2$ h $x^3 - 16x^2$

13 Factorise these algebraic expressions.
   a $x^2 + 4xy + 4y^2$ b $16x^2 - 15xy + 9y^2$
   c $5xy + 10x^2y^2$ d $10 - 3x - x^2$
   e $10x + 3 - 8x^2$ f $15xy + 2y - 3x^2y$
   g $x^4 - 7x^2 + 12$ h $2x^4y^4 + x^2y^2 - 6$

APPLICATIONS

6.4 Expanding and factorising 2

RECAP
- To expand double brackets multiply each term in the first bracket by each term in the second bracket.
- To expand three brackets, first multiply two brackets together and then multiply each term of the new quadratic expression by each term in the third bracket.
- To factorise an expression you reverse the process and rewrite it using brackets.

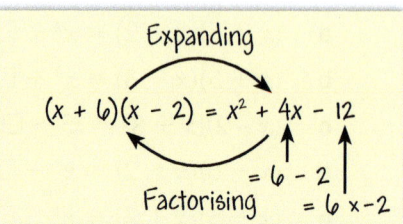
Expanding
$(x + 6)(x - 2) = x^2 + 4x - 12$
Factorising $= 6 - 2$
$= 6 \times -2$

HOW TO
To solve a problem involving quadratic expressions
1. Read the question, give any unknown a letter and decide if you need to expand or factorise
2. Collect like terms. Check your answer by expanding.
3. Answer the question

EXAMPLE

Use this diagram to create an algebraic identity.

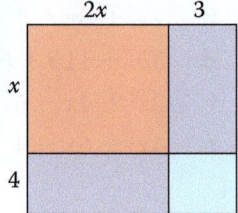

① Calculate the area as a whole or as the sum of parts.
Whole area = $(2x + 3) \times (x + 4)$
Red area = $2x \times x = 2x^2$
② Purple area = $x \times 3 + 4 \times 2x = 11x$
Blue area = $4 \times 3 = 12$
③ Both ways of calculating give the same area.
$(2x + 3)(x + 4) \equiv 2x^2 + 11x + 12$

EXAMPLE

a Prove this identity. $(a + b)(a - b) \equiv a^2 - b^2$
b Hence or otherwise evaluate these expressions *without* using a calculator.
   i 107×93 **ii** $78^2 - 22^2$

a $(a + b)(a - b) \equiv a \times a - a \times b + b \times a - b \times b$ ② Using FOIL.
$\equiv a^2 - b^2$

b **i** $107 \times 93 = (100 + 7)(100 - 7)$ **ii** $78^2 - 22^2 = (78 + 22)(78 - 22)$
$= 100^2 - 7^2$ $= 100 \times 56$
$= 10000 - 49$ $= 5600$ ③
$= 9951$ ③

EXAMPLE

Simplify these algebraic fractions.

a $\dfrac{x^2 + 5x + 6}{x + 2}$ **b** $\dfrac{x^2 + 2x - 15}{x^2 - 4x + 3}$ **c** $\dfrac{p^2 + 8p + 15}{p^2 - 9}$

① Factorise the quadratic expressions and look for terms to cancel.

a $= \dfrac{(x + 3)(x + 2)^1}{x + 2^1}$ ②   b $= \dfrac{(x + 5)(x - 3)^1}{(x - 1)(x - 3)^1}$ ②   c $= \dfrac{(p + 3)(p + 5)}{(p + 3)(p - 3)}$ ②

$= x + 3$ ③ $= \dfrac{x + 5}{x - 1}$ ③ $= \dfrac{p + 5}{p - 3}$ ③

Algebra Formulae and functions

Reasoning and problem solving

Exercise 6.4A

1 Use each diagram to create a quadratic identity.

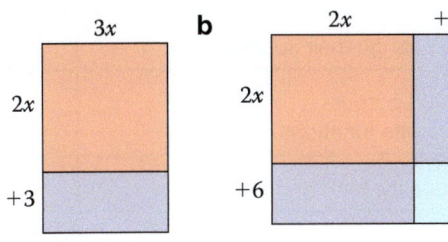

2 Martina has been given a set of 1×1 square tiles arranged into a rectangle and a square.

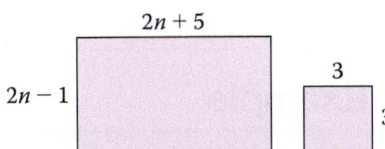

Can she rearrange all the tiles to make a square? Give your reason.

3 Carry out these calculations without using a calculator.

- **a** $67^2 - 33^2$
- **b** $34^2 - 26^2$
- **c** $61^2 - 59^2$
- **d** $55^2 - 45^2$
- **e** 62×58
- **f** 47×33
- **g** 47×53
- **h** 54×46

4 a Write an identity for $(a + b)^2$.

b Hence or otherwise, evaluate these expression *without* using a calculator.

- **i** $1.32^2 + 2 \times 1.32 \times 2.68 + 2.68^2$
- **ii** $21^2 - 2 \times 21 \times 11 + 11^2$

5 Prove that the difference of two consecutive square numbers is the sum of their square roots.

6 Prove that the difference of two consecutive even square numbers is a multiple of 4.

7 For any three consecutive integers prove that the square of the middle number is always greater than the product of the other two numbers.

8 Prove that every expression in this sequence evaluates to 2.

$2 \times 3 - 1 \times 4$

$3 \times 4 - 2 \times 5$

$4 \times 5 - 3 \times 6$

*9 Find the values of these expressions.

- **a** $(4 - \sqrt{2})(4 + \sqrt{2})$
- **b** $(2 + \sqrt{3})(2 - \sqrt{3})$
- **c** $(5 + \sqrt{6})(5 - \sqrt{6})$
- **d** $(4 - \sqrt{3})(4 + \sqrt{3})$
- **e** $(5 - \sqrt{11})(5 + \sqrt{11})$
- **f** $(6 + \sqrt{5})(6 - \sqrt{5})$

p.264

*10 Factorise fully $a^4 - b^4$.

*11 a Calculate

- **i** $(a + b)^2$ **ii** $(a + b)^3$ **iii** $(a + b)^4$

b Can you find a pattern in your results?

12 Prove these identities.

- **a** $(x + 1)(x - 1)(x + 1) \equiv x^3 + x^2 - x - 1$
- **b** $(x + 2)(x - 1)(x + 2) \equiv x^3 + 3x^2 - 4$
- **c** $(x - 1)(x^2 + x + 1) \equiv x^3 - 1$

13 Find the values of a, b and c such that

- **a** $x^2 + 6x - 2 \equiv a(x + b)^2 - c$
- **b** $x^2 + 10x + 3 \equiv a(x + b)^2 - c$
- **c** $-x^2 + 6x + 14 \equiv a(x + b)^2 + c$
- **d** $2x^2 + 6x + 10 \equiv a(x + b)^2 + c$
- **e** $-6x^2 + 36x + 24 \equiv a(x + b)^2 + c$

14 Using your answers to question **13**

- **i** say whether the expression has a largest or smallest value.
- **ii** At what value of x does this occur?
- **iii** What is the value of the expression at the **turning point**?

*15 Simplify fully these expressions.

- **a** $\dfrac{x^2 + 8x + 15}{4x + 12}$
- **b** $\dfrac{2y^2 + 17y - 30}{y^2 - 100}$
- **c** $\dfrac{3x^2 - 7x - 6}{15x + 10}$
- **d** $\dfrac{4x^2 - 64}{2x^2 - 2x - 24}$
- **e** $\dfrac{x^2 + 10x + 24}{x^2 + 4x - 12}$
- **f** $\dfrac{3x^2 + 5x - 2}{9x^2 - 9x + 2}$

16 Simplify fully this expression.

$$\dfrac{p^2 - 4p - 12}{p^2 - 36} \times \dfrac{p + 6}{p^2 - 2p - 8}$$

1150, 1151, 1156, 1157

Summary

Checkout
You should now be able to...

Test it
Questions

✓		
✓	Substitute values into formulae and rearrange formulae to change their subject.	1, 2
✓	Write an equation to represent a function and find inputs and outputs. Find the inverse of a function and construct and use composite functions.	3
✓	Use the terms expression, equation, formula, identity, inequality, term and factor.	4, 5
✓	Construct proofs of simple statements using algebra.	6, 7
✓	Expand brackets to get a quadratic expression and factorise quadratics into brackets.	8 – 11
✓	Multiply three binomials.	12

Language	Meaning	Example
Equation	An algebraic expression containing an = sign and at least one unknown. Only true for specific values of the unknown.	$x^2 - 5x + 6 = 0$ Only true for $x = 2$ or $x = 3$.
Formula	An equation linking two or more variables.	$V = IR$
Subject	The variable before the = sign in a formula.	V is the subject.
Rearrange	Rewrite a formula with a different variable as the subject.	Rearranged to make I the subject. $I = \dfrac{V}{R}$
Function	A rule that links each input value with *one* output value.	Domain: 0, 1, 2, 3 → $f(x) = 2x$ → Range: 0, 2, 4, 6 Inputs / Outputs
Domain	The set of input values for the function.	
Range	The set of output values for the function.	
Composite function	A two-step function. The output of the first step is used as the input for the second step.	$f(x) = 2x$ $g(x) = x + 1$ $fg(x) = 2x + 2$ $gf(x) = 2x + 1$
Identity	An equation that is true for every possible value for the variables.	$\dfrac{a}{4} \equiv 0.25 \times a$
Proof	A series of logical statements that show that if certain facts are true then something else must always be true.	The difference of the squares of two consecutive integers is always odd. $(n + 1)^2 - n^2 = n^2 + 2n + 1 - n^2$ $= 2n + 1$
Counter-example	An example which shows that a statement can be false, that is, not always true.	Statement: All primes are odd. Counter-example: 2 is a prime.
Expand	Remove the brackets in an expression by multiplying.	$(3x + 1)(2x - 3) = 6x^2 - 9x + 2x - 3$ $= 6x^2 - 7x - 3$
Factorise	Write an expression as a product of terms.	$6x^2 - 7x - 3 = (3x + 1)(2x - 3)$
Quadratic	An expression of the form $ax^2 + bx + c$ where a, b and c are integers and $a \neq 0$.	$6x^2 - 7x - 3$ $a = 6, b = -7, c = -3$
Binomial	A polynomial with two terms.	$(x + 2)$ $(2x - 3)$ $(y + 4)$

Algebra   Formulae and functions

Review

1. Use the formula $v = u + at$ to calculate
   a. v if $u = 5$, $a = -2$ and $t = 8$
   b. u if $v = 60$, $a = 4$ and $t = 10$
   c. a if $v = 36$, $u = 0$ and $t = 4$
   d. t if $v = 20$, $u = 50$ and $a = -6$

2. Rearrange each formula to make X the subject.
   a. $3 + 2X = A$
   b. $AX - B = 3C$
   c. $\dfrac{3X + Y}{4Z} = 5$
   d. $\sqrt{X + 4} = 2Y$
   e. $X^2 - 2K = L^2$

3. The functions f and g are defined as $f(x) = 5x$ and $g(x) = x + 3$.
   a. Write down the value of
      i. $f(7)$
      ii. $g(-5)$.
   b. Write down the inverse of
      i. $f(x)$
      ii. $g(x)$.
   c. Write down these composite functions.
      i. $gf(x)$
      ii. $fg(x)$

4.
   > $5x^2 + 3$　　$v^2 = u^2 + 2as$
   > $7a + 5 = 19$　　$3(x + 2) \equiv 3x + 6$
   > $2y + 9 > 37$

   From the box chose an example of
   a. a formula
   b. an identity
   c. an expression
   d. an equation.

5. Write an inequality for each of these statements.
   a. x is greater than -2.
   b. y is less than or equal to 0.

6. Use algebra to show that this identity is true.
   $5(2x + 3) + 2(4 - 5x) \equiv 23$

7. Prove that the sum of any three consecutive integers is a multiple of three.

8. Expand and simplify these expressions.
   a. $(x + 9)(x + 2)$
   b. $x(x - 7)(x - 6)$
   c. $(3x - 2)(x + 11)$
   d. $(4x + 2)(3x - 1)$

9. Factorise these quadratic expressions.
   a. $x^2 - 81$
   b. $16x^2 - 49$
   c. $x^2 - 7x$
   d. $21x^2 + 28x$

10. Factorise these quadratic expressions.
    a. $x^2 - 3x - 4$
    b. $x^2 - 7x + 10$
    c. $x^2 + 8x - 9$
    d. $2x^2 + 7x + 3$
    e. $6x^2 + 7x - 3$
    f. $8x^2 - 16x + 6$

11. Simplify fully these algebraic fractions.
    a. $\dfrac{x^2 + 3x + 2}{x + 1}$
    b. $\dfrac{x^2 - 16}{x^2 + 8x + 16}$
    c. $\dfrac{2x^2 - 3x + 1}{2x^2 + x - 1}$

What next?

Score		
0 – 4		Your knowledge of this topic is still developing. To improve, look at MyiMaths: 1150, 1151, 1156, 1157, 1159, 1170, 1171, 1186, 1247
5 – 9		You are gaining a secure knowledge of this topic. To improve your fluency look at InvisiPens: 06Sa – p
10 – 11		You have mastered these skills. Well done you are ready to progress! To develop your problem solving skills look at InvisiPens: 06Aa – f

Assessment 6

1 a Erin and Stina both substitute the values $u = -4$, $v = 12$ and $t = 8$ into the expression $s = \frac{1}{2}(u + v)t$. Erin says that this gives $s = 64$. Stina says that this gives $s = 32$.
Who is correct? Give your reason. [1]

b Erin and Stina now both substitute the values $u = 0$, $a = 2$ and $s = 16$ into the expression $v^2 = u^2 + 2as$. Erin says that this gives $v = 64$. Stina says that this gives $v = 8$.
Who is correct? Give your reason. [2]

2 Each formula has been rearranged into the given form.

a $a = -n^2\left(x - \dfrac{h}{n^2}\right)$ into $n = \sqrt{\dfrac{h-a}{x}}$ [3]

b $t = \left(\sqrt{\dfrac{1-e}{1+e}}\right)x$ into $e = \dfrac{x^2 - t^2}{x^2 + t^2}$ [3]

c $\dfrac{1}{S} = \dfrac{1}{R} + \dfrac{1}{r}$ into $S = \dfrac{rR}{r+R}$ [2]

Suggest one way to do each rearrangement.

3 A mobile phone provider 'TextUnending' charges 15p each minute if you call between the peak times of 09:00 and 20:00 and 12p per minute at other times, called off-peak. Texts cost 14 pence per text at all times.

a Using the letters C for the cost (in pence), p for the number of peak minutes, q for the number of off-peak minutes and t for the number of text messages sent, write down a formula to work out the cost. [2]

b Hugo made one call for 3 minutes 20 seconds at 13:15 and another lasting from 20:32 to 21:07 exactly. He also sent 8 text messages. How much did this cost? [3]

c Cheng made three calls lasting from 9:39 to 10:02, 11:53 to 12:17 and 19:42 to 20:15. She also sent 25 text messages. Work out the total cost. [4]

4 The formula for the length of a skid, S m, for a vehicle travelling at v km/h is $S = \dfrac{v^2}{170}$

a A car skids while travelling at 100 km/h. How long is the skid? [1]

b A car and a lorry are travelling head-on towards each other on a narrow road.

They see each other and start to skid at the same instant.

The car is travelling at 70 km/h and the lorry at 52 km/h.

Calculate the distance they are apart if they just stop in time. [3]

c Find the speed of a vehicle which skidded for 60 m. Give your answer to the nearest km/h. [2]

5 Stellio is sitting on a cliff ledge above the sea.

When he is x metres above sea level, the horizon is y miles away. Variables, y and x, are connected by the formula $y = 3.57\sqrt{x}$.

a How far out to sea can Stellio see when he is 8.5 m above the sea? [1]

b Calculate Stellio's height above the sea when the distance to the horizon is 8.55 km. [2]

c A pirate ship sails past the cliff 33 km offshore when Stellio is 85 m above the sea. Can Stellio see the pirate ship? [2]

6 You score a total of x marks in 3 tests. In another x tests you score another 27 marks but your mean score remains the same. How many marks did you score in the first 3 tests? [4]

7 a Are these identities correct? If not, then give the correct answer.
    i $(p-4)(p-7) = p^2 + 28$ [2]
    ii $(v+9)(v-7) + (4-5v)^2 = 6v^2 - 47$ [4]

  b Are these identities correct? If not, then give the correct answer.
    i $z^2 + 13z + 36 = (z+4)(z+9)$ [2]
    ii $v^2 - 100 = (v-10)^2$ [2]
    iii $30x^2 + 13xy - 3y^2 = 3x(5x + 3y)$ [2]

8 a i Factorise $2.4^2 - 2 \times 2.4 \times 3.6 + 3.6^2$ [2]
    ii Evaluate your factorised expression. [1]

  b Work out these without using a calculator.
    i $89^2 - 11^2$ [3]
    ii $6.89^2 - 3.11^2$ [3]

9 Write $x^2 - 8x + 30$ in the form $(x-a)^2 + b$. [3]

10 For each statement either prove that it is always true or find a counter-example to show that it is false.
  a An odd number times an even number is always odd. [1]
  b Prime numbers have one factor. [1]
  c The sum of two even numbers is always even. [2]
  d Any number squared is more than 0. [1]
  e Two prime numbers multiplied together are always odd. [1]
  f The sum of three consecutive even numbers is always divisible by 6 [2]
  g The square of any number is never a prime number. [2]

11 Work out the rule that turns *each* input in these function machines into its corresponding output.

  a 6, 10, 12 → □ → □ → 25, 41, 49 [2]
  b 6, 15, 36 → □ → □ → 0, 3, 10 [2]

12 a Hassan says that the inverse of $g(x) = 2x + 1$ is $2x - 1$.
    Is he correct? If not, what is the correct inverse of $g(x) = 2x - 1$? [3]

  b For the functions $f(x) = 3x - 1$ and $g(x) = x^2$. Find
    i $f(6)$ [2]   and   **ii** $fg(6)$. [2]

Revision 1

1. Jenni gets the bus to work. The fare is $1.75 for a single journey and no return tickets are issued. She works Monday to Wednesday, Friday and Saturday. Explain if Jenni would save money if she bought a weekly ticket costing $15. [4]

2. Ankit buys petrol and collects 'Frequent Traveller Miles' (FTM) vouchers that give him free flights. He gets 5 FTMs for every 30 litres of petrol that he buys. The petrol for Ankit's car costs 136.9p per litre. He needs 600 FTMs for a free trip to Nice.

   a. How much does it cost Ankit to collect enough vouchers? [5]

   Ankit's car averages 38 miles per gallon. Take 1 litre as being 0.22 gallons.

   b. How many miles does Ankit have to drive to get his free trip? [5]

3. Taci is lining the base and sides of a drawer with paper. The drawer has the dimensions shown.

   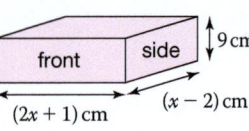

   a. Give the following in terms of x. Expand all brackets.
      i. The area of the front of the drawer. [2]
      ii. The area of a side of the drawer. [2]
      iii. The total area of paper Taci uses. [6]
      iv. The volume of the drawer. [2]

   b. The total area of paper Taci uses is $51x$ cm². Find the numerical volume of the drawer. Show your working. [5]

4. a. A hoist is made up of 4 metal girders PQ, QR, QS and RS. The girders are attached to the horizontal ground PST. P, Q and R lie on a straight line. The distances PS and SR are equal, RQS is a right angle and angle QPS is 28°.

   i. Find angles QSR, RST and PSR. Explain your answers. [9]
   ii. Explain why $PQ = QR$. [2]

   b. Another hoist is made up of 5 girders, AB, BC, CD, DB and AD. AB is parallel to DC and the girders AB, BD and BC are all the same length. Angle BAD is 54°.

   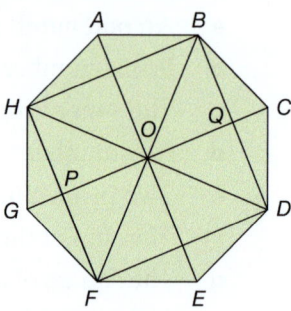

   i. What sort of triangles are ABD and BCD? [1]
   ii. Find angles ABD and DBC. [8]
   iii. Is triangle ABD similar to triangle BCD? Explain your answer. [1]

5. $ABCDEFGH$ is a regular octagon.

   a. Are the following statements correct or not. Fully describe the shape in each case.
      i. HBD is an isosceles triangle. [4]
      ii. $HBDF$ is a trapezium. [2]
      iii. $HBCG$ is a parallelogram. [3]

   b. Using the letters in the diagram, name
      i. a triangle congruent to HOP [1]
      ii. a triangle similar to HOP. [1]

   c. Find the value of these angles.
      i. ABC [4]   ii. POF [1]
      iii. HOC [1]   iv. ABH [2]

6. Stars in the night sky are grouped by their surface temperatures, measured in thousands of degrees.

Group	O	B	A	F	G	K	M
Surface Temperature (1000 °C)	40	15	8	6	5	4	2

   To illustrate this data draw a
   a. bar chart [5]   b. pie chart. [5]

7 The data shows the number of nights and number of guests staying in a hotel.

No. of nights (x)	1	2	3	4	5	6	7	8	9	10	11	12
No. of guests (f)	4	5	3	6	8	3	9	7	5	4	2	6

  a Calculate the
    i mean [4]
    ii median length of stay. [3]
  b Find the
    i range [1]
    ii interquartile range. [5]

8 a In an election Jordin Trusain, polled 8082 votes and won 36% of all the votes cast. How many votes were cast altogether? [2]

  b Ali Imyur polled 7010 votes. What percentage of the total vote did she poll? [2]

  c The number of votes cast represented a turnout of 63.7% to the nearest 0.1%. What are the minimum and maximum number of voters in the total electorate? [4]

9 a Solomon sits 6 exam papers and obtains a mean mark of 57.5. He then sits another paper and scores 45. What is his mean score now? [5]

  b Sabina sits some exam papers and obtains a mean mark of 32, with 288 marks altogether. She takes another test and her new mean is 35.7. What did she score in the last test? [5]

10 Flask A contains 80 ml of sulphuric acid and flask B contains 100 ml of water. Brianna transfers 20 ml of water from flask B into flask A.

  a What fraction of the liquid in flask A is water? [2]

The sulphuric acid and water in flask A are *thoroughly mixed* together. A 20 ml spoonful is transferred to flask B.

10 b How much
    i sulphuric acid does the spoon hold [1]
    ii water does the spoon hold? [1]
  c Give the new volumes of sulphuric acid and water in
    i flask A [2]
    ii flask B. [2]

11 Matilya says '$10x$ is always bigger than x'. Explain whether she is right. [2]

12 The overtaking distance D, when one vehicle passes another, is given by the formula $D = \dfrac{V(L + 130)}{V - U}$ where U is the speed of the slower vehicle, V the speed of the faster vehicle and L the length of the slower vehicle. V and U are in mph, D and L are in feet.

  a A car at 70 mph, overtakes a van of length 12 ft, travelling at 30 mph. What is the overtaking distance? [2]

  b Later, travelling at 55 mph, the car passes a car travelling at 25 mph. The overtaking distance is 400 ft. Find the length of the car being overtaken. [5]

  c The law in a particular country says that all overtaking must be done in a maximum distance of 850 ft. Use your formula to calculate the minimum speed Peter, driving a car, must be travelling to overtake a coach of length 41 ft travelling at 60 mph. [6]

  d A motorway runs parallel to a railway line. A train of 8 coaches, each of length 65 feet including engines, is travelling at 49 mph. A car is travelling at 68 mph. Can the car pass the train in under a mile? (1 mile = 5280 ft) [5]

7 Working in 2D

Introduction

Self-similarity is the property in which an entire shape is mathematically similar to a part of itself. This means is that if you 'zoom in' on a small part of the shape, you get an exact replica of the original shape itself. Self-similarity is used in fractal images, like the one you can see here. It has real-world use in describing the structure of coastlines, as well as the natural growth of plants such as ferns and the formation of crystals and snowflakes.

What's the point?

The real world, being mathematically untidy, is never exactly self-similar. However self-similarity provides a very useful model in understanding the complex geometries seen in nature, which can't usually be reduced to simple rectangles and circles.

Objectives

By the end of this chapter, you will have learned how to …
- Measure line segments and angles accurately.
- Use scale drawings and bearings.
- Calculate the areas of triangles, parallelograms, trapeziums and composite shapes.
- Describe and transform shapes using reflections, rotations, translations (described as 2D vectors) and enlargements (including fractional and negative scale factors).
- Identify what changes and what stays the same under a combination of transformations.

Check in

1 These shapes are drawn on a centimetre square grid.

a 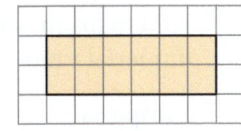 b 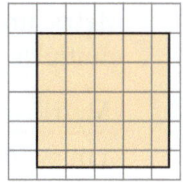 c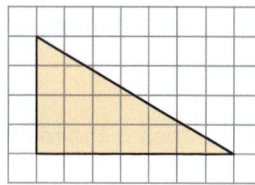

For shapes **a–c**
  i find the perimeter in **i** cm, **ii** mm.
  iii calculate the area of the shape in cm².

2 Write the coordinates of the points **A–E**.

3 Write the equations of the lines **a–e**.

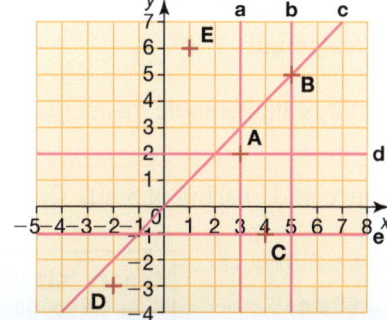

Chapter investigation

Create a snowflake!

Step 1: Draw an equilateral triangle.

Step 2: Draw equilateral triangles on each of the three sides (carefully – you'll have to divide each side into three equal parts).

Step 3: You have now got 12 sides. Draw equilateral triangles on each of these.

If it's still not snowflaky enough, try once more – but you'll find it starts getting very fiddly!

SKILLS

7.1 Measuring lengths and angles

Distances are usually measured in **metric** units.
Maps and plans use **scales**.

- The ratio scale 1 : a means the real length is $a \times$ the length on the map.

For example, the scale 1 : 200 means
real length = 200 × length on the map
map length = real length ÷ 200

Metric units use multiples of 10.

Metric
1 m = 1000 mm
1 m = 100 cm
1 km = 1000 m

Old maps may use imperial units

Imperial
1 foot = 12 inches
1 yard = 3 feet
1 mile = 1760 yards
5 miles ≈ 8 km
1 foot ≈ 30 cm

- A bearing is a three-figure angle measured *clockwise from north*.

EXAMPLE

Find the distance and bearing of Santa Cruz de Tenerife from Arrecife.

1 cm represents 5 000 000 cm
= 50 000 m = 50 km
Distance on map = 5.4 cm
Real distance = 5.4 × 50
= 270 km
Bearing = 360° − 103°
= 257°
Sum of angles at a point is 360°.

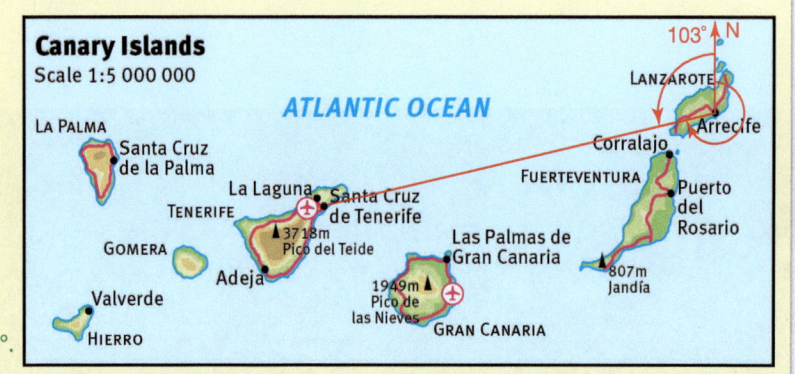

EXAMPLE

A sports hall is to include a badminton court.
The sketch shows the dimensions.
Draw a scale diagram using a scale of 1 : 200.

You can draw the accurate scale diagram on plain paper, squared paper or graph paper.

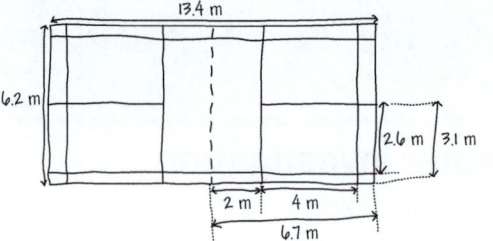

× 100 to change the real lengths in m to cm then ÷ 200 to find the length on the plan.

Real (m)	Real (cm)	Plan (cm)
13.4 m	1340 cm	6.7 cm
6.2 m	620 cm	3.1 cm
6.7 m	670 cm	3.35 cm
2.6 m	260 cm	1.3 cm
3.1 m	310 cm	1.55 cm

÷ 200 =

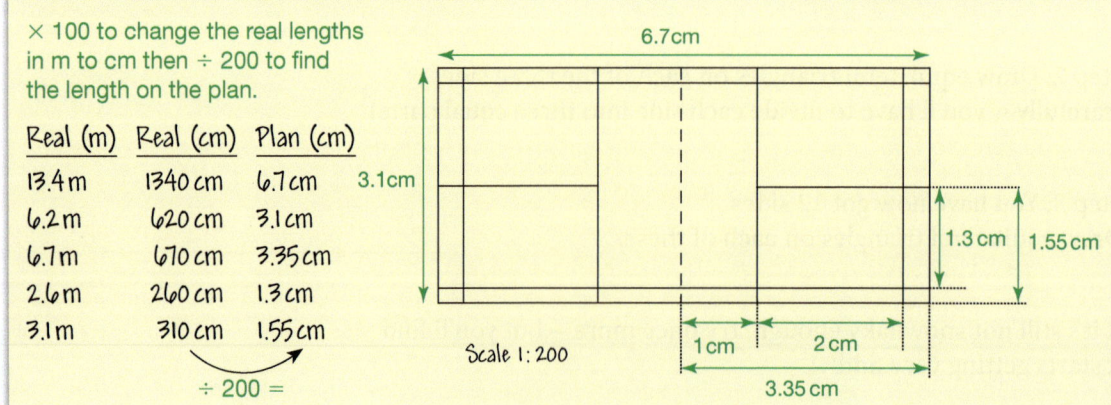

Scale 1 : 200

Geometry Working in 2D

Fluency

Exercise 7.1S

1 Use tracing paper and the map of the Canary Isles to find the distance and bearing of

   a Arrecife from Las Palmas
   b Valverde from Adeja
   c Las Palmas from Santa Cruz de Tenerife
   d Santa Cruz de Tenerife from Las Palmas
   e Puerto del Rosario from Arrecife.

2 The scale of the map of England is 1 : 5 000 000.
   Find the distance and bearing of

   a Cambridge from London
   b Oxford from Manchester
   c Bristol from Birmingham
   d Birmingham from London
   e Oxford from Cambridge.

3 The scale of the map of Grasmere is 1 : 25 000.

   a Find the real distance that is represented by 1 cm on the map.
   b Find the actual length and width of the island in Grasmere.
   c i   Use string to find the distance around Grasmere on the map.
      ii Find the actual distance around Grasmere.

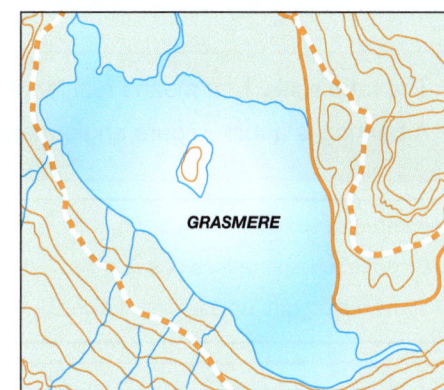

4 The sketch shows the dimensions of a student flat.

   Draw an accurate scale diagram of the flat.

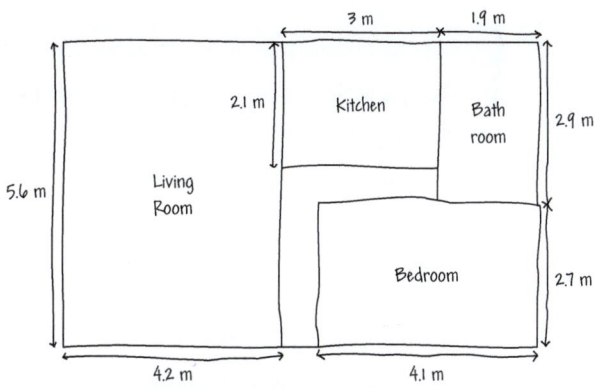

Did you know…

You can colour any map using no more than four colours, such that neighbouring regions always have different colours. The proof of this theorem, by Appel and Haken in 1976, was the first to rely on using computers.

APPLICATIONS

7.1 Measuring lengths and angles

RECAP
- The ratio scale 1 : *a* means real length = *a* × length on the map.
- A bearing is a three-figure angle measured *clockwise from north*.

HOW TO
To construct a scale drawing
1. Choose a scale, unless one is given.
2. Work out or measure the lengths or angles required.
3. Draw a diagram (unless one is given).
4. Give the answers including units.

EXAMPLE

A boat travels for 260 miles on a bearing of 115°, then 340 miles on a bearing of 263°.
a i   Find the direction in which the boat must travel to return directly to the start.
   ii Find the length of the return journey.
b Explain how the accuracy of the answers depends on the scale you have used.

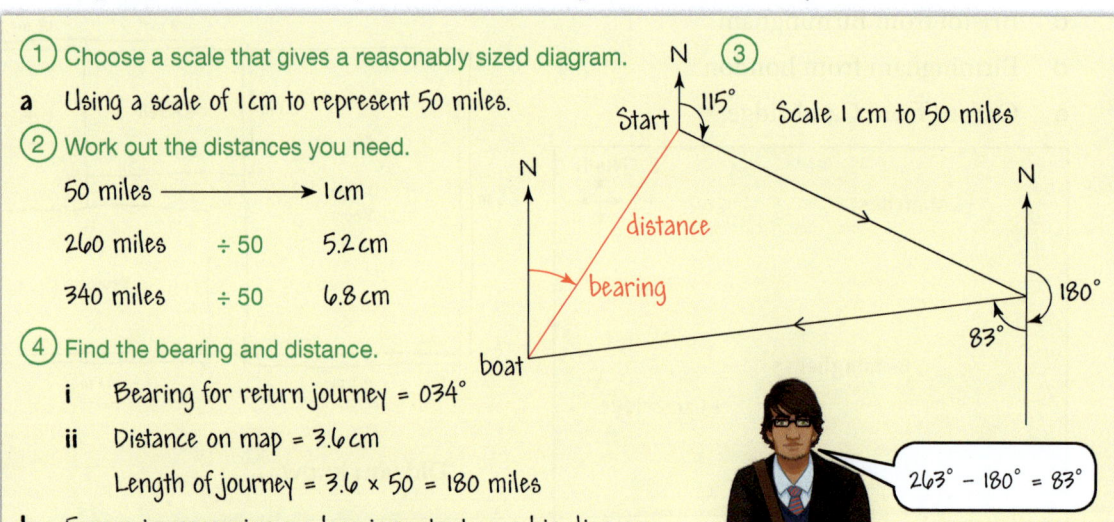

① Choose a scale that gives a reasonably sized diagram.
a Using a scale of 1 cm to represent 50 miles.
② Work out the distances you need.

50 miles ⟶ 1 cm
260 miles ÷ 50 5.2 cm
340 miles ÷ 50 6.8 cm

④ Find the bearing and distance.
  i   Bearing for return journey = 034°
  ii Distance on map = 3.6 cm
      Length of journey = 3.6 × 50 = 180 miles
b Errors in measuring are less important on a big diagram.
   The larger the scale, the more accurate the answers will be.

263° − 180° = 83°

EXAMPLE

This is part of a plan of Seth's kitchen.
Seth wants a cooker to fit in the space.
The widths of cookers are 900 mm, 1000 mm and 1100 mm.
Which width should Seth choose? Explain your answer.

| scale 1 : 50 | wall |
| kitchen units | space for cooker |

①②③ Measure the space on the diagram.
Width of space = 19 mm

Measure in mm so that you don't need to change units.

④ Compare the real size of the space with the cooker widths.
Real width of space = 19 × 50 = 950 mm
Seth should choose the 900 mm cooker because this will fit into the space.
The widths of the other cookers are more than 950 mm, so they will not fit.

Geometry Working in 2D

Reasoning and problem solving

Exercise 7.1A

1. A plane leaves airport *A* and flies 420 miles on a bearing of 067°.

   The plane then changes course to 312° and travels a further 480 miles to land at airport *B*.

   a. Find the distance and bearing for a direct return journey from *B* to *A*.

   b. How will the distance change if the initial bearing, 067°, increases?

2. Liam sees a lifeboat on a bearing of 138° from Mevagissey. Dina sees the same lifeboat on a bearing of 215° from the Rame Head chapel.

   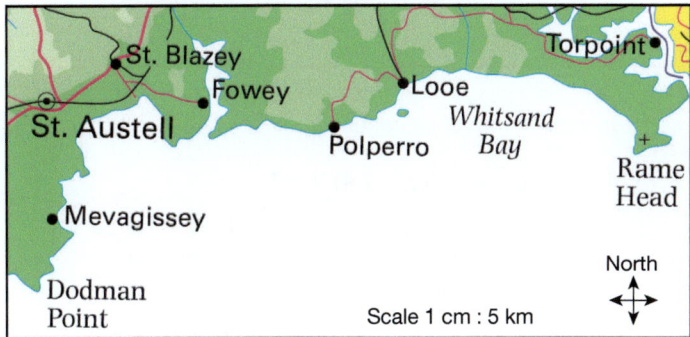

   Dina says the lifeboat is about 5 km nearer to Liam than to her.

   Use tracing paper to copy the coastline and work out if Dina is correct.

3. The measurements in the table are taken from a plot of land.

Vertex	Distance from *A* in direction of *AD*	Perpendicular distance from *AD*
B	$AX = 120$ m	$BX = 185$ m
C	$AZ = 370$ m	$CZ = 240$ m
D	$AD = 580$ m	0 m
E	$AY = 240$ m	$EY = 175$ m

   The farmer wants to know the area of the land and how much it will cost to surround it with fencing. The fencing costs $75 per 50 metre roll.

   a. Use a scale diagram to estimate
      i. the cost of the fencing
      ii. the area of the land.

   b. You are asked to make your estimates ten times more accurate. Describe how you can do this.

4. The scale of this map of Sicily is 1 : 500 000

   a. A poster advertises a cruise around the coast of the island.

      Estimate the minimum distance the boat will travel.

   b. A tourist wants to travel along the coastal road from Palermo to Messina.

      Estimate how long this will take at an average speed of 50 km/hr.

   *c. Estimate the area of Sicily in km².

5. Aldo is 8 km west of Bea.
   Cecile is 7 km from Aldo and 5.5 km from Bea.

   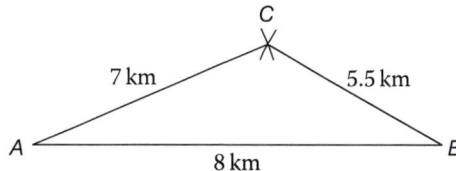

   a. Draw a scale drawing to show this information.

   b. Use your scale drawing to find the bearing of
      i. Cecile from Aldo
      ii. Cecile from Bea.

SKILLS

7.2 Area of a 2D shape

Area is the amount of space inside a 2D shape.

In the metric system, area is measured in **mm²**, **cm²**, **m²** or **km²**

- Area of a rectangle = length × width
- Area of a parallelogram
  = base × perpendicular height
- Area of a triangle
  = ½ base × perpendicular height
- Area of a trapezium
  = ½ sum of the parallel sides × perpendicular height

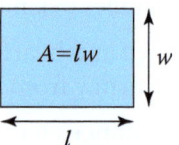
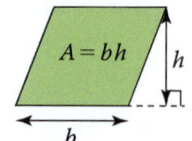
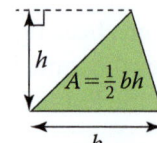
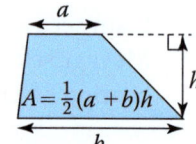

EXAMPLE

A hole is cut in a trapezium.

Find the percentage of the trapezium that is left.

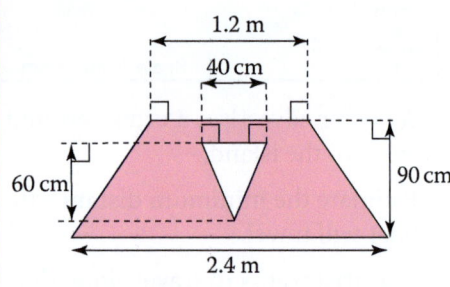

You can work in metres or centimetres.

Area of trapezium = ½ (1.2 + 2.4) × 0.9
                  = ½ × 3.6 × 0.9
                  = 1.62 m²

Area of hole = ½ × 0.4 × 0.6
             = 0.12 m²

Area of trapezium left = 1.62 − 0.12
                       = 1.5 m²

% of shape that is left = $\frac{1.5}{1.62}$ × 100%
                        = 92.59
                        = 92.6% (3 sf)

Units must be consistent
90 cm = 0.9 m
40 cm = 0.4 m
60 cm = 0.6 m

EXAMPLE

The area of a trapezium is 650 cm².

The lengths of its parallel sides are 20 cm and 30 cm.

Calculate the height of the trapezium.

Substitute the known values into the formula, then rearrange.

A = ½ (a + b) h

650 = ½ (20 + 30) h

650 = ½ × 50h

25h = 650

h = 650 ÷ 25 = 26

The height of the trapezium = 26 cm

Sketch the trapezium if it helps.

Geometry Working in 2D

Fluency

Exercise 7.2S

1. Find the area of each trapezium
   i using the trapezium formula
   ii by splitting it into other shapes.

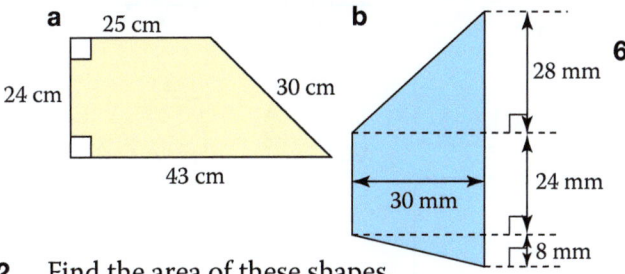

2. Find the area of these shapes.

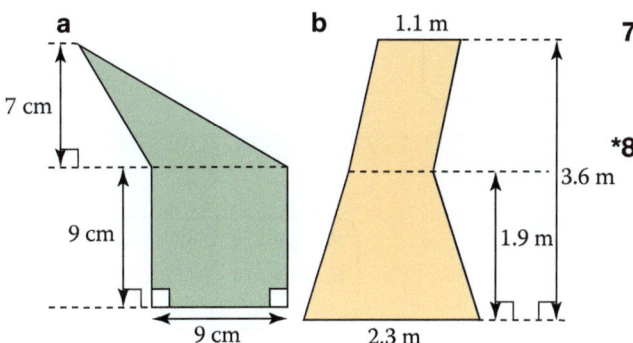

3. Copy the table. Find the missing values.

Shape	Base	Height	Area
Rectangle	16 mm		240 mm²
Parallelogram		45 cm	1800 cm²
Triangle	4 m		5 m²

4. A hole is cut in each shape.
   Find the percentage of each shape that is left.

   a
   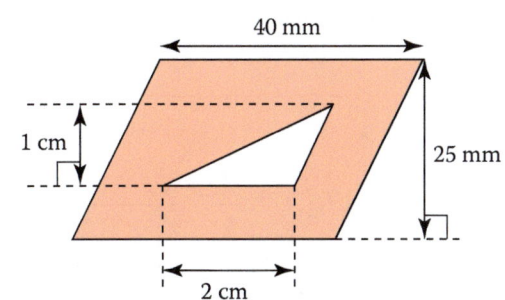

   b
   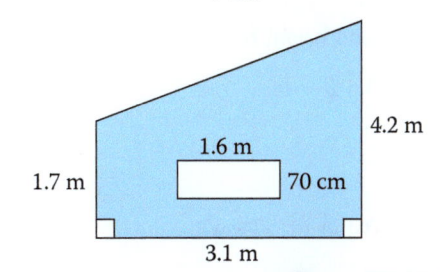

5. The area of a trapezium is 132 cm².
   The height of the trapezium is 12 cm and the length of one parallel side is 15 cm.
   Find the length of the other parallel side.

6. A shape consists of a rectangle surmounted by a parallelogram. The total area of the shape is 112 cm². Calculate the height, h, of the rectangle.

   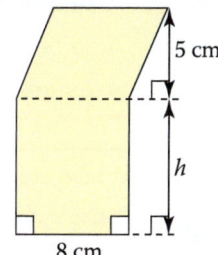

7. Find the length of the side of a square whose area is equal to that of a triangle with base 27 cm and height 24 cm.

*8. A regular pentagon and a regular hexagon each have sides of length 4 cm.
   The other dimensions are as shown.

   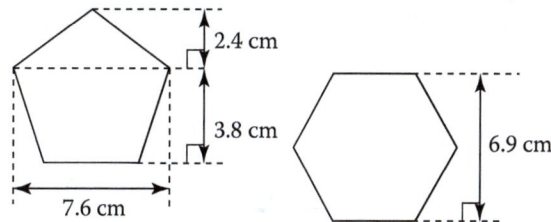

   Write the area of the pentagon as a percentage of the area of the hexagon.

   p.60

9. The diagram shows a window in the side of a house.
   The window is 80 cm wide and 1.5 m high.
   Find the area of this side of the house excluding the window.

   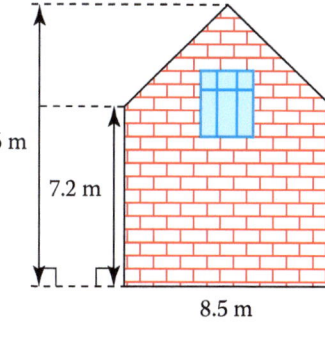

10. A square sheet of card has sides of length 1.2 m.
    a How many rectangles, with sides of length 6 cm and 8 cm can you cut out of the square?
    b How many right angled triangles with sides of length 3 cm, 4 cm and 5 cm can you cut out of the square?

1108, 1128, 1129

137

APPLICATIONS

7.2 Area of a 2D shape

RECAP
- Area of a rectangle = length × width
- Area of a parallelogram = base × perpendicular height
- Area of a triangle = $\frac{1}{2}$ base × perpendicular height
- Area of a trapezium
  = $\frac{1}{2}$ sum of the parallel sides × perpendicular height

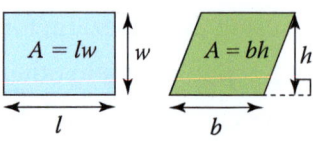
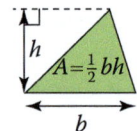
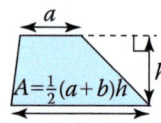

HOW TO

To find the area of a shape
1. Draw a diagram (if needed) and identify simple shapes in it.
2. Decide which formula or formulae to use.
3. Find the areas of the simple shapes, including the units.
4. State your conclusion and, when required, your reason.

EXAMPLE

This tunnel must have an area of at least 28 m². Calculate the minimum height.

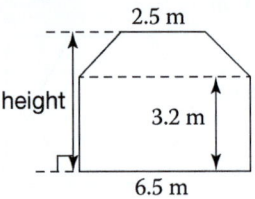

1. Given
2. For a rectangle, $A = lw$.
3. Area of rectangle = 6.5 × 3.2
   = 20.8 m²
2. For a trapezium, $A = \frac{1}{2}(a + b)h$
   Given A use the formula to find h.
3. Minimum area of trapezium = 28 − 20.8
   = 7.2

Substituting

$7.2 = \frac{1}{2} \times (2.5 + 6.5) \times h$

$4.5h = 7.2$

$h = 7.2 \div 4.5 = 1.6$

4. Minimum height = 3.2 + 1.6 = 4.8 metres

You can check the answer by using it to work out the area of the tunnel.

EXAMPLE

What fraction of this regular hexagon is shaded blue?

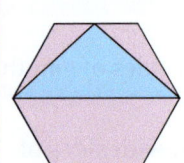

1. A regular hexagon can be split into 6 congruent triangles.
2. Use $A = \frac{1}{2}bh$ for these triangles and then the blue triangle.

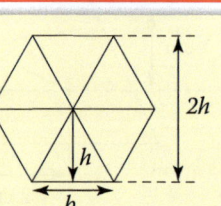

3. Total area of hexagon = $6 \times \frac{1}{2}bh = 3bh$
   The blue triangle has base $2b$ and height h.
   Area of blue triangle = $\frac{1}{2} \times 2b \times h = bh$

   Fraction that is blue = $\frac{bh}{3bh} = \frac{1}{3}$

   The shaded area is $\frac{1}{3}$ of the area of the hexagon.

Geometry   Working in 2D

Reasoning and problem solving

Exercise 7.2A

1. Erica says the area of this plot is over $\frac{1}{4}$ hectare. (1 hectare = $10\,000\,\text{m}^2$)

   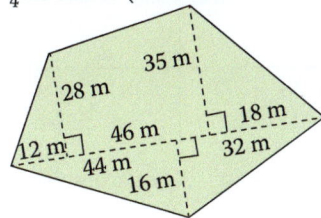

   a Is Erica correct?

   b What assumptions could affect the answer to **a**?

2. The area of this water channel must be at least $240\,\text{cm}^2$.

   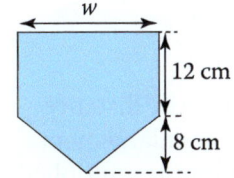

   a Find the minimum width, w.

   b What will happen to the minimum width if the 12 cm is increased and the 8 cm reduced by the same amount?

3. What fraction of each shape is shaded?

   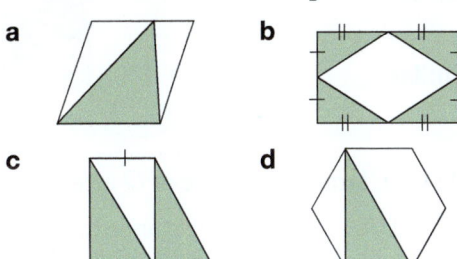

4. Use x and y axes from 0 to 6 on centimetre square grid paper.

   a Plot and join these points to make an arrow.
   (0, 3), (3, 6), (5, 6), (3, 4), (6, 4)
   (6, 2), (3, 2), (5, 0), (3, 0), (0, 3)

   b Show how to find the area of the arrow by three different methods.

5. This shape is made from two overlapping squares. What is its area?

   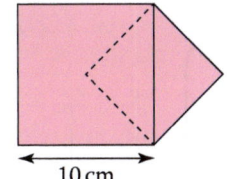

6. The diagram shows the dimensions of Rami's bathroom floor.

   Rami wants to cover the floor with square tiles with sides of length 400 mm.

   The tiles cost €29.50 for each pack of 5 tiles.

   Rami says it will cost less than €200 to tile the floor. Is Rami correct? Show your working.

   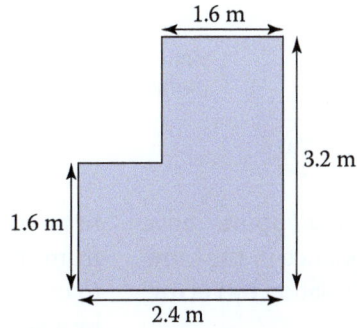

7. Amy has a rectangular map 30 cm long and 20 cm wide. Amy says 'Over half of this map lies within 4 cm of the edge.'

   Is Amy correct? Explain your answer.

*8. a How many badges like this can Zosia cut from a square sheet of metal with sides of length $\frac{1}{4}$ metre?

   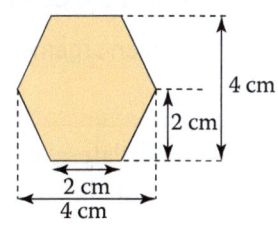

   b Find the percentage of metal that is wasted.

*9. PQRS is a parallelogram.
   X is the mid-point of PQ.
   Y is the point of intersection of PR and QS.
   Show that the area of triangle PXY is one eighth of the area of parallelogram PQRS.

10. Find the area of these shapes.

    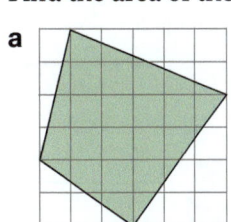 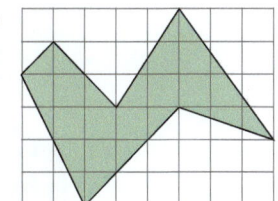

     = $1\,\text{cm}^2$

SKILLS

7.3 Transformations 1

A **transformation maps** points in the **object** to points in the **image** and causes the position, and for enlargements the size, of shapes to change.

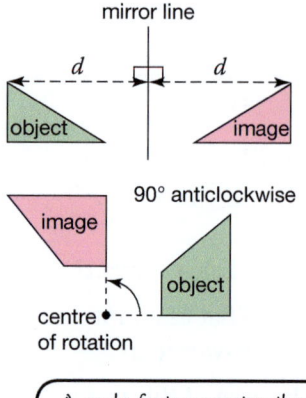

- All points on the **mirror line** do not move in a **reflection**.
- The **centre of rotation** does not move in a **rotation**.
- The **centre of enlargement** does not move in an **enlargement**.

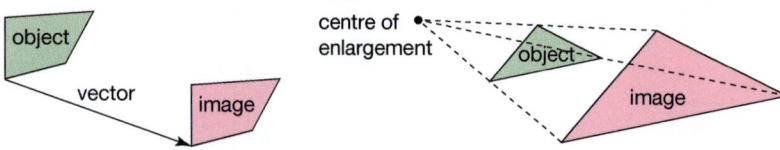

In a **translation**, *all* points move by the same distance in the same direction. A translation by **vector** $\binom{3}{-1}$ moves every point 3 units to the right and 1 unit down.

In an enlargement the distance from the centre of enlargement to every other point is multiplied by a **scale factor** (SF).

A scale factor greater than 1 enlarges the shape. A scale factor between 0 and 1 makes the shape smaller.

- In a reflection, rotation or translation the image and object shapes are **congruent**.
- In an enlargement the image and object shapes are **similar**.

EXAMPLE

Quadrilateral **Q** has vertices at (1, 1), (3, 1), (5, 3) and (1, 5). Draw the image of **Q** after

a reflection in the line $y = -1$
b rotation of 90° anti-clockwise about $(-1, 1)$
c translation by vector $\binom{-7}{-8}$
d enlargement, centre (3, 3) scale factor $\frac{1}{2}$.

Transform each vertex of Q and join them to find the image.
a Each vertex has an image on the opposite side of the mirror line, an equal distance from it.
b Rotate each vertex 90° anti-clockwise about (−1, 1). Use tracing paper if you wish.
c Each vertex moves 7 units left and 8 units downwards.
d Each image point is half as far from the centre as the corresponding object point.

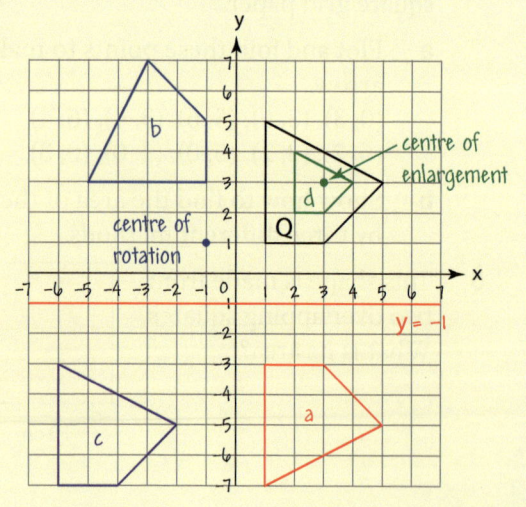

The line $y = -1$ goes through points like (1, −1), (2, −1) and (3, −1).

Geometry   Working in 2D

Fluency

Exercise 7.3S

For questions **1** and **2** use *x* and *y* axes from −4 to 4.

1. Triangle *T* has vertices at (1, 1), (4, 0) and (2, 3). Draw *T* and its image after
   a. reflection in the *x* axis
   b. reflection in the *y* axis.

2. Parallelogram *P* has vertices (1, 2), (4, 1), (4, 3) and (1, 4). Draw *P* and its image after
   a. rotation of 90° clockwise about (0, 0)
   b. rotation of 180° clockwise about (0, 0)
   c. rotation of 90° anti-clockwise about (0, 0).

For questions **3** and **4** use *x* and *y* axes from −6 to 6.

3. a. Draw and name shape *A* with vertices (−2, 1), (−2, 4), (−1, 4), (−1, 3) and (0, 1).
   b. Draw the image of *A* after each of these translations.
      i. $\binom{4}{2}$ ii. $\binom{-1}{-7}$ iii. $\binom{3}{-6}$ iv. $\binom{-3}{1}$

4. a. Draw and name the shape *X* with vertices at (1, 0), (2, 2), (3, 0) and (2, −2).
   b. Draw the image of *X* after each of these enlargements.
      i. Centre (0, 0), scale factor 2
      ii. Centre (4, 0), scale factor 3
      iii. Centre (2, −6), scale factor $\frac{1}{2}$

5. Copy the diagram onto isometric paper. Draw the image of flag *F* after
   a. enlargement, scale factor 3, centre *C*
   b. rotation of 120° clockwise about *D*
   c. rotation of 120° anti-clockwise about *D*.

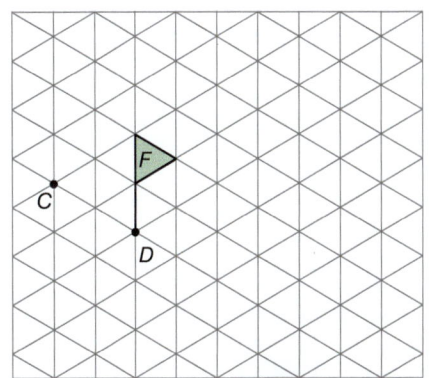

6. a. For each part use a copy of this pentagon.

   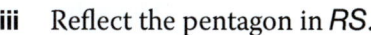

   i. Reflect the pentagon in *PQ*.
   ii. Reflect the pentagon in *QR*.
   iii. Reflect the pentagon in *RS*.
   iv. Reflect the pentagon in *TR*.

   b. i. Describe the position of the mirror line of the reflection that maps the pentagon onto itself.
   ii. What is the image of *P* in this reflection?

For questions **7** and **8** use *x* and *y* axes from −7 to 7.

7. a. Join points $A(-4, 2)$, $B(-3, 4)$, $C(-4, 5)$ and $D(-5, 4)$ to form a quadrilateral.
   What type of quadrilateral is *ABCD*?
   b. Reflect *ABCD* in the line $y = -1$. Label the image $A_1B_1C_1D_1$.
   c. Rotate *ABCD* 90° clockwise about (0, 1). Label the image $A_2B_2C_2D_2$.
   d. Translate *ABCD* by vector $\binom{9}{-7}$. Label the image $A_3B_3C_3D_3$.
   e. Enlarge *ABCD* using centre (−4, 4), scale factor 2. Label the image $A_4B_4C_4D_4$.

8. a. Join points $P(1, -5)$, $Q(4, -5)$, $R(4, -2)$ and $S(1, 1)$ to form a quadrilateral.
   What type of quadrilateral is *PQRS*?
   b. Reflect *PQRS* in the line $y = x$. Label the image $P_1Q_1R_1S_1$.
   c. Translate *PQRS* by vector $\binom{-8}{1}$. Label the image $P_2Q_2R_2S_2$.
   d. Rotate *PQRS* 180° clockwise about (4, −1). Label the image $P_3Q_3R_3S_3$.
   e. Enlarge *PQRS* using centre (1, −5), scale factor $\frac{1}{3}$. Label the image $P_4Q_4R_4S_4$.

1099, 1113, 1115, 1127   SEARCH

APPLICATIONS

7.3 Transformations 1

RECAP

To describe a	give
• Reflection	The position of the mirror line
• Rotation	The angle of rotation
	The direction (clockwise or anti-clockwise)
	The centre of rotation
• Translation	The vector or the distance and direction
• Enlargement	The scale factor
	The centre of enlargement

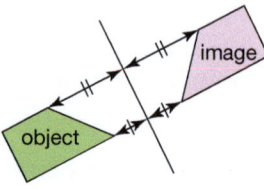

▲ The mirror line is halfway between the object and image.

HOW TO

To identify a transformation
1. Draw a diagram (unless one is given).
2. Decide which type of transformation is involved and find the information needed. Use tracing paper if you wish.
3. Give a full description of the transformation.

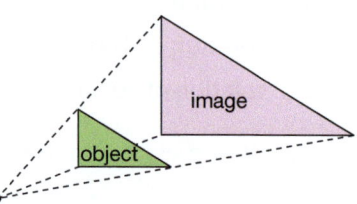

▲ To find the centre of enlargement, draw lines between corresponding points.

EXAMPLE

Write a full description of the transformation that maps the flag F onto

a A **b** B **c** C **d** D.

a ② A has changed size, it is half as tall and half as wide as F ⇒ enlargement.
Join corresponding points to find the centre of the enlargement.

③ Enlargement, scale factor $\frac{1}{2}$, centre $(9, 1)$

b ② B is in the same orientation but a different position ⇒ translation. F moves right by 2 squares and down by 9 squares.

③ Translation by vector $\begin{pmatrix} 2 \\ -9 \end{pmatrix}$

c ② C is 'flipped' ⇒ reflection. Points like $(-2, 2)$, $(-1, 1)$ and $(1, -1)$ lie halfway between F and C.

③ Reflection in mirror line $y = -x$

d ② D is turned, 90° anti-clockwise ⇒ rotation. Use tracing paper to find the centre of rotation. (You will use perpendicular bisectors later.)

③ Rotation 90° anti-clockwise about $(0, -2)$.

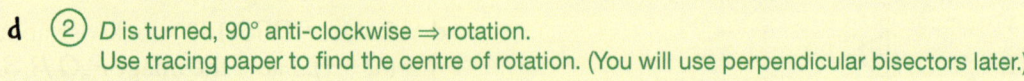

Geometry Working in 2D

142

Reasoning and problem solving

Exercise 7.3A

1. Describe fully the transformation that maps
   - **a** A onto B
   - **b** A onto C
   - **c** C onto B
   - **d** D onto E
   - **e** E onto F
   - **f** D onto F
   - **g** G onto H
   - **h** I onto triangle J.

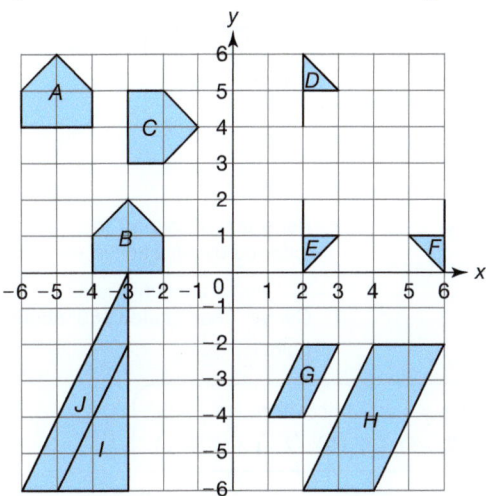

2. The table gives the vertices of a kite and its image after a transformation.

Kite	(2, 0)	(3, 2)	(2, 3)	(1, 2)
Image	(6, 0)	(9, 6)	(6, 9)	(3, 6)

   - **a** Describe what happens to the coordinates.
   - **b** Give a full description of the transformation.

3. Use x and y axes from -6 to 6.
   - **a** Draw triangle T with vertices at $(-6, -4)$, $(6, -4)$ and $(2, 4)$ and its image, S, with vertices at $(0, -1)$, $(3, -1)$ and $(2, 1)$.
   - **b** Give a full description of the transformation that maps T onto S.

4. Rohini says a rotation of 90° clockwise about $(1, 4)$ maps A onto B.
   Is Rohini correct? Give your reason.

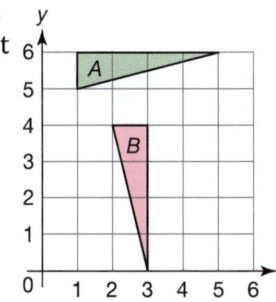

5. **a** Describe three *different* transformations that map A onto B.
   b Which point(s) do not move under each transformation?

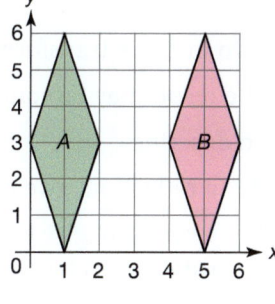

*6. Describe the transformation that maps
   - **a** A onto P, B onto Q, C onto R and D onto S
   - **b** A onto Q, B onto P, C onto S and D onto R
   - **c** A onto R, B onto S, C onto P and D onto Q.

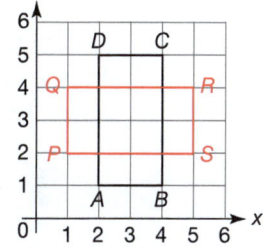

*7. ABCD is a square. Give full descriptions of each of the following transformations in which ABCD is mapped onto itself.
   - **a** A maps onto C and C onto A whilst B and D do not move.
   - **b** B maps onto D and D onto B whilst A and C do not move.
   - **c** A maps onto C, B onto D, C onto A and D onto B.
   - **d** A maps onto B, B onto C, C onto D and D onto A.

*8. Describe the transformation that maps
   - **a** $A(1, 5)$ onto $A'(3, 3)$ and $B(-1, -3)$ onto $B'(-5, 1)$
   - **b** $P(1, -1)$ onto $P'(-2, -2)$ and $Q(3, 2)$ onto $Q'(1, -4)$.

9. What happens to coordinates when
   - **a** a shape is reflected in
     - **i** the x axis
     - **ii** the y axis
     - **iii** $y = x$
     - **iv** $y = -x$
     - *v lines like $x = 3$ or $y = 3$
   - **b** a shape is rotated through
     - **i** 90° clockwise about $(0, 0)$
     - **ii** 180° clockwise about $(0, 0)$
     - **iii** 90° anti-clockwise about $(0, 0)$.

SKILLS

7.4 Transformations 2

Two or more transformations may be combined.

The result may be equivalent to a single transformation.

To describe

- a reflection, give the mirror line.
- a rotation, give the centre and angle and say whether the rotation is clockwise or anti-clockwise.
- a translation, give the vector (or distance and direction).
- an enlargement, give the centre of enlargement and the scale factor.
  - A scale factor greater than 1 makes the shape bigger
  - A scale factor between 0 and 1 makes the shape smaller.
  - A negative scale factor inverts the shape.

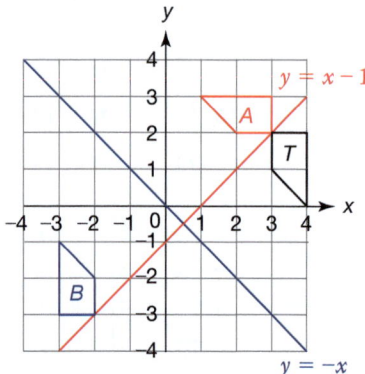

▲ Reflection of shape T in $y = x - 1$ followed by reflection in $y = -x$.

T, A and B are congruent.
Overall T has rotated 180° about the point $(\frac{1}{2}, -\frac{1}{2})$.

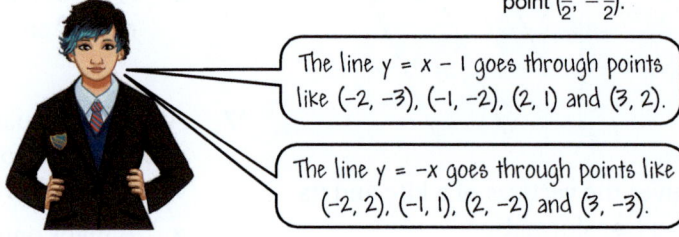

The line $y = x - 1$ goes through points like $(-2, -3)$, $(-1, -2)$, $(2, 1)$ and $(3, 2)$.

The line $y = -x$ goes through points like $(-2, 2)$, $(-1, 1)$, $(2, -2)$ and $(3, -3)$.

EXAMPLE

Use the triangle with vertices $A(1, 1)$, $B(7, 1)$ and $C(1, 5)$ to find the single transformation that is equivalent to enlargement, centre $(-3, 3)$, scale factor $-\frac{1}{2}$, followed by translation by vector $\begin{pmatrix} 12 \\ -9 \end{pmatrix}$.

To do the enlargement: measure from the centre to A then go half as far from the centre in the opposite direction to give the image A_1.

Repeat for B_1 and C_1 to obtain $A_1B_1C_1$.

To do the translation: move each vertex of $A_1B_1C_1$ 12 units right and 9 units downwards to give the final image $A_2B_2C_2$.

The sides of triangle $A_2B_2C_2$ are half as long as those of $ABC \Rightarrow$ enlargement. To find the centre of this enlargement join AA_2, BB_2 and CC_2.

The single transformation that maps ABC onto $A_2B_2C_2$ is an enlargement, centre $(5, -3)$, scale factor $-\frac{1}{2}$.

Remember to give all three pieces of information.

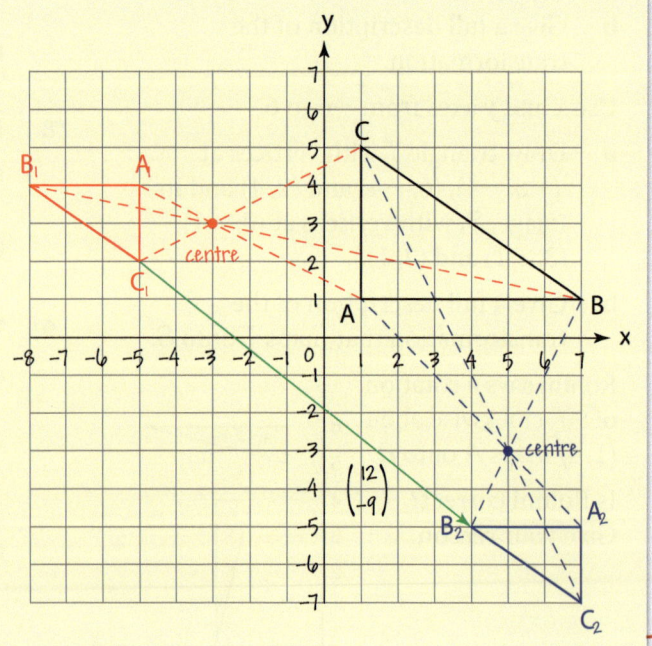

Geometry   Working in 2D

Fluency

Exercise 7.4S

1. **a** Copy trapezium, T, and lines $y = x - 1$ and $y = -x$ from the top of the previous page.
   b Reflect T in $y = -x$, then reflect the image in $y = x - 1$
   c Does the order of the reflections affect the final result? If it does say how.
   d Which point(s) do not move?

   For questions **2** to **6** use x and y axes from –6 to 6.

2. Use the quadrilateral with vertices at $P(0, 4)$, $Q(5, 4)$, $R(5, 5)$ and $S(2, 6)$ to find the single transformation that is equivalent to
   a reflection in the x axis followed by reflection in the y axis
   b reflection in $y = 2$ followed by reflection in $y = -3$.

3. Use the triangle with vertices at $A(-5, 1)$, $B(-3, 1)$ and $C(-1, 4)$ to find the single transformation that is equivalent to
   a rotation through 90° clockwise about the origin, (0, 0) followed by rotation through 180° clockwise about the point (2, 0).
   b rotation through 180° clockwise about the origin, (0, 0) followed by rotation through 180° clockwise about the point (2, 0).

4. Use the trapezium with vertices $P(4, 3)$, $Q(6, 3)$, $R(5, 5)$ and $S(4, 5)$ to find the single transformation that is equivalent to translation by vector $\begin{pmatrix} -3 \\ -6 \end{pmatrix}$ followed by enlargement, centre (0, 0), scale factor -2.

5. **a** Use the quadrilateral with vertices $A(-5, -3)$, $B(-2, -6)$, $C(-2, -2)$, $D(-4, -2)$ to find the single transformation that is equivalent to rotation of 90° clockwise about B followed by translation by vector $\begin{pmatrix} 1 \\ 9 \end{pmatrix}$.
   b What is the single transformation that returns the final image to ABCD?

6. **a** Draw triangle $K(-1, 4)$, $L(-3, -2)$, $M(3, 2)$ and its image after enlargement, centre (0, 2), scale factor 2. Label the image $K_1L_1M_1$.
   b Draw the image of $K_1L_1M_1$ after enlargement, centre (2, -2), scale factor $-\frac{1}{4}$. Label the image $K_2L_2M_2$.
   c Describe the single transformation that maps KLM onto $K_2L_2M_2$.
   d **i** Describe the single transformation that maps $K_2L_2M_2$ onto KLM.
       ii Which point(s) do not move?

*7 **a** **i** On a copy of the diagram below, show the result when P is rotated by 120° anti-clockwise about C, then the image is reflected in M.

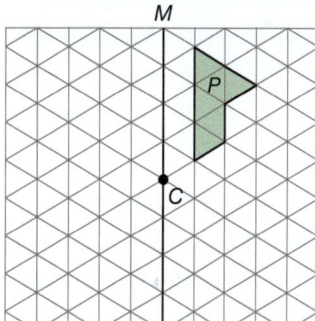

   ii Describe the single transformation that maps P onto the final image.
   b Repeat part **a** on a new copy of the diagram, but this time reflect P in M before rotating the image by 120° anti-clockwise about C.

*8 **a** **i** Draw the triangle with vertices at $A(2, 2)$, $B(3, 2)$ and $C(2, 4)$ and its image after reflection in $y = x - 1$ followed by translation by the vector $\begin{pmatrix} -4 \\ 4 \end{pmatrix}$.
   ii Find the single transformation that maps ABC onto the final image.
   b Repeat part **a** but translate ABC first.

APPLICATIONS

7.4 Transformations 2

RECAP

To identify the type of transformation compare the object and the image.

- Congruent shapes, same orientation ⇒ translation
  Give vector (or distance and direction)
- Congruent shapes, image 'flipped' ⇒ reflection
  Give mirror line
- Congruent shapes, image 'turned' ⇒ rotation
  Give, centre, angle and direction
- Similar shapes ⇒ enlargement
  – image same orientation and enlarged scale factor > 1
  – image same orientation and reduced scale factor between 0 and 1
  – image inverted scale factor negative
  Give centre and scale factor

▲ Geometric patterns on tiles, wallpaper and fabric are often made by combining or repeating transformations.

HOW TO

To identify a transformation
1. Draw a diagram (unless one is given).
2. Decide which type of transformation is involved and find the information needed.
3. Give a full description of the transformation.

EXAMPLE

a **i** Reflect *T* in *M*, then reflect the image in *N*.

ii Describe the single transformation that maps *T* onto the final image and find a connection with the position of *M* and *N*.

b What happens if the order of reflections is reversed?

c Continue to reflect the images in *M* and *N* until the results give a symmetrical pattern and describe the symmetry.

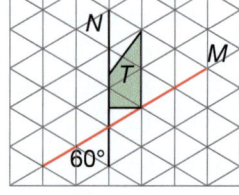

a **i** ① The first diagram shows the images.

ii ② The trapezium has rotated.
Use tracing paper to find the centre.

③ Rotation of 120° anti-clockwise about the point of intersection of the mirror lines. Angle of rotation = 2 × angle between the mirror lines.

b The second diagram shows *T* after reflection in *N* then *M*.
The direction of rotation is reversed to clockwise.

c Adding one more reflection gives a symmetrical pattern.
The pattern has three lines of symmetry (two of them being the mirror lines) and rotational symmetry of order 3.

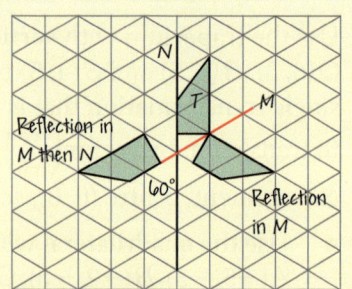

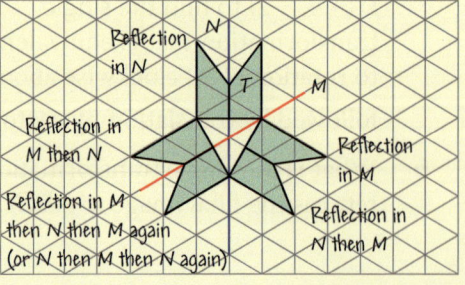

Geometry   Working in 2D

Reasoning and problem solving

Exercise 7.4A

1. **a** **i** On a copy of this diagram, reflect T in M, then reflect the image in N.

   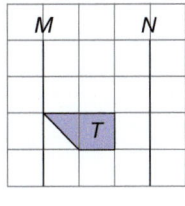

   ii Describe the single transformation that maps T onto the final image.

   b What happens if the order of reflections is reversed?

2. **a** **i** On a copy of this diagram, reflect T in M, then reflect the image in N.

   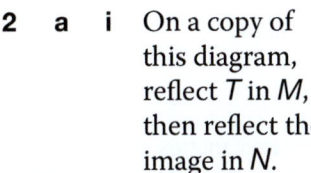

   ii Describe the single transformation that maps T onto the final image and describe the connection with the position of the mirror lines.

   b What happens if the order of reflections is reversed?

   c Continue to reflect T until you have a symmetrical pattern and describe the symmetry.

3. A single transformation is applied twice to a hexagon.

   The result is as shown.

   Describe two possible transformations.

   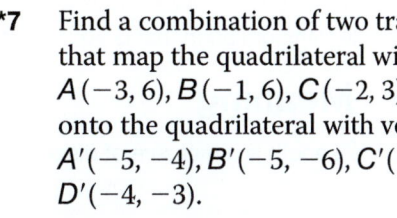

4. **a** On a copy of this diagram, rotate the flag by 180° about P, then rotate the image by 180° about Q. What is the equivalent single transformation?

   b Repeat part **a** with anti-clockwise rotations of 90°.

5. **a** Describe two translations that are together equivalent to translation by vector $\begin{pmatrix} 5 \\ -1 \end{pmatrix}$.

   b How many possible answers are there to part **a**? What do they have in common?

6. Find another transformation that has the same effect as an enlargement, centre (0, 0), scale factor -1.

*7. Find a combination of two transformations that map the quadrilateral with vertices $A(-3, 6)$, $B(-1, 6)$, $C(-2, 3)$, $D(-4, 5)$ onto the quadrilateral with vertices $A'(-5, -4)$, $B'(-5, -6)$, $C'(-2, -5)$, $D'(-4, -3)$.

*8. The diagonals of square PQRS intersect at T. Describe the transformations that can be used to map triangle PQT onto each of the other congruent triangles.

   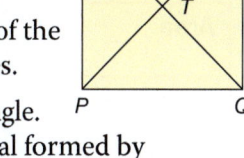

9. ABC is an isosceles triangle. Describe the quadrilateral formed by reflecting ABC in its longest side when

   a $\angle B = 90°$ **b** $\angle B$ is obtuse.

10. Investigate how transformations are used in patterns on wallpapers, fabric, tiles and kaleidoscopes.
    Make your own patterns.

11. Investigate whether each statement is always true, sometimes true or never true. Explain and illustrate each answer.

    a One translation followed by another is equivalent to a translation.

    b One reflection followed by another is equivalent to a reflection.

    c One rotation followed by another is equivalent to a rotation.

    d A translation followed by a reflection is equivalent to a reflection.

    e A translation followed by a rotation is equivalent to a rotation.

Summary

Checkout
You should now be able to...

Test it
Questions

	Checkout	Test it Questions
✓	Measure line segments and angles accurately.	1
✓	Use scale drawings and bearings.	1
✓	Calculate the areas of triangles, parallelograms, trapeziums and composite shapes.	2
✓	Describe and transform shapes using reflections, rotations, translations (described as 2D vectors) and enlargements (including fractional and negative scale factors).	3 – 5
✓	Identify the single transformation that is equivalent to a combination of transformations.	6

Language	Meaning	Example
Length	Length is a measure of distance.	Millimetres, centimetres, metres and kilometres are all measures of length. Length can be measured with a ruler.
Angle	The amount that one straight line is turned relative to another that it meets or crosses.	Angles are measured in degrees. One degree is $\frac{1}{360}$th of a complete turn. Use a protractor to measure an angle.
Area	The amount of space occupied by a 2D shape.	Area = 12 units^2 Perimeter = 14 units
Perimeter	The total distance around the edges that outline a shape.	
Transformation	A geometric mapping that takes the points in an **object** to points in an **image**.	Rotation, reflection, translation, enlargement.
Translation	A transformation in which all the points in the object are moved the same distance and in the same direction.	Translation $\binom{4}{2}$
Reflection / Mirror line	A transformation that moves each point to a point on the other side of a mirror line and an equal distance from it.	Mirror line
Rotation / Centre of rotation	A transformation that turns points through a fixed angle whilst keeping their distance from the centre of rotation fixed.	anticlockwise; Centre of rotation
Enlargement / Scale factor / Centre of Enlargement	A transformation that moves points a fixed multiple, the scale factor, of their distance from the centre of enlargement.	Centre of enlargement; Scale factor = $\frac{1}{3}$

Geometry Working in 2D

Review

1. a What is the bearing of B from A?
   b What is the bearing of A from B?
   c What is the length of AB in real life?

   Scale 1 cm = 5 km

2. Work out the area of these shapes.

   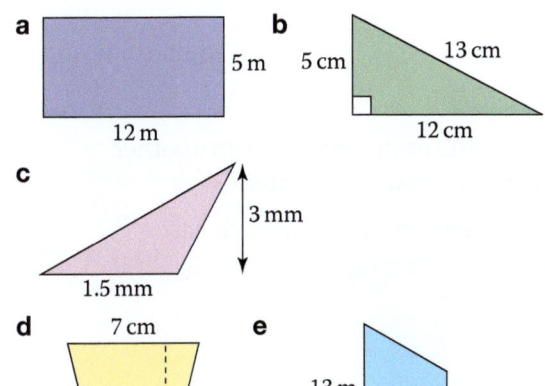

   a 12 m × 5 m rectangle
   b triangle with 5 cm, 13 cm, 12 cm
   c triangle with 1.5 mm base, 3 mm height
   d trapezium: 7 cm top, 11 cm side, 2 cm bottom
   e parallelogram: 13 m side, 2.5 m height

3. Copy this diagram.
   a Reflect triangle A in the line y = 1 and label the image B.
   b Rotate triangle A 90° anti-clockwise about (0, 1) and label the image C.
   c Describe the transformation A to D.

   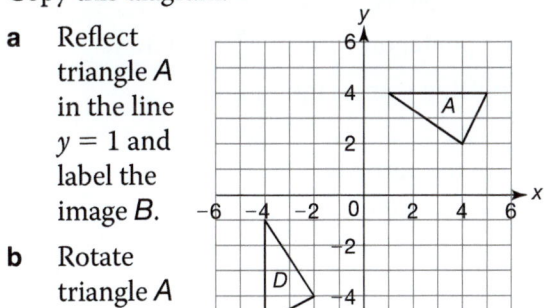

4. Copy this diagram.

   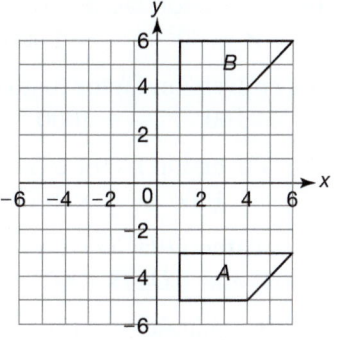

   a Translate shape A by the vector $\begin{pmatrix} -5 \\ 2 \end{pmatrix}$ and label the image C.
   b Describe the transformation A to B.

5. Make a copy of this diagram.

   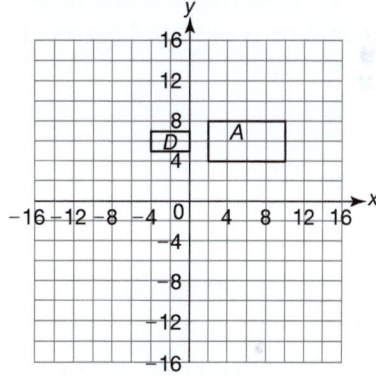

   a Enlarge A by scale factor 0.25 from centre of enlargement (6, 0) and label the new rectangle B.
   b Enlarge A by scale factor −1.5 using the origin as the centre of enlargement and label the image C.
   c Describe the transformation A to D.

6. a Use the triangle with vertices at (4, 0), (4, 4) and (4, 6) to find the single transformation that is equivalent to reflection in y = x followed by rotation by 90° clockwise about (0, 0).
   b Is the result the same if the order of the two transformations is reversed?

What next?

Score		
0 – 2		Your knowledge of this topic is still developing. To improve, look at MyiMaths: 1086, 1099, 1103, 1108, 1113, 1115, 1117, 1125, 1127, 1128, 1129
3 – 5		You are gaining a secure knowledge of this topic. To improve your fluency look at InvisiPens: 07Sa – r
6		You have mastered these skills. Well done you are ready to progress! To develop your problem solving skills look at InvisiPens: 07Aa – f

Assessment 7

1. **a** What is the conversion factor to go from centimetres to kilometres?
   ÷ 1000   × 1000   ÷ 1 000 000   × 100 000   ÷ 100 000 [1]

   b What is the conversion factor to go from litres to mililitres?
   ÷ 100   × 100   ÷ 1000   × 1000   ÷ 1 000 000 [1]

   c What is the conversion factor to go from grams to tonnes?
   × 1000   ÷ 1000   × 100 000   ÷ 1 000 000   × 1 000 000 [1]

2. **a** Draw a quadrilateral $ABCD$ with sides $BC = 6.5$ cm, $AD = DC = 7.5$ cm and $\angle ADC = 105°$ and $\angle DCB = 60°$. [2]

   b Measure   **i** side AB [1]   **ii** $\angle BAD$ [1]   **iii** $\angle ABC$ [1]

   c Give the mathematical name of quadrilateral $ABCD$. [1]

   d Use a ruler to find the midpoint of each side. Join the midpoints to form another quadrilateral. What is the mathematical name of this new quadrilateral? [2]

3. Suha makes the following statements about the angles shown below. In each case state whether Suha is right and correct her answer if she is wrong.

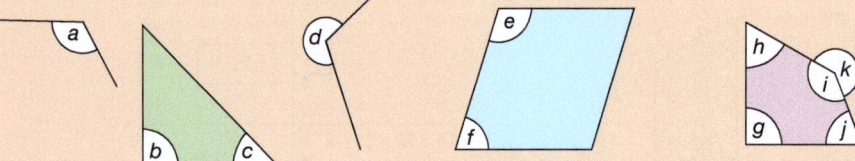

   a a is obtuse   **b** b is acute   **c** c is acute   **d** d is reflex   **e** e is obtuse   **f** f is reflex
   g g is obtuse   **h** h is reflex   **i** i is obtuse   **j** j is obtuse   **k** k is acute [11]

4. **a** Use a protractor and ruler to construct a regular pentagon with side length 3 cm. [3]

   b Find the area of the shape drawn in part **a** by splitting your diagram into appropriate shapes. [4]

5. Three buildings, a chemist's, a shop and a hospital are the vertices of a triangle. The bearing of the shop from the chemist's is 065°.

   a Work out and write down the bearing of
   i the hospital from the shop [1]
   ii the hospital from the chemist's [1]
   iii the shop from the hospital [2]
   iv the chemist's from the hospital [2]
   v the chemist's from the shop [2]

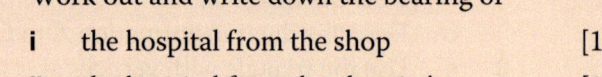

   b Look at the 'reverse' bearings of the chemist's from the shop and the shop from the chemist's. Repeat this for the other two sets of 'reverse' bearings. Write down the connection between a bearing and its 'reverse' bearing. [1]

6. Square patio slabs come in 3 sizes: 1 m × 1 m, 2 m × 2 m and 3 m × 3 m. Farakh wants to build a square patio of side 7 m.

   a How many of the 1 m × 1 m slabs would Farakh need? [1]

   b Explain whether Farakh can build the patio using
   i just 2 m × 2 m slabs [1]   **ii** just 3 m × 3 m slabs. [1]

6 c Can Farakh build a patio using just 2 m × 2 m *and* 3 m × 3 m slabs?
      Draw a diagram to illustrate your answer. [4]

7 What fraction of each regular polygon is shaded?

   a   b

[4]

8 The diagram shows the dimensions of a garden in the shape of a trapezium. The garden contains a triangular flowerbed and the rest is lawn. Each square metre of the lawn needs 35 grams of fertiliser. Is a $1\frac{1}{2}$ kilogram bag of fertiliser enough for this lawn? Explain your answer.

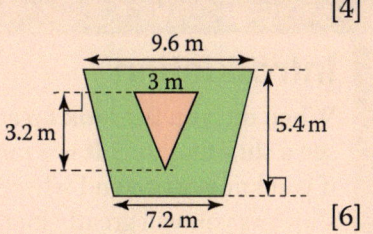

[6]

9 Copy these diagrams and draw enlargements of the shapes with scale factors and centres of enlargement, O, as shown.

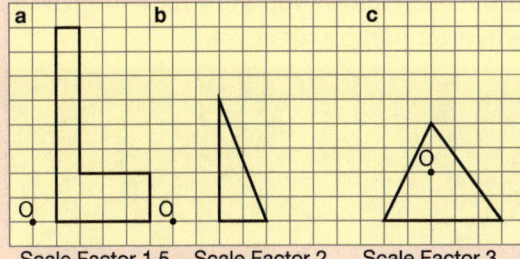

Scale Factor 1.5   Scale Factor 2   Scale Factor 3   [6]

10 In a computer game, a rocket takes off from its launching pad and travels 45 miles vertically upwards and 25 miles right.

   a Show the path of the rocket on a squared grid.
      Take 1 square width/length as representing 10 miles. [1]
   b Write the vector which represents this translation. [1]

   At this position part of the rocket is ejected and the remainder travels a further 15 miles right and 20 miles up.

   c Add this new translation to your diagram for part **a**, attaching it to the end of the first translation. [1]
   d Write this new translation as a vector. [1]
   e Using your diagram, find the *single* translation vector that represents the complete journey. [1]
   f How are the vectors in parts **b**, **d** and **e** related? [1]

11 a Jo transforms the triangle, A, into the triangle B.
       Describe this transformation fully. [2]
    b Rotate triangle A 90° clockwise, about the origin. Label the image C. [2]
    c Reflect triangle C in the line $y = -x$. Label the image D. [2]
    d Fully describe the single transformation that maps A on to D. [2]

8 Probability

Introduction

When did you last look up at the night sky and see a shooting star? It is a rare event, although if you know when and where to look for meteor showers you will greatly increase your chances of seeing one. The world is full of uncertainty, from the unpredictable appearance of shooting stars or earthquakes, to manmade events like the result of a hockey match. Probability is the branch of mathematics that deals with the study of uncertainty and chance.

What's the point?

You apply probability whenever you weigh up everyday risks. For example, should I take an umbrella today? A basic understanding of probability allows you to be more prepared for whatever life throws at you!

Objectives

By the end of this chapter, you will have learned how to …

- Use experimental data to estimate probabilities of future events.
- Calculate theoretical probabilities using the idea of equally likely events.
- Compare theoretical probabilities with experimental probabilities.
- Recognise mutually exclusive events and exhaustive events and know that the probabilities of mutually exclusive exhaustive events sum to 1.

Check in

1. Work out the value of these expressions.

   a $1 - \frac{2}{5}$
   b $1 - \frac{4}{7}$
   c $1 - \frac{3}{8}$
   d $1 - \frac{4}{11}$
   e $\frac{2}{3} + \frac{1}{6}$
   f $\frac{1}{5} + \frac{1}{4}$
   g $\frac{1}{3} + \frac{5}{8}$
   h $\frac{1}{4} + \frac{3}{8}$

2. Calculate these fractions of an amount.

   a $\frac{2}{5} \times 100$
   b $\frac{1}{9} \times 360$
   c $\frac{3}{8} \times 56$
   d $\frac{7}{30} \times 240$

3. Change these fractions to decimals.

   a $\frac{7}{10}$
   b $\frac{3}{4}$
   c $\frac{3}{8}$
   d $\frac{2}{5}$
   e $\frac{1}{3}$
   f $\frac{1}{16}$
   g $\frac{1}{9}$
   h $\frac{5}{8}$

Chapter investigation

Two people can play an old game called 'Rock, Paper, Scissors'.

In the game, you make a shape with your hand.

- Rock is a closed fist.
- Paper is an open palm with closed fingers.
- And scissors is two fingers held like scissor blades.

The players reveal their 'hand' simultaneously.

Rock beats scissors, paper beats rock, and scissors beats paper.

Is there a best strategy for playing this game?

SKILLS

8.1 Probability experiments

- Probability measures how likely an **event** is to happen.
- All probabilities have a value from 0 (impossible) to 1 (certain).

You can use repeated experiments to estimate probabilities.

- A probability based on data from an experiment is called the **relative frequency**.
- Relative frequency = $\dfrac{\text{Number of favourable trials}}{\text{Total number of trials}}$

You can give probabilities as fractions, decimals or percentages.

EXAMPLE

A drawing pin is thrown 10 times and lands point up in 7 cases.

a Estimate the probability that a drawing pin lands point up.

The drawing pin is now thrown another 90 times and lands point up another 67 times.

b Calculate the relative frequency of the drawing pin landing point up using all the observations made so far.

a P(point up) = $\dfrac{7}{10}$ = 0.7 = 70%

b P(point up) = $\dfrac{7 + 67}{10 + 90}$ = $\dfrac{74}{100}$ = 0.74 = 74% There were a total of 74 out of 100.

0.74 is a more reliable estimate because it is based on more information.

- If you increase the number of trials, the outcome of an experiment is likely to become closer to the **theoretical** (expected) outcome.

If you know the probability of an event, you can calculate how many times you **expect** the outcome to happen.

- Expected frequency = number of trials × probability of the event

EXAMPLE

8% of men have red-green colour-blindness. Women are rarely colour-blind. A group of 65 men and 40 women are attending a meeting. Estimate how many people in the group suffer from red-green colour-blindness.

65 × 0.08 = 5.2 or about 5 are likely to be colour-blind.
Only the men are likely to be colour-blind.

Probability   Probability

Fluency

Exercise 8.1S

1. **a** Rafael and Sebastian have played 12 tennis matches against each other in the last year. Rafael won 6 out of 12 matches. Estimate the probability Sebastian wins the next match between the two players.
   b A piece of toast is dropped twenty times. It lands butter side down 14 times. Estimate the probability that toast will land butter side up when dropped again.

2. Hari works at Quick-Fix garage.
   Over the last month, 40 customers have come to the garage with a flat tyre caused by a puncture.
   Hari records which tyre was punctured.

Tyre	Frequency
Front left	8
Front right	7
Back left	13
Back right	12

   Estimate the probability that the next customer with a flat tyre has a puncture on their
   a front left tyre   **b** front right tyre
   c back left tyre   **d** back right tyre.
   Give your answer as a
   i fraction   **ii** decimal   **iii** percentage.

3. The probability of a drawing pin landing point up when dropped is $\frac{3}{4}$. Estimate how many times it will land point up if it is dropped
   a 20 times   **b** 7 times.

4. Hazem wants to pick a red ball and can choose bag A or B to pick a ball at random from.

Bag A	Colour	Red	Blue
	Frequency	1	2

Bag B	Colour	Red	Blue
	Frequency	2	8

   a Which bag should Hazem choose? Give the reason for your answer.
   b Hazem takes out two balls without replacing. Hazem wants to pick two reds. Which bag should he choose? Give the reason for your answer.

5. There are 10 coloured balls in a bag. One ball is taken out and then replaced in the bag. The colours of the ball are shown in the table.

Colour	Red	Green	Blue
Frequency	9	14	27

   a How many times was a ball taken out of the bag?
   b Estimate the probability of taking out
   i a red ball   **ii** a green ball
   iii a blue ball.
   c How many balls of each colour do you think are in the bag?
   d How could you improve this guess?

6. Weather records show that approximately 30% of April days in a certain village have some rain. Estimate the number of days the village will have some rain in April next year.

7. A financial advisor claims that, on average, 70% of his recommended shares have increased in value. He has recommended 20 shares in the current year.
   How many shares would you expect to increase in value over this year?

8. Throw an ordinary dice until you get a six, counting how many throws you make, including the one on which the six appears.
   a Do this twenty times and calculate the relative frequency from your observations for each of the values you have recorded.
   b Often when this experiment is done, the outcome with the highest relative frequency is one throw.
   Give a reason why that happens.

9. Lorna rolls a dice and gets a six on the first four rolls.
   She thinks that the dice may be biased. Describe how you would test Lorna's dice to see if it was biased.

1211, 1264

APPLICATIONS

8.1 Probability experiments

RECAP
- You can estimate the probability of an event by conducting an experiment.
- Probability data from an experiment is called the **relative frequency**.
  $$\text{Relative frequency} = \frac{\text{Number of favourable trials}}{\text{Total number of trials}}$$
- When you repeat an experiment, you may get a different outcome.
- The expected frequency = number of trials × probability of the event.

HOW TO
1. Think first in terms of words – likely / evens / unlikely.
2. Calculate the relative or expected frequency of the event.
3. Answer the question. Remember that the estimated probability becomes more reliable as you increase the number of trials.

```
0           0.5           1
|  Unlikely  |  Likely  |
Impossible  Even chance  Certain
```

EXAMPLE

Aleesha takes a ball out of a bag, records the colour and places the ball back in the bag. She repeats this 40 times and records her results in a table.

Colour	Yellow	Green	Blue
Frequency	16	4	
Relative frequency			

a Complete Aleesha's table.

b Which of these bags *could be* Aleesha's bag? Explain your answer.

A, B, C (bags of coloured balls shown)

a
② Use the formula for relative frequency.
Blue frequency = 40 − (16 + 4) = 20
Yellow: 16 ÷ 40 = 0.4
Green: 4 ÷ 40 = 0.1
Blue: 20 ÷ 40 = 0.5

Colour	Yellow	Green	Blue
Frequency	16	4	20
Relative frequency	0.4	0.1	0.5

b ① In Bag A the chance of blue is a little more likely than the chance of yellow. Green is impossible.
③ A is not her bag because bag A has no green balls.
① Bag B has an even chance of yellow, and blue is more likely than green.
③ B could be her bag. The probabilities are close to the relative frequencies from the experiment.
① In Bag C yellow is likely, blue and green are unlikely.
③ C could be her bag. Aleesha should do more trials to get more reliable results.

EXAMPLE

Katrina plays a game at a fair. Each try costs $1.
If the spinner lands on the WIN zone you win $2.
If the spinner lands on the SAFE zone you get your $1 back.
Katrina plays the game 10 times.
How much prize money should she expect to win or lose?

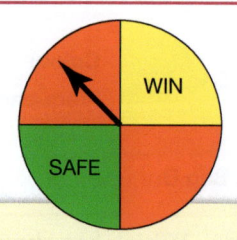

① There are 10 trials.
② Probability of winning $2 = $\frac{1}{4}$ ➡ ③ Times she wins £2 = $10 \times \frac{1}{4} = 2.5$

Probability of winning $1 back = $\frac{1}{4}$ ➡ Times she wins £1 = $10 \times \frac{1}{4} = 2.5$

Expected prize money = (2.5 × $2) + (2.5 × £1) = $5 + $2.50 = $7.50
Playing 10 games costs $10. Katrina should expect to lose $2.50.

Probability   Probability

Reasoning and problem solving

Exercise 8.1A

1. Sumbula spins this spinner 4 times. She says that if the spinner is fair then the spinner will land on each colour once. Is Sumbula correct?

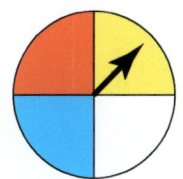

2. A bag contains 100 coloured balls. Xavier, Yvonne and Zoe each select a ball from the bag, record if the ball is red and replace the ball. The table shows their results

	Number of trials	Number of red balls
Xavier	5	4
Yvonne	20	16
Zoe	100	95

   a Xavier says that the probability of choosing a red ball is $\frac{4}{5}$.
   Criticise his statement.

   b Zoe says that the bag must contain 95 red balls. Is she correct?

   c i Explain why the most reliable estimate of the relative frequency of choosing a red ball is 0.92.
   ii Does the bag contains 92 red balls?

3. Alik spins a spinner and records the result in a table. He adds each column when he lands on a new colour.

Colour	Red	White	Blue
Frequency	10	5	
Relative freq.		0.2	0.4

   a Copy and complete Alik's table. What do you notice about the sum of the relative frequencies?

   b How many of these spinners could be Alik's?

   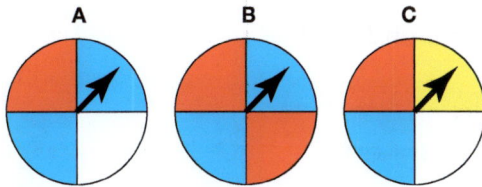

   c Draw three possible spinners that could be Alik's spinner. You can divide your spinners into as many sectors as you like.

*4 Jareer thinks that fewer people are born on a Saturday or Sunday than on a weekday because any planned births will be scheduled during the week.

He collects information from a hundred friends of different ages about the day their birthday falls on this year and finds that Saturday and Sunday seem to be as common as other days.

Suggest why the information he has collected does not mean that his idea about the days on which people are born is wrong.

*5 A red bag has 5 white and 10 black balls in it, and a blue bag has 5 white and 5 black balls in it. All the balls are identical except for their colour.

   a You choose a ball from the red bag at random and note its colour before returning it to the bag.
   If you do this 30 times, estimate how many times you will see a white ball.

   b If you have used the blue bag instead, would you expect to see a white more often, less often or about the same?

*6 The coloured sections on this spinner are all equal.
Labeeb is playing a game which charges 50p a go.
He wins a prize if the spinner lands on a blue sector.

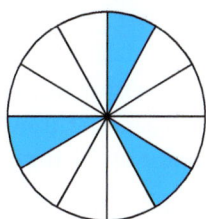

   a If Labeeb has 24 goes, how often would he be likely to win?

   b If the prize is £1, how much profit does the game make on average?

SKILLS

8.2 Theoretical probability

- A **trial** is an activity or experiment.
- An **outcome** is the result of a trial.
- An **event** is one or more outcomes of a trial.

> Rolling a dice is a trial.
> The outcomes of rolling a dice are 1, 2, 3, 4, 5, 6.
> Rolling an even number is an event.

The outcomes of some experiments are equally likely, for example, throwing a fair coin or rolling a fair dice.

- The **theoretical probability** of an event is

$$\text{Probability of event happening} = \frac{\text{Number of outcomes in which event happens}}{\text{Total number of all possible outcomes}}$$

EXAMPLE

A fair dice is rolled and the number showing on the top is scored. Find the probability of these events.

a 3 is scored.
b A factor of 12 is scored.
c A prime number is scored.

> There are 6 possible outcomes which are equally likely.
> a $P(3) = \frac{1}{6}$ There is one 3.
> b $P(\text{Factor of 12}) = \frac{5}{6}$ Five outcomes are factors of 12: 1, 2, 3, 4 and 6.
> c $P(3) = \frac{3}{6} = \frac{1}{2}$ Three outcomes are prime: 2, 3 and 5.

The probability of an event can be written as P(Event).

Theoretical probability is based on equally likely outcomes.

- You use experimental probability when the event is unfair or biased or when the theoretical outcome is unknown.

EXAMPLE

Sami threw a drawing pin 10 times and it landed point up 7 times. Cristiano threw a similar drawing pin 43 times and it landed point up 31 times.

a Calculate the relative frequency of the pin landing point up for each person.
b Which of these is the better estimate of the probability?

> a Sami $\frac{7}{10} = 0.7$ Cristiano $\frac{31}{43} = 0.72$ (2 dp)
> b Cristiano used the larger sample so he has the better estimate.

The closer experimental probability is to theoretical probability the less likely it is that there is bias.

- The experimental probability becomes a better estimate for the theoretical probability as you increase the number of trials.

Probability   Probability

Fluency

Exercise 8.2S

1 A fair dice is rolled and the number showing on the top is scored.
Find these probabilities

   a P(6)   **b** P(less than 3)
   c P(factor of 10)   **d** P(square number)

2 Are the outcomes in each case equally likely? If they are, give the probability of each outcome. If they are not, explain why.

   a The score showing when a fair die is thrown.

   b Whether a drawing pin lands point up or down when it is thrown in the air.

   c The day of the week a baby is born on.

   d The number of times I have to toss a fair coin until I see a head.

   e What the letter is, if a letter is chosen at random from a book.

3 All the counters in a bag are either green or black. At the start the probability a counter chosen at random from the bag is green is $\frac{1}{4}$. Counters are not replaced when they are taken out.

The first two counters taken out of the bag are green.

What is the least number of black counters that must still be in the bag?

4 A class of 24 students are each given a similar biased dice and asked to record how many times they saw a six when they threw their dice 20 times.

8	9	5	7	6	8	7	8
6	7	4	9	7	7	6	7
8	6	9	5	3	6	6	7

   a Give the highest and lowest relative frequencies found by any of the students.

   b Find the best possible estimate of the probability of a six showing on the biased dice from the results given.

5 For the following events give a value between 0 and 1 for the probability of the event described happening.

   a You score a 4 when you throw a fair dice

   b A set of cards has the numbers 1-9 written on one of each of the cards.
You pick a card at random and it is an even number.

   c You score less than 7 when you throw a fair dice.

6 Naseefa wonders how often there is a difference of more than 2 when you throw a pair of fair dice. She does an experiment, noting down how often it happens after each block of 10 throws.
Her results are

5	4	5	3	2	4	6
3	3	3	5	4	2	4

   a Give the relative frequencies at the end of each block of 10 throws.
The first two are $\frac{5}{10} = 0.5$, and $\frac{9}{20} = 0.45$.

   b What is the best estimate Naseefa has of the probability of getting a difference of more than 2?

*7 Three friends were investigating how often a biased dice showed a six.
Sergio got 10 sixes from 40 throws.
Annalise got 8 sixes from 40 throws.
Dominika got 21 sixes from 80 throws.

- Sergio says they should take his as it is the median of the three estimates.
- Annalise says they should take the mean of the three.
- Dominika says they should combine all the results and use that relative frequency as the estimate.

Which method is the best?

APPLICATIONS

8.2 Theoretical probability

RECAP
- You use theoretical probability for fair activities, for example rolling a fair dice.
- For unfair or biased activities, or where you cannot predict the outcome, you use experimental probability (relative frequency).
- Relative frequency is the proportion of successful trials in an experiment.

Theoretical probability
$$= \frac{\text{Number of favourable outcomes}}{\text{Total number of outcomes}}$$

Relative frequency
$$= \frac{\text{Number of successful trials}}{\text{Total number of trials}}$$

HOW TO
To compare theoretical probability with relative frequency
1. Start by assuming that the activity is fair and calculate the theoretical probability of the event.
2. Use the results of the experiment to calculate the relative frequency.
3. If the relative frequency and theoretical probability are very different, then the activity could be biased.

The more trials you carry out, the more reliable the relative frequency will be.

EXAMPLE

Kaseem rolled a dice 50 times and in 14 of those he scored a 2.
Is the dice biased towards 2? Explain your answer.

① Theoretical probability 2 $= \frac{1}{6} = 0.166\ldots$ If the dice is fair, then every outcome is equally likely.

② Relative frequency of rolling a 2 $= \frac{14}{50} = 0.28$ Converting to decimals makes it easier to compare.

③ The relative frequency is higher than the theoretical probability.
The dice appears to be biased towards 2.

EXAMPLE

A fair spinner is divided into blue, red and yellow sectors.
Valerie knows that the blue sector is one quarter of the spinner.
Valerie can't remember the size of angle x, but she knows that it is either 150° or 160°.
She spins the spinner 100 times and records her results in a table.

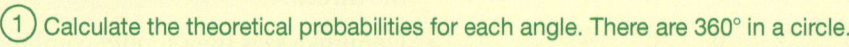

Colour	Blue	Yellow	Red
Frequency	23	42	35

Find the size of angle x. Explain your answer.

① Calculate the theoretical probabilities for each angle. There are 360° in a circle.

$x = 150°$ P(Yellow) $= \frac{150}{360} = 0.416\ldots$ $x = 160°$ P(Yellow) $= \frac{160}{360} = 0.444\ldots$

Angle for red $= 360 - (150 + 90) = 120$ Angle for red $= 360 - (160 + 90) = 110$

If $x = 150°$, P(Red) $= \frac{120}{360} = 0.333\ldots$ P(Red) $= \frac{110}{360} = 0.305\ldots$

② Use the information in the table to find the relative frequency.

Relative frequency of yellow $= \frac{42}{100} = 0.42$ Relative frequency of red $= \frac{35}{100} = 0.35$

③ As the spinner is fair, the relative frequency and theoretical probabilities should be similar.

The relative frequencies are closest to the probabilities for $x = 150°$.

Probability Probability

Reasoning and problem solving

Exercise 8.2A

1. The table below shows the gender and hair colour of 30 students in a class.

	Fair	Dark
Boy	7	9
Girl	8	6

   A student is chosen at random from the class. What is the probability that the student is

   a a girl   b a dark-haired boy?

2. Devi rolled a dice 100 times and in 71 of those rolls her number was a factor of 12.

   Is the dice biased towards 5?

   Give your reason.

3. ABCD is a square and E and F are the midpoints of sides AD and BC.

   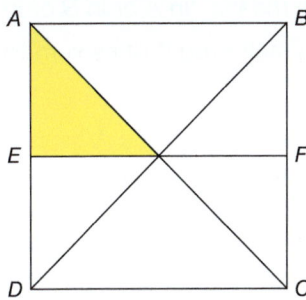

   If a point is selected at random in the square what is the probability it is in the shaded area?

4. The square with sides 4 cm shown has four quarter circles shaded.
   A point is chosen at random in the square. Calculate the probability the point is in the shaded area.

   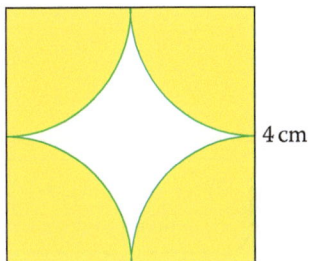

5. Bag 1 contains 8 red balls and 12 green balls. Bag 2 contains 5 red balls and 7 green balls. The balls are identical apart from colour.

   a You choose a ball from bag 1 at random. What is the probability you choose a green ball?

   All the balls are now put into one bag and you choose a ball at random.

   b What is the probability you choose a red ball?

6. The coloured sections on the spinner are all equal.

   If the spinner lands on a green section, the score is doubled.

   If the spinner lands on a blue section, the score is halved.

   Shameem is playing a game with his friend Andrea who goes first, with the result shown.

   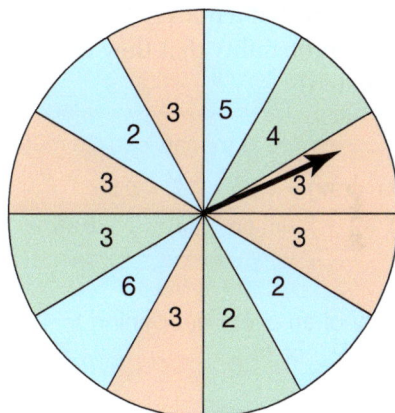

   What is the probability that Shameem scores more than Andrea?

7. Iveta has a spinner divided equally into 19 sectors.

   The sectors are white, black and red.

   There are the same number of black and red sectors.

   Iveta spins the spinner 80 times and records her results in a table.

Colour	White	Black	Red
Frequency	17	32	31

   Explain why the spinner could be biased.

SKILLS

8.3 Mutually exclusive events

- Two events are **mutually exclusive** if they cannot both happen at the same time.

EXAMPLE

You throw an ordinary dice. Here are three possible events.
A a prime number   B an odd number   C a square number
Which of these pairs of events are mutually exclusive?
a A and B   b A and C   c B and C

A set of events is exhaustive if they include all possible outcomes.

A 2, 3 and 5   B 1, 3 and 5   C 1 and 4
a A and B both contain 3 and 5, so not mutually exclusive
b A and C have nothing in common, so they are mutually exclusive
c B and C both contain 1, so not.

- If A and B are mutually exclusive then P(A ∪ B) = P(A) + P(B) means, 'in A or in B or in both'.
- If a set of mutually exclusive events are also **exhaustive** then their probabilities sum to 1.

EXAMPLE

The probability that a hockey team wins a match is 0.65. The probability of a draw is 0.2. What is the probability that they win or draw the match? What is the probability that they lose the match?

The outcomes win, draw and lose are mutually exclusive and exhaustive.
P(win or draw) = 0.65 + 0.2 = 0.85
P(lose) = 1 − P(win or draw) = 1 − 0.85 = 0.15

- Probability of an event happening = 1 − Probability of event not happening
  P(A) = 1 − P(not A)

EXAMPLE

The probability that you are selected for a random body search by an electronic security monitoring system is 0.06. What is the probability you will not be selected?

1 − 0.06 = 0.94   The probabilities add to 1.

EXAMPLE

The probability that a child in a nursery class has a cat is 0.3.
a Find the probability that the child does not have any pets.
b Find the probability that the child has a hamster.

These events may not be mutually exclusive..

a Not known, they can have other animals as pets.
b Not known, a child may have either a cat or a hamster, both or neither!

Probability Probability

Fluency

Exercise 8.3S

1. For each of the following pairs of events say whether or not they are mutually exclusive.
   a. *P*: a girl has red hair
      Q: she has brown eyes
   b. *F*: a boy gets the same score on 2 dice
      G: his total score is odd
   c. *M*: a girl gets the same score on 3 dice
      N: her total score is odd.
   d. *X*: a girl gets the same score on 3 dice
      Y: her total score is a multiple of 6.

2. A pack of cards contains 60 cards in four colours: black, red, green and blue. There are 15 of each colour.
   The black cards carry the numbers 1 to 15.
   The red cards are multiples of 2.
   The green cards are multiples of 3.
   The blue cards are multiples of 6.
   The top card is turned over.

   For each pair of events, say whether or not they are mutually exclusive.
   a. *A*: the card is black, and
      B: it is an even number
   b. *C*: the card is red, and
      D: it is an odd number
   c. *E*: the card is green, and
      F: it is a factor of 20
   d. *G*: the card is blue, and
      H: it is a prime number

3. All the counters in a bag are either green or black. The probability a counter chosen at random from the bag is green is $\frac{1}{3}$.
   a. Find the probability that a counter chosen at random is blue.
   b. Find the probability that a counter chosen at random is black.

4. All the counters in a bag are either green, white or black.
   The probability a counter chosen at random from the bag is green is $\frac{1}{3}$, and for white it is $\frac{1}{2}$.
   What is the probability a counter chosen is
   a. black
   b. not white?

5. A fair dice has a triangle, square, circle, rectangle, sphere and a cylinder showing on the six sides. The dice is rolled.
   Find the probability that
   a. it shows a 3D shape.
   b. it does not show a 2D shape with straight edges.

6. Seamons has two packs of 12 cards, each labelled 1 to 12.
   He removes all the multiples of 4 from one of the packs, combines the two packs and shuffles them.
   He turns over the top card.
   Find the probability that the card is
   a. a multiple of 6
   b. an even number
   c. not a prime number.

7. A biased dice has the probabilities shown in the table.

Score	1	2	3	4	5	6
Probability	0.1	0.2	*x*	0.3	0.1	0.1

   Find the probability of getting
   a. a three
   b. an even number
   c. not a prime number

*8. A bag contains at least green, blue and white balls. The probability a ball chosen at random is green is $\frac{1}{7}$. Blue is twice as likely to be chosen as green and white is twice as likely to be chosen as blue. Explain why you know there are no other colours in the bag.

*9. A bag contains just green, blue and white balls. Blue is three times as likely to be chosen as green, and white is twice as likely to be chosen as blue.
   What is the probability a ball chosen at random from the bag is
   a. green
   b. white?

APPLICATIONS

8.3 Mutually exclusive events

RECAP
- Events are mutually exclusive if they cannot happen at the same time.
- If the events are also exhaustive, then the probabilities sum to 1.
- P(A) = 1 − P(not A).

You could list the outcomes in a table.

HOW TO
1. Read the question carefully and think about the outcomes possible. It often helps to list them.
2. Think what you know about mutually exclusive and exhaustive events to help you calculate probabilities.

EXAMPLE

A red and a blue dice are thrown together and the higher score is recorded.
Find the probability that the score recorded is 5 or less.

① There are 36 possible outcomes, and these can be shown in a table.

	1	2	3	4	5	6
1	1	2	3	4	5	6
2	2	2	3	4	5	6
3	3	3	3	4	5	6
4	4	4	4	4	5	6
5	5	5	5	5	5	6
6	6	6	6	6	6	6

The complementary event is just another way of saying 1 minus the event.

② It is easier to find the **complementary** event

P(5 or less) = 1 − P(6)
$= 1 - \frac{11}{36}$
$= \frac{25}{36}$

EXAMPLE

There are a number of balls in a bag, identical except for their colour.
A ball is taken from the bag at random.
The probability of taking
- green is $\frac{1}{6}$
- blue is $\frac{1}{4}$
- black is $\frac{1}{2}$.

Are any other colours in the bag?

② If the events are exhaustive then their probabilities add up to 1.

$\frac{1}{6} + \frac{1}{4} + \frac{1}{2} = \frac{2}{12} + \frac{3}{12} + \frac{6}{12} = \frac{11}{12}$ There must be at least one other colour.

Probability Probability

164

Reasoning and problem solving

Exercise 8.3A

1. A red and a blue dice are thrown together and the difference between the scores is recorded.

   a. Draw a table to show the possible outcomes.

   b. What is the probability of a difference of
      i exactly one   ii zero   iii more than 1?

2. There are a number of balls in a bag, identical except for their colour. A ball is taken from the bag at random. Serena says that the probability of taking a green is $\frac{1}{4}$, taking a blue is $\frac{1}{3}$ and taking a black is $\frac{1}{2}$.
   Can Serena be correct?
   Give the reason for your answer

3. There are a number of balls in a bag, identical except for their colour.
   A ball is taken from the bag at random. Arinda says that the probability of taking a green is $\frac{1}{6}$, taking a blue is $\frac{1}{3}$, taking a white is $\frac{1}{8}$, and taking a black is $\frac{3}{8}$.

   a. Can you tell if Arinda is correct? Give the reason for your answer.

   b. Can you tell if there are any other colours? Give the reason for your answer.

4. A bag of sweets contains 4 vanilla fudge pieces, 3 white chocolate caramels, 3 chocolate covered fudge pieces, and 2 caramel fudge pieces.
   What is the probability that a sweet chosen at random contains

   a chocolate   b fudge   c caramel?

5. A set of cards with the numbers 1 to 10 is shuffled and a card chosen at random. Here are four possible events.
   A A prime number   B A factor of 36
   C An even number   D An odd number

   a. List any pairs of mutually exclusive events.

   b. List any pairs of exhaustive events.

   c. Explain why A and C are not mutually exclusive.

6. The table below shows the gender and hair colour of the 28 students in a class.

	Fair	Dark	Red
Male	6	9	3
Female	7	6	0

   A student is chosen at random from the class, and here are four possible events.
   M The student is male.
   F The student is female.
   R The student has red hair.
   D The student has dark hair.
   Giving a reason, explain whether the following statements are true or false.

   a. M and F are mutually exclusive.

   b. M and F are exhaustive.

   c. M and D are mutually exclusive.

   d. F and R are mutually exclusive.

7. A red and a blue dice are thrown together and the blue score is subtracted from the red score. Both dice are unbiased.

   a. Draw up a table like the one in the first example to show the possible outcomes.

   b. What is the probability the score is
      i −2   ii 5   iii less than 5?

8. The probabilities for a biased dice are shown in the table.

Score	1	2	3	4	5	6
Probability	0.1	0.2			0.2	0.1

   Getting a 4 is three times as likely as getting a 3.
   Find the probability of getting a

   a 3   b 4   c prime number.

*9. A spinner is constructed so that the numbers 1 to 10 are possible, and the probability of seeing a number $k + 1$ is twice as likely as seeing the number k (for $k = 1, 2, \ldots 9$).
   Find the probability of getting a

   a 1   b 10.

   Give your answers as decimals correct to 3 decimal places.

Summary

Checkout
You should now be able to...

	Test it Questions
✓ Use experimental data to estimate probabilities and expected frequencies.	1, 2, 7
✓ Use tables to represent the outcomes of probability experiments.	1, 2
✓ Calculate theoretical probabilities and expected frequencies using the idea of equally likely events.	3, 4
✓ Recognise mutually exclusive events and exhaustive events and know that the probabilities of mutually exclusive exhaustive events sum to 1.	3, 5, 6
✓ Compare theoretical probabilities with experimental probabilities.	7

Language	Meaning	Example
Trial	An activity or experiment.	Rolling a single dice.
Outcome	The result of a trial.	Rolling a single dice and getting a 2.
Event	One or more outcomes of a trial.	Getting a prime number when you roll a conventional six-sided dice.
Impossible	It cannot happen. The probability of it happening is 0.	Rolling a conventional six-sided dice and getting the number 7.
Certain	It must happen. The probability of it happening is 1.	Rolling a conventional single dice and getting a number less than 7.
Relative frequency	The **experimental probability** of an outcome after several trials is $$\frac{\text{Number of times the outcome happened}}{\text{Number of times the activity was done}}$$	A drawing pin is thrown 100 times. It lands point down 72 times. The relative frequency of the pin landing point down is $\frac{72}{100} = 0.72$
Expected frequency	How many times you expect the outcome to happen. Number of trials × probability of the event	The expected frequency of getting a 5 when you roll a fair dice 40 times is $\frac{1}{6} \times 40 = 6.666... = 6.\dot{6}$ about 7 times.
Theoretical probability	A value between 0 and 1 describing the likelihood of an event determined by logical consideration of the trial. If all possible outcomes are equally likely this is given by $$\frac{\text{Number of ways the outcome could happen}}{\text{Number of possible outcomes}}$$	There is 1 way of rolling a two on a fair dice. There are 6 different possible outcomes. The theoretical probability of rolling a two is $\frac{1}{6}$.
Bias/Biased	All outcomes are not equally likely.	A fair dice has the numbers 1, 1, 2, 3, 4, 5 on its faces. It is more likely that you will roll a 1 than any of the other outcomes. The dice is biased.
Equally likely	All outcomes have the same probability of happening.	When you roll a fair dice all the outcomes {1, 2, 3, 4, 5, 6} are equally likely. They all have a probability of $\frac{1}{6}$.

Review

1 Adina records the ages of children in a play area.

Age (in years)	Frequency
Younger than 2	6
2 – 4	16
5 – 7	10
8 – 11	7
Older than 11	1

a What is the relative frequency of children younger than 2 years old?

b On a different day there are 25 children in the park. Use Adina's data to estimate the number of children who are in the 2–4 age range.

2 A spinner with five colours on it is repeatedly spun and the results recorded.

Outcome	Relative frequency
Blue	0.2
Red	a
Green	a
Yellow	$3a$
Purple	0.05

a What is the value of a?

b Copy and complete the table to show all the relative frequencies.

c Use these results to estimate how many times the spinner would land on blue if it were spun 300 times.

3 A bag contains 12 blue and 26 green counters. A counter is selected at random.
What is the probability the counter is

a blue **b** red

c green or blue?

4 A fair dice is rolled twice. What is the probability that the two numbers add up to

a 5 **b** 11?

5 The probability that a tennis player wins a point is $\frac{11}{20}$. What is the probability that they lose a point?

6 Waleed always chooses either a biscuit, an apple or a banana to eat with a cup of tea. The probability he chooses fruit is $\frac{18}{25}$ and he is exactly twice as likely to pick an apple as a banana.

Calculate the probability of Waleed choosing

a a biscuit **b** a banana

c an apple.

7 An unbiased dice is thrown 60 times and the number 3 occurs 14 times.

a What is the relative frequency of a 3?

b What is the theoretical probability of a 3?

The dice is now thrown an additional 100 times.

c What would you expect to happen to the relative frequency of a 3?

What next?

Score		
	0 – 3	Your knowledge of this topic is still developing. To improve, look at MyiMaths: 1211, 1262, 1263, 1264
	4 – 6	You are gaining a secure knowledge of this topic. To improve your fluency look at InvisiPens: 08Sa – e
	7	You have mastered these skills. Well done you are ready to progress! To develop your problem solving skills look at InvisiPens: 08Aa – e

Assessment 8

1. Amanda says that it is impossible to thrown a coin five times and see tails every time. Is she correct? Give a reason for your answer. [1]

2. A fair coin is thrown and lands on heads twice. What is the probability of a head on the third throw? Explain your answer. [2]

3. A dice is thrown 100 times and the following results are obtained.

Score	1	2	3	4	5	6
Frequency	14	17	22	18	15	14

   Calculate the relative frequency of scoring an odd number. [2]

4. One in every eight children in a town has asthma. How many children would you expect to have asthma

   a in a school of 1600 children [1]

   b in a road where 14 children live [1]

   c in a class of 34 children? [1]

5. The sides of a biased spinner are labelled 1, 2, 3, 4 and 5. The probability that the spinner will land on each of the numbers 2, 3 and 4 is given in the table.

Number	1	2	3	4	5
Probability	x	0.14	0.26	0.22	x

   The probability that the spinner lands on either of the numbers 1 or 5 is the same.

   a It is claimed that $x = 0.2$. Explain why this is wrong and find the correct value of x. [3]

   The spinner is spun 200 times.

   b How many times should you expect the spinner to land on 3? [1]

   c How many times should you expect the spinner to land on an odd number? [2]

   d How many times should you expect the spinner to land on a prime number? [2]

6. Wissam and Ben buy identical packets of mixed vegetable seeds. They pick out the pea seeds from their packets. Their results are shown in the tables.

   Wissam's results

Number of packets	Number of seeds in each packet	Number of pea seeds in each packet
5	30	5, 5, 8, 3, 6

   Ben's results

Number of packets	Number of seeds in each packet	Total number of pea seeds
10	30	51

   a Write down an estimate of the probability of selecting a pea seed at random from a packet using

   i Wissam's results ii Ben's results. [4]

   b Whose results are likely to give the better estimate of the probability? Explain your answer. [2]

7. Lena buys 20 bottles of juice.
   There were four bottles each of apple, blueberry, grape, orange and pineapple.

   a Lena takes a bottle from the fridge at random.
   What is the probability that it is apple juice? [1]

   b Lena's son drinks all of the orange juice. Lena takes one bottle at random from the fridge.
   What is the probability that it is either grape of pineapple juice?
   Write your answer in its simplest form. [2]

Probability Probability

8 A box of cat food contains 30 individual packets.
The probability that a tuna packet is chosen from the box is 0.3
How many packets of tuna packets are in the box? [2]

9 The probability that a train arrives early is 0.18
The probability that the train arrives on time is 0.31

    a A passenger complains that the train is late more often than it is on time or early. Is the passenger correct? [2]

    b If you take 225 journeys on this train during one year, how many days can you expect to be on time? [1]

10 The table below shows the probabilities of selecting raffle tickets from a drum.
The tickets are coloured red, white or blue and numbered 1, 2, 3 or 4.
A raffle ticket is taken at random from the drum.

    a Calculate the probability that
        i it is white and numbered 3 [1]
        ii it is numbered 1 [2]
        iii it is red [2]
        iv it is either blue or numbered 2. [3]

    b What raffle ticket is impossible to draw? Explain your answer. [1]

		Number			
		1	2	3	4
Colour	Red	$\frac{2}{25}$	$\frac{1}{25}$	$\frac{3}{50}$	$\frac{1}{10}$
	White	$\frac{9}{50}$	$\frac{7}{50}$	$\frac{3}{25}$	$\frac{1}{50}$
	Blue	$\frac{3}{50}$	$\frac{4}{25}$	$\frac{1}{25}$	0

11 At a busy road junction, the traffic lights operate as follows
red: 40 seconds   red and amber: 10 seconds   green: 40 seconds   amber: 10 seconds

Work out the probability of each event. Give your answer as a simplified fraction.

    a the lights show green [1] **b** there is an amber light showing [2]
    c there is not a red light showing [2] **d** either green or amber lights are showing [2]

12 a Are these pairs of events mutually exclusive? Give reasons for your answers.
        i Red section and multiples of 3. [2]
        ii Blue section and prime numbers. [2]
        iii White section and multiples of 7. [2]

    b Every number on the spinner is increased by 1.
How does this change your answers to part **a**? [6]

13 During any day in December the probability that it will snow is 0.1. The probability that it will be sunny is 0.3
Is the probability that it will snow or be sunny on any given day in December 0.4?
You must give a reason for your answer. [2]

14 A regular pentagon, two squares and a triangle are joined to create a spinner.
Assume that the spinner is unbiased.

    a Find the probability of landing on the blue region of the spinner. [4]

    b How does the assumption that the spinner is unbiased affect your answer to part **a**? [1]

9 Measures and accuracy

Introduction

During the 1936 Olympic Games held in Berlin in Nazi Germany, the US athlete Jesse Owens broke the 100 m men's world record, with a time of 10.2 s, a record that remained unbroken for 20 years afterwards. He also took the gold medal in the long jump in the same games and he is widely believed to have been the greatest athlete ever. It is said that Hitler was not impressed by Owens's victories, as he had hoped that the games would be a validation of Aryan supremacy.

The current 100 m world record is held by Usain Bolt of Jamaica, with a time of 9.58 s in 2009. Note the increased accuracy in the reporting of the time in 2009 as compared with 1936.

What's the point?

Being able to rely on accurate decimal measurements is vital in athletics. In a 'photo-finish' with two competitors crossing the finishing line apparently together, 0.01 seconds can be the difference between gold and silver!

Objectives

By the end of this chapter, you will have learned how to …
- Use approximate values to estimate calculations.
- Use an estimate to check an answer obtained using a calculator.
- Solve problems involving speed and density.
- Look at a value that has been rounded and work out upper and lower bounds for the original value.

Check in

1 Write these numbers to **i** 1 decimal place (1dp)
ii 2 significant figures (2sf).

  a 38.5   **b** 16.08   **c** 103.88   **d** 0.082   **e** 0.38

2 For each calculation **i** write down a mental estimate
ii use a calculator to give an exact answer.

  a 18×53 **b** 3.77×89.5 **c** $3870 \div 79$

  d $642 \div 28.7$ **e** $\dfrac{101 \times 23}{17.1 + 4.9}$ **f** $\dfrac{37 - 84}{0.13 \times 8}$

Chapter investigation

Most of the quantities you can measure, such as length and mass, are based on the decimal number system; however time is not. During the French Revolution, it was proposed to replace the standard system of hours and minutes with French Revolutionary Time, in which time would be measured using powers of 10.

Investigate French Revolutionary Time and write a short report.

At what times would your school day start and finish using this system?

SKILLS

9.1 Estimation and approximation

- An **approximate** value can be found by rounding.

A useful first approximation is to 1 significant figure but other approximations may be better.

EXAMPLE

Find approximate values for these calculations.

a $76.5 + 184.2$ **b** $12.3 - 8.9$ **c** $183.2 \div 17.6$ **d** 22×14.53

Round each number appropriately before doing the calculation.

a $76.5 + 184.2 \approx 80 + 200 = 280$
b $12.3 - 8.9 \approx 12 - 9 = 3$
c $183.2 \div 17.6 \approx 200 \div 20 = 10$
d $22 \times 14.53 \approx 20 \times 15 = 300$

- Approximations can be used to **estimate** an answer before doing a calculation. Estimates are useful for checking your answer or adjusting place value.

You need to be careful when estimating powers or subtracting numbers that are close. For example, 1.3 is quite close to 1, but 1.3^7 is not close to 1^7.

EXAMPLE

Estimate the value of these calculations.

a $\dfrac{563 + 1.58}{327 - 4.72}$

b $\dfrac{3.27 \times 4.49}{1.78^2}$

c $\dfrac{\sqrt{2485}}{1.4^3}$

d $\dfrac{2.45^3}{2.44 - 2.31}$

a Ignore the relatively small amounts added and subtracted.
Estimate $= 600 \div 300 = 2$

b $\sqrt{3} = 1.73$ to 2 dp.
1.78^2 is 'a bit more than 3', so it cancels with 3.27
Estimate $= 4.5$

c $2485 \approx 2500$ and $\sqrt{2500} = 50$
$1.4^2 \approx 2$, so $1.4^3 \approx 1.4 \times 2 = 2.8 \approx 3$
Estimate $= 50 \div 3 \approx 17$

d $2.45 \approx 2.5$ and $2.5^2 = 6.25$, so $2.5^2 \approx 6$
$2.45^3 \approx 6 \times 2.5 = 15$
$2.44 - 2.31 \approx 0.1$ Beware when approximating the denominator that it is not 0.
Estimate $= 15 \div 0.1 = 150$

You might be able to use some number facts to help you get a closer approximation.

EXAMPLE

Estimate $\sqrt{5}$ to 1 dp.

'$\Rightarrow$' means 'implies'.
'$\approx$' means 'approximately equal to'.

$\sqrt{4} = 2, \sqrt{9} = 3 \Rightarrow 2 < \sqrt{5} < 3$.

As 5 is closer to 4 than 9, $\sqrt{5}$ is going to be closer to 2 than 3. So an estimate would be a little more than 2.

```
  2.5          6.25 > 5              2.2
× 2.5       ⇒ 2 < √5 < 2.5         × 2.2         4.84 < 5                   2.25        5.0625 > 5
 500           Try 2.2              440         ⇒ 2.2 < √5 < 2.5          × 2.25        ⇒ 2.2 < √5 < 2.25
+125                                + 44         Try 2.25                  45000
6.25                                4.84                                    4500
                                                                           +1125
√5 ≈ 2.2 (1 dp)                                                            5.0625
```

Number Measures and accuracy

Fluency

Exercise 9.1S

1. Round each of these numbers to
   i 1 sf   ii 1 dp   iii the nearest integer.
   a 8.3728   b 18.82   c 35.84
   d 278.72   e 1.3949   f 3894.79
   g 0.008 372 h 2399.9   i 8.9858
   j 14.0306   k 1403.06   l 140 306

   When asked to estimate an answer you should show how you obtained your estimate.

2. Estimate the answer to each calculation by rounding to the degree of accuracy given.
   a $37.43 \div 3.52$ (1 sf)
   b 2.497×1.99 (1 dp)
   c $6342 \div 897$ (2 sf)

3. Write a suitable estimate for each of these calculations.
   a $4.88 + 3.07$
   b $216 + 339$
   c $0.0049 + 0.00302$
   d $43.89 - 28.83$
   e 3.77×0.85
   f $44.66 \div 0.89$
   g 3.76×4.22
   h 17.39×22.98
   i $\dfrac{4.59 \times 7.9}{19.86}$
   j $54.31 \div 8.8$
   k 4.98×6.12
   l $17.89 + 21.91$
   m $\dfrac{5.799 \times 3.1}{8.86}$
   n $34.8183 - 9.8$
   o $\dfrac{32.91 \times 4.8}{3.1}$
   p $272.701 - 43$
   q $(9.8^2 + 9.2 - 0.438)^2$

4. Estimate these square roots.
   Try estimating to 1 dp, then check your estimates with a calculator.
   a $\sqrt{2}$ b $\sqrt{8}$ c $\sqrt{10}$
   d $\sqrt{15}$ e $\sqrt{20}$ f $\sqrt{26}$
   g $\sqrt{32}$ h $\sqrt{45}$ i $\sqrt{70}$

5. Estimate each of these calculations. Use a calculator to check your estimates.
   a $\dfrac{29.91 \times 38.3}{3.1 \times 3.9}$
   b $\dfrac{16.2 \times 0.48}{0.23 \times 31.88}$
   c $(4.8^2 + 4.2 - 0.238)^2$
   d $\dfrac{63.8 \times 1.7^2}{1.78^2}$
   e $\sqrt{(2.03 \div 0.041)}$
   f $\sqrt{(27.6 \div 0.57)}$

6. Explain why approximating the numbers in these calculations to 1 significant figure would *not* be an appropriate method for estimating the results of the calculations.
   a $\dfrac{5.39 + 4.72}{0.53 - 0.46}$
   b $\sqrt{(1.52 - 1.49)}$

7. Use approximations to estimate the value of each of these calculations.
   a $\dfrac{317 \times 4.22}{0.197}$
   b $\dfrac{4.37 \times 689}{0.793}$
   c $\dfrac{4.75 \times 122}{522 \times 0.38}$
   d $4.8^3 - 8.5^2$
   e $\dfrac{9.32 - 3.85}{0.043 - 0.021}$
   f $7.73 \times \left(\dfrac{0.17 \times 234}{53.8 - 24.9}\right)$
   g $\dfrac{48.75 \times 4.97}{10.13^2}$
   h $\sqrt{\dfrac{305.3^2}{913}}$
   i $\dfrac{\sqrt{9.67 \times 8.83}}{0.087}$
   j $\dfrac{6.8^2 + 11.8^2}{\sqrt{47.8 \times 52.1}}$
   k $\dfrac{(23.4 - 18.2)^2}{3.2 + 1.8}$
   l $\sqrt{\dfrac{2.85 + 5.91}{0.17^2}}$

8. Estimate these calculations and say whether your estimate will be larger or smaller than the exact answer.
   a $129.3 - 74.6$
   b $2.612 + 0.77$
   c 65×63
   d 65×71
   e $93.6 \div 5.8$
   f $11 \div 5.4$

9. Check your answers to questions **5**, **7** and **8** with a calculator. How do your estimates compare?

10. Irwin estimates
    $\dfrac{47.3 \times 18.9}{8.72} \approx 100.$
    What approximations could he have used to get his estimate?

APPLICATIONS

9.1 Estimation and approximation

RECAP
- Rounding can be used to approximate numbers.
- Approximation can be used to estimate answers to complex calculations.
- You can use estimates to check answers and to adjust place values.

Rounding to 1 sf is usually helpful!

HOW TO
To solve problems using approximations
1. Read the question and decide what you need to do.
2. Estimate using appropriate approximations.
3. Answer the question and round your answer to a suitable degree of accuracy.

EXAMPLE

Philip says 22 feet is 70 m to 1 sf. One foot = 12 inches and 1 inch = 25.400 cm.

a Use an estimate to show he is wrong.
   Suggest a mistake that he might have made.
b Comment on the accuracy of your estimate.

a ① Using 1 dp should give a good estimate.
   1 inch ≈ 2.5 cm   ② Round to 1 dp.
   1 foot ≈ 30 cm   12 × 2.5
   22 feet ≈ 22 × 0.3 = 6.6 m
                      = 7 m (1 sf)
   ③ Philip was wrong by a power of 10, he may have incorrectly thought 30 cm equals 3 m.

b ① Compare to accurate value.
   2.54 × 12 × 22 = 6.7056   ② Calculator
                 = 6.71 m (2 dp)
   My estimate was 11 cm (6.71 − 6.6) too small. This is less than a 2% error, so the estimate is a good one.
   ③ Make sure you justify your reason mathematically.

EXAMPLE

Jehan is planting a new rectangular lawn. He is told that he needs between 45 and 60 grams of seed per square metre of lawn. The space he has prepared is 4.8 m × 4.9 m. Seed can be bought in 800 g bags. How many bags should he buy? Give your reason.

It is better to have too much than too little seed.
Area of lawn = 4.8 × 4.9
            ≈ 5 × 5 = 25 m²   ② Round up to 1 sf.
Seed required ≤ 25 × 60 = 1500 g ① Use the larger amount (60 > 45).
Number of bags ≤ 1500 ÷ 800 < 2   ② Over estimate.
Jehan should buy 2 bags. This gives some spare. ③

EXAMPLE

Bo wants to hire a car to travel from Hutang to Shanghai, a distance of 118 miles. He will return the car to Hutang. The hire charge for a small car is $39.98 and for a larger car is $68.92. It costs 12 cents per mile to run the large car and 15 cents per mile to run the small car. Which car would you advise Bo to choose. Give your reason.

① Estimate the cost of fuel there and back and add it to the hire fee.
Large 120 × 0.12 × 2 + 70 ≈ 14 × 2 + 70 ≈ 30 + 70 = $100   ② 12² = 144 ≈ 140 and 14 × 2 = 28 ≈ 30
Small 120 × 0.15 × 2 + 40 ≈ 120 × 0.3 + 40 ≈ 40 + 40 = $80 ② 15 × 2 = 30 and 12 × 3 = 36 ≈ 40
③ The small car. It is ≈ $20 cheaper.

Number   Measures and accuracy

Reasoning and problem solving

Exercise 9.1A

1. Eka wants to estimate how many centimetres are in a yard. He knows that 1 inch = 2.5400 cm and 1 yard = 36 inches.

   a. Approximately how many cm are in a yard by rounding 2.5400 cm to 1 sf.

   b. Was it a good estimate? Give your reason.

2. Seth is staining a fence. The tin says one litre covers 6 m². Stain comes in 5 litre pots.

   Seth's fence is 1.8 m high and 48 m long.

   How many pots of stain should he buy? Show your working.

3. a. Estimate mentally these calculations.

   i. $258 + 362$
   ii. $64 \div 27$
   iii. 62.7×211.8
   iv. $96.7 - 64.8$

   b. Explain if each answer in part **a** is an overestimate or an underestimate.

4. Mark writes

   $$\frac{19.3 \times 204}{3.8} \approx 100$$

   Is he correct? Give your reason.

5. Gosia is making jam to fill some 4 fluid ounce jars she bought at a car boot sale. She knows that 1 fl oz = 28.3495 ml.

   a. Estimate how many jars she can *fill* if she makes 550 ml of jam.

   b. About how much jam would she need to make in order to fill one more jar?

6. Caroline has 490 MB of space left on her camera's SD card. She is taking photographs which are 6.448 MB in size.

   a. Approximately how many can she take?

   b. If 1GB = 1000 MB and she has a 1.5 GB memory card, about how many photos has she taken when the card is full?

7. Paul's mobile phone gives a summary of his bank account to date. On 28th February he had about £2000 in his account.

06/03	ATM TOWN	−£100.00
06/03	DC DIRECT DEBIT	−£156.45
06/03	DDEBIT	−£99.30
06/03	CASH TRM MAR 06	−£80.00
04/03	CHQ IN	+£84.37
02/03	ABC & G	+£59.21
01/03	SALARY	+£1758.64
28/02	BALANCE	£2058.63

   Approximately how much money is in his account now?

8. Chen does the following calculations in his homework.

   i. $1.86 \times 5.432 = 10.10$ (2 dp)
   ii. $17.543 \times 543.25 = 90\,000$ (1 sf)
   iii. $12.15 \div 17.55 = 1.7$ (1 dp)

   a. Using estimates only, explain if Chen's answers are reasonable.

   b. For any answers that seem incorrect based on your estimate, say what Chen might have done wrong.

9. Greta is driving from Hambery to Herselt, a distance of 278 miles. Her diesel car does 43 mpg (miles per gallon). 1 gallon is about 5 litres.

   a. Estimate how many litres she will use on her journey to Herselt.

   b. In fact 1 gallon = 4.54609 litres. Will she use more or less diesel than you estimated?

   c. Greta thinks that her average speed will be 55 miles per hour. If she leaves at 10:30 estimate when will she arrive?

*10. Estimate the number of jelly beans in a 1 litre jar.

SKILLS

9.2 Calculator methods

You can use a scientific calculator to carry out more complex calculations that involve decimals.

EXAMPLE

Use your calculator to work out $3.46 + 2.9 \times 4.8$
Give your answer to one decimal place.

Estimate $= 3 + 3 \times 5 = 3 + 15 = 18$ BIDMAS

You type 3 . 4 6 + 2 . 9 × 4 . 8 =

The calculator should display $3.46+2.9\times4.8$
                                 17.38

$17.38 = 17.4$ (1 dp)   Check ≈ 18 ✓

Modern calculators have the BIDMAS rules built into them. Typing $2 + 3 \times 4$ gives 14 not 20!

A scientific calculator has **bracket** keys.

EXAMPLE

Use your calculator to work out $(3.9 + 2.2)^2 \times 2.17$
Write all the figures on your calculator display.

Estimate $= (4 + 2)^2 \times 2 = 6^2 \times 2 = 36 \times 2 = 72$ BIDMAS

You type (3 . 9 + 2 . 2) x^2 × 2 . 1 7 =

The calculator should display $(3.9+2.2)2\times2.17$
                                 80.7457

80.7451 Check ≈ 72 ✓

You can use the bracket keys to do calculations where the **order of operations** is not obvious.

EXAMPLE

Use a calculator to work out the value of this expression.
Write all the figures on the calculator display.

$$\frac{21.42 \times (12.4 - 6.35)}{(63.4 + 18.9) \times 2.83}$$

Estimate $= \dfrac{20 \times (12 - 6)}{(60 + 20) \times 3} = \dfrac{120}{240} = 0.5$

Rewrite the calculation as
$(21.42 \times (12.4 - 6.35)) \div ((63.4 + 18.9) \times 2.83)$
Type this into the calculator.

$(21.42\times(12.4-6.35))\div((63.4+18.9)\times2.83)$ → $(21.42\times(12.4-$
$0.5564 0 1856$

$0.556\ 401\ 856$ Check ≈ 0.5 ✓

EXAMPLE

Write 10 000 seconds in hours, minutes and seconds.

Do the calculation in stages.
Convert 10 000 s to minutes, then convert the remainder back to seconds.

$10000\div60$ Ans-166   Ans×60
166.6666667 0.666666667 40

$10\ 000\ s = 166.\dot{6}\ min = 166\ min, 40\ s$

Convert 166 mins to hours, then convert the remainder back to minutes.

$166\div60$ Ans-2   Ans×60
2.766666667 0.766666666 46

$166\ min = 2.7\dot{6}\ hrs = 2\ hrs, 46\ min$

$10\ 000\ s = 2\ hrs, 46\ min, 40\ s$ Check $40 + 60 \times (46 + 60 \times 2) = 10000$ ✓

Number Measures and accuracy

Fluency

Exercise 9.2S

For each question first estimate the answer and then use your calculator to find an accurate answer.

1. Calculate these giving your answers to one decimal place.

   a $3.4 + 6.2 \times 2.7$

   > Estimate $= 3 + 6 \times 3$
   > $\qquad\quad = 3 + 18$
   > $\qquad\quad = \underline{\quad\quad}$
   > Exact $\;\;= \underline{\quad\quad}$

   b $1.98 \times 11.7 - 4.6$

   c $7.8 + 19.3 \div 4.12$

   d $2.09 \times 2.87 + 3.25 \times 1.17$

   e $13.67 \div 1.75 + 3.24$

   f $1.2 + 3.7 \times 0.5$

   g $802.6 \times 2.014 \div 3.92$

2. Calculate these giving all the figures on your calculator display.

   a $(2.3 + 5.6) \times 3^2$

   b $2.3^2 \times (12.3 - 6.7)$

   c $(2.8^2 - 2.04) \div 2.79$

   d $7.2 \times (4.3^2 + 7.4)$

   e $11.33 \div (6.2 + 8.3^2)$

   f $(2.5^2 + 1.37) \times 2.5$

3. Calculate these, first giving all the digits on your calculator display, then rounding your answers to one decimal place. Proof 1 comment suggested 2 significant figures.

   a $\dfrac{5.4 + 3.8}{4.5 - 2.9}$

   b $\dfrac{3.8 - 1.67}{4.3 - 2.68}$

   c $\dfrac{12.4 + 5.8}{14.5 - 3.9}$

   d $\dfrac{13.08 - 2.67}{2.13 + 2.68}$

4. Convert these times into weeks, days, hours, minutes and seconds.

   a 50 000 s

   b 100 000 s

   c 500 000 s

   d 1 million s

   e 10 million s

   f 30 million s

5. Calculate these, first giving all the digits on your calculator display, then rounding your answers to two significant figures.

   a $3.2 \times (2.8 - 1.05)$

   b $2.8^2 \times (9.4 - 0.083)$

   c $16 \div (5.1^2 \times 7.2)$

   d $(3.8 + 8.9) \times (2.2^2 - 7.6)$

   e $1.8^3 + 4.7^3$

   f $52 \div (4.6 - 1.8^2)$

6. Calculate these giving all the figures on your calculator display.

   a $\dfrac{462.3 \times 30.4}{(0.7 + 4.8)^2}$

   b $\dfrac{13.58 \times (18.4 - 9.73)}{(37.2 + 24.6) \times 4.2}$

7. Calculate these giving all the figures on your calculator.

   a $\dfrac{165.4 \times 27.4}{(0.72 + 4.32)^2}$

   b $\dfrac{(32.6 + 43.1) \times 2.3^2}{173.7 \times (13.5 - 1.78)}$

   c $\dfrac{24.67 \times (35.3 - 8.29)}{(28.2 + 34.7) \times 3.3}$

   d $\dfrac{1.45^2 \times 3.64 + 2.9}{3.47 - 0.32}$

   e $\dfrac{12.93 \times (33.2 - 8.34)}{(61.3 + 34.5) \times 2.9}$

   f $\dfrac{24.7 - (3.2 + 1.09)^2}{2.78^2 + 12.9 \times 3}$

8. Find the power key on your calculator, $\wedge$ or $x^{\blacksquare}$ or y^x or ….

   a Use this key to help you investigate these sequences

   i $\left(1 + \dfrac{1}{n}\right)^n$ ii $\left(1 + \dfrac{2}{n}\right)^n$

   for increasing values of the number $n = 2, 3, 10, 100, 1000, \ldots$.

   b What is the relationship between your results for parts **i** and **ii**?

APPLICATIONS

9.2 Calculator methods

RECAP
- Know how to use *your* calculator to work out complex calculations.
- Know how to apply the order of operations with your calculator.

HOW TO

To solve problems using your calculator
1. Read the question and decide what to do. Estimate the answer.
2. Use your calculator and check the answer agrees with your estimate.
3. Answer the question – interpret your calculator display in the context of the question.

EXAMPLE

Put brackets into each of these calculations to make them correct.

a $11.3 + 7.2 \times 2.5 = 46.25$
b $21.6 - 3 \times 4 + 3.2 = 0$
c $11 \div 5.5 - 4 = -2$
d $11 \div 5.5 - 5 = 22$

① If you cannot see where the brackets should go, work systematically through the different positions. Estimating first can help.

a $(11 + 7) \times 3 = 54$ ① Estimate. Using $\times 3$ for $\times 2.5$ will give overestimates.
$11 + (7 \times 3) = 32$
$(11.3 + 7.2) \times 2.5 = 46.25$ ② ③

b $22 - (3 \times 4) + 3 = 13$
$(22 - 3) \times 4 + 3 = 79$ ①
$22 - 3 \times (4 + 3) = 1$
$21.6 - 3 \times (4 + 3.2) = 0$ ② ③

c $11 \div (6 - 4) = 5.5$ ① Estimate
$(12 \div 6) - 4 = -2$
$(11 \div 5.5) - 4 = -2$ ② ③

d $11 \div (5.5 - 5) = 22$ ③
Notice $11 \times 2 = 22$ and $5.5 - 5 = 0.5$
$(11 \div 0.5 = 22)$ so no estimates needed.

EXAMPLE

Joseph books four tickets online for a show, at €25.90 each. He has to pay a €2.90 booking fee per ticket, and a 'single transaction fee' of €2.85. Tickets 'on the door' cost €30 each.

Did he save money by booking online? How much?

Estimate $= (30 \times 4) - (25 \times 4 + 3 \times 4 + 3) = €120 - €115 = €5$ ①   Exact $= 120 - 118.05 = 1.95$ ②
Yes, he saved €1.95 ③

EXAMPLE

A daytime taxi rate is set at £2.60 standing charge, plus £1.80 per mile, plus 20p for 40 seconds waiting time. 30p is added for every extra adult.

Nasser and Denise hire a taxi for a 4 mile journey, with 2 minutes waiting time. Nasser gives the driver £15 and says 'keep the change.' What tip did he give?

① Total cost = standing charge + mileage cost + waiting time + extra adult
Total cost $= 2.60 + 4 \times 1.80 + (2 \times 60 \div 40) \times 0.20 + 0.30$
Estimate $= 3 + 4 \times 2 + 3 \times 0.2 + 0.3 \approx 12$
Exact $= £10.70$ ② ≈ 12 ✓
Harry gives a tip of £4.30   ③ $15.00 - 10.70$

Number Measures and accuracy

Reasoning and problem solving

Exercise 9.2A

1 i Estimate first, then use your calculator to decide which of these calculations are correct.

    a $7.2 + (8.3 \times 3.4) = 35.42$

    b $36 \div (2.5 + 5.5) = 19.9$

    c $(36 \div 2.5) + 5.5 = 4.5$

    d $(21 - 3) \times (4 + 3) = 126$

ii Move the brackets where necessary to make all the answers correct.

2 The night rate for taxi fares is set at $3.40 standing charge, plus $2.20 per mile, plus 20 cents for every 30 seconds. An extra 40 cents is added for every additional adult.

  a Amali takes a taxi for 8 miles, which involves 5 minutes waiting. How much is the fare for this journey?

  b Robyn and her two sisters share a taxi for a 3 mile journey, with 3 minutes of waiting time. She gives the driver a tip which is 10% of the fare. How much does she pay?

3 Carpet tiles are sold in boxes of 12. Style A is €2.25 per tile, Style B is €3.95, and Style C is €4.49. They are only sold in full boxes.

Chand creates a pattern of tiles that requires $\frac{1}{3}$ of a box of Style C, $\frac{3}{4}$ of a box of Style A and $\frac{1}{2}$ box of Style B.

  a Estimate the cost of his pattern.

  b Use your calculator to find the exact cost in one calculation.

He sells the leftover tiles for a profit of €5 total.

  c How much did he sell the spare tiles for?

  d If his pattern was twice the size, could he still buy only three boxes of tiles?

4 Which calculation, **A – D**, gives an answer that is

  a zero   **b** a whole number

  c more than 100   **d** negative?

Estimate first, then use your calculator.

  A $97.31 \times 2 - 26.3 \times 7.5$

  B $(15.2^2 + 7) \div 2$

  C $5.7 \times 16 - 11 \times 3.2$

  D $\dfrac{2.3^2 - (3.2 + 2.09)}{0.7 \times 5 - 1.4 \times 3.8}$

5 Luciana sees the following offer.

> **T2 Phone deal**
> Monthly Fee £12.99
>
> FREE 200 texts every month
> FREE 200 minutes every month
> Extra text messages 3.2p each
> Extra minutes 5.5p each

Her current mobile phone costs £22.99 per month but she gets unlimited free texts and voice minutes.

What would the new offer cost her

  a in February, when she used 189 texts and 348 voice minutes?

  b in March, when she used 273 texts and 219 voice minutes?

  c Is the new offer a good deal for Luciana? Give your reason.

*6 The speed of sound depends, amongst other things, on the temperature of the air it travels through. If T stands for temperature in degrees Celsius, then the speed of sound, in feet per second, is $\dfrac{\sqrt{(273 + T)} \times 1087}{16.52}$

  a What is the difference in the speed of sound at 0 °C compared to 18 °C?

  b To convert feet per second to miles per hour, you multiply by 15, then divide by 22. Calculate the speed of sound in miles per hour at 11.5 °C using a single calculation.

SKILLS

9.3 Measures and accuracy

You can measure **length**, **mass** and **capacity** using metric units.

- Length is a measure of distance.
  Standard units are mm, cm, m and km.

- Mass is a measure of the amount of matter in an object. Standard units are g, kg and tonnes.

- Capacity measures the amount of fluid that a container will hold. Standard units are ml, cl and litres.

- Volume is the amount of space a solid shape takes up. Standard units are cm³ and m³.

The weight of an object is actually a measure of the force of gravity on that object.

Compound measures such as speed and density describe how one quantity changes in proportion to another.

- **Speed** = $\dfrac{\text{Distance}}{\text{Time}}$
  Units such as m/s, km/h

- **Density** = $\dfrac{\text{Mass}}{\text{Volume}}$
  Units such as g/cm³, kg/m³

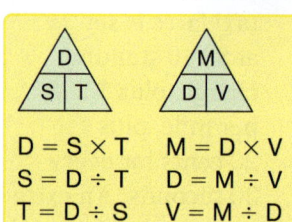

$D = S \times T$ $M = D \times V$
$S = D \div T$ $D = M \div V$
$T = D \div S$ $V = M \div D$

Density is a measure of the amount of mass per unit of volume.

EXAMPLE (p.300)

Find the density of a piece of wood with cross-sectional area 42 cm², length 12 cm and mass 693 g.

Volume of a prism = cross-sectional area × length
Volume = 42 × 12 = 504 cm³
Density = mass ÷ volume
Density = 693 ÷ 504 = 1.375 g/cm³

Measurements are not exact. Their accuracy depends on the precision of the measuring instrument, the skill of the person making the measurement and whether they have been rounded.

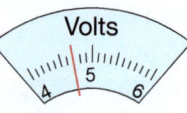

◀ To the nearest volt the meter reads 5 volts. $4.5 \leq x < 5.5$

- If a quantity is measured to within a given **error interval** then
  measurement $-\frac{1}{2}$ error interval $\leq$ true value $<$ measurement $+\frac{1}{2}$ error interval

EXAMPLE

The mass of a meteorite is given as 235.6 g. Find the lower and upper bounds of the mass.

The **implied accuracy** is one decimal place, that is, the mass is given correct to the nearest 0.1 g.

$235.55 \leq \text{mass} < 235.65$ $\frac{1}{2} \times 0.1 = 0.05$

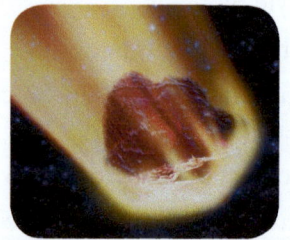

Number   Measures and accuracy

Fluency

Exercise 9.3S

> You need to learn these equivalences.
>
> | Length | Mass | Capacity |
> |---|---|---|
> | 10 mm = 1 cm | 1000 g = 1 kg | 1000 ml = 1 litre |
> | 100 cm = 1 m | 1000 kg = 1 tonne | 100 cl = 1 litre |
> | 1000 m = 1 km | | 1000 cm³ = 1 litre |

1 Give the name and abbreviation of an appropriate metric unit to measure these quantities.
   a Mass of a whale.
   b Mass of an ant.
   c Width of a book.
   d Amount of petrol in a car.
   e Amount of tea in a mug.

2 Convert these measurements to the units shown.
   a 20 mm = ___ cm   b 400 cm = ___ m
   c 450 cm = ___ m   d 4000 m = ___ km
   e 0.5 cm = ___ mm   f 4.5 kg = ___ g
   g 6000 g = ___ kg   h 6500 g = ___ kg
   i 2500 kg = ___ t   j 3 litres = ___ ml

3 Calculate these average speeds.
   a 10 miles in 4 hours
   b 16 km in 30 minutes
   c 1 hour to travel 80 km

4 Calculate the time taken if the average speed is 60 mph and the distance travelled is 150 miles.

5 Calculate the distance travelled if the average speed is 15 mph and the journey takes $2\frac{1}{2}$ hours.

6 A 420 km journey takes 6 hours and uses 30 litres of petrol. Calculate
   a the average speed
   b the petrol consumption in km per litre.

7 Calculate these densities.
   a Mass = 480 kg   Volume = 1 m³
   b Mass = 250 g   Volume = 6 cm³
   c Mass = 1 tonne   Volume = 5 m³
   d Mass = 7.8 kg   Volume = 2000 mm³

8 If 1500 m is travelled in 4.5 mins, what is the speed in
   a metres per minute
   b metres per second?

9 A cube of side 2 cm has a mass of 40 grams.
   a Find the density, in g/cm³, of the material from which the cube is made.
   b A cube of side length 2.6 cm is made from the same material. Find the mass of this cube, in grams.

10 Emulsion paint has a density of 1.95 kg/litre. Find
    a the mass of 4.85 litres of the paint.
    b the number of litres of the paint that would have a mass of 12 kg.

11 Each of these measurements was made correct to one decimal place. Write the upper and lower bounds for each measurement.
    a 5.8 m   b 16.5 litres
    c 0.9 kg   d 6.3 N
    e 10.1 s   f 104.7 cm
    g 16.0 km   h 9.3 m/s

12 Find the upper and lower bounds of these measurements, which were made to varying degrees of accuracy.
    a 6.7 m   b 7.74 litres
    c 0.813 kg   d 6 N
    e 0.001 s   f 2.54 cm
    g 1.162 km   h 15 m/s

13 Find the upper and lower bounds of these measurements, which are correct to the nearest 5 mm.
    a 35 mm   b 40 mm
    c 110 mm   d 45 mm
    e 22 cm   f 1 m
    g 0.5 m   h 0.003 km

*14 Find the maximum and minimum possible total mass of
    a 12 boxes, each of which has mass 14 kg, to the nearest kilogram
    b 8 parcels, each of mass 3.5 kg to the nearest tenth of a kilogram.

APPLICATIONS

9.3 Measures and accuracy

RECAP
- You can measure length, mass and capacity using metric units.
- A compound measure, such as speed or density, connects two different measurements.
- Most measures are approximate due to rounding and their true value lies within an error interval
measurement $-\frac{1}{2}$ error $\leq$ true value $<$ measurement $+\frac{1}{2}$ error.

$$\text{Speed} = \frac{\text{Distance}}{\text{Time}}$$

$$\text{Density} = \frac{\text{Mass}}{\text{Volume}}$$

HOW TO

To solve problems involving measure and accuracy
1. Read the question and decide what to do.
2. Convert measurements to the same units and use your knowledge of measures and accuracy.
3. Answer the question, use suitable units and give your answer to an appropriate degree of accuracy.

EXAMPLE

Stuart puts his honey into one pound jars. One pound is 454 grams.

a How many jars can he fill with 10 kg of honey?

b He sells the jars at £4.50 each. How much money does he make?

a ① How many lots of 1 lb in 10 kg.
10 kg = 10 000 g   ②
10000 ÷ 454 = 22.02...
He can fill 22 jars.   ③ Need full jars.
b 22 × 4.50 = £99   ② ③

EXAMPLE

Matija is laying a driveway using gravel. He measures it to be approximately 20 m long by 5 m wide, to the nearest metre. A company says he needs 1 bulk bag for every 10 m², at $88 per bag.

a How much will the driveway cost?

b What is the largest area that could be required and how much extra would this cost?

① Work out the area, work out how many bags are required and then multiply to get the cost.

a 20 × 5 ÷ 10 × 88 = $880   ③   ② Area ÷ 10 × price per bag.

b 20.5 × 5.5 = 112.75 m²   ② Use upper limits for maximum area.
112.75 ÷ 10 = 11.275, so 12 bags needed.   Need to buy full bags.
Extra cost = 88 × 2 = $176   ③   Two extra bags.

EXAMPLE

A toy car travels 4.5 m in 2.3 s. Both measurements are given to one decimal place. Find upper and lower bounds for the car's speed in m/s.

	Maximum	Minimum
X + Y	$X_{Max} + Y_{Max}$	$X_{Min} + Y_{Min}$
X − Y	$X_{Max} − Y_{Min}$	$X_{Min} − Y_{Max}$
X × Y	$X_{Max} × Y_{Max}$	$X_{Min} × Y_{Min}$
X ÷ Y	$X_{Max} ÷ Y_{Min}$	$X_{Min} ÷ Y_{Max}$

① Speed = distance ÷ time
② 4.45 ≤ distance < 4.55, 2.25 ≤ time < 2.35
③ Upper bound = 4.55 ÷ 2.25 = 2.022... m/s
Furthest distance in shortest time.
Lower bound = 4.45 ÷ 2.35 = 1.893... m/s
Shortest distance in longest time.

Number Measures and accuracy

Reasoning and problem solving

Exercise 9.3A

1. A river is normally 28 cm deep. After a storm it rises by $3\frac{1}{2}$ metres. At 3.89 m there is danger of flooding.

   a How much more could the river rise before it floods?

   b The record height is 5.18 m. How much above the normal level is this?

2. Sunita is travelling. It takes her 25 minutes to walk to the station, 8 minutes waiting for her train, 1 hour 42 minutes on the train, 5 minutes waiting for a taxi, and 25 minutes in the taxi.

   a If she leaves at 11 am, what time does she arrive?

   The train costs £25.70, the taxi costs £8.65

   b She gave the taxi driver £10 and said 'keep the change' for a tip. How much was the tip worth?

3. The maximum load a lift can carry is 450 kg. 6 passengers get in. Their masses, correct to the nearest kg, are 82, 73, 80, 56, 80 and 78 kg.

   a Is the maximum load exceeded?

   b Is it certain that it will be safe? Explain your answers.

4. The side length of a cube is measured to be 3.4 cm (1 dp). Jayne says the volume of the cube is over 40 cm³. Is Jayne correct? Show your working.

5. Sachin is competing in a cycle race over four stages, two in Britain and two in France.

   Stage 1   100 miles   Stage 2   75 miles
   Stage 3   232 km   Stage 4   84 km

   He knows that 5 miles is roughly 8 kilometres. His average speed is 40 km/h.

   a How long does the race take him?

   b What is his speed in miles per hour (mph)?

6. Andreas ran 50 m in 6.4 s. Ben ran 80 m in 10.3 s. All measurements are given to one decimal place. Cai says 'Ben was faster'. Is this possible? Show your working.

7. Craig is building a garage.
   The base measures 8 m by 5 m to the nearest metre and is 15 cm deep to the nearest centimetre.

   The density of concrete is 2400 kg/m³ and the cost is $55 per tonne.

   a What volume of concrete does he need?

   b How much will the mass be?

   c How much will this cost?

   d How much concrete short could he be?

   e How much spare could he have?

*8. A car travels 315 miles in 7 hours 30 mins. It uses 50 litres of petrol.

   a What is its average speed?

   b An advert for this type of car says it does 45 mpg (miles per gallon). A gallon is approximately 4.5 litres. Is the advert reasonably accurate?

9. A metal rod in the shape of a cuboid is measured to have a mass of 3925 kg. The dimensions of the rod are measured as 25 cm × 25 cm × 8.0 m.

   Marina is trying to decide what metal the rod is made from. The densities of five different possible metals are shown in the table.

Metal	Density kg/m³
Titanium	4500
Tin	7280
Steel	7850
Brass	8525
Copper	8940

   Marina claims that the rod must be made of steel. Do you agree? Explain your answer fully.

*10. A gold cylinder has a mass of 1.2 kg (1 dp) The density of gold is 19.3 g/cm³ (1 dp)

   a If the radius of the cylinder is 2.0 cm (1 dp) what is the upper bound on the height of the cylinder?

   b If the height of the cylinder is 6.5 cm (1 dp) what is the lower bound on the radius of the cylinder?

   Volume of cylinder = π × (radius)² × height

Summary

Checkout
You should now be able to...

	Checkout	Test it Questions
✓	Use approximate values obtained by rounding to estimate calculations.	1, 2
✓	Use an estimate to check an answer obtained using a calculator.	2
✓	Use, and convert between, standard units of length, mass, capacity and other measures including compound measures.	3 – 7
✓	Solve problems involving compound measure such as speed and density.	6, 7
✓	Find upper and lower bounds of a quantity that has been rounded.	8, 9
✓	Find upper and lower bounds of expressions that involve quantities that have been rounded.	10, 11

Language	Meaning	Example
Approximation	A stated value of a number that is close to but not equal to the true value of a number. Usually obtained by rounding.	$\pi = 3.141592654...$ $= 3.1$ (1 dp) $= 3$ (1 sf)
Estimate	A simplified calculation based on approximate values or a judgement of a quantity.	$1.9^2 \times \pi \simeq 2^2 \times 3 \simeq 12$ Exact $= 11.34114....$
Length	Length is a measure of linear extent.	A standard ruler is 30 cm long.
Mass	A measure of the amount of matter in an object. Weight measures the force of gravity acting on a mass.	The mass of a bag of sugar is 1 kg.
Volume	A measure of the amount of 3D space occupied by an object.	Volume $= 2000$ cm^3
Capacity	A measure of the amount of fluid that a 3D shape will hold.	Capacity $= 2$ litres
Speed	A measure of the distance travelled in a unit of time.	The speed of a car is 65 km/h.
Density	A measure of the amount of mass in a unit of volume.	The density of iron is 7.87 g/cm^3.
Accuracy	How close a measured or calculated quantity is to the true value.	$1.239648 = 1.2$ to 1 dp $= 1.24$ to 2 dp
Implied accuracy	The accuracy of a value implied by the number of significant figures or decimal places given.	1.3 has implied accuracy of 1 dp. 1.30 has implied accuracy of 2 dp
Upper bound	The largest value a quantity can take.	$1.45 \leq 1.5$ (1dp) < 1.55
Lower bound	The lowest value a quantity can take.	Lower bound Upper bound
Error interval	The range of values between the lower and upper bounds.	

Number   Measures and accuracy

Review

1. Round each of these numbers to
   i 2 decimal places
   ii 2 significant figures.
   a 93 021
   b 27.941
   c 0.006 25
   d 0.895

2. a Estimate the answers to these calculations.
   i $3.8 + 27.3 \times 2.1$
   ii $\dfrac{9.81^2 + 112}{\sqrt{3.7}}$
   iii $\dfrac{2.2 + 0.6 \times 7.1}{11.2 - 3.6^2}$
   iv $\sqrt{\dfrac{43 + 67}{7.6 - 3.4}} - 2.9$

   b Find the answers to the calculations in part a using a calculator. Give your answers to 2 decimal places.

3. State two metric units for each of these quantities.
   a length   b mass
   c volume   d capacity.

4. Convert these measurements to litres.
   a 672 cl   b 205 ml
   c 3500 cm³

5. Vikentiy took 97 minutes to complete his homework. He finished at 21:52, at what time did he start?

6. Scott leaves Kent town at 13:52. He travels at 15 m/s to Clare Valley which is 137.7 km away. When does he arrive in Clare Valley?

7. A block of silver has mass 3.147 kg and density 10.49 g/cm³. Find its volume.

8. For each of these measurments give its
   i upper bound   ii lower bound.
   a 12.5 cm (1 dp)
   b 11.5 kg (3 sf)
   c 1.00 m (2 dp)
   d 0.025 km (2 sf)

9. Tsui cuts a piece of wood of length x which is measured as 85.0 cm to the nearest millimetre. Write an inequality to describe the range of values x could take.

10. $X = 10$ (1 sf) $Y = 3.5$ (1 dp)
    Find the maximum and minimum values of these expressions.
    a $X - Y$ b $\dfrac{Y}{X}$
    c X^2 d $\sqrt{X + Y}$

11. The triangle shown below has been measured to the nearest cm.
    a What is the lower bound for the area of the triangle?
    b What is the upper bound for the area of the triangle?

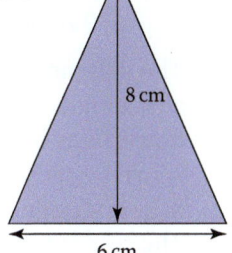

8 cm

6 cm

What next?

Score		
0 – 4		Your knowledge of this topic is still developing. To improve, look at MyiMaths: 1005, 1006, 1043, 1057, 1067, 1121, 1246, 1932, 1933
5 – 9		You are gaining a secure knowledge of this topic. To improve your fluency look at InvisiPens: 09Sa – f
10 – 11		You have mastered these skills. Well done you are ready to progress! To develop your problem solving skills look at InvisiPens: 09Aa – d

Assessment 9

1. The following answers are incorrect. Find the correct answers.

   a. 3.234 to 3 sf is 3.234 [1]
   b. 29.482 to 2 sf is 29.48 [1]
   c. 0.203762 to 1 sf is 0 [1]
   d. 312 to the nearest 10 is 31 [1]
   e. 5678 to the nearest 100 is 5600 [1]
   f. 256 133 to the nearest 1000 is 256 [1]

2. Ahmed, Bart and Christian make these estimates for the following calculations. For each calculation, decide who makes the best estimate.

		Ahmed	Bart	Christian	
a	5.8 × 6.5	30	36	40	[1]
b	20.2 × 5.6	100	110	120	[1]
c	7.9 × 8.8	56	66	72	[1]
d	6.6 ÷ 1.3	5	6	9	[1]
e	7.9 ÷ 0.91	7	8	9	[1]

3. On average, a person blinks about 6 times per minute. Esther says that a good estimate for the number of times a person blinks in one day is 9000. Use your calculator to explain how she got that estimate. [2]

4. Estimate the square root of 186. Give your answer to the nearest integer. [2]

5. a. Patrick estimates the value of the following calculation to be $1\frac{1}{3}$. Find his mistakes and give a sensible estimation of the value.

   $$\frac{5.8^2 - 1.5 \times 1.8^2}{4.59 + 9.21} \approx \frac{6^2 - 2 \times 2^2}{5 + 10} = \frac{36 - 16}{15} = 1\frac{1}{3}$$ [3]

   b. Aleksei estimates that $\frac{28.9 + \sqrt{0.51}}{(25.2 - 14.7)^2} \approx 0.0775$

   Use your calculator to find the difference between the exact value and Aleksei's estimate. Say why this difference is large. [2]

6. At a wedding reception, guests ate from a chocolate fountain. Everyone ate some chocolate and none was left over! The chocolate in the fountain had a mass of 3.56 kg and each person ate, on average, 88 g of chocolate. Estimate the number of people at the reception. [2]

7. The diagram shows the plan of a garden in the shape of a rectangle 19.5 m by 10.5 m. Four identical triangular flowerbeds have been cut from the corners, each with shorter edges 5.6 m and 2.8 m. There is a square fishpond, side length 6.6 m, in the centre of the garden. The rest of the garden is grass.
   Estimate the percentage of the garden that is taken up by the grass. Give your answers to the nearest 1%. [5]

   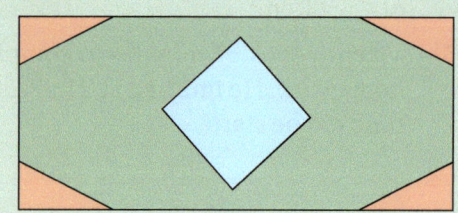

8. Put brackets into each of the expressions to make them correct.

   a. 4.6 + 4.1 + 1.2 × 2.6 = 18.38 [1]
   b. 14.9 − 6.8 ÷ 3.7 − 1.2 = 12.18 [1]
   c. 3.4 × 1.6 + 5.9 − 2.8 = 8.54 [1]
   d. 2.6 + 7.56 ÷ 1.8 − 0.72 = 6.08 [1]
   e. 12.3 − 5.2 × 1.6 + 3.4 × 2 = 14.76 [1]

9 | tonne | kilogram | gram | kilometre | metre | centimetre | millimetre | litre | centilitre | millilitre |

Estimate the following values using an appropriate unit from the list in each case.
- **a** The capacity of a paddling pool. [1]
- **b** The mass of a full suitcase going on holiday. [1]
- **c** The mass of an eyeliner pencil. [1]
- **d** The distance from your classroom to the headteacher's room. [1]
- **e** The length of a rugby pitch. [1]
- **f** The mass of an aeroplane. [1]
- **g** The amount in a dose of cough mixture. [1]
- **h** The thickness of a DVD case. [1]
- **i** The amount of coffee consumed in a café during a day. [1]
- **j** The distance run in a marathon race. [1]
- **k** The length of a hairbrush. [1]
- **l** The amount of tea in one cup. [1]

10
- **a** Hamil cycles 165 km in $2\frac{1}{2}$ hours. What is his average speed? [2]
- **b** Lewis' racing car travels 9500 metres in 3.8 minutes. What is its average speed in km/h? [2]
- **c** A train travels for $2\frac{3}{4}$ hours at an average speed of 175 km/h. How far does it go? [2]
- **d** An aeroplane travels for $7\frac{1}{3}$ hours at an average speed of 465 mph. How many miles does it fly? [2]
- **e** A snail travels 14 m at 0.7 m/h. How long does his journey take? [2]

11
- **a** Water has a density of 1 g/cm³. A cricket ball has a mass of 158 grams and volume of 195 cm³. Would it float in water? Explain your answer. [3]
- **b** A gold bracelet has a mass of 44 g and a density of 19.3 g/cm³. What is its volume? [2]
- **c** A concrete girder measuring 25 m by 15 m by 6 m has a density of 2.4 g/cm³. What is the mass of the bar in tonnes? [3]

12 Write down the upper and lower limits of the following measurements:
- **a** The shortest distance from the moon to the earth is 221 460 miles to 5 sf. [1]
- **b** Farakh's leg is 90 cm long to the nearest 10 cm. [1]
- **c** A box of chocolates weighs 455 g to the nearest 5 g. [1]
- **d** In July 2013 Mo Farah beat Steve Cram's 28 year-old British 1500 m record with a time of 3 minutes 28.8 s to the nearest 0.1 s. [1]
- **e** The number of tea bags in a box is 240 to the nearest 4 bags. [1]
- **f** A lorry weighs 28 tonnes to the nearest tonne. [1]
- **g** A walking stick has a length of 586 mm to the nearest mm. [1]

13 A rectangular tray measures 45 cm by 24 cm by 0.3 cm correct to the nearest cm. Its mass is 76 g correct to the nearest gram.
- **a** Calculate the biggest and smallest possible values of its volume. [3]
- **b** Calculate the biggest and smallest possible values of its density. [3]

10 Equations and inequalities

Introduction
Maths is not all about whether something is equal to something else. Sometimes it can be useful to know when a quantity needs to be less than or more than a particular value. An example could be a recipe for a soft drink, where the proportion of cane sugar needs to be within certain bounds to conform to standards and to customers' appetites. Some companies are very secretive about the formula for their particular branded soft drink. All you are allowed to know is that the proportions of ingredients lie within a certain range.

Inequalities are statements that allow us to work mathematically with this kind of restriction.

What's the point?
In the food and drink industries it is important to be able to work within restrictions on particular ingredients, which are often imposed by health legislation. In the pharmaceutical industry, it can be critical as too much of one particular chemical can have harmful effects.

Objectives
By the end of this chapter, you will have learned how to …
- Solve linear equations including when the unknown appears on both sides.
- Solve quadratic equations using factorisation, completing the square and the quadratic formula.
- Solve a pair of linear or linear plus quadratic simultaneous equations.
- Use iterative processes to find approximate solutions to equations.
- Solve inequalities and display your solution on a number line or a graph.

Check in

1. Evaluate these, expressing your answers in their simplest form.

   a $\frac{7}{30} + \frac{1}{6}$ b $\frac{1}{5} + \frac{2}{9}$ c $\frac{3}{4} - \frac{1}{4}$ d $2\frac{1}{6} + 3\frac{2}{5}$

   e $\frac{5}{6} \times \frac{7}{10}$ f $\frac{5}{8} \div \frac{1}{5}$

2. What value must the □ represent in each case?

   a □ + 3 = 12 b □ × 5 = 20 c □ − 11 = 19 d □² = 100

3. Rewrite these expressions, by expanding any brackets and collecting like terms.

   a $5x + 7y - 4x + 9y$ b $11x + 9x^2 - 8x$ c $7p - 9$

   d $3(2x - 1) + 6(3x - 4)$ e $2(4y - 8) - 3(y - 9)$ f $x(3x - 2y)$

   g $(x + 9)(x - 7)$ h $(2w - 8)(3w - 4)$ i $(p - q)^2$

4. Factorise these expressions.

   a $x^2 + 5x + 6$ b $x^2 - 2x - 24$ c $x^2 - 6x + 9$ d $x^2 - 100$

Chapter investigation

This L shape is drawn on a 10 × 10 grid numbered from 1 to 100. It has 5 numbers inside it. We can call it L_{35} because the largest number inside it is 35.

What is the sum total of the numbers inside L_{35}?

Find a connection between n and L_n. Can you write a formula to describe the relationship? What if the L shape is extended in each direction?

1	2	3	4	5	6	7	8	9	10
11	12	13	14	15	16	17	18	19	20
21	22	23	24	25	26	27	28	29	30
31	32	33	34	35	36	37	38	39	40
41	42	43	44	45	46	47	48	49	50
51	52	53	54	55	56	57	58	59	60
61	62	63	64	65	66	67	68	69	70
71	72	73	74	75	76	77	78	79	80
81	82	83	84	85	86	87	88	89	90
91	92	93	94	95	96	97	98	99	100

SKILLS

10.1 Solving linear equations

- A **linear** equation is an equation of degree one.

$2x + 1 = 83 \qquad 3(x - 2) = 23$

A linear equation can be solved by algebraic methods. To do this you *must* do the same to the expressions on *both* sides of the = sign in order to keep the equation balanced.

$\dfrac{x}{2} + 3 = 31 \qquad 3x - 2 = 5x + 4$

EXAMPLE (p.30)

Solve these equations.

a $3a + 5 = 23$ **b** $4(b - 5) = -24$ **c** $4c - 5 = 1$ **d** $\dfrac{x}{3} - 2 = 7$

a
$3a + 5 = 23 \quad -5$
$3a = 18 \quad \div 3$
$a = 6$

b Expand the bracket.
$4b - 20 = -24 \quad +20$
$4b = -4 \quad \div 4$
$b = -1$

c
$4c - 5 = 1 \quad +5$
$4c = 6 \quad \div 4$
$c = \dfrac{6}{4} = \dfrac{3}{2}$

d
$\dfrac{x}{3} - 2 = 7 \quad +2$
$\dfrac{x}{3} = 9 \quad \times 3$
$x = 27$

For part **a**

a	a	a	5
	23		

→

a	a	a
	18	

→

a
6

In **b** you could first divide by 4.
$4(b - 5) = -24$
$b - 5 = -6$
$b = -1$

To solve a linear equation you may first have to rearrange it so that the unknown is only on one side.

EXAMPLE

Solve these equations.

a $4x + 3 = 2x + 12$ **b** $10 - 5x = 6 + 3x$ **c** $\dfrac{x - 3}{2} = \dfrac{2x - 1}{3}$

a
$4x + 3 = 2x + 12 \quad -2x$
$2x + 3 = 12 \quad -3$
$2x = 9 \quad \div 2$
$x = \dfrac{9}{2}$

b
$10 - 5x = 6 + 3x \quad +5x$
$10 = 6 + 8x \quad -6$
$4 = 8x \quad \div 8$
$x = \dfrac{4}{8} = \dfrac{1}{2}$

c
$\dfrac{x - 3}{2} = \dfrac{2x - 1}{3} \quad \times 2 \times 3$
$3(x - 3) = 2(2x - 1) \quad \text{Expand}$
$3x - 9 = 4x - 2 \quad -3x$
$-9 = x - 2 \quad +2$
$x = -7$

Linear equations can also be solved approximately using graphs.

EXAMPLE

Use this graph of $y = 3x - 2$ to solve these equations.

a $3x - 2 = 7$
b $3x - 2 = 12$
c $3x - 2 = -2$

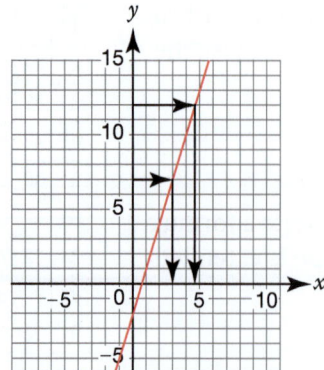

Read off where the graph takes the given y value.

a $x = 3 \qquad y = 7$
b $x \approx 4.5 \qquad y = 12$
c $x = 0 \qquad y = -2$

Algebra   Equations and inequalities

Fluency

Exercise 10.1S

1 Solve these equations.
- **a** $4x + 7 = 71$
- **b** $5x + 13 = 53$
- **c** $8x - 3 = 37$
- **d** $7x - 15 = 20$
- **e** $20 = 2x + 6$
- **f** $25 = 4x - 7$

2 Solve these equations.
- **a** $4(m + 7) = 44$
- **b** $5(m + 6) = 45$
- **c** $8(p - 3) = 48$
- **d** $7(p - 15) = 63$
- **e** $20 = 2(f + 6)$
- **f** $36 = 4(h - 7)$
- **g** $-10 = 2(w + 6)$
- **h** $-12 = 4(x - 7)$

3 Solve these equations.
- **a** $\dfrac{x}{2} + 3 = 13$
- **b** $\dfrac{m}{3} + 6 = 21$
- **c** $\dfrac{p}{4} - 3 = 12$
- **d** $\dfrac{2(m + 4)}{5} = 8$

4 Using this graph of $y = -5 - 2x$ to solve these equations.
- **a** $-5 - 2x = 12$
- **b** $-5 - 2x = -4$
- **c** $-5 - 2x = 4$
- **d** $11 = -5 - 2x$

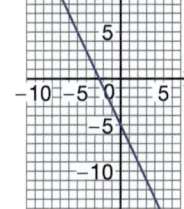

5 a Draw the graph of $y = 2x + 3$, using axes $0 \leq x \leq 10$.
  b Using the graph, solve these equations.
  - **i** $2x + 3 = 13$
  - **ii** $2x + 3 = 12$
  - **iii** $2x + 3 = 19$
  - **iv** $7 = 2x + 3$

6 a Draw the graph of $y = \dfrac{x}{3} - 1$ using axes $-10 \leq x \leq 10$.
  b Using the graph, solve these equations.
  - **i** $\dfrac{x}{3} - 1 = 2$
  - **ii** $\dfrac{x}{3} - 1 = -4$
  - **iii** $-1 = \dfrac{x}{3} - 1$
  - **iv** $\dfrac{x}{3} - 1 = 0$

7 Solve these equations.
- **a** $5x + 6 = 2x + 27$
- **b** $3y + 2 = y + 12$
- **c** $4p + 10 = p + 4$
- **d** $5f + 6 = 2f - 3$
- **e** $2x + 14 = 4x + 4$
- **f** $8 + x = 15 + 2x$
- **g** $2b + 4 = 6b + 12$
- **h** $3p = 3 + 9p$

8 Solve these equations.
- **a** $11 - 3x = 2x - 4$
- **b** $15 - 4x = x + 10$
- **c** $6 - 2x = 8 - x$
- **d** $12 - 2x = 18 - 5x$
- **e** $10 - 4p = 2p - 8$
- **f** $20 - 4h = 10 - 2h$
- **g** $2(3 - 8x) = 3(3 - 6x)$
- **h** $6 - (3y - 9) = -2(5 - y)$

9 Solve these equations.
- **a** $\dfrac{x + 1}{3} = \dfrac{x - 1}{4}$
- **b** $\dfrac{2y - 1}{3} = \dfrac{y}{2}$
- **c** $\dfrac{p + 7}{3} = \dfrac{2p - 4}{5}$
- **d** $\dfrac{5q - 9}{2} = \dfrac{3 - 2q}{6}$
- **e** $\dfrac{3s + 1}{5} = \dfrac{2s}{3}$
- **f** $\dfrac{t}{2} + \dfrac{t}{4} = 7$
- **g** $\dfrac{4u}{3} - \dfrac{2u}{5} = 9$
- **h** $\dfrac{3v}{4} + \dfrac{7v}{6} = \dfrac{1}{3}$
- **i** $\dfrac{5}{w + 5} = \dfrac{15}{w + 7}$
- **j** $\dfrac{3}{z - 1} = \dfrac{9}{2z - 1}$

*10 Solve these equations.
- **a** $(x + 3)(x + 4) = (x + 7)(x - 2)$
- **b** $(y - 7)^2 = (y + 5)^2$

*11 Solve these equations.
- **a** $\dfrac{5}{x + 1} - \dfrac{2}{x + 2} = \dfrac{3}{x + 3}$
- **b** $\dfrac{4}{x - 1} - \dfrac{2}{x - 3} = \dfrac{2}{x + 3}$
- **c** $\dfrac{3}{x + 1} + \dfrac{1}{x - 3} = \dfrac{4}{x - 1}$
- **d** $\dfrac{4}{x + 1} - \dfrac{3}{x + 2} = \dfrac{2}{2x - 1}$

APPLICATIONS

10.1 Solving linear equations

RECAP
- To solve an equation do the *same* operation to *both* sides.
- Work systematically: remove fractions by multiplying throughout by their denominators, expand brackets, collect like terms, rearrange to get the unknown on one side.

$$\frac{4-x}{3} = \frac{3-x}{2}$$
$$2(4-x) = 3(3-x)$$
$$8 - 2x = 9 - 3x$$
$$8 + x = 9$$
$$x = 1$$

HOW TO

To solve problems involving finding an unknown
① Read the question and decide what mathematics you need to form an equation.
② Simplify the equation by
  • clearing fractions   • expanding brackets   • collecting like terms.
③ Solve the equation and give your answer in the context of the question.

EXAMPLE

The square and the rectangle have the same area.
Find the value of x.

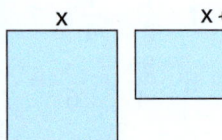

All lengths are given in metres.

Area of square = Area of rectangle ①
$x^2 = (x+5)(x-2)$ ② Remember brackets first (FOIL).
$x^2 = x^2 + 5x - 2x - 10$ Collect like terms.
$x^2 = x^2 + 3x - 10$ Subtract x^2 from each side.
$0 = 3x - 10$ Add 10 to both sides.
$10 = 3x$ Divide by 3.
$x = 3\frac{1}{3}$ m ③

EXAMPLE

I divide 15 by a number and get $\frac{3}{4}$.
What is the number?

Let x = the number $\frac{15}{x} = \frac{3}{4}$ ①
$15 \times 4 = 3 \times x$ ② **Cross multiply.**
$60 = 3x$ $\div 3$
$x = 20$ ③

EXAMPLE

Work out the value of these three expressions if the value of the expressions is in ascending order and their mean is equal to their median.

$5(x-13) \quad x+1 \quad 2(x-4)$

$\frac{1}{3}[5(x-13) + x + 1 + 2(x-4)] = x + 1$ ① Mean = median.
$5(x-13) + x + 1 + 2(x-4) = 3(x+1)$ ② Multiply both sides by 3.
$5x - 65 + x + 1 + 2x - 8 = 3x + 3$ Expand brackets.
$8x - 72 = 3x + 3$ Collect like terms.
$5x - 72 = 3$ $-3x$
$5x = 75$ $+72$
$x = 15$

Substitute $x = 15$ into the three expressions.
$5(x-13) = 5(15-13) = 10 \qquad x + 1 = 15 + 1 = 16 \qquad 2(x-4) = 2(15-4) = 22$
The three numbers are 10, 16 and 22.   ③

Algebra   Equations and inequalities

Reasoning and problem solving

Exercise 10.1A

All distances in this exercise are given in metres.

1. The perimeter of this shape is 30 m. Work out the length of each side.

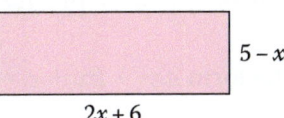

2. The diagram shows an isosceles triangle. The equal angles are 10° less than double the third angle.

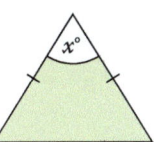

Work out the size of each angle.

3. The angles in a triangle are $x°$, $x + 20°$ and $x + 40°$.

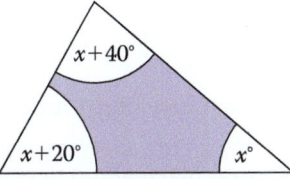

Work out the angles of the triangle.

4. The perimeters of these two shapes are equal.

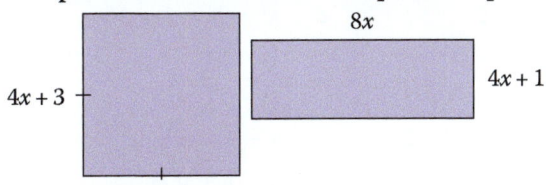

What are the dimensions of each shape?

5. The sides of an isosceles triangle are $3x + 4$, $5x − 8$ and $x + 10$.
What are the possible lengths of the sides of the triangle?

6. Work out these numbers.

    a. I think of a number, divide it by 16 and I get 10.

    b. I think of a number, add 4, divide it by 12 and I get 7.

    c. I think of a number, take 3 and divide it by 11. This gives me the same answer as when I take the same number and divide it by 8.

7. The mean values of each set of expressions are equal. Find x and the numbers in each set.

    Set 1
    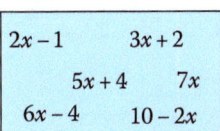

    Set 2

    $3x − 7$ $5x + 8$

    $13 − x$

    $2(3x + 1)$ $12 + 4x$

8. The areas of the square and rectangle are equal.

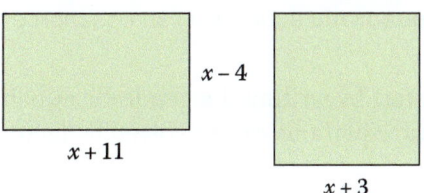

What are the dimensions of each shape?

9. Work out the length of this rectangle.

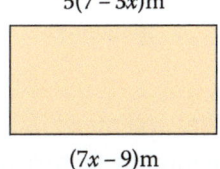

10. In 10 years' time, my age will be double what it was 11 years ago. How old am I?

11. Work out the length of a square of side $2x + 3$ with an area equal to the area of a rectangle measuring $x + 6$ by $4x − 2$.

12. A number is doubled and divided by 5. Three more than the number is divided by 8. Both calculations give the same answer. What is the number?

13. The sum of one-fifth of Lucy's age and one-seventh of her age is 12.
Use algebra to work out Lucy's age.

14. These trapeziums have equal areas. What are the lengths of the parallel sides?

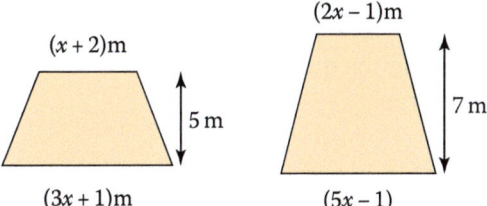

10.2 Quadratic equations

> A **quadratic** equation is an equation of degree two. $x^2 = 12$, $3x^2 + 8x + 12 = 0$

Always start by writing the quadratic equation in the form $ax^2 + bx + c = 0$.
Quadratic equations can be solved by factorising.

> If $p \times q = 0$, then $p = 0$ and/or $q = 0$.

EXAMPLE

Solve these equations. Give the solutions to 1 decimal place when appropriate.

a $x^2 = 30$ **b** $x^2 + 3x = 0$

a $x = \pm\sqrt{30}$ The inverse of 'square' is 'square root'.
$x = 5.5$ (1 dp) or
$x = -5.5$ (1 dp)

b $x(x + 3) = 0$ Factorise.
$x = 0$ or $x + 3 = 0$
$x = 0$ or $x = -3$

If you cannot solve the equation by factorising then you can use **completing the square**.

EXAMPLE

Solve the equations, giving the solutions to 1 decimal place when appropriate.

a $x^2 + 7x + 10 = 0$ **b** $2x^2 - 3x = 2$ **c** $x^2 - 4x - 2 = 0$

a $x^2 + 7x + 10 = 0$
$(x + 5)(x + 2) = 0$
$x + 5 = 0$ or $x + 2 = 0$
$x = -5$ or $x = -2$

b $2x^2 - 3x - 2 = 0$
$(2x + 1)(x - 2) = 0$
$2x + 1 = 0$ or $x - 2 = 0$
$x = -\tfrac{1}{2}$ or $x = 2$

c $x^2 - 4x - 2 = 0$
$(x - 2)^2 - 4 - 2 = 0$
$(x - 2)^2 = 6$
$x - 2 = \pm\sqrt{6}$
$x = 2 \pm \sqrt{6}$
$x = 4.4$ or $x = -0.4$ (1 dp)

You can always use the **quadratic formula**.

> If $ax^2 + bx + c = 0$ then $x = \dfrac{-b \pm \sqrt{b^2 - 4ac}}{2a}$

EXAMPLE

Solve the equation $5x^2 + 3x - 10 = 0$ using the quadratic formula.

$a = 5, b = 3, c = -10$

$x = \dfrac{-3 \pm \sqrt{3^2 - 4 \times 5 \times -10}}{2 \times 5} = \dfrac{-3 \pm \sqrt{9 - (-200)}}{10} = \dfrac{-3 \pm \sqrt{209}}{10}$

$x = -1.746$ or $x = 1.146$ (3 dp)

Approximate solutions to quadratic equations can also be found using graphs.

EXAMPLE

Use this graph of $y = x^2 + 5x$ to find approximate solutions for the equation $x^2 + 5x = 2$.

Find the intersection points of $y = 2$ and $y = x^2 + 5x$.

$x \approx 0.4$ and $x \approx -5.4$

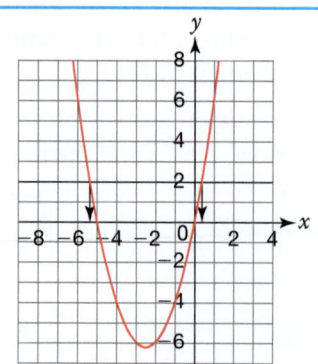

Algebra   Equations and inequalities

Fluency

Exercise 10.2S

When appropriate give your answers to 2 decimal places.

1. Solve these quadratic equations by factorising.
   a. $x^2 + 5x + 6 = 0$
   b. $x^2 + 7x + 12 = 0$
   c. $x^2 + 8x + 15 = 0$
   d. $x^2 + 8x + 16 = 0$
   e. $0 = x^2 + 18x + 17$
   f. $0 = x^2 + 15x + 26$
   g. $x^2 + x - 6 = 0$
   h. $x^2 + x - 12 = 0$
   i. $0 = x^2 + 3x - 18$
   j. $0 = x^2 + 9x - 22$
   k. $x^2 - x - 6 = 0$
   l. $x^2 - x - 12 = 0$
   m. $x^2 - 2x - 15 = 0$
   n. $0 = x^2 - 6x - 16$
   o. $0 = x^2 - 13x - 30$
   p. $0 = x^2 - 3x - 28$
   q. $0 = x^2 - 10x + 24$
   r. $0 = x^2 - 12x + 35$
   s. $2x^2 + 12x + 10 = 0$
   t. $3x^2 + 14x + 8 = 0$
   u. $3x^2 - 18x + 27 = 0$
   v. $4x^2 - 11x - 3 = 0$

2. Solve these quadratic equations by completing the square.
   a. $x^2 + 10x + 25 = 0$
   b. $x^2 + 6x + 9 = 0$
   c. $x^2 - 8x + 16 = 0$
   d. $x^2 - 16x + 64 = 0$
   e. $x^2 + 18x + 12 = 0$
   f. $x^2 + 10x + 18 = 0$
   g. $x^2 + 3x - 6 = 0$
   h. $x^2 + 7x - 12 = 0$
   i. $0 = x^2 + 12x - 18$
   j. $0 = x^2 + 11x - 8$
   k. $2x^2 - 10x - 6 = 0$
   l. $3x^2 - 6x - 12 = 0$

3. Li is solving the equation $2x^2 + 6x - 3 = 0$ using the method of completing the square.

   Her first step is
   $2x^2 + 6x - 3 = (2x + 3)^2 - 9 - 3$

   Li has made a mistake, what is it?

4. Solve these quadratic equations.
   a. $x^2 - x = 20$
   b. $x^2 + 12x = 13$
   c. $8 = x^2 + 2x$
   d. $x^2 + 4x = 21$
   e. $x^2 + 45 = 14x$
   f. $13x = x^2 + 30$
   g. $2x^2 = x + 6$
   h. $3x^2 - 6 = 7x$
   i. $x(x + 10) = -21$
   j. $(x + 2)^2 = 9x$

5. Solve these equations by using the quadratic formula.
   a. $x^2 + 9x - 25 = 0$
   b. $x^2 + 3x - 11 = 0$
   c. $x^2 - 8x + 6 = 0$
   d. $x^2 - 7x + 3 = 0$
   e. $x^2 + 20x = 45$
   f. $x^2 + 13x = 2$
   g. $3x^2 + 3x - 2 = 0$
   h. $5x^2 + 7x - 12 = 0$
   i. $0 = 6x^2 + 12x - 15$
   j. $0 = 4x^2 + 11x - 18$

6. Explain why $x^2 + 4x + 20 = 0$ does not have a solution.

7. Use this graph of $y = x^2 + 3x$ to find approximate solutions, to one decimal place when appropriate, for these equations.

   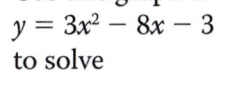

   a. $x^2 + 3x = 0$
   b. $x^2 + 3x = 2$
   c. $x^2 + 3x = 4$
   d. $x^2 + 3x = -2$

8. Use this graph of $y = 3x^2 - 8x - 3$ to solve

   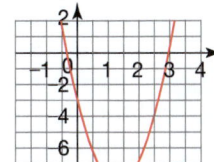

   a. $3x^2 - 8x - 3 = 1$
   b. $3x^2 - 8x = 3$
   c. Explain why $3x^2 - 8x - 3 = -9$ does not have a solution.

*9. Solve $x^3 + 3x^2 - 6x = 0$.

10. a. Show that $x^2 - 30x + 195 = 0$
    b. Solve the equation to find the length of the longest side.

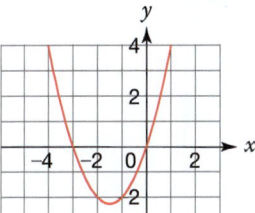

APPLICATIONS

10.2 Quadratic equations

RECAP
- A quadratic equation has the form $ax^2 + bx + c = 0$.
- You can solve a quadratic equation by factorising, completing the square or using the quadratic formula.
- Factorising usually gives integer or fraction answers. Completing the square and the formula usually give answers involving surds.

$x^2 - 6x + 5 = 0$
- $(x - 1)(x - 5) = 0$
  $\Rightarrow x - 1 = 0, x = 1$
  or $x - 5 = 0,$
  $x = 5$
- $(x - 3)^2 - 9 + 5 = 0$
  $(x - 3)^2 = 4$
  $x - 3 = \pm 2 \Rightarrow x = 3 \pm 2$
  $= 1$ or 5
- $x = \dfrac{6 \pm \sqrt{6^2 - 4 \times 1 \times 5}}{2 \times 1}$
  $= \dfrac{6 \pm \sqrt{16}}{2}$
  $= 3 \pm 2 = 1$ or 5

HOW TO

To solve a problem involving a quadratic equation

① Use the information in the question to form a quadratic equation.

② Rearrange the quadratic so that it equals zero.

③ Solve the quadratic by factorising, completing the square or using the quadratic formula.

④ Check that your answers make sense in context. You may need to reject one solution.

EXAMPLE

Two numbers have a product of 105 and a difference of 8.

Find the two numbers.

① Let the two numbers be x and $x - 8$.
$x(x - 8) = 105$

② $x^2 - 8x - 105 = 0$

③ $(x + 7)(x - 15) = 0$
$x + 7 = 0$ or $x - 15 = 0$
$x = -7$ and $x - 8 = -15$ or $x = 15$ and $x - 8 = 7$.

④ Two answers for x lead to two answers for $x - 8$.
The two numbers are -7 and -15 or 7 and 15.

EXAMPLE

If the area of the trapezium is 400 cm^2, show that $x^2 + 10x = 400$ and find the value of x correct to 3 dp.

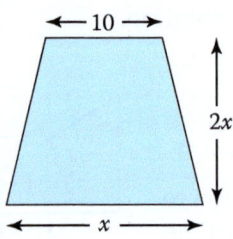

① Area of trapezium $= \dfrac{(a + b)}{2} h$
$400 = \dfrac{1}{2}(x + 10) \times 2x$
$400 = x(x + 10)$
$x^2 + 10x = 400$ (as required)

② $x^2 + 10x - 400 = 0$

③ Use the formula $x = \dfrac{-b \pm \sqrt{b^2 - 4ac}}{2a}$

$x = \dfrac{-10 \pm \sqrt{10^2 - 4 \times 1 \times -400}}{2}$

$= 15.61552...$ or $-25.61552...$

④ Since x is a length, it must be positive,
$x = 15.616$ (3 dp)

Algebra   Equations and inequalities

Reasoning and problem solving

Exercise 10.2A

1. A rectangle has a length that is 7 cm more than its width, w. The area of the rectangle is 60 cm².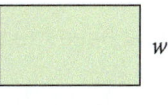

   a Show that $w^2 + 7w - 60 = 0$.

   b Find the dimensions of the rectangle.

2. Prove that $p^2 + 6p + 9$ can never be negative.

3. Use the quadratic formula to solve $2x^2 + 7x + 9 = 0$.
   What do you notice?

4. Give answers to 2 decimal places where appropriate.

   a Two numbers which differ by 4 have a product of 117. Find the numbers.

   b The length of a rectangle exceeds its width by 4 cm.
   The area of the rectangle is 357 cm².
   Find the dimensions of the rectangle.

   c The square of three less than a number is ten. What is the number?

   d Three times the reciprocal of a number is one less than ten times the number. What is the number?

   e The diagonal of a rectangle is 17 mm. The length is 7 mm more than the width. Find the dimensions of the rectangle.

   f If this square and rectangle have equal areas, find the side length of the square.

   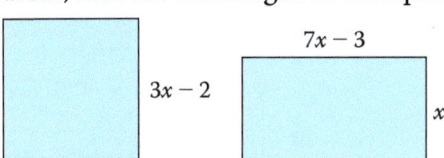

5. a The height of a closed cylinder is 5 cm and its surface area is 100 cm².
   Given that the radius is r, show that
   $\pi r^2 + 5\pi r - 50 = 0$

   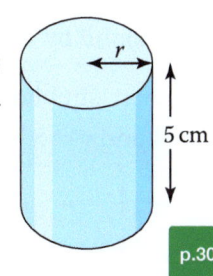

   b Hence, find the diameter of the base of the cylinder.

6. a The perimeter of this rectangle is 46 cm and its diagonal is 17 cm.

   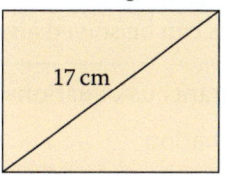

   Find the dimensions of the rectangle.

   b The diagonal of this rectangle is 13 cm. What is the perimeter of the rectangle?

   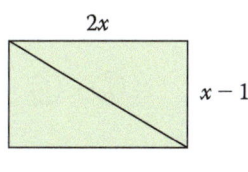

   c The area of this hexagon is 25 m². Find the perimeter of the hexagon.

   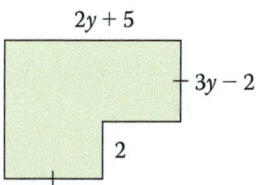

   d The surface area of the sphere is equal to the curved surface area of the cylinder. Which has the larger volume?

   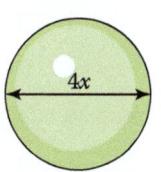
   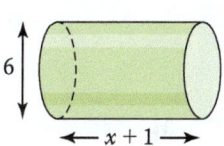

7. Solve these equations.

   a $10x + 7 = \dfrac{3}{x}$ b $10x = 1 + \dfrac{3}{x}$

   c $\dfrac{x^2 + 3}{4} + \dfrac{(2x - 1)}{5} = 1$

   d $\dfrac{3}{(x + 1)} + \dfrac{4}{(2x - 1)} = 2$

   e $x^4 - 13x^2 + 36 = 0$

*8. By completing the square on $ax^2 + bx + c = 0$, prove the quadratic equation formula.
   $$x = \dfrac{-b \pm \sqrt{b^2 - 4ac}}{2a}$$

SKILLS

10.3 Simultaneous equations

- **Simultaneous** equations are a set of equations that have the same solution.

Simultaneous equations can be solved algebraically using the methods of **elimination** or **substitution**.

EXAMPLE

Solve these simultaneous equations $x + 2y = 17$ and $3x + 2y = 19$

a using elimination b using substitution.

For both methods, it's useful to label the equations for reference purposes.

a
$x + 2y = 17$ ①
$3x + 2y = 19$ ②
② − ① $\quad 2x = 2 \quad$ Eliminate y.
$x = 1$
$1 + 2y = 17 \quad x = 1$ in ①
$2y = 16$
$y = 8$
$x = 1$ and $y = 8$

b
$x + 2y = 17$ ①
$3x + 2y = 19$ ②
From ① $\quad x = 17 - 2y$ ③
③ in ② $\quad 3(17 - 2y) + 2y = 19 \quad$ Simplify.
$51 - 6y + 2y = 19$
$51 - 4y = 19$
$32 = 4y \quad 51 - 19 = 32$
$y = 8$
$x = 17 - (2 \times 8) = 1 \quad y = 8$ in ③

Elimination is good for simultaneous linear equations.
Substitution is good for simultaneous linear and non-linear equations.

EXAMPLE

Solve these simultaneous equations.

a $\quad 2x - y = 9$ ①
$\quad 5x - y = 21$ ②

b $\quad 3x + 2y = 20$ ①
$\quad 2x - 3y = 22$ ②

c $\quad x + 3y = 2$ ①
$\quad y = x^2$ ②

a Eliminate y.
② − ① $\quad 3x = 12$
$x = 4$
$2 \times 4 - y = 9 \quad x = 4$
$8 - y = 9 \quad$ in ①
$8 = 9 + y$
$y = -1$

b Eliminate y.
$3 \times$ ① $\quad 9x + 6y = 60$ ③
$2 \times$ ② $\quad 4x - 6y = 44$ ④
③ + ④ $\quad 13x = 104$
$x = 8$
$24 + 2y = 20 \quad x = 8$ in ①
$2y = -4$
$y = -2$

c Substitute for x.
$x = 2 - 3y \quad$ Rearrange ①
$y = (2 - 3y)^2 \quad$ Sub in ②
$y = 4 - 12y + 9y^2$
$9y^2 - 13y + 4 = 0$
$(9y - 4)(y - 1) = 0$
$y = 1$ and $x = -1 \quad$ Sub in ①
or $y = \frac{4}{9}$ and $x = \frac{2}{3}$

Approximate solutions to simultaneous equations can be found by drawing graphs.

EXAMPLE

Use the graphs to find an approximate solution for these simultaneous equations.

a $\quad y = 2x - 7$ and $2y + x = 3$

b $\quad x + 2y = 4$ and $y = x^2 - 2$

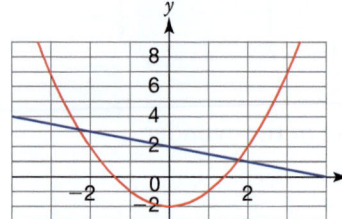

The solution is the **intersection** of the graphs.

a $\quad x \approx 3.3, y \approx -0.3$
b $\quad x \approx -2.3, y \approx 3.1$
$\quad x \approx 1.8, y \approx 1.1$

Fluency

Exercise 10.3S

1 a Find the pairs of solutions for these equations.
   i $2x + y = 11$ **ii** $6x + y = 27$

b Solve these simultaneous equations.
   $2x + y = 11$ $6x + y = 27$

2 Solve these simultaneous equations.

a $4x + 4y = 16$ **b** $3x + 2y = 19$
   $x + 4y = 13$ $x + 2y = 9$

c $4m + 4n = 24$ **d** $3x + 2y = 16$
   $m + 2n = 8$ $2x + y = 9$

e $5x + 3y = 17$ **f** $4e + 3f = 13$
   $x + 6y = -2$ $3e + 5f = 18$

g $2m + 3n = 14$ **h** $2y + 3x = 5$
   $m = 14 - 5n$ $y = 7 - 3x$

3 Solve these simultaneous equations.

a $4x - 4y = 20$ **b** $x - 2y = 11$
   $x - 4y = 2$ $3x - 2y = 25$

c $5p - 3q = 27$ **d** $a - 2b = 5$
   $5p - q = 29$ $4a - 5b = 23$

4 Solve these simultaneous equations.

a $4x + 2y = 26$ **b** $x + 3y = 13$
   $x - 2y = 4$ $3x - 3y = 15$

c $5p + 3q = 7$ **d** $3a + 2b = 9$
   $2p - q = 11$ $4a - 5b = 35$

5 Use the graphs to find approximate solutions for these simultaneous equations.
   $x + 3y = 4$
   $y = 2x + 2$

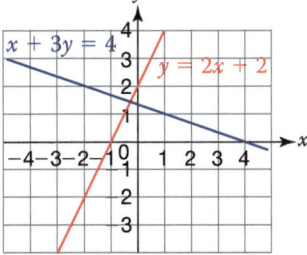

6 By drawing graphs, find approximate solutions for these simultaneous equations.

a $x + 3y = 4$ **b** $x + 2y = 6$
   $y = 2x - 2$ $y = 3x + 1$

c $y = 3x - 1$ **d** $y = 3x - 7$
   $y = x - 2$ $x = y + 2$

*7 Solve these simultaneous equations.

a $\dfrac{x}{3} - \dfrac{y}{4} = \dfrac{3}{2}$ **b** $\dfrac{a}{2} + 3b = 1$
   $2x + y = 14$ $5a - 7b = 47$

c $p - \dfrac{2q}{3} = \dfrac{26}{3}$ **d** $\dfrac{5s}{6} + \dfrac{t}{4} = 8$
   $\dfrac{p}{4} + 3q - 1 = 0$ $\dfrac{2s}{5} + \dfrac{t}{10} = 4$

8 Solve these simultaneous equations, giving your answer to 2 decimal places where appropriate.

a $x + y = 8$ **b** $x + y = 9$
   $y = x^2$ $y = x^2 - 2$

c $x + 2y = 7$ **d** $3x + 2y = 8$
   $y = 2x^2$ $y = 2x^2$

e $x + 2y = 3$ **f** $x + 2y = 4$
   $x^2 + y^2 = 3$ $x^2 + y^2 = 4$

g $x^2 - y^2 = 2$ **h** $xy = 3$
   $x + 2y = 10$ $x + 2y = 8$

i $y = x^2 - 3x + 7$
   $y - 5x + 8 = 0$

j $x^2 + 3xy = 10$
   $x = 2y$

9 Explain why these simultaneous equations do not have a solution.
   $x - y = 8$ $y = x^2 - 6$

10 a These simultaneous equations have one solution, find it.
   $x^2 + y^2 = 5$ $y = 2x + 5$

b Explain why there is only one solution using graphs.

*11 How many solutions do these simultaneous equations have? Give your reason.

a $x - y = 2$ **b** $x + y = 5$
   $y = x^3$ $y = \dfrac{1}{x}$

12 The sum of the ages of Bob's grandparents is 135 years. The difference between their ages is 11.

What are the possible ages of Bob's grandparents?

APPLICATIONS

10.3 Simultaneous equations

RECAP

- You can solve simultaneous equations by
  - Elimination – good for pairs of linear equations
  - Substitution – good for simultaneous linear and non-linear equations
  - Drawing graphs – any crossing points give approximate solutions.
- Two linear equations can have no, one or infinitely many pairs of solutions.
- A linear and a quadratic equation can have no, one or two pairs of solutions.

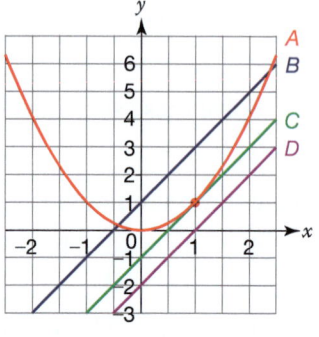

A $y = x^2$ B $y = 2x + 1$
C $y = 2x - 1$ D $y = 2x - 2$
A & B have 2 solutions
A & C have 1 solution
A & D have 0 solutions

HOW TO

To solve a simultaneous equation problem

① Use the information in the question to form a pair of equations.

② Solve the equations using elimination, substitution or by drawing a graph.

③ Give your answers and check that they make sense.

EXAMPLE

The perimeter of this isosceles triangle is 50 cm. Find the length of its base.

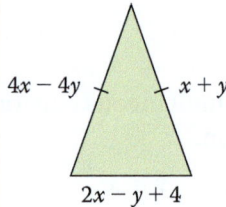

① There are two unknowns, so you need two equations to find them.
Isosceles: $4x - 4y = x + y \Rightarrow 3x - 5y = 0$ (1)
Perimeter: $7x - 4y + 4 = 50 \Rightarrow 7x - 4y = 46$ (2)

② $7 \times$ (1) $21x - 35y = 0$ (3)
$3 \times$ (2) $21x - 12y = 138$ (4)
(4) − (3) $23y = 138$
$y = 6$
Substitute in (1) $3x - 30 = 0$
$3x = 30$
$x = 10$

③ The base is $(2 \times 10) - 6 + 4 = 18$ cm.

You might remember same signs so subtract.

EXAMPLE

Solve the equation $5x - x^2 = 2x - 1$ graphically.

① Split the problem into two simultaneous equations.
② Plot the graphs of $y = 5x - x^2$ and $y = 2x - 1$.

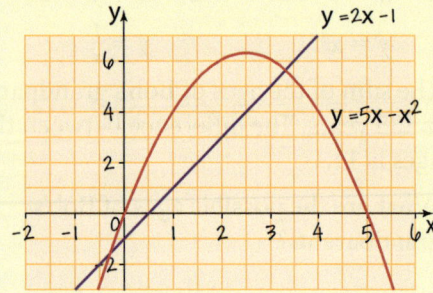

The solutions are $x \approx -0.3$ and $x \approx 3.3$.

Graphs can only give you an approximate solution. To get an exact solution you need to use algebra.

Algebra   Equations and inequalities

Reasoning and problem solving

Exercise 10.3A

1. Use the information given to solve these problems.

   a. How much does a lemon cost?

3 lemons	4 oranges
4 oranges	5 lemons
$1.27	$1.61

   b. The perimeter of this triangle is 30 cm. How long is the base?

   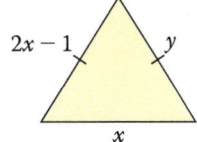

2. Use the information given to solve these problems.

   a. Two numbers have a sum of 41 and a difference of 7. What are the numbers?

   b. One number is 6 more than another. Their mean average is 20. What are the numbers?

   c. 230 students and 29 staff are going on a school trip. They travel by large and small coaches. The large coaches seat 55 and the small coaches seat 39. If there are no spare seats and five coaches are to make the journey, how many of each coach are used?

   d. In an isosceles triangle, the largest angle is 30° more than double the equal angles. Find the angles in the triangle.

   e. Uncle Jack gave me a €25 book token at Christmas. At Firestone's Bookshop I can use the token to buy exactly 3 paperbacks and 1 hardback book or 1 paperback and 2 hardbacks.

   Find the cost of a paperback and a hardback book.

3. Use the given information to find each pair of numbers.

   a. Two numbers have a sum of 23 and a difference of 5. What numbers are they?

   b. Two numbers have a difference of 6. Twice the larger plus the smaller number also equals 6. What are the numbers?

4. Find the value of each symbol in this puzzle.

5. Use a graphical method to solve these problems.

   a. Twice one number plus three times another is 4. Their difference is 2. What are the numbers?

   b. The sum of the ages of James and Isla is 4. The difference between twice Isla's age and treble James' age is 3. How old are they?

6. a. Explain why the simultaneous equations $y = 2x - 1$ and $y = 2x + 4$ have no solution.

   b. Is it possible to have a pair of simultaneous equations with more than one solution? Give your reason.

7. Two lines intersect. One has gradient 4 and y-axis intercept 3. The other has gradient 6 and cuts the y-axis at (0, 1). What is the point of intersection of the lines?

8. Use graphs to find the approximate solutions of these equations.

   a. $x^2 - x - 2 = 2$
   b. $x^2 - x - 2 = -1$
   c. $x^2 - x - 2 = x + 1$
   d. $x^2 - x - 2 = 0$
   e. $2 - x = x^2 - x - 2$

9. By finding the prime factor decompositions of the numbers on the right hand side, or otherwise, solve these simultaneous equations.

   a. $xy^5 = -96$
      $2xy^3 = -48$

   b. $2^{p+q} = 32$
      $3^{q-2p} = 6561$

SKILLS

10.4 Approximate solutions

Not all equations are easily solved.
One method of finding an approximate solution is by bracketing the solution.

EXAMPLE

Find a solution to the equation $f(x) = x^3 - 2x + 5 = 0$ to 2 significant figures.

Work out $f(x)$ to find a pair of positive and negative values.
$f(0) = 5$, $f(2) = 9$, $f(-2) = 1$, $f(-3) = -16$

x	f(x)	Bounds
-2	+1	
-3	-16	-3 < soln < -2
-2.1	-0.061	-2.1 < soln < -2.0
-2.08	+0.161088	-2.10 < soln < -2.08
-2.09	+0.050671	-2.10 < soln < -2.09

$f(x)$ crosses the x-axis between −3 and −2.
The solution is closer to −2.
$x = -2.1$ (2 sf)

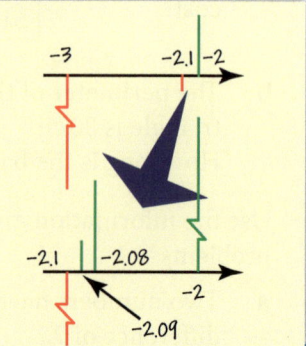

Approximate solutions to equations can also be found using **iteration**.
This means repeating a process in order to obtain a sequence of approximate solutions: $x_1, x_2, x_3, \ldots x_n, \ldots$ Often one value is used to obtain the next value.

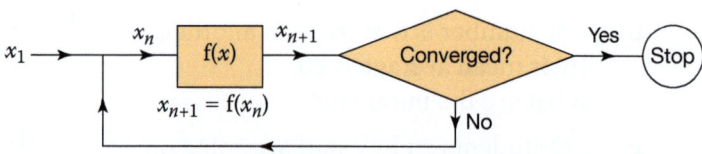

▲ The iterates have converged when the output equals the input: $x_{n+1} = x_n$.

EXAMPLE

a Iterate this function six times.
Start with $x_1 = 1$. $f(x) = \dfrac{2}{3}\left(x + \dfrac{1}{x^2}\right)$

b What number do the iterates converge to?

a $x_2 = \dfrac{2}{3}\left(1 + \dfrac{1}{1^2}\right) = \dfrac{4}{3}$ = 1.3333...

$x_3 = \dfrac{2}{3}\left(1.333\ldots + \dfrac{1}{1.333\ldots^2}\right)$ = 1.263888...

$x_4 = \dfrac{2}{3}\left(1.263888\ldots + \dfrac{1}{1.263888\ldots^2}\right)$ = 1.259933493...

$x_5 = 1.25992105\ldots$ $x_6 = 1.25992105\ldots$

b let $x = x_n = x_{n+1}$

$x = \dfrac{2}{3}\left(x + \dfrac{1}{x^2}\right) \Rightarrow 3x = 2x + \dfrac{2}{x^2} \Rightarrow$

$x = \dfrac{2}{x^2} \Rightarrow x^3 = 2 \Rightarrow x = \sqrt[3]{2}$

The process finds the cube-root of 2.

Investigate doing these iterations using the ANS key on your calculator.

EXAMPLE

a Use the iterative formula $x_{n+1} = \sqrt[3]{2x_n - 5}$ to find a solution of $x^3 - 2x + 5 = 0$ accurate to four decimal places. Start with $x_1 = -2$.

b If the iterative formula converges, show that it gives a solution to the equation $x^3 - 2x + 5 = 0$.

a $x_2 = \sqrt[3]{2 \times (-2) - 5}$ = -2.0800838
$x_3 = \sqrt[3]{2 \times (-2.0800838) - 5}$ = -2.0923507
$x_4 = \sqrt[3]{2 \times (-2.0923507) - 5}$ = -2.0942170
$x_5 = -2.0945007$ $x_6 = -2.0945438$
$x = -2.0945$ (4 dp)

b If the formula converges then.
$x = \sqrt[3]{2x - 5}$
$x^3 = 2x - 5$
$x^3 - 2x + 5 = 0$

Algebra   Equations and inequalities

Fluency

Exercise 10.4S

1. Use a bracketing method to find a positive solution to these equations to four decimal places.

   a $x^2 - 2 = 0$
   b $x^2 - x - 1 = 0$
   c $x^3 - 2x^2 = 1$
   d $x^3 + x = 100$
   e $10 = \sqrt[3]{2x^3 + 1}$
   f $2^x - 10 = 0$

2. a Iterate these functions, starting with $x_1 = 1$.

   i $f(x) = \dfrac{1}{2}\left(x + \dfrac{2}{x}\right)$

   ii $f(x) = \dfrac{1}{4}\left(3x + \dfrac{10}{x^3}\right)$

   b Find the exact numbers that these iterative sequences converge to.

3. Tien is finding an approximate solution for the equation $x^3 - 5x + 5 = 0$ using this iterative formula.

   $x_{n+1} = \sqrt[3]{5x_n - 5}$

   Tien starts with $x_1 = -2$ and uses a spreadsheet to generate $x_2, x_3, x_4, \ldots$

x_1	-2
x_2	-2.466212074
x_3	-2.587865565
x_4	-2.617793522
x_5	-2.625052098
x_6	-2.62680652
x_7	-2.627230218
x_8	-2.627332522

   a Use Tien's results to find a solution to 3 decimal places of $x^3 - 5x + 5 = 0$.

   b By substituting your answer to part **a** into $x^3 - 5x + 5 = 0$, comment on the accuracy of Tien's solution to $x^3 - 5x + 5 = 0$.

4. Sara is trying to find an approximate solution for the equation $x^3 - 4x + 1 = 0$. She has rearranged the equation to form this iterative formula.

   $x_{n+1} = (4x_n - 1)^{\frac{1}{3}}$

   She starts with

   $x_1 = -2$ $x_2 = ((4 \times -2) - 1)^{\frac{1}{3}}$
   $= -2.0800838$

   a Find x_3, x_4, x_5, x_6 and x_7.

   b Comment on your results to part **a**.

   c Using the same iterative formula $x_{n+1} = (4x_n - 1)^{\frac{1}{3}}$ but starting with $x_1 = 2$, find another solution of $x^3 - 4x + 1 = 0$.

5. a Show that if this iterative formula converges

   $x_{n+1} = \sqrt[3]{6x_n - 3}$

   then it gives a solution to the equation $x^3 - 6x + 3 = 0$.

   b Starting with $x_1 = 2$ and using the recursive formula, find an approximate solution of $x^3 - 6x + 3 = 0$.

6. *a Rearrange the equation $x^3 + 2x - 5 = 0$ to form an iterative formula.

   b Starting with $x_1 = 1$ and using your recursive formula, find an approximate solution of $x^3 + 2x - 5 = 0$.

7. a Show that the equation $x^4 - 2x - 8 = 0$ can be rearranged to form this iterative formula.

   $x_{n+1} = \sqrt[4]{2x_n + 8}$

   b Using the graph of $y = x^4 - 2x - 8$, select a suitable value for x_1.

   c Using your value of x_1 and the iterative formula, find an approximate solution of $x^4 - 2x - 8 = 0$.

   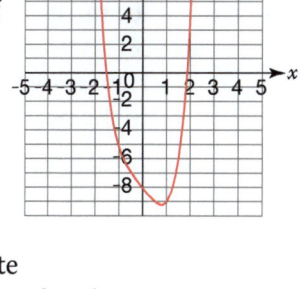

*8 Use iteration to find an approximate solution of $x^2 + \cos x = 2$

9. Ellie is programming a spreadsheet to solve an equation iteratively. The iterative formula is $x_{n+1} = \sqrt[3]{7x_n + 1}$ and $x_1 = 3$.
   What does Ellie need to type into cell A2?

	A	B	C	D
1				
2	=			
3				

APPLICATIONS

10.4 Approximate solutions

RECAP

- You find solutions to equations by using an iterative method.
- Iteration involves repeating the same instructions until you find an accurate solution.
- A formula of the form $x_{n+1} = f(x_n)$ can often be used to obtain an approximate solution of an equation.

HOW TO

① Read the question and obtain an equation which needs to be solved.
② Use an approximate method of solution if the equation is too difficult to solve algebraically.
③ Answer the question

EXAMPLE

Loans are compared using an interest rate known as the annual percentage rate (APR). For a loan of £L that is repaid in 3 equal annual payments of £R, the APR is $100A$% where

$L(1 + A)^3 = R(3 + 3A + A^2)$.

A car manufacture offers a loan of £10 000 which has to be repaid in 3 annual payments of £4000.

a Prove that that the APR formula for the loan can be rearranged to $5A^3 + 13A^2 + 9A - 1 = 0$ Hence, show that the APR is 10% to the nearest percentage.

b Afraah, Brian and Carly each suggest an iterative formala to find a more accurate value for the APR.

Afraah $\quad A_{n+1} = \sqrt[3]{\dfrac{1 - 13A_n^2 - 9A_n}{5}}$

Brian $\quad A_{n+1} = \sqrt{\dfrac{1 - 5A_n^3 - 9A_n}{13}}$

Carly $\quad A_{n+1} = \dfrac{1 - 5A_n^3 - 13A_n^2}{9}$

Subsutitute $A_1 = 0.1$ into each formula. Use your answers to decide which of the three formulae can be used to find a more accurate value for the APR.

a ① Substitute the values into the APR equation.
$10\,000(1 + A)^3 = 4000(3 + 3A + A^2)$
$5(1 + A)^3 = 2(3 + 3A + A^2) \quad \div 2000$
Expand the brackets in two stages.
$5(1 + A)(A^2 + 2A + 1) = 6 + 6A + 2A^2$
$5(A^2 + 2A + 1 + A^3 + 2A^2 + A) = 6 + 6A + 2A^2$
$5(A^3 + 3A^2 + 3A + 1) = 6 + 6A + 2A^2$
$5A^3 + 15A^2 + 15A + 5 = 6 + 6A + 2A^2$
$5A^3 + 13A^2 + 9A - 1 = 0$

② Show that $100A$ lies between 9.5 and 10.5 by showing that the equation has a solution between 0.095 and 0.105
$A = 0.095$, LHS $= -0.02$
$A = 0.105$, LHS $= 0.09$
③ The APR is 10% to the nearest percentage.

b Afraah $\quad A_2 = \sqrt[3]{\dfrac{1 - 13(0.1)^2 - 9(0.1)}{5}} = -0.18171...$

Brian $\quad A_2 = \sqrt{\dfrac{1 - 5(0.1)^3 - 9(0.1)}{13}} = 0.08548...$

Carly $\quad A_2 = \dfrac{1 - 5(0.1)^3 - 13(0.1)^2}{9} = 0.09611...$

The formula should give a value between 0.095 and 0.105.
③ Carly's formula can be used to find a more accurate value.

Algebra   Equations and inequalities

Reasoning and problem solving

Exercise 10.4A

1. A loan of $12000 has to be repaid in 3 annual payments of $6000.
   For a loan of L that is repaid in 3 equal annual payments of £R, the APR is $100A\%$ where $L(1+A)^3 = R(3 + 3A + A^2)$.

   a. Show that the APR formula for the loan can be written as
   $2A^3 + 5A^2 + 3A - 1 = 0$.
   Hence, show that the APR is 23% to the nearest one per cent.

   b. Aria tries to use the iterative formula
   $$A_{n+1} = \sqrt{\frac{1 - 2A_n^2 - 3A_n}{5}}$$
   to find a more accurate value for the APR starting with $A_1 = 0.23$
   Will Aria's iterative formula converge on the solution? Explain your answer.

   c. Use the iterative formula
   $$A_{n+1} = \frac{1 - 2A_n^3 - 5A_n^2}{3}$$
   to find the value of the APR to 1 dp.

2. Karlie and Taylor each put £1000 into saving accounts.

   After n years, Karlie's simple interest account contains £$1000(1 + 0.04n)$ whilst Taylor's compound interest account contains £$1000(1.03)^n$.

   a. Show that both accounts have the same amount of money when $1 = 1.03^n - 0.04n$

   b. Use a bracketing method with starting values for $n = 8$ and $n = 40$ to find the numbers of years when both accounts have the same amount of money.

   *c. Draw a flow chart, containing only questions with yes/no answers, to describe how to find a solution using bracketing.

3. This 'Babylonian' iterative formula
   $$x_{n+1} = \frac{x_n}{2} + \frac{2}{2x_n}$$
   can be used to find a fraction approximation to $\sqrt{2}$.

   a. Starting with $x_1 = 1$ work out x_4.

   b. What is your x_4^2?

4. a. This 'Babylonian' iterative formula
   $$x_{n+1} = \frac{x_n}{2} + \frac{3}{2x_n}$$
   can be used to find a fraction approximation to $\sqrt{3}$.
   Starting with $x_1 = 1$ find x_3.

   b. What happens if you use the much simpler formula $x_{n+1} = \frac{3}{x_n}$?

5. In 2010, a survey of the birds on an island counted approximately 200 kittiwakes.

   A conservationist used the logistic equation
   $P_{n+1} = P_n(1.4 - 0.001P_n)$
   to predict the expected population, P_n, n years later.

   a. What did the equation predict for the population of kittiwakes each year from 2011 to 2015?

   b. Describe how the population was predicted to change.

   c. Assuming that the population reaches a roughly constant value, find this population.

Did you know...

Some historians believe that 4000 years ago Babylonian mathematicians used an iterative formula to find the square roots of numbers.

SKILLS

10.5 Inequalities

- **Inequalities** are expressions involving these symbols.
  - $<$ less than $\leq$ less than or equal to
  - $>$ greater than $\geq$ greater than or equal to

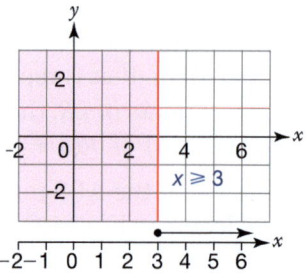

Linear and quadratic inequalities can be solved algebraically using the same techniques you use to solve equations. Solutions can be represented using a number line or graphically.

EXAMPLE

Solve the linear inequalities and represent the solution on a number line.

a $3x < 15$ b $4x + 3 \geq 21$ c $6 < 3(x - 2) < 18$ d $\dfrac{5x - 2}{3} \leq 6$

a $3x < 15$ $\div 3$
   $x < 5$

b $4x + 3 \geq 21$ -3
   $4x \geq 18$ $\div 4$
   $x \geq 4.5$

c $6 < 3(x - 2) < 18$
   $6 < 3x - 6 < 18$ $+6$
   $12 < 3x < 24$ $\div 3$
   $4 < x < 6$

d $\dfrac{5x - 2}{3} \leq 6$ $\times 3$
   $5x - 2 \leq 18$ $+2$
   $5x \leq 20$ $\div 5$
   $x \leq 4$

The only integer solution is 5.

EXAMPLE (p.274)

Show the region defined by these three inequalities.

$x \geq -2$
$y > x$ and $x + y < 3$

The required region is left unshaded at each stage.

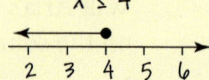

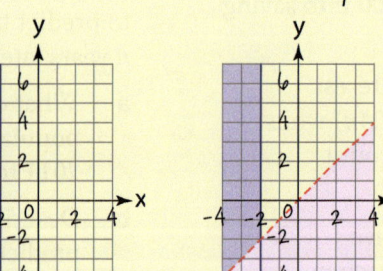
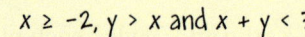
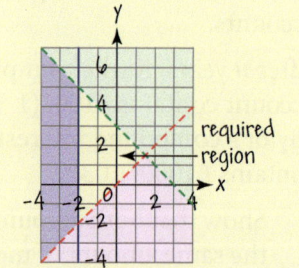

EXAMPLE

Solve the quadratic inequalities.

a $x^2 < 9$ b $x^2 > 25$ c $x^2 + 5x + 6 \leq 0$

Replace the inequality by an equation, find its roots and inspect values of x on each side of the roots.

a $x < 3$ and $x > -3$
   or $-3 < x < 3$

b $x > 5$ and $x < -5$

c $(x + 2)(x + 3) \leq 0$
   $-3 \leq x \leq -2$

It is useful to sketch the graph of the quadratic function to solve the inequality.

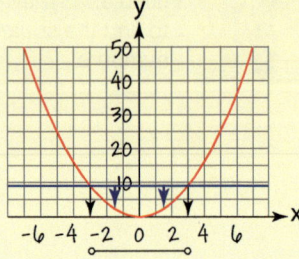
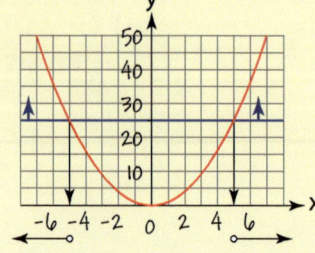
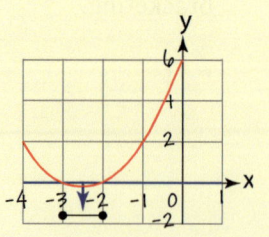

Algebra   Equations and inequalities

Fluency

Exercise 10.5S

1. List all the integer values of n such that $-2 < n < 5$.

2. Solve these linear inequalities.
   a $2x > 12$
   b $2x + 1 < 17$
   c $3x - 2 < 19$
   d $5x - 2 \leq 28$
   e $3(x - 2) < 36$
   f $5(x - 2) \leq 20$
   g $27 \leq 9(x - 2)$
   h $-5 < 5(1 + 3x)$
   i $10 > 2x > 2$
   j $1 < 2x + 1 < 17$
   k $7 \leq 3x - 2 < 19$
   l $3 < 5x - 2 \leq 28$
   m $-2 \leq 5x + 3 \leq 23$
   n $-5 < 8x + 3 < 19$
   o $-20 < 3x - 5 < 19$
   p $-5 < 10 + 3x < -2$

3. Find the smallest or largest integer that satisfies these inequalities.
   a $\frac{x}{2} + 3 > 7$
   b $\frac{x}{2} - 5 < 4\frac{1}{2}$
   c $\frac{3x - 6}{4} > 3$
   d $\frac{5x - 2}{3} \leq 6$
   e $3 < \frac{9x - 12}{5}$
   f $8 > \frac{4(x + 3)}{3}$

4. Is it possible to write a single linear inequality to represent this solution?

   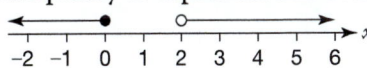

   Give your reason.

5. Show each of these inequalities
   i   using a number line
   ii graphically.
   a $2 \leq x \leq 5$
   b $-1 \leq x < 3$
   c $-4 < y \leq -2$
   d $-3 < y < 4$

6. Use inequalities to describe the unshaded regions in these graphs.

   a 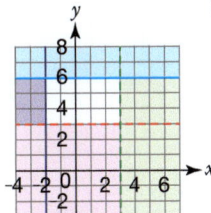   b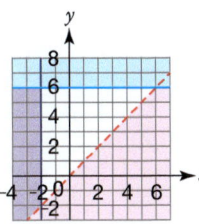

6. c 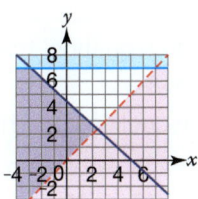   d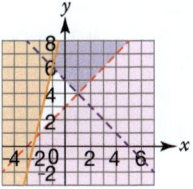

   e 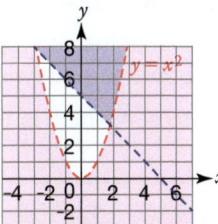   f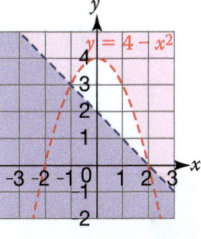

7. Show the regions defined by these inequalities.
   a $y > x, x < 3$
   b $y \geq x, x > -2, y \leq 3$
   c $y > x + 2, x + y > 2$
   d $x \geq 0, y \geq 0, x + 2y < 6$

8. Solve these quadratic inequalities.
   a $x^2 < 64$
   b $x^2 > 1$
   c $x^2 + 2x > 0$
   d $x^2 - 6x \leq 0$
   e $x^2 + 6x + 8 < 0$
   f $x^2 + x < 12$
   g $2x^2 - 5x - 3 \leq 0$
   h $3x^2 + 2 \leq 0$

9. Find a quadratic inequality that is represented by this solution set.

10. Saqr thinks the solution for the quadratic inequality $x^2 \geq 49$ is $-7 \leq x \geq 7$.

    Do you agree with Saqr?

    Explain your reasoning.

11. Explain why the inequality $3x^2 + 2 \leq 0$ does not have a solution set.

*12. Show the region defined by the inequalities $y > x^2$ and $y + x^2 < 8$.

13. In 15 years time, Patricia will be more than double the age she was 10 years ago.

    What is the oldest age Patricia could be?

1161, 1162, 1163, 1189

207

APPLICATIONS

10.5 Inequalities

RECAP

- An inequality is an 'equation' that uses one of the symbols <, ≤, ≥ or >.
- You can show inequalities graphically
  - For a single variable use a number line with o for strict inequalities (< or >, the end point is *not* included) otherwise use •.
  - For a pair of variables use a graph with dashed lines for strict inequalities otherwise solid lines.
  - Test a point to see which is the required region.

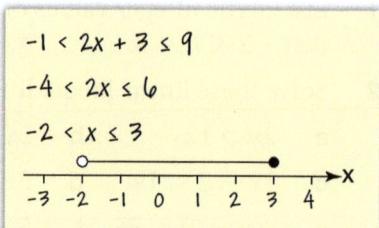

HOW TO

To solve a problem involving inequalities

① Read the question. You may need to write inequalities to model the situation.

② Simplify the inequalities and show them graphically if needed.

③ Choose a point in your solution area and check it obeys all the inequalities.

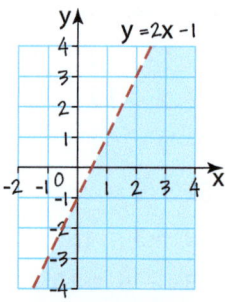

▲ The region $y > 2x - 1$.
(0, 0) is in the region $0 > 2 \times 0 - 1$.

EXAMPLE

Neha is pricing items to sell in her cafe. She wants the cost of 2 sandwiches to be less than £5. Neha wants the cost of 2 sandwiches and 3 cupcakes to be less than £9.
x is the price of a sandwich and y is the price of a cupcake.
Show the region on a graph that meets Neha's requirements.

① $2x < 5 \Rightarrow x < \dfrac{5}{2}$ 2 sandwiches cost less than £5.
   $2x + 3y < 9$ 2 sandwiches and 3 cupcakes cost less than £9.
   $x > 0$ $y > 0$ The sandwiches and cupcakes are not free.

② All strict inequalities so draw dashed lines.

   $2x + 3y = 9 \Rightarrow y = \dfrac{9 - 2x}{3}$

   $y = 0$ (x-axis) shade below.
   $x = 0$ (y-axis) shade to the left.
   $x = \dfrac{5}{2}$ shade to the right.
   $3y + 2x = 9$ shade above.

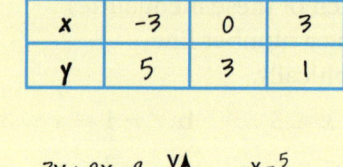

State which region is the required region.
The required region is unshaded.

③ Check that a point in your unshaded solution area obeys all the inequalities.
   (1, 1) $y = 1$ and $1 > 0$ ✓
   $x = 1$ and $1 > 0$ ✓
   $x = 1$ so $2 \times 1 = 2$ and $2 < 5$ ✓
   $x = 1, y = 1$ so $(3 \times 1) + (2 \times 1) = 5$ and $5 < 9$ ✓
   (1,1) obeys all the required inequalities.

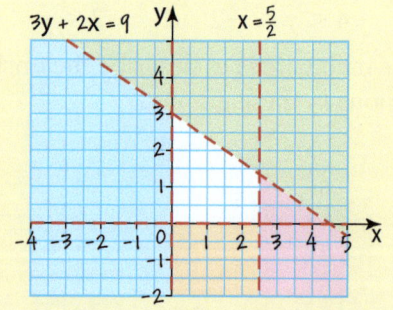

Algebra   Equations and inequalities

Reasoning and problem solving

Exercise 10.5A

1 a The numerical value of the area of this rectangle exceeds the numerical value of its perimeter. Write an inequality and solve it to find the range of values of x.

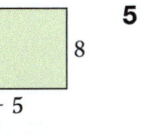

b Given that x is an integer, find the smallest possible value that x can take.

2 Explain why it is not possible to find a value of y such that $3y \leq 18$ and $2y + 3 > 15$.

3 Draw graphs to show the regions satisfied by these inequalities.

- **a** $y \leq x + 5$
- **b** $y > 2x + 1$
- **c** $y \geq 1 - 3x$
- **d** $y < \frac{1}{4}x - 2$
- **e** $2y \geq 3x + 4$
- **f** $3y < 9x - 7$
- **g** $x + y < 9$
- **h** $2x + 4y > 7$

4 What inequalities are shown by the *shaded* region in each of these diagrams.

a
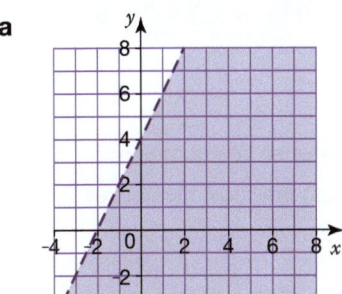

b
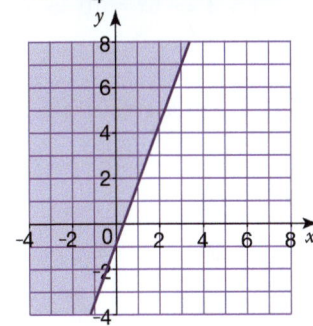

5 Here are two graphs of quadratic equations, with horizontal lines drawn as shown. Write the inequalities represented by the *shaded* regions.

a
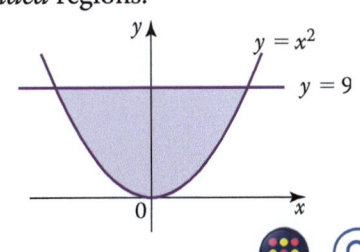

b
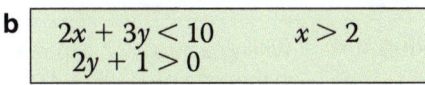

6 On one diagram, show the region satisfied by the three inequalities. List all the integer coordinates that satisfy all three inequalities.

a
$x \geq -1$	$y < 2$
$y \geq 3x - 1$	

b
$2x + 3y < 10$	$x > 2$
$2y + 1 > 0$	

7 Draw the region satisfied by these three inequalities
$$y \geq x^2 - 4, \quad y < 2, \quad y + 1 < x.$$

8 300 students are going on a school trip and 16 adults will accompany them. The head teacher needs to hire coaches to take them on the trip. She can hire small coaches that seat 20 people or large coaches that seat 48 people. There must be at least 2 adults on each coach to supervise the students.

a If x is the number of small coaches hired and y is the number of large coaches, explain why these inequalities model the situation.
$x \geq 0, y \geq 0, 5x + 12y \geq 79, x + y \leq 8$

b Show the region on a graph that satisfies all four inequalities.

9 Draw diagrams and write inequalities that when shaded would give

a a shaded, right-angled triangle with right angle at the origin

b an unshaded trapezium.

Summary

Checkout
You should now be able to...

	Test it Questions
✓ Solve linear equations including when the unknown appears on both sides.	1, 2
✓ Solve quadratic equations using factorisation, completing the square and the quadratic formula.	3 – 7
✓ Solve a pair of linear or linear plus quadratic simultaneous equations.	7
✓ Use iterative processes to find approximate solutions to equations.	8
✓ Solve inequalities and display your solution on a number line or a graph.	9, 10

Language	Meaning	Example
Completing the square	Writing a quadratic expression $ax^2 + bx + c$ in the form $p(x + q)^2 + r$.	$2x^2 - 4x - 3 = 2(x - 1)^2 - 5$ $= 0 \Rightarrow x = 1 \pm \sqrt{\frac{5}{2}}$
Quadratic formula	A formula for the solutions of the quadratic equation $ax^2 + bx + c = 0$. $x = \dfrac{-b \pm \sqrt{b^2 - 4ac}}{2a}$	$2x^2 - 5x + 2 = 0, a = 2, b = -5, c = 2$ $x = \dfrac{5 \pm \sqrt{5^2 - 4 \times 2 \times 2}}{2 \times 2} = \dfrac{5 \pm 3}{4} = 2, \dfrac{1}{2}$
Simultaneous equations	Two or more equations that are true at the same time for the same values of the variables.	① $3x - y = 2$ ② $2x + y = 8$ Both true when $x = 2$ and $y = 4$. On a graph the lines intersect at (2, 4).
Elimination	A method of solving simultaneous equations by removing one of the variables.	① + ② $5x = 10 \Rightarrow x = 2$
Substitution	Replacing one of the variables in a simultaneous equation with an expression found by rearranging the other equation.	① $\Rightarrow y = 3x - 2$ into ② $2x + (3x - 2) = 8 \Rightarrow 5x = 10 \Rightarrow x = 2$ ① $y = 4$
Inequality	A comparison of two quantities that may not be equal.	$5x - 1 < 9$ $5x - 1$ is strictly less than 9.
Iteration	A procedure which is repeated in order to generate a sequence of numbers which give more accurate solutions to a problem. Usually the output of the procedure is used as the input for the next iteration.	$x_{n+1} = \dfrac{1}{2}\left(x_n + \dfrac{2}{x_n}\right)$ $x_1 = 1 \to x_2 = 1.5 \to x_3 = 1.41\dot{6} \to$ $x_4 = 1.4142... \to \sqrt{2}$

Algebra   Equations and inequalities

Review

1 Solve these equations.

　a $3a - 7 = 35$　**b** $\frac{b}{4} + 3 = 28$

　c $5c + 23 = 8$　**d** $4x + 7 = 3x + 13$

　e $3x - 9 = 8x - 69$

　f $10 - 4x = 24 - 8x$

2 Shakeel takes x hours to run a marathon and Andy takes 36 minutes longer. The sum of their times is 7 hours.

　a Write an equation in x for the sum of their times.

　b Solve your equation to find Shakeel's and Andy's times. Give your answers in hours and minutes.

3 Solve these quadratic equations by factorising.

　a $x^2 - x - 12 = 0$

　b $x^2 - 12x + 35 = 0$

　c $x^2 + 12x + 36 = 0$

　d $16x^2 - 121 = 0$

　e $6x^2 - 9x = 0$

　f $3x^2 + 7x + 4 = 0$

　g $10x^2 - 19x + 6 = 0$

4 Solve this quadratic equation.

$2x^2 - 7x = 15$

5 a Complete the square for these quadratic expressions.

　　i $x^2 - 4x - 1$　**ii** $x^2 + 3x + 2$

　b Solve these quadratic equations by completing the square.

　　i $x^2 - 10x + 16 = 0$

　　ii $2x^2 + 8x = 24$

6 Use the quadratic formula to solve these equations giving your answers to 3 significant figures.

　a $x^2 - 5x - 7 = 0$　**b** $3x^2 + 9x + 5 = 0$

7 Solve these pairs of simultaneous equations.

　a $3x + 4y = 5$　**b** $5v - 3w = -2$
　　　$5x - 5y = 20$　　　$4v - 7w = -8.5$

　c $y - x = 2$　**d** $y = x^2 + 5x$
　　　$y = x^2$　　　　$y = x + 5$

8 a Starting with $x_1 = 2$ use this iterative formula to find x_5.

$x_{n+1} = \sqrt[3]{27 - x_n}$

　b Show that if this iterative formula converges then it gives a solution to the equation $x^3 + x - 27 = 0$.

9 Solve these inequalities and display the results on a number line.

　a $5x - 12 > 13$　**b** $5x + 17 \leq 2x + 8$

　c $17 - 2x > 9$　**d** $3(10 - 3x) < -24$

　e $x^2 + 8x + 7 < 0$　**f** $x^2 + 3x - 18 \geq 0$

10 Draw suitable diagrams to show these inequalities. You should leave the required region *unshaded* and label it **R** in part **c**.

　a $-1 \leq x < 3$　**b** $y \leq 4$ or $y \geq 5$

　c $1 < x < 4, y > 0$ and $x + y \leq 4$

What next?

Score		
0 – 4		Your knowledge of this topic is still developing. To improve, look at MyiMaths: 1057, 1160, 1161, 1162, 1163, 1169, 1174, 1177, 1181, 1182, 1185, 1189, 1236, 1319, 1928, 1929
5 – 9		You are gaining a secure knowledge of this topic. To improve your fluency look at InvisiPens: 10Sa – r
10		You have mastered these skills. Well done you are ready to progress! To develop your problem solving skills look at InvisiPens: 10Aa – g

Assessment 10

1. **a** A bucket of water has a mass of 27.5 kg. The mass of the water is 10 kg greater than the mass of the bucket. How heavy is the bucket? [2]

   b Oliver is three times as old as his brother Albert. Oliver is 6 years older than Albert. How old are Oliver and Albert? [2]

   c Sajaa cut a cake into 3 pieces. Jack's piece was 35 g heavier than Seamus's. Seamus's piece was 22 g lighter than Sajaa's. The total mass was 454 g. Calculate the mass of Sajaa's piece. [4]

   d Brendan, Arsene and José go on holiday. Brendan takes x Euros, Arsene takes half as much as José, and José takes 150 Euros more than Brendan. Altogether, they took 2000 Euros. How much money did each person take? [3]

   e The ages of Milo and Fizz are in the ratio $3:4$. In 6 years time they will be in the ratio $9:10$. How old are they now? [4]

   f Lewis drove on the motorway, averaging 70 mph. He then left the motorway and drove on another road averaging 40 mph. He drove a total of 163 miles in 2.5 hours. How far did he drive along the motorway? [4]

2. Mario attempted to kayak 30 km for charity. Sophia gave him €5 for each *complete* kilometre. Mario gave Sophia €2 for each *complete* kilometre not covered. The balance was given to charity. Mario gave €115 to the charity.

   a Mario kayaked k kilometres. Write down an equation for k and solve it. [4]

   b Find the minimum number of kilometres that Mario needed to complete to be sure of making a contribution to charity. [3]

3. **a** A square has side length s cm. Another square has a side 2 cm shorter than the first. The total area of the squares is 200 cm². Find the *exact* side length of the first square. [4]

   b Two consecutive prime numbers have a difference of 6 and a product of 3127. Find the numbers. [4]

   c The formula $S = \dfrac{n(n+1)}{2}$ represents the sum of the numbers $1 + 2 + 3 + \cdots + n$. The sum of the numbers 1 to n is 5050. Find n. [3]

4. A cricket ball is hit straight upwards. The formula $h = 20t - 5t^2$ represents its height above the ground, t seconds after he hits it.

   a Find the times when the height of the ball is 15 m above the ground. Say why there are two possible answers. [3]

   b Find the time when the height of the ball is 20 m above the ground. Say why there is only one answer. [3]

   c Form a quadratic equation to find the time when the ball is 25 m above the ground. Give two reasons why there is no solution to this equation. [3]

   d Find the time when the ball next hits the ground. Explain your answer. [2]

5. The formula $2d = n(n - 3)$ represents the number of diagonals, d, in a polygon with n sides. A polygon has 54 diagonals. How many sides are there in the polygon? [3]

6. Greg says that the solutions to the equation $b^2 + 9b + 4 = 0$ are $b = 13$ or $b = 5.04$. Solve the equation by using the quadratic formula. Give your answers to 2 dp. Was Greg correct? [4]

Algebra   Equations and inequalities

7 Solve these pairs of simultaneous equations.

   a $y = x + 2$ and $y = x^2 + 3x - 22$ [5]

   b $x + y = 1$ and $x = (y - 1)^2 - 12$ [6]

8 **a** On a coach trip, an adult ticket is £a and a child ticket £c. Tickets for 2 adults and 3 children cost £26.50; tickets for 3 adults and 5 children cost £42.

     How much is   **i** an adult's ticket [4]   **ii** a child's ticket? [1]

   b A magic goose lays both brown and golden eggs. 5 brown eggs and 2 golden eggs have a total mass of 200 g, while 8 brown and 3 golden eggs have a total mass of 310 g. Find the mass of a golden egg. [5]

   c 5 packets of 'Doggibix' and 3 packets of 'Cattibix' cost £5.49
      3 packets of 'Doggibix' and 1 packet of 'Cattibix' cost £2.79
      Find the cost of one packet of   **i** Doggibix [4]   **ii** Cattibix? [1]

   d A DVD costs $\$d$ and a CD costs $\$c$. 3 DVDs and 2 CDs cost $45.83 in total.
      1 DVD and 3 CDs cost $26.92 in total.
      What is the cost of
      i a CD [4]   **ii** a DVD? [1]

   e Farmer Scott can buy 2 sheep and 6 goats, or 10 sheep and 2 goats, for £3500.
      What is the price of   **i** a sheep [4]   **ii** a goat? [1]

   f Titus Lines is going fishing and needs bait. He can buy 5 maggots and 6 worms for 38 pence or 6 maggots and 12 worms for 60 pence.
      What is the price of
      i a maggot [4]   **ii** a worm? [1]

9 Rasheeq represents the inequality $-1 < x \leqslant 3$ on a number line as shown.

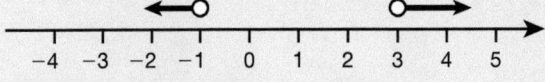

Correct his mistakes. [3]

10 Megan says that the region indicated by the inequalities $2y - x < 5$, $x + y > 4$ and $x \leqslant 5$ is shown on the diagram.
Correct her diagram and shade the correct region. [4]

11 **a** A lorry averages 22 mpg. It will travel for 345 miles on a full tank of diesel. The driver has g gallons in his tank. Write down an appropriate inequality for g and solve it. [2]

   b David and Kaatima are getting married. Food has been ordered for 100 guests. 120 guests have agreed to come but at least 15% of these will not come. On the day, there is not enough food for everyone. How many guests attend the wedding? Find all possible solutions. [4]

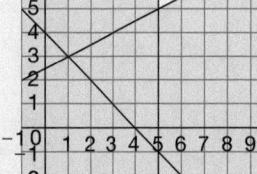

   c A busy station has 12 platforms. There are f freight trains and p passenger trains at platforms. There are 2 more passenger trains than freight trains. There are at least 3 freight trains at platforms.

      i Find all possible solutions for f and p. [4]

      ii Are there any times when all 12 platforms are in use? [1]

      iii Find the greatest possible number of empty platforms. [1]

11 Circles and constructions

Introduction
The invention of the wheel was most definitely a landmark event in human technological development, giving people the ability to travel at speed. However it was the use of gears and cogs on a massive scale during the Industrial Revolution that really accelerated advancement, not just in technology but also in social and economic development.

What's the point?
Without mankind's understanding of circles, and how their properties can be exploited in marvellous ways, we would still be living in largely agricultural communities in a pre-industrial state, with no computers, mobile phones, cars, …

Objectives
By the end of this chapter, you will have learned how to …
- Find the area and circumference of a circle and composite shapes involving circles.
- Calculate arc lengths, angles and areas of sectors.
- Prove and apply circle theorems.
- Use standard ruler and compass constructions and solve problems involving loci.

Check in

1 Work out the missing angles.

a 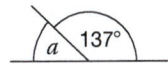 b 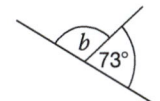 c 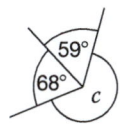 d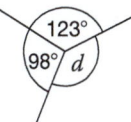

2 Work out the missing angles in these shapes.

a 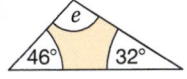 b 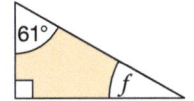 c 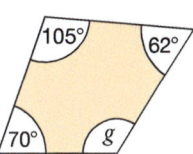 d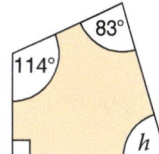

3 Find the area of these shapes.

a 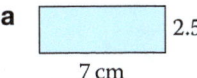 b 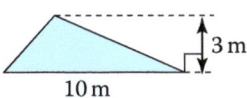 c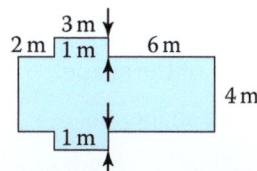

Chapter investigation

Sketch a circle, and draw a straight line through it. How many pieces have you divided the circle into?

Draw a second straight line through the circle. What is the maximum number of pieces you can divide the circle into?

Continue drawing straight lines through the circle (the lines can cut at any angle, and the circle can be as big as you want it to be). Is there a relationship between the number of cuts and the maximum number of pieces? Investigate.

SKILLS

11.1 Circles 1

Perimeter is the total distance around the sides of a 2D shape.
Diameter of a **circle** = 2 × **radius**

- **Circumference** of a circle, $C = \pi d = 2\pi r$ where $\pi = 3.14159...$
  Rearranging gives $d = \dfrac{C}{\pi}$ or $r = \dfrac{C}{2\pi}$
- Area inside a circle, $A = \pi r^2$
  Rearranging gives $r = \sqrt{\dfrac{A}{\pi}}$

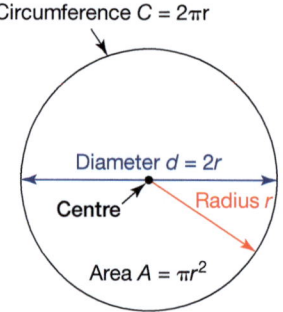

EXAMPLE

This shape is made from a semi-circle and an isosceles triangle.
Find, in terms of π,
a the perimeter of the shape.
b the area of the shape.

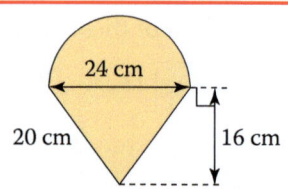

a Arc length of semi-circle = $\dfrac{1}{2} \times \pi d$ A semi-circle is half of a circle.

$= \dfrac{1}{2} \times \pi \times 24$

$= 12\pi$ cm   *An arc is a part of a circle's circumference.*

Total perimeter = $12\pi + 20 + 20$ Add the other sides of the shape.

$= (12\pi + 40)$ cm   Or $4(3\pi + 10)$ by taking out the common factor.

b Area of triangle = $\dfrac{1}{2} \times 24 \times 16 = 192$ cm² $\dfrac{1}{2} \times$ base $\times$ *perpendicular* height

Area of semi-circle = $\dfrac{1}{2} \times \pi r^2$

$= \dfrac{1}{2} \times \pi \times 12^2$ Radius = diameter ÷ 2

$= 72\pi$ cm²

Total area = $(72\pi + 192)$ cm²   Or $24(3\pi + 8)$ cm² by taking out the common factor.

Sometimes you may be given the circumference or area and be asked to find the radius or diameter.

EXAMPLE

The area of a circle is 200 m². Calculate the diameter of the circle.

Put 200 into the area formula.

$200 = \pi r^2$

Rearrange to find r.

$r^2 = \dfrac{200}{\pi}$

$= 63.6619...$

$r = \sqrt{63.6619...}$

Diameter = $7.9788... \times 2 = 15.9576...$

The diameter of the circle is 16.0 m (3 sf)

Carry out the working on your calculator — just round the final answer.

Geometry   Circles and constructions

216

Fluency

Exercise 11.1S

1. **a** Estimate the value of
   - **i** 27^2
   - **ii** 71.5^2
   - **iii** $\sqrt{8900}$
   - **iv** $\sqrt{43600}$

   b Use a calculator to find each value to 3 sf.

2. Find the circumference and area of each circle.
   Use approximations to check each answer.

   a 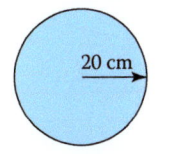 20 cm

   b 64 mm

   c 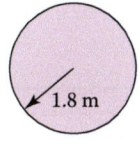 1.8 m

   d 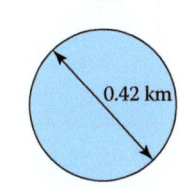 0.42 km

3. The circumference of a circle is 88 cm.
   Find the diameter of the circle.

4. The area of a circle is $1.72\,m^2$.
   Find the radius of the circle.

5. Copy and complete the table for four circles.

	radius	diameter	circumference	area
a	6 cm			
b		4.6 m		
c			98 mm	
d				$15.2\,m^2$

6. Each shape is made from a rectangle and semi-circle.

   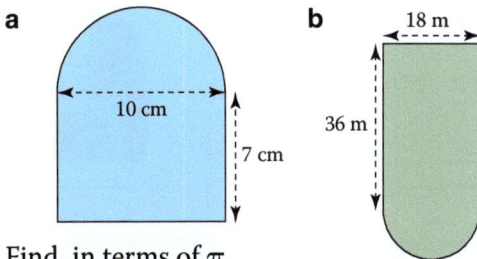

   a 10 cm, 7 cm

   b 18 m, 36 m

   Find, in terms of π,
   - **i** the perimeter of each shape
   - **ii** the area of each shape.

7. A circular hole is cut in a square card.
   Find the area of card that is left.
   Give your answer to 3 significant figures.

   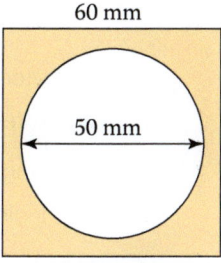
   60 mm, 50 mm

8. A symmetrical pond is in the shape of two semi-circles.
   Find, to 3 significant figures,

   a the perimeter of the pond

   b the area of the pond.

   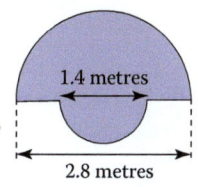
   1.4 metres, 2.8 metres

9. For each shape, calculate to 3 sf
   - **i** the perimeter
   - **ii** the area

   a This shape is made from a right-angled triangle and a semicircle.

   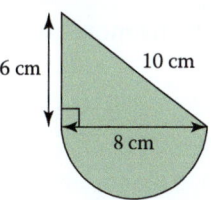
   6 cm, 10 cm, 8 cm

   b This shape is made from a parallelogram and a semicircle.

   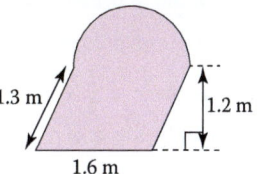
   1.3 m, 1.2 m, 1.6 m

*10 This pendant is made from an isosceles trapezium and a semi-circle.
    Each circular hole has diameter 4 mm.
    Find the area of the pendant.
    Give your answer in terms of π.

    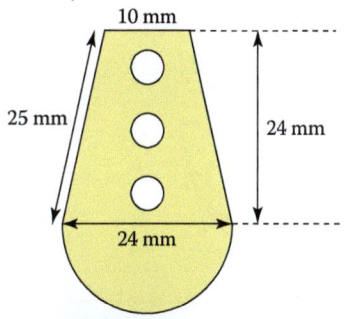
    10 mm, 25 mm, 24 mm, 24 mm

APPLICATIONS

11.1 Circles 1

RECAP
- Diameter of a circle = 2 × radius
- Circumference of a circle, $C = \pi d$ or $2\pi r$
  Rearranging gives $d = \dfrac{C}{\pi}$ or $r = \dfrac{C}{2\pi}$
- Area inside a circle, $A = \pi r^2$
  Rearranging gives $r = \sqrt{\dfrac{A}{\pi}}$

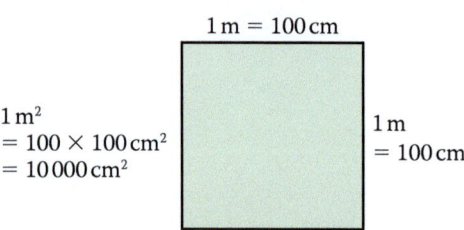

HOW TO

To calculate perimeters or areas involving circles
1. Decide which formulae and units to use.
2. Calculate the perimeters or areas needed.
3. Answer the question, giving units where necessary.

$1\,m^2 = 100 \times 100\,cm^2 = 10\,000\,cm^2$

$1\,m = 100\,cm$

To convert area units use $1\,m^2 = (100\,cm)^2 = 10\,000\,cm^2$

EXAMPLE

Ditmir wants to give the front of his dog's kennel two coats of paint.

He has enough paint to cover $\tfrac{1}{2}\,m^2$.

Does Ditmir need any more paint? Show your working.

p.136

1. Use the area formulae for a triangle, rectangle and semi-circle.
   Work in metres.
2. Find the area of the shapes on the front, ignoring the hole.

   Area of triangle $= \tfrac{1}{2} \times 0.75 \times 0.3 = 0.1125\,m^2$
   Area of rectangle $= 0.75 \times 0.35 = 0.2625\,m^2$

   > Area of triangle $= \tfrac{1}{2}bh$

2. Find the area of the shapes making the hole.

   Area of semi-circle $= \tfrac{1}{2} \times \pi r^2 = \tfrac{1}{2} \times \pi \times 0.2^2 = 0.02\pi\,m^2$
   Area of rectangle $= 0.4 \times 0.15 = 0.06\,m^2$ Height $= 0.35 - 0.2$
   Area to be painted $= 0.1125 + 0.2625 - 0.02\pi - 0.06 = 0.252168...\,m^2$
   Area for 2 coats of paint $= 2 \times 0.252168... = 0.504336...\,m^2$

   > Or you could work in cm and then convert to m using $10\,000\,cm^2 = 1\,m^2$.

3. $0.5043...\,m^2$ is more than $\tfrac{1}{2}\,m^2$, so Jake does need more paint.

EXAMPLE

Show that the perimeter of this shape is equal to πL.

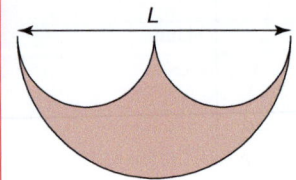

1. A semi-circle is half of a circle, so its arc length is $\tfrac{1}{2}\pi d$.
2. The perimeter consists of one large semi-circle, diameter L and and two small semi-circles, diameter $\tfrac{1}{2}L$.

   Total perimeter $= \tfrac{1}{2}\pi L + \tfrac{1}{2}\pi(\tfrac{1}{2}L) + \tfrac{1}{2}\pi(\tfrac{1}{2}L)$

   $= \tfrac{1}{2}\pi L + \tfrac{1}{4}\pi L + \tfrac{1}{4}\pi L$

3. $= \pi L$ as required.

Geometry Circles and constructions

Reasoning and problem solving

Exercise 11.1A

1. The diagram shows the dimensions of a flowerbed in Yusuf's garden.

   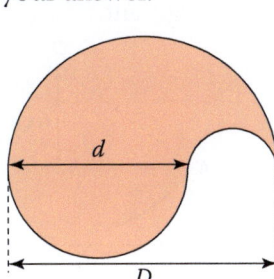

   a Yusuf says that 12 metres of edging will be enough to go around the flowerbed. Is Yusuf correct? Show your working.

   b Yusuf also wants to buy fertiliser to feed the flowers six times in the summer. He uses 35 grams per square metre each time. Is a 2 kilogram bag of fertiliser enough? Explain your answer.

2. This shape is formed by 3 semi-circles. Show that

   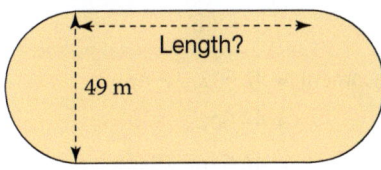

   a the perimeter of the shape $= \pi D$

   *b the area of the shape $= \frac{1}{4}\pi Dd$

3. The diagram shows the inner perimeter of a running track.

   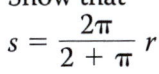

   a The total length must be 400 m. Find the length of the straight sides.

   b The track has 6 lanes, each 1.5 m wide. Find the difference in length between the outer and inner perimeters.

4. The diameter of the wheels on Nicole's bike is 26 inches. Nicole says the wheels go round over 1000 times for every mile she travels. Is Nicole correct? Show how you decide. (1 mile = 1760 yards, 1 yard = 36 inches)

5. A semicircle has radius s.
   A circle has radius r.
   The perimeter of the semicircle equals the circumference of the circle.
   Show that
   $$s = \frac{2\pi}{2 + \pi} r$$

6. The diagram shows a regular hexagon inscribed in a circle. Write the perimeter of the hexagon as a fraction of the circumference of the circle. Give your answer in terms of π.

   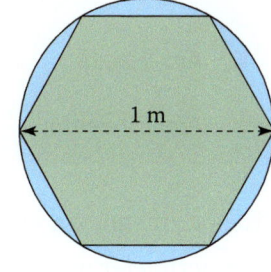

7. A rectangular sheet of metal is 1.2 metres long and 80 centimetres wide.

   A badge-making machine cuts circles of metal from this sheet.

   a i Find the maximum number of circles of diameter 4 cm that can be cut from the sheet.

   ii Sally says that less than 25% of the metal is wasted. Is Sally correct? Show your working.

   *b The diameter 4 cm is only correct to the nearest millimetre.
   What difference could this make to the answers to part **a**?

*8 The area of this aircraft window must be less than 100 square inches.
   Find the maximum value for the height h.

   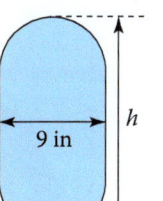

*9 a Show that the shaded area is given by the formula
   $A = r^2(4 - \pi)$

   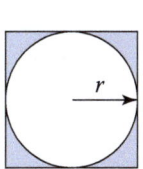

   b Without using a calculator, show that this is less than 25% of the area of the square.

10. Both these curves are made from semicircles. Which is longer? Give your reason.

    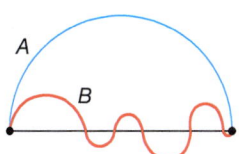

SKILLS

11.2 Circles 2

- **Arc** length $\quad s = \dfrac{\theta}{360°} \times 2\pi r \quad$ or $\quad \dfrac{\theta}{360°} \times \pi d$
- Area of **sector** $\quad A = \dfrac{\theta}{360°} \times \pi r^2$

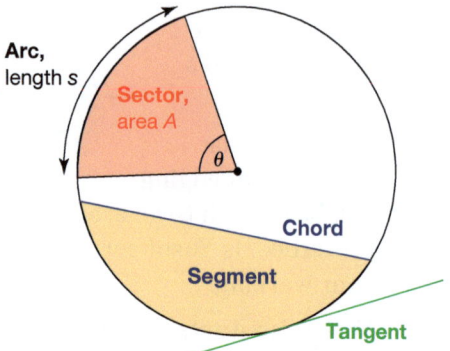

The perimeter of a sector consists of an arc and two radii.
The perimeter of a segment consists of an arc and a chord.
To find the area of a segment, you will need to use the area of a sector and subtract the area of a triangle.

EXAMPLE

a Find the area of this sector.

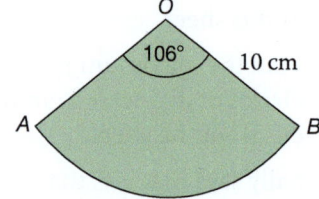

b Triangle *OAB* is removed from the sector. Find the area of the remaining segment.

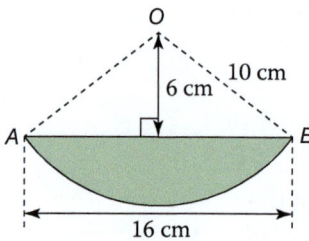

a Area of sector = $\dfrac{106°}{360°} \times \pi \times 10^2$

= 92.502...

= 92.5 cm² (3 sf)

You can store 92.502.... in a calculator memory to use later.

b Area of triangle $OAB = \dfrac{1}{2} \times 16 \times 6$

= 48

Area of segment = 92.502... − 48

= 44.502...

= 44.5 cm² (3 sf)

EXAMPLE

Find the perimeter of this sector.
Give your answer in terms of π.

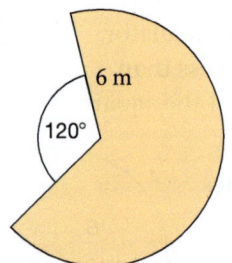

The perimeter includes the arc and the two radii.

Angle in the sector = 360° − 120°   *Angles at a point.*

= 240°

Arc length = $\dfrac{\cancel{240}^{2}°}{\cancel{360}_{3}°} \times 2\pi \times \cancel{6}^{2}$

Simplify by cancelling.

= 8π

Perimeter = 8π + 6 + 6

= (8π + 12) m or 4(2π + 3) m

Fluency

Exercise 11.2S

1. Simplify, leaving each answer in terms of π
   a $\frac{60°}{360°} \times 2\pi \times 24$ b $\frac{120°}{360°} \times 2\pi \times 15$
   c $\frac{80°}{360°} \times \pi \times 6^2$ d $\frac{135°}{360°} \times \pi \times 12^2$

2. Copy and complete the table for four sectors.

	angle	diameter	arc length	area of sector
a	40°	18 mm		
b	135°	1.6 m		
c	252°	3.5 cm		
d	312°	0.46 km		

3. For each sector, find in terms of π,
   i the perimeter   ii the area.
   a 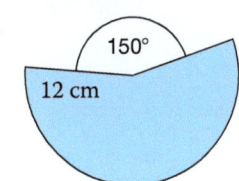   b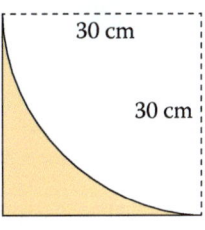

4. A **quarter-circle** of radius 30 cm is cut from a square with sides of length 30 cm. Find the area that is left. Give your answer in terms of π.
   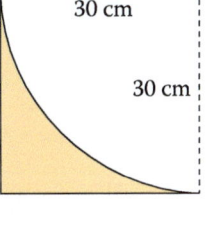

5. a Find the area of this sector.
   b Find the area of the segment that remains when triangle *OPQ* is removed.
   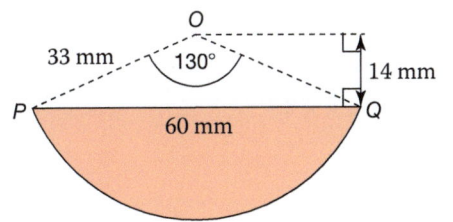

6. Find, in terms of π,
   a the shaded area
   b the perimeter of shaded area.
   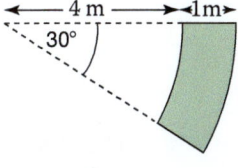

7. A shape is made from a quarter-circle and a rectangle. Find, in terms of π,
   a the perimeter of the shape
   b the area of the shape.
   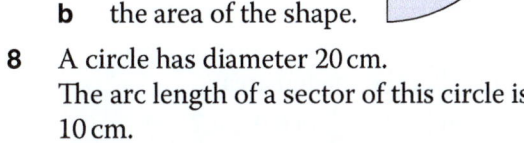

8. A circle has diameter 20 cm. The arc length of a sector of this circle is 10 cm. Find the area of the sector.

9. A sector of a circle has radius 30 cm and area 500 cm². Find the perimeter of the sector.

*10. The sectors in this pattern all have the same angle. The radius of the red sectors is twice as long as the radius of the black sectors. What percentage of the pattern is red?
   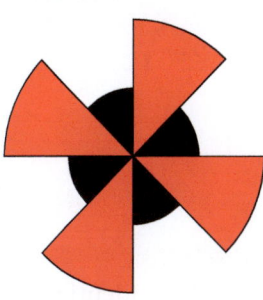

11. The diagram shows the shape of a flowerbed.
   a Find the length of edging needed to go around the perimeter of the flowerbed.
   b Find the area of the flowerbed.
   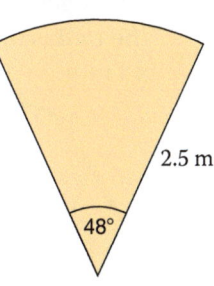

Did you know…

A circle has the largest area for a given perimeter. Tree trunks minimise the risk of damage by being round.

1118, 1952

APPLICATIONS

11.2 Circles 2

RECAP
- Arc length $\quad s = \dfrac{\theta}{360°} \times 2\pi r \quad$ or $\quad \dfrac{\theta}{360°} \times \pi d$
- Area of **sector**, $A = \dfrac{\theta}{360°} \times \pi r^2$

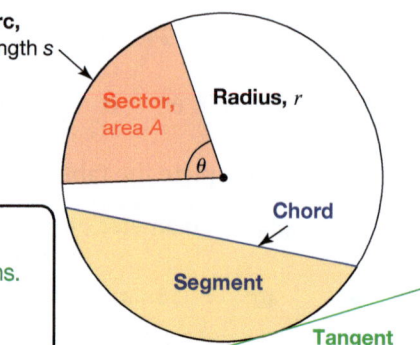

HOW TO

To solve problems involving circles and parts of circles
1. Draw a diagram if it helps. Label useful points and given dimensions.
2. Decide which formula to use.
3. Work out the answer to the question, giving units where necessary.

EXAMPLE

Show that the shaded area is given by $A = \dfrac{1}{8}\pi(R + r)(R - r)$.

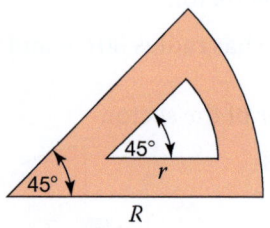

p.34
p.120

② Use area of sector, $A = \dfrac{\theta}{360°} \times \pi r^2$

Area of large sector $= \dfrac{45°}{360°} \times \pi \times R^2 \quad$ Cancel the fraction.

$\qquad\qquad\qquad\quad = \dfrac{1}{8}\pi R^2$

Area of small sector $= \dfrac{1}{8}\pi r^2 \quad$ The fraction is the same.

Shaded area, $A = \dfrac{1}{8}\pi R^2 - \dfrac{1}{8}\pi r^2$

$\qquad\qquad A = \dfrac{1}{8}\pi(R^2 - r^2) \quad$ Taking out the common factor.

③ $A = \dfrac{1}{8}\pi(R + r)(R - r) \quad$ Factorising the difference of 2 squares.

EXAMPLE

p.48

Three identical pencils are held together with a band. The radius of each pencil is 4 mm.

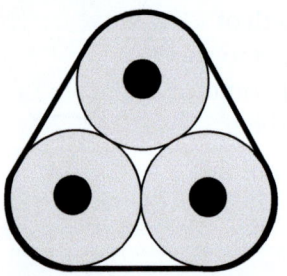

Show that the length of the band is $8(\pi + 3)$ mm.

① Joining the pencil centres gives an equilateral triangle. Drawing other radii from the centres to the ends of the arcs gives 3 congruent rectangles.

② The arc length formula needs an angle

$\angle PAU = 360° - 60° - 90° - 90°$
$\qquad\quad = 120°$

Arc $PU = \dfrac{120°}{360°} \times 2\pi \times 4 = \dfrac{1}{3} \times 8\pi$

③ Length of a straight section, $UT = AC = 8$ mm

Total length $= 3 \times \dfrac{1}{3} \times 8\pi + 3 \times 8$

$\qquad\qquad\quad = 8\pi + 24$

$\qquad\qquad\quad = 8(\pi + 3)$ mm

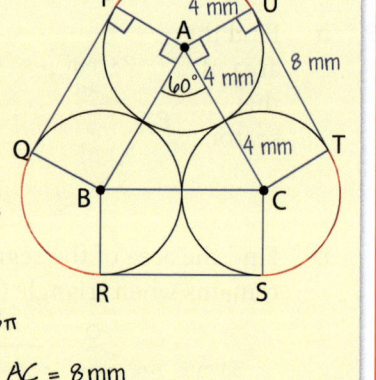

Geometry   Circles and constructions

Reasoning and problem solving

Exercise 11.2A

1. A continuous belt is needed to go around four identical oil drums as shown.
   The diameter of each drum is 24 inches.
   Work out the length of the belt. Give your answer in terms of π.

   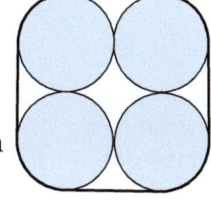

2. A sector of radius 3 cm is cut from one corner of an equilateral triangle.
   Find the area of the remaining shape.

   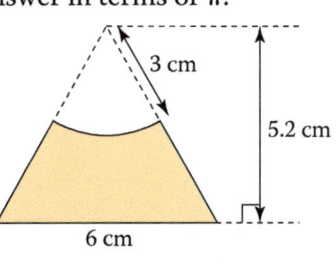

3. Two quarter-circles are used to draw the leaf on this square tile. Show that

   a the perimeter of the leaf is 20π cm

   b the area of the leaf is $200(\pi - 2)$ cm².

   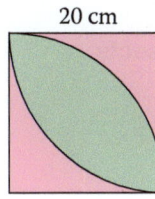

4. The diagrams show the water in two pipes.

   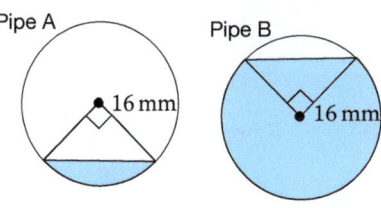

   Show that the difference between the cross-sectional area taken up by the water in these pipes is $128(\pi + 2)$ mm².

5. a Show that the area of this shape is given by $A = \frac{1}{12}\pi(R + r)(R - r)$

   b Find the area when $R = 4\frac{1}{2}$ m and $r = 1\frac{1}{2}$ m.
   Give your answer in terms of π.

   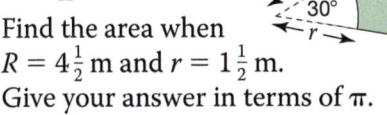

6. Find the *exact* area of this shape.

   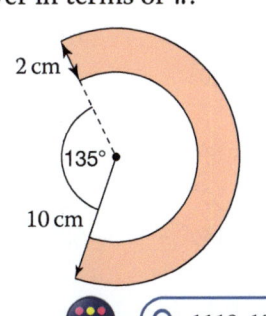

7. a Through what angle does the large cog turn when the small cog rotates once?

   b Ali says that the angle turned by the small cog is always $2\frac{1}{2}$ times the angle turned by the large cog. Is this true? Show how you decide.

   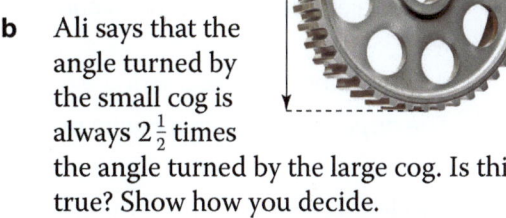

*8 The diagram shows the entrance to a road tunnel.

   a Calculate the perimeter.

   b The area needs to be at least 24 m².
   Does the area satisfy this requirement? Explain your answer.

   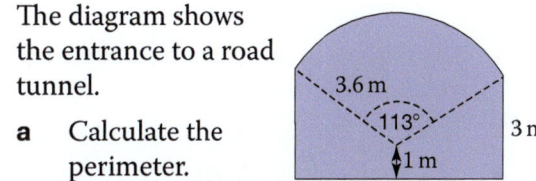

*9 Two sectors are cut from an isosceles triangle as shown. Find, in terms of π, the area of the remaining shape.

Did you know…

The diameter of the London Eye is 135 m. It has 32 capsules spaced equally along its circumference and takes 30 minutes to complete 1 revolution.

Can you work out the length of the arc between each capsule and the next?

How far does a capsule travel in ten minutes?

1118, 1952

SKILLS

11.3 Circle theorems

Some **circle theorems** involve facts about angles in a circle.

- The angle **subtended** by an arc at the centre of a circle is equal to twice the angle subtended at the circumference.
- The angle subtended at the circumference by a semi-circle is 90°.
- Angles in the same segment are equal.
- The opposite angles in a **cyclic quadrilateral** sum to 180°.

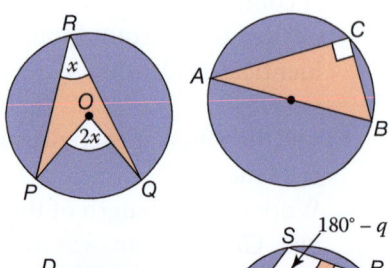

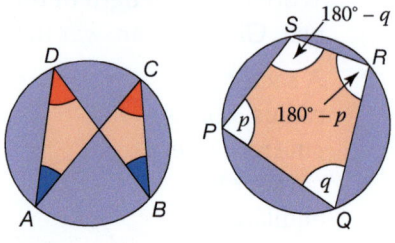

Some circle theorems are about radii, tangents and chords.

- The tangent at any point on a circle is perpendicular to the radius at that point.
- Tangents from an external point are equal in length.
- The perpendicular from the centre to a chord **bisects** the chord.
- The angle between a tangent and a chord is equal to the angle in the **alternate segment**.

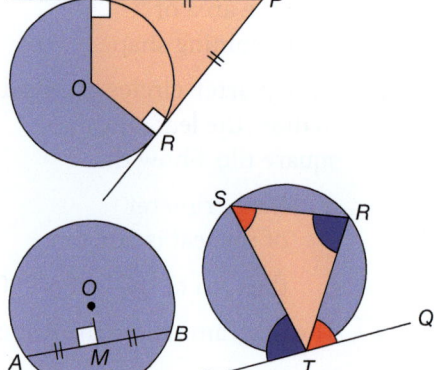

p.44 You can use these and other geometrical facts from earlier chapters to find angles.

EXAMPLE

Angle $CBE = 65°$

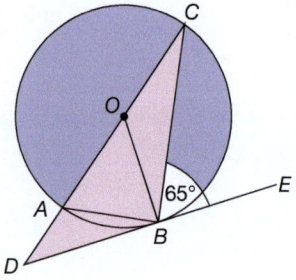

Find these angles, giving reasons.
a angle OBC   b angle BOC
c angle BAC   d angle ODB

Always say which geometrical fact you are using.

a $\angle OBC = 90° - 65°$ Angle between tangent BE and
         $= 25°$ radius OB is 90°.

b $\angle BOC = 180° - 25° - 25°$ OBC is an isosceles triangle
         $= 130°$ with equal base angles.

Radii often give useful isosceles triangles.

c $\angle BAC = 130° \div 2$ Angle subtended by BC at centre is
         $= 65°$ twice the angle at the circumference.

Or $\angle BAC$ is 65° because of the alternate segment theorem.
Or $\angle OCB = 25° = \angle OBC$ $\angle CBA = 90°$ angle in semi circle
$\angle CAB = 180 - 90° - 25° = 65°$

d In triangle ODB

$\angle DBO = 90°$ Angle between tangent DE
                         and radius OB.

$\angle BOD = 180° - \angle BOC$ Angles at O on a straight line.
        $= 50°$

$\angle ODB = 180° - 90° - 50°$ Angle sum of a triangle.
        $= 40°$

There are often many different ways. For example, you could find $\angle BOD$ from isosceles triangle BOA or by multiplying $\angle BCA$ by 2.

Geometry   Circles and constructions

Fluency

Exercise 11.3S

1. Work out *a*, *b*, *c* and *d* giving your reason for each answer.

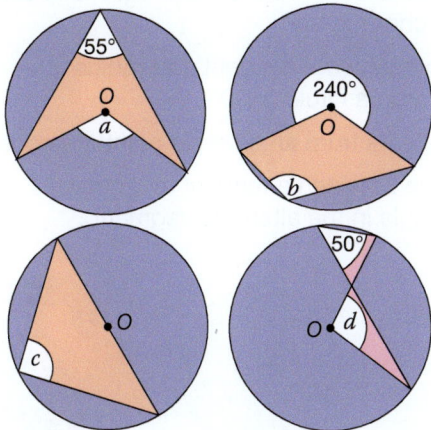

2. Work out the marked angles, giving reasons.

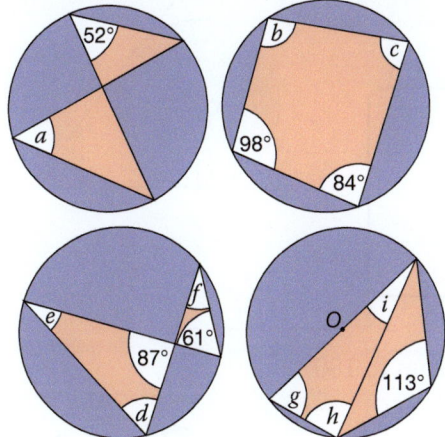

3. Work out *x* and *y*, giving your reasons.

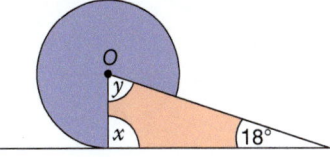

4. Work out *p*, *q* and *r* giving your reasons.

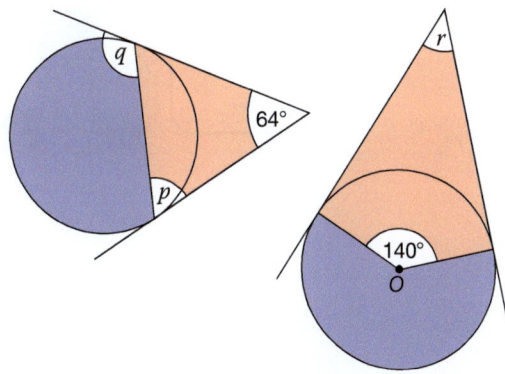

5. Use the alternate segment theorem to write down the values of *a* and *b*.

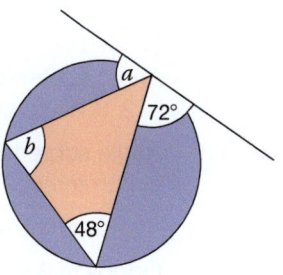

6. Work out angle *QOS*, angle *QRS* and angle *OQR*.

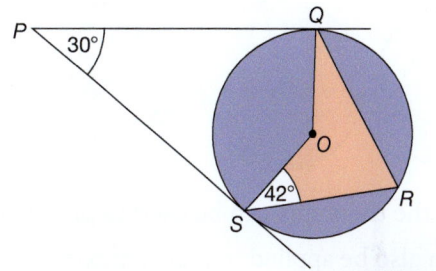

7. Angle $ABD = 64°$

   Work out angle *ACD*, angle *CAD*, angle *ADF* and angle *CDE*.

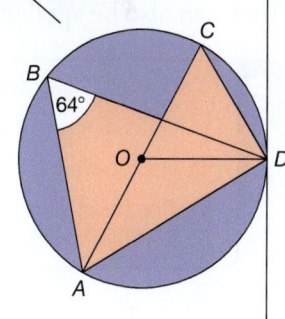

*8. Angle $PST = 53°$ and angle $SPT = 49°$. Work out all the angles of cyclic pentagon *PQRST*.

Check that the angle sum is correct for a pentagon.

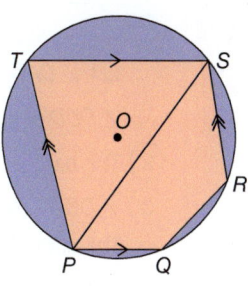

Did you know...

Until the work of Johannes Kepler and Tycho Brahe in the seventeenth century, for aesthetic reasons, astronomers thought that heavenly bodies followed paths created by combinations of circular motions.

1087, 1142, 1321

APPLICATIONS

11.3 Circle theorems

RECAP

- Angle subtended by arc at centre = 2 × angle subtended at circumference
- Angle subtended by semi-circle at circumference = 90°
- Angles in the same segment are equal.
- Opposite ∠s in a cyclic quadrilateral sum to 180°.
- Perpendicular from centre to a chord cuts the chord in half.
- Angle between tangent and radius at a point on a circle = 90°
- Tangents from an external point are equal.
- Angle between a tangent and chord = angle in the alternate segment.

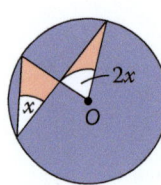 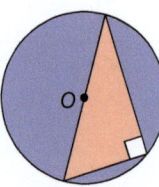 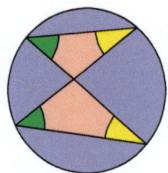 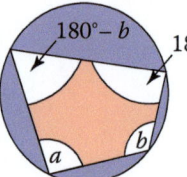 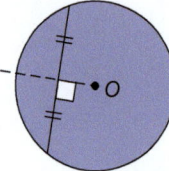 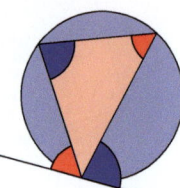

Some of the theorems can be used to prove others.

They can also be applied in real contexts.

HOW TO

To solve an angle problem involving circles

1. Draw a diagram (or use one that is given), add radii if useful.
2. Decide which theorems may apply.
3. Construct a chain of reasoning to find the angles you need or to prove the required results.

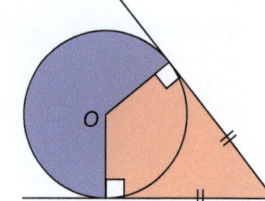

EXAMPLE

Prove that the opposite angles of a cyclic quadrilateral sum to 180°.

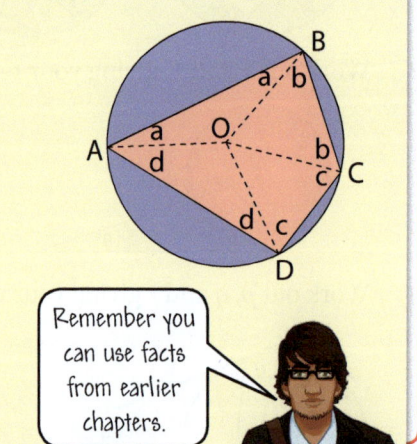

① Add radii to give isosceles triangles OAB, OBC, etc.
  Let $\angle OAB = a$, $\angle OBC = b$,
  $\angle OCD = c$ and $\angle ODA = d$.
② The base angles of an isosceles triangle are equal.
③ $\angle OBA = \angle OAB = a$, etc.
  The sum of the interior angles in quadrilateral $ABCD$ is 360°.

$$360° = (a + b) + (b + c) + (c + d) + (d + a)$$
$$= 2(a + b + c + d) \quad \text{Each letter appears twice.}$$
$$180° = a + b + c + d \quad ÷ 2$$
$$= \angle ABC + \angle CDA \quad \angle ABC = a + b, \angle CDA = c + d$$
$$= \angle BAD + \angle DCB \quad \angle BAD = a + d, \angle DCB = b + c$$

Remember you can use facts from earlier chapters.

Geometry Circles and constructions

Reasoning and problem solving

Exercise 11.3A

1. **a** **i** Use triangle AOC to write angle AOC in terms of x.

   ii Hence, write angle AOD in terms of x.

   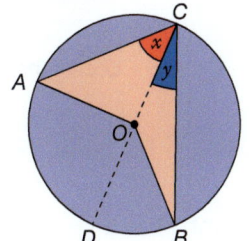

   b Use similar reasoning to write angle BOD in terms of y.

   c State the circle theorem that your working proves.

   d **i** Show how this theorem can be used to prove that angles in the same segment are equal.

   ii What special case of the theorem gives the result that the angle subtended by a semi-circle is 90°?

2. Prove that the exterior angle PST of cyclic quadrilateral PQRS is equal to the interior opposite angle PQR.

   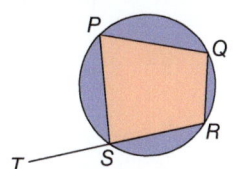

3. Any tangent is perpendicular to the radius drawn at its point of contact with a circle.

   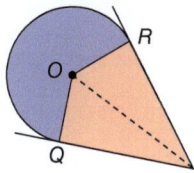

   a Use this result to prove that triangle PQO is congruent to triangle PRO and hence that PQ = PR.

   b What can you say about the line joining the centre to the point of intersection of the tangents and the angle between the tangents?

4. M is the mid-point of chord AB.

   Prove that OM is perpendicular to AB.

   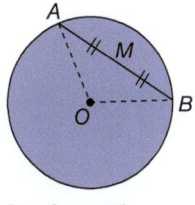

   This is the converse of the theorem that the perpendicular from the centre to a chord cuts the chord in half.

5. **a** Copy and complete the following to give each angle in terms of x.

   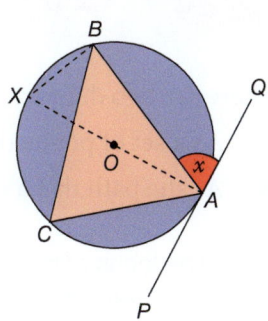

   Give reasons.
   $\angle BAX = \square$
   $\angle BXA = 180° - \square - \square = \square$
   $\angle BCA = \angle BXA = \square$

   b Part **a** is a proof of the alternate segment theorem. Use this theorem to find two other angles on the diagram that are equal.

6. The sides of quadrilateral ABCD touch a circle at P, Q, R and S.

   Prove that
   $AB + CD = BC + DA$

   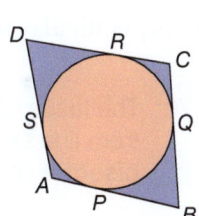

*7 Two chords AB and AC in a circle are equal. Prove that the tangent at A is parallel to BC.

*8 Points P, Q and R lie on a circle.

   The tangent at P is parallel to QR and the tangent at Q is parallel to PR.

   Prove that triangle PQR is equilateral.

9. Angle POB = 112°. Angle CAB = 48°
   PT is a tangent. AB is a diameter.
   Find $\angle PTB$.
   Show your working.

   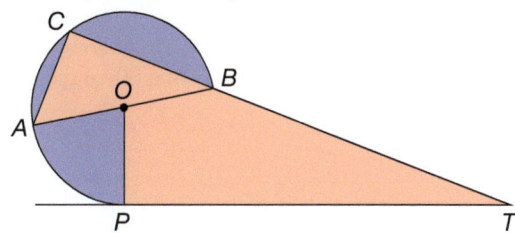

10. Prove that $\angle BOC = 4 \times \angle OAC$

    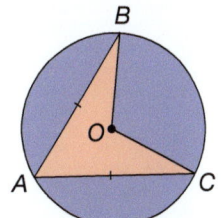

SKILLS
11.4 Constructions and loci

There are two different ways to think of a **locus**.
- As a set of points that follow one or more rules.
- As the path that a moving point follows.

Loci is the plural of locus.

- The locus of points that are **equidistant** from a point, P, is a circle with centre P.

- The locus of points that are equidistant from two given points, A and B, is the **perpendicular bisector** of AB.

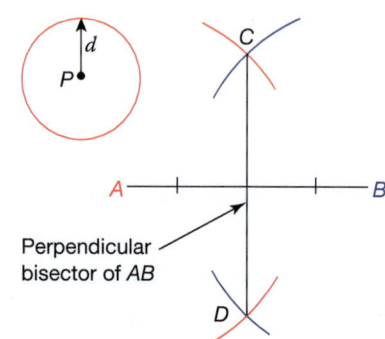

You can use a pencil, ruler and compasses to **construct** the perpendicular bisector of a **line segment**. You can also construct a perpendicular to a line from or at a given point, P.

- The locus of points that are equidistant from line segments PQ and PR is the **angle bisector** of angle P.

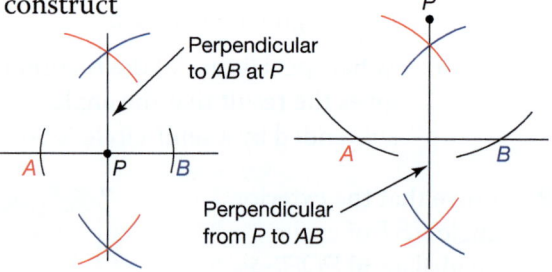

The example below shows how to bisect an angle.

A locus may be one or more points, a line or a region.

Sometimes the rules for a locus give overlapping regions.

The perpendicular is the shortest distance from a point to a line.

EXAMPLE

Using a *pencil, ruler and compasses only*,
a construct angle $ABC = 60°$
b on your diagram, shade the locus of points that are nearer to AB than BC.

a Draw a line AB.
With centre A and radius equal to AB, draw an arc above AB. With centre B and the *same* radius, draw another arc to intersect the first arc at C. Join BC to give a 60° angle at B.

b With centre B draw arcs at X and Y.
With centre X, then Y draw arcs to meet at Z.
BZ bisects the angle at B.
The points below BZ are nearer to AB than BC.

Draw the angle bisector as a dotted line to show that it is *not* part of the locus.

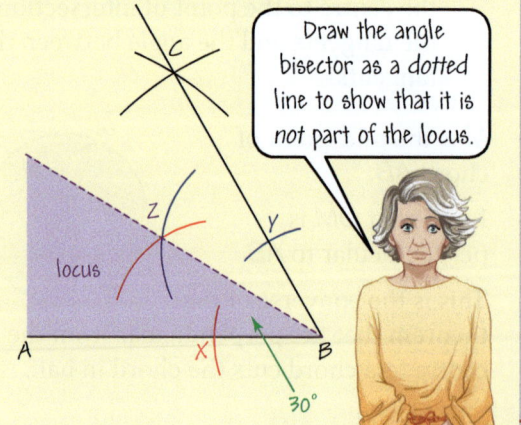

Geometry Circles and constructions

Fluency

Exercise 11.4S

Use a pencil, ruler and compasses only.

1. **a** Sketch the locus of points that are 2 cm from a fixed point *P*.
   b Shade the locus of points that are less than 2 cm from *P*.

2. Draw 2 parallel lines *PQ* and *RS*.

   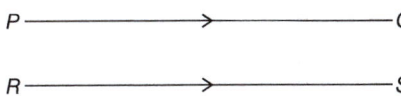

   On your diagram sketch the locus of a point that moves so that its distance from *PQ* is equal to its distance from *RS*.

3. Label two points, *A* and *B*, 7.8 cm apart. Construct the locus of points that are equidistant from *A* and *B*.

4. **a** Draw any vertical line *QR*. Mark a point *P* at one side as shown.
   b Construct the shortest line from *P* to *QR*.
   c Use a protractor to check that the angle formed is 90°.

5. **a** Construct an angle of 90°.
   b Bisect your angle to give an angle of 45°.

6. Construct each of the following angles. Use a separate diagram for each part.
   a 120° **b** 15° ***c** 75°

7. Draw a line, *AB*, of length 6.4 cm and the locus of points that are 3 cm from *AB*.

8. Using a line *XY*, 7 cm long, show the locus of points that are less than 4 cm from *X* and less than 5 cm from *Y*.

9. Draw an accurate diagram to show the locus of points that are more than 3.5 cm, but less than 4.5 cm from a fixed point *P*.

10. Draw separate diagrams to show the locus of a point *P* which moves so that it is
    a outside a 5 cm by 7 cm rectangle and 2 cm from it
    b outside an equilateral triangle with sides of length 4 cm and 1 cm from it.

11. **a** Draw a square with sides of length 6 cm. Label the vertices *A*, *B*, *C* and *D*.
    b On your diagram show the locus of points inside *ABCD* that are less than 3 cm from *A* and more than 4 cm from *BC*.

12. **a** Draw triangle *ABC* with *AB* = 6 cm, *BC* = 8 cm and angle *ABC* = 90°.
    b On your diagram show the locus of points inside *ABC* that are more than 7 cm from *C* and nearer to *A* than to *B*.

*13. **a** Construct an equilateral triangle, *ABC*, with sides of length 10 cm.
    b Find and mark the point that is equidistant from *AB*, *BC* and *AC*. Label this point *X*.

*14. **a** Construct quadrilateral *ABCD* with the dimensions shown.
    b Find and mark the point that is equidistant from *B*, *C* and *D*. Label this point *Y*.

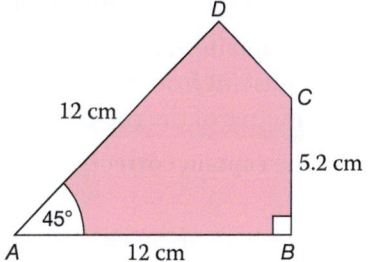

15. Construct a triangle *ABC* with angle *CAB* = 30°, *AB* = 8 cm and *BC* = 6 cm.

16. Draw sketches to show
    a the locus of a point *P* on the circumference of a coin as it rolls along the ruler
    b the locus of a vertex *Q* of the square card as it is rotated against the ruler.

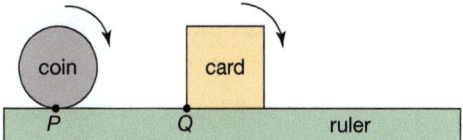

APPLICATIONS

11.4 Constructions and loci

RECAP

- The locus of points that are equidistant from a point, P, is a circle with centre P.
- The locus of points that are equidistant from two given points, A and B, is the **perpendicular bisector** of AB.
- The locus of points that are equidistant from line segments PQ and PR is the **angle bisector** of angle P.

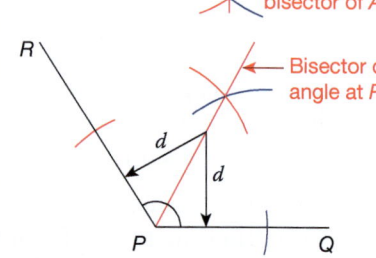

HOW TO

To solve a loci construction problem

① Draw a sketch or a scale diagram.
   If a scale is not given, choose one that is easy to use.
② Use a pencil, pair of compasses and ruler to carry out constructions. Leave all construction lines on your diagram and label clearly.
③ Answer any questions asked.

EXAMPLE

A large rock, R, lies 36 metres due North of a lighthouse, L.

A wreck, W, lies 90 metres from the lighthouse on a bearing of 300°.

a Using a *pencil, ruler and compasses only*, draw a scale diagram to show the positions of the lighthouse, wreck and rock. Use a scale of 1 cm to represent 20 metres.

b A ship approaches the lighthouse from the South West.

The captain wishes to sail between the lighthouse and wreck, on a course that is equidistant from them. He thinks he can stay on this course without coming within 30 metres of the rock.

Is the captain correct? Show how you decide.

a ① Distance LR on the scale diagram = 36 ÷ 20 = 1.8 cm

② A bearing of 300° is 60° to the west of north.
Construct a 60° angle at L.

Distance LW on the scale diagram = 90 ÷ 20 = 4.5 cm

The first diagram shows the position of L, W and R.

Leave all construction lines on the diagram.

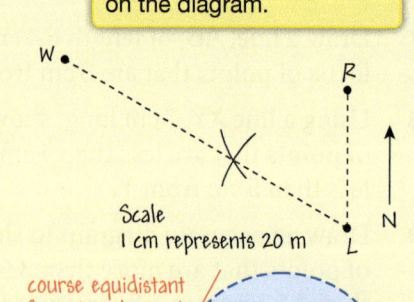

b The course that is equidistant from L and W is the perpendicular bisector of LW. Construct this on the diagram.
The points within 30 metres of the rock lie inside a circle.

30 metres on the scale diagram = 30 ÷ 20
= 1.5 cm

③ The shaded circle shows the points that are within 30 metres of the rock.

Some of the points on the captain's course lie inside this circle, so the captain is wrong.

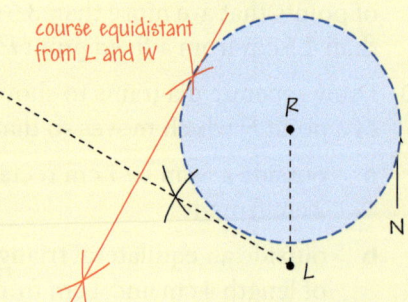

Geometry Circles and constructions

Reasoning and problem solving

Exercise 11.4A

1. A lifeboat L is 10 km from another lifeboat K on a bearing of 045°. They both receive a distress call from a ship. The ship is within 7 km of K and within 5 km of L.

   Draw a scale drawing to show the positions of K and L. Shade on your diagram the area in which the ship could be.

2. Ayton is 23 km west of Bramley and Collingford is 29 km from Bramley on a bearing of 210°.

   A mobile phone mast is planned to be located equidistant from the three villages.

   Use a scale diagram to show the location of the mobile phone mast and give a reasonable estimate of how far it is from each village.

3. Meera's lawnmower cable can reach 12 m from the power point at P.

   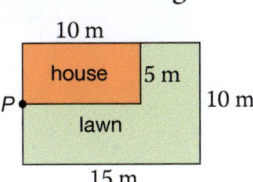

   a Use a scale diagram to show which parts of the lawn Meera can mow.

   b What is the minimum length of cable Meera needs to reach all of the lawn?

4. Ann sets off on a bearing of 060° from A to walk across this rectangular plot of land.

   At the same time, Barid starts from B and keeps the same distance from AB and BC as he walks across the plot.

   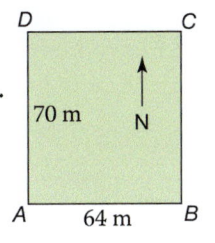

   Ann and Barid meet a few minutes later.

   a Use a scale drawing to show where Ann and Barid meet.

   b Barid says that he walks faster than Ann.

   Do you think this is true? Give your reason.

5. A caretaker is polishing the floor of a 2 metre wide corridor. The polisher is in the shape of a circle, diameter 50 cm. Calculate the area of the floor that is polished each time the caretaker moves the polisher perpendicular to the walls from one side of the corridor to the other.

   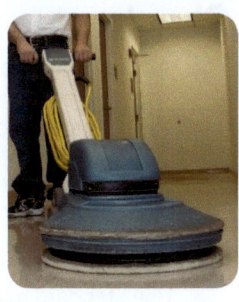

6. Genca wants to tether her goat on this grass using a 5 foot long chain. One end of the chain will be attached to the goat and the other end to a ring that can slide along an 18 foot long rail.

   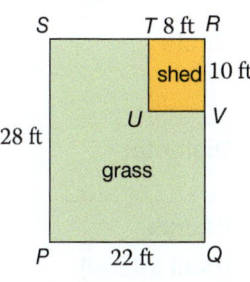

   a Calculate the area of the grass that the goat can reach if Genca puts the rail along the sides TU and UV of the shed.

   b Where on the perimeter of the grass should Genca put the rail to allow the goat to reach the greatest area of grass? Explain your answer.

*7 a Jamie says that the construction of an angle bisector works because of congruent triangles.

      Is this true? Give your reason.

   b Give geometric reasons why each of the following constructions work.

      i   60° angle

      ii Perpendicular bisector of a line segment.

Did you know…

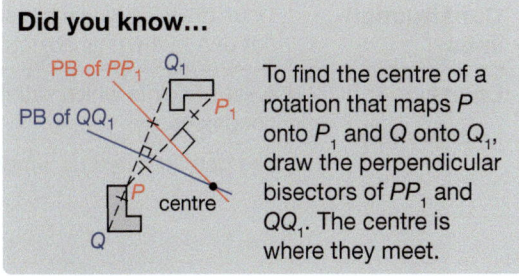

To find the centre of a rotation that maps P onto P_1 and Q onto Q_1, draw the perpendicular bisectors of PP_1 and QQ_1. The centre is where they meet.

Summary

Checkout
You should now be able to...

Test it
Questions

✓	Find the area and circumference of a circle and composite shapes involving circles.	1 – 3
✓	Calculate arc lengths, angles and areas of sectors.	4
✓	Prove and apply circle theorems.	5, 6
✓	Use standard ruler and compass constructions and solve problems involving loci.	7 – 9

Language	Meaning	Example
Circle	A closed curve in a flat surface which is everywhere the same distance from a single fixed point.	
Diameter	A chord that passes through the centre of the circle.	
Radius Radii (plural)	A straight line segment drawn from the centre of the circle to the perimeter.	
Circumference	The distance around the edge of a circle.	
Arc	A continuous section of the circumference of a circle.	
Chord	A straight line segment with endpoints on the circumference of a circle.	
Tangent	A straight line which touches the circle at one point only and does not cut the circle.	
Segment	The 2D shape enclosed by an arc and a chord.	
Sector	The shape enclosed by two radii and an arc.	
Bisect	Cut into two parts of the same shape and size.	
Perpendicular bisector	A line which bisects another line at right angles.	
Construct	Draw something accurately using compasses and a ruler.	
Construction lines	Lines drawn during a construction that are not part of the final object.	
Locus Loci (plural)	A set of points which satisfy a given set of conditions. The path followed by a moving point.	The set of points less than 1 cm from P is the interior of a circle.

Geometry   Circles and constructions

Review

1 For this circle calculate the
   a area
   b circumference.

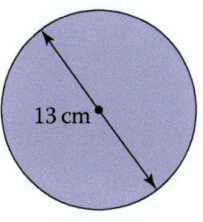

2 The area of a circle is 160 cm². What is the diameter of the circle to 1 decimal place?

3 The shape of a swimming pool consists of a rectangle with a semi-circle at one end as shown.

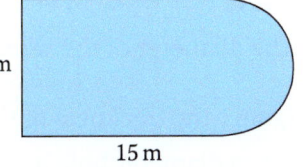

   Calculate
   a the area
   b the perimeter of the swimming pool.

4 Calculate
   a the area
   b the arc length
   c the perimeter of the sector.

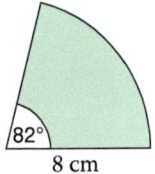

5 Calculate the size of angles a, b, c and d. Give reasons for your answers.

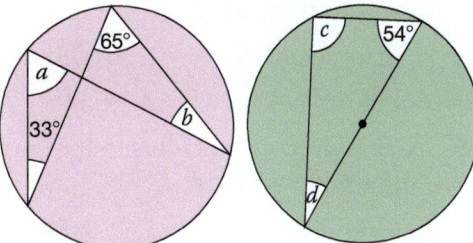

6 Calculate the size of angles a and b. Give reasons for your answers.

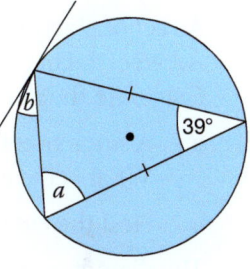

7 a Draw a line exactly 6.5 cm long.
   b Use a ruler and pair of compasses to accurately construct the perpendicular bisector of this line.

8 A dog is attached to a lamp post by a rope which is 2.4 m long and can rotate freely around the post.

   Use a scale of 1 : 60 to draw to scale the locus of the furthest points the dog can reach.

9 Use a pencil, ruler and pair of compasses to construct a triangle with sides of lengths 8 cm, 7 cm and 5 cm.

What next?

Score		
0 – 4		Your knowledge of this topic is still developing. To improve, look at MyiMaths: 1083, 1087, 1088, 1089, 1090, 1118, 1142, 1147, 1321
5 – 8		You are gaining a secure knowledge of this topic. To improve your fluency look at InvisiPens: 11Sa – i
9		You have mastered these skills. Well done you are ready to progress! To develop your problem solving skills look at InvisiPens: 11Aa – g

Assessment 11

1. **a** Match the correct radius to the correct circumference and find the missing radius. Show your working.

   Circumferences   **i** 61.58 cm   **ii** 314.16 cm   **iii** 2.01 cm

   Radii   **A** x cm   **B** 0.320 cm   **C** 50.0 cm   [3]

   b Match the correct radius to the correct area and find the missing radius. Show your working.

   Areas   **i** 176.7 m²   **ii** 81.1 m²   **iii** 0.407 m²

   Radii   **A** 5.08 m   **B** 7.50 m   **C** y m   [3]

2. A reel of electrical cable contains 50 m of cable. The diameter of the wheel is 25 cm. An electrician completely unwinds the reel.
   Calculate the number of times the reel rotates. (Ignore the thickness of the cable). [3]

3. **a** A mint sweet is 19 mm in diameter with an 8 mm diameter hole in the middle. What is the area of the face to the nearest mm²? [3]

   b Find the circumferences of the outer part and inner holes. [2]

4. The diameter of a bike's wheels is 27.56 inches. 1 mile = 63 360 inches. How many times, to the nearest whole number, do the wheels rotate on a 50 mile ride? [4]

5. The London Eye's wheel has a diameter of 120 m. The wheel rotates at 26 cm per second. How long, to the nearest minute, does the wheel take to make a complete revolution? [4]

6. The diagram shows a heart shape.

   There are two semicircles at the top each 6 cm in diameter.

   The sides are made of two circular arcs. The right hand arc, YZ has centre X and the left hand arc, XZ has centre Y. The distance from the middle of the two semicircles at the top to the point Z is 10.39 cm.
   Calculate

   a the total perimeter of the heart diagram [6]

   b the area of the heart diagram. [8]

7. A sector has a radius equal to its arc length.

   Show that the angle at the centre is 57.3° to 1 dp. [3]

8. This is a diagram of the centre circle of an ice hockey pitch. There are players at the centre point O and on the circumference at points A, B and C. Angle AOB = 69°

   Write down the values of the angles p, q and r.
   Give reasons for your answers. [6]

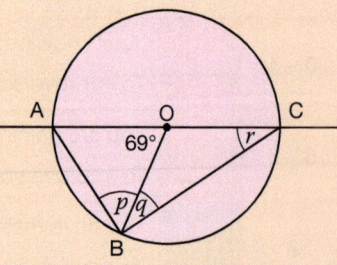

9 The diagram shows a surveyor's trundle wheel.
The handle is inclined at 34° to the vertical.
O is the centre of the circle and the diameter, AC, through O is horizontal.
Write down the values of the angles BOC, BCO and CAB.
Give reasons for your answers. [6]

10 The diagram shows the framework housing a ball on a tenpin bowling rink.
O is the centre of the circle and the diameter through it is vertical.
The line at the bottom of the framework is the horizontal runner on which the ball rests. Write down the values of the angles v, w and x. Give reasons for your answers. [6]

11 *Without* using the Alternate Segment Theorem prove that x = y. [6]

O is the centre of the circle.

The diagram includes the tangent to the circle at B.

A, B and C are points on the circumference.

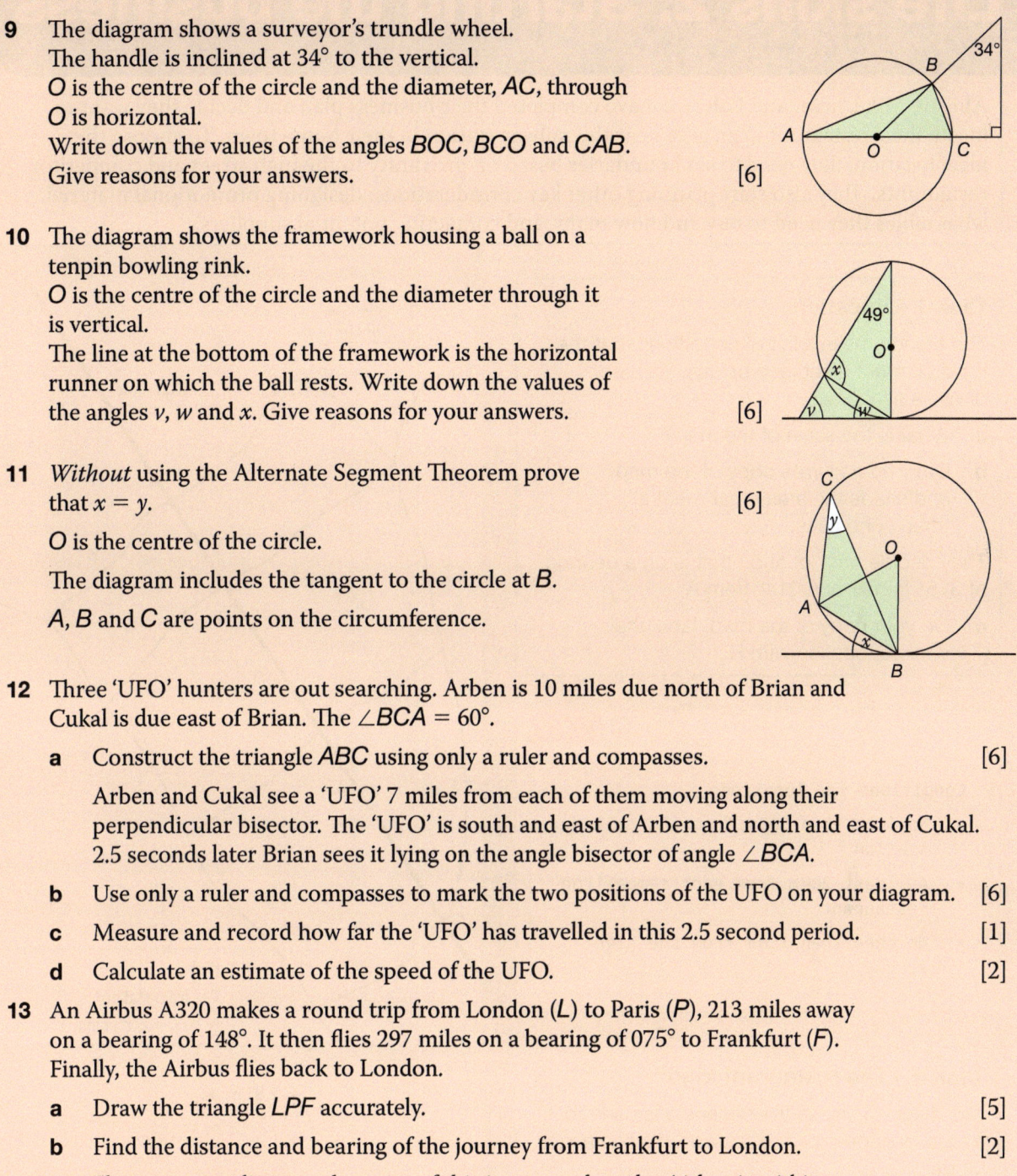

12 Three 'UFO' hunters are out searching. Arben is 10 miles due north of Brian and Cukal is due east of Brian. The ∠BCA = 60°.

  a Construct the triangle ABC using only a ruler and compasses. [6]

  Arben and Cukal see a 'UFO' 7 miles from each of them moving along their perpendicular bisector. The 'UFO' is south and east of Arben and north and east of Cukal. 2.5 seconds later Brian sees it lying on the angle bisector of angle ∠BCA.

  b Use only a ruler and compasses to mark the two positions of the UFO on your diagram. [6]

  c Measure and record how far the 'UFO' has travelled in this 2.5 second period. [1]

  d Calculate an estimate of the speed of the UFO. [2]

13 An Airbus A320 makes a round trip from London (L) to Paris (P), 213 miles away on a bearing of 148°. It then flies 297 miles on a bearing of 075° to Frankfurt (F). Finally, the Airbus flies back to London.

  a Draw the triangle LPF accurately. [5]

  b Find the distance and bearing of the journey from Frankfurt to London. [2]

  c Show on your diagram the parts of this journey when the Airbus is within 50 miles of London. [2]

Life skills 2: Starting the business

Abigail, Mike, Juliet and Raheem, have completed their business plan and decide they want to locate their restaurant in an area near the railway station in their home town. To choose the ideal location, they need to set boundaries based on proximity to the high street and competitor restaurants. They also start planning other key considerations: designing promotional material, what tables they need to buy and how many, and contacting potential suppliers.

Task 1 – Location

The friends make a list of conditions that the location must meet (see below). Raheem draws a scale map.

a   What is the scale of the map?

b   Draw an accurate copy of the map and shade the areas that meet all three conditions.

They decide on a location that is on a bearing of 315° from T and 014° from A.

c   On your copy of the map, label their desired location with R.

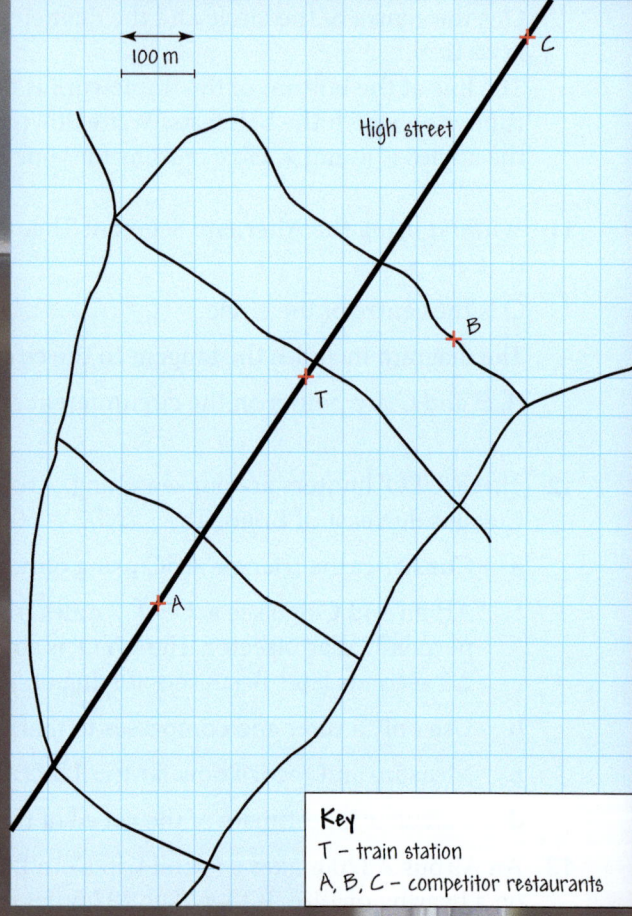

```
Conditions for location
 • No more than 500m from the railway
   station
 • At least 200m from each competitor
   restaurant
 • No more than 150m from the high
   street
```

Task 2 – The restaurant logo

The diagram shows the dimensions used for a logo to appear on the restaurant's business cards.

a   Find the area of the logo.

For use on A5 flyers, the logo is enlarged by a scale factor of 2.5.

b   Find the area of the enlarged logo.

The logo is the only part of the business card and the flyer that uses coloured ink.

c   A colour ink cartridge lasts long enough to print 4000 business cards. How many A5 flyers would you expect to print using one colour ink cartridge?

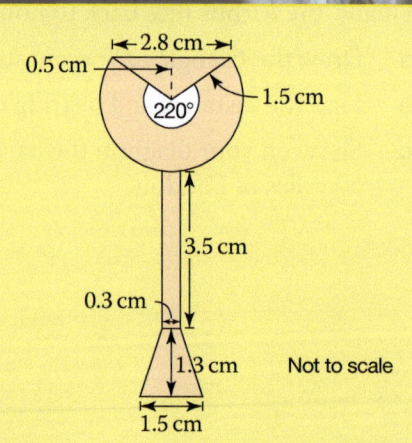

▲ Juliet's logo for use on their promotional material. It consists of a triangle, a circular sector, a rectangle and a trapezium, as shown in the second figure.

Life skills

Task 3 – Safety regulations

The restaurant must adhere to fire safety regulations.

a x is the number of customers (non-staff) in the restaurant and y is the number of staff in the restaurant. Write down two inequalities in x and y to express the fire safety regulations.

b On graph paper, draw x and y axes from $x = 0$ to 40 and from $y = 0$ to 40. Shade the region where both your inequalities are satisfied.

c Use your inequalities and/or graph to find the maximum number of customers allowed in the restaurant at any time.

FIRE

Fire regulations

- Total number of staff and non-staff must not exceed 32.
- Ratio of non-staff to staff must not exceed 8 : 1.

Table	Shape of top	Dimensions
Style A	Circular	Diameter = 1.4 m
Style B	Regular octagon	Side length = 58 cm
		Length between opposite sides = 1.4 m

▲ Possible styles of table for restaurant.

Task 4 – Choosing tables

They narrow down their choice of table based on the maximum number of customers they could have.

a Without doing any calculations, which table top (style A or style B) has the larger area?

b Calculate the area of each table top.

Style A is cheaper, so they decide to buy that one.

c Is this a reasonable decision? Give the reason for your answer.

Task 5 – Number of tables

The dining space is rectangular, and measures 12 m by 8 m. Abigail draws a sketch to indicate the gap required from each wall, and between each table.

a How many tables of Style A can fit in the dining space?

b If the measurements of 12 m and 8 m are accurate to the nearest 10 cm, find upper and lower bounds for the area of the rectangular space.

c Do you still agree with your answer to part **a**? Why?

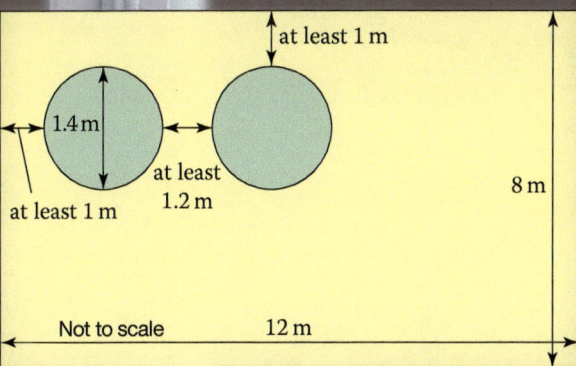

Task 6 – Calling potential suppliers

Each member of the team makes telephone calls to potential suppliers. They logged how many calls they made in the first week, and how many went unanswered.

a Which person had the highest proportion of unanswered calls?

b Estimate how many unanswered calls you would expect out of the next 50 calls made by the team.

c What assumptions have you made in answering part **b**?

Caller	Total number of calls made	Number of unanswered calls
Abigail	30	10
Mike	22	8
Juliet	40	18
Raheem	18	8

▲ Number of calls made in the first week.

12 Ratio and proportion

Introduction

Colour theory is based on the idea that all possible colours can be created from three 'primary colours'. Video screens use the additive colours red, green and blue. Artists often use red, yellow and blue as the basis for mixing paints. The printing industry uses the subtractive colours cyan, magenta and yellow. The ratio in which these primary colours are mixed determines the colour of the result.

What's the point?

An understanding of ratio is essential for artists in mixing colours on a palette to achieve a desired result. Furthermore, an understanding of proportion is essential for artists in ensuring that the elements of their artwork are in the 'right proportions'.

Objectives

By the end of this chapter, you will have learned how to …
- Express proportions of amounts as fractions or percentages.
- Divide a quantity in a given ratio.
- Use scale factors to convert between lengths on maps and scale diagrams and the distances they represent.
- Calculate percentage increases and decreases using multiplication.
- Find the original value of a quantity that has undergone a percentage increase or decrease.

Check in

1. Calculate each of these expressions.

   a. $\frac{2}{3}$ of 120
   b. 0.35×60
   c. 28% of 70
   d. $\frac{3}{5}$ of 280

2. Carry out each of these conversions.

   a. Convert $\frac{3}{5}$ to a percentage.
   b. Convert 0.35 to a fraction.
   c. Convert 65% to a fraction.
   d. Convert $\frac{7}{20}$ to a decimal.

3. Write these ratios in their simplest form.

   a. 4 : 12
   b. 5 : 30
   c. 6 : 9
   d. 15 : 6

Chapter investigation

The ratio of height to head circumference for the average person is said to be around 3 : 1. Investigate.

SKILLS

12.1 Proportion

- A **proportion** is a part of the whole. It can be written using a fraction, a decimal or a percentage.

EXAMPLE

Find   **a** $\frac{3}{5}$ of 85   **b** 120% of 45   **c** $\frac{7}{8}$ of 86

a $\frac{3}{5} \times 85^{17} = \frac{3}{1} \times 17 = 51$ Cancel the 5s before multiplying.

b Work out 20% of 45 and add it to the original amount.

120% of 45 = (100% of 45) + (20% of 45)

20% of 45 = $\frac{1}{5} \times 45 = 9$ ⟹ 120% of 45 = 45 + 9 = 54

c Work out $\frac{1}{8}$ of 86 and then subtract from the original amount.

$\frac{1}{8}$ of 86 = 86 ÷ 8 = 10 + $\frac{6}{8}$ = 10$\frac{3}{4}$ ⟹ $\frac{7}{8}$ of 86 = 86 − 10$\frac{3}{4}$ = 76 − $\frac{3}{4}$ = 75$\frac{1}{4}$

You can express one quantity as a percentage of another in three steps:
1. Write the first quantity as a fraction the second.
2. Convert the fraction to a decimal by division.
3. Convert the decimal to a percentage by multiplying by 100.

> Percentage is per-cent which means parts per hundred.

EXAMPLE

What proportion is   **a** 4 of 32   **b** 5 of 35   **c** 12 of 4.8?
Write your answers as percentages.

a $\frac{4}{32} = \frac{1}{8}$

so 4 is $\frac{1}{8}$ of 32

$\frac{1}{8} = \frac{1}{8} \times 100\%$

= 12$\frac{1}{2}$%

b $\frac{5}{35} = \frac{1}{7}$

so 5 is $\frac{1}{7}$ of 35

$\frac{1}{7} = \frac{1}{7} \times 100\%$

= 14.3%

c $\frac{12}{4.8} = \frac{120}{48} = \frac{5}{2}$

so 12 is 5.2 × 4.8

$\frac{5}{2} \times 100\% = 250\%$

EXAMPLE

Skye took two tests. In German she scored 35 out of 50, and in French she scored 60 out of 80. Write these scores as percentages.

German

35 out of 50 = $\frac{35}{50}$

= 35 ÷ 50

= 0.7

= 70%

French

60 out of 80 = $\frac{60}{80}$

= 60 ÷ 80

= 0.75

= 75%

- You can compare proportions by converting them to percentages or decimals.

Ratio and proportion   Ratio and proportion

Fluency

Exercise 12.1S

1 12.5% of the mass of a soft drink is sugar. Find the amount of sugar in these amounts of drink.
 a 750 g
 b 45 g
 c 1 kg
 d 1250 g

2 Three-fifths of the volume of a fruit punch is orange juice. Find the amount of orange juice in these volumes of fruit punch. Show all of your working.
 a 150 cm³
 b 380 cm³
 c 2 m³
 d 280 cm³

3 Calculate
 a 120% of 50 g
 b 90% of 40 mm
 c 95% of 400 g
 d 80% of 39 km

4 What proportion of
 a 15 is 5
 b 20 is 10
 c 10 cm is 3 cm
 d 12 kg is 3 kg
 e €30 is €12
 f $8 is $16
 g 200 is 25
 h 6.4 is 0.4
 i £1.44 is 12p
 j 1.2 m is 30 cm

 Write your answers as
 i fractions in their lowest terms
 ii percentages to 1 dp.

5 Write the proportion of each of these shapes that is shaded. Write each of your answers as
 i a fraction in its simplest form
 ii a percentage (to 1 dp as appropriate).

 a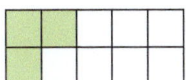
 b
 c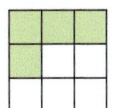
 d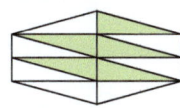

6 Work out these proportions, giving your answers as
 i fractions in their lowest terms
 ii percentages.
 a 7 out of every 20
 b 8 parts in a hundred
 c 6 out of 20
 d 75 in every 1000
 e 18 parts out of 80
 f 9 parts in every 60
 g 20 out of every 3
 h 20 out of 8
 i 412 parts in a hundred
 j 6400 in every thousand

7 Use an appropriate method to work out these percentages. Show your method each time.
 a 50% of 270 kg
 b 27.9% of 115 m
 c 37.5% of $280
 d 25% of 90 cm³
 e 19% of 2685 g
 f 27.5% of €60.00

8 Four candidates stood in an election.
 A received 19 000 votes
 B received 16 400 votes
 C received 14 800 votes
 D received 13 200 votes

 Write each of these results as a percentage of the total number of votes.

 Give your answers to 1 decimal place.

APPLICATIONS

12.1 Proportion

RECAP
- A proportion is part of the whole.
- You can use percentages, fractions and decimals to describe proportion.
- If you know what proportion of the total a part is, you can find the actual size of the part by multiplying the proportion by the total.
- To calculate a proportion, write the part as a fraction of the total.
- You can compare proportions by converting them to percentages.

HOW TO
1. Decide which value is the 'whole'.
2. Express the part as a fraction of the whole, or multiply by the fraction to find the part.
3. Convert to decimals or percentages if necessary for comparison.

EXAMPLE

A 1 kg bag of 'Grow Up' fertiliser contains 45 grams of phosphate. A 500 gram packet of 'Top Crop' fertiliser contains 20 grams of phosphate. What is the proportion of phosphate in each fertiliser?

1. Write the whole.

   The whole for 'Grow Up' fertiliser is 1 kg = 1000 g.
   The whole for 'Top Crop' fertiliser is 500 g.

2. Express the proportions as fractions.

   Proportion of phosphate in 'Grow Up'
   $= \dfrac{45}{1000} = \dfrac{9}{200}$

   Proportion of phosphate in 'Top Crop'
   $= \dfrac{20}{500} = \dfrac{4}{100}$

3. Convert the fractions to percentages.

   Percentage of phosphate in 'Grow Up'
   $= \dfrac{45}{1000} \times 100 = 4.5\%$

   Percentage of phosphate in 'Top Crop'
   $= \dfrac{20}{500} \times 100 = 4\%$

EXAMPLE

Marvin has $2800. He gives $\dfrac{1}{5}$ to his son and $\dfrac{1}{4}$ to his daughter. How much does Marvin keep himself? You must show all of your working.

1. Write the whole.

   The whole is $2800.

2. Multiply by the fraction to find the size of the part.

   Marvin's son receives
   $\dfrac{1}{5}$ of $2800 = \dfrac{2800}{5} = \$560$

   Marvin's daughter receives
   $\dfrac{1}{4}$ of $2800 = \dfrac{2800}{4} = \$700$

   Marvin keeps $2800 − ($560 + $700) = $1540

Ratio and proportion   Ratio and proportion

Reasoning and problem solving

Exercise 12.1A

1. A 250 ml glass of fruit drink contains 30 ml of pure orange juice. What proportion of the drink is pure orange juice?
   Give your answer as

   a a fraction in its lowest terms

   b a percentage.

2. Samantha wins £4500 in a competition. She gives $\frac{1}{3}$ to her mother and $\frac{1}{5}$ to her sister.

   a How much does she keep? Show your working.

   b What proportion of the prize money does she give away? Give your answer as a fraction and as a percentage.

3. A dairy farmer takes 200 cheeses to sell at a market. In the first hour she sells 20% of the cheeses. In the second hour she sells 15% of those that are left.

   a How many cheeses has she sold in total?

   b What percentage of the original number of cheeses did she sell in the first 2 hours?

4. Sunita took three tests. In Maths she scored 48 out of 60, in English she scored 39 out of 50 and in Science she scored 55 out of 70.

   a In which subject did she do

      i   the best

      ii the worst?

   b Suggest a way in which you could give Sunita a single overall score for her three tests.

5. 20% of the cars in a car park are red.
   40 cars are not red.
   How many cars are in the car park?

6. Rita is driving 220 miles from Belo Horizonte to Minas Gerais.
   Rita drives at an average speed of 45 mph for the first hour.
   She drives at an average speed of 60 mph for the next two hours.
   Rita says that
   "After three hours I will have completed three-quarters of the journey."
   Do you agree with Rita? Explain your answer.

7. Jada compares the amount of sugar in three different types of fruit juice.
   - Tropical 22.4 g of sugar in 180 ml
   - Pineapple 28.8 g of sugar in 250 ml
   - Blackcurrant 23.4 g of sugar in 200 ml

   Jada says that tropical juice has the highest proportion of sugar. Is she correct? Show your working.

8. a Complete these equivalent fractions.

      $\frac{16}{?} = \frac{18}{45}$ $\frac{18}{45} = \frac{20}{?}$

   b A company has 45 employees at the start of 2015.
      18 employees are female.
      By 2016, two more female employees have joined the company.
      The **proportion** of female employees has not changed.
      Use your answer to part **a** to find the number of total employees in 2016.

9. Jenni compares the proportion of the shaded areas of two shapes.

   Shape A   Shape B

   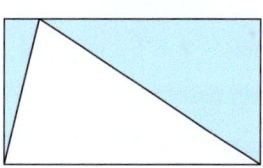

   She says that shape A has the larger proportion of shaded area.
   Do you agree with Jenni? Explain your answer.

10. What proportion of this shape is shaded?

    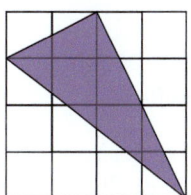

SKILLS

12.2 Ratio and scales

You can compare the size of two or more quantities using a **ratio**.
2 : 3 means '2 parts of one quantity, compared with 3 parts of another'.

- You can simplify a ratio by cancelling common factors. When a ratio cannot be simplified any further it is in its **simplest form**.

Two ratios are equivalent if they have the same simplest form.

EXAMPLE

Chuang the snake is only 30 cm long. George the snake is 90 cm long. Express this as a ratio.

Compare the two lengths.
= Chuang's length : George's length
= 30 cm : 90 cm
= 30 : 90

Express the ratio 30 : 90 in its simplest form.
30 : 90
÷10 → = 3 : 9 ← ÷10
÷3 → = 1 : 3 ← ÷3

Check that the measurements have the same units.

The ratio 1 : 3 means that 90 is three times bigger than 30.

You can divide a quantity in a given ratio.

EXAMPLE

Shuaib and Patrick share $355 in the ratio 3 : 7. How much money do they each receive?

Find the total number of parts of the whole amount.
Total number of parts = 3 + 7 = 10 parts

Find the value of one part.
Each part = $355 ÷ 10 = $35.50
Shuaib receives 3 parts = 3 × $35.50 = $106.50
Patrick receives 7 parts = 7 × $35.50 = $248.50

Check your answer by adding up all the parts. They should add up to the amount being shared!
$106.50 + $248.50 = $355

A simplified ratio does not contain any units in the final answer.

- You can express a **ratio** in the form 1 : n using division. This is often called a **scale**.

Maps and plans are drawn to scale. You can solve problems involving scales by multiplying or dividing by the scale.

In this scale diagram, 1 cm on the drawing represents 100 cm on the real house.

Corresponding lengths are multiplied by the same **scale factor**. You can write the scale factor as a ratio.

You can write this scale factor as 1 cm represents 100 cm or 1 : 100.

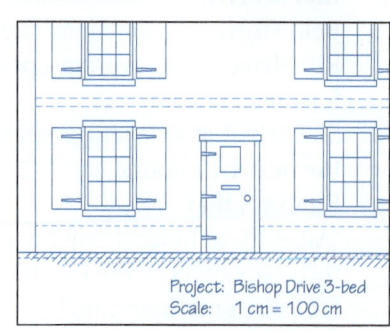

Project: Bishop Drive 3-bed
Scale: 1 cm = 100 cm

Ratio and proportion   Ratio and proportion

Fluency

Exercise 12.2S

1 Write each of these ratios in its simplest form.
- **a** 2:6
- **b** 15:5
- **c** 6:18
- **d** 98:28
- **e** 0.5:50
- **f** 30:625
- **g** 24:156
- **h** 96:120
- **i** 115:184
- **j** 1369:111
- **k** 16.8:45.6
- **l** 3.36:3.99

2 Write each of these ratios in its simplest form.
- **a** 40 cm : 1 m
- **b** 55 mm : 8 cm
- **c** 3 km : 1200 m
- **d** 4 m : 240 cm
- **e** 700 mm : 42 cm
- **f** 12 mins : 450 secs

3 Match the pairs of equivalent ratios.
- **a** 4:2
- **A** 36:48
- **b** 21:14
- **B** 48:24
- **c** 45:60
- **C** 32:48
- **d** 74:111
- **D** 51:34

4 Solve each of these problems.
- **a** Divide $90 in the ratio 3:7.
- **b** Divide 369 kg in the ratio 7:2.
- **c** Divide 103.2 tonnes in the ratio 5:3.
- **d** Divide 35.1 litres in the ratio 5:4.
- **e** Divide €36 in the ratio 1:2:3.

5 Solve each of these problems.
Give your answers to 2 decimal places where appropriate.
- **a** Divide £75 in the ratio 8:7.
- **b** Divide $1000 in the ratio 7:13.
- **c** Divide 364 days in the ratio 5:2.
- **d** Divide 500 g in the ratio 2:5.
- **e** Divide 600 m in the ratio 5:9.

6 Divide the amounts in the ratios indicated.
- **a** $500   2:5:3
- **b** 360°   4:1:3
- **c** 2 km   9:4:7

Hint for c: Convert to metres.

7 Express each of these ratios as a ratio in the form $1:n$ (a scale).
- **a** 2:6
- **b** 3:12
- **c** 10:20
- **d** 8:40
- **e** 3:6
- **f** 5:15
- **g** 4:20
- **h** 12:36
- **i** 30:60
- **j** 9:45
- **k** 45:90
- **l** 20:120

8 A map has a scale of 1:50 000 or 1 cm represents 50 000 cm. Calculate in metres the actual distance represented on the map by these measurements.
- **a** 2 cm
- **b** 8 cm
- **c** 10 cm
- **d** 0.5 cm
- **e** 14.5 cm

9 A map has a scale of 1:5000.
- **a** What is the distance in real life of a measurement of 6.5 cm on the map?
- **b** What is the distance on the map of a measurement of 30 m in real life?

10 A map has a scale of 1:500.
- **a** What is the distance in real life of a measurement of 10 cm on the map?
- **b** What is the distance on the map of a measurement of 20 m in real life?

11 A map has a scale of 1:5000.
- **a** What is the distance in real life of a measurement of 4 cm on the map?
- **b** What is the distance on the map of a measurement of 600 m in real life?

12 A map has a scale of 1:2000.
- **a** What is the distance in real life of a measurement of 5.8 cm on the map?
- **b** What is the distance on the map of a measurement of 3.6 km in real life?

1036, 1038, 1039, 1103

APPLICATIONS

12.2 Ratio and scales

RECAP
- A ratio allows you to compare two amounts.
- 5:2 means 5 parts of one thing compared with 2 parts of another.
- You can simplify a ratio by diving both parts by the same number.
- You can divide a quantity in a given ratio.
- Real-life lengths are reduced in proportion on a scale diagram or map. The scale of the diagram gives the relationship between the drawing and real life.

You can write a map scale as a ratio 1:n.

HOW TO
1. Set out the problem to show the ratio or proportion.
2. Multiply or divide to keep the ratios the same.
3. Write the answer.

EXAMPLE

A paint mix uses red and white paint in a ratio of 1:12.
How much white paint will be needed to mix with 1.8 litres of red paint?

① Set out the problem as equivalent ratios.

    red : white
      1 : 12
1.8 litres : 1.8 × 12 = 21.6 litres

② 3 × 0.3 = 0.9

③ 0.9 ml of red paint is needed.

EXAMPLE

Kara and Mia share money in the ratio 3:7.
Mia receives £168 more than Kara.
How much money did they share?

① Mia gets 4 more parts than Kara.

    4 parts = £168

② 1 part = £42

    Total number of parts = 3 + 7 = 10
    10 parts = £420

③ They shared £420.

EXAMPLE

In this scale drawing 1 cm represents 3 m.
Calculate the area of the front.
State your units.

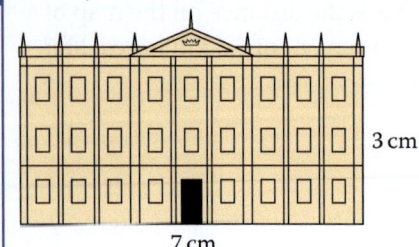

① Change 3 m to cm.   3 m = 300 cm
   Write the scale.

        ×300
   1 cm   300 cm
        ÷300

② Multiply by the scale.   3 cm × 300 = 900 cm = 9 m
                              7 cm × 300 = 2100 cm = 21 m

③ Calculate the answer.   Area of front = 9 m × 21 m = 189 m^2

Ratio and proportion   Ratio and proportion

Reasoning and problem solving

Exercise 12.2A

1 The ratio of the length of a car to the length of a van is 2:3.
The car has a length of 240 cm.
  a Express the length of the car as a percentage of the length of the van.
  b Calculate the length of the van.

2 The ratio of the mass of Faraj to Morgan is 6:5. Morgan has a mass of 85 kg.
  a Express the mass of Faraj as a percentage of the mass of Morgan.
  b Calculate the mass of Faraj.

3 A metal alloy is made from copper and aluminium.
The ratio of the mass of copper to the mass of aluminium is 5:3.
  a What mass of the metal alloy contains 45 grams of copper?
  b Work out the mass of copper and the mass of aluminium in 184 grams of the metal alloy.

4 Siobhan and Silmi shared $700 in the ratio 2:3.
Siobhan gave a quarter of her share to Karen. Silmi gave a fifth of his share to Karen.
What fraction of the $700 did Karen receive?

5 Using this scale drawing of the Eiffel Tower, calculate
  a the height
  b the width of the base.

Scale: 1 cm represents 60 m

6 On the plan of a house, a door measures 3 cm by 8 cm.
If the plan scale is 1 cm represents 25 cm.
Calculate the dimensions of the real door.

7 This is a scale drawing of a volleyball court. The drawing has a scale of 1:200. Calculate
  a the actual distances marked x, y and z
  b the area of the court.
State the units of your answers.

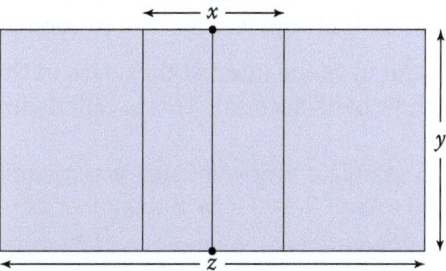

8 The ratio of red counters to blue counters in a bag is 3:7. What is the probability of choosing a blue counter?

9 Nick and Safiyya share money in the ratio 2:5. Nick gets £120 less than Safiyya.
  a How much money did they share?
  b How much money did they both receive?

10 Marnie, Hana and Jessa share money between them.
Hannah gets twice as much as Marnie. Jessa gets one-third of Hana's amount.
  a Write the ratio of Marnie: Hana: Jessa in the form $m:h:j$ where m, h and j are whole numbers and the ratio is in its simplest form.
  b Hana receives $120. How much more money did Marnie receive than Jessa?

11 The ratio of the internal angle to the external angle in a regular polygon is 3:1.
How many sides does the polygon have?

12 The angles P and Q in this trapezium are in the ratio 11:25.
Find the size of angles P and Q.

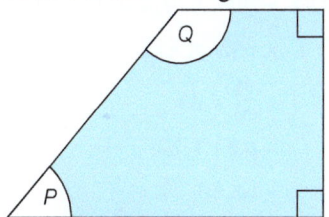

SKILLS

12.3 Percentage change

You often need to calculate a percentage of a quantity.

- A quick method, especially when using a calculator, is to multiply by the appropriate decimal number.

To find 38% of a quantity, multiply by 0.38.

Banks and building societies pay **interest** on money in an account. The interest is always written as a **percentage**.

People can have the interest they earn at the end of each year paid out of their bank account. This is called **simple interest**.

- To calculate simple interest you multiply the interest earned at the end of the year by the number of years.

EXAMPLE

Calculate the simple interest on $3950 for 4 years at an interest rate of 5%.

> Calculate the interest for one year.
> Don't forget to estimate.
> 5% of $3950
> ≈ 5% of $4000
> = $200

Interest each year = 5% of $3950
$$= \frac{5}{100} \times 3950$$
$$= 0.05 \times 3950 = \$197.50$$

Total amount of simple interest after 4 years = 4 × $197.50
= $790

- To calculate a **percentage increase**, work out the increase and add it to the original amount.
- To calculate a **percentage decrease**, work out the decrease and subtract it from the amount.

Percentages are used in real life to show how much an amount has increased or decreased.

You can calculate a percentage increase or decrease in a single calculation.

EXAMPLE

a In a sale all prices are reduced by 16%. A pair of trousers normally costs £82. What is the sale price of the pair of trousers?

b Last year, Leanne's Tax bill was €968. This year the local council have raised the bill by 16%. How much is Leanne's new bill?

a Sale price = (100 − 16)% of the original price
= 84% of €82
$$= \frac{84}{100} \times 82$$
= 0.84 × 82
= 68.88
= €68.88

−16%   +16%
84%   100%   116%

b New bill = (100 + 16)% of the original bill
= 116% of €968
$$= \frac{116}{100} \times €968$$
= 1.16 × €968
= €1122.88

Ratio and proportion Ratio and proportion

Fluency

Exercise 12.3S

1. Write a decimal number equivalent to each percentage.
   - a 50%
   - b 60%
   - c 25%
   - d 8.5%
   - e 0.15%
   - f 0.01%

2. Calculate these amounts without using a calculator.
   - a 10% of $400
   - b 10% of 2600 cm
   - c 5% of 64 kg
   - d 25% of 80 m
   - e 50% of 380p
   - f 5% of $700
   - g 25% of 12 kg
   - h 20% of €31

3. Calculate these percentages, giving your answer to 2 decimal places where appropriate.
   - a 45% of 723 kg
   - b 25% of $480
   - c 23% of 45 kg
   - d 21% of 28 kg
   - e 17.5% of €124
   - f 34% of 230 m

4. Calculate these percentages, giving your answer to two decimal places where appropriate.
   - a 7% of $3200
   - b 12% of £3210
   - c 27% of €5400
   - d 3.5% of $2200
   - e 0.3% of €4450
   - f 3.7% of £12 590

5. Which is larger: 10% of £350 or 15% of £200?

6. Which is larger: 5% of $40 or 8% of $28?

7. Calculate the simple interest paid on $4580
   - a at an interest rate of 4% for 3 years
   - b at an interest rate of 11% for 5 years
   - c at an interest rate of 4.6% for 4 years
   - d at an interest rate of 8.5% for 3 years.

8. Calculate the simple interest paid on
   - a an amount of £3950 at an interest rate of 10% for 2 years
   - b an amount of $6525 at an interest rate of 8.5% for 2 years
   - c an amount of €325 at an interest rate of 2.4% for 7 years
   - d an amount of $239.70 at an interest rate of 4.25% for 13 years.

9. Write the decimal number you must multiply by to find these percentage increases.
   - a 20%
   - b 30%
   - c 45%

10. Write the decimal number you must multiply by to find these percentage decreases.
    - a 40%
    - b 60%
    - c 35%

11. Calculate these percentage changes.
    - a Increase £450 by 10%
    - b Decrease 840 kg by 20%
    - c Increase $720 by 5%
    - d Decrease 560 km by 30%
    - e Increase €560 by 17.5%
    - f Decrease 320 m by 20%

12. Calculate these amounts.
    - a Increase £250 by 10%
    - b Decrease $2830 by 20%
    - c Increase €17 200 by 5%
    - d Decrease €3600 by 30%
    - e Increase $3.60 by 17.5%
    - f Decrease £2500 by 20%

13. Calculate each of these amounts using a mental or written method.
    - a Increase £350 by 10%
    - b Decrease 74 kg by 5%
    - c Increase $524 by 5%
    - d Decrease 756 km by 35%
    - e Increase 960 kg by 17.5%
    - f Decrease €288 by 25%

14. Calculate these percentage changes. Give your answers to 2 decimal places as appropriate.
    - a Increase €340 by 17%
    - b Decrease 905 kg by 42%
    - c Increase £1680 by 4.7%
    - d Decrease 605 km by 0.9%
    - e Increase $2990 by 14.5%
    - f Decrease 2210 m by 2.5%

1060, 1073, 1237, 1302, 1934

APPLICATIONS

12.3 Percentage change

RECAP
- You can find a percentage of an amount by multiplying the quantity by an equivalent decimal or fraction.
- You can calculate a percentage increase or decrease by calculating the percentage increase or decrease and then adding it to, or subtracting it from, the original amount.

HOW TO

To write one number as a percentage of another
1. Write the first number as a fraction of the other.
2. Convert the fraction to a percentage.

EXAMPLE

Find
a 13 as a percentage of 20   b 8 as a percentage of 23

a (1) $\frac{13}{20} = \frac{13 \times 5}{20 \times 5} = \frac{65}{100} = 65\%$ (2) Convert to a fraction with a denominator of 100.

b (1) $\frac{8}{23} = (8 \div 23) \times 100\% = 34.8\%$ (2) Use a calculator to convert to a percentage.

If a price was reduced from £20 to £13, then the discount would be (100 − 65)% = 35%.

- In a **reverse percentage** problem, you are given an amount after a percentage change, and you have to find the original amount.

HOW TO

To calculate the original amount after a percentage change
1. Write the percentage change as a decimal.
2. Find the original amount by dividing by the decimal.
3. You can test your answer.

EXAMPLE

Find the original price of a denim jacket reduced by 15% to £32.30.

(1) 100% − 15% = 85%
85% = 0.85
(2) original price = £32.30 ÷ 0.85 = £38
(3) You can test your answer by finding 85% of £38
85% of £38 = 0.85 × £38 = £32.30

EXAMPLE

Following a 5% price increase, a car radio costs $168.
How much did it cost before the increase?

(1) 100% + 5% = 105%
105% = 1.05
(2) original price = $168 ÷ 1.05 = $160
(3) 105% of $160 = 1.05 × $160 = $168

Ratio and proportion   Ratio and proportion

Reasoning and problem solving

Exercise 12.3A

1. Use a written method to find
   - **a** 12 as a percentage of 20
   - **b** 36 as a percentage of 75
   - **c** 24 as a percentage of 40

2. Use a calculator to find
   - **a** 19 as a percentage of 37
   - **b** 42 as a percentage of 147
   - **c** 8 as a percentage of 209

3. **a** A t-shirt is reduced from £25 to £20. Find the percentage discount.
   b A coat is reduced from $80 to $64. Find the percentage discount.

4. **a** Jenna's savings increase from £400 to £450. Calculate the interest rate.
   b Mu'mina's antique vase increases in value from €600 to €720. Calculate the percentage increase in value.

5. A car costs $9750 when new. 5 years later it is sold for $4500. What is the average percentage loss each year?

6. Carys is trying to find the original price of an item that is on sale. Its sale price is £56 and it was reduced by 20%.

   > 20% of 56 = 11.2
   > 56 + 11.2 = 67.2
   > The answer is £67.20

   Carys' working is shown here.
   What is wrong with her working?

7. A book costs $4 after a 20% price reduction. How much did it cost before the reduction? Show your working.

8. These numbers are the results when some amounts were increased by 10%. For each one, find the original number.
   - **a** 55   **b** 44
   - **c** 88   **d** 121

9. Find the original cost of the following items.
   - **a** A vase that costs £7.20 after a 20% price increase.
   - **b** A table that costs €64 after a 20% decrease in price.

10. **a** Francesca earns £350 per week. She is awarded a pay rise of 3.75%. Maazin earns £320 per week. He is awarded a pay rise of 4%. Who gets the bigger pay increase? Show all your working.
    b Bertha's pension was increased by 5.15% to $82.05. What was her pension before this increase?

11. Calculate the original cost of these items, before the percentage changes shown. Show your method. You may use a calculator.
    - **a** A hat that costs $46.50 after a 7% price cut.
    - **b** A skirt that costs €32.80 after a price rise of 6%.

12. A car manufacturer increases the price of a Sunseeker sports car by 6%. The new price is $8957. Calculate the price before the increase.

13. During 2005 the population of Camtown increased by 5%. At the end of the year the population was 14 280.
    What was the population at the beginning of the year?

*14. To decrease an amount by 8%, multiply it by 0.92.
    For a further decrease of 8%, multiply it by 0.92 again, and so on.
    Use this idea to calculate
    - **a** the final price of an item with an original price of €380, which is given two successive price cuts of 8%.
    - **b** the final price of an item with an original price of €2400, which is given three successive price cuts of 10%.

Summary

Checkout

You should now be able to...

	Test it Questions
✓ Find fractions and percentages of amounts and express one number as a fraction or percentage of another.	1, 2, 11
✓ Divide a quantity in a given ratio and reduce a ratio to its simplest form.	2 – 6
✓ Use scale factors, scale diagrams and maps.	6 – 8
✓ Solve problems involving percentage change.	9 – 12

Language	Meaning	Example
Proportion	A proportion is a part of the whole. Two quantities are in proportion if one is always the same multiple of the other.	If there are 6 eggs in a carton Total number of eggs = 6 × number of full cartons
Ratio	A ratio compares the size of one quantity with the size of another.	Ratio of blue squares to yellow squares = 2 : 6 = 1 : 3
Simplify (ratio)	Divide both parts by common factors.	
Scale	The ratio of the length of an object in a scale drawing to the length of the real object.	The Ordnance survey produce maps with scales such as: 1 : 100 000, 1 : 50 000 and 1 : 25 000.
Scale drawing	An accurate drawing of an object to a given scale.	
Percentage	A type of fraction in which the value given is the number of parts in every hundred.	$\frac{33}{100} = 33\%$
Interest	A fee paid to somebody for the use of their money. This is a percentage of the loan amount that must be paid to the lender in addition to the loan itself.	A loan with an interest rate of 4% means that the borrower has to pay back an extra 4% of the amount.
Simple interest	Interest that is calculated on the original amount only and not on any extra interest that has built up.	£100 saved in a bank account at 4% Amount after 3 years = 100 × (1 + 3 × 0.04) = £112
Percentage increase/ decrease	An increase/decrease by a percentage of the original amount.	4.20 increased/decreased by 25% = 1.25 × 4.20 = 0.75 × 4.20 = 5.25 = 3.15
Reverse percentage problem	Calculating the original value of a quantity using the value after a percentage change.	Quantity after a 15% decrease = 544 85% of original quantity = 544 100% of original quantity = $\frac{544}{0.85}$ = 640

Ratio and proportion   Ratio and proportion

Review

1. A class has 16 boys and 9 girls.
   a. What fraction of the class are boys?
   b. What percentage of the class are girls?
   c. This class has girls and boys in the same proportion as the overall year group. There are 125 students in the year group, how many of them are girls?

2. Ghaaliya spends 8 hours at work each day.
   a. What fraction of the day does she spend at work?
   b. Write the ratio of the amount of time she spends at work to not at work.

3. Write these ratios in their simplest form.
   a. $22:33$ b. $45:27$ c. $2\,cm:5\,m$

4. Share
   a. £42 in the ratio $5:1$
   b. $99 in the ratio $2:4:3$

5. Palben is twice as old as Kao.
   a. Write a ratio of Palben's age to Kao's age.
   b. What fraction of Palben's age is Kao?

6. The ratio of the mass of milk to flour in a recipe for batter is $2:3$.
   a. How much milk and how much flour will be needed to make 1 kg of batter?
   b. How much milk will be needed to mix with 450 g of flour?
   c. What fraction of the mass of the mixture is milk?

7. This circle is enlarged by scale factor 5.
   a. What is the diameter of the enlarged circle?

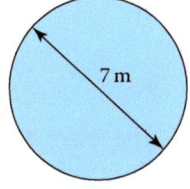

   b. What is the area of the enlarged circle?

8. A map is drawn using the scale $1:400\,000$.
   a. How far in real life is a length of 2 cm on the map?
   b. A road is 12.5 km long, how long will it be on the map?

9. a. Increase 60 by 34%.
   b. Decrease 192 by 58%.

10. A bank pays 1.9% per annum interest on a savings account. £6000 is paid in at the beginning of the year and no further deposits or withdrawals are made.
    How much is in the account after 1 year?

11. Aslam has a mortgage of $200 000. He pays $1100 a month. This covers only the interest on his mortgage so he still owes the bank $200 000.
    a. What is the annual simple interest rate?
    After 20 years of paying the interest of $1100 each month the mortgage ends and he pays back the $200 000.
    b. How much money has he paid in total over the 20 years?
    c. What is the total amount paid as a percentage of the original amount borrowed?

12. a. The price of a loaf of bread rose 8% to £1.34, what was it before the increase?
    b. A car is reduced by 17% and now costs €7520, what was the original price of the car?

What next?

Score			
	0 – 5		Your knowledge of this topic is still developing. To improve, look at MyiMaths: 1015, 1036, 1037, 1038, 1039, 1060, 1073, 1103, 1237, 1934
	6 – 10		You are gaining a secure knowledge of this topic. To improve your fluency look at InvisiPens: 12Sa – j
	11 – 12		You have mastered these skills. Well done you are ready to progress! To develop your problem solving skills look at InvisiPens: 12Aa – c

Assessment 12

1. There are some Brazil nuts, walnuts and hazelnuts in a bowl. 3 out of 5 of the nuts are Brazils, W out of N are walnuts and 3 out of every 20 are hazelnuts.
   a. Work out the values of W and N in their simplest form. [2]
   b. Write all these proportions as percentages. [3]
   c. Which nut represents the smallest proportion of the bowl's contents? [1]
   d. Which nut represents the biggest proportion of the bowl's contents? [1]
   e. What percentage of the nuts in the bowl are either Brazil nuts or chestnuts? [1]
   f. What fraction of the nuts are *not* walnuts? [1]
   g. What fraction of the nuts in the bowl are neither walnuts nor Brazil nuts? [1]

2. a. i. Nomsah received $\frac{13}{25}$ of the prize money from a competition.
         Write this proportion as a percentage and a decimal. [1]
      ii. The total prize was £225 000 rand. How much did Nomsah win? [1]
   b. A car was bought for $17 250. A year later its value had decreased by $\frac{6}{25}$.
      How much was the car worth a year later? [2]

3. A tin contains 63 biscuits, some made with chocolate and the rest 'plain', in the ratio 3:4. Calculate the numbers of chocolate and plain biscuits in the tin. [2]

4. Cheese straws, are made using flour, butter and cheese in the ratio 24:6:9.
   a. Write this ratio in its simplest form. [1]
   b. A cheese straw mixture uses 160 g of flour.
      Work out the number of grams of butter and cheese used. [2]

5. A café serves cups of coffee in three sizes: small, medium and large.
   The volumes are in the ratio 2:5:8.
   a. A medium coffee is 375 cm³. Find the volumes of the other two sizes. [2]
   b. The café serves pots of tea in three sizes, the ratio of volumes small:medium:large is the same as it is for coffee. The capacity of a large pot of tea is 1 litre.
      Find the capacity of the medium and small pots. [2]

6. a. The cast of a school play has 24 boys and 20 girls.
      Write this ratio in its simplest form. [1]
   b. 1 more boy joins the cast but 5 girls leave. Write the new ratio in the form $n:1$. [2]

7. The ratio of girls to boys in a cycling squad is 4:7. There are 88 people in the squad.
   a. How many of each gender are in the squad? [2]
   b. The number of girls goes up by 12 and the number of boys goes up by 37.5%.
      Find the number of girls and boys in the squad after the change and show that the ratio of girls to boys stays the same. [4]

8. Two circles have diameters of 15 m and 20 m. Find the ratio of their
   a. circumferences [2]   b. areas. [2]

   Write all answers in their simplest form.

9 It takes Dwain 24.5 seconds to run 200 m. At the same pace, how long will Dwain take to run
   a   74 m   [2]   **b**   172 m?   [2]
Give both answers to the nearest 0.1 s.

10 A road atlas' scale is 1 inch to 5 miles.
   a   Gatwick airport to Heathrow is 8.1 inches on a map. How far apart are they?   [1]
   b   The straight-line distance between Liverpool FC and Newcastle United FC is 128 miles. How far apart are they on the map?   [1]
   c   The straight-line distance from John O' Groats to Lands End on a map is 120.4 inches. The actual distance by road is 838 miles. Calculate the difference in mileage between the actual distance and the straight-line distance.   [3]

11 On a street map of London the scale is written as 1 : 20 000.
   a   How many metres in London is represented by 1 cm on the map?   [1]
   b   On the map, Buckingham Palace is 5.85 cm from the statue of Eros in Piccadilly Circus. How far apart are they in reality? Give your answer in km.   [1]
   c   Trafalgar Square is 0.96 km from Big Ben. How far apart are they on the map?   [1]
   d   A sick child is taken from Paddington Station to Great Ormond Street Children's Hospital. On the map, the distance is represented by 20.75 cm.
The ambulance travels at an average speed of 80 km/h.
Show that the ambulance journey takes less than 200 seconds.   [4]

12 a   Zhan says that to increase a value by 29.1% you multiply by 29.1. Felix says you multiply by 1.291 Who is correct?   [1]
   b   Explain how to decrease 550 ml by 45.4%.   [2]

In the following questions give your answers to 2 decimal places wherever appropriate.

13 a   A station has 45 steps up to the platform bridge. The first section has 19 steps. What percentage of the total number of steps does the first section take?   [1]
   b   Josefina buys material for her wedding dress. The shop assistant cut her material from a 25 m roll, leaving 17.25 m on the roll. What percentage of the roll has Josefina bought?   [2]

14 a   A painting is bought for $2 500 000. Some years later it was revalued as being worth 8.5% more than originally. What is its new value?   [2]
   b   In 2000 the average price of a new house was €86 100. In 2010 this average price was €164 300. By what percentage had the 2010 value risen above the 2000 value?   [2]
   c   'FALSEPRINT', a film processor, sells prints in various sizes. They claim that the area 7″ by 5″ print is more than 45% bigger than their 6″ by 4″ size. Are they correct? Give your reasons.   [3]

15 A river contains 250 crocodiles. After two years of hunting, the number of crocodiles decreases by 8% in the first year and 5.5% in the second.
How many crocodiles, to the nearest crocodile, are there in the river after
   a   1 year   [2]   **b**   2 years?   [2]

16 a   A bike was bought for $4 776. The bike was sold at a loss of 11.3%. How much was the bike sold for?   [2]
   b   After a pay rise of 3.5% Osborne's monthly salary is £2 566.80 What was his monthly salary before his increase?   [2]
   c   A shop sells a fishing rod for €251.34 The shop makes a profit of 6%. How much did the shop pay for the rod?   [2]

Revision 2

1. Andrea draws a quadrilateral with vertices at the coordinates (1,2), (3,4), (4,7) and (3,10). Find the area of this quadrilateral. [4]

2. The diagram shows a company logo.
   $BC = CD = 8$ cm,
   $AC = 15$ cm and
   $AE = 9$ cm.
   Calculate the area of the logo. [4]

   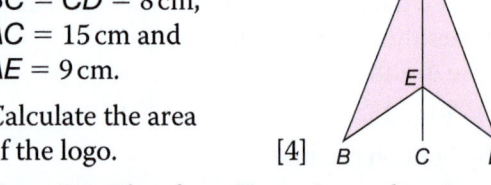

3. Town B is 5 km from Town A on a bearing of 135°. Town C is 12 km from Town A on a bearing of 045°.
   a. Sketch the triangle ABC. [2]
   b. Find angle BAC. [1]
   c. Calculate the area of the triangle enclosed by the three towns. [1]

4. A field in the shape of a trapezium has two parallel sides, 150 m and 176 m. The perpendicular distance between them is 243 m. Calculate the area of the field in
   a. m² [2]
   b. hectares. (1 hectare = 10 000 m²) [1]

5. a. Triangle ABC has been transformed to triangle PQR as shown. Describe this transformation fully. [3]

   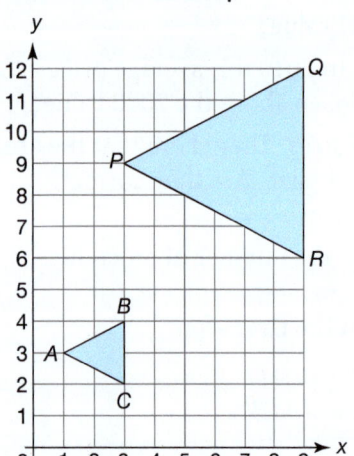

   b. Reflect triangle ABC in the line $y = -x$ and label the image A'B'C'. [2]

6. Triangle A has vertices at (2, 2), (6, 2) and (4, 6). Triangle B has vertices at (3, −3), (9, −6) and (3, −9). Draw both triangles on a coordinate grid and fully describe the two transformations needed to map A onto B. [6]

7. Zarita's bookshelf can hold 38 books of an average thickness of 2.9 cm.
   a. Estimate the length of the bookcase. [1]
   The average mass of a book is 1100 g.
   b. Estimate the total mass of the books on Zarita's bookshelf. [1]
   c. Use your calculator to find exact answers to parts a and b. [2]
   d. Find your percentage error for the estimates in parts a and b. [4]

8. A shape is made up of a cylinder, radius 1.5 m with height 2.3 m, and a cuboid measuring 5 m × 3 m × 4 m. All measurements are correct to the nearest cm.
   a. Find the upper and lower bounds for the total volume of the shape. [6]
   b. Calculate the upper and lower bounds of the mass of the cylinder if the density of the material is 2.8 g/cm³ [2]

9. Rose is planting flowers, Purple Lupins which cost $3 each and White Stocks which cost $4 each. She must spend at least $10 and has room for not more than 6 Lupins and 4 Stocks. She wants at least twice as many Stocks as Lupins.
   a. Taking the y axis as Lupins and the x axis as Stocks, draw four straight lines to illustrate the information above. [4]
   b. By shading *out* inappropriate areas identify the area which satisfies these conditions. [1]
   c. Use your graph to find the minimum number of each flower Rose can buy given that she wants at least one flower of each type. [2]

10. A length of wire is 58 cm long. It can be bent into a rectangle x cm long. The area of the rectangle is 100 cm². Find the dimensions of the rectangle. [5]

11 A coffee machine takes only 10p and 50p coins. When emptied, it had 41 coins in it totalling £10.10 How many of each value coin did the machine have? [4]

12 A triangle has angles p, q and r, where $p > 21$, $q > 59$. Find an appropriate inequality for r. [2]

13 A drinks manufacturer makes circular bottle tops. A rectangular metal sheet, of dimensions 35 cm by 30 cm is cut into 42 squares. A circle of the largest possible size is then cut from each square.

   a Find the percentage of metal sheet discarded in this process. [4]
   b The remainder of the sheet is recast. How many bottle tops could be made from it
      i   using the arrangement described in part a
      ii using any arrangement? [4]

14 A man is trying to move a heavy roller of radius r, using a straight crowbar, BDF. The angle between the crowbar and the ground, ABC, is 60°, AE is the diameter of the circle and O is the centre.

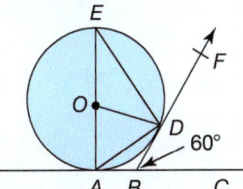

   a Find angles DEA and EDF. Explain your answers. [6]
   b Prove that $AD = r$. [3]
   c Calculate, in terms of r, the length of arc AD and the area of the sector AOD. [3]

15 A house has a rectangular garden, $ABCD$, which is 12 m by 9 m, CD is the side of the house and is 9 m long. A window at the centre of CD has a modem in it which gives a WiFi signal range of 7.5 m.

   a Use a ruler and compasses to construct a diagram of the garden. [5]
   b Shade in the area of the garden within range of the WiFi. [2]
   c Ida has a deckchair 7.2 m from A and 5.2 m from B. Is she within WiFi range? Explain your answer. [3]
   d Draw and measure a line on your diagram to find the WiFi range needed to make all of the garden accessible. [1]

16 A tennis club has the ratio men : women as 7 : 6. There are 104 people in the club.

   a How many of each sex are there in the club? [2]

The number of men and the number of women each rises by 25%.

   b Does the ratio of men to women change? Explain your answer. [2]
   c What are the new numbers of women and men in the squad? [1]
   d The same number of men leave as women join. The ratio is now 1 : 1. How many people of each gender are in the tennis club now? [1]

17 On a map 1 inch represents 3 miles on the ground.

The straight-line distance from Milan to Magenta is 22 miles. How far apart are they on the map? [1]

18 a A daily newspaper claims an average daily sale of 1 456 739. What percentage increase would take its daily average sales to 1 500 000? [2]
    b The value of a new car is £18 000. It depreciates by 16% yearly. How much is the car worth at the end of 3 years? [3]

19 Sunetra has 6 cards: 1, 3, 5, 7, 9 and 11. She selects two cards and puts the smaller number on top of the larger to make a fraction.

   a Show there are 15 possible outcomes. [3]
   b Work out the probability her number is
      i $\frac{1}{3}$ [1]   ii less than $\frac{1}{4}$ [3]

13 Factors, powers and roots

Introduction
Cryptography is the study of codes, with the aim of communicating in a secure way without messages being deciphered. Modern cryptography is often based on prime numbers. In particular finding very large numbers that can be written as the product of two not-quite-as-large prime factors. Finding what these two prime numbers are is a challenge, even with modern computers; but you need to find them in order to crack the code.

What's the point?
In the modern day, illegal computer-based syndicates use increasingly sophisticated techniques to access sensitive digitally-held information, including bank accounts. Prime number encryption increases the security of stored data, making it harder for the hackers.

Objectives
By the end of this chapter, you will have learned how to …
- Know and use the language of prime numbers, factors and multiples.
- Write a number as a product of its prime factors.
- Find the HCF and LCM of a pair of integers.
- Estimate the square or cube root of an integer.
- Find square and cube roots of numbers and apply the laws of indices.
- Simplify expressions involving surds including rationalising fractions.

Check in

1. Write all the factors of each number.
   - **a** 12
   - **b** 30
   - **c** 120
   - **d** 360
2. Write a list of all of the prime numbers up to 100. (There are 25 of them.)
3. Find the results of these calculations, giving the answers in index form.
   - **a** $2^3 \times 2^4$
   - **b** $3^5 \div 3^2$
   - **c** $5^2 \times 5^3 \times 5^2$
   - **d** $6^6 \div 6^4$
   - **e** $7^8 \div (7^2 \times 7^3)$
   - **f** $(4^6 \times 4^2) \div (4^3 \times 4)$
4. Write the value of each number.
   - **a** 4^0
   - **b** 6^0
   - **c** 5^1
   - **d** 2^{-1}

Chapter investigation

77 is the product of two prime numbers: 7 and 11.

That is, $7 \times 11 = 77$

Is 702 the product of two prime numbers? If not, why not?

What about 703?

SKILLS

13.1 Factors and multiples

You can write a number as a **product** of **factors** in different ways.

- A **prime** number is a number with only *two* factors – itself and 1.

This means 1 is not a prime number.

- Any integer can be written as a unique product of its prime factors. This is called the **prime factor decomposition** of the number.

EXAMPLE

Write 990 as a product of prime factors.

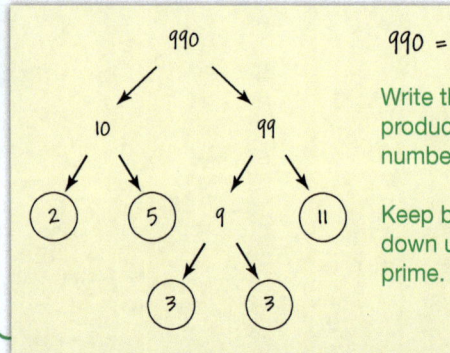

$990 = 2 \times 3^2 \times 5 \times 11$

Write the number as the product of two smaller numbers.

Keep breaking the numbers down until you reach a prime.

Try starting the factor tree a different way. What do you notice?

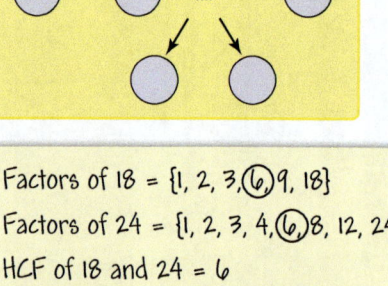

- The **highest common factor (HCF)** of two numbers is the largest number that is a factor of both of them.

- The **lowest common multiple (LCM)** of two numbers is the smallest number that is a **multiple** of both of them.

Factors of 18 = {1, 2, 3, ⑥, 9, 18}
Factors of 24 = {1, 2, 3, 4, ⑥, 8, 12, 24}
HCF of 18 and 24 = 6
Multiples of 18 = 18, 36, 54, ⑦②, 90, ...
Multiples of 24 = 24, 48, ⑦②, 96, ...
LCM of 18 and 24 = 72

You can find the HCF and LCM of two numbers by writing their **prime factors** in a Venn diagram.

p.410

- The HCF is the product of the numbers in the **intersection**.
- The LCM is the product of all the numbers in the diagram.

The intersection is the place where the two sets cross over.

EXAMPLE

Find the HCF and LCM of 60 and 280.

Find the prime factorisation of each number.

$60 = 2^2 \times 3 \times 5$ $280 = 2^3 \times 5 \times 7$

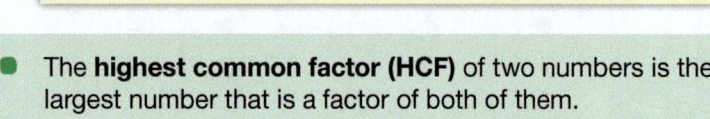

An alternative is to use the highest power of each factor.
LCM = $2^3 \times 3 \times 5 \times 7 = 840$
     = 3×280
     = $60 \times 2 \times 7$

The LCM is the product of all the numbers in the diagram.
LCM = $2^3 \times 3 \times 5 \times 7 = 840$

The HCF is the product of the numbers in the intersection.
HCF = $2^2 \times 5 = 20$

Number Factors, powers and roots

Fluency

Exercise 13.1S

1 Copy and complete these calculations to show the different ways that 24 can be written as a product of its factors.
   a $24 = \square \times 2$
   b $24 = 3 \times \square$
   c $24 = 2 \times 3 \times \square$
   d $24 = 4 \times \square$

2 Each of these numbers has just two prime factors, which are not repeated. Write each number as the product of its prime factors.
   a 77
   b 51
   c 65
   d 91
   e 119
   f 221

3 Copy and complete the factor tree to find the prime factor decomposition of 18.

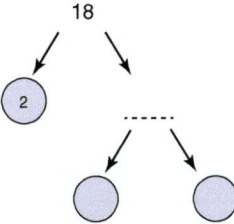

4 Work out the values of these expressions.
   a $3^2 \times 5^2$
   b $2^3 \times 5$
   c $3^2 \times 7^2$
   d $2^2 \times 3 \times 5^2$

5 Write each number as the product of powers of its prime factors.
   a 36
   b 120
   c 34
   d 48
   e 27
   f 105
   g 99
   h 37
   i 91

6 Write the prime factor decomposition for each of these numbers.
   a 1052
   b 2560
   c 825
   d 715
   e 1001
   f 219
   g 289
   h 2840
   i 2695
   j 1729
   k 3366
   l 9724
   m 11830
   n 2852
   o 10179

7 The diagram below shows how Azza found the LCM of 8 and 6.

   Multiples of 8 = 8, 16, (24), 32, 40, 48 ...
   Multiples of 6 = 6, 12, 18, (24), 30, 36 ...

   Use Azza's method to find the LCM of 12 and 9.

8 Using the method from question 7, find the LCM of these pairs of numbers.
   a 4 and 5
   b 12 and 18
   c 5 and 30
   d 12 and 30
   e 14 and 35
   f 8 and 20

9 Find the HCF of each pair of numbers, by drawing a Venn diagram or otherwise.
   a 35 and 20
   b 48 and 16
   c 21 and 24
   d 25 and 80
   e 28 and 42
   f 45 and 60

10 Find the LCM of each pair of numbers, by drawing a Venn diagram or otherwise.
   a 24 and 16
   b 32 and 100
   c 22 and 33
   d 104 and 32
   e 56 and 35
   f 105 and 144

11 Find the LCM and HCF of these numbers.
   a 180 and 420
   b 77 and 735
   c 240 and 336
   d 1024 and 18
   e 762 and 826
   f 1024 and 1296

12 Find the HCF and LCM of these numbers.
   a 30, 42, 54
   b 90, 350, 462
   c 462, 510, 1105
   d 44, 57, 363

13 a Show that 33 105 is divisible by 15.
   b Show that 262 262 is divisible by 1001.

14 A number is **abundant** if its factors, excepting the number itself, sum to a total greater than the number. If this total is less than the number it is **deficient** and if the total equals the number it is **perfect**.

   For example 12 has factors 1, 2, 3, 4, 6 (and 12) $1 + 2 + 3 + 4 + 6 = 16$ and $16 > 12$ so 12 is abundant. Complete the table below by classifying the following numbers, the first one has been done for you.

   12, 72, 40, 86, 30, 50, 64, 6, 27, 28

Abundant	12
Perfect	
Deficient	

APPLICATIONS

13.1 Factors and multiples

RECAP
- The **highest common factor** (HCF) of two numbers is the largest number that is a factor of them both.
- The **lowest common multiple** (LCM) of two numbers is the smallest number that they both divide into.
- You can find the HCF and LCM of two numbers by writing their **prime factors** in a Venn diagram.
  - The HCF is the product of the numbers in the intersection.
  - The LCM is the product of all the numbers in the diagram.

HOW TO
To solve problems involving factors or multiples
1. Read the question – decide how to use your knowledge of factors and multiples.
2. Be systematic – use listing, factor trees and/or Venn diagrams to help you to find multiples and factors.
3. Answer the question – make sure that you explain your answers fully.

▲ John Venn invented Venn diagrams as a way to represent sets. They are used extensively in probability. This stained glass window celebrating Venn can be round in the dining hall of Gonville and Caius College at Cambridge University.

EXAMPLE
Explain why the square of a prime number has exactly three positive factors.

1. A prime number, p, has two factors, 1 and p (itself).
2. List the factors of the square of a prime number.
   $p^2 = 1 \times p^2$ and $p^2 = p \times p$
3. The only factors are: 1, the prime number, p, and the number itself, p^2.

EXAMPLE
The highest common factor of two numbers is 6.
The lowest common multiple is 1080.
Zeena says that the two numbers must be 54 and 120.
Show that there are other possibilities.

1. Find another pair of numbers with the same HCF and LCM.
2. Write the prime factors of 54 and 120 in a Venn diagram.
   $54 = 2 \times 3^3$ $120 = 2^3 \times 3 \times 5$

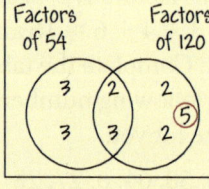 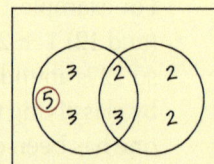

You can move the factors around to find another pair of numbers with the same HCF and LCM, but the numbers in the intersection have to stay the same.

3. Find another pair of numbers with the same HCF and LCM.
   A possible pair of numbers is
   $2 \times 3^3 \times 5 = 270$ and $2^3 \times 3 = 24$
   Other possible pairs are $2 \times 3 = 6$ and $2^3 \times 3^3 \times 5 = 1080$
   or $3^3 \times 2^3 = 216$ and $2 \times 3 \times 5 = 30$.

Number   Factors, powers and roots

Reasoning and problem solving

Exercise 13.1A

1. Explain why the cube of a prime number has exactly four factors.

2. Recall the definition of a deficient number from exercise **13.1S**. Use algebra to show that the square of a prime must be deficient.

3. The highest common factor of two numbers is 30. The lowest common multiple is 900. Omar says that the two numbers must be 150 and 180. Show that there is another possibility.

4. Two numbers have HCF = 15 and LCM = 90. One of the numbers is 30. What is the other number?

5. Look at this statement.

   > The product of any two numbers is equal to the product of their HCF and their LCM.

   a. Test the statement for three pairs of numbers. Do you think it is true?

   b. Use a Venn diagram to justify your answer to part **a**.

6. a. Amos says 'all odd numbers are prime numbers'. Give two examples that show he is wrong.

   b. Aya says 'all prime numbers are odd numbers'. Give one example to show that she is wrong.

   c. Arik says 'take one off any multiple of 6 and you always get a prime number'.
      i. Give 3 examples where this is true.
      ii. Give 1 example where it is false.

7. Emily checks her phone every 10 minutes to see if her friend Maleeka is available to chat. Maleeka checks hers every 8 minutes. They both decide to check at 9:00 am. When is the first time after this that they are both online?

8. Two needles move around a dial. The faster needle moves around in 24 seconds and the slower needle in 30 seconds. If the two needles start together at the top of the dial, how many seconds does it take before they are next together at the top?

9. A wall measures 234 cm by 432 cm. What is the largest size of square tile that can be used to cover the wall, without needing to cut any of the tiles?

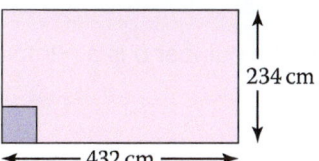

10. The number 18 can be written as $2 \times 3 \times 3$. You can say that 18 has three prime factors.

    a. Find three numbers with exactly three prime factors.

    b. Find five numbers with exactly four prime factors.

    c. Find four numbers between 100 and 300 with exactly five prime factors.

    d. Find a two-digit number with exactly six prime factors.

11. A cuboid has a volume of 1815 cm^3. Each side of the cuboid is a whole number of centimetres and each side is longer than 1 cm. Find all the possible dimensions of the cuboid.

12. Find the HCF and LCM of $x^2 - 1$ and $2x^2 - x - 1$.

13. The ratio $x : y$ is $3 : 7$. Find the following in terms of x.

    a. The HCF of $9y + x$ and $3y + 4x$.

    b. The LCM of $9y + x$ and $3y + 4x$.

14. How many factors do these numbers have?

    a. 2^3 b. $2^2 \times 3^3$ c. $5^2 \times 7^3$
    d. 625   e. 45   f. 484

15. The product of each pair of circled expressions is given on the edge between them. Copy the diagrams and fill in the missing expressions.

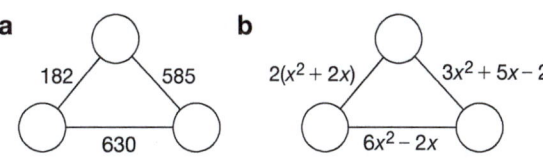

13.2 Powers and roots

- The **square root** of a number a is a number that when squared equals a.

- The **cube root** of a number b is a number that when cubed equals b.

Identify the buttons on your calculator which compute the square root, cube root and powers of a number. They may look like this

$\sqrt{25} = 5$ and $5 \times 5 = 25$ $\sqrt[3]{64} = 4$ and $4 \times 4 \times 4 = 64$

EXAMPLE

Estimate the value of $\sqrt{30}$ to 1 decimal place.

$5^2 = 25$, 5 is too small.
$6^2 = 36$, 6 is too big.
The answer must lie between 5 and 6.
$5.5^2 = 30.25$, 5.5 is too big.
$5.4^2 = 29.16$, 5.4 is too small.
The answer lies between 5.4 and 5.5, test their midpoint to see which is closer.
$5.45^2 = 29.7025$, 5.45 is too small.
The solution lies closer to 5.5 so $\sqrt{30} = 5.5$ (1 dp)

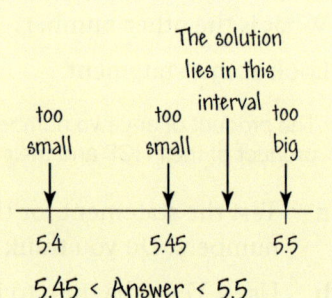

$5.45 <$ Answer < 5.5
Both bounds = 5.5 (1 dp)

You can use **index** notation to describe **powers** of any number.
$4.6^5 = 4.6 \times 4.6 \times 4.6 \times 4.6 \times 4.6$

Powers of the same base number can be multiplied and divided.

- To multiply powers of the same base, add the indices. $5^3 \times 5^2 = 5^5$
- To divide powers of the same base, subtract the indices. $5^8 \div 5^2 = 5^6$
- To raise a power to a power, multiply the indices. $(5^2)^4 = 5^8$
- For all values of x, except $x = 0$, $x^0 = 1$. $5^0 = 1$

EXAMPLE

Simplify these expressions, giving your answers in index form.

a $7^2 \times 5^3 \times 7^3 \times 5^4$
b $(2^5 \times 3^4) \div (2^3 \times 3^2)$

a $7^2 \times 5^3 \times 7^3 \times 5^4 = 7^{2+3} \times 5^{3+4}$
   $= 7^5 \times 5^7$
b $(2^5 \times 3^4) \div (2^3 \times 3^2) = 2^{5-3} \times 3^{4-2}$
   $= 2^2 \times 3^2$

EXAMPLE

Write down the value of these expressions.

a $(16^3 - 81 \times 17)^0$
b $(4.8)^1$

a There is no need to evaluate the expression in the brackets (except to note that it is non-zero).
   $(16^3 - 81 \times 17)^0 = 1$
b $x^1 = x$ for any value of x (including decimal numbers).
   $(4.8)^1 = 4.8$

Number   Factors, powers and roots

Fluency

Exercise 13.2S

1. **a** Find the square root of 20 to 1 decimal place.
   b Find.
   i $\sqrt{40}$ **ii** $\sqrt{60}$ **iii** $\sqrt{95}$

2. Find the cube root of these numbers to 1 decimal place.
   a $\sqrt[3]{20}$ **b** $\sqrt[3]{50}$
   c $\sqrt[3]{80}$ **d** $\sqrt[3]{150}$

3. Solve each of these equations, giving your answer to 2 dp when appropriate. Remember to give both solutions.
   a $x^2 = 1345$ **b** $x^2 = 38.6$
   c $x^2 = 7093$ **d** $x^2 = 234.652$

4. Calculate these, giving your answers to 2 dp when appropriate.
   a $\sqrt[3]{12\,167}$ **b** $\sqrt[3]{-216}$
   c $\sqrt[3]{-70}$ **d** $\sqrt[3]{0.015\,625}$

For questions 5 to 12 give your answers in index form.

5. Simplify these expressions.
   a $6^2 \times 6^3$ **b** $4^5 \times 4^4$ **c** $11^5 \times 11^2$
   d $1^{17} \times 1^{13}$ **e** $3^6 \times 3^6$ **f** $9^9 \times 9$

6. Simplify these expressions.
   a $7^8 \div 7^6$ **b** $8^6 \div 8^2$ **c** $3^3 \div 3^2$
   d $4^7 \div 4$ **e** $2^9 \div 2^9$ **f** $12^8 \div 12^6$

7. Simplify these expressions.
   a $(3^2)^2$ **b** $(3^3)^4$ **c** $(3^4)^3$
   d $(3^9)^0$ **e** $(3^6)^3$ **f** $(3^2)^9$

8. Simplify these expressions.
   a $8^6 \times 8^2 \div 8^3$ **b** $5^7 \times 5^2 \div 5^4$
   c $2^8 \times 2^3 \div 2^5$ **d** $9^6 \times 9^3 \div 9^7$
   e $(8^5)^2 \times 8^5 \div 8^2$ **f** $(7^6 \times 7^5)^3 \div 7^4$
   g $4^6 \times (4^8 \div 4^4)^3$ **h** $(6^2 \times 6^2 \div 6^3)^4$

9. Simplify these expressions.
   a $3^4 \times 3^2 \div (3^3 \times 3^2)$
   b $(5^6 \div 5^2) \times 5^4 \times 5^2$
   c $(4^5 \div 4^2) \div (4^6 \div 4^5)$
   d $(7^9 \div 7^2) \div (7^2 \times 7^3)$
   e $(8^7 \div 8^4)^0 \times 8^5 \times 8^3$
   f $9^3 \times (9^5 \div 9^2)^4 \times 9^4$

10. Simplify these expressions.
    a $\dfrac{4^2 \times 4^2}{4^2}$ **b** $\dfrac{9^8}{9^2 \times 9^4}$ **c** $\dfrac{8^6 \div 8^3}{8^2}$
    d $\left(\dfrac{5^9 \times 5^4}{5^3 \times 5^7}\right)^2$ **e** $\dfrac{6^3 \times 6^4}{6^5 \div 6^3}$ **f** $\dfrac{(8^9 \div 8^2)^3}{8^7 \div 8^2}$

11. Simplify these expressions as far as possible.
    a $4^2 \times 3^3 \times 4^2$ **b** $8^5 \times 7^2 \div 8^2$
    c $6^2 \times 5^3 \times 6^2 \times 5^3$ **d** $5^4 \times 2^3 \div 5^2$
    e $9^5 \times 7^2 \times (7^2 \times 9^2)^3$ **f** $8^2 \times 5^6 \times (8^3 \div 5^3)^4$
    g $(9^3 \times 2^5 \div 2^3 \times 9^2)^2$ **h** $3^4 \times 8^5 \times 3^4 \times 8^2$

12. Simplify these expressions.
    a $\dfrac{5^2 \times 8^5}{8^2}$ **b** $\dfrac{6^5 \times 7^2}{6^3}$
    c $\dfrac{6^4 \times 5^4}{6^2 \times 5^2}$ **d** $\dfrac{(7^8 \times 5^6)^3}{5^3 \times 7^2}$
    e $\dfrac{8^7 \times 8^5}{3^2 \times 8^5}$ **f** $4^3 \times \left(\dfrac{4^5 \times 5^9}{4^3 \times 5^7}\right)^2$
    g $\dfrac{6^9 \times 7^5}{6^7 \times 7^3} \times 6^2$ **h** $4^3 \times \dfrac{7^6 \times 4^5}{4^4 \times 7^3} \times 7^2$

13. Write down the answer on your calculator display for these expressions. Then round your answer to 2 decimal places.
    a $(-3.9)^2$ **b** 2.1^2 **c** $(-0.7)^2$
    d 13.25^2 **e** $(-5.4)^3$ **f** 9.9^3
    g $(-0.1)^3$ **h** 16.85^3 **i** π^3

14. Find the value of each of these expressions. For example, $5^3 = 5 \times 5 \times 5 = 125$.
    a 4^2 **b** 4^3 **c** 2^5 **d** 10^2
    e 10^3 **f** 3^3 **g** 2^3 **h** 3^2

15. Write each number as a power of the given number.
    a $81 = 9^\square$ **b** $125 = 5^\square$
    c $128 = 2^\square$ **d** $100\,000 = 10^\square$
    e $81 = 3^\square$ **f** $343 = 7^\square$

16. Find the value of x, y and z.
    a $5^x = 625$ **b** $4^y = 64$ **c** $z^4 = 81$

*17 Estimate the value of x to 1 decimal place if $6^x = 40$.

APPLICATIONS

13.2 Powers and roots

RECAP
- Add indices when multiplying powers of the same number.
- Subtract indices when dividing powers of the same number.
- Multiply indices when finding the power of a power.
- Any number (except 0) to the power of 0 is 1.

> The expression $3^4 \times 3^1 \div 3^2$ can be simplified to 3^3. The expression $2^3 \times 5^4$ can't be simplified in the same way as the base numbers are different.

HOW TO

To solve problems involving powers and roots

1. Read the question – decide how to use your knowledge of powers and roots.
2. Apply the index laws and look out for chances to use known facts about squares and cubes.
3. Answer the question – make sure that you leave your answer in index form if the question asks for it.

EXAMPLE

Find three whole numbers n, such that $\sqrt{5 + 4n}$ is a whole number.

1. If $\sqrt{5 + 4n}$ is a whole number, then $5 + 4n$ must be a square number.

   $5 + 4n$ = square number

2. Find a square number that is 5 more than a multiple of 4.

   Square numbers: 1, 4, 9, 16, 25, 36, 49, 64, 81, 100 ...
   Multiples of 4: 4, 8, 12, 16, 20, 24, 28, 32, 36, 40, 44, 48 ...
   $5 + 4 \times 1 = 9$ $5 + 4 \times 5 = 25$ $5 + 4 \times 11 = 49$

3. Three possible values for n are 1, 5 and 11.

> This method is more efficient than trying different values of n until you find a correct solution.

EXAMPLE

a Victoria tosses a coin and throws a dice. List the different ways they can land.
b She then tosses three coins and throws two dice. How many different ways can they land?

1. When listing, work systematically and use abbreviations or symbols.

a $2 \times 6 = 12$ ways   2. 2 ways a coin can land, and 6 ways for a dice.
   H1, H2, H3, H4, H5, H6, T1, T2, T3, T4, T5, T6

b $2^3 \times 6^2$ ways = 288 ways   2. 2 ways for each coin and 6 ways for each dice.

EXAMPLE

The coordinates (x, y) of points on the circle C are given by the equation $x^2 + y^2 = 25$.
Write down all the points on this circle where both coordinates are integers.

1. Find pairs of square numbers that add to 25.

   Square number 0, 1, 4, 9, 16, 25   2. Remember to include 0.
   $0 + 25 = 25$ and $9 + 16 = 25$
   $9 = 3^2 = (-3)^2$ $16 = 4^2 = (-4)^2$ $25 = 5^2 = (-5)^2$

3. Include all the possible pairs (x, y) remembering that each combination can be written in two ways.

   Possible pairs are (0, 5), (5, 0) (0, -5), (-5, 0)
   (3, 4), (4, 3), (-3, 4), (4, -3), (3, -4), (-4, 3), (-3, -4), (-4, -3)

Number Factors, powers and roots

Reasoning and problem solving

Exercise 13.2A

1. Find a pair of prime numbers a, b such that $\sqrt{a^2 - b}$ is a whole number.

2. Find a pair of whole numbers p, q such that $\sqrt{p^3 - 3q}$ is a whole number.

3. An exam has 5 multiple choice questions each with 4 choices of answer.
   a. How many answers are there on the paper?
   b. Kiera chooses one answer at random for every question. How many different ways could she have done this? Give your answer in index form.
   c. Haneef sits a multiple choice test with 7 questions, each with 3 possible answers. He selects one answer for every question at random. Which of the two students has a higher chance of getting full marks? p.154

4. The coordinates (x, y) of points on a circle C satisfy $(x - 1)^2 + (y - 4)^2 = 20$
   Find the points on this circle where both coordinates are integers.

5. Solve these problems.
   a. $\sqrt{7} = 2.645751$
      Why does $(2.645751)^2 \neq 7$?
   b. The product of two consecutive numbers is 3192.
      What are the two numbers?

6. A scale is split into five sections, each section starting on a multiple of 2.

   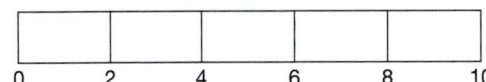

   Copy the diagram, and put these numbers into the correct sections.

   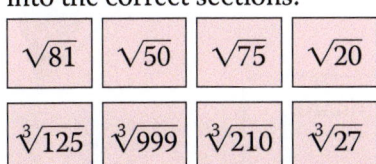

7. a. Find the number that is
   - a multiple of 9
   - a cube number
   - less than 40.
   b. Can you work out the number from two clues? Show your working.

8. Some cube numbers are also square numbers.
   a. Show that 64 is one of these.
   b. Show that 3^6 is another.
   c. What is the next number which is a square and a cube?

9. | | | |
   |---|---|---|
   | $6^0 = 1$ | $6^1 = 6$ | $6^2 = 36$ |
   | $6^3 = 216$ | $6^4 = 1296$ | $6^5 = 7776$ |
   | $6^6 = 46656$ | | |

   Use the table to
   a. explain why $36 \times 216 = 7776$
   b. work out $46656 \div 36$.

10. Computers work in powers of two. Four common units are shown in this table.

Unit	Approximate	Exact
Kilobyte (kB)	A thousand	2^{10}
Megabyte (MB)	A million	2^{20}
Gigabyte (GB)	1000 million	2^{30}
Terabyte (TB)	A million million	2^{40}

    a. What are the percentage errors in each approximation?
    b. Talaal writes, 1 megabyte $=$ (1 kilobyte)^2. Is he correct? Give your reason.

11. Find the value of the letter in each equation.
    a. $3^a \times 9 = 3^{18}$
    b. $5^b = 25^{b-1}$
    c. $4^c = 16^{2c+1}$
    d. $9^{2d-1} = 27^d$

12. Show that $\dfrac{2 \times 4^{x+3}}{2^{2x+4}} = 8$.

13. Small cubes, each with volume $40\,\text{cm}^3$, are stacked to make a larger cube.
    a. What is the volume of the whole stack?
    b. Estimate the length of the side of one small cube to 1 decimal place.
    c. Use your answer to estimate $\sqrt[3]{320}$.

*14. The graphs of $y = \dfrac{x^4 \times x}{x^{2+a}}$ and $y = x^5 \div x^a$, where a is a constant, cross at a point P. Show that the x-coordinate of P must be 1 or 0.

SKILLS

13.3 Surds

The square root of two cannot be written exactly as a fraction, a finite decimal or a recurring decimal. Instead a special notation is used to represent the exact value.

$\sqrt{2} = 1.4142135623730950488\ldots$

- Numbers like $\sqrt{2}$, $\sqrt{3}$ and $\sqrt{5}$ are called **surds**.

Surds can be simplified by looking at their factors.
$\sqrt{20} = \sqrt{4 \times 5} = \sqrt{4} \times \sqrt{5} = 2\sqrt{5}$.

- The square root of a number equals the product of the square roots of the number's factors.

Did you know…

A dangerous number…

Numbers that cannot be written as fractions are called irrational numbers. Their discovery is usually attributed to Hippasus, a member of Pythagoras' school of Mathematicians. Legend has it that Hippasus' revelation that $\sqrt{2}$ could not be written as a fraction was so threatening to Pythagoras' work that Pythagoras had Hippasus drowned to conceal the truth.

EXAMPLE

a Simplify $\sqrt{96}$.
b Simplify $\sqrt{12} + \sqrt{48}$.
c Write $6\sqrt{2}$ in the form $\sqrt{n}$, where n is an integer.
d Simplify $(5 - 2\sqrt{5})(3 + 2\sqrt{5})$.

a $\sqrt{96} = \sqrt{16 \times 6}$
    $= \sqrt{16} \times \sqrt{6} = 4\sqrt{6}$
b $\sqrt{12} + \sqrt{48} = \sqrt{4} \times \sqrt{3} + \sqrt{16} \times \sqrt{3}$
    $= 2\sqrt{3} + 4\sqrt{3} = 6\sqrt{3}$ *Collect like terms.*
c $6\sqrt{2} = \sqrt{6^2 \times 2} = \sqrt{36 \times 2} = \sqrt{72}$
d $(5 - 2\sqrt{5})(3 + 2\sqrt{5}) = 15 + 10\sqrt{5} - 6\sqrt{5} - 20$ *Using FOIL.*
    $= -5 + 4\sqrt{5}$ *Collecting like terms.*

To simplify a surd look for square factors of the number under the square root.

Decimal **approximations** are useful in practical contexts but leaving an answer in **surd form** gives an exact answer.

If you have an expression with a surd in the **denominator**, you should **rationalise** it.

If you wanted to mark out a square of area 5 m², you would give the side length as 2.24 m, not $\sqrt{5}$ m.

- To rationalise the denominator in a fraction, multiply numerator and denominator by the denominator. $\dfrac{\sqrt{n}}{\sqrt{n}} = 1$

EXAMPLE

Rewrite each of these expressions without surds in the denominator.

a $\dfrac{5}{\sqrt{2}}$ b $\dfrac{7}{\sqrt{8}}$

a $\dfrac{5}{\sqrt{2}} = \dfrac{5}{\sqrt{2}} \times \dfrac{\sqrt{2}}{\sqrt{2}}$
    $= \dfrac{5\sqrt{2}}{2}$

b $\dfrac{7}{\sqrt{8}} = \dfrac{7}{\sqrt{8}} \times \dfrac{\sqrt{8}}{\sqrt{8}}$
    $= \dfrac{7\sqrt{8}}{8}$

Number   Factors, powers and roots

Fluency

Exercise 13.3S

1 Simplify these expressions.
  a $\sqrt{3} + \sqrt{3}$ b $\sqrt{5} + \sqrt{5}$
  c $\sqrt{9} + \sqrt{4}$ d $\sqrt{49} + \sqrt{2} - \sqrt{16}$
  e $3\sqrt{7} + \sqrt{49} - \sqrt{7}$
  f $17 + \sqrt{17} - \sqrt{9}$

2 Write each of these expressions as the square root of a single number.
  a $\sqrt{2} \times \sqrt{3}$ b $\sqrt{5} \times \sqrt{3}$
  c $\sqrt{11} \times \sqrt{13}$ d $\sqrt{3} \times \sqrt{7} \times \sqrt{11}$

3 Write these expressions in the form $\sqrt{a} \times \sqrt{b}$, where a and b are prime numbers.
  a $\sqrt{14}$ b $\sqrt{33}$ c $\sqrt{21}$
  d $\sqrt{35}$ e $\sqrt{44}$ f $\sqrt{51}$

4 Write these expressions in the form $a\sqrt{b}$, where a is an integer and b is a prime number.
  a $\sqrt{20}$ b $\sqrt{27}$ c $\sqrt{98}$
  d $\sqrt{48}$ e $\sqrt{28}$ f $\sqrt{45}$
  g $\sqrt{63}$ h $\sqrt{363}$ i $\sqrt{512}$
  j $\sqrt{192}$ k $\sqrt{500}$ l $\sqrt{845}$

5 Write these expressions in the form $\sqrt{n}$, where n is an integer.
  a $4\sqrt{3}$ b $5\sqrt{2}$ c $4\sqrt{5}$
  d $10\sqrt{2}$ e $9\sqrt{5}$ f $7\sqrt{7}$

6 Without using a calculator, evaluate these expressions.
  a $\sqrt{12} \times \sqrt{3}$ b $\sqrt{18} \times \sqrt{8}$
  c $\sqrt{33} \times \sqrt{132}$ d $\sqrt{63} \times \sqrt{28}$
  e $\sqrt{77} \times \sqrt{33}$ f $\sqrt{35} \times \sqrt{56}$
  g $\sqrt{30} \times \sqrt{150}$ h $\sqrt{128} \times \sqrt{8}$
  i $\sqrt{14} \times \sqrt{15} \times \sqrt{42}$
  j $\sqrt{39} \times \sqrt{6} \times \sqrt{208}$

7 Simplify these expressions.
  a $3\sqrt{5} + \sqrt{20}$ b $\sqrt{28} + 5\sqrt{7}$
  c $7\sqrt{12} - 2\sqrt{27}$ d $4\sqrt{28} - 3\sqrt{63}$
  e $\sqrt{50} + \sqrt{8} - \sqrt{72}$
  f $3\sqrt{56} - 7\sqrt{126} + \sqrt{1400}$

8 Simplify these expressions, giving your answers in surd form where necessary.
  a $\sqrt{2} \times \sqrt{2}$ b $\sqrt{5} \times \sqrt{5}$
  c $\sqrt{3}(\sqrt{3} + 3)$ d $\sqrt{4}(\sqrt{3} + 4)$
  e $\sqrt{15}(3 + \sqrt{5})$ f $\pi(2^3 - \sqrt{20})$
  g $(1 + \sqrt{5})(2 + \sqrt{5})$
  h $(4 - \sqrt{7})(6 - 2\sqrt{7})$
  i $(2 - \sqrt{3})^2$ j $(3\sqrt{5} + 2\sqrt{7})^2$

9 Simplify these expressions.
  a $\dfrac{\sqrt{18} \times \sqrt{27}}{\sqrt{54}}$ b $\dfrac{\sqrt{98} \times \sqrt{12}}{\sqrt{48} \times \sqrt{128}}$
  c $\dfrac{\sqrt{8}}{\sqrt{10}} \div \dfrac{2\sqrt{45}}{\sqrt{18}}$ d $\dfrac{3\sqrt{2}}{10} + \dfrac{\sqrt{8}}{\sqrt{50}}$
  e $\dfrac{4\sqrt{63}}{15} - \dfrac{3\sqrt{42}}{10\sqrt{24}}$ f $\dfrac{\sqrt{65}}{\sqrt{28}} \div \dfrac{\sqrt{39}}{2\sqrt{35}}$

10 Use a calculator to find an approximate decimal value for each of these expressions. Give your answers to 2 decimal places.
  a $4\sqrt{2}$ b $\sqrt{5} + 1$
  c $2 + \sqrt{5}$ d $36\pi - 7$

11 Rationalise the denominator of each of these fractions.
  a $\dfrac{1}{\sqrt{2}}$ b $\dfrac{1}{\sqrt{3}}$ c $\dfrac{3}{\sqrt{7}}$
  d $\dfrac{5}{\sqrt{6}}$ e $\dfrac{5}{\sqrt{10}}$ f $\dfrac{9}{\sqrt{15}}$

12 Rewrite each of these fractions without roots in the denominator.
  a $\dfrac{2}{\sqrt{8}}$ b $\dfrac{2}{\sqrt{10}}$ c $\dfrac{3}{\sqrt{12}}$
  d $\dfrac{5}{\sqrt{30}}$ e $\dfrac{8}{\sqrt{40}}$ f $\dfrac{3}{\sqrt{120}}$

APPLICATIONS

13.3 Surds

RECAP
- Numbers such as $\sqrt{2}$ are called surds.
- For any pair of surds $\sqrt{a} \times \sqrt{b} = \sqrt{a \times b}$.
- Surds are written in the simplest form when the smallest possible integer is written inside the square root sign.

$\sqrt{54} = \sqrt{9 \times 6} = 3\sqrt{6}$
$(1 + \sqrt{3})^2 = 1 + 2\sqrt{3} + (\sqrt{3})^2 = 4 + 2\sqrt{3}$
Write the rational number first then the surd.

$\dfrac{1}{\sqrt{7}} = \dfrac{1}{\sqrt{7}} \times \dfrac{\sqrt{7}}{\sqrt{7}} = \dfrac{\sqrt{7}}{7}$

HOW TO

To manipulate surds
1. Read the question – decide how to use your knowledge of surds.
2. Use $\sqrt{a} \times \sqrt{b} = \sqrt{a \times b}$ and algebraic techniques to simplify expressions containing surds.
3. Answer the question and leave your answer in simplified surd form.

EXAMPLE

Find the area and perimeter of this rectangle.

$3\sqrt{3}$ cm
$2 + \sqrt{3}$ cm

1. The area is the product of the height and the width, the perimeter is the sum of the four sides.
2. Area $= 3\sqrt{3} \times (2 + \sqrt{3}) = 6\sqrt{3} + 9$ cm^2
   Perimeter $= 2(2 + \sqrt{3}) + 2(3\sqrt{3}) = 4 + 8\sqrt{3}$ cm

EXAMPLE

The diagram shows two similar rectangles.
Find the length of side x.

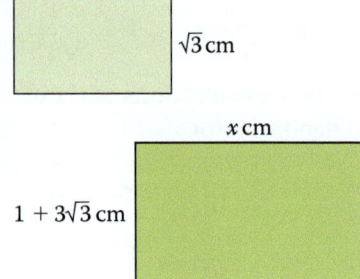

$3 + \sqrt{3}$ cm
$\sqrt{3}$ cm
x cm
$1 + 3\sqrt{3}$ cm

1. The rectangles are similar so $\dfrac{\text{width}}{\text{height}}$ is the same.

$\dfrac{x}{1 + 3\sqrt{3}} = \dfrac{3 + \sqrt{3}}{\sqrt{3}}$

2. Solve for x by multiplying both sides by $1 + 3\sqrt{3}$.

$x = \dfrac{(3 + \sqrt{3})}{\sqrt{3}} \times (1 + 3\sqrt{3})$

$= \dfrac{3 + 9\sqrt{3} + \sqrt{3} + 9}{\sqrt{3}} = \dfrac{12 + 10\sqrt{3}}{\sqrt{3}}$

$= \dfrac{12}{\sqrt{3}} \times \dfrac{\sqrt{3}}{\sqrt{3}} + 10$ 3. Rationalise.

$= 4\sqrt{3} + 10$ cm

EXAMPLE

Find the area of this equilateral triangle. Give your answer in surd form.

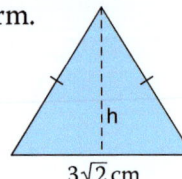

$3\sqrt{2}$ cm

1. Use Pythagoras' theorem to find the height of the triangle, then use the formula for the area of a triangle.

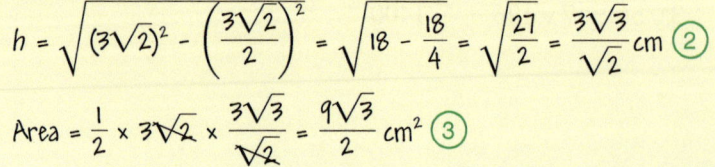

Number   Factors, powers and roots

Reasoning and problem solving

Exercise 13.3A

1 Find the perimeter and area of each shape. Give your answers in simplified surd form.

a   **b**

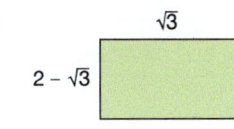

2 The diagram shows two similar triangles.

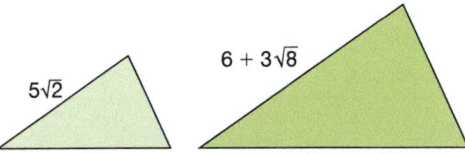

a Find the scale factor of enlargement.

b Find the scale factor for the areas of the two triangles.

Give your answers in simplified surd form.

3 An isosceles triangle has base $\sqrt{175}$ cm as shown. Its area is 10 cm².

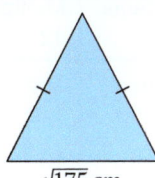

a Find its height.

b Find its perimeter.

4 You are told that $\sqrt{2} \approx 1.414$, $\sqrt{3} \approx 1.732$ and $\sqrt{5} \approx 2.236$. Use this information to estimate the value of each of these surds. Show your working, and give your answers to 4 significant figures.

a $\sqrt{2} + \sqrt{3}$ **b** $\sqrt{10}$

c $\sqrt{125}$ **d** $\sqrt{120}$

5 Write down four more surds whose value can be estimated using the information given in question **4**, and estimate the values of your surds.

6 Put these expressions in ascending order. Give your reasons.

A $\dfrac{6 + \sqrt{27}}{3} - \dfrac{2 + \sqrt{3}}{4}$

B $\dfrac{20 + 3\sqrt{7}}{3} - (5 + \sqrt{28})$

C $\dfrac{8 + \sqrt{45}}{3} - \dfrac{4 - \sqrt{20}}{9}$

7 Find the size of the shaded area. Give your answer in the form $p + q\sqrt{3}$ where p and q are integers.

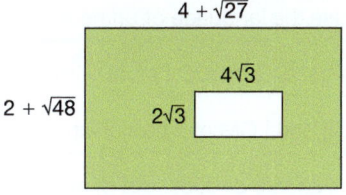

8 A square, area 36 cm², is cut into 4 triangles and rearranged to make a rectangle.

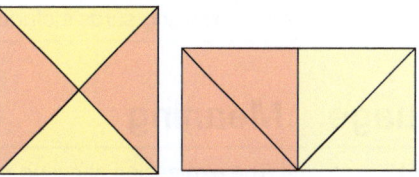

a How long is the side of the red square?

b What is the perimeter of the rectangle?

*9

$\sqrt{5}$	$2 - \sqrt{5}$	$\sqrt{64}$	$8\sqrt{2}$
$\sqrt{2}$	$\sqrt{20}$	$3 + \sqrt{5}$	$2 + \sqrt{5}$
$3 - \sqrt{5}$	$3\sqrt{2}$	$\sqrt{10}$	$\sqrt{100}$

From this table, find

a two numbers which can be written as integers

b a pair which add up to give 4

c a pair where one number is double the other

d a pair where one is the other squared

e a pair which multiply to give 4

f a pair which multiply to give 48

g the largest value

h the smallest value.

10 The numbers X, Y and Z satisfy $Y^2 = XZ$.

a Find Z if $X = \sqrt{3}$ and $Y = 1 - \sqrt{3}$. Give your answer in the form $p + q\sqrt{3}$, where p and q are integers or fractions.

b Explain why it would not be possible to find Y if $X = \sqrt{5}$ and $Z = 1 - \sqrt{5}$.

Summary

Checkout
You should now be able to...

Test it
Questions

✓	Know and use the language of prime numbers, factors and multiples.	1
✓	Write a number as a product of its prime factors.	2, 3
✓	Find the HCF and LCM of a pair of integers.	3
✓	Estimate the square or cube root of an integer.	4
✓	Find square and cube roots of numbers and apply the laws of indices.	5, 6
✓	Simplify expressions involving surds including rationalising fractions.	7, 8

Language	Meaning	Example
Product	The expression showing multiplication.	$15 = 3 \times 5$
Multiple	The original number multiplied by an integer.	Multiples of 6 include $12 \ (2 \times 6)$ and $18 \ (3 \times 6)$.
Factor	A whole number that divides exactly into a given integer.	The factors of 18 are 1, 2, 3, 6, 9 and 18.
Prime	A number that has only two factors, itself and one.	$13 = 1 \times 13$
Prime factor decomposition	Writing a number as a product of its prime factors.	$52 = 2^2 \times 13$
Highest common factor (HCF)	The largest factor that is shared by two or more numbers.	Factors of 30 / Factors of 70 Venn diagram: 3 | 2, 5 | 7 $HCF = 2 \times 5 = 10$ $LCM = 3 \times 2 \times 5 \times 7 = 210$
Lowest common multiple (LCM)	The smallest multiple that is shared by two or more numbers.	
Square root	A number that when multiplied by itself is equal to a given number.	$\sqrt{64} = 8 \quad (8 \times 8 = 64)$
Cube root	A number that when multiplied by itself two times is equal to a given number.	$\sqrt[3]{64} = 4 \quad (4 \times 4 \times 4 = 64)$
Surd	The root of a number which cannot otherwise be written exactly.	$\sqrt{2}$ $\sqrt[3]{5}$
Rationalise	To rewrite a fraction so that it does not contain any surds in the denominator.	$\dfrac{1}{\sqrt{2}} = \dfrac{\sqrt{2}}{2}$

Review

1 | 1 | 2 | 3 | 15 | 37 | 63 | 101 | 105 |

List the numbers in the panel that are
- **a** prime
- **b** factors of 105
- **c** multiples of 21.

2 Write each number as the product of its prime factors. Use index notation where appropriate.
- **a** 105
- **b** 37
- **c** 300
- **d** 126

3 For each pairs of numbers find the
- **i** lowest common multiple
- **ii** highest common factor.
- **a** 5 and 7
- **b** 13 and 39
- **c** 60 and 36
- **d** 30 and 108

4 Estimate the value of these roots. Check your estimate using your calculator.
- **a** $\sqrt{30}$
- **b** $\sqrt[3]{45}$

5 Calculate the value of these expressions.
- **a** $\sqrt[3]{64}$
- **b** $\sqrt[3]{125}$
- **c** 4^3
- **d** 3^4

6 Simplify these expressions giving your answer in index form.
- **a** $7^2 \times 7^5 \div 7^3$
- **b** $(3^5 \div 3^2)^3$
- **c** $\dfrac{3^{11} \div 3^2}{3^6}$
- **d** $(7^{12} \div 7^3) \times 7^4 \times 7^8$
- **e** $3^4 \times 5^3 \times 3^{-6} \div 5^2 \times 3^2$

7 Simplify these expressions involving surds.
- **a** $\sqrt{108}$
- **b** $5\sqrt{3} - \sqrt{27} + \sqrt{8}$
- **c** $2\sqrt{5} \times \sqrt{5}$
- **d** $\sqrt{6} \times \sqrt{2}$
- **e** $3\sqrt{8} \div \sqrt{2}$
- **f** $5\sqrt{10} \div 10\sqrt{5}$
- **g** $(\sqrt{3} - 2)(\sqrt{3} - 5)$
- **h** $\dfrac{\sqrt{8} + \sqrt{2}}{\sqrt{8} - \sqrt{2}}$

8 Rationalise these fractions.
- **a** $\dfrac{1}{\sqrt{7}}$
- **b** $\dfrac{3}{\sqrt{6}}$
- **c** $\dfrac{5}{2\sqrt{5}}$
- **d** $\dfrac{4 + \sqrt{6}}{\sqrt{6}}$
- **e** $\dfrac{3\sqrt{2} - 5}{\sqrt{128} - 4\sqrt{2}}$
- **f** $\dfrac{4\sqrt{3} + 5\sqrt{2}}{6\sqrt{3} + \sqrt{27}}$

What next?

Score		
	0 – 3	Your knowledge of this topic is still developing. To improve, look at MyiMaths: 1032, 1033, 1034, 1044, 1053, 1064, 1065, 1924
	4 – 7	You are gaining a secure knowledge of this topic. To improve your fluency look at InvisiPens: 13Sa – f
	8	You have mastered these skills. Well done you are ready to progress! To develop your problem solving skills look at InvisiPens: 13Aa – h

Assessment 13

1. Isa and Josh both try to write 19 800 as a product of its prime factors.

    Isa writes $19\,800 = 2^3 \times 3^2 \times 5^2 \times 11$
    Josh writes $19\,800 = 2 \times 3^3 \times 5 \times 11$

    Who is correct? Write 19 800 as a product of its prime factors using index notation. Show your working. [2]

2. All square numbers over 1 can be written as the sum of two prime numbers.

    a Show there are two ways of doing this for the number 16. [2]

    b Show there are five ways of doing this for the number 64. [3]

    c Show that there is only one possible way in which any *odd* square number can be written as the sum of two prime numbers. [2]

3. Ms Connell took her class to the monkey house at the zoo, but a naughty child opened the cage door and let the monkeys out with the children. Ms Connell counted about 70 heads and tails and the ratio of heads to tails was roughly 3 : 2. She also knew that the number of children was a prime number.
   How many children were in Ms Connell's class? [4]

4. It is claimed that all prime numbers can be found by substituting $p = 0, 1, 2$, etc into the formula $P = 41 - p + p^2$.

    a Confirm the claim for $p = 0, 3$ and 6. [3]

    b Write down one value of p which shows that the claim is *not* correct. Explain your reasoning. [2]

5. a Is 27 the highest common factor of 54 and 270?
       If not, show how to work out the highest common factor of these two numbers. [2]

    b Show that 4320 is the lowest common multiple of 96 and 270 by using prime factors. [2]

6. MegaBurgers sell frozen burgers. The burgers are first packed into boxes of a uniform size and then the boxes are packed into crates of various sizes.
   A large crate contains 48 burgers, a medium crate contains 30 burgers and a small crate contains 18 burgers.
   What is the largest number of burgers that could be in a box? [3]

7. This is an alternative method to find the HCF of two numbers.

    1. Write down the numbers side by side.
    2. Cross out the largest and write underneath it the 'difference' between the largest and smallest.
    3. Repeat step 2 until the numbers left are the same.
    4. The remaining number is the HCF of the two original numbers.

    a Use this method with these numbers.

       i 30 and 66   [2]   ii 252 and 588   [2]

    b Does this method work with more than two numbers? [3]

Number Factors, powers and roots

8 Trains from Birmingham go to Glasgow, Liverpool and Leeds. Trains to Glasgow leave once an hour, to Liverpool every 45 minutes and to Leeds every 25 minutes. At 6 am three trains leave for all these destinations. When is the next time trains to all these destinations will leave simultaneously? [4]

9 a Zainab says that to evaluate $3^3 \times 3^2$ you multiply the indices. Saqib says you add the indices. Who is correct? Use the correct rule to evaluate $3^3 \times 3^2$. [2]

b Gino says that to evaluate $14^8 \div 14^2$ you subtract the indices. Jonas says you divide the indices. Who is correct? Use the correct rule to evaluate $14^8 \div 14^2$. [2]

10 Solve these equations for p and q.

a $6^{(p+5)} = 36^p$ [2]   **b** $8^{(2q-4)} = 4^{(q-2)}$ [3]

11 A cuboid gift box has length $5 + 2\sqrt{7}$ width $5\sqrt{3}$ and depth $\sqrt{27}$ as shown.

a Find the area of

  i face A [2]   **ii** face B [2]

  iii face C. [2]

b Find the total surface area. [2]

c A piece of ribbon will be used to wrap around the gift box in both directions.
What is the minimum length of ribbon needed? [2]

d Find the volume of the gift box. [2]

12 The answer to each of these simplifications is wrong.
In each case explain the error and give the correct solution.

a $\sqrt{3} + \sqrt{3} = \sqrt{6}$ [1]   **b** $4\sqrt{3} \times 5\sqrt{12} = 20\sqrt{36}$ [2]

c $(\sqrt{3} + 4)^2 = 3 + 16 = 19$ [3]   **d** $(\sqrt{6} - 2)^2 = 6 - 4 = 2$ [3]

e $(3 + \sqrt{3})(7 - \sqrt{3}) = 21 - 3 = 18$ [3]

f $(\sqrt{7} + \sqrt{5})(\sqrt{7} - \sqrt{5}) = 12$ [2]

g $(4\sqrt{8} + 2\sqrt{11})(3\sqrt{8} - 5\sqrt{11}) = 12\sqrt{64} - 110$ [3]

13 Rationalise the denominator in each of the following expressions.
Leave the fraction in its simplest form.

a $\dfrac{4}{\sqrt{5}}$ [2]   **b** $\dfrac{8}{\sqrt{8}}$ [2]

c $\dfrac{1 - \sqrt{2}}{3\sqrt{2}}$ [2]   **d** $\dfrac{12 - 3\sqrt{18}}{\sqrt{45}}$ [2]

e $\dfrac{15 - \sqrt{21}}{\sqrt{28} + \sqrt{63}}$ [3]   **f** $\dfrac{\sqrt{44}}{\sqrt{704} - \sqrt{176}}$ [3]

14 A right-angled triangle has sides as shown.
Show that both these statements are correct.

a $h = 6$ [3]

b The perimeter of the triangle is numerically double its area. [4]

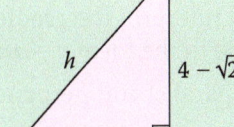

14 Graphs 1

Introduction
Kinematics is the topic within maths that deals with motion. By writing equations and drawing graphs that describe the relationship between distance, speed and acceleration, it is possible to calculate things such as the speed of a vehicle at a particular time during its journey.

What's the point?
Having the mathematical tools to describe how objects move means that we can better understand our world, which is in continual motion.

Objectives
By the end of this chapter, you will have learned how to …
- Find and interpret the gradient and y-intercept of a line and relate these to the equation of the line in the form $y = mx + c$.
- Identify parallel and perpendicular lines using their equations.
- Draw line graphs and quadratic curves.
- Identify roots, intercepts and turning points of quadratic curves using graphical and algebraic methods.
- Use graphs to solve problems involving distance, speed and acceleration.

Check in

1. Evaluate these expressions for i $x = 3$ ii $x = -2$.
   a x^2
   b x^3
   c $2x^2$
   d $x^3 + x$
   e $3x^2 - x$
   f $2x^3 + 2x$
   g $x^3 + 4x^2$
   h $2x^3 - 4x^2 - x$

2. Plot these coordinate groups on one set of axes. Join each group of coordinates with a straight line.
   a $(3, 1), (3, 2), (3, 3)$
   b $(1, -2), (2, -2), (3, -2)$
   c $(1, 3), (2, 5), (3, 7)$
   d $(1, 4), (2, 3), (3, 2)$

3. Find the value of the unknown in each equation.
   a $9 = 3n + 3$
   b $6 = 1 + 5m$
   c $10p - 4 = 1$

4. For each line give its i gradient ii y-axis intercept iii direction.
   a $y = 3x + 4$
   b $y = 10 - 4x$
   c $2y = 8x + 10$
   d $2y - 4x = 15$
   e $y = 7$
   f $x = 2y - 4$

Chapter investigation

The points $(2, 4)$ and $(5, 13)$ both lie on the same straight line.

Find the equation of this line.

The point $(3, -2)$ lies on a line that is perpendicular to this line.
Find the equation of the perpendicular line.

SKILLS

14.1 Equation of a straight line

You need two pieces of information to specify a straight line. This is usually the **gradient** and **y-intercept**.

The gradient defines the steepness of a straight line.

The y-intercept gives the value of y where the line crosses the y-axis.

- The equation $y = mx + c$ describes a straight line with gradient m and y-intercept c.
- Gradient of a line = $\dfrac{\text{Change in } y}{\text{Change in } x}$

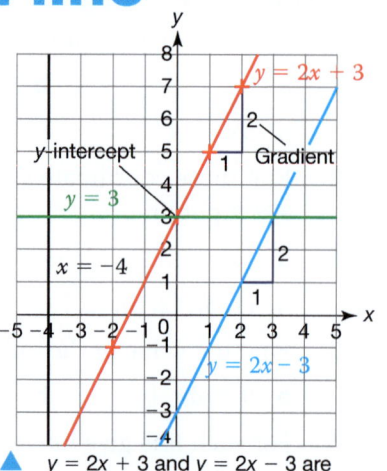

▲ $y = 2x + 3$ and $y = 2x - 3$ are parallel because they both have the same gradient, 2.

EXAMPLE

Find the equation of the line that passes through the points (1, 3) and (5, 11)

Find the gradient of the line: $\dfrac{(11 - 3)}{(5 - 1)} = \dfrac{8}{4} = 2$

Therefore the equation of the line is $y = 2x + c$

Choose one of the points to work with: (1, 3) is on the line, so
$3 = 2 \times 1 + c$
$3 = 2 + c$
$c = 1$
$y = 2x + 1$

EXAMPLE

a Using a ruler and protractor draw the line that passes through the point (4, 1) and is perpendicular to the line $y = \tfrac{1}{2}x - 1$ shown below.

b Find the equation of the perpendicular line.

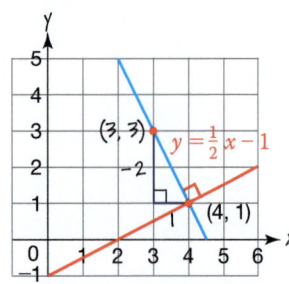

b To calculate the gradient compare two points on the line.

Gradient $= \dfrac{1 - 3}{4 - 3} = \dfrac{-2}{1}$
$= -2$

The change in y is negative. The line goes down 2 units for every 1 unit right.

To calculate the y-intercept substitute into $y = -2x + c$.

(4, 1) lies on $y = -2x + c$
$\Rightarrow 1 = -2 \times 4 + c$
$c = 9$
$y = -2x + 9$

Check using (3, 3): $-2 \times 3 + 9 = 3$

The equation of a line with gradient m passing through the point (a, b) can be written
$\dfrac{y - b}{x - a} = m$ or $y = m(x - a) + b$

- If a line L has gradient m then any line perpendicular to L has gradient $-\dfrac{1}{m}$.

Algebra   Graphs 1

Fluency

Exercise 14.1S

1 A line has gradient 3 and y-intercept -5.
  a Draw the graph of this function.
  b Write the equation of the line.

2 A line has gradient $\frac{3}{4}$ and y-intercept 2.
  a Draw the graph of this function.
  b Write the equation of the line.

3 The graphs of six functions are plotted.

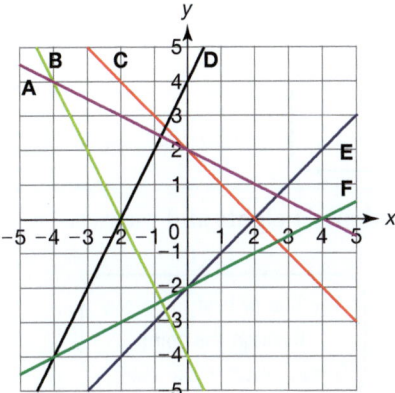

Match each function to a graph.
  a $-2x - y = 4$ **b** $x + y = 2$
  c $y - 2x = 4$ **d** $x + 2y = 4$
  e $x - 2y = 4$ **f** $x - y = 2$

4 For each graph
  i write its gradient and y-intercept
  ii write the equation of the line.

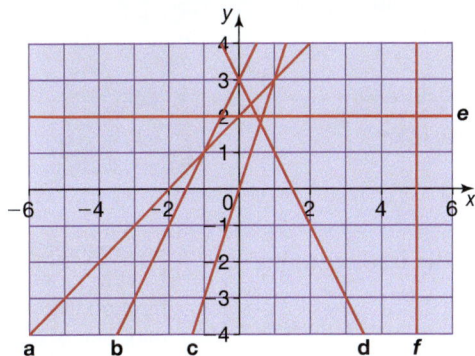

5 Copy and complete this table.

Gradient	y-intercept	Equation
3	5	
5	-2	
-2	7	
$\frac{1}{2}$	9	
$-\frac{1}{4}$	-3	
0	4	
1	0	

6 The points $(-1, -4)$ and $(3, 6)$ are opposite vertices of a rectangle. If the sides of the rectangle are parallel to the x- and y-axes write down their equations.

7 Find the equations of these lines.
  a Gradient 6,   passes through $(0, 2)$
  b Gradient -2, passes through $(0, 5)$
  c Gradient -1, passes through $(0, \frac{1}{2})$
  d Gradient -3, passes through $(0, -4)$

8 For each of these lines, give the equation of a line parallel to it.
  a $y = 2x - 1$ **b** $y = -5x + 2$
  c $y = -\frac{1}{4}x + 2$ **d** $y = 7 - 4x$
  e $y = 6 + \frac{3}{4}x$ **f** $2y = 9x - 1$

9 Find the equations of these lines.
  a A line parallel to $y = -4x + 3$ and passing through $(-1, 2)$.
  b A line parallel to $2y - 3x = 4$ and passing through $(6, 7)$.

10 Give the equation of a line perpendicular to each of these lines.
  a $y = 2x - 1$ **b** $y = -5x + 2$
  c $y = -\frac{1}{4}x + 2$ **d** $y = 7 - 4x$
  e $y = 6 + \frac{3}{4}x$ **f** $2y = 9x - 1$

11 Find the equation of a line
  a perpendicular to $y = 2x + 4$ and passing through $(3, 7)$
  b perpendicular to $3y = 8 - 2x$ and crossing the y-axis at the same point.

12 Find the equations of the line segments joining each of these pairs of points.
  a $(4, 9)$ to $(8, 25)$ **b** $(5, 6)$ to $(10, 16)$
  c $(2, 1)$ to $(5, 2)$ **d** $(-2, 5)$ to $(3, -5)$

13 Identify which lines form perpendicular pairs and which is the odd one out.

C $y = \frac{1}{2}x + 7$	**F** $y = 9 - \frac{1}{4}x$
A $y = 9 - \frac{1}{2}x$	**D** $y = \frac{5}{8}x$	**G** $5y = 6 - 8x$
B $2y = 8x - 1$	**E** $y = 2x + 4$

1153, 1311, 1314, 1957   SEARCH

279

APPLICATIONS

14.1 Equation of a straight line

RECAP
- The equation of a straight line is $y = mx + c$, where m is the gradient and c is the y-intercept of the line.
- Parallel lines have the same gradient.
- The gradients of two perpendicular lines multiply together to give -1.

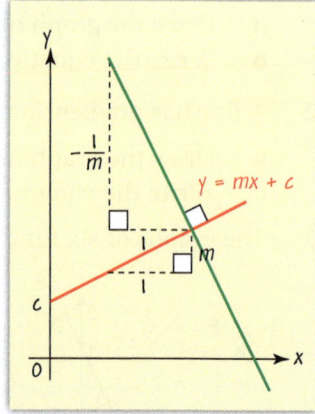

HOW TO

To find the equation of a straight line
1. Find the gradient of the line.
   Use the graph in the question or draw a sketch.
2. Find the y-intercept either using the graph or by substituting a known point into the equation $y = mx + c$.
3. Give your answer in the context of the question.

EXAMPLE (p.324)

Interpret the line of best fit on this scatter diagram of 18 students' heights and masses.

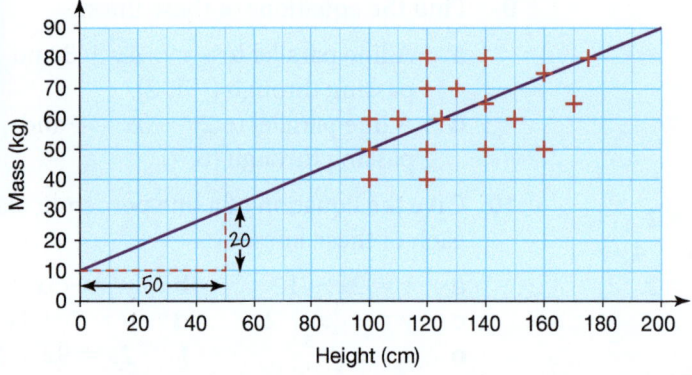

The line of best fit passes through the point (mean height, mean mass).

1. The gradient, $m = \dfrac{20}{50} = \dfrac{2}{5}$ 2. The y-axis intercept, $c = 10$
   The equation of the graph is $y = \dfrac{2}{5}x + 10$.
3. The gradient tells you that for every 5 cm you grow you gain 2 kg.
   The intercept tells you that at 0 cm height, you have mass 10 kg.
   It does not always make sense to interpret the intercept!

EXAMPLE (p.226)

What is the equation of the perpendicular bisector of the line segment passing through (4, 8) and (6, 16)?

A perpendicular bisector goes through the midpoint of a line segment and is at right angles to it.

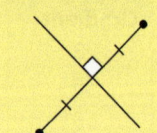

1. Gradient $= \dfrac{16 - 8}{6 - 4} = 4$
2. The midpoint is (mean of x-coordinates, mean of y-coordinates).
   Midpoint is $\left(\dfrac{4+6}{2}, \dfrac{8+16}{2}\right) = (5, 12)$
   Let $y = -\dfrac{1}{4}x + c$ Gradient of perpendicular is $-\dfrac{1}{m}$.
   At (5, 12) $12 = \left(-\dfrac{1}{4}\right) \times 5 + c$
   $c = 13\dfrac{1}{4}$
3. The equation is $y = -\dfrac{1}{4}x + 13\dfrac{1}{4}$

Algebra   Graphs 1

Reasoning and problem solving

Exercise 14.1A

1 Here are the equations of several lines.

  A $y = 3x - 2$
  B $y = 4 + 3x$
  C $y = x + 3$
  D $y = 5$
  E $2y - 6x = -3$
  F $y = 3 - x$

  a Which three lines are parallel to one another?
  b Which two lines cut the y-axis at the same point?
  c Which line has a zero gradient?
  d Which line passes through (2, 4)?
  e Which pair of lines are reflections of one another in the y-axis?

2 For each graph, find its equation in the form $y = mx + c$ and interpret the meaning of m and c. State if it is sensible to interpret c.

  a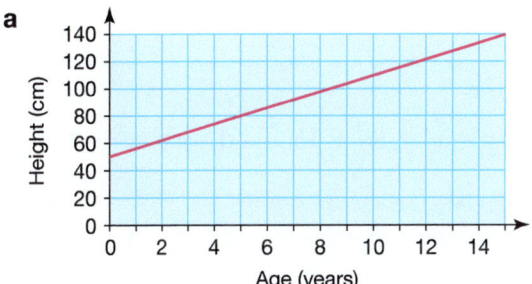

  b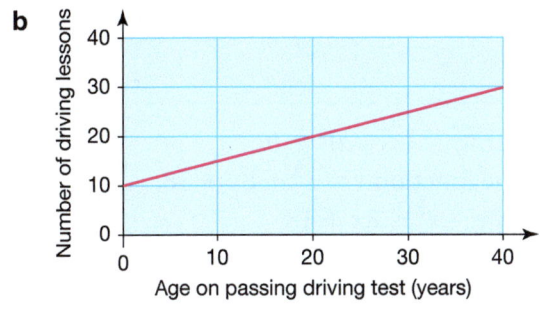

3 Mrs Lui gives her class a geography test. The results (%) for Paper 1 and Paper 2 for 10 students, together with a line of best fit, are shown on a scatter diagram. The points (50, 40) and (60, 50) lie on the line of best fit.

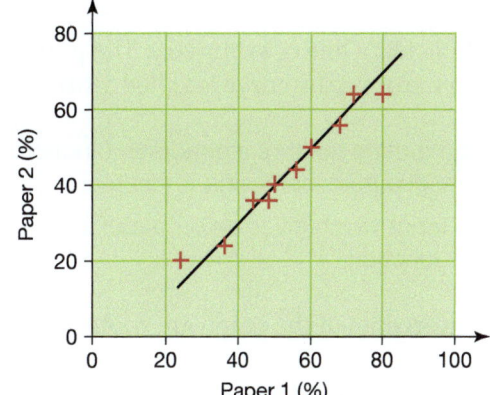

  Interpret the line of best fit.

4 Find the equation of the perpendicular bisector of the segment joining each pair of points.

  a (3, 10) and (7, 12)
  b (2, 20) and (5, 18)
  c (−2, 7) and (4, −10)

5 Write an expression for the gradient of a line perpendicular to the line segment joining (3t, 9) to (2t, 12).

6 The triangle formed by joining the points (5, 12), (14, 24) and (2, 40) is right-angled. True or false?

7 Jawhar solved a pair of simultaneous equations graphically. His solution was $x = 2$ and $y = 3$. Use this diagram to help you find the equation of the second line.

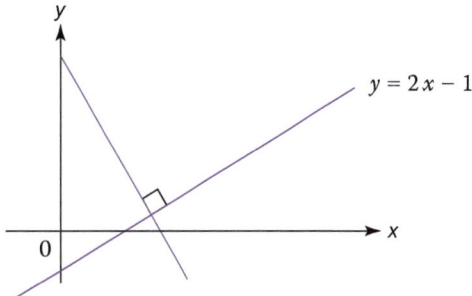

*8 The point P (3, 4) lies on a circle with centre (0, 0).
  Find the equation of the tangent at P.

14.2 Properties of quadratic functions

SKILLS

A parabola has a line of symmetry. The point where the line of symmetry crosses the curve is called a **turning point**.

- At a turning point of a quadratic function, the function changes from increasing to decreasing or vice versa.
- An input which gives an output of zero is called a **root** of the function.

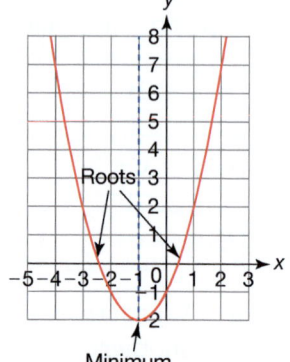

The x-**intercepts** of the graph are roots of the function.

EXAMPLE

For the function $y = x^2 - 3$, find
a the coordinates of the turning point
b the value of the y-intercept
c the roots.

Plot the function $y = x^2 - 3$.
a $(0, -3)$
b -3
c Using the graph.
$x = -1.8$ and $x = 1.8$
Using algebra.
$x^2 - 3 = 0$
$x^2 = 3$
$x = \sqrt{3}$ or $-\sqrt{3}$ $\pm\sqrt{3}$

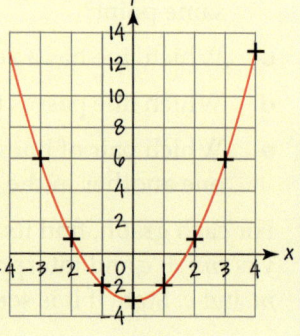

EXAMPLE p.116

Use algebra to find the roots of these functions.
a $y = x^2 - 2x$ b $y = (x + 1)(x - 7)$ c $y = x^2 + 8x + 15$ d $y = x^2 - 4$

The roots of the function are values that have an output of 0.

a $x^2 - 2x = 0$
$x(x - 2) = 0$
$x = 0$ and $x = 2$

b $(x + 1)(x - 7) = 0$
$x = -1$ and $x = 7$

c $x^2 + 8x + 15 = 0$
$(x + 3)(x + 5) = 0$
$x = -3$ and $x = -5$

d $x^2 - 4 = 0$
$(x + 2)(x - 2) = 0$
$x = -2$ and $x = 2$

Quadratic functions can have 0, 1 or 2 roots.

EXAMPLE

a Write the function $y = x^2 + 8x + 18$ in the form $y = (x + a)^2 + b$.
b Find the coordinates of the turning point of the function $y = x^2 + 8x + 18$.
c Explain why there are no roots of the function $y = x^2 + 8x + 18$.

$y = (x + a)^2 + b$ is known as **completed square form**.

a $y = x^2 + 8x + 18 = (x + 4)^2 - 16 + 18$ $x^2 + 8x = (x + 4)^2 - 16$
$y = (x + 4)^2 + 2$

b The x^2 term of the function is positive. Therefore the turning point is a smallest value.
$y = (x + 4)^2 + 2$ has a lowest output when $x = -4$ and $y = 2$.
The turning point of the function is at $(-4, 2)$.

c The quadratic has a lowest value at $(-4, 2)$.
Therefore the curve does not cross the x-axis and has no x-intercepts, that is, there are no roots.
This information can be used to sketch the graph.

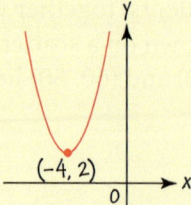

Algebra   Graphs 1

Fluency

Exercise 14.2S

1 a Sketch the graph of these functions on the same axes and label each line.

$y = x^2$ $y = x^2 - 2$ $y = -x^2$

b Which function has a maximum?

c Which function has two roots?

2 For each of these functions find
  i the equation of the line of symmetry
  ii the number of roots
  iii the coordinates of the turning point.
  a $y = x^2 - 8x + 20$
  b $y = 7x - x^2$
  c $y = x^2 - 4$
  d $y = -x^2 - 10x - 27$
  e $y = -x^2 + 3x + 4$
  f $y = x^2 - 8x + 16$

3 Use your results from question **2** to match each function to the sketch of its graph.

A 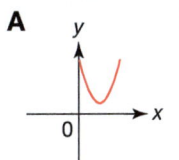 B 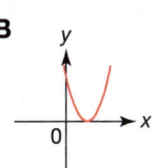 C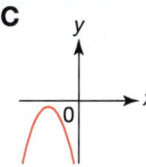

D 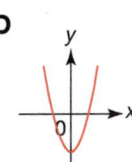 E 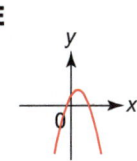 F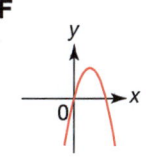

4 Use algebra to find the roots of these functions.
  a $y = x^2 + 7x$
  b $y = x^2 - 8x$
  c $y = x^2 + 5x$
  d $y = x^2 + 2x$
  e $y = 2x^2 + 8x$
  f $y = 2x^2 + 5x$
  g $y = 4x^2 + 12x$
  h $y = 5x^2 + 3x$

5 Use algebra to find the roots of these functions.
  a $y = x^2 + 7x + 6$
  b $y = x^2 - 3x + 2$
  c $y = x^2 + 5x + 6$
  d $y = x^2 + x - 12$
  e $y = x^2 - x - 6$
  f $y = x^2 - 5x + 4$
  g $y = x^2 - 2x - 15$
  h $y = x^2 + 9x + 14$

6 Use algebra to find the roots of these functions.
  a $y = x^2 + 4x + 4$
  b $y = x^2 - 16$
  c $y = x^2 - 4$
  d $y = x^2 + 10x + 25$

7 a Express the function $y = x^2 + 6x + 3$ in the form $y = (x + a)^2 + b$.
  b Write the coordinates of the y-intercept.
  c Find the coordinates of the turning point.
  d Sketch the graph of $y = x^2 + 6x + 3$.
  e How many roots does the function have?

8 For each of these functions
  i express the function in completed square form
  ii find the coordinates of the turning point
  iii sketch the graph of the function
  iv write down the number of roots of the function.
  a $y = x^2 + 2x - 3$
  b $y = x^2 + 2x + 3$
  c $y = x^2 + 8x + 12$
  d $y = x^2 - 6x + 12$
  e $y = x^2 + 5x - 3$
  f $y = x^2 - 6x + 9$
  g $y = -x^2 + 10x - 1$
  h $y = -x^2 + 3x + 7$

*9 a Write the function $y = 3x^2 + 6x - 2$ in the form $y = p(x + q)^2 + r$.
  b Write the coordinates of the y-intercept.
  c Find the coordinates of the turning point.
  d Sketch the graph of $y = 3x^2 + 6x - 2$.
  e How many roots does the function have?

*10 Calculate $b^2 - 4ac$ for the function $y = x^2 + 4x + 9$. Explain how this shows that the function has no roots.

*11 Use $b^2 - 4ac$ to show that the function $y = 3x^2 + 2x - 8$ has two roots.

*12 A cricket ball is hit into the air. Its height (in metres) above ground after t seconds is given by the formula $h = 1 + 8t - t^2$. What is the greatest height that the ball reaches?

APPLICATIONS

14.2 Properties of quadratic functions

RECAP
- A parabola has a line of symmetry. The point where the line of symmetry crosses the curve is called a turning point.
- The turning point of a quadratic function will either be a maximum or a minimum.
- An input which gives an output of zero is called a root of the function.

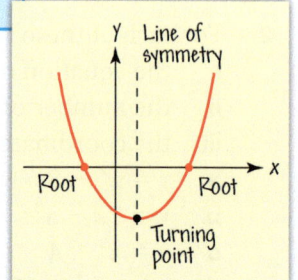

HOW TO
To solve problems involving quadratic graphs
1. Identify the quadratic function.
2. Sketch a graph.
3. Identify whether the roots or turning point help solve the problem.
4. Complete the necessary algebra and interpret the solution.

EXAMPLE

A company manufactures and sells packs of batteries. If the selling price of their batteries is too low they make no profit. If the selling price is too high they do not sell enough to make a profit. The company works out that that profit is a function of price:
$P = -s^2 + 7s - 10$ where $P =$ profit (p) and $s =$ selling price ($)

a At what selling price will the company start losing money?
b What selling price will maximise their profit? How much profit will they make?

1. The quadratic function will form a 'negative parabola' since the s^2 term is negative. The graph could be plotted, but it is quicker to factorise.
2. $P = -s^2 + 7s - 10 \Rightarrow P = -(s^2 - 7s + 10) \Rightarrow P = -(s - 2)(s - 5)$
   The roots of the function are $s = 2$ and $s = 5$.
3. Since a parabola is symmetrical the maximum occurs halfway between the roots: $(5 + 2) \div 2 = 3.5$
4. a More than $5.
   b $3.50
      $P = -3.5^2 + 7 \times 3.5 - 10 = 2.25$. Their profit is 2.25 cents.

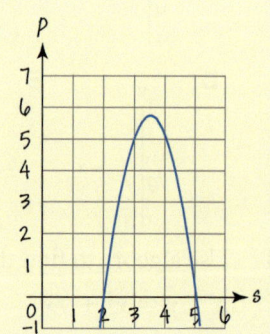

EXAMPLE

A second company also sells batteries. Their model uses the profit function $P = -s^2 + 8s - 6$ where $P =$ profit (p) and $s =$ selling price (£).
a What selling price will maximise their profit?
b How much profit will they make in this case?
c Comment on the profit for a selling price of £0.

a $P = -s^2 + 8s - 6 \Rightarrow P = -(s^2 - 8s + 6)$
This does not factorise easily so complete the square instead.
$s^2 - 8s + 6 = (s - 4)^2 - 16 + 6 = (s - 4)^2 - 10 \Rightarrow P = -(s - 4)^2 + 10$
The optimum selling price is £4.
b $s = 4 \Rightarrow P = -4^2 + 8 \times 4 - 6 = 10$. Their profit is 10p.
c $P = -0^2 + 8 \times 0 - 6 = -6$. The company loses 6p per battery.
   The model does not make sense at this point.

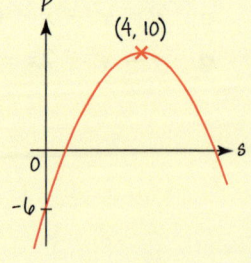

Algebra   Graphs 1

Reasoning and problem solving

Exercise 14.2A

1. A company establishes a profit function of $P = 6s - s^2 - 5$ where P = profit (cents) and s = selling price ($).
   a. Sketch the graph of the profit function.
   b. At what prices would the profit be zero?
   c. What selling price will maximise their profit? How much profit will they make?

2. A company uses a profit function of $P = -2s^2 + 900s - 100\,000$ where P = profit (€) and s = selling price (€).
   a. What selling price will maximise their profit? How much profit will they make?
   b. At what prices would the profit be zero?

3. The height above ground of a javelin is modelled by the function $h = 100 + 48x - x^2$ where h = height in centimetres and x = horizontal distance in metres.
   a. Sketch the graph of the function.
   b. Find the greatest height of the javelin.
   c. State the length of the throw.

4. The jet of water in a fountain is modelled by the function $y = -\frac{1}{10}x(x - 50)$ where x = distance from source (cm) and y = height (cm).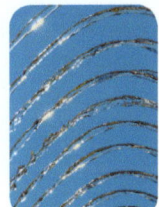
   a. At what distance from the source does the jet enter the water again?
   b. What is the greatest height reached by the jet?

5. Shot balls used to be made by dropping molten lead from the top of a tower into a pool of water. After lead is dropped, its height (in metres) above ground is given by the formula $h = 44.1 - 4.9t^2$, where t is time in seconds.
   a. Sketch the graph of the function
   b. What height is the lead dropped from?
   c. After how many seconds does the lead land in the pool of water?

6. A quadratic function has roots at $x = -2$ and $x = 8$ and a turning point at $(3, 5)$.
   Sketch the graph of this function.

7. A quadratic function has roots $x = -3$ and $x = 1$ and intercept $y = 6$.
   Sketch the graph of this function.

8. A quadratic function has a turning point at $(-3, 0)$ and a y-intercept at $(0, 4)$
   How many roots does the function have?

9. A quadratic function has a lowest point at $(4, -3)$ and a y-intercept at $y = 1$.
   a. Sketch the graph of this function on squared paper.
   b. Use your sketch to estimate the roots of the function.

*10. A parabolic trough solar collector captures energy from sunlight. Every cross-section focuses the sun's rays onto a single point.
A parabolic trough is designed using the function $y = \frac{1}{20}x^2 + 5$. The single point is $(0, 10)$. A feature of this point is that every position on the parabola is equidistant from it and the x-axis.
   a. Verify that the point $(10, 10)$ is on the curve.
   b. Find another point with integer coordinates that lies on the curve.
   c. Use graphing software to verify your solution.
   d. Investigate this feature of the parabola.

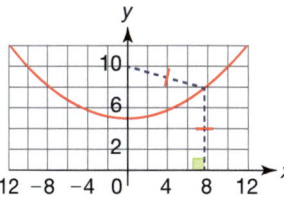

*11. Write each function in question **2** in Exercise 14.2S on p. 283 in the form $y = p(x - q)^2 + r$.
Explain how the values of p, q and r relate to the answers to question 2 parts **i–iii**.

SKILLS

14.3 Kinematic graphs

A **distance-time graph** shows information about a journey.

The gradient of a straight line in a distance-time graph is the **speed** of the object.

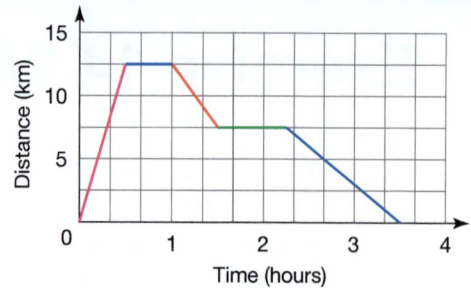

EXAMPLE

Shao is running.

At the end of her training session her GPS app shows this graph.

a What was Shao's speed during the first 45 minutes?

b What is Shao's average speed for the whole journey?

> To get answers in km/h use 45 mins = 0.75 hrs.
>
> a speed = $\frac{\text{distance}}{\text{time}} = \frac{5}{0.75}$ = 6.666... km/h
>   = 6.7 km/h (1 dp)
>
> b For average speed use total distance and total time.
>
>   average speed = $\frac{\text{total distance}}{\text{total time}} = \frac{12.5 \text{ km}}{1.75 \text{ hours}}$ = 7.142857... km/h
>   = 7.1 km/h (1 dp)

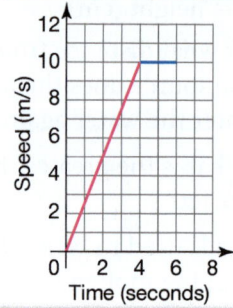

Remember the gradient is 'change in y' over 'change in x'.

Speed-time graphs also give information about a journey.

- The gradient of a straight line in a speed-time graph is the **acceleration** of the object.

- The area under a line in a speed-time graph is the **distance** travelled by the object.

Velocity is a vector. It is speed in a certain direction.

EXAMPLE

The speed-time graph describes the journey of a radio-controlled car.

a Describe the car's journey in words.

b What is the acceleration of the car during the first four seconds?

c What is the distance travelled over the first six seconds?

> a The car starts at rest and reaches a speed of 10 m/s after 4 seconds.
>   Initially the car accelerates steadily.   *The graph is a straight line.*
>   Then, between 4 and 6 seconds, the car travels at a constant speed.   *The line is horizontal.*
>
> b Calculate the gradient.
>
>   Acceleration = $\frac{\text{change in y}}{\text{change in x}} = \frac{10 \text{ m/s}}{4 \text{ seconds}}$ = 2.5... m/s²
>
> c Calculate the area below the line.
>
>   From $t = 0$ to $t = 4$ Area = $\frac{1}{2} \times 4 \times 10 = 20$ Or as a trapezium
>   From $t = 4$ to $t = 6$ Area = $2 \times 10 = 20$ Area = $\frac{1}{2}(2 + 6) \times 10$
>   The distance travelled is 40 metres.   = 40 m

Algebra Graphs 1

Fluency

Exercise 14.3S

1 This distance-time graph shows Lisa's coach journey.

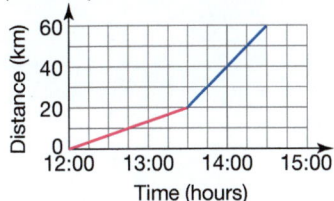

   a What is the speed between 12:00 and 13:30?

   b What is the speed between 13:30 and 14:30?

2 Janu is riding a bike. Information about her journey is shown in this graph.

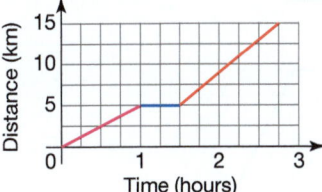

Janu's journey starts at 8 am.

   a What is her average speed for the whole ride?

   b What is her speed between 10:00 and 10:30?

3 Mark sets off from home in his car at 2 pm. He stops to get petrol and then continues on his journey.
   Mark returns home later in the afternoon.
   The distance-time graph shows more information about his journey.

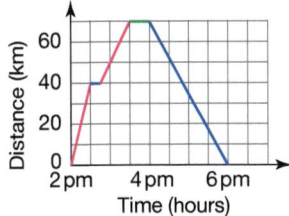

   a Between which times was Mark travelling fastest? Explain how you know.

   b What was Mark's average speed on the outward journey?

   c What was Mark's average speed on the homeward journey?

   d The gradient of the line after 4 pm is negative. What does this tell you about the speed?

4 The speed-time graph shows information about a runner in a race.

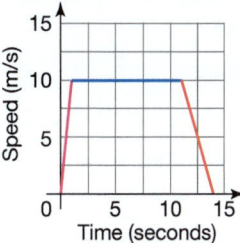

   a What is the speed at 12 seconds?

   b At what times is the speed 6 m/s?

   c What is the acceleration during the first second?

   d Find the overall distance travelled.

5 The diagram shows a speed-time graph.

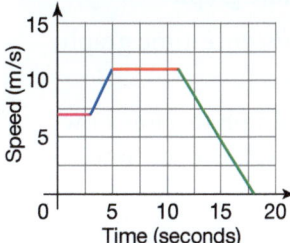

   a What is the acceleration between 3 and 5 seconds?

   b What is the distance travelled between 3 and 5 seconds?

   c What is the overall distance travelled?

*6 Describe the journey shown by this speed-time graph.

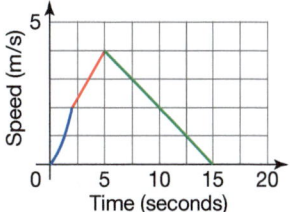

7 A train leaves a station at 11 am and travels at a constant speed of 65 km/h.
   A second train leaves the same station 30 minutes later. It travels at 150 km/h.

   a Construct a distance-time graph to show this.

   b At what time are the two trains the same distance from the station?

APPLICATIONS

14.3 Kinematic graphs

RECAP
- In a distance–time graph the gradient of a straight line is the speed of the object.
- In a speed–time graph
  - the gradient of the straight line is the acceleration of the object
  - the area under the line is the distance travelled by the object.

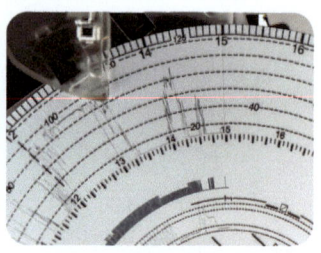

The tachograph on a lorry plots a distance–time graph and a speed–time graph every 24 hrs.

HOW TO

To sketch a distance–time or acceleration–time graph given a speed–time graph
1. Split the total time into distinct sections.
2. Consider how the speed varies in each section.
3. Sketch each section of the distance–time graph in turn.

EXAMPLE

For this speed-time graph
a sketch the *distance*-time graph for this data
b construct a graph to show how acceleration varies over time.

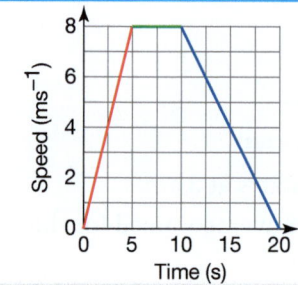

1. Consider 0 – 5 s, 5 – 10 s and 10 – 20 s separately.
2. Distance = area under graph
   Acceleration = gradient of graph

<u>0 to 5</u> Speed: steady increase 0 → 8 m/s
Total distance = $\frac{1}{2} \times 5 \times 8 = 20$ m

3. Rate of covering distance increases as speed increases.

Acceleration = $\frac{8}{5}$ = 1.6 m/s²

<u>5 to 10</u> Speed: constant at 8 m/s
Total distance = 5 × 8 = 40 m

Distance increases at a steady rate: straight line, positive gradient.

Acceleration = 0 m/s²

<u>10 to 20</u> Speed: steady decrease, 8 → 0 m/s
Total distance = $\frac{1}{2} \times 10 \times 8 = 40$ m

3. Rate of covering distance decreases as speed decreases.

Acceleration is $\frac{-8}{10}$ = −0.8 m/s²
(Deceleration is 0.8 m/s²)

a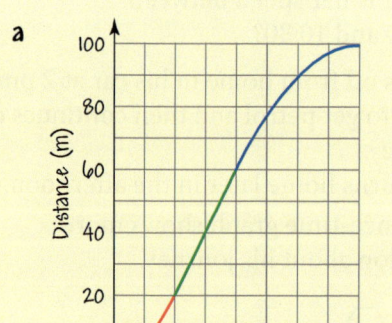

b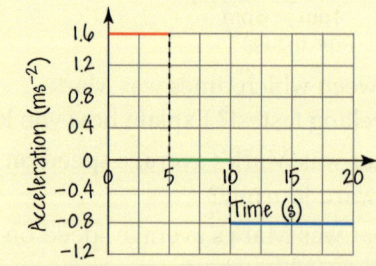

Algebra   Graphs 1

Reasoning and problem solving

Exercise 14.3A

1   For each of the speed-time graphs
    i   sketch a distance-time graph
    ii construct an acceleration-time graph.

    a
    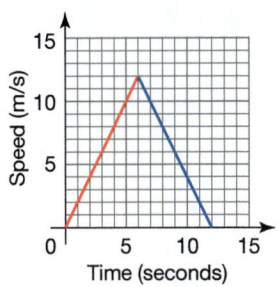

    b
    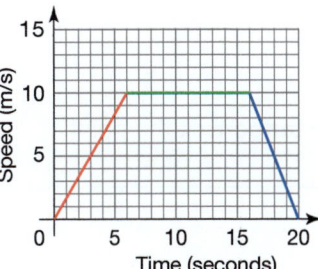

    c
    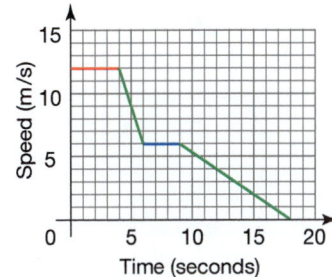

2   Two cyclists start a training ride at 7:30 am.

    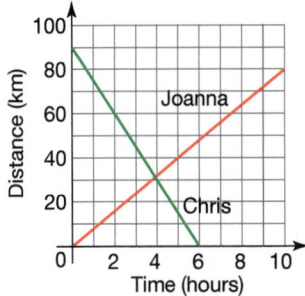

    a   What happens at 11:20 am?
    b   What is the average speed of each cyclist?

3   This speed-time graph gives information about two cars.

    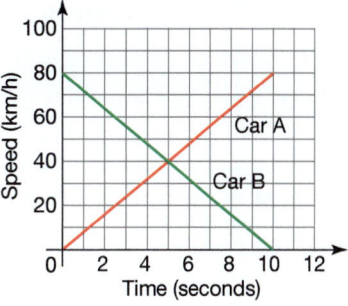

    a   What is the distance travelled by each car?
    b   Sketch, on the same axes, a distance-time graph for each car.
    c   Plot, on the same axes, an acceleration-time graph for each car. Clearly state the units.

4   Explain why each of these graphs are impossible.

    a   b
    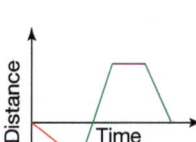

    c   d
    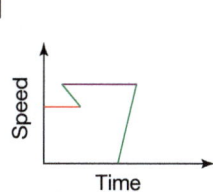

5   Nima and Sara run a 400 m race. This graph shows the race.

    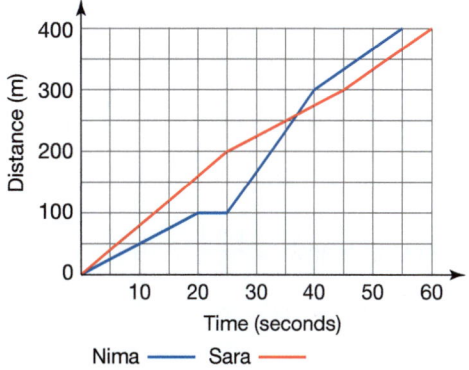

    Describe the race.
    Ensure that you state who won.

SKILLS

14.4 Gradients and areas under graphs

The gradient of a straight line, $\frac{\text{change in } y}{\text{change in } x}$, is the **rate of change** of y with respect to x.

- If two points on a straight line are (x_1, y_1) and (x_2, y_2) then gradient $= \frac{y_2 - y_1}{x_2 - x_1}$.

EXAMPLE

Find the gradient of the line that passes through the points

a $(-2, 3)$ and $(7, 6)$ b $(3, 5)$ and $(6, -7)$.

a Gradient $= \frac{y_2 - y_1}{x_2 - x_1} = \frac{6 - 3}{7 - -2} = \frac{3}{9} = \frac{1}{3}$ b Gradient $= \frac{-7 - 5}{6 - 3} = \frac{-12}{3} = -4$

The gradient, or rate of change, varies for a curve, such as $y = x^2$.

- The gradient at a given point on a curve is the gradient of the **tangent** at the point.

EXAMPLE

Find the gradient of the curve $y = x^2$ when $x = 2$.

Draw an accurate graph.
Use your ruler to draw a line that just touches the curve at $(2, 4)$.
Find two points on the tangent line.

$(1, 0)$ and $(2, 4)$ are points on the tangent.

Gradient $= \frac{4 - 0}{2 - 1} = \frac{4}{1} = 4$

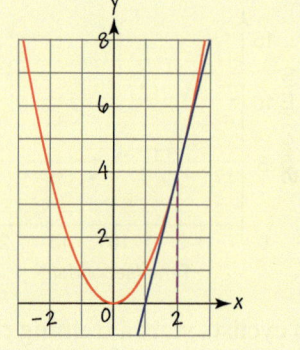

It is useful to be able to find the area under a curve between two points.
The area under a speed–time graph gives the distance travelled.

EXAMPLE

Estimate the area under the curve $y = x^2$ between the points when $x = -3$ and $x = 0$.

You can estimate this area by splitting it into two trapeziums and a triangle.

Area of large trapezium $= \frac{9 + 4}{2} \times 1 = 6.5$

Area of small trapezium $= \frac{4 + 1}{2} \times 1 = 2.5$

Area of triangle $= \frac{1 \times 1}{2} = 0.5$

Estimate for whole area $= 6.5 + 2.5 + 0.5 = 9.5$
This is an over-estimate of the actual area as each of the three shapes is slightly too large.

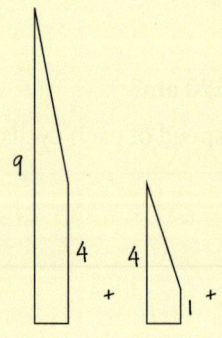 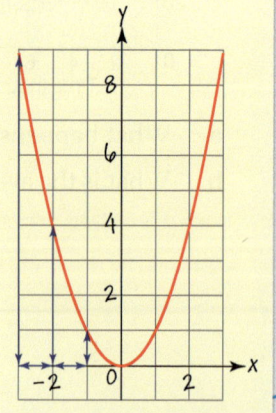

Fluency

Exercise 14.4S

1 Find the gradient of the straight line through these points.
  a (1, 1) and (5, 9)
  b (2, 6) and (8, 2)
  c (−2, −1) and (1, 5)
  d (−2, 1) and (1, −2)
  e (1, −2) and (2, 4)
  f (0, −3) and (−2, 5)
  g (−4, 3) and (2, 1)
  h (3, −1) and (7, 1)
  i (−5, −2) and (3, 6)
  j (−1, 5) and (7, 2)

> You may use graph plotting software to help with questions **2** and **3**.

2 a Plot the graph of $y = x^2$ for values of x from −4 to 4. If using pencil and paper, aim to use a single side of A4 graph paper.
  b Find the gradient of the tangent to the curve $y = x^2$ at these points.
    i $x = 1$ ii $x = 3$ iii $x = -1$
    iv $x = -2$ v $x = 0$ vi $x = -3$

3 a Plot the graph of the function $y = \frac{1}{2}x^2 + 3$ for $-4 \leq x \leq 4$.
  b Find the gradient of the tangent to the curve $y = \frac{1}{2}x^2 + 3$ at these points.
    i (2, 5)   ii (−3, 7.5)   iii $x = -2$
    iv $x = 0$ v $x = 1$ vi $x = 3$
  c Use your results to predict the gradient at the point $x = 5$. Give your reason.

4 The graph shows the function $y = x^3 + 2$.

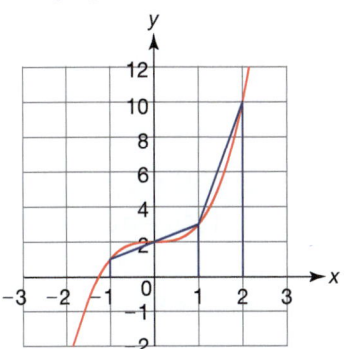

4 a Use the trapeziums to estimate the area under the curve between the points where $x = -1$ and $x = 2$.
  b Is your answer an over-estimate or an under-estimate? Give your reason.

5 The graph shows the function $y = x^2 - x + 2$.

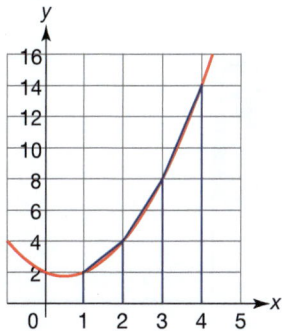

  a Use the trapeziums to estimate the area under the curve between $x = 1$ and $x = 4$.
  b A more accurate estimate would be found if each trapezium had a 'width' of $\frac{1}{2}$. Work out the value of this estimate.

6 Estimate the distance travelled by an object in this velocity-time graph.

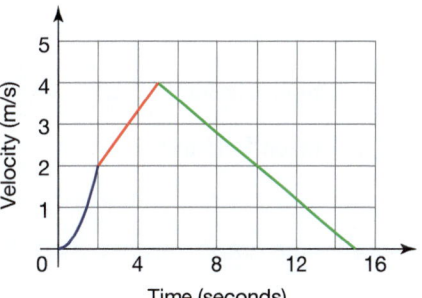

7 a Find the area under the line $y = 2x$ for values of x between
    i 0 and 1   ii 0 and 2
    iii 0 and 3   iv 0 and 4.
  b Plot the areas you found in part **a** as a function of the upper bound of the x range and join them by a smooth curve. Comment on what you notice.

8 Draw the graph of $y = 3x^2$ for $0 \leq x \leq 4$. Repeat question **7** finding the areas under the curve $y = 3x^2$.

1128, 1312, 1944, 1953

APPLICATIONS

14.4 Gradients and areas under graphs

RECAP
- The gradient of a straight line is the **rate of change** of y with respect to x.
- If two points on a straight line are (x_1, y_1) and (x_2, y_2) then the gradient $= \dfrac{y_2 - y_1}{x_2 - x_1}$
- The gradient at a point on a curve is the gradient of the **tangent** at that point.
- The area under a curve can be estimated by splitting it into simple shapes.

HOW TO
To solve problems involving gradients and areas of curves
1. Use a clear diagram that has large enough scales to read values accurately.
2. Draw a tangent at the relevant point or split the area into simple shapes.
3. Choose suitable points to substitute into the formula or calculate the total area of the simple shapes.
4. Interpret the solution in the context of the question.

EXAMPLE

The speed–time graph shows information about a runner during the first 2 seconds of a race.

What was the runner's **acceleration** at 1.25 seconds?

Acceleration is the rate of change of speed with respect to time.

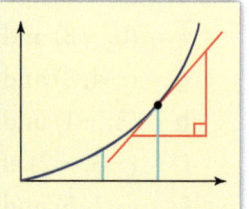

3. Use the points (2.25, 8) and (1, 2).

Gradient of tangent $= \dfrac{8 - 2}{2.25 - 1} = \dfrac{6}{1.25} = 4.8$

4. Acceleration $= 4.8 \, m/s^2$.

- In a speed–time graph the area under the curve gives the distance travelled.

EXAMPLE

Estimate the distance travelled by the runner in the first two seconds.

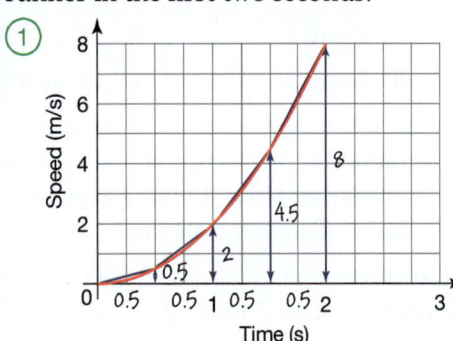

2. Split the area at every 0.5 seconds.
3. Read the required lengths off the graph.

Area of large trapezium $= \dfrac{8 + 4.5}{2} \times 0.5 = 3.125$

Area of medium trapezium $= \dfrac{4.5 + 2}{2} \times 0.5 = 1.625$

Area of small trapezium $= \dfrac{2 + 0.5}{2} \times 0.5 = 0.625$

Area of triangle $= \dfrac{1}{2} \times 0.5 \times 0.5 = 0.125$

Estimate for whole area $= 3.125 + 1.625 + 0.625 + 0.125$
$= 5.5$

4. The estimated distance travelled is 5.5 metres.

Algebra   Graphs 1

Reasoning and problem solving

Exercise 14.4A

1 Here is a speed–time graph.

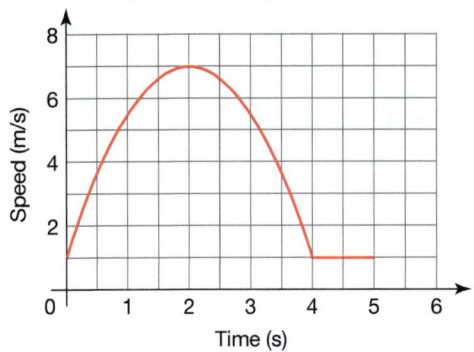

  a Work out the acceleration at 1 second.
  b Find the acceleration at 4.5 seconds.
  c Work out the deceleration at 3.5 seconds.
  d Estimate the distance travelled over the 5 second period.

2 Rifaah's grandmother put £1000 into a savings account when he was born. She adds money every year so that the total increases by 10%. The (red) graph shows this information.

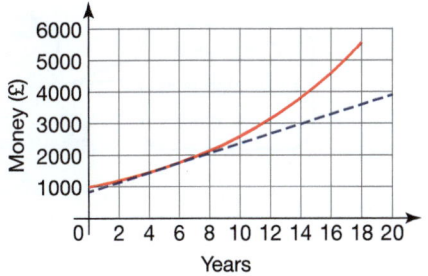

  a How much money will be in the account when Rifaah is 18?
  b The tangent at 5 years is shown. The tangent goes through the points (14, 3000) and (−3, 400). Find the gradient of the tangent.
  c What is the rate of change of money in the account at 5 years? State the units of your answer.
  d Work out the rate of change of money in the account at
     i 10 years   ii 15 years.

3 Rory borrows $150 from a short-term loan company. The money should be paid back in 18 days with interest of $28. This is equivalent to an interest rate of 365% per annum.

If Rory does not pay the money back, the total amount to be paid back grows very quickly as shown in the graph.

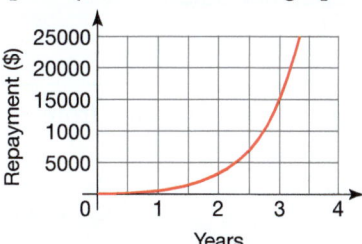

The function for the graph is
$y = 150 \times 4.65^x$ where x = years and y = repayment.

Work out
  a the repayment after 3 years
  b the rate of change of the repayment at 3 years.

4 Georgina is conducting an experiment. She measures the temperature of a substance once a minute for 10 minutes. The results are shown in the table.

Time	0	1	2	3	4	5	6	7	8	9	10
°C	100	50	33	26	20	17	14	13	11	10	9

Estimate the rate of change in the temperature at 3 minutes.

5 A radio-controlled model car travels a distance of 14 metres, and its speed–time graph looks like this.

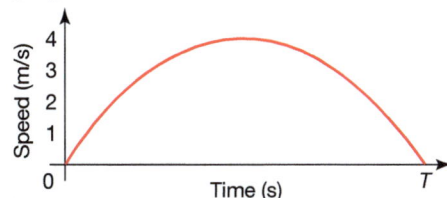

Estimate the time T that the car takes to come to rest again.

*6 Estimate the area between these graphs.
  a $x \geq 0$, $y = 8 - x^2$ and $y = 2x$
  b $y = -x^2$, $y = x^3$ and $x = 2$
  c $x^2 + y^2 = 8$ and $y \geq x^2 - 2$

SKILLS

14.5 Exponential and trigonometric functions

The **trigonometric** functions $y = \sin x$ and $y = \cos x$ produce **waves**.

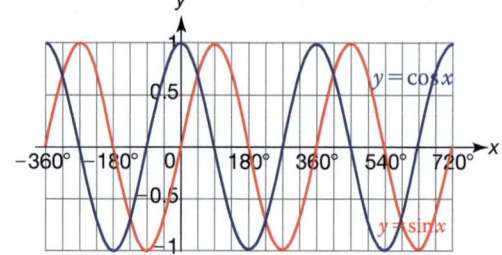

▲ The functions $y = \sin x$ and $y = \cos x$ both have **period** 360°.

- A function which repeats after a fixed interval is called **periodic**.

EXAMPLE

Plot the graph of $y = \sin x$ for x from $-45°$ to $450°$.

Create a table of values using steps of 45°.
Round answers to two decimal places when necessary.

x	-45	0	45	90	135	180
y	-0.71	0	0.71	1	0.71	0

x	225	270	315	360	405	450
y	-0.71	-1	-0.71	0	0.71	1

Draw a smooth curve through the points.

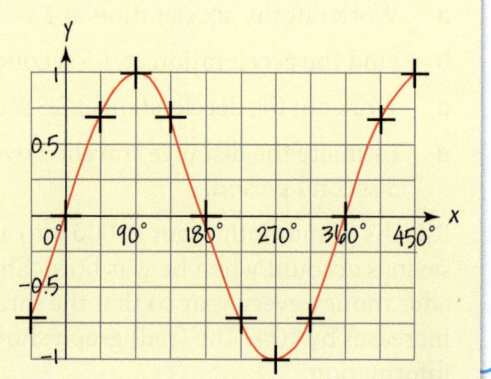

- An **exponential** function has the form $y = a^x$.

The x-axis is a line that the function $y = a^x$ gets as close as you like to but never actually touches.

Check your calculator is in degree mode!

EXAMPLE

Plot the graph of $y = 2^x$.

Create a table of values including positive and negative numbers.

x	-3	-2	-1	0	1	2	3
y	$\frac{1}{8}$	$\frac{1}{4}$	$\frac{1}{2}$	1	2	4	8

Draw a smooth curve through the points.
The curve never touches the x-axis.
An exponential curve will continue to increase at a faster rate.

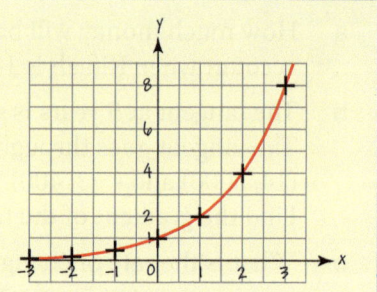

Algebra   Graphs 1

Fluency

Exercise 14.5S

1. Plot the graph of the function $y = 3^x$ for $-4 \leqslant x \leqslant 4$.

2. Plot the graph of the function $y = \left(\frac{1}{2}\right)^x$ for $-4 \leqslant x \leqslant 4$.

3. a Create a table of values for the function $y = \cos x$. Use values of x from $-45°$ to $450°$, increasing in steps of $45°$.
   b Plot the graph of the function $y = \cos x$.
   c What is the period of the function?
   d Write the maximum and minimum values of the function.

4. a Create a table of values for the function $y = \tan x$. Use values of x from $-30°$ to $390°$, increasing in steps of $30°$.
   b Plot the graph of the function $y = \tan x$.
   c Draw and label the lines which are asymptotes of the function.
   d What is the period of the function?

5. a Plot the graphs of these functions on the same axes.
   $y = 2^x \qquad y = 4^x \qquad y = 5^x$
   b Compare the three functions, noting similarities and differences.

6. Plot the graphs of these functions on the same axes.
   $y = \left(\frac{1}{3}\right)^x \qquad y = \left(\frac{1}{4}\right)^x \qquad y = \left(\frac{1}{6}\right)^x$
   Compare the three functions, noting similarities and differences.

7. Describe the graph of the function $y = 1^x$.

8. Match the four graphs to the four equations.

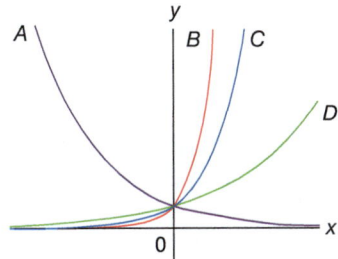

   a $y = 4^x$ b $y = 1.8^x$
   c $y = 12^x$ d $y = \left(\frac{1}{2}\right)^x$

9. a Sketch the graphs of these functions on the same axes.
   Label the y-intercept in each case.
   $y = 3^x \qquad\qquad y = 3^x + 2$
   $y = 3^x - 5 \qquad y = 3^x - 1$
   b Write about what you notice.

10. a Plot the graph of these functions on the same set of axes.
    Label the minimum point in each case.
    $y = x^2 \qquad\qquad y = (x+1)^2$
    $y = (x+4)^2 \qquad y = (x-3)^2$
    b Write about what you notice.

11. Use your observations from question 9 to help with this question. Sketch the graph of these functions for values of x from $0°$ to $360°$. Label all the intersections with the axes in each case.
    a $y = \cos x$ b $y = \cos x - 5$
    c $y = \cos x - 1$ d $y = \cos x + 2$

*12. Use your observations from question 10 to help with this question. Sketch the graph of these functions for values of x from $0°$ to $360°$. Label all the intersections with the axes in each case.
    a $y = \sin x$ b $y = \sin(x + 30°)$
    c $y = \sin(x + 120°)$ d $y = \sin(x - 90°)$

Use graphing software for questions 13 and 14.

13. Plot the graphs of
    a $y = 7^x$ and $y = -7^x$
    b $y = \tan x$ and $y = -\tan x$
    c $y = x^2 + 2$ and $y = -(x^2 + 2)$
    d Write about what you notice.

14. Elijah writes, 'cos $(x + 90) = \sin x$'. Do you agree with Elijah? Explain your answer.

APPLICATIONS

14.5 Exponential and trigonometric functions

RECAP

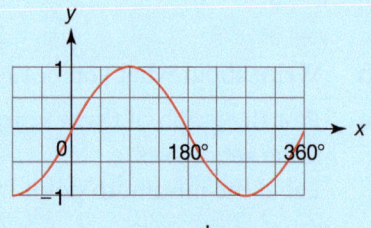

$y = \sin x$

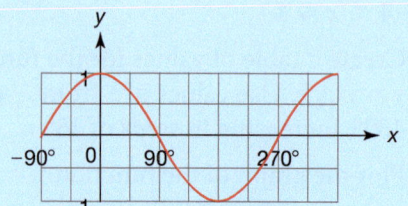

$y = \cos x$

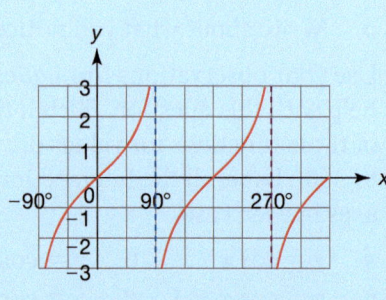

$y = \tan x$

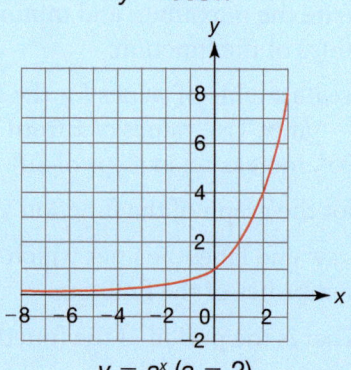

$y = a^x$ ($a = 2$)

HOW TO

To use a trigonometrical graph to solve an equation
1. Extract the relevant trigonometrical function from the equation.
2. Plot the graph of this function.
3. Draw a line using the relevant value from the equation.
4. Use the points of intersection to read off the solution to the equation.

EXAMPLE

The graph shows $y = \sin x$

a How many solutions to the equation $\sin x = 0.3$ lie between 0 and 360?

b Estimate the value of these two solutions.

c Use your calculator to find a solution to the equation $\sin x = 0.3$. Write down all the figures on the display.

d Using your answers to a, b and c, find the second solution to $\sin x = 0.3$ correct to five decimal places.

a ① There are two points of intersection of $y = \sin x$ and $y = 0.3$ between 0 and 360, therefore there are two solutions

b ② The x values at the points of intersection provide the required solutions. These values are about 17.5° and 162.5°.

c ③ Check that the calculator is in degrees mode. $x = 17.45760312°$

d ④ The second solution is 180° − the first solution. 180° − 17.45760312° = 162.54239688°

Algebra Graphs 1

Reasoning and problem solving

Exercise 14.5A

1 The graph shows $y = \tan x$.

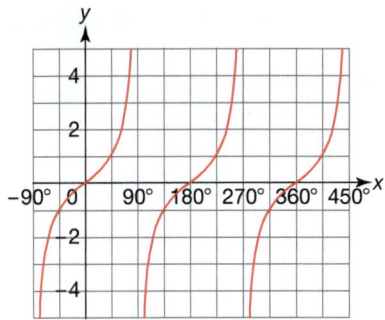

  a Use the graph to find three angles such that $\tan x = 1$.
  b Explain how you could continue finding angles with a tangent of 1.

2 The graph shows $y = \sin x$

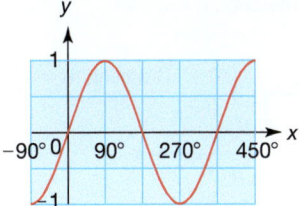

  a Use the graph to find three angles such that $\sin x = 0.5$
  b Explain how you could continue finding angles with a sine of 0.5

3 The graph shows $y = \cos x$

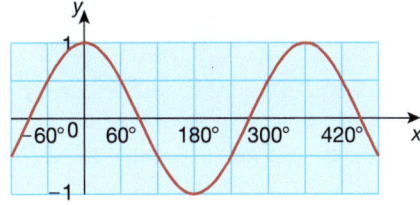

  a Use the graph to estimate three solutions to the equation $\cos x = 0.8$
  b How many solutions to the equation $\cos x = 0.8$ lie between -180 and 720?

4 Bethia is a scientist. She uses the graph of $y = 2^x$ to model the growth of microorganisms in a culture.

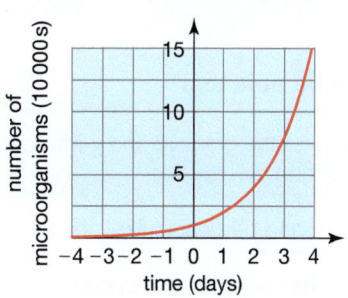

  a Use the graph to find the number of microorganisms at the start of day 4.
  b Use the graph to estimate the time at which the number of microorganisms passes 100 000.

5 The graph shows the function $y = \sin(x - p)$.

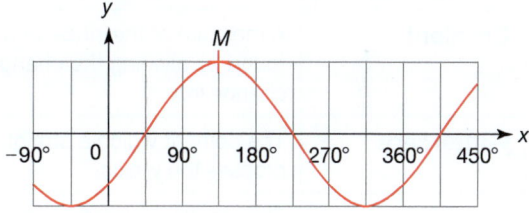

  a What are the coordinates of the point M?
  b Work out the value of p.
  c Josh says, 'The graph could be of another function $y = \cos(x - q)$ instead.'
     Josh is correct.
     Work out a possible value of q.

  > There are lots of possible answers to part **b**. Keep it simple.

*6 Are these statements true or false. In each case give your reason.
  a $\cos(x) = \cos(-x)$
  b $\sin(x) = \sin(x + 180°)$
  c $\tan(x) = \tan(x + 180°)$
  d $\sin(x + 90°) = \cos(x)$
  e $\cos(90° - x) = \sin(x)$
  f $\tan(x) = -\tan(-x)$

7 a Investigate what happens with $y = a \sin x$, and $y = \sin(ax)$, when a is varied.
  b Research to find out where these kinds of waves are used in the real world.

> Use graphing software for question **7**.

Summary

Checkout
You should now be able to...

Test it
Questions

✓ Find and interpret the gradient and y-intercept of a line and relate these to the equation of the line in the form $y = mx + c$.	1 – 3
✓ Identify parallel and perpendicular lines using their equations.	4
✓ Draw line graphs and quadratic curves.	5, 6
✓ Identify roots, intercepts and turning points of quadratic curves using graphical and algebraic methods.	7, 8
✓ Use graphs to solve problems involving distance, speed and acceleration.	9

Language	Meaning	Example
Gradient	A measure of the slope of a line on a graph found by dividing the change in y by the change in x.	y-intercept, $c = 3$; Gradient, $m = -\frac{1}{2}$; $y = -\frac{1}{2}x + 3$
y-intercept	The point at which a straight line or curve crosses the y-axis.	
$y = mx + c$	The standard form for a straight line graph. m = gradient, c = y-intercept	
Quadratic function	A function of the form $f(x) = ax^2 + bx + c$ where a, b and c are integers and $a \neq 0$.	$f(x) = 6x^2 - 7x - 3$ $a = 6, b = -7, c = -3$
Parabola	The shape of a quadratic curve.	Turning point $(-2,3)$; Turning point $(2,0)$; Root; $y = 3 - (x + 2)^2$; $y = 4(x - 2)^2$
Turning point	A point on a curve where the gradient changes from increasing to decreasing, ∩ decreasing to increasing, ∪.	
Root	An x-intercept of a graph. If $y = f(x)$, a root is a solution to the equation $f(x) = 0$, that is, an input to the function $f(x)$ that gives output 0.	
Kinematics	A branch of mathematics relating to the motion of objects.	Acceleration = $\frac{\text{change in velocity}}{\text{change in time}} = \frac{10}{5} = 2 \text{ m/s}^2$
Speed	The gradient of a distance–time graph.	
Acceleration	The gradient of a speed–time graph.	

Review

1. **a** What is the gradient of this line?
   b What is the equation of this line?

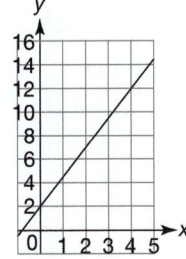

2. What is the equation of a line that has gradient -3 and passes through $(0, 5)$?

3. State the gradients and y-intercepts of the lines with these equations.
   a $y = 7 - 2x$ **b** $y = x + 9$
   c $x + y = 3$ **d** $2x + 3y - 5 = 0$

4. Given a line with the equation $y = 5x + 2$, write down the equation of the line that is
   a parallel and passes through $(2, 16)$
   b perpendicular and passes through $(10, -1)$.

5. Draw the graphs of these functions.
   a $y = 5x + 2$ **b** $y = 14 - 3x$

6. Draw the graphs of these functions.
   a $y = 4x^2 - 16$ **b** $y = x^2 + 10x + 25$
   c $y = 2x^2 - 5x - 3$
   d $y = -x^2 + x + 6$

7. Here is a graph of the function $y = f(x)$. Write down
   a the y-intercept
   b the x-intercepts
   c the roots
   d the x-coordinate of the turning point

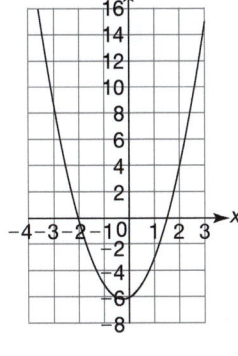

8. **a** Complete the square for each of these functions.
   i $f(x) = x^2 + 2x - 5$
   ii $f(x) = 2x^2 - 12x + 5$
   iii $f(x) = 5x - x^2$
   b Work out the coordinates of the turning point for each of the functions.

9. The graph shows the distance covered by a sprinter.

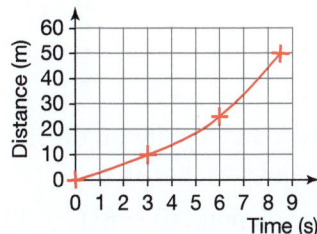

   a What is the total distance run?
   b What was the speed of the sprinter in the first 3 seconds?
   c What is the average speed, in km/h, of the sprinter?
   d What is happening to the speed of the sprinter during the run?

   At the end of the run the sprinter takes 5 seconds to slow down to a complete stop.

   e What is the sprinter's deceleration, assuming it is constant?

What next?

Score		
0 – 4		Your knowledge of this topic is still developing. To improve, look at MyiMaths: 1153, 1169, 1180, 1185, 1311, 1312, 1314, 1322, 1323, 1396
5 – 8		You are gaining a secure knowledge of this topic. To improve your fluency look at InvisiPens: 14Sa – k
9		You have mastered these skills. Well done you are ready to progress! To develop your problem solving skills look at InvisiPens: 14Aa – f

Assessment 14

1. **a** State the gradient of each line (A–D) on the grid. [2]
   b Match each equation to one of the lines A, B, C and D drawn on the graph. [3]
      i $5x + 3y = 20$ ii $2y - 3x = 9$
      iii $2x - y = 0$ iv $2x + 3y = 0$
   c State the y-intercept for each equation. [2]

2. Work out the gradient and y-intercept for each of the following straight lines.
   a $y = 2x + 7$ [1]
   b $y = 9 + 4x$ [1]
   c $y = 6x - 11$ [1]
   d $y = 12 - 4x$ [1]
   e $4x - 7y = 14$ [2]
   f $15x + 14y = 35$ [2]

3. Given the gradient and a point on the line, find the equation of each line in the form $y = mx + c$.
   a Gradient $= -5$, point $(0, -61)$ [2]
   b Gradient $= 3$, point $(1, 2\frac{1}{2})$ [2]
   c Gradient $= \frac{1}{3}$, point $(3, -1)$ [2]
   d Gradient $= \frac{1}{4}$, point $(0, -\frac{3}{4})$ [2]

4. Calculate the gradients of the straight lines which pass through each pairs of points.
   a $(1, 5)$ and $(5, 9)$
   b $(6, 7)$ and $(8, -9)$
   c $(-3, 5)$ and $(-4, 6)$
   d $(-11, 0)$ and $(5, -8)$ [4]

5. Find the equation of each line in the form $y = mx + c$.
   a Line parallel to $y - 7x - 9 = 0$ that intercepts the y-axis at $(0, 5)$. [2]
   b Line parallel to $3x + y - 77 = 0$ that intercepts the y-axis at $(0, -7.25)$. [2]
   c Line parallel to $2x + 6y = 15$ that intercepts the y-axis at $(0, 2\frac{2}{3})$. [2]

6. Here are the equations of ten lines.

   A $y = 4x - 8$ **B** $x + 1 = 0$ **C** $4x = 3y$ **D** $4y + x - 15 = 0$ **E** $y = x$

   F $3x + y + 8 = 0$ **G** $4 - x = 0$ **H** $2y - 8x + 1 = 0$ **I** $y - 7 = 0$ **J** $x - \frac{1}{4}y + 3 = 0$

   a Find the three lines that are parallel to one another. [2]
   b Find two pairs of lines that are perpendicular to one another. [2]
   c Which line has a zero gradient? [1]
   d Which two lines are parallel to the y-axis? [2]
   e Which two lines pass through the point $(1, 3\frac{1}{2})$? [2]
   f Which two lines pass through the origin? [1]

7. **a** Find the mid-point and gradient of the line segment joining $(-1, 5)$ and $(7, 3)$. [3]
   b Find the equation of the perpendicular bisector of this line segment. [3]

Algebra   Graphs 1

8 A town planner investigated the correlation between the population density and distance from the urban centre in a large city. These are the results.

Distance (km)	2	4	6	8	10	12	14	16	18	20	22	24	26
Population density (people per square km)	95	93	90	33	78	66	92	59	48	40	33	27	27

   a The line of best fit passes through (6, 90) and (20, 41). Find the equation of the line of best fit. [4]

   b Complete this sentence.
As the distance from the urban centre increases by 1 km, the population density _____ by _____ people per square km. [2]

   c Is the result at 8 km an outlier? Explain your answer. [1]

9 **a** Draw $y = x^2 - x - 6$ and $y = x$ on the same grid for values of x from -4 to 3. [5]

   b Estimate the coordinates of the points where the graphs intersect. [1]

   c Use your graph to estimate the solutions of the equation $x^2 - x - 6 = 0$. [2]

   d By factorising, solve the equation $x^2 - x - 6 = 0$. Compare your answer with your estimate in part **c**. [2]

10 A metal spring stretches when a mass of m grams is hung on the end. The distance stretched, d cm, is given by the formula $d = 8 + \dfrac{m}{15}$.

   a Complete the table of values for m and d. Write your answers to 1 dp where appropriate. [2]

m	10	20	30	40	50	60	70	80	90	100
d										

   b Plot the graph of d against m for values of m from 0 to 100. [3]

   c Use your graph to
     i find the stretch, d, when a mass of 75 g is hung on the spring [1]
     ii find the mass hung on the spring when $d = 9$ cm. [1]

   d What is the length of the spring when no weights are hung on it? [1]

11 The average safe stopping distance for cars d metres, is given by the equation $d = \dfrac{3v^2}{125} + \dfrac{v}{2.7}$, where v is the speed of the car in km/h.

   a Complete the table of values for d and v. Write your distances to the nearest metre. [4]

v	0	10	20	30	40	50	60	70	80	90	100
d											

   b Draw the graph of $d = \dfrac{3v^2}{125} + \dfrac{v}{2.7}$ [4]

   c Use your graph to find the safe stopping distance when a car travels at
     i 25 km/h   **ii** 42 km/h   **iii** 77 km/h. [3]

15 Working in 3D

Introduction
In the Cave of Crystal Giants in Mexico, vast crystals of selenite grow to lengths of up to 11 metres, weighing up to 55 tons – the largest crystals to have yet been discovered. The cave was only discovered because there was a mine nearby. It is likely that there are even more awesome geological structures lying somewhere undiscovered beneath our feet!

What's the point?
We live in a three-dimensional world. 3D geometry allows us to describe the wonderful things that we can see in the natural world, as well as helping us to devise increasingly sophisticated structures in the manmade world.

Objectives
By the end of this chapter, you will have learned how to …
- Draw and interpret plans and elevations of 3D shapes.
- Calculate the volume of cuboids and prisms.
- Calculate the surface area and volume of spheres, pyramids, cones and composite shapes.
- Know and apply the relationship between lengths, areas and volumes of similar shapes.

Check in

1. **a** Draw a net of this cube.
   b Hence calculate its surface area.
   c Calculate its volume.

2. For each circle work out its
   i area
   ii circumference.

   a 5 cm   **b** 7.4 cm

Chapter investigation

A manufacturer wants to create a drinks container with a capacity of 1 litre. In order to minimise the cost, they also want the container to have the least surface area.

Advise the manufacturer on suitable designs for the container.

SKILLS

15.1 3D shapes

A **solid** is a **three-dimensional** shape. The diagrams show some examples. You may find it easier to draw them on square or **isometric** paper.

A **prism** has the same **cross-section** throughout its length. **Cubes**, **cuboids** and **cylinders** are all prisms.
Other prisms are named according to the shape of their cross-sections.

The base of a **pyramid** is a polygon. The other faces are triangles.

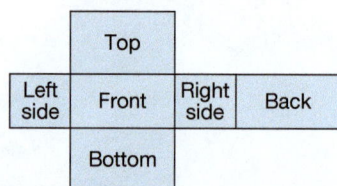
▲ Cuboid ▲ Cube
▲ Triangular prism ▲ Hexagonal prism
▲ Square-based pyramid ▲ Triangle-based pyramid or tetrahedron
▲ Cylinder ▲ Cone ▲ Sphere

● A **net** is a 2D shape that can be folded to make a 3D shape.

Imagine cutting along some of the edges of the cuboid and opening it out to give this net.

	Top		
Left side	Front	Right side	Back
	Bottom		

● 3D shapes can be represented by **plans** and **elevations**.

EXAMPLE

The sketch shows a 3D shape on dotty isometric paper.

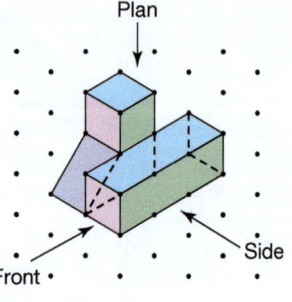

Draw
a a **plan** of the shape
b a **front elevation**
c a **side elevation**.

a The plan is the view from above.
   You see the blue and purple faces.

Plan
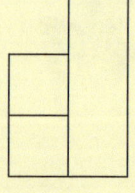

Imagine standing at the front and looking down.
You don't see any of the vertical front or side faces.

b The front elevation is the view from the front.
   You see the pink and purple faces.

Front elevation
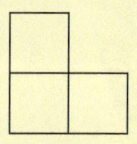

c The side elevation is the view from the side.
   You see the green faces.

Side elevation
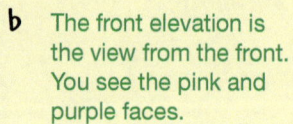

Use dotted lines to show edges that you can't see.

Geometry   Working in 3D

Fluency

Exercise 15.1S

1 a Copy and complete the table.

3D solid	Number of		
	Edges E	Vertices V	Faces F
Cube			
Triangular prism			
Hexagonal prism			
Triangle-based pyramid			
Square-based pyramid			

b Write down an expression for E in terms of V and F. Check that your answer works for other prisms and pyramids.

2 Which of these are possible nets for a cube?

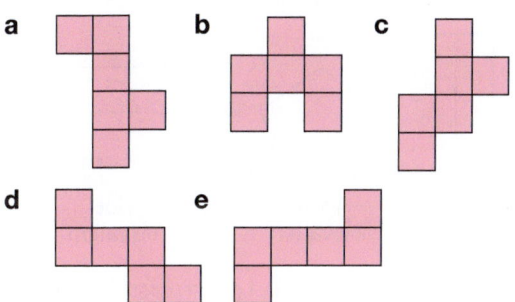

3 Sketch and name the 3D shape that can be made from each net.

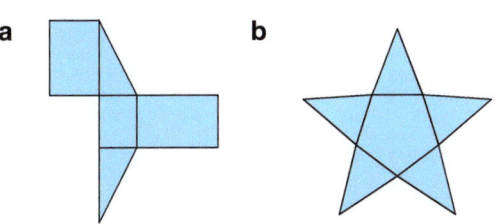

4 Draw an accurate net of each shape.

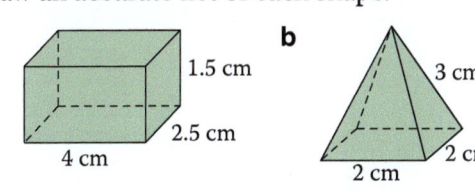

5 Draw the plan, front and side elevation of
  a a cylinder
  b a square-based pyramid
  c a cone
  d a tetrahedron
  e These shapes made from
      i 7 cubes   **ii** 10 cubes.

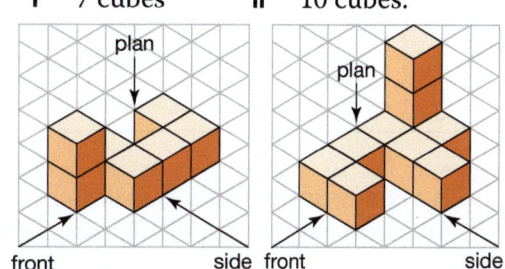

  ***f** These isometric drawings.

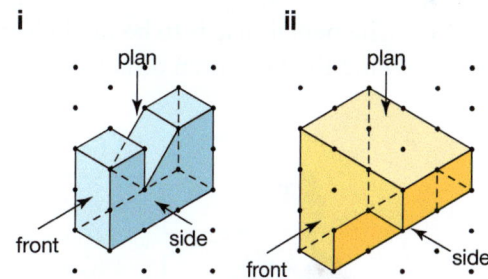

6 The table gives the plans and elevations of some solids. Draw a 3D sketch of each solid.

	plan	front elevation	side elevation
a			
*b			
*c			

***7** Use card to make accurate models of some 3D shapes.

305

APPLICATIONS

15.1 3D shapes

RECAP
- A net is a 2D shape that can be folded to make a 3D shape.
- 3D shapes can be represented by plans and elevations.

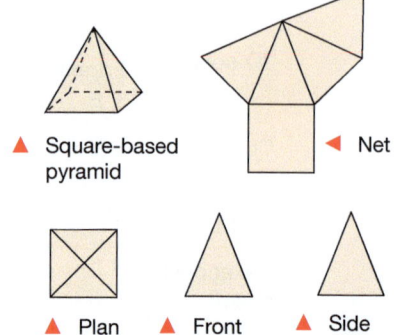

▲ Square-based pyramid ◀ Net

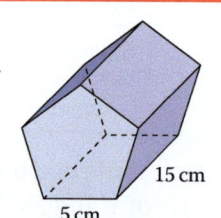

▲ Plan ▲ Front elevation ▲ Side elevation

HOW TO

When asked to create a 2D representation of a 3D shape
1. Think about which type of diagram you need to draw.
2. When asked for a scale diagram, choose a scale that is easy to use and work out the dimensions you need.
3. Use the diagram to answer the questions.

EXAMPLE

A manufacturer wants to make a pencil box in the shape of a prism. The cross-section is a regular pentagon. The diagram gives the dimensions.

a Sketch a plan, front and side elevation of the box.
b Draw a scale diagram of the net.
c The pencil case is to be made from a rectangular sheet of material. Find the minimum dimensions of the sheet needed for your net.

a ① Imagine looking at the prism from above, then from the front, then from the side.

Use dotted lines for edges that you can't see.

plan   front elevation   side elevation

b ② Choose a scale that is easy to use. (Yours could be larger.)

Scale 1 cm to represent 5 cm
Length of each side of the pentagon = 1 cm
Length of prism = 15 ÷ 5 = 3 cm.

A pentagon can be split into 3 triangles.
Angle sum = 3 × 180° = 540°.
Each angle of pentagon = 540° ÷ 5 = 108°.

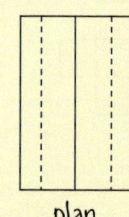

scale 1 cm represents 5 cm

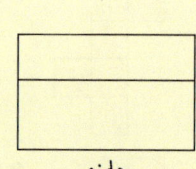

c ③ Measure the smallest rectangle that fits round the net.

Measured length = 6.1 cm
Actual length = 6.1 × 5 = 30.5 cm
Measured width = 5 cm
Actual width = 5 × 5 = 25 cm

The minimum dimensions are length 31 cm and width 25 cm (to nearest cm).

Geometry Working in 3D

Reasoning and problem solving

Exercise 15.1A

1. Draw sketches to show the plan, front elevation and side elevation of these steps and ramp.

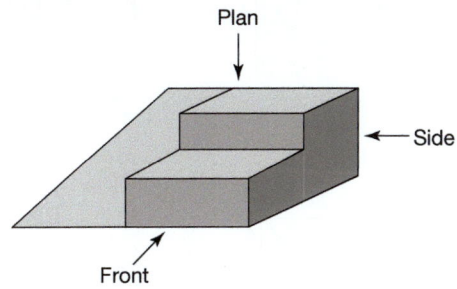

2. Draw scale diagrams to show the plan, front elevation and side elevation of this church.

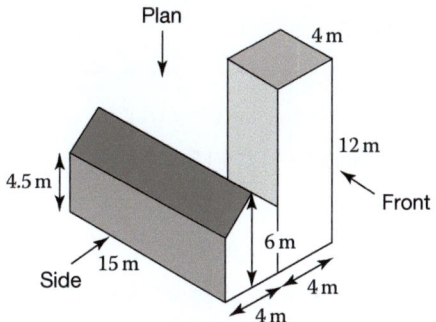

3. Tina wants to make a box to hold a perfumed soap to give as a present. The box will be in the shape of a prism with a trapezium cross-section as shown.

   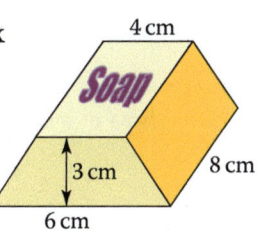

   a. Draw a net that Tina could use.

   b. Tina wants to decorate all the edges of the box with lace and glue a flower to each vertex.

   She buys $\frac{1}{2}$ metre of lace and 10 flowers. Is this enough? Explain your answer.

4. The sketch shows a cylindrical can that has been opened out. The height of the can is 11.5 cm. The radius of the can is 3.8 cm

   Find the total area of metal used to make the can.

   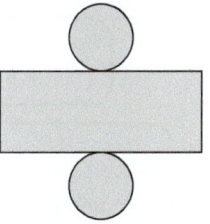

5. A manufacturer wants a box for matches to have an open tray and a sleeve to go around it.

   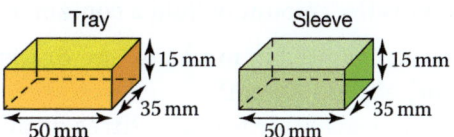

   a. Draw accurate nets for the tray and sleeve.

   b. How would the thickness of the card used to make the net affect your answer?

6. Sofia uses this card to make the biggest possible open-topped box in the shape of a cube. How long are the sides of Sofia's box?

   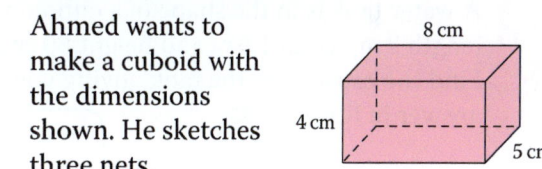

*7. Ahmed wants to make a cuboid with the dimensions shown. He sketches three nets.

   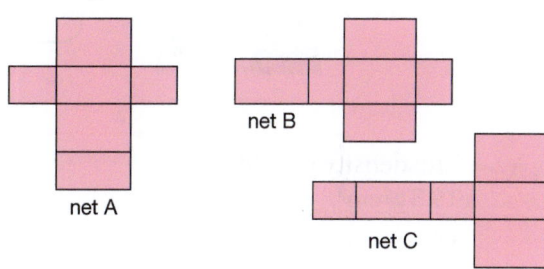

   Ahmed plans to cut a net from a rectangular card. Which net wastes the smallest area of card?

   State two assumptions you make that would affect your answer.

8. Sonja makes nets for pyramids by using a circle to draw the triangular faces. Use Sonja's method to construct

   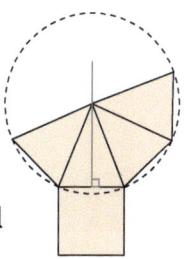

   a. a square-based pyramid

   *b. a pentagonal-based pyramid

   *c. a hexagonal-based pyramid.

1078, 1098, 1106

SKILLS

15.2 Volume of a prism

Volume is the amount of space occupied by a 3D shape.
Capacity is the amount of fluid a container can hold.

Volume is measured in **mm³**, **cm³**, **m³** or **km³**.
$1\,m^3 = 1\,000\,000\,cm^3$
Liquids are usually measured in **litres** or **millilitres**.
$1\,m^3 = 1000$ litres, 1 litre $= 1000\,cm^3$ and $1\,ml = 1\,cm^3$

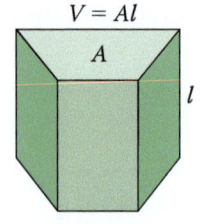
▲ Prism

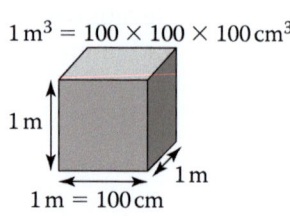

- Volume of a prism = area of cross-section × length
- Volume of a cuboid = length × width × height
- Volume of a cylinder = π × radius² × height
- Mass = volume × density

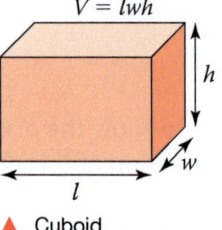
▲ Cuboid

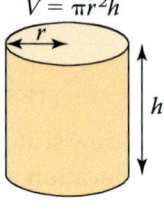
▲ Cylinder

EXAMPLE

A water tank is in the shape of a cuboid with length 2 m, width 1.5 m and height 80 cm. Find the capacity of the tank, giving your answer in litres.

Make sure your units are consistent $80\,cm = 0.8\,m$

For a cuboid $V = lwh$
Volume $= 2 \times 1.5 \times 0.8$
$= 3 \times 0.8$
$= 2.4\,m^3$
Capacity in litres $= 2.4 \times 1000$
$= 2400$ litres

EXAMPLE

The density of gold is $19.3\,g/cm^3$.
Find the mass of this gold bar.

Work in cm to give the volume in cm³

Area of triangular cross-section
$= \frac{1}{2} \times 8 \times 6$
$= 24\,cm^2$
Volume of prism $= 24 \times 20$
$= 480\,cm^3$
Mass $= 480 \times 19.3$
$= 9264\,g$ or $9.264\,kg$

EXAMPLE

The volume of a cylinder is $550\,cm^3$. The height is $12\,cm$. Calculate the radius.

Substitute the values into $V = \pi r^2 h$, then rearrange.
$550 = \pi \times r^2 \times 12$
$r^2 = \dfrac{550}{12\pi} = 14.589...$
$r = \sqrt{14.589...} = 3.819...$
The radius of the cylinder $= 3.82\,cm$ (3 sf)

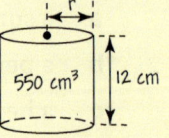

Sketch the cylinder if it helps.

When calculating $\dfrac{550}{12\pi}$ take care with the denominator.

Geometry   Working in 3D

Fluency

Exercise 15.2S

1 Find the volume of these solids.

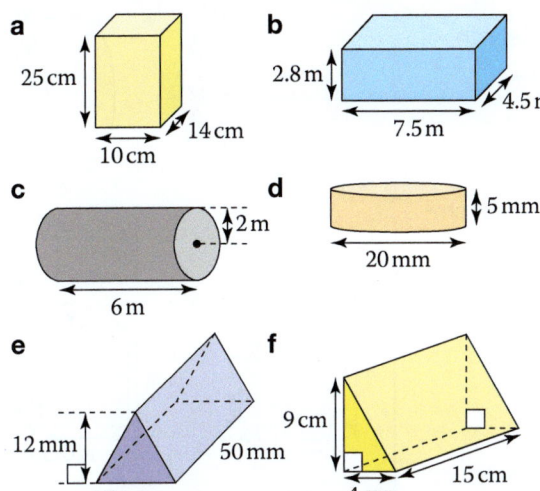

a (cuboid) 25 cm, 14 cm, 10 cm
b (cuboid) 2.8 m, 4.5 m, 7.5 m
c (cylinder) 2 m, 6 m
d (cylinder) 5 mm, 20 mm
e (triangular prism) 12 mm, 16 mm, 50 mm
f (triangular prism) 9 cm, 4 cm, 15 cm

5 The diagram shows the cross-section of a prism. The prism is 1.4 m long. The density is 0.7 g/cm³. Find the mass.

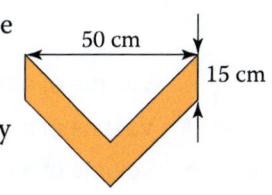
50 cm, 15 cm

6 The concrete in this prism has a density of 2400 kg/m³. Find the prism's mass.

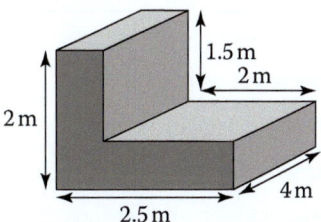
1.5 m, 2 m, 2 m, 2.5 m, 4 m

2 The table gives dimensions of some cuboids. Copy the table. Find the missing values.

Length	Width	Height	Volume
12 mm	7 mm	9 mm	
16 cm	5 cm		240 cm³
	4 m	2 m	64 m³
14 mm		5 mm	490 mm³

7 a A cylindrical water tank has diameter $\frac{1}{2}$ metre and height 64 cm. Find the capacity in litres.

b The volume of a cylinder is 510 cm³. The radius is 45 mm. Calculate the height.

c A cylindrical can has a capacity of 440 ml. The height is 10 cm. Calculate the diameter.

3 Find the volume of these solids that have holes running through their whole length.

a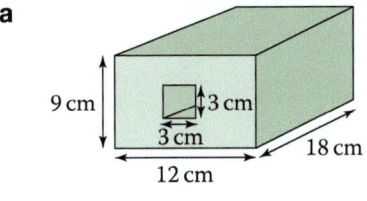
9 cm, 3 cm, 3 cm, 12 cm, 18 cm

b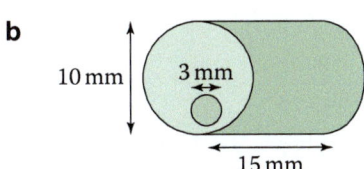
10 mm, 3 mm, 15 mm

8 The diagram shows the cross-section of a prism. The volume of the prism is 750 cm³. Find the length.

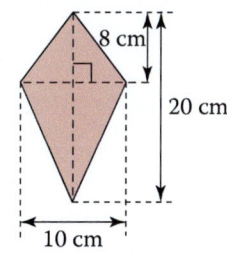
8 cm, 20 cm, 10 cm

*9 The cross-section of this prism is a regular hexagon with sides of length 22 mm. The density is 8.5 g/cm³. Work out the mass.

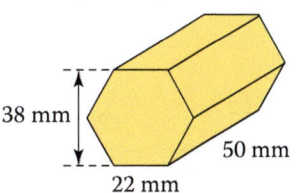
38 mm, 22 mm, 50 mm

4 The density of silver is 10.5 g/cm³. Find the mass of this silver prism.

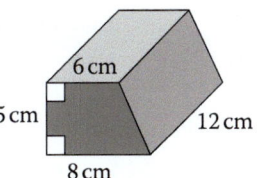
6 cm, 5 cm, 8 cm, 12 cm

10 The diagram shows a planter. How many litres of compost does it take to fill it?

30 cm, 25 cm, 24 cm, 16 cm, 1.2 m

APPLICATIONS

15.2 Volume of a prism

RECAP
- Volume of a prism = area of cross-section × length
- Volume of a cuboid = length × width × height
- Volume of a cylinder = π × radius² × height
- Mass = volume × density
- $1 m^3 = 1\,000\,000 \, cm^3 = 1000$ litres   1 litre = $1000 \, cm^3$

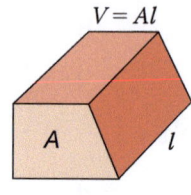
▲ Prism

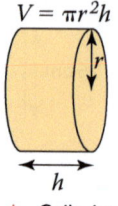
▲ Cylinder

HOW TO
To solve problems involving the volumes of prisms
1. Draw a diagram (if needed).
2. Decide which formula (or formulae) and units to use.
3. Find the answer, including the units.
4. State your conclusion and reasons clearly (where necessary).

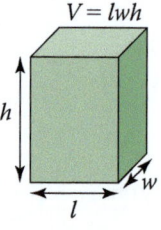
▲ Cuboid

$1 m^3 = 100 \times 100 \times 100 \, cm^3$

EXAMPLE
20 kg of molten metal is made into prisms. The density of the metal is $8 \, g/cm^3$. How many prisms are made?

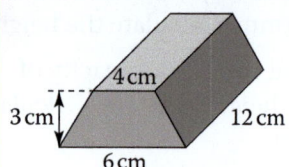

② Volume = mass ÷ density
Volume of metal = 20 000 ÷ 8 = 2500 cm³
② Area of trapezium $A = \frac{1}{2}(a + b)h$
Area of trapezium = $\frac{1}{2}(6 + 4) \times 3 = 15 \, cm^2$
② Volume of prism $V = Al$
Volume of prism = 15 × 12 = 180 cm³
③ Number of prisms = 2500 ÷ 180 = 13.88...
④ Number of prisms = 13   Round down.

EXAMPLE
A drinks manufacturer wants a can to hold 330 ml of cola. The diameter of the can must be 7 cm.

a How tall does the can need to be?

The cans are to be packed in boxes of 12.

b Give the dimensions of a box that could be used.

c What advice can you give about the use of your answers in practice?

a ①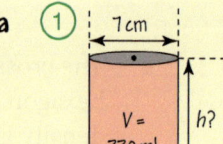

② For a cylinder $V = \pi r^2 h$ 1 ml = 1 cm³
③ Substitute the values.
330 = π × 3.5² × h   Radius = 7 ÷ 2
330 = 12.25π × h   Rearrange to find h.
$h = \frac{330}{12.25\pi} = 8.57 \, cm$ (3 sf)

b ① A possible box.   ② Length of box = 3 × 7 cm = 21 cm
Width of box = 2 × 7 cm = 14 cm
Height of box = 2h = 17.1 cm (3 sf)

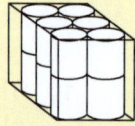

c ④ h and the box dimensions will need to be a little larger. This is because the answers have been rounded downwards and the thickness of the can and box have been neglected.

Geometry   Working in 3D

Reasoning and problem solving

Exercise 15.2A

1 a How many cups of coffee can you fill from a 2 litre coffee pot?
   b If the cup is only filled to a depth of 6 cm, how many cups can you now fill?

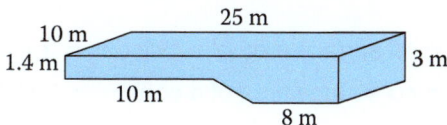

2 This metal cuboid is melted and made into triangular prisms.

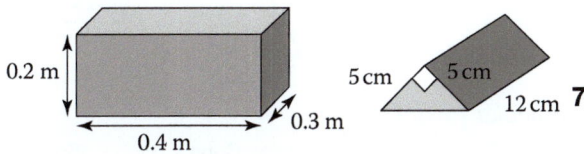

   a How many prisms are made?
   b How would your answer change if 5% of the original metal was wasted?

3 How many tins of soup can you pour into the pan?

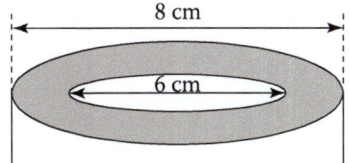

4 A trough with this cross-section must hold 50 litres of water. Find the shortest possible length of the trough.

5 Plastic pipe has the cross section shown. The density of the plastic is 600 kg/m³.

   a What length of pipe is made from 1.2 tonnes of plastic? (1 tonne = 1000 kg)
   b If the dimensions of the cross-section were doubled, say how your answer would change.

6 Water flows into this swimming pool at a rate of 500 litres per minute.

   a Estimate how many hours it takes to fill the swimming pool. State any assumptions you make.
   b What extra information would allow you to find a more accurate estimate?

7 How many of these CD boxes is it possible to fit into a box measuring 75 cm by 60 cm by 30 cm?

8 Water is poured from A into B until the depth of the water in each container is the same. Find this depth.

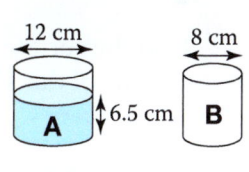

*9 Unsharpened pencils have a cylinder of graphite surrounded by wood. The diameter of the graphite is 2 mm. The density of the graphite is 640 kg/m³. The density of the wood is 420 kg/m³.

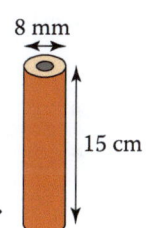

   a Find the mass of 500 pencils.
   b i Suggest dimensions for a box to pack the 500 pencils in.
      ii Find the percentage of the box that is filled by the pencils.

*10 Hsia measures a £1 coin and writes down
     Diameter = 22 mm   Thickness = 3 mm
     The table gives the metals used to make the coin.

Metal	Copper	Nickel	Zinc
Density (kg/m³)	8930	8800	7135
% of coin	70%	5.5%	24.5%

Estimate the mass of a £1 coin.

1137, 1138, 1139, 1246

311

15.3 Volume and surface area

Surface area is the total area of the faces of a 3D shape.

- Curved surface area of a cylinder = $2\pi \times$ radius $\times$ height
- Curved surface area of a cone = $\pi \times$ radius $\times$ *slant* height
- Surface area of a sphere = $4 \times \pi \times$ radius^2

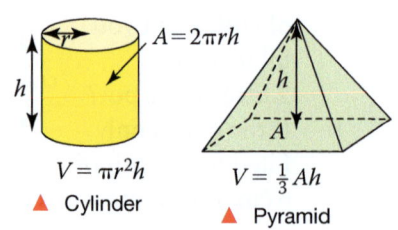
▲ Cylinder   ▲ Pyramid

When shapes are *similar*, the *area* scale factor is the *square* of the linear scale factor.

- Volume of a pyramid = $\frac{1}{3} \times$ area of base $\times$ *perpendicular* height
- Volume of a cone = $\frac{1}{3} \times \pi \times$ radius$^2 \times$ *perpendicular* height
- Volume of a sphere = $\frac{4}{3} \times \pi \times$ radius^3

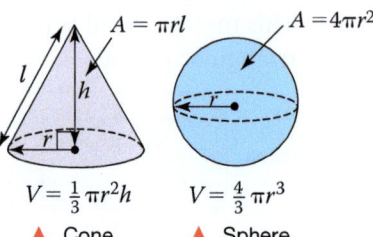
▲ Cone   ▲ Sphere

When shapes are *similar*, the *volume* scale factor is the *cube* of the linear scale factor.

EXAMPLE

Find the surface area and volume of this solid.

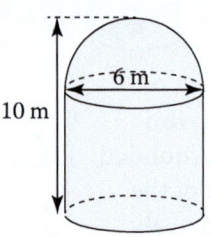

Add the area of the hemisphere, the cylinder and the circular base.
Surface area = $2\pi r^2 + 2\pi rh + \pi r^2 = 3\pi r^2 + 2\pi rh$
$= 3 \times \pi \times 3^2 + 2 \times \pi \times 3 \times 7$
$= 27\pi + 42\pi$
Surface area = 69π or $217\,m^2$ (3 sf)

A **hemisphere** is half of a sphere.

Add the volume of the hemisphere and cylinder.
Volume = $\frac{2}{3}\pi r^3 + \pi r^2 h$
$= \frac{2}{3} \times \pi \times 3^3 + \pi \times 3^2 \times 7$
$= 18\pi + 63\pi$
Volume = 81π or $254\,m^3$ (3 sf)

The answers in terms of π are more accurate.

EXAMPLE

Find the volume of this **frustum** of a cone.

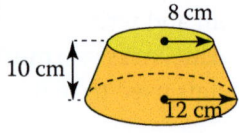

The volume scale factor is the cube of the linear scale factor

A small cone has been removed from a larger cone.
Using corresponding sides of the similar triangles
$\frac{h}{h+10} = \frac{8}{12} = \frac{2}{3}$
$3h = 2h + 20$
$h = 20$

Volume of large cone = $\frac{1}{3} \times \pi \times 12^2 \times 30$
$= 1440\pi\,cm^3$

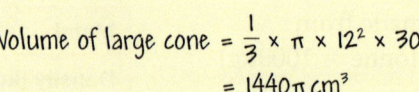

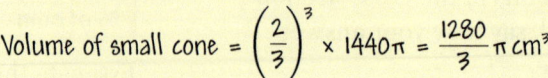

Geometry   Working in 3D

Fluency

Exercise 15.3S

1. Calculate the surface area and volume of each solid. Give your answers in terms of π.

   a 6 cm

   b 48 mm

   c 12 cm, 15 cm, 9 cm

   d 3.9 m, 3 m, 3.6 m

2. Work out the volume and surface area of each solid.

   a 6.4 cm, 13 cm, 8 cm, 8 cm, 8 cm

   b 7.2 m, 11 m, 5 m, 8 m

3. Calculate the volume of each frustum.

   a 3 cm, 2 cm, 6 cm

   b 0.8 m, 2.4 m, 3 m

4. **a** Find the volume in cm³ of this frustum of a square-based pyramid.

   36 mm, 42 mm, 60 mm

   b Four of the faces are trapeziums. The perpendicular height of the trapeziums is 43.7 mm. Find the total surface area in cm².

5. Find the height of each shape.

   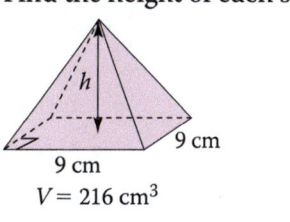
   $V = 216$ cm³, 9 cm, 9 cm, h

   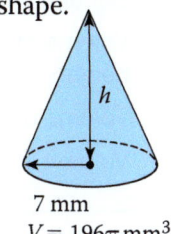
   7 mm, $V = 196\pi$ mm³, h

6. **a** The surface area of a cube is 294 cm². Find the volume of the cube.

   b The surface area of a sphere is 784π cm². Find the radius of the sphere.

   c The volume of a sphere is 4500π cm³. Find the diameter of the sphere.

7. Two square-based pyramids are joined. The total volume is 2700 mm³. Find x.

   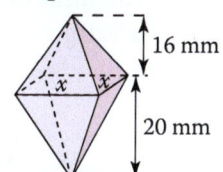
   16 mm, 20 mm, x

8. A hemisphere is cut from a cylinder. For the remaining solid, find

   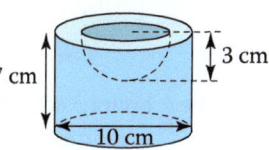
   7 cm, 3 cm, 10 cm

   a the volume
   b the surface area.

*9. **a** Find, in terms of π, the surface area and volume of this solid.

   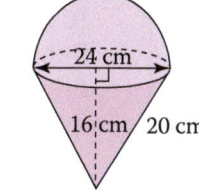
   24 cm, 16 cm, 20 cm

   b A similar solid is 7 cm tall. Find its surface area and volume.

*10. This container consists of a frustum of a cone and a hemisphere. Find its capacity in litres.

   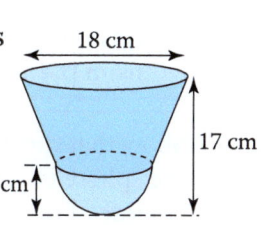
   18 cm, 17 cm, 5 cm

11. The capacity of a hemispherical ladle is $\frac{1}{4}$ litre. Calculate the diameter of the ladle.

APPLICATIONS

15.3 Volume and surface area

RECAP

- Curved surface area of a cylinder = $2\pi \times$ radius $\times$ height
- Curved surface area of a cone = $\pi \times$ radius $\times$ slant height
- Surface area of a sphere = $4 \times \pi \times$ radius^2
- Volume of a pyramid = $\frac{1}{3} \times$ area of base $\times$ *perpendicular* height
- Volume of a cone = $\frac{1}{3} \times \pi \times$ radius$^2 \times$ *perpendicular* height
- Volume of a sphere = $\frac{4}{3} \times \pi \times$ radius^3

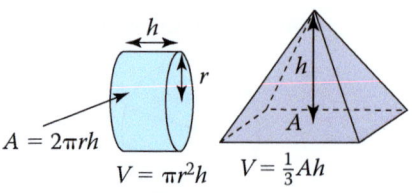

▲ Cylinder   ▲ Pyramid

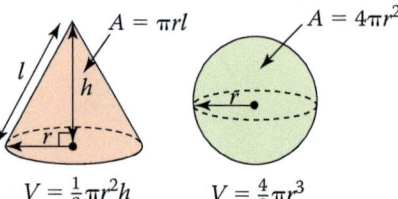

▲ Cone   ▲ Sphere

HOW TO

To solve volume problems
1. Draw a diagram (if needed or helpful).
2. Decide which formula(e) and units to use.
3. Find the answer, including the units.
4. State your conclusion and reasons clearly (where necessary).

EXAMPLE

The mass of this trophy is 920 g.

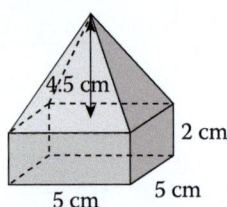

a Javiero says the trophy is made from silver. Rosie says it is nickel silver. Who is correct?
Density of silver = 10 490 kg/m^3
Density of nickel silver = 8900 kg/m^3.

b A similar trophy has a square base with edges of length 4 cm. Assuming it is made from the same metal, what will be its mass?

a ② Mass = volume × density
For a cuboid $V = lwh$. For a pyramid $V = \frac{1}{3}Ah$

③ Volume of trophy = $5 \times 5 \times 2 + \frac{1}{3} \times 25 \times 4.5$
= 50 + 37.5 = 87.5 cm^3
Volume of trophy in m^3 = 87.5 ÷ 1 000 000 = 8.75×10^{-5} m^3
Mass of trophy in kg = 920 ÷ 1000 = 0.92 kg
Density = $\frac{mass}{volume}$ = $\frac{0.92}{8.75 \times 10^{-5}}$ = 10 500 kg/m^3 (3 sf)
This is nearer to 10 490 kg/m^3 than 8900 kg/m^3

④ Javiero is correct — the trophy is made of silver.

b ③ Find the scale factor, for length, then volume.

Length scale factor = $\frac{4}{5}$ Volume scale factor = $\left(\frac{4}{5}\right)^3 = \frac{64}{125}$

Mass depends on volume, so it has the same scale factor.

Mass of similar trophy = $\frac{64}{125} \times 920 = 471$ g (3 sf)

EXAMPLE

Show that the volume of the sphere is $\frac{2}{3}$ of the volume of the cylinder that surrounds it.

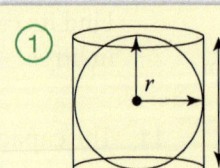

② Volume of cylinder = $\pi r^2 \times 2r$ $V = \pi r^2 h$
= $2\pi r^3$
Volume of sphere = $\frac{4}{3}\pi r^3$ $V = \frac{4}{3}\pi r^3$
= $\frac{2}{3} \times 2\pi r^3$

④ Volume of the sphere = $\frac{2}{3}$ of the volume of the cylinder.

Geometry   Working in 3D

Reasoning and problem solving

Exercise 15.3A

1. A conical hole is drilled in a metal hemisphere. The mass of the object is 120 g.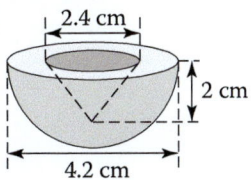
   a. Is it more likely that the object is made from aluminium with density $2.6\,g/cm^3$ or steel with density $7.5\,g/cm^3$?
   b. A similar object is 6.3 cm wide. Find the mass of this object.

2. Show that the surface area of the sphere is $\frac{2}{3}$ of the surface area of the cylinder that surrounds it.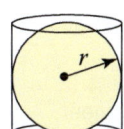

3. a. This glass holds 125 ml when full. Find h.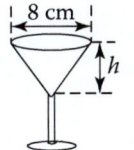
   b. Find the diameter of a hemispherical glass that holds 125 ml.

4. a. i. What height must the can be to hold 1 litre of oil?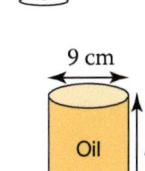
      ii. Calculate the surface area.
   b. Find the diameter, height and surface area of a similar can that will hold 5 litres.

5. a. OA and OB are joined to make a cone. Find the diameter of the cone.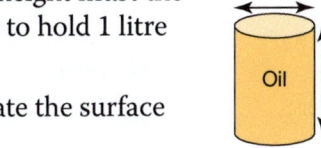
   b. Prove that the curved surface area of any cone is given by $A = \pi r l$.

6. Three tennis balls, each of diameter 6.8 cm fit tightly into a closed cylindrical tube.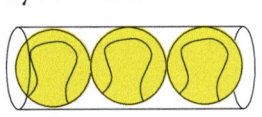
   a. Calculate the total surface area of the tube.
   b. Find the percentage volume of the tube that is filled by the tennis balls.

7. a. Draw a net of this frustum and find its surface area.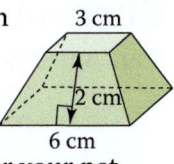
   b. Find the dimensions of the smallest rectangular card that you could use for your net.
   c. What percentage of the card is wasted?

8. Thirty ball bearings of diameter 8 mm are put into this glass of water. Find the increase in the depth of the water.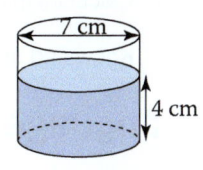

9. This container needs to hold 2 tonnes of liquid.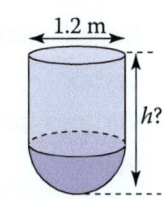
   The density of the liquid is $800\,kg/m^3$.
   Find h.

10. The density of marble used for this pedestal is $2750\,kg/m^3$.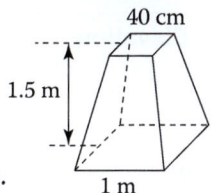
    A crane can lift a maximum load of 2 tonnes.
    Can the crane lift this pedestal?

*11. A food manufacturer wants to sell ice cream in tubs shaped as frustums of cones.
   a. Find dimensions for a tub that will have a capacity of 1 litre.
   b. Find dimensions of a similar tub that will have a capacity of 2 litres.

*12. A frustum of a cone has a base radius R, a top radius of r and height h.
   a. Prove that the height of the cone from which the frustum is cut is $H = \dfrac{Rh}{R-r}$
   b. Show that the volume of the frustum is $V = \frac{1}{3}\pi h(R^2 + Rr + r^2)$

*13. A frustum of a pyramid has a square base with sides of length X, a square top with sides of length x and height h.
   Show that the volume of the frustum is $V = \frac{1}{3}h(X^2 + Xx + x^2)$

Summary

Checkout
You should now be able to...

Test it
Questions

✓ Draw and interpret plans and elevations of 3D shapes.	1, 2
✓ Calculate the volume of cuboids and prisms.	3, 4
✓ Calculate the surface area and volume of spheres, pyramids, cones and composite shapes.	5, 6
✓ Know and apply the relationship between lengths, areas and volumes of similar shapes.	7

Language	Meaning	Example
Face	A flat surface of a solid enclosed by edges.	Vertex, Edge, Face
Edge (solid)	A line along which two faces meet.	
Vertex / **Vertices (plural)**	A point at which two or more edges meet.	
Plan	A drawing of a 3D object looking straight down at the object from directly overhead.	Plan, Front, Side
Elevation	A 2D drawing of a 3D object looking straight at the object from the front or side.	
Net	A 2D shape that can be folded to make a 3D solid.	4 cm cube; Surface area of cube = 6 × 4 × 4 = 96 cm²
Surface area	The total area of all the faces of a 3D solid.	
Volume	The amount of space occupied by, or inside, a 3D shape.	Volume of cube = 4 × 4 × 4 = 64 cm³
Cross-section	The 2D shape formed when a solid shape is cut through in a specified direction, usually parallel to one of its faces.	Cross section; Cylinder, Triangular prism
Prism	A 3D solid with a constant cross-section.	
Pyramid	A 3D solid with a polygon as its base. All the other faces are triangular in shape and meet at a single vertex.	Regular tetrahedron, Square-based pyramid, Irregular pyramid
Cylinder	A prism with a circular cross-section.	Cylinder, Cone, Sphere
Cone	A solid with a circular base and one vertex.	
Sphere	A 3D shape with every point on its surface the same distance from the centre.	
Frustum	The part of a cone (or pyramid) which remains when the top part is cut off with a cut parallel to the base	Frustum

Geometry   Working in 3D

Review

1. Draw
   a. the front elevation
   b. the plan view of the solid.

   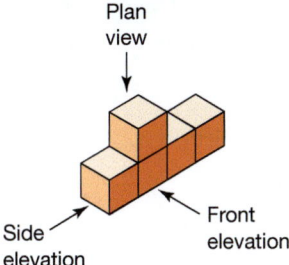

2. Here are three elevations of a 3D solid made from cubes.

   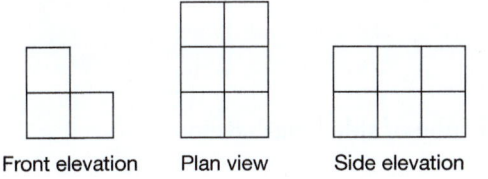

   Front elevation   Plan view   Side elevation

   Draw the 3D solid these views come from.

3. A cuboid has dimensions 14 m, 3.5 m and 5 m.

   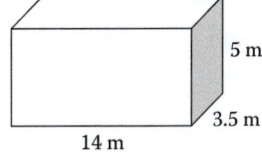

   Calculate
   a. the volume
   b. the surface area of the cuboid.

   The cuboid is made from a material with a density of 1.5 kg/m³.

   c. What is the mass of the cuboid?

4. A cylinder has a radius of 13 cm and a length of 18 cm.

   Calculate
   a. the volume
   b. the surface area assuming the cylinder is closed at both ends.

5. Calculate the volume of these 3D shapes.

   a.   b.

   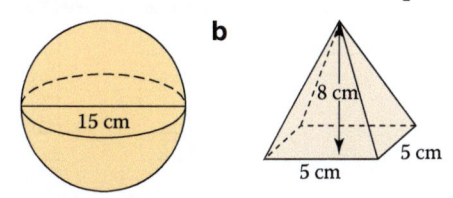

6. Calculate the volume and surface area of this shape.

   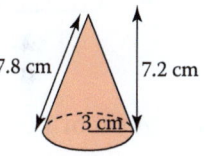

7. A prism of volume 20 cm³ is enlarged to a solid of volume 160 cm³.

   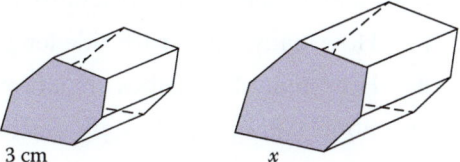

   a. What will be the length x?

   The area of the cross section of the larger solid is 40 cm².

   b. What is the area of the cross section of the smaller solid?

What next?

Score		
	0 – 3	Your knowledge of this topic is still developing. To improve, look at MyiMaths: 1078, 1098, 1106, 1107, 1122, 1136, 1137, 1138, 1139, 1246
	4 – 6	You are gaining a secure knowledge of this topic. To improve your fluency look at InvisiPens: 15Sa – j
	7	You have mastered these skills. Well done you are ready to progress! To develop your problem solving skills look at InvisiPens: 15Aa – g

Assessment 15

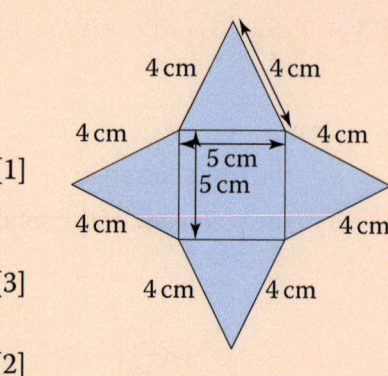

1. **a** A shape has four faces that are all triangles.
   What is the mathematical name of the shape? [1]

   b The diagram shows an incorrect net of a square based
   pyramid with base length 4 cm and longest edge 5 cm.
   Draw the correct net for this shape. [3]

2. Match the two isometric drawings, **A** and **B**, to the two sets
   of plans and elevations. [2]

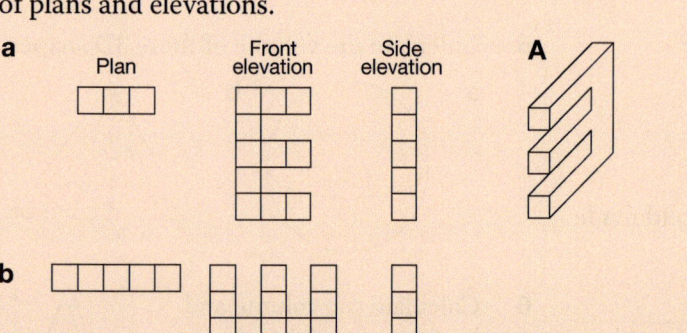

3. A set of 1 cm cubes is exactly enough to make a solid cube of side length 8 cm.

   a How many 1 cm cubes are there? [1]
   b How many cubes with side length 2 cm can be made? [2]
   c How many cubes with side length 4 cm can be made? [2]
   d **i** How many cubes with side length 3 cm can be made? [2]
   ii How many 1 cm cubes are left over? [1]

4. A large cuboid storage box has dimensions, 24 cm × 30 cm × 36 cm.

   a How many small boxes, with dimensions 6 cm × 5 cm × 2 cm, will fit into the large
   storage box? [3]

   b Is this statement correct? Give reasons for your answer.
   The ratio of the surface area of the large box to the surface area of the small box
   equals the ratio of the volume of the large box to the volume of the small box. [5]

5. Chips are sold in two sizes, Thin chips are 9.6 cm long with a square cross section
   of side 10 mm. Fat chips are 6.7 cm long with a square cross section of side 12 mm.

   a Show that the two sizes of chips have approximately the same volume. [2]
   b Healthier chips have a smaller surface area so they absorb less fat.
   Which of the two types of chip is healthier to eat? [3]

6. The chocolate bar in the diagram consists of a slab of milk
   chocolate with a cylinder of cream filling.
   The whole bar is then covered with chopped nuts.

   a Calculate the volume of
   i cream filling [2] **ii** chocolate. [3]
   b Calculate the surface area of the chocolate bar. [5]

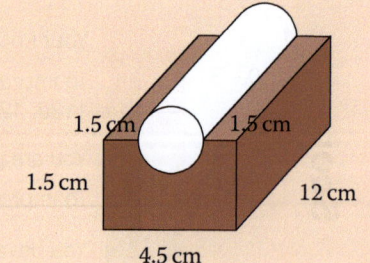

Geometry   Working in 3D

7 a A slab of cheese is in the shape of a prism with a right-angled triangular cross section. The triangle has sides of 1.5, 3.6 and 3.9 inches. The length of the prism is 5 inches. Calculate **i** its total surface area [4] **ii** its volume. [2]

b A pyramid made of gold has a rectangular base with dimensions 55 cm × 40 cm. The perpendicular height of the pyramid is 65 cm.

    i Calculate its volume. [3]

    ii The gold is melted down and recast as a number of spheres of diameter 10 cm. How many complete spheres can be made? [4]

c A glass tube is 8 cm long, with an outer radius of 0.5 cm and a central hole of diameter 0.5 cm. Calculate

    i the volume of the glass [4]

    ii the mass of the tube if the density of the glass is 2.6 g/cm^3. [1]

8 A toy consists of a hemisphere and a cone with the same radius.

Calculate the **a** volume [5] **b** surface area of the toy. [5]

9 An unsharpened wooden pencil is in the shape of a hexagonal prism. Each side of the hexagon is 4 mm and the pencil is 6.5 cm long. The graphite core is a cylinder of radius 0.6 mm. Calculate the

a cross sectional area [2] **b** total volume [2]

c volume of wood in the pencil. [4]

10 A café sells ice cream and has an advertising stand outside made of a cylinder, a cone and a hemisphere. The plastic has thickness 1.25 inches.

a Calculate the volume of the complete model. [6]

b Calculate the volume of the plastic in the hemisphere. [2]

c Calculate the total surface area. [8]

11 The clepsydra, or water clock, was invented in the 14th century BC. Water slowly drips out of the hole at the bottom and time is measured by the level of water remaining.

The clepsydra is filled at midnight and water drips out a a rate of 3 ml/min. Find the following.

a The capacity, to the nearest ml, of the full cone. [2]

b The time taken, to the nearest minute, to empty the clepsydra. [1]

c The radius of the water's surface when its height is 15 cm. [2]

d The volume of water when the height is 15 cm. [2]

e The volume of water lost before its height drops to 15 cm. [1]

f The actual time when the height of water is 15 cm. [1]

The 'hour' marks on the side of the clepsydra were evenly spaced.

g Explain why this was not sensible and suggest a better shape. [1]

Life skills 3: Getting ready

Abigail, Mike, Juliet and Raheem are now ready to start making the building suitable for their restaurant. They have also chosen some suppliers to work with, and are considering what stocks of non-perishable food and drink they will need. They analyse the data that they have gathered from their market research to help set their menu and price list.

Task 1 – The ingredients

One of the starters on the menu is an old family recipe that Mike borrowed from his grandmother: *Artichokes with anchovies*.

The friends are keen to avoid waste, and want to use ingredients efficiently. Each tin of anchovy fillets contains 24 anchovy fillets.

Each jar of artichoke hearts contains 20 artichoke hearts.

How many portions of *Artichokes with anchovies* will they have to make in order to use up complete tins of anchovy fillets and complete jars of artichoke hearts with no leftovers?

Artichokes with anchovies – serves 1
2 artichoke hearts
3 anchovy fillets
1 tablespoon of olive oil
8 tablespoons of butter, softened
2 garlic cloves
1 pinch each, salt and pepper
Juice of 1 lemon

Task 2 – Price list

They decide to look again at the data they collected from their small pilot survey (*Life skills 1: The business plan*).

p.100

a On graph paper, plot separate male and female scatter graphs of amount prepared to spend against age.

b For each of these graphs, describe the correlation and explain what it suggests in each case.

c Should they use these graphs alone to decide a specific age group to target for their restaurant? Why?

Task 3 – Supplier meeting

Raheem drives from the restaurant to the vegetable suppliers, 9 miles away. Each friend keeps a log of all business-related travel and time spent in meetings so that they can claim back costs on expenses and keep a record of how much time they are investing.

a Draw a distance-time graph showing Raheem's journey.

b At what time does he arrive back at the restaurant?

c How long was he away from the restaurant?

```
Veggie-r-us supplier meeting 3,
journey log
Total distance to supplier, 9 miles.
1. Departed restaurant at 1 pm.
2. 6-minute drive down high street
   (2 miles).
3. 7 miles at 60mph (miles per hour)
   on motorway.
4. Reach supplier, stay there for
   10 minutes.
5. 3 miles on motorway at 60 mph.
6. Stop at motorway services (5 minutes).
7. 4 miles on motorway at 40mph
   (bad traffic).
8. 2 miles through town (4 minutes).
```

Life skills

Alan Plug
Call-out charge: £40
Hourly rate: £35
Tel: 3141 592654

Bill Tap
Call-out charge: £60
Hourly rate: £32
Tel: 1414 213562

Task 4 – Maintenance

They need to hire a plumber to fix some pipes in the restaurant.

There are two local plumbers who have been recommended by friends.

a On the same axes, draw graphs showing the cost (excluding parts) of hiring each plumber against the time taken.

b For what range of times is Bill cheaper than Alan?

Both quote that the job will take about 6.5 hours.

c Who should they get to do the job? Give your reason.

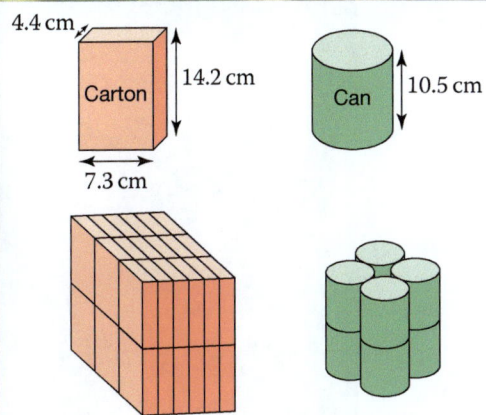

▲ Top: dimensions of one carton and one can. Bottom: stacked arrangement of cartons and an example of how cans are able to stack.

Annual amount spent eating out (£)	Number of people
0–400	36
Over 400–600	21
Over 600–800	38
Over 800–1000	35
Over 1000–1200	26
Over 1200–1400	18
Over 1400–1700	17
Over 1700–2000	9
Total	200

▲ Annual spend on eating out in the area.

Task 5 – Stock keeping

The friends need to decide between buying cartons or cans of tomatoes. The cartons are stacked in 2 layers in a box. Each layer is 6 cartons deep and 3 cartons across.

a Find the volume of a box.
Add an extra 5% to the volume for packaging.

The boxes are stored in a cuboid cupboard. Its dimensions are:

70 cm deep, 115 cm wide and 180 cm high.

b Calculate the volume of the cupboard in cm^3 and m^3. Explain how these figures are related.

c What percentage of the space in the cupboard do 30 boxes take up?

d Find the surface area of one carton.

The cans are cylindrical and have the same volume as a carton.

e What is the radius of the can?

f The cans can be stacked as shown. Work out how many cans could fit in the same space taken up by one cardboard box.

g If the cans and cartons cost the same which would you buy? Give your reasons.

Task 6 – Repeat business

The friends decide to investigate the results of a larger survey into peoples' spending on eating out.

a Calculate an estimate of the mean for the annual amount spent eating out.

b Draw a histogram showing the frequency density.

c Draw a cumulative frequency graph showing the data.

d Find the median and interquartile range of the annual amount spent eating out.

e Draw a box plot showing the data.

f What conclusions can you make from the different representations of the data? Do they support the mean?

Juliet assumes that a given person will spend the same amount at each of the four local restaurants.

g If Juliet is right, how much should the friends expect the average person to spend at their restaurant in a year?

16 Handling data 2

Introduction
In many countries, literacy is almost taken for granted. By the time people reach adulthood, they can generally read and write. In many other countries, this is not the case. Statistics show that there is a correlation, or link, between the wealth per person of a country (as measured in 'GDP per capita') and the adult literacy rate. In 2013, the UK GDP and literacy rate were $36 000 and 99% respectively; by comparison, in Sierra Leone they were $2000 and 35%.

The branch of statistics that deals with the relationship between variables is called correlation.

What's the point?
Understanding the relationship between quantities helps us to make informed decisions on a global scale. Literacy problems will not be resolved effectively unless poverty is also tackled.

Objectives
By the end of this chapter, you will have learned how to …
- Calculate summary statistics from a grouped frequency table.
- Construct and interpret cumulative frequency curves and box plots.
- Plot scatter graphs and recognise correlation.
- Use tables and line graphs to represent time series data.

Check in

1. Work out:
   a. $\frac{1}{2}$ of 124
   b. 50% of 120
   c. $\frac{1}{2}$ of (36 + 1)
   d. 25% of 60
   e. $\frac{1}{4}$ of (27 + 1)
   f. 25% of 120
   g. $\frac{3}{4}$ of 144
   h. 75% of 200

2. This graph shows the cost per day to hire a power tool.

   a. How much does it cost to hire the power tool for
      i. 3 days
      ii. 5 days?

   b. Gautam has £40.
      What is the maximum number of days he can hire the power tool?

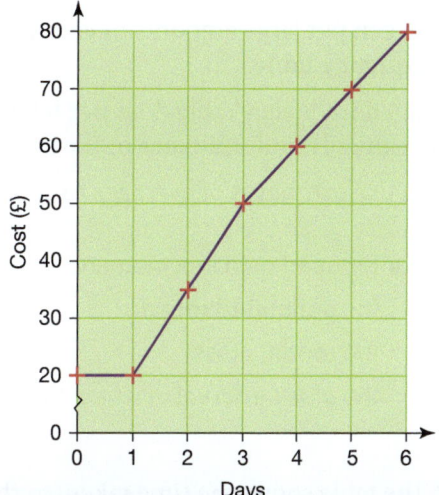

Chapter investigation

Imogen and Nilaksh are trying to work out if there is a correlation between the size of an animal and how long it lives.

Nilaksh says, 'Larger animals live longer than smaller ones. This must be true because elephants live much longer than mice.'

Imogen replies, 'Well, I read that the oldest animal ever found was a small clam that lived for over 400 years. Imagine living 400 years as a clam!'

Is there any correlation between animal size and longevity.

SKILLS

16.1 Averages and spread 2

- You can put large amounts of continuous data into a **grouped frequency table**.

A grouped frequency table does not tell you the actual data values so you can only find estimates of the averages.

- You use **estimates** of averages to summarise the data.

> - For grouped data in a frequency table, you can calculate
>   - the **estimated mean**
>   - the **modal class**
>   - the **class interval** in which the median lies.

The modal class is the class with the greatest frequency.

EXAMPLE

The table shows the time taken, to the nearest minute, by a group of students to solve a crossword puzzle.

Time, t (minutes)	Frequency
$5 < t \leq 10$	2
$10 < t \leq 15$	14
$15 < t \leq 20$	13
$20 < t \leq 25$	6
$25 < t \leq 30$	1

For these data, work out
a the **modal class**
b the class containing the **median**
c an estimate for the **mean**

a Modal class is $10 < t \leq 15$

b Class containing median is $15 < t \leq 20$

c

Time, t (minutes)	Frequency	Midpoint	Midpoint × frequency
$5 < t \leq 10$	2	7.5	7.5 × 2 = 15
$10 < t \leq 15$	14	12.5	12.5 × 14 = 175
$15 < t \leq 20$	13	17.5	17.5 × 13 = 227.5
$20 < t \leq 25$	6	22.5	22.5 × 6 = 135
$25 < t \leq 30$	1	27.5	27.5 × 1 = 27.5
Total	**36**		**580**

Total number of students: 36
Total time: 580

Add two extra columns in the table.

Find the midpoint of each interval.

Find the totals of the Frequency and Midpoint × frequency columns.

Estimated mean = $\dfrac{\text{Estimated total time}}{\text{Total number of students}}$

Estimated mean = $\dfrac{580}{36}$ = 16.1 (1 dp)

Statistics   Handling data 2

Fluency

Exercise 16.1S

1 The masses, to the nearest kilogram, of 25 men are shown.

69	82	75	66	72
73	79	70	74	68
84	63	69	88	81
73	86	71	74	67
80	86	68	71	75

You can use a scientific calculator to work out the mean of grouped data. You should find out how to do this on your calculator.

a Copy and complete the frequency table.

Mass (kg)	Tally	Number of men
60 to 64		
65 to 69		
70 to 74		
75 to 79		
80 to 84		
85 to 89		

b State the modal class.

c Find the class interval in which the median lies.

2 The speeds of 10 cars in a 20 mph zone are shown in the frequency table.

Speed (mph)	Mid-value	Number of cars	Mid-value × Number of cars
11 to 15		1	
16 to 20		6	
21 to 25		2	
26 to 30		1	

a Calculate the number of cars that are breaking the speed limit.

b Copy the frequency table and calculate the mid-values for each class interval.

c Complete the last column of your table and find an estimate of the mean speed.

3 The grouped frequency tables give information about the time taken to solve four different crosswords.

For each table, copy the table, add extra working columns and find

i the modal class

ii the class containing the median

iii an estimate of the mean.

a

Time, t (minutes)	Frequency
$5 < t \leq 10$	2
$10 < t \leq 15$	14
$15 < t \leq 20$	13
$20 < t \leq 25$	6
$25 < t \leq 30$	1

b

Time, t (minutes)	Frequency
$0 < t \leq 10$	3
$10 < t \leq 20$	6
$20 < t \leq 30$	4
$30 < t \leq 40$	5
$40 < t \leq 50$	2

c

Time, t (minutes)	Frequency
$5 < t \leq 10$	8
$10 < t \leq 15$	5
$15 < t \leq 20$	7
$20 < t \leq 25$	4
$25 < t \leq 30$	0
$30 < t \leq 35$	1

d

Time, t (minutes)	Frequency
$5 < t \leq 15$	3
$15 < t \leq 25$	9
$25 < t \leq 35$	7
$35 < t \leq 45$	8
$45 < t \leq 55$	2
$55 < t \leq 65$	1

4 The heights of 50 Year 10 students were measured. The results are shown in the table.

Height, h, cm	Number of students
$150 \leq h < 155$	3
$155 \leq h < 160$	5
$160 \leq h < 165$	15
$165 \leq h < 170$	25
$170 \leq h < 175$	2

a What is the modal group?

b Estimate the mean height.

c Which class interval contains the median?

1201, 1202, 1254, 1255　SEARCH

APPLICATIONS

16.1 Averages and spread 2

RECAP

- For large amounts of data presented as grouped data in a frequency table, you cannot calculate the exact mean, mode or median. Instead, you can calculate
  - the estimated mean
  - the modal class
  - the class interval in which the median lies.

The mean, mode and median are measures of average. The range is a measure of spread.

HOW TO

Compare grouped data
1. Compare a measure of average such as the modal class, median class or estimated mean.
2. Compare the ranges of the data sets.

EXAMPLE

The tables show the ages of people attending concerts to see the bands Badness and Cloudplay.

Badness	
Ages, a (years)	Frequency
$20 \leq a < 30$	1600
$30 \leq a < 40$	4300
$40 \leq a < 50$	2100
$50 \leq a < 60$	1000
$60 \leq a < 70$	300

Cloudplay	
Ages, a (years)	Frequency
$15 \leq a < 20$	2800
$20 \leq a < 30$	4600
$30 \leq a < 40$	3300

Compare the ages of the people attending the two concerts.

① Compare a measure of average.

The modal class of the ages is greater at Badness concerts then Cloudplay concerts.

The highest frequency for Badness is in the class interval 30–40 years old, whereas the highest frequency for Cloudplay is in the interval 20–30 years old.

② Compare the range in ages.

There is more variation in the ages of the people at Badness concerts than Cloudplay converts.

The largest possible range for Cloudplay is 39 − 15 = 24 years.
The smallest possible range for Badness is 69 − 20 = 49 years.

You could also compare the median class or the estimated mean.

Statistics   Handling data 2

326

Reasoning and problem solving

Exercise 16.1A

1. Jayne kept a daily record of the number of miles she travelled in her car during two months.

	December	January
Miles travelled, m	Frequency	Frequency
$0 < m \leq 20$	3	0
$20 < m \leq 40$	8	5
$40 < m \leq 60$	10	12
$60 < m \leq 80$	6	8
$80 < m \leq 100$	4	6

   a Estimate the mean number of miles for each month.

   b Find the modal class and median class for each month.

   c Compare the number of miles Jayne travelled in December and January.

2. Jafar carried out a survey to find the time taken by 120 teachers and 120 office workers to travel home from work.

   Teachers

Time taken, t (minutes)	Frequency
$0 < t \leq 10$	12
$10 < t \leq 20$	33
$20 < t \leq 30$	48
$30 < t \leq 40$	20
$40 < t \leq 50$	7

   Office workers

Time taken, t (minutes)	Frequency
$10 < t \leq 20$	2
$20 < t \leq 30$	21
$30 < t \leq 40$	51
$40 < t \leq 50$	28
$50 < t \leq 60$	18

   a Estimate the mean number of minutes for the teachers and office workers.

   b Find the modal class and median class for the teachers and office workers.

   c Make comparisons between the time taken by the teachers and office workers to travel home from work.

3. The masses of some apples are shown in the table.

Mass, w (g)	Frequency
$30 \leq w < 40$	6
$40 \leq w < 50$	28
$50 \leq w < 60$	21
$60 \leq w < 70$	25

   Granny Smith apples have a mean mass of 45 g and a range of 39 g.

   Is the data in the table about Granny Smith apples?

4. A company produces three million packets of crisps each day. It states on each packet that the bag contains 25 g of crisps. To test this, a sample of 1000 bags are weighed.

   The table shows the results.

Mass, w (g)	Frequency
$23.5 \leq w < 24.5$	20
$24.5 \leq w < 25.5$	733
$25.5 \leq w < 26.5$	194
$26.5 \leq w < 27.5$	53

   Is the company justified in stating that each bag contains 25 g of crisps?

   Show your working and justify your answer.

5. Two machines are each designed to produce paper 0.3 mm thick.

   The tables show the actual output of a sample from each machine.

	Machine A	Machine B
Thickness, t (mm)	Frequency	Frequency
$0.27 \leq t < 0.28$	2	1
$0.28 \leq t < 0.29$	7	50
$0.29 \leq t < 0.30$	32	42
$0.30 \leq t < 0.31$	50	5
$0.31 \leq t < 0.32$	9	2

   Compare the output of the two machines using suitable calculations.

   Which machine is producing paper closer to the required thickness?

SKILLS

16.2 Box plots and cumulative frequency graphs

EXAMPLE

The heights of 120 boys are given in the table.

Height, h (cm)	$145 \leqslant h < 150$	$150 \leqslant h < 155$	$155 \leqslant h < 160$	$160 \leqslant h < 165$	$165 \leqslant h < 170$
Frequency	8	27	48	31	6

a Use the table to draw a cumulative frequency diagram.

b Given that the tallest and shortest boys are 167 cm and 146 cm respectively, draw a box plot for the data.

c Estimate the number of boys under 152.5 cm (5 feet).

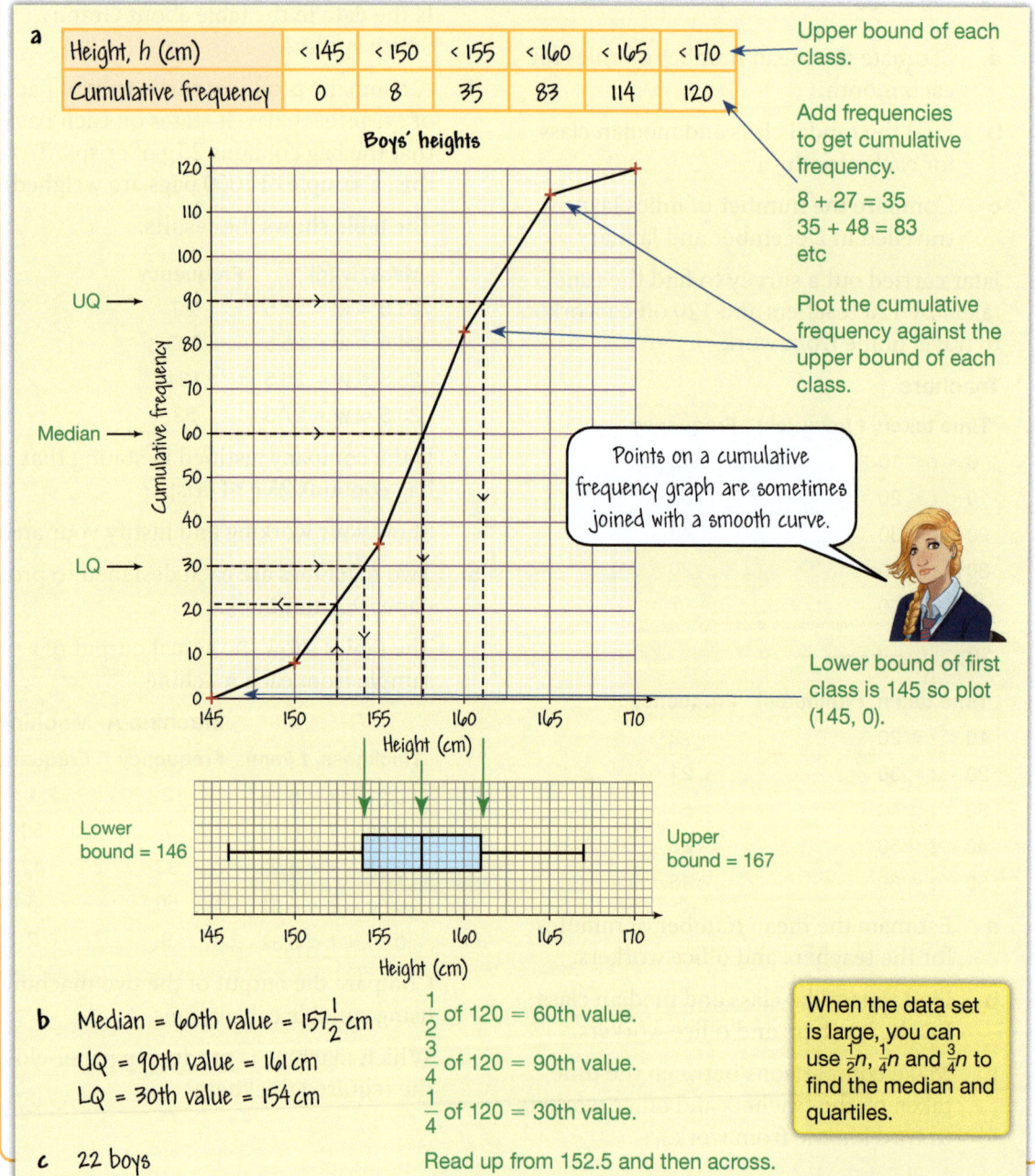

a

Height, h (cm)	< 145	< 150	< 155	< 160	< 165	< 170
Cumulative frequency	0	8	35	83	114	120

Upper bound of each class.

Add frequencies to get cumulative frequency.
8 + 27 = 35
35 + 48 = 83
etc

Plot the cumulative frequency against the upper bound of each class.

Points on a cumulative frequency graph are sometimes joined with a smooth curve.

Lower bound of first class is 145 so plot (145, 0).

b Median = 60th value = $157\frac{1}{2}$ cm
UQ = 90th value = 161 cm
LQ = 30th value = 154 cm

$\frac{1}{2}$ of 120 = 60th value.
$\frac{3}{4}$ of 120 = 90th value.
$\frac{1}{4}$ of 120 = 30th value.

When the data set is large, you can use $\frac{1}{2}n$, $\frac{1}{4}n$ and $\frac{3}{4}n$ to find the median and quartiles.

c 22 boys

Read up from 152.5 and then across.

Statistics   Handling data 2

Fluency

Exercise 16.2S

1 Mr Cheong summarised the test results for a group of students.

Lowest mark 41%   Lower quartile 54%
Median 60%   Highest mark 84%
Upper quartile 70%

Draw a box plot for these results.

2 In a science experiment, Davinder recorded reaction times to the nearest tenth of a second. He summarised the data.

Quickest time 3.2   Lower quartile 3.7
Median 4.4   Slowest time 9.6
Upper quartile 7.5

Draw a box plot to represent these results.

> For each of the data sets in questions 3–7
> **a** draw a cumulative frequency diagram
> **b** draw a box plot.
> For each part **c** use the cumulative frequency diagram to make each estimate.

3 The heights of 100 girls are shown in the grouped frequency table.

Height h (cm)	Frequency
$145 \leq h < 150$	7
$150 \leq h < 155$	25
$155 \leq h < 160$	46
$160 \leq h < 165$	17
$165 \leq h < 170$	5

b The shortest girl is 149 cm.
The tallest girl is 166 cm.

c Estimate the number of girls with height
i < 152 cm   **ii** > 163 cm.

4 The ages of teachers in a school are shown in the grouped frequency table.

Age, A (years)	Frequency
$20 \leq A < 30$	18
$30 \leq A < 40$	37
$40 \leq A < 50$	51
$50 \leq A < 60$	28
$60 \leq A < 70$	16

b The youngest teacher is 22.
The oldest teacher is 65.

c Estimate the number of teachers who are
i < 35 years old   **ii** > 55 years old.

5 The times taken to complete a crossword puzzle are shown in the grouped frequency table.

Time t (minutes)	Frequency
$0 \leq t < 10$	4
$10 \leq t < 20$	11
$20 \leq t < 30$	29
$30 \leq t < 40$	37
$40 \leq t < 50$	27
$50 \leq t < 60$	12

b The shortest time is 6 mins.
The longest time is 59 mins.

c Estimate the number of people who took
i < 25 minutes
ii > 45 minutes to complete the puzzle.

6 The masses of a sample of cats and kittens are shown in the grouped frequency table.

Mass, w (grams)	Frequency
$1500 \leq w < 2000$	9
$2000 \leq w < 2500$	22
$2500 \leq w < 3000$	37
$3000 \leq w < 3500$	20
$3500 \leq w < 4000$	12

b The lightest cat is 1600 g.
The heaviest cat is 3745 g.

c Estimate the number of cats and kittens that weighed
i < 2200 g   **ii** > 3600 g.

7 The heights of sunflowers growing in one field are shown in the grouped frequency table.

Height, h (cm)	Frequency
$40 \leq h < 60$	2
$60 \leq h < 80$	17
$80 \leq h < 100$	28
$100 \leq h < 120$	39
$120 \leq h < 140$	24
$140 \leq h < 160$	10

b The shortest plant is 45 cm.
The tallest plant is 155 cm.

c Estimate the number of sunflowers that were
i < 130 cm   **ii** > 90 cm.

1194, 1195, 1333

APPLICATIONS

16.2 Box plots and cumulative frequency graphs

RECAP
- You can represent grouped data on a cumulative frequency diagram.
- You use a box plot to show the range, median and interquartile range (IQR) of a set of data.

HOW TO

To compare two distributions
1. Compare a measure of average, such as the median.
2. Compare a measure of spread, such as the range or IQR.

EXAMPLE

These cumulative frequency graphs summarise the masses of a sample of 100 men and 100 women.

Make two comparisons between the masses of the men and women.

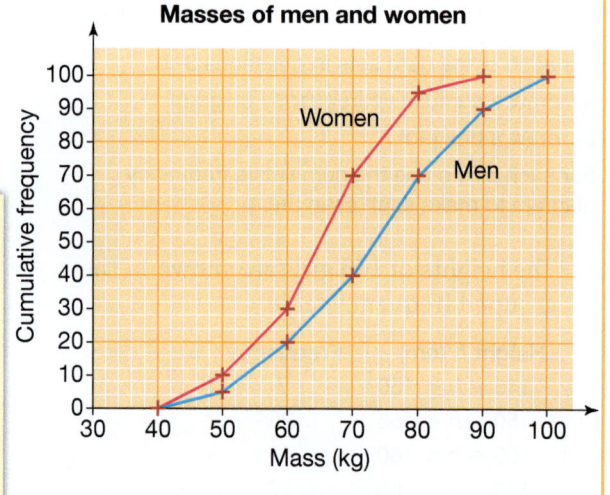

1. Median mass of women = 65 kg
   Median mass of men = 73 kg
   On average, the women are lighter than the men.
2. Range of women's masses = 90 − 40 = 50 kg
   Range of men's masses = 100 − 40 = 60 kg
   The men's masses vary more than the women's masses.

EXAMPLE

These box plots summarise the heights of samples of 13- and 14-year-old boys and girls.

Compare like measures such as the medians, IQR, etc.

Write two comparisons between the heights of the boys and the girls.

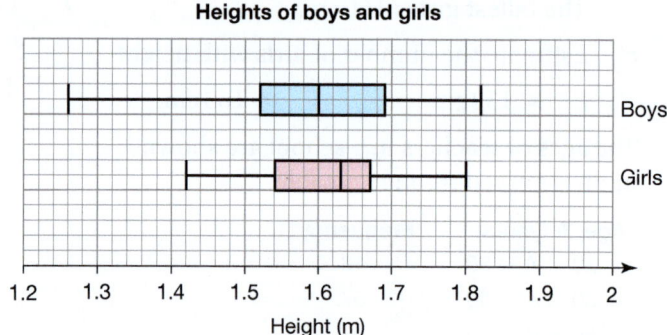

1. **Median** height of girls = 1.63 m   Median height of boys = 1.60 m
   On average, the girls are taller than the boys.
2. **IQR** for girls = 1.67 − 1.54 = 0.13 m   IQR for boys = 1.69 − 1.52 = 0.17 m

   The IQR for the boys is greater than for the girls so the middle 50% of the boys have more variation in their heights.

Statistics   Handling data 2

Reasoning and problem solving

Exercise 16.2A

1. Write three comparisons between the test results of a group of girls and a group of boys.

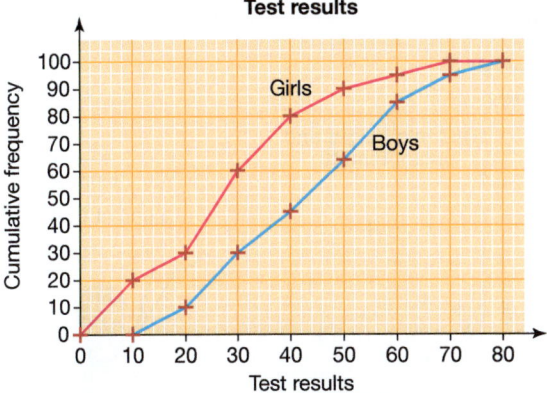

2. Write three comparisons between the heights of samples of sunflowers grown by two farmers.

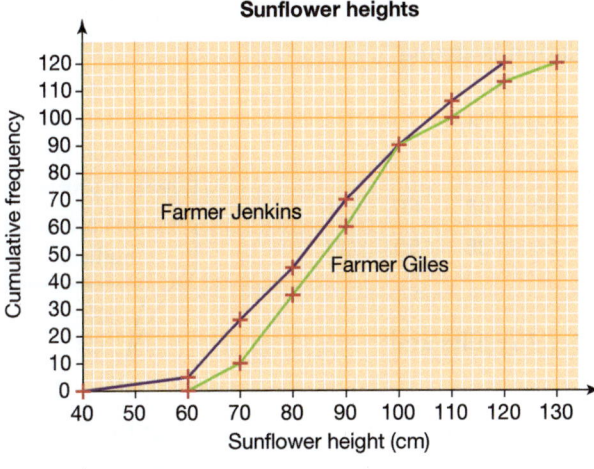

3. Write three comparisons between the mobile phone bills paid by samples of boys and girls.

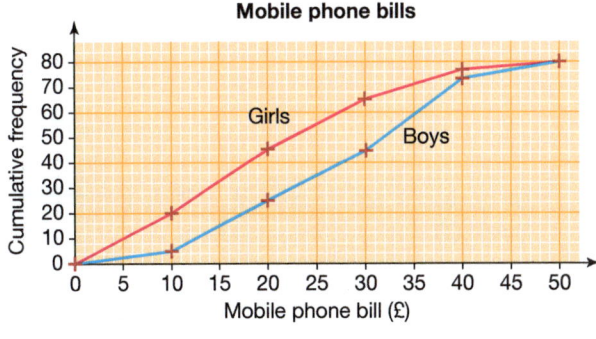

For questions **4** to **7**, write two comparisons between each pair of box plots.

4. The box plots summarise the waiting times, to the nearest minute, of a group of patients at the doctor and the dentist.

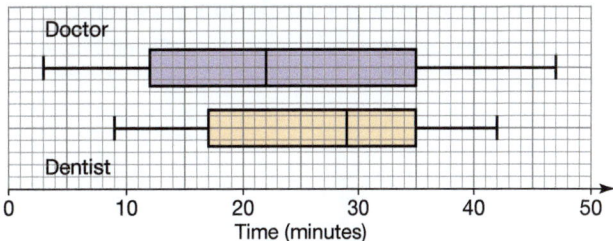

5. The box plots summarise the reaction times, of a group of boys and girls.

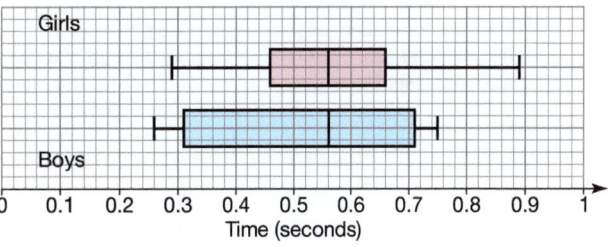

6. The box plots summarise the French and English test results of a group of students.

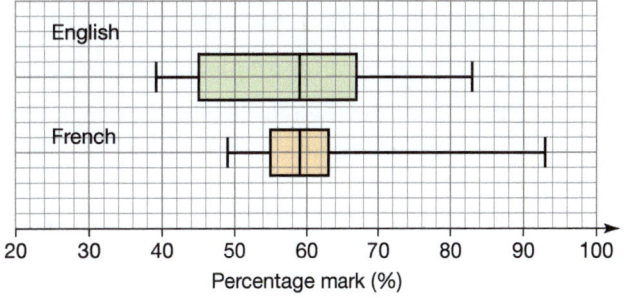

7. The box plots summarise the average length of a phone call, to the nearest minute, made by two groups of girls aged 13 and 17.

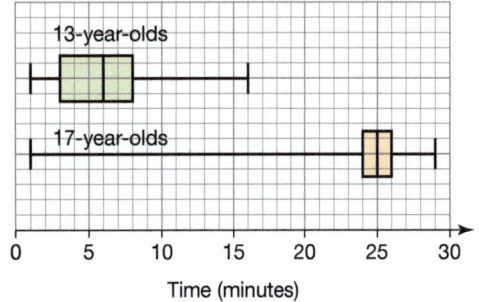

SKILLS

16.3 Scatter graphs and correlation

You can use a **scatter graph** to compare two sets of data, for example, height and mass.

- The data is collected in pairs and plotted as coordinates.
- If the points lie roughly in a straight line, there is a **linear relationship** or **correlation** between the two **variables**.

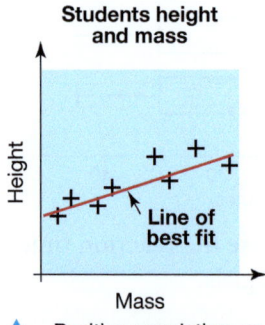

▲ Positive correlation: as height increases, mass also increases.

▲ Negative correlation: as the age of a car increases, the price decreases.

▲ No correlation: there is no relationship between height and exam mark.

> You cannot draw a line of best fit for no correlation.

If the points lie close to the line of best fit, the correlation is strong.

EXAMPLE

The table shows the exam results (%) for Paper 1 and Paper 2 for 10 students.

Paper 1	56	72	50	24	44	80	68	48	60	36
Paper 2	44	64	40	20	36	64	56	36	50	24

a Draw a scatter graph and line of best fit.

b Describe the relationship between the Paper 1 results and Paper 2 results.

> Points that are an exception to the general pattern of the data are called **outliers**.

a Plot the exam marks as coordinates. The line of best fit should be close to all the points, with approximately the same number of crosses on either side of the line.

b There is a positive correlation. The higher the mark on paper 1, the higher the mark on paper 2.

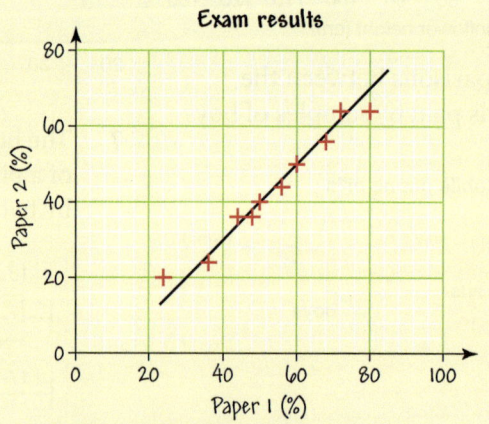

> Performing well on Paper 1 does not cause you to perform well on Paper 2. The tests could be RE and sport.

- If two variables are correlated then it does not always mean that one **causes** the other.

The events could both be a result of a common cause, or there could be no connection at all between the events.

Statistics   Handling data 2

Fluency

Exercise 16.3S

1 Match each scatter graph to the correct type of correlation.

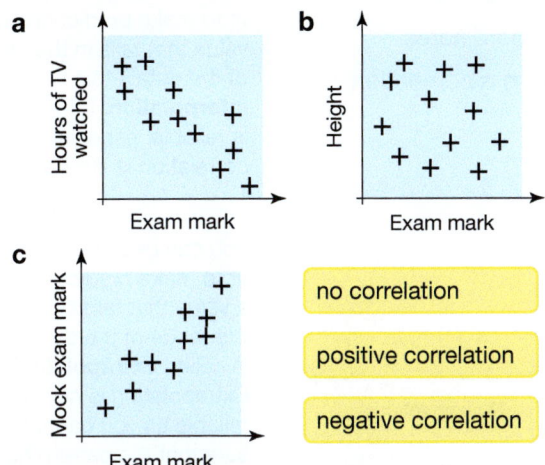

no correlation

positive correlation

negative correlation

2 Describe the points A, B, C, D and E on each scatter graph.

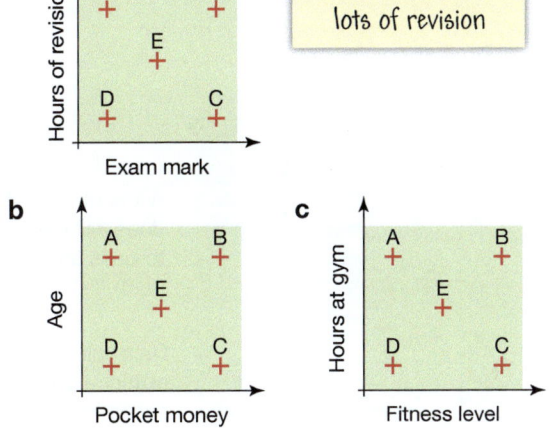

A low exam mark, lots of revision

3 Match each scatter graph to the correct type of correlation.

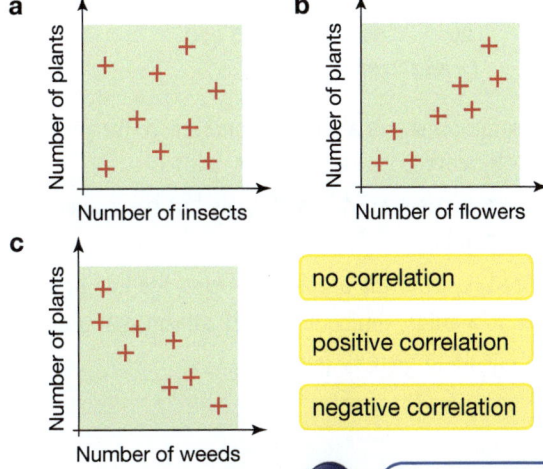

no correlation

positive correlation

negative correlation

4 The table shows the amount of water used to water plants and the daily maximum temperature.

Water (litres)	25	26	31	24	45	40	5	13	18	28
Maximum temperature (°C)	24	21	25	19	30	28	15	18	20	27

a Copy and complete the scatter graph for this information.

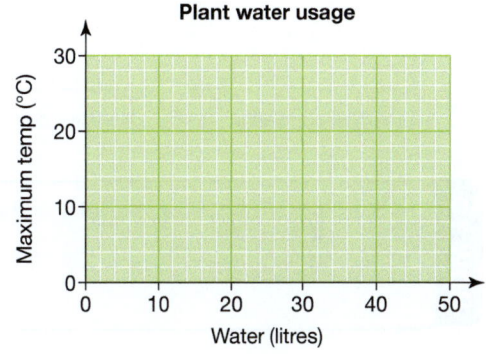

b State the type of correlation shown in the scatter graph.

c Copy and complete these sentences:

i As the temperature increases, the amount of water used _____.

ii As the temperature decreases, the amount of water used _____.

5 The times taken, in minutes, to run a mile and the shoe sizes of ten athletes are shown in the table.

Shoe size	10	$7\frac{1}{2}$	5	9	6	$8\frac{1}{2}$	$7\frac{1}{2}$	$6\frac{1}{2}$	8	7
Time (mins)	9	8	8	7	5	13	15	12	5	6

a Draw a scatter graph to show this information.

Use 2 cm to represent 1 shoe size on the horizontal axis.

Use 2 cm to represent 5 minutes on the vertical axis.

b State the type of correlation shown in the scatter graph.

c Describe, in words, any relationship that the graph shows.

1213, 1250 SEARCH

APPLICATIONS

16.3 Scatter graphs and correlation

RECAP

- You can compare two sets of data on a scatter graph.
- The data is collected in pairs and plotted as coordinates.
- If the points lie roughly in a straight line, there is a correlation between the two variables.

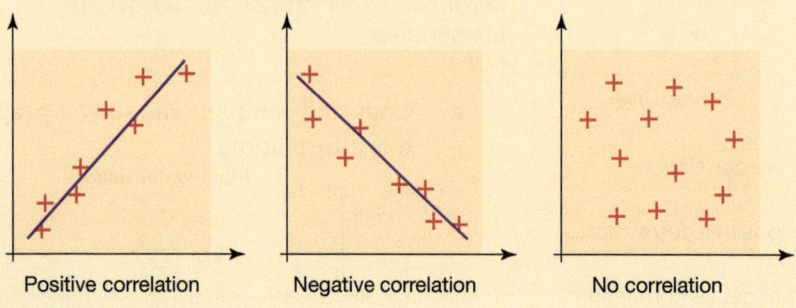

You can use the line of best fit to make predictions for a value that falls in the range of the data – this is called **interpolation**. Interpolation is reliable, especially if the correlation is strong.

You can use the line of best fit to make predictions for a value that falls outside the range of the data – this is called **extrapolation**. Extrapolation is not always reliable as you cannot be sure that the pattern holds for values outside of the data collected.

HOW TO

① Describe and interpret the relationship shown by the diagram.

② Make predictions based on the correlation shown.

EXAMPLE

The scatter graph shows the number of goals scored by 21 football teams in a season plotted against the number of points gained.

a Describe the relationship between the goals scored and the number of points.

b Describe the goals and points for team A.

c If a team scored 45 goals, how many points would you expect it to have?

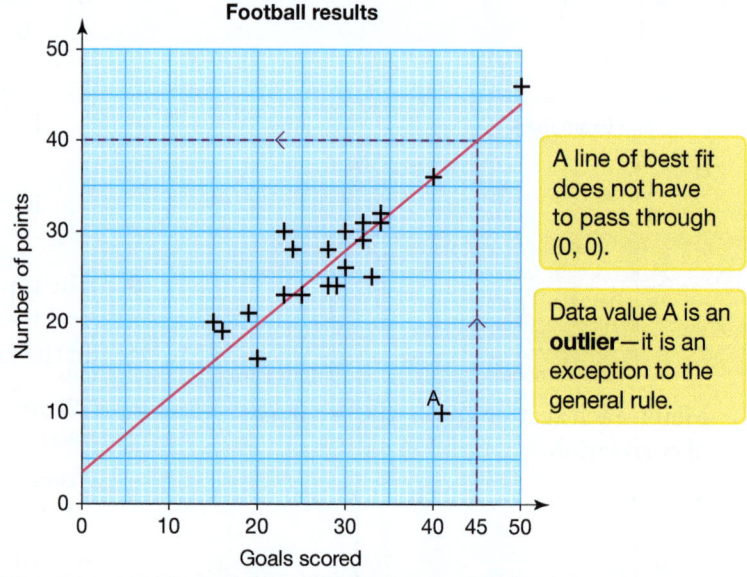

A line of best fit does not have to pass through (0, 0).

Data value A is an **outlier**—it is an exception to the general rule.

① Describe and interpret the relationship shown by the diagram.

② Make predictions based on the correlation shown.

a The graph shows a positive correlation: the more goals scored, the more points gained.

b Team A has scored lots of goals but has gained very few points.

c Reading from the graph, and using the line of best fit as a guide, you could expect a team that scored 45 goals to gain 40 points.

Statistics   Handling data 2

Reasoning and problem solving

Exercise 16.3A

1   The scatter graph shows the exam results (%) for Paper 1 and Paper 2 for 11 students.

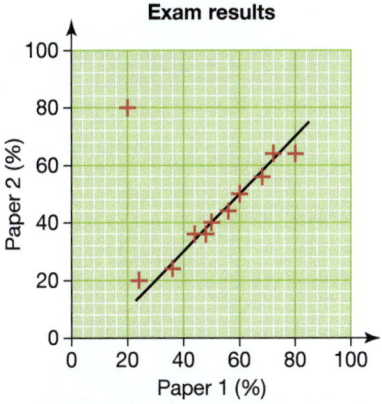

   a Describe the relationship between Paper 1 and Paper 2 results.

   b Identify an outlier and describe how this student performed in the papers.

   c If a student scored 40 in Paper 1, what score would you expect them to gain in Paper 2?

   d Jenna extends the line of best fit on the graph and tries to predict the score that a student who scored 100 on Paper 2 will score on Paper 1.
   Will her estimate make sense?

2   The graph shows the marks in two papers achieved by ten students.

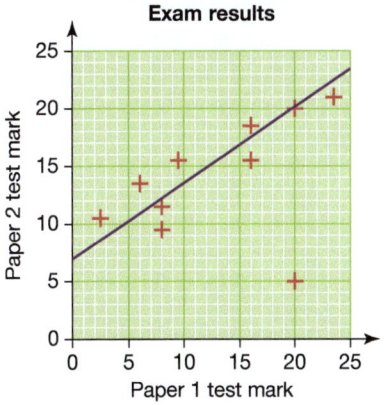

   Use the line of best fit to estimate

   a the Paper 2 mark for a student who scored 13 in Paper 1

   b the Paper 1 mark for a student who scored 23 in Paper 2.

   c Describe the outlier in the scatter graph. How well did this student perform on the two papers?

3   The table shows the age and diameter, in centimetres, of trees in a forest.

Age (years)	10	27	6	22	15	25	11	16	21	19
Diameter (cm)	20	78	9	65	38	74	25	44	59	50

   a Draw a scatter graph for the data.

   b State the type of correlation between the age and diameter of the trees.

   c Draw a line of best fit.

   d If the diameter of a tree is 55 cm, estimate the age of the tree.

   e Use your graph to estimate the diameter of a tree that is one year old.

   f Explain why the estimate in part **d** is more reliable than the estimate in part **e**.

4   The table shows the number of hot water bottles sold per month in a chemist's shop and the average temperature for each month.

Month	Average monthly temperature °C	Sales of hot water bottles
Jan	2	32
Feb	4	28
Mar	7	10
April	10	4
May	14	6
June	19	0
July	21	2
Aug	20	3
Sept	18	7
Oct	15	15
Nov	11	22
Dec	5	29

   a Draw a scatter graph for the data.

   b Describe the correlation shown and the relationship between the results.

   c Draw in a line of best fit.

   d Predict the average temperature in a month 20 hot water bottles were sold.

   e The weather forecast predicts an average January temperature of −5°C. Could you use your graph to find how many hot water bottles would be sold?

1213, 1250

SKILLS

16.4 Time series

You can use a **line graph** to show how data changes as time passes.

The data can be discrete or continuous.

The temperature of a liquid is measured every minute.

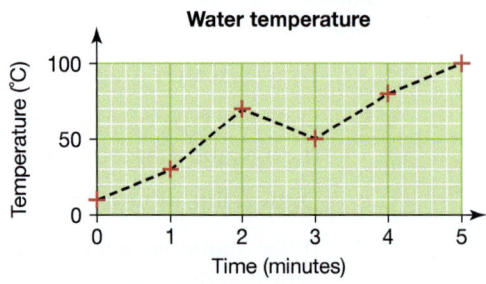

This is an example of a **time series graph**.

- A time series graph shows
  - how the data changes over time, or the **trend**
  - each individual value of the data.

EXAMPLE

The table shows the average monthly rainfall, in centimetres, in Vienna over the last 30 years.

Month	Jan	Feb	Mar	Apr	May	Jun	Jul	Aug	Sep	Oct	Nov	Dec
Rainfall (cm)	8.7	6.3	6.8	6.3	5.6	6.7	5.1	6.4	6.4	7.4	7.8	9.2

Draw a time series graph to show this information.

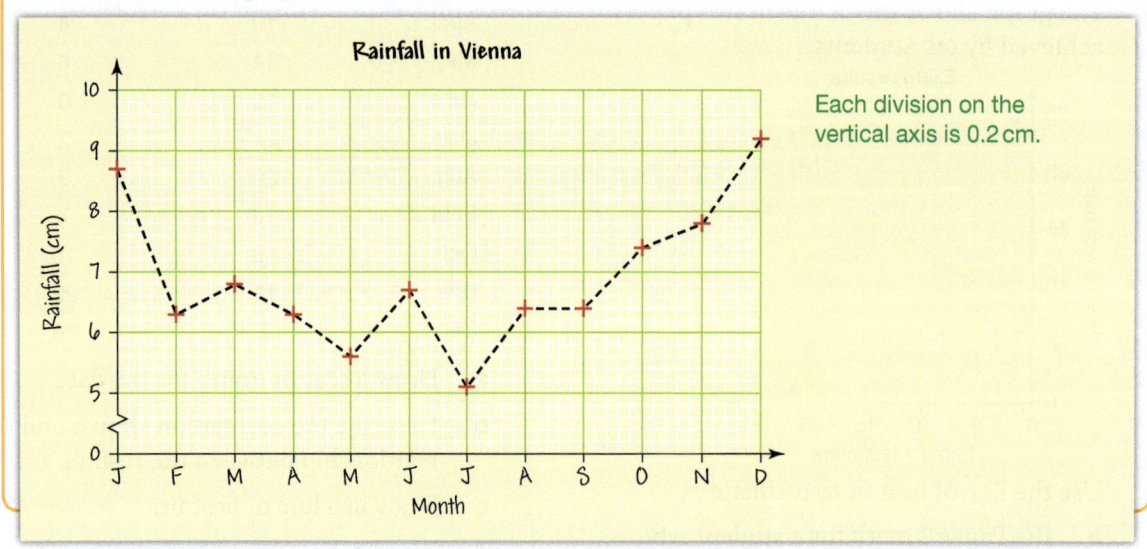

Each division on the vertical axis is 0.2 cm.

Statistics Handling data 2

Fluency

Exercise 16.4S

1 The number of photographs taken each day during a 7-day holiday is given.

Sunday	Monday	Tuesday	Wednesday	Thursday	Friday	Saturday
8	12	11	16	19	2	13

Copy and complete the line graph to show this information.

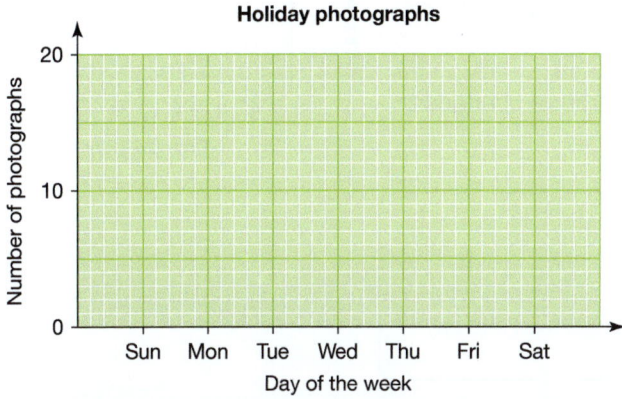

2 The numbers of DVDs rented from a shop during a week are shown.

Sunday	Monday	Tuesday	Wednesday	Thursday	Friday	Saturday
18	9	7	11	15	35	36

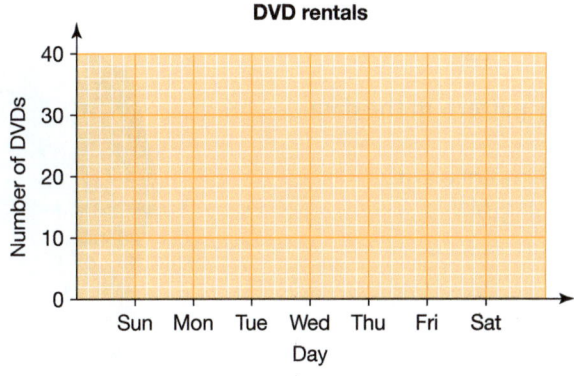

Copy and complete the line graph, choosing a suitable vertical scale.

3 Every year on his birthday, Paco's mass in kilograms is measured.

Age	2	3	4	5	6	7	8	9	10	11	12	13
Mass (kg)	14	16	18	20	22	25	28	31	34	38.5	40	45

Draw a line graph to show the masses.

4 The hours of sunshine each month are shown in the table.

Jan	Feb	Mar	Apr	May	Jun	Jul	Aug	Sep	Oct	Nov	Dec
43	57	105	131	185	176	194	183	131	87	53	35

Draw a line graph to show the hours of sunshine.

5 The daily viewing figures, in millions, for a reality TV show are shown.

Day	Sat	Sun	Mon	Tue	Wed	Thu	Fri
Viewers (in millions)	3.2	3.8	4.3	4.5	3.1	5.2	7.1

Draw a line graph to show the viewing figures.

1198, 1939

APPLICATIONS

16.4 Time series

RECAP

- You can use a line graph to show how data changes over time.
- This is sometimes called a time series graph. It shows:
  - how the data changes over time, or the trend
  - each individual value of the data.
- This graph shows the how the temperature changes over time.

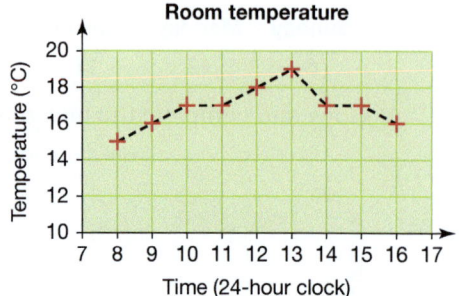

HOW TO

1. Draw axes on graph paper and label with suitable scales.
2. Plot the data from the table as coordinate pairs. Join the plotted points with straight lines.
3. Discuss any short term trends, seasonal variation and any longer term trends.

EXAMPLE

Ceyda's quarterly gas bills over a period of two years are shown in the table.

	Jan–March	April–June	July–Sept	Oct–Dec
2003	£65	£38	£24	£60
2004	£68	£42	£30	£68

Plot the data on a graph and comment on any pattern in the data.

1. Draw axes on graph paper with time on the horizontal axis. Plot time on the horizontal axis, J–M means Jan–March.
2. Plot the coordinates as crosses on the grid. Join them up with straight lines.

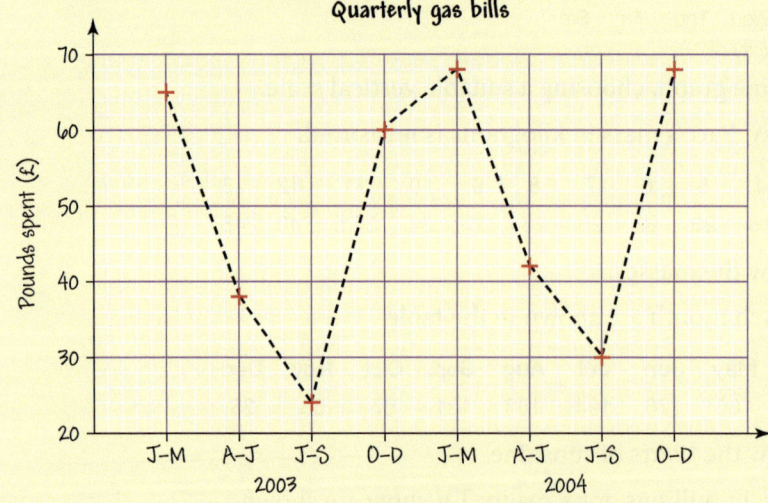

The graph shows seasonal variation.

3. Gas bills are highest in the winter months and lowest in the summer months. This annual pattern appears to repeat itself.

There is a slight trend for the bills to rise from year to year.

Statistics   Handling data 2

Reasoning and problem solving

Exercise 16.4A

For each of questions **1–6**

    a Plot the data on a graph

    b Comment on any patterns in the data.

1 The table shows Abdul's monthly mobile phone bills.

Jan	Feb	Mar	April	May	June	July	Aug	Sept	Oct	Nov	Dec
$16	$12	$15	$18	$16	$18	$12	$10	$12	$15	$16	$20

2 The table shows Egle's quarterly electricity bills over a two-year period.

	Jan–March	April–June	July–Sept	Oct–Dec
2004	€45	€20	€15	€48
2005	€54	€24	€18	€50

3 The table shows monthly ice-cream sales at Angelo's shop during one year.

Jan	Feb	Mar	April	May	June	July	Aug	Sept	Oct	Nov	Dec
£16	£12	£15	£18	£38	£48	£52	£58	£18	£15	£16	£40

4 A town carried out a survey over a number of years to find the percentage of local teenagers who used the town's library. The table shows the results.

year	1998	1999	2000	2001	2002	2003	2004	2005
%	14	18	24	28	25	20	18	22

5 Christabel kept a record of how much money she had earned from babysitting during three years.

	Jan–April	May–August	Sept–Dec
2001	£12	£18	£30
2002	£21	£33	£60
2003	£39	£42	£72

6 Steve kept a record of his quarterly expenses over a period of two years.

	Jan–March	April–June	July–Sept	Oct–Dec
2003	$35	$56	$27	$12
2004	$39	$68	$29	$18

Summary

Checkout
You should now be able to...

	Test it Questions
✓ Calculate summary statistics from a grouped frequency table.	1
✓ Construct and interpret cumulative frequency curves and box plots.	2
✓ Plot scatter graphs and recognise correlation.	3
✓ Use tables and line graphs to represent time series data.	4

Language	Meaning	Example
Box plot	A drawing which displays the median, quartiles and greatest and least value of a set of data.	See page 328.
Cumulative frequency	The total of all the frequencies of a set of data up to a particular value of data.	See page 328.
Scatter graph	A graph that displays bivariate data – data points which involve two variables.	See page 334.
Line graph	A graph that uses points connected by lines to show how something changes in value (as time goes by, or as something else happens).	
Line of best fit	A straight line through the points on a scatter graph which shows the trend of the data.	
Correlation	A measure of how strongly two variables appear to be related.	
Time series graph	A graph showing a sequence of values at different times.	
Trend (time series)	The long term behaviour of the data, 'averaging out' short term fluctuations.	

Statistics   Handling data 2

Review

1 The table gives data about the annual salaries in a company.

Salary, s ($1000s)	Frequency
$10 < s \leq 20$	5
$20 < s \leq 30$	25
$30 < s \leq 40$	26
$40 < s \leq 50$	17
$50 < s \leq 60$	5
$60 < s \leq 70$	2

  a What is the modal class?
  b In which class does the median lie?
  c Estimate the mean salary.

2 Use the data given in question **1** to answer the following questions.
  a Draw a cumulative frequency graph for this data.
  b Use your graph to estimate the
   i lower quartile **ii** upper quartile
   iii median **iv** IQR
   v percentage of people who earn less than $45,000 a year.
  c Draw a box plot for this data.

3 The table shows the engine size and average CO_2 emissions of a number of cars.

Engine Capacity (cc)	CO_2 g/km
2200	175
1950	160
1900	150
1400	125
1500	145
1350	130
1200	125
900	100

  a Draw a scatter diagram to display this data.
  b Describe the correlation between engine capacity and CO_2 emissions.
  c Draw a line of best fit on your diagram.
  d Estimate the emissions for a car with an engine size of 2000 cc.
  e Can you use this graph to estimate the engine size of a car which emits 200 g/km?

4 a Draw a line graph for this time series data.

Time	Temperature (°C)
06:00	15
09:00	19
12:00	20
15:00	24
18:00	25
21:00	20

  b Comment on any pattern(s) present in the data.

What next?

Score		
0 – 1		Your knowledge of this topic is still developing. To improve, look at MyiMaths: 1194, 1195, 1198, 1201, 1202, 1213, 1250, 1254, 1255, 1333
2 – 3		You are gaining a secure knowledge of this topic. To improve your fluency look at InvisiPens: 16Sa – k
4		You have mastered these skills. Well done you are ready to progress! To develop your problem solving skills look at InvisiPens: 16Aa – f

Assessment 16

1 The table shows the heights of boys and girls in a swimming club.

For each of these groups

a boys

b girls

c boys and girls combined

find

Height, h (cm)	Boys	Girls
$150 \leq h < 155$	3	9
$155 \leq h < 160$	7	11
$160 \leq h < 165$	14	12
$165 \leq h < 170$	8	6
$170 \leq h < 180$	4	1

    i the modal class [6]

    ii the class interval containing the median height [6]

    iii an estimate of the mean height. [12]

2 The histogram shows the number of hours spent revising for an exam by a group of 48 students.

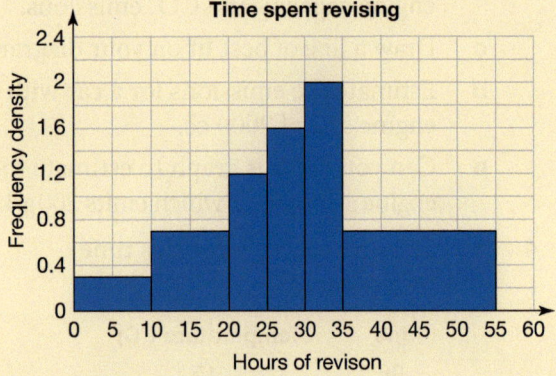

  a What is the modal class? [2]

  b Which class interval contains the median? [2]

  c Estimate the mean time spent revising. [4]

3 The following table gives the waiting time for a fairground ride.

Waiting time, t (mins)	$0 < t \leq 5$	$5 < t \leq 10$	$10 < t \leq 15$	$15 < t \leq 20$	$20 < t \leq 25$	$25 < t \leq 30$
Frequency	3	5	12	16	7	2

  a Draw the cumulative frequency diagram for this information. [6]

  b Use your diagram to answer the following.

    i How many people waited less than 10 minutes or less? [1]

    ii How many people waited more than 15 minutes? [1]

  c Use your cumulative frequency diagram to estimate the values of

    i the median [1]

  and **ii** the interquartile range. [2]

  d Given that the shortest wait was 0 mins and the longest wait was 27 mins, draw a box plot for the data. [4]

Statistics   Handling data 2

4 The table shows the results of a group of Key stage 2 children's English and Science tests.

Science	22	25	31	33	37	45	48	54	56	61	68	70	75	80	86	89	90	93	96	98
English	44	93	47	58	12	48	76	55	82	60	75	38	59	52	71	66	80	15	72	68

   a   Plot a scatter diagram to show this data. [4]
   b   Draw the line of best fit. [1]
   c   Comment on the type of correlation. [1]
   d   Comment on the two outliers. [1]
   e   Use your line to estimate
      i   a Science Mark corresponding to an English mark of 52. [1]
      ii an English Mark corresponding to a Science mark of 52. [1]

5 The number of patients in the waiting room of a health centre was recorded at hourly intervals and the results recorded on this time graph.

   a   How many patients were in the waiting room at 10:00? [1]
   b   How many patients were in the waiting room at 12:00? [1]
   c   How many patients were in the waiting room at 14:30? [1]
   d   Why is your answer to **c** only an approximation? [1]
   e   What is the maximum number of patients recorded in the waiting room? [1]
   f   Give a possible reason for the three peaks in the time series. [1]

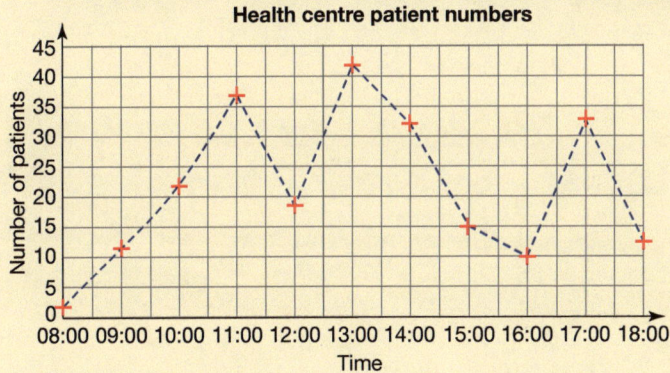

6 An ice cream maker has issued its sales figures for the past two years.
(Figures are to the nearest $1000)

	Jan	Feb	Mar	Apr	May	Jun	Jul	Aug	Sep	Oct	Nov	Dec
Year 1	100	95	125	150	176	203	251	266	204	131	101	88
Year 2	70	70	105	135	165	176	215	189	217	145	133	110

   a   Draw a time series diagram of these figures with Year 2 superimposed on top of Year 1. [6]
   b   Compare the summer and winter sales for these periods. [2]

17 Calculations 2

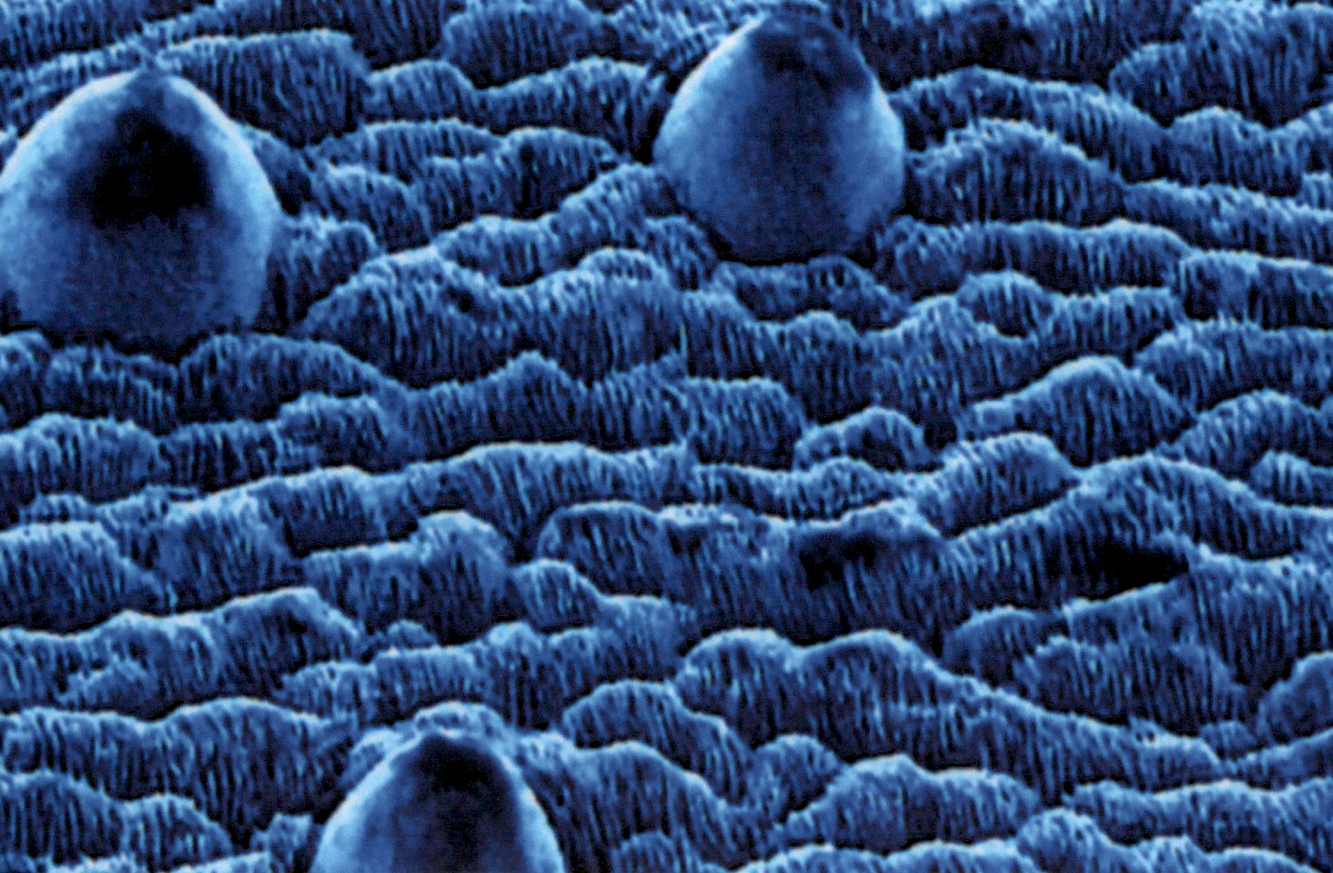

Introduction
An electron microscope is much more powerful than a normal microscope, and is used to look at very small objects. Biologists use electron microscopes to look at micro-organisms such as viruses. Materials scientists might use one to analyse crystalline structures. Nowadays there are electron microscopes that can detect objects that are smaller than a nanometre (0.000000001 m), so they can 'see' molecules and even individual atoms.

What's the point?
Mathematics needs to be able to describe very small quantities, such as lengths measured in nanometres. Without this ability, researchers could not analyse microscopic organisms like viruses.

Objectives
By the end of this chapter, you will have learned how to …
- Perform calculations involving roots and indices, including negative and fractional indices.
- Perform exact calculations involving fractions, surds and π.
- Work with numbers in standard form.

Check in

1 Simplify each expression.

   a $4\pi + 3\pi$ b $\pi(3^2 + 4)$ c $4\pi(5^2 - 4^2)$

   d $\sqrt{(5^2 - 3^2)}$ e $\sqrt{2} + 2\sqrt{2}$ f $5\sqrt{3} - \sqrt{3}(4^2 - 3 \times 4)$

2 a List all of the factors of these numbers.

   i 6   ii 12   iii 28   iv 36

   b List all of the prime numbers between 1 and 50.

3 Write these multiplications in power form. For example, $4 \times 4 \times 4 = 4^3$.

   a 3×3 b $4 \times 4 \times 4 \times 4 \times 4$ c $6 \times 6 \times 6$ d $5 \times 5 \times 5 \times 5$

4 Evaluate these expressions.

   a 2^4 b 3^3 c 4^2 d 5^3 e 2^7 f 7^1

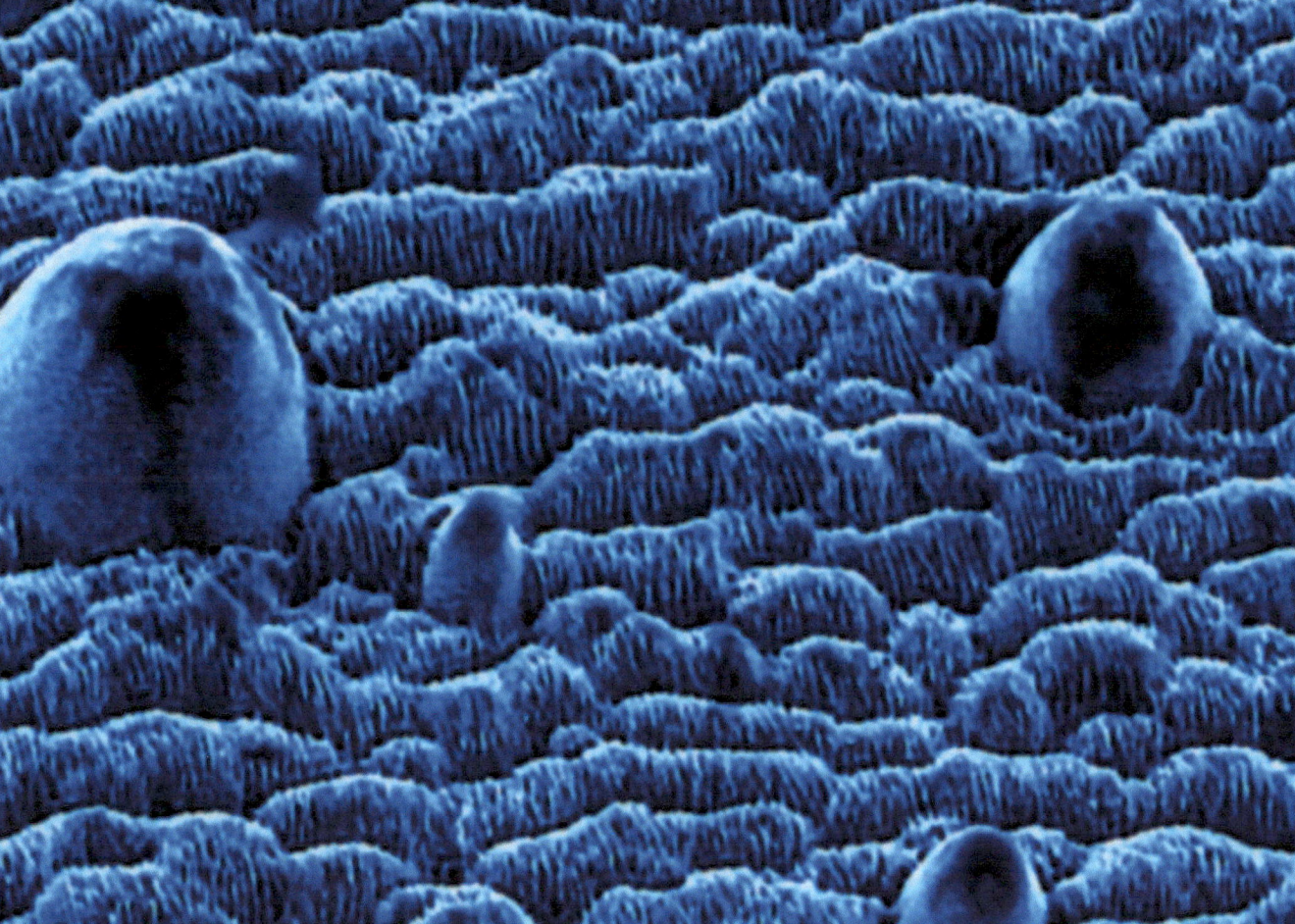

Chapter investigation

A picometre is a unit of length that is used to describe atomic distances. It is 0.000 000 000 001 m in length.

See if you can find out about the smallest metric unit of length that has a name, and if it has any possible real-world uses. What about the largest unit of length?

SKILLS

17.1 Calculating with roots and indices

Indices can be fractions as well as whole numbers.

They obey the same index laws as positive integer powers.

- $x^{\frac{1}{2}} = \sqrt{x}$ for any value of x.
  In general, $x^{\frac{1}{n}} = \sqrt[n]{x}$ (the nth root of x).
- $x^{-1} = \dfrac{1}{x}$, for any value of $x \cdot 0$.
  This is the **reciprocal** of x.
- In general, $x^{-n} = \dfrac{1}{x^n}$
  This is the reciprocal of x^n.

$5^{\frac{1}{2}} \times 5^{\frac{1}{2}} = 5^{\frac{1}{2}+\frac{1}{2}} = 5^1 = 5.$
but $\sqrt{5} \times \sqrt{5} = 5$,
so $5^{\frac{1}{2}}$ means $\sqrt{5}$.

$\dfrac{1}{5} = 1 \div 5 = 5^0 \div 5^1 = 5^{0-1} = 5^{-1}$,
so 5^{-1} means $\dfrac{1}{5}$.

$\dfrac{1}{5^2} = 1 \div 5^2 = 5^0 \div 5^2 = 5^{0-2} = 5^{-2}$,
so 5^{-2} means $\dfrac{1}{5^2}$.

Index laws
$x^a \times x^b = x^{a+b}$
$x^a \div x^b = x^{a-b}$
$(x^a)^b = x^{a \times b}$

Zero has no reciprocal so 0^{-1} is not defined.

EXAMPLE

Evaluate these expressions.

a $16^{\frac{1}{2}}$ b 8^{-1} c 10^{-2}

a $16^{\frac{1}{2}} = \sqrt{16} = 4$ b $8^{-1} = \dfrac{1}{8}$ c $10^{-2} = \dfrac{1}{10^2} = \dfrac{1}{100} = 0.01$

The positive square root.

A **fractional** index means a root.

A **negative** index means a reciprocal.

You can give meaning to more complex fractional powers using the index laws.

- $x^{\frac{m}{n}} = (x^{\frac{1}{n}})^m = (\sqrt[n]{x})^m$ or $x^{\frac{m}{n}} = (x^m)^{\frac{1}{n}} = \sqrt[n]{x^m}$

EXAMPLE

Find the value of these expressions.

a $4^{-\frac{1}{2}}$ b $9^{\frac{3}{2}}$ c $8^{\frac{2}{3}}$ d $4^{-\frac{5}{2}}$

a $4^{-\frac{1}{2}} = \dfrac{1}{4^{\frac{1}{2}}} = \dfrac{1}{\sqrt{4}} = \dfrac{1}{2}$

b $9^{\frac{3}{2}} = (9^{\frac{1}{2}})^3 = (\sqrt{9})^3 = (3)^3 = 27$

c $8^{\frac{2}{3}} = (8^{\frac{1}{3}})^2 = (\sqrt[3]{8})^2 = 2^2 = 4$

d $4^{-\frac{5}{2}} = \dfrac{1}{4^{\frac{5}{2}}} = \dfrac{1}{(4^{\frac{1}{2}})^5} = \dfrac{1}{(\sqrt{4})^5} = \dfrac{1}{(2)^5} = \dfrac{1}{32}$

It is usually easier to find roots first, then powers.

You can use the index laws with fractional and negative indices.

EXAMPLE

Simplify these expressions.

a $2^{\frac{3}{4}} \times 2^{\frac{1}{2}}$ b $2^{\frac{1}{2}} \div 2^{\frac{1}{6}}$ c $(2^{-\frac{1}{2}})^4$ d $(2^{-2} \times 2^{\frac{3}{2}})^2$

a $2^{\frac{3}{4}} \times 2^{\frac{1}{2}} = 2^{\frac{3}{4}+\frac{1}{2}}$
   $= 2^{\frac{5}{4}}$

b $2^{\frac{1}{2}} \div 2^{\frac{1}{6}} = 2^{\frac{1}{2}-\frac{1}{6}}$
   $= 2^{\frac{1}{3}}$

c $(2^{-\frac{1}{2}})^4 = 2^{-\frac{1}{2} \times 4}$
   $= 2^{-2}$

d $(2^{-2} \times 2^{\frac{3}{2}})^2 = (2^{-\frac{1}{2}})^2$
   $= 2^{-\frac{1}{2} \times 2} = 2^{-1}$

Number Calculations 2

Fluency

Exercise 17.1S

1 Evaluate these expressions.
   a $100^{\frac{1}{2}}$ b $16^{0.5}$ c $49^{\frac{1}{2}}$
   d $4^{0.5}$ e $121^{\frac{1}{2}}$ f $144^{0.5}$
   g $8^{\frac{1}{3}}$ h $27^{\frac{1}{3}}$ i $100^{0.5}$
   j $81^{\frac{1}{2}}$ k $9^{\frac{1}{2}}$ l $125^{\frac{1}{3}}$
   m $0^{\frac{1}{2}}$ n $1000^{\frac{1}{3}}$ o $64^{\frac{1}{3}}$

2 Write these numbers in index form.
   a $\frac{1}{2}$ b $\frac{1}{5}$ c $-\frac{1}{7}$
   d 0.5 e 0.1 f $0.\dot{3}$

3 Write these expressions in index form.
   a $\frac{1}{7^2}$ b $\frac{1}{9^2}$ c $\frac{1}{2^2}$
   d $\frac{1}{2^5}$ e $\frac{1}{3^4}$ f $\frac{1}{6^4}$

4 Write these expressions as fractions.
   a 8^{-2} b 7^{-3} c 5^{-2}
   d 9^{-4} e 3^{-2} f 9^{-3}

5 Write these expressions
   i as fractions
   ii as decimals.
   a 3^{-2} b 2^{-3} c 10^{-5}
   d 1^{-7} e 8^{-1} f 2^{-4}

6 Evaluate these expressions.
   a 4^{-2} b 4^{-1} c 4^0
   d $4^{0.5}$ e 4^1 f 4^2
   g 4^3 h $4^{-0.5}$ i 4^{-3}

7 Write these expressions in index form.
   a $\frac{1}{\sqrt{3}}$ b $\frac{1}{\sqrt{5}}$ c $\frac{1}{\sqrt{11}}$

8 Evaluate these expressions.
   a $25^{\frac{1}{2}}$ b $25^{-\frac{1}{2}}$ c 25^0
   d $25^{\frac{3}{2}}$ e 25^{-1} f $25^{-\frac{3}{2}}$
   g $16^{\frac{3}{4}}$ h $27^{\frac{2}{3}}$ i $4^{-\frac{3}{2}}$
   j $81^{-0.25}$ k $125^{-\frac{2}{3}}$ l $100^{-\frac{7}{2}}$
   m $4^{\frac{3}{2}}$ n $8^{\frac{2}{3}}$ o $9^{\frac{5}{2}}$
   p $100^{-\frac{1}{2}}$ q $16^{-\frac{3}{2}}$ r $1000^{\frac{2}{3}}$
   s $400^{-\frac{1}{2}}$ t $169^{\frac{3}{2}}$

9 Simplify these expressions.
   a $2^{\frac{1}{3}} \times 2^{\frac{1}{6}}$ b $2^{\frac{1}{3}} \div 2^{\frac{1}{6}}$
   c $(2^{\frac{3}{2}})^3$ d $2^{\frac{2}{5}} \times 2^{\frac{3}{5}} \div 2$
   e $(2^{\frac{3}{4}} \times 2^{\frac{1}{2}})^2$ f $(2^3 \times 2^{\frac{5}{2}}) \div (2^{\frac{7}{2}} \times 2^2)$

10 Simplify these expressions.
   a $3^{-2} \times 3^3$ b $3^{-4} \div 3^2$
   c $(3^{-1})^2$ d $3^{-2} \times 3^{-1} \div 3^3$
   e $(3^{-1} \div 3^{-2})^{-1}$ f $(3^{-2} \div 3^{-1})^2 \div 3^{-6}$

11 Simplify these expressions.
   a $(5^{\frac{1}{2}})^2$ b $(5^{\frac{1}{3}})^2$ c $(5^2)^{\frac{1}{3}}$
   d $(5^{\frac{1}{3}})^{\frac{1}{2}}$ e $(5^{-2})^2$ f $(5^3)^{-2}$
   g $(5^{-1})^{-3}$ h $(5^{-\frac{5}{4}})^{-2}$ i $(5^{-2})^{\frac{3}{4}}$
   j $(5^{-\frac{2}{5}})^{\frac{3}{2}}$ k $(5^{\frac{6}{5}})^{-\frac{1}{3}}$ l $(5^{-\frac{2}{3}})^{\frac{5}{4}}$

12 Simplify these expressions.
   a $2^{-\frac{3}{2}} \times 2^{\frac{1}{4}}$ b $2^{-\frac{1}{6}} \div 2^{-\frac{1}{3}}$
   c $(2^{\frac{1}{8}} \div 2^{-\frac{1}{2}})^3$ d $2^{\frac{3}{2}} \times 2^{-2} \div 2^{\frac{1}{4}}$
   e $(2^{\frac{5}{4}} \times 2^{-1})^{-\frac{3}{2}}$ f $(2^3 \div 2^{-\frac{1}{2}})^{-2} \times (2^{\frac{3}{2}} \div 2^{-1})$

13 Write these expressions as
   i powers of 4 ii powers of 16.
   a $\frac{1}{4}$ b 16 c $\frac{1}{16}$ d $\frac{1}{2}$

14 Write these expressions as powers of 10.
   a $\frac{1}{10}$ b $\sqrt{10}$ c $(\sqrt{10})^3$ d $\frac{1}{(\sqrt{10})^5}$

15 Write these expressions as powers of 5.
   a 0.04 b 0.008 c $\frac{1}{\sqrt[3]{5}}$
   d $\sqrt[3]{25}$ e $\frac{1}{\sqrt{125}}$ f $\frac{1}{\sqrt[3]{625}}$

16 Fill in the missing numbers.
   a i $125 = 25^{\square}$
     ii $25 = 125^{\square}$
   b i $32 = 4^{\square}$
     ii $4 = 32^{\square}$
   c i $81 = 27^{\square}$
     ii $27 = 81^{\square}$
   d i $0.125 = 16^{\square}$
     ii $16 = 0.125^{\square}$

*17 Evaluate $\left(2\frac{1}{4}\right)^{-\frac{1}{2}}$

APPLICATIONS

17.1 Calculating with roots and indices

RECAP

- Indices are defined for fractional and negative powers.
  $$x^{\frac{1}{n}} = \sqrt[n]{x} \qquad x^{\frac{m}{n}} = (\sqrt[n]{x})^m \qquad x^{-n} = \frac{1}{x^n} \qquad x^0 = 1$$
- The same index laws we use for positive whole number powers also apply to fractional and negative powers.
  $$x^a \times x^b = x^{a+b} \qquad x^a \div x^b = x^{a-b} \qquad (x^a)^b = x^{a \times b}$$

$$27^{-\frac{2}{3}} = \frac{1}{27^{\frac{2}{3}}} = \frac{1}{(\sqrt[3]{27})^2} = \frac{1}{3^2} = \frac{1}{9}$$

$$2^{\frac{2}{3}} \times 2^{\frac{1}{6}} = 2^{\frac{2}{3} + \frac{1}{6}} = 2^{\frac{5}{6}}$$

$$2^{\frac{2}{3}} \div 2^{\frac{1}{6}} = 2^{\frac{2}{3} - \frac{1}{6}} = 2^{\frac{1}{2}}$$

HOW TO

To solve problems involving roots and indices
1. Read the question – decide the mathematics required.
2. Apply the index laws as appropriate.
3. Answer the question – leave the answer in index form if asked to do so.

A formula where a variable appears in a power is said to be exponential. For example $y = 2^x$.

EXAMPLE

Given favourable conditions, the number of bacteria cells in an infected area can double every 20 minutes.

a Starting with one cell, how many exist after
   i 1 hour   ii 2 hours   iii 3 hours   iv 4 hours.
b How long will it be until there are over a million cells?

p.442

① Doubling every 20 minutes is to do with powers of 2. 1 hour is 3 × 20 minutes.

a i $2 \times 2 \times 2 = 8$ Doubles 3 times.   ii $2^6 = 64$ Doubles 6 times.
  iii $2^9 = 512$ Doubles 9 times.   iv $2^{12} = 4096$ Doubles 12 times.

b $2^{20} = 1048576$ 2^{20} is over a million. Check that 2^{19} is not over a million.
  $2^{19} = 524288$ 2^{19} is less than a million, so 2^{20} is the first power above a million.
  6 hr, 40 min. 20×20 mins

EXAMPLE

Find the missing powers in these equations.

a $(2^{-2x} \div 2^{-5})^2 = 2^{14}$ b $2^{\frac{1}{3}} \times 2^{\frac{1}{2} - x} = \frac{1}{2^{\frac{1}{6}}}$ c $\dfrac{2^{\frac{1}{2}} \times 4^x \div 2^{-\frac{1}{4}}}{8^{2x}} = 2$

① Simplify the expressions using the index laws.

a $(2^{-2x} \div 2^{-5})^2 = 2^{14}$ ②
$(2^{-2x - -5})^2 = 2^{14}$
$2^{2(5 - 2x)} = 2^{14}$
Compare indices.
$2(5 - 2x) = 14$
$5 - 2x = 7$
$x = -1$ ③

b $2^{\frac{1}{3}} \times 2^{\frac{1}{2} - x} = \frac{1}{2^{\frac{1}{6}}}$ ②
$2^{\frac{1}{3} + \frac{1}{2} - x} = 2^{-\frac{1}{6}}$
$2^{\frac{5}{6} - x} = 2^{-\frac{1}{6}}$
$\frac{5}{6} - x = -\frac{1}{6}$
$x = 1$ ③

c First write 4 and 8 as powers of 2.
$\dfrac{2^{\frac{1}{2}} \times (2^2)^x \div 2^{-\frac{1}{4}}}{(2^3)^{2x}} = 2$
$2^{\frac{1}{2} + 2x - -\frac{1}{4} - 6x} = 2$ ②
$2^{\frac{3}{4} - 4x} = 2^1$
$\frac{3}{4} - 4x = 1 \Rightarrow 4x = -\frac{1}{4}$
$x = -\frac{1}{16}$

Number   Calculations 2

Reasoning and problem solving

Exercise 17.1A

1 Jack's beanstalk doubles in height every 24 hours. It is measured every day at 9 am. On Monday it was 2 cm high.

   a **i** How high was it at 9 am on the Friday after?

      ii How high was it at 9 am on the following Monday?

   b **i** On which day was it measured to be above 1 metre high?

      ii On which day was it measured to be above 10 metres?

   c Write your answers to part **a** as powers of 2.

2 The Richter scale measures the power of an earthquake. Each level is ten times more powerful than the previous value. So level 4 is 10 times as powerful as level 3.

   a How many times more powerful is

      i level 4 compared to level 1?

      ii level 9 (Complete devastation) compared to level 5 (Buildings shake)?

   b What level is an earthquake 10^5 times more powerful than level 3?

3 The volume of a cube is x cm³.

   a Find an expression for the surface area of the cube in the form ax^b, where a and b are fractions or integers.

   b The surface area of a cube is 54 cm². Find the volume of the cube.

   c A cube with volume 216 cm³ has surface area 216 cm². Are there any other cubes that have the same numerical value for the volume and surface area? Give your reason.

4 Arrange these numbers in ascending order.

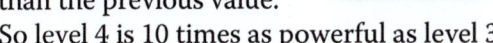

$27^{-\frac{2}{3}} \quad 9^{\frac{3}{2}} \quad 16^{\frac{3}{4}} \quad \left(\frac{1}{2}\right)^{-4} \quad \left(\frac{1}{8}\right)^{-\frac{5}{3}} \quad 25^0$

5 Find the missing powers in these equations.

   a $2^3 \times 2^a = 2^7$ **b** $4^5 \div 4^b = 4^2$

   c $5^c \times 5^4 = 5^2$ **d** $1259^d = 1$

   e $3^{\frac{3}{2}} = (3^{\frac{1}{2}})^e$ **f** $5^{\frac{1}{3}} \times 5^f = 5$

6 Find the value of the letter in each of these equations.

   a $\sqrt{5} = 5^a$ **b** $\sqrt[3]{600} = 600^b$

   c $\sqrt{100} = 1000^c$ **d** $2 = 16^d$

7 Solve these equations.

   a $(2^{-3x} \div 2^{-2})^2 = 2^{10}$

   b $\dfrac{2^{3x} \times 2^4}{2^3 \times 2^{4x}} = \dfrac{1}{2^2}$

   c $(2^{4+x} \div 2^{2+2x})^2 = (2^{3x} \times 2^{x-2})^{-1}$

   d $2^{\frac{3}{2}} \times 2^x \div 2^{\frac{2}{3}} = 2$

   e $(2^{\frac{1}{2}} \times 2^x)^{\frac{1}{2}} = \dfrac{1}{(2^{\frac{1}{3}})^3}$

8 Copy and complete these number pyramids. The number in each cell is the product of the numbers in the two cells immediately below it.

   a 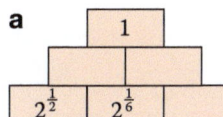   **b**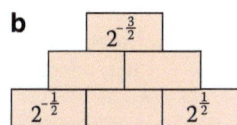

9 **a** Copy and complete this table of values for 9^x.

x	-1	$-\frac{1}{2}$	0	$\frac{1}{2}$	1
9^x					

   b Plot a graph of 9^x, for $-1 \leqslant x \leqslant 1$.

   c On the same set of axes, plot the graph of 4^x for the same values of x. Describe any similarities and differences between the two graphs.

p.360

SKILLS

17.2 Exact calculations

Using a calculator does not always give you the most accurate answer.

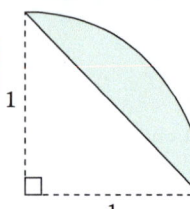

The perimeter of this shape is given by

$$\sqrt{1^2 + 1^2} + \frac{1}{4} \times 2\pi \times 1 = \begin{cases} 2.9850098... & \text{Using a calculator} \\ \sqrt{2} + \frac{1}{2}\pi & \text{Exact} \end{cases}$$

Some calculators may give answers with surds and pi in them.

In practice the decimal approximation might be more useful than the exact answer!

Surds (like $\sqrt{2}, \sqrt{3}, \sqrt{5}, ...$) and π cannot be written as fractions and have decimal expansions which go on for ever.
You have to truncate or round the decimals.

$\sqrt{2} = 1.414\,213\,562... = 1.414$ (4 sf) $\pi = 3.141\,592\,654... = 3.142$ (4 sf)

- You can write an **exact answer** using π and surds.

EXAMPLE

Simplify these expressions.

a $2(\pi - 3) + 4(\sqrt{3} + 2)$ **b** $(2\pi + 1)^2 - (\pi + 2)(\pi - 3)$ **c** $\sqrt{8} - 3\sqrt{2} + \sqrt{45}$

a $= 2\pi - 6 + 4\sqrt{3} + 8$
$= 2 + 2\pi + 4\sqrt{3}$

Treat π and $\sqrt{3}$ like two variables 'x and y'.

b $= 4\pi^2 + 4\pi + 1 - (\pi^2 - \pi - 6)$
$= 7 + 5\pi + 3\pi^2$

π and π^2 are unlike terms, like 'x and x^2'.

c $= \sqrt{4 \times 2} - 3\sqrt{2} + \sqrt{9 \times 5}$
$= 2\sqrt{2} - 3\sqrt{2} + 3\sqrt{5}$
$= -\sqrt{2} + 3\sqrt{5}$

Use $\sqrt{ab} = \sqrt{a} \times \sqrt{b}$ and look for square number factors.

If a calculation involves fractions it is best to use them as far as possible.

EXAMPLE

Calculate

a $\frac{1}{2} + \frac{2}{3}$

b $\frac{3}{4} \times \frac{2}{9}$

c $\frac{4}{5} \div \frac{3}{10}$

a $\frac{1}{2} + \frac{2}{3} = \frac{3 + 4}{6} = \frac{7}{6} = 1\frac{1}{6}$

b $\frac{\cancel{3}^1}{\cancel{4}_2} \times \frac{\cancel{2}^1}{\cancel{9}_3} = \frac{1}{2} \times \frac{1}{3} = \frac{1}{6}$

c $\frac{4}{5} \div \frac{3}{10} = \frac{4}{\cancel{5}_1} \times \frac{\cancel{10}^2}{3} = \frac{4 \times 2}{1 \times 3} = \frac{8}{3} = 2\frac{2}{3}$

You can always write a fraction as a decimal but it may not be clear that it hasn't been truncated or it may not look as nice.
$\frac{1}{8} = 0.125$
$\frac{1}{7} = 0.142\,857$

You may have to work with combinations of fractions, integers, surds and numbers such as π.

EXAMPLE

Calculate this expression
a exactly
b to 3 dp.

$\frac{7 + \sqrt{20}}{4} - \frac{3 + \sqrt{5}}{5}$

a $= \frac{5(7 + 2\sqrt{5}) - 4(3 + \sqrt{5})}{4 \times 5}$ $\sqrt{20} = \sqrt{4 \times 5} = 2\sqrt{5}$

$= \frac{35 + 10\sqrt{5} - 12 - 4\sqrt{5}}{20}$

$= \frac{23 + 6\sqrt{5}}{20}$

b $= 1.820\,820\,3...$ Using a calculator.
$= 1.821$ (3 dp)

Number Calculations 2

Fluency

Exercise 17.2S

1 Simplify these expressions.
  a $4(\pi + 2) - 3(\pi - 1)$
  b $5(3 - \pi) + 4(2\pi - 3)$
  c $\pi(\pi + 1) - 2(3 - \pi)$
  d $(2\pi + 3)(\pi - 1)$
  e $(3\pi - 4)^2$
  f $(2\pi - 1)(2\pi + 1) - (\pi - 1)^2$

2 Simplify these expressions.
  a $\sqrt{20} + 3\sqrt{5}$
  b $\sqrt{27} + 2\sqrt{3} - \sqrt{24}$
  c $(5 + 3\sqrt{3}) - (2 + 4\sqrt{3})$
  d $3\sqrt{28} + 4\sqrt{12} + \sqrt{14}$
  e $\sqrt{6}(3 - 2\sqrt{6})$
  f $\sqrt{3}(4 - 2\sqrt{6} + \sqrt{12})$
  g $(2 + \sqrt{3})(3 + \sqrt{3})$
  h $(3 - \sqrt{2})^2$
  i $(5 - \sqrt{7})(5 + \sqrt{7})$
  j $(4 + 3\sqrt{5})(6 - \sqrt{5})$

3 Evaluate these calculations exactly.
  a $\frac{2}{3} \times \frac{3}{4}$ **b** $\frac{5}{9} \div \frac{1}{3}$ **c** $2\frac{1}{2} \times \frac{5}{8}$
  d $\frac{8}{9} \div 1\frac{2}{3}$ **e** $3\frac{1}{2} \div 2\frac{1}{4}$ **f** $5\frac{1}{5} \times 2\frac{3}{4}$

4 Evaluate these calculations exactly.
  a $\frac{1}{3} + \frac{1}{5}$ **b** $\frac{2}{5} + \frac{3}{7}$ **c** $\frac{3}{8} - \frac{2}{7}$
  d $\frac{8}{15} + \frac{4}{9}$ **e** $6\frac{1}{2} - 1\frac{5}{8}$ **f** $\frac{2}{5} + \frac{1}{3} + \frac{1}{4}$

5 Evaluate these calculations exactly.
  a $\frac{2}{5} \times \left(\frac{3}{4} + \frac{2}{3}\right)$ **b** $\left(\frac{2}{3} + \frac{4}{5}\right) \div \left(\frac{2}{5} + \frac{3}{7}\right)$
  c $\frac{5}{6} \div \frac{3^2 + 4^2}{10}$ **d** $\left(\frac{3}{7}\right)^2 \times \left(\frac{4}{5} - \frac{1}{7}\right)$
  e $\left(\frac{1}{2} + \frac{5}{9}\right)^2 + \left(\frac{2}{3}\right)^3$

6 Simplify these expressions.
  a $\frac{4}{3}\pi \times 6^3 + \frac{2}{3}\pi \times 3^2$
  b $\frac{1}{3}\pi \times 9^2 \times 5 - \frac{1}{3}\pi\left(\frac{18}{5}\right)^2 \times 2$
  c $\frac{1}{2}(4\pi \times 11^2 + 4\pi \times 9^2)$
  d $16\pi + 4\pi\sqrt{6^2 + 4^2}$
  e $\frac{8\pi}{5} + \frac{3\pi}{4}$
  f $2\frac{1}{4}\pi + 1\frac{1}{2} \times 4\frac{2}{3}\pi$
  g $\frac{5}{6}\pi - 2\sqrt{2} + \frac{7}{9}\pi + \sqrt{28}$
  h $\frac{\sqrt{15\pi} \times 4 \times \sqrt{3\pi}}{6\sqrt{3}}$

7 Simplify these expressions.
  a $\frac{6 + \sqrt{27}}{3} - \frac{2 + \sqrt{3}}{4}$
  b $\frac{20 + 3\sqrt{7}}{3} - (5 + \sqrt{28})$
  c $\frac{6 + \sqrt{8}}{2} + \frac{3 + \sqrt{2}}{9}$
  d $\frac{8 + \sqrt{45}}{3} - \frac{4 + \sqrt{20}}{9}$

8 Rationalise these expressions.
  a $(1 + \sqrt{5}) \div \sqrt{5}$
  b $(2 + \sqrt{7}) \div \sqrt{28}$
  c $(5 - \sqrt{11}) \div 3\sqrt{7}$
  d $(1 + 2\sqrt{3}) \div \sqrt{48}$
  e $(5 + 2\sqrt{3}) \div \sqrt{24}$
  f $(7 - 2\sqrt{5}) \div \sqrt{50}$

9 Find an approximate decimal value for each of these expressions. Give your answers to 2 decimal places.
  a $4\sqrt{2}$ **b** $\sqrt{5} + 1$
  c $2 + \sqrt{5}$ **d** $36\pi - 7$

10 Ayben says that all these six calculations give the same answer. Is she correct? Show your working.
  a $\frac{3}{4} + \frac{1}{2}$ **b** $\frac{1}{2} \times 2\frac{1}{2}$
  c $2\frac{3}{4} - 1\frac{1}{2}$ **d** $3 \div 2\frac{2}{5}$
  e $\sqrt{50} \div \sqrt{32}$ **f** $\frac{1}{4}(3 + \sqrt{2})(3 - \sqrt{2})$

APPLICATIONS

17.2 Exact calculations

RECAP (p.88, p.214, p.264)
- To find an exact answer, do *not* use decimals. Instead use fractions in their lowest terms, simplified surds and multiples of π, as appropriate, throughout the calculation.
- A surd is in its simplest form when the smallest possible integer appears inside the square root.
- To manipulate an expression treat π and any surds as variables and collect like terms, etc.

$$1\tfrac{1}{4} + 2\tfrac{5}{6} = \tfrac{5}{4} + \tfrac{17}{6}$$
$$= \tfrac{15 + 34}{12}$$
$$= \tfrac{49}{12} = 4\tfrac{1}{12}$$
$$\sqrt{98} = \sqrt{49} \times \sqrt{2} = 7\sqrt{2}$$

HOW TO
To find an exact answer to a question
1. Read the question – decide which formulae and mathematical techniques you will need to use.
2. Calculate with fractions, surds and multiples of π.
3. Answer the question – collect like terms, simplify any surds and write fractions in their lowest terms.

$$1 + \tfrac{5}{2}\sqrt{2} - 2(\pi + \sqrt{2})$$
$$= 1 + \tfrac{5}{2}\sqrt{2} - 2\pi - 2\sqrt{2}$$
$$= 1 + \tfrac{1}{2}\sqrt{2} - 2\pi$$

EXAMPLE (p.300)

The blue cylinder has diameter 4 cm and height 3 cm. The red cylinder has diameter 3 cm and height 4 cm. The cuboid has a square base with side 3 cm, and is 5 cm high.

a What is the exact total volume of the shapes?
b The blue cylinder is enlarged by scale factor $\tfrac{3}{2}$, what is the new total of the volumes?

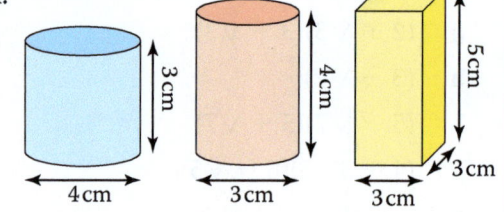

a Blue volume $= \pi \times \left(\tfrac{4}{2}\right)^2 \times 3 = 12\pi$
   Red volume $= \pi \times \left(\tfrac{3}{2}\right)^2 \times 4 = \pi \times \tfrac{9}{4} \times 4 = 9\pi$
   Yellow volume $= 3 \times 3 \times 5 = 45$
   Total volume $= 45 + 21\pi$ cm³

   ① Volume of cylinder $= \pi r^2 h$.
   ① Volume of a cuboid $= l \times w \times h$.
   ② Terms involving π are like terms.

b Volume scale factor $= \left(\tfrac{3}{2}\right)^3 = \tfrac{27}{8}$
   New blue volume $= \tfrac{27}{8} \times {}^3 12\pi = \tfrac{81}{2}\pi = 40\tfrac{1}{2}\pi$
   New total volume $= 40\tfrac{1}{2}\pi + 9\pi + 45$
   $= 49\tfrac{1}{2}\pi + 45$ cm³

   ③ Cancel common factors.

Leaving π in your answers actually makes working easier!

EXAMPLE (p.194)

Solve these simultaneous equations.
$2p + q = 6\sqrt{3}$ (A)
$2p - 2q = 3\pi$ (B)

① Eliminate p to find q.
$2p + q - (2p - 2q) = 6\sqrt{3} - 3\pi$
$3q = 6\sqrt{3} - 3\pi$
$q = 2\sqrt{3} - \pi$ ③

② Subtract (A) – (B).

① Substitute q into (A).
$2p + 2\sqrt{3} - \pi = 6\sqrt{3}$ ②
$2p = 4\sqrt{3} + \pi$
$p = 2\sqrt{3} + \tfrac{\pi}{2}$ ③

Number   Calculations 2

Reasoning and problem solving

Exercise 17.2A

1. A bathroom window is a semicircle with an internal diameter of 80 cm. Work out the exact area and perimeter of the glass in the window.

    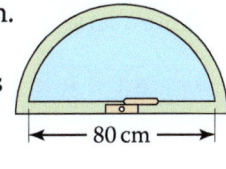

2. Shamin is cutting out circles for an art project. She has squares of card that are 4 cm wide.
    If Shamin cuts out the largest possible circle, what is the exact area of the card remaining?

    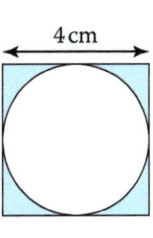

3. Find the exact solutions to these simultaneous equations.

    a $x + y = 3\sqrt{5}$
       $2x - y = 3\pi$

    b $3x + y = 9\sqrt{5} + \pi$
       $x + y = 5\sqrt{5} - \pi$

    c $3x + 2y = 2(\pi - \sqrt{5})$
       $2x + 3y = 3\sqrt{5} - 2\pi$

4. A circle of radius 10 cm has the same area as a square with sides x cm.
    Find the exact value of x.

5. A cube with side length 5 cm has the same volume as a cone with vertical height 3 cm.
    Show that the exact radius of the cone is $\dfrac{5\sqrt{5}}{\sqrt{\pi}}$ cm.

6. The diagram shows a cuboid with volume 25 cm³.
    What is the exact surface area of the cuboid?

    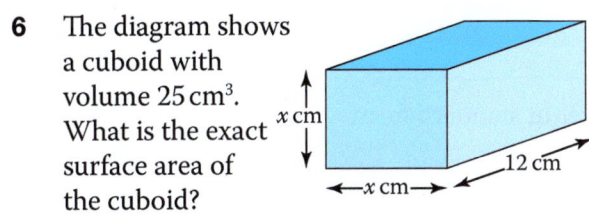

7. A spherical orange has diameter 12 cm. Once peeled, its volume is $\dfrac{500\pi}{3}$ cm³. How thick was the skin?

8. A hemisphere has surface area 60π mm². Find its volume.

9. The areas of two squares are in the ratio 1:3. The side of the larger square is 12 cm. What is the side of the smaller square? Give your answer in the form $a\sqrt{b}$ where a and b are integers.

10. An equilateral triangle has sides 2 cm.

    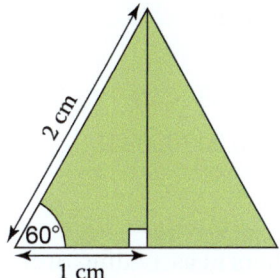

    a Show that the height of the triangle is $\sqrt{3}$ cm.

    The triangle is enlarged and the area of the image is $3\sqrt{3}$ cm².

    b Find the height of the enlarged triangle.

    c Find the side length of the enlarged triangle.

11. Rachel bought a cylindrical tube containing three exercise balls. Each exercise ball is a sphere of radius 5 cm.
    The exercise balls touch the sides of the tube.
    The exercise balls touch the top and bottom of the tube.

    Work out the exact volume of empty space in the tube.

    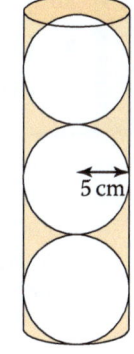

*12 Lindsey and Neiva are arguing about the square root of two. Lindsey says,
    'you can write it as a fraction, $\sqrt{2} = \dfrac{m}{n}$'.
    Neiva says,
    'that means both m and n must be even so your fraction can never be in its lowest terms'.

    Who is right? Give your reason.

SKILLS

17.3 Standard form

You can use **standard form** to represent large and small numbers.

- In standard form, a number is written as $A \times 10^n$, where $1 \leqslant A < 10$ and n is an integer.

EXAMPLE

Write these numbers in standard form.
a 235
b 0.23×10^6
c 0.45
d 0.000 000 416

a 2.35×10^2
b 2.3×10^5
c 4.5×10^{-1}
d 4.16×10^{-7}

EXAMPLE

Write these numbers in ascending order.
6.35×10^4, 5.44×10^4, 6.95×10^3, 7.075×10^2, 9.9×10^{-1}

9.9×10^{-1}, 7.075×10^2, 6.95×10^3, 5.44×10^4, 6.35×10^4

First order the powers of 10. $10^{-1} < 10^2 < 10^3 < 10^4$.

Then compare numbers with the same power of 10. $5.44 < 6.35$.

- To multiply or divide numbers given in standard form, multiply or divide the numbers and then add or subtract the powers.

EXAMPLE

Simplify these expressions.
a $(3.6 \times 10^5) \div (1.2 \times 10^3)$
b $(5.4 \times 10^4) \times (2 \times 10^{-3})$

a $\dfrac{3.6 \times 10^5}{1.2 \times 10^3} = (3.6 \div 1.2) \times (10^5 \div 10^3) = 3 \times 10^{5-3} = 3 \times 10^2$

b $(5.4 \times 10^4) \times (2 \times 10^{-3}) = (5.4 \times 2) \times (10^4 \times 10^{-3})$
$= 10.8 \times 10^{4-3} = 1.08 \times 10^2$

- To add or subtract numbers given in standard form, first rewrite the numbers with the same power of 10 and then add or subtract the numbers.

EXAMPLE

Calculate $3.2 \times 10^5 + 7.1 \times 10^4$. Give your answer in standard form.

$3.2 \times 10^5 + 7.1 \times 10^4 = 3.2 \times 10^5 + 0.71 \times 10^5$
$= (3.2 + 0.71) \times 10^5 = 3.91 \times 10^5$

You need to know how to enter standard form calculations into your calculator and how to interpret the display.

EXAMPLE

Use your calculator to work out $(6.43 \times 10^6) \div (4.21 \times 10^{-2})$.

$\boxed{6.43 \times 10^6 \div 4.21 \times 10^{-2}}$
$\boxed{1.527315914 \times 10^8}$ 1.53×10^8 (3 sf)   Round up

Number Calculations 2

Fluency

Exercise 17.3S

1. Write these powers as ordinary numbers.
   a 10^3 b 10^6 c 10^5

2. Write these powers as ordinary numbers.
   a 10^0 b 10^{-2} c 10^{-5}

3. Write these numbers in standard form.
   a 9000   b 650   c 6500
   d 952   e 23.58   f 255.85

4. Write these numbers in standard form.
   a 0.00034   b 0.1067
   c 0.0000091   d 0.315
   e 0.0000505   f 0.0182
   g 0.00845   h 0.000000000306

5. Write these numbers in ascending order.
   4.05×10^4 4.55×10^4 9×10^3
   3.898×10^4 1.08×10^4 5×10^4

6. Write these numbers as ordinary numbers.
   a 6.35×10^4 b 9.1×10^{17}
   c 1.11×10^2 d 2.998×10^8

7. Convert these standard form numbers to ordinary numbers.
   a 4.5×10^{-3} b 3.17×10^{-5}
   c 1.09×10^{-6} d 9.79×10^{-7}

8. These numbers are not in standard form. Rewrite each of them correctly in standard form.
   a 60×10^1 b 45×10^3
   c 0.65×10^1 d 0.05×10^8
   e 21.5×10^3 f 0.7×10^{14}
   g 122.516×10^{18} h 0.015×10^9
   i 28×10^{-2} j 0.4×10^{-1}
   k 13.5×10^{-4} l 12×10^{-8}

For questions **9** to **13**, do not use a calculator and give your answers in standard form.

9. Evaluate these calculations.
   a $(5 \times 10^3) \times (5 \times 10^4)$
   b $(8 \times 10^7) \times (3 \times 10^5)$
   c $(2.5 \times 10^{-3}) \times (2 \times 10^2)$
   d $(4.6 \times 10^{-6}) \times (2 \times 10^{-2})$

10. Evaluate these calculations.
    a $(2 \times 10^6) \div (4 \times 10^4)$
    b $(3 \times 10^5) \div (4 \times 10^2)$
    c $(4 \times 10^4) \div (2 \times 10^6)$
    d $(8.4 \times 10^{-2}) \div (2 \times 10^6)$

11. Evaluate these calculations.
    a $(4 \times 10^6) \div (2 \times 10^8)$
    b $5 \times 10^4 \times 2 \times 10^{-6}$
    c $(3 \times 10^{-3}) \div (2 \times 10^5)$
    d $(4 \times 10^5) \div (2 \times 10^2)$
    e $5 \times 10^{-3} \times 5 \times 10^{-4}$
    f $(9.3 \times 10^{-2}) \div (3 \times 10^{-6})$

12. Evaluate these calculations.
    a $(5 \times 10^{-1}) + (2 \times 10^{-2})$
    b $(4 \times 10^{-2}) + (6 \times 10^{-3})$
    c $(2 \times 10^{-2}) + (9 \times 10^{-4})$
    d $(1.5 \times 10^{-2}) - (2 \times 10^{-3})$
    e $(7 \times 10^4) - (5 \times 10^5)$
    f $(6 \times 10^2) - (7 \times 10^{-1})$

13. Evaluate these calculations.
    a $(6.4 \times 10^{-4}) + (7.1 \times 10^{-3})$
    b $(9.9 \times 10^5) - (2.7 \times 10^4)$
    c $(4.8 \times 10^{-6}) + (3.9 \times 10^{-5})$
    d $(3.3 \times 10^2) - (7.5 \times 10^1)$
    e $(9.8 \times 10^5) - (6.4 \times 10^5)$
    f $(3.5 \times 10^{-2}) + (9.7 \times 10^{-3})$

14. Use your calculator to check your answers to questions **9–13**.

15. a Re-write the following expressions using standard form.
    i 3000×210 ii $0.0042 \div 0.6$
    iii 1200×0.09 iv $0.049 \div 70$
    b Complete the calculations without using a calculator, giving your answer in standard form.

1049, 1050, 1051

APPLICATIONS

17.3 Standard form

Did you know…

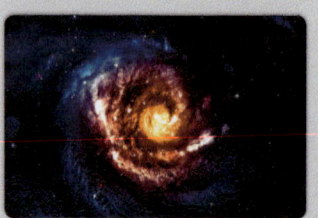

Standard form helps scientists to represent very small quantities, like the volume of an atom, and very large quantities, like the number of atoms in the universe. It is estimated that there are roughly 1×10^{80} fundamental particles in the known universe.

RECAP

- A number in standard form is written $A \times 10^n$, where n is a positive or negative integer and $1 \leq A < 10$.

HOW TO

To use standard form in calculations
1. Read the question – decide which calculation you need to carry out.
2. Calculate using your knowledge of standard form.
3. Give your answer in standard form and check it is sensible.

EXAMPLE

Palmira wrote
$$(7.2 \times 10^{-4}) \div (3.6 \times 10^{-8}) = 2 \times 10^{-12}$$
Is she correct? Explain your answer.

1. No. She has multiplied 10^{-4} and 10^{-8}, instead of dividing.
2. The correct answer is
$$(7.2 \div 3.6) \times (10^{-4} \div 10^{-8}) = 2 \times 10^{-4--8} = 2 \times 10^4$$
A quick check shows Palmira's answer is incorrect. Since you are dividing one number by another smaller number, the answer must be **greater** than 1.

EXAMPLE

The mass of a carbon atom is 2×10^{-23} g. How many atoms are there in one kilogram of carbon?

$$1000 \div (2 \times 10^{-23}) = (1 \times 10^3) \div (2 \times 10^{-23})$$
$$= (1 \div 2) \times (10^3 \div 10^{-23})$$
$$= 0.5 \times 10^{26}$$
$$= 5 \times 10^{25}$$

1. Divide 1 kg by the mass of one carbon atom.
3. Answer in standard form.

There are 5×10^{25} atoms in one kilogram of carbon.
A very large number as you would expect!

EXAMPLE

Mercury is 5.79×10^7 km from the Sun.
Venus is 1.08×10^8 km from the Sun.
Find the minimum distance between Mercury and Venus.

1. Draw a sketch of the orbits.
   Minimum distance = $(1.08 \times 10^8) - (5.79 \times 10^7)$
2. Write both numbers with the same power of 10.
   $= (10.8 \times 10^7) - (5.79 \times 10^7)$
   $= (10.8 - 5.79) \times 10^7$
3. $= 5.01 \times 10^7$ km   Include the units.

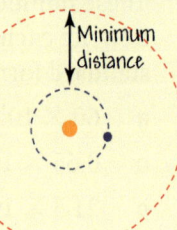

EXAMPLE

An HD video camera captures frames at 1920 × 1080 pixels. It records 50 frames per second.
a   How many pixels are stored for 3 minutes of video? (Answer in standard form to 2 dp.)
b   The camera's LCD viewing screen is 8.75 cm wide. How wide is one pixel?

1. **a** involves multiplication and a large answer, **b** requires division and a very small answer.
a $1920 \times 1080 \times 50 \times 60 \times 3 = 18\,662\,400\,000$
    $= 1.87 \times 10^{10}$ pixels
b $8.75 \div 1920 = 0.00455729$
    Width of a pixel = 4.56×10^{-3} (3 sf)

1. Size × fps × seconds in a minute × 3 min
2. Answer in standard form.
1. Screen width ÷ number of pixels across.
2. Answer in standard form.

Number Calculations 2

Reasoning and problem solving

Exercise 17.3A

Use the information in the table in the questions that follow.

	Mass (kg)	Diameter (km)
Sun	2×10^{30}	–
Moon	–	3.5×10^3
Mercury	3.30×10^{23}	–
Venus	4.87×10^{24}	–
Earth	5.97×10^{24}	1.3×10^4
Mars	6.42×10^{23}	–
Jupiter	1.90×10^{27}	1.398×10^6
Saturn	5.68×10^{26}	1.164×10^5
Uranus	8.68×10^{25}	–
Neptune	1.02×10^{26}	–

1. Approximately how many times heavier than the Earth is the Sun?

2. The mass of one atom of the element mercury is 3.3×10^{-22} g.
   How many mercury atoms have the same mass as the planet Mercury?

3. The width of a plant cell is 60 micrometres.
   A micrometre is 1×10^{-6} m.
   The diagram of a plant cell in a science textbook has width 3 cm.
   How many times bigger is the diagram than the real plant cell?

4. A bumblebee has mass 5.2×10^{-5} kg.
   An adult man has mass 70 kg.
   A bumblebee can carry 75% of its mass.
   How many bumblebees would it take to lift the man?
   Give your answer in standard form to 3 sf.

5. The diameter of the Sun is 400 times greater than the diameter of the Moon.
   How many times smaller is the diameter of the Earth than the diameter of the Sun?

6. The Earth is 1.496×10^8 km from the Sun.
   Mars is 2.279×10^8 km from the Sun.
   Find the minimum and maximum distances between Earth and Mars.

7. Light travels about 3×10^8 metres per second.
   a. How long does it take for light to travel 1 m?
   b. Find the distance light travels in 1 year.
   Give your answers in standard form.

8. The mass of a proton is approximately 1800 times greater than the mass of an electron. The mass of an electron is 9.11×10^{-31} kg. A hydrogen atom contains one proton and one electron. Find the mass of a hydrogen atom. Give your answer in standard form.

9. Carrie says the total mass of the eight planets given in the table is $7.686\,12 \times 10^{26}$ kg.
   a. Which planet did Carrie forget to include in her total?
      What is the correct total mass?
   b. What percentage of the total mass does Earth account for?
      Give your answer to 2 sf.

10. Digital televisions are available in a range of different screen resolutions. They can be described by their width and height in pixels.

    | SD 576p | 720×576 |
    | HD 1080p | 1920×1080 |
    | UHD 2160p | 3840×2160 |
    | UHD 4320p | 7680×4320 |
    | UHD 8640p | $15\,360 \times 8640$ |

    a. Yvonne says, 'the difference between the highest number of pixels and the second highest is larger than the difference between the smallest number of pixels and the second smallest'. Is she correct?
    b. How many times more pixels does an HD 1080p have compared to a SD 576p?
    c. Which resolution is best described as '2 million pixels'?

11. Lamar claims that Jupiter is twice as dense as Saturn. Do you agree?
    Explain your answer.

*12. Venus travels around the Sun at an speed of 1.26×10^5 km/hr in a circular orbit, radius 1.08×10^8 km.
    A Venutian year is the time it takes to complete one orbit.
    A Venutian day is 243 Earth days.
    Is a Venutian day longer than its year?
    Show your working.

Summary

Checkout
You should now be able to...

Test it
Questions

✓ Perform calculations involving roots and indices, including negative and fractional indices.	1 – 4
✓ Perform exact calculations involving fractions, surds and π.	5 – 8
✓ Work with numbers in standard form.	9 – 11

Language	Meaning	Example
Index **Base** **Power**	In index notation, the index or power shows how many times the base has to be multiplied by itself. The plural of index is indices.	Power or index $5^3 = 5 \times 5 \times 5$ Base
Fractional index	A fractional index indicates that a root needs to be found.	$64^{\frac{1}{3}} = \sqrt[3]{64} = 4$
Negative index	The reciprocal of the base to the positive power of the index.	$2^{-3} = \frac{1}{2^3} = \frac{1}{8}$ $\quad 64^{-\frac{1}{3}} = \frac{1}{\sqrt[3]{64}} = \frac{1}{4}$
Reciprocal	The reciprocal of a number is one divided by that number. A number multiplied by its reciprocal equals 1.	The reciprocal of 5 is $\frac{1}{5} = 1 \div 5 = 0.2$ $5 \times \frac{1}{5} = 1$
Root	The *n*th root of a number *x* is the number which to the power *n* equals *x*: $(\sqrt[n]{x})^n = x$	$\sqrt[5]{243} = 3$ $(\sqrt[5]{243})^5 = 3^5 = 3 \times 3 \times 3 \times 3 \times 3 = 243$
Surd	Surds are roots that can only be written accurately using fractional indices or a root symbol.	$\sqrt{2}$ and $\sqrt{3}$ are surds. $\sqrt{4} = 2$ so $\sqrt{4}$ is not a surd.
Approximation	A stated value of a number that is close to but not equal to the true value of a number.	π = 3.14 (3 sf)
Surd form	Writing an exact answer using surds instead of finding a decimal approximation.	$\sqrt{3} + \sqrt{3} = 2\sqrt{3}$
Exact calculation	A calculation that does not involve truncated decimals or other approximations. Exact answers are given in terms of integers, fractions, surds and π.	$\frac{4}{3}\pi \times \left(1\frac{1}{2}\right)^3 = \frac{4}{3} \times \pi \times \frac{3^{3\,2}}{2^{3\,1}} = \frac{9}{2}\pi$ $\sqrt{7^2 - 3^2} = \sqrt{40} = 2\sqrt{10}$
Standard form	Writing a number in the form of a number greater than or equal to 1 and less than 10 multiplied by an integer power of 10.	$498\,000 = 4.98 \times 10^5$ $0.0056 = 5.6 \times 10^{-3}$

Number   Calculations 2

Review

1. Write these numbers in index form
   a $0.\dot{1}$ b $\dfrac{1}{5^7}$ c $\sqrt[3]{12}$ d $\dfrac{1}{\sqrt[3]{8}}$

2. Write down the value of these expressions.
   a $10^5 \times 10^2$ b $5^6 \div 5^3$ c 7^0
   d $(3^2)^3$ e $\sqrt{100}$ f $64^{\frac{1}{2}}$
   g $125^{-\frac{1}{3}}$ h 6^{-2} i $4^{\frac{3}{2}}$
   j $(13^{\frac{1}{2}})^2$ k $27^{\frac{2}{3}}$ l $121^{-0.5}$

3. Simplify these expressions.
   a $2^{\frac{3}{7}} \times 2^{\frac{2}{7}}$ b $2^{\frac{3}{7}} \div 2^{\frac{2}{7}}$
   c $(2^{\frac{3}{7}})^2$ d $2^{-3} \times 2^{-4}$
   e $2^{-3} \div 2^{-4}$ f $(2^{-3})^{-4}$
   g $2^{-\frac{5}{4}} \times 2^{\frac{1}{2}} \div 2^{-\frac{3}{4}}$ h $(2^{-2} \times 2^{\frac{5}{6}})^2 \div 2^{-\frac{2}{3}}$

4. Find the value of the letter in each of these equations.
   a $27 = 3^a$ b $\sqrt{6} = 6^b$
   c $\sqrt[3]{4} = 2^c$ d $\dfrac{1}{1000} = 10^d$

5. Evaluate these expressions exactly.
   a $\dfrac{5}{14} \times 8$ b $\dfrac{5}{9} \times \dfrac{3}{20}$ c $\dfrac{7}{15} + \dfrac{2}{3}$
   d $\dfrac{6}{7} - \dfrac{3}{8}$ e $5\dfrac{1}{5} - 1\dfrac{1}{3}$ f $8\dfrac{1}{3} \times \dfrac{3}{5}$
   g $\dfrac{6}{11} \div 9$ h $\dfrac{3}{14} \div \dfrac{9}{7}$ i $2\dfrac{1}{4} \div 3\dfrac{1}{3}$

6. Simplify these expressions.
   a $7 + 2\pi - 3 - 5\pi$
   b $1 - \sqrt{12} + \sqrt{3}$
   c $11\pi(\sqrt{64} - 5)$
   d $\dfrac{6 + \sqrt{27}}{2} + \dfrac{8 + \sqrt{75}}{3}$
   e $(9 + 2\sqrt{7})(1 - \sqrt{7})$
   f $\dfrac{5 - \sqrt{5}}{3\sqrt{5}}$

7. Find the missing sides in these shapes.
   a 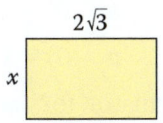 b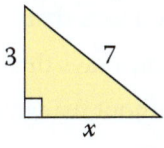

   Perimeter of rectangle = $\sqrt{3}(4 + \sqrt{2})$

8. a Calculate the exact diameter of the circle and the semi-circle.

   i 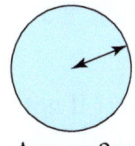 ii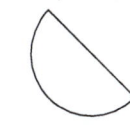
   Area = 2π Area = 18π

   b Calculate the exact circumference of the circle and the perimeter of the semi-circle.

9. Write these approximations in standard form.
   a The UK population in 2014: 63 million.
   b The average distance from Earth to the Sun: 149 600 000 km.
   c The width of a transistor: 0.000 022 mm.
   d 30 g in kg.

10. Write these as ordinary numbers.
    a 2.18×10^9 b 3.1×10^5
    c 5×10^{-7} d 9.92×10^{-5}

11. Calculate these and leave your answer in standard form.
    a $(4 \times 10^3) \times (2.1 \times 10^8)$
    b $(2.4 \times 10^8) \div (5 \times 10^{-3})$
    c $(4.5 \times 10^7) + (3 \times 10^5)$
    d $(8.4 \times 10^{-6}) - (2 \times 10^{-4})$

What next?

Score		
	0 – 4	Your knowledge of this topic is still developing. To improve, look at MyiMaths: 1033, 1045, 1049, 1050, 1051, 1065, 1074, 1301, 1924
	5 – 9	You are gaining a secure knowledge of this topic. To improve your fluency look at InvisiPens: 17Sa – f
	10 – 11	You have mastered these skills. Well done you are ready to progress! To develop your problem solving skills look at InvisiPens: 17Aa – g

Assessment 17

1. Leon says that   **a** $0^2 = 1$ [1]   **b** $49^{\frac{1}{2}} = 7$ [1]   **c** $(-3)^2 = -9$ [1]
   Soraya says that   **a** $0^2 = 0$ **b** $49^{\frac{1}{2}} = \frac{1}{49}$ **c** $(-3)^2 = 9$
   Without using your calculator, for each value who is correct?

2. Are these statements correct? If not, give the correct answer.
   a To work out $15^7 \times 15^5$ you multiply the indices. [1]
   b To work out $(3^4)^5$ you multiply the indices. [1]
   c **i** $(3^4)^0 = 3^4$ **ii** $(3^4)^0 = 1$ **iii** $(3^4)^0 = 3$. [1]
   d $7^7 \times 7^2 \div 7^6 = 7^8$. [1]

3. Are these answers correct? If not give the correct answer.
   a $25^{\frac{3}{2}} = 125$ [2] **b** $216^{\frac{2}{3}} = 144$ [2] **c** $49^{-\frac{1}{2}} = -7$ [1]
   d $196^{\frac{3}{2}} = 941\,192$ [2] **e** $9^{-\frac{5}{2}} = 243$ [2] **f** $156^0 = 156$ [1]

4. Find the values of a, b and c that make these expressions correct.
   a $\sqrt{7} = 7^a$ [1]   **b** $\sqrt[5]{40} = 40^b$ [1]
   c $\sqrt{81} = 9^c$ [2]   **d** $5 = 625^d$ [2]

5. Write the following expressions as powers of **i** 3   **ii** 9.
   a 3 [2]   **b** 9 [2]   **c** 81 [2]
   d $\frac{1}{3}$ [2]   **e** $\frac{1}{9}$ [2]   **f** $\frac{1}{81}$ [2]
   g 27 [2]   **h** $\frac{1}{27}$ [2]   **i** $\sqrt{3}$ [2]

6. **a** The volume of a cone, whose radius is the same as its vertical height, x, is given by the formula $V = kx^3$. Show that $k = \frac{\pi}{3}$. [2]
   b The curved surface area of this cone, A, is given by the formula $A = px^q$ where q can be written as a fraction or integer and p cannot. Find the values of p and q. [3]
   c The curved surface area of a similar cone is $16\pi\sqrt{2}\,\text{m}^2$.
   Find the volume of this cone. [2]

7. An unsharpened wooden pencil is in the shape of a hexagonal prism.
   The side of the hexagon is $\sqrt{3}$ cm. The pencil is 2π cm long.
   The graphite core is a cylinder with radius $0.1\sqrt{2}$ cm. Calculate the following exact values.
   a The area of the hexagonal cross section. [3]
   b The total volume of pencil and core. [2]
   c The volume of the core. [2]
   d The volume of wood in the pencil. [1]

8. **a** The areas of six oceans and seas are given below in miles² (mi²).
   Convert them to ordinary numbers and write them in increasing order of size.

Malay Sea	3.14×10^6 miles²	Bering Sea	8.76×10^5 miles²
Indian Ocean	2.84×10^7 miles²	Caribbean Sea	1.06×10^6 miles²
English Channel	2.9×10^4 miles²	Baltic Sea	1.46×10^5 miles²

   [4]

Number   Calculations 2

8 b Which is bigger: 1×10^9 or $999\,999\,999$ and by how much? [2]

  c Find the value of n in each of the following equations.

    i $4.7 \times 10^n = 47\,000$ [1]   **ii** $6.81 \times 10^n = 681$ [1]

    iii $3.467 \times 10^n = 3\,467\,000$ [1]   **iv** $27.5 \times 10^n = 0.0275$ [1]

  d Rewrite each of these sentences using ordinary numbers.

    i The distance from Mexico City to Moscow is 1.0763×10^4 km. [1]

    ii The energy released by the wingbeat of a honey bee is 8×10^{-4} Joules/second. [1]

  e i Rewrite the following sentence using standard form.

    The average length of a bedbug is $\frac{4}{1\,000\,000}$ of a kilometre. [1]

    ii Write the distance given in part **i** in millimetres. [1]

9 a The Wright Brothers Flyer I, the world's first aircraft, had a mass of 3.4×10^2 kg.
The Saturn V Rocket had a mass of 2.96×10^6 kg.
How many times heavier is a Saturn V than a Flyer I? [2]

  b Flyer I attained a speed of 3.04×10^0 ms^{-1} on its first flight and the supersonic airliner, Concorde, attained a speed of 2.179×10^3 kmh^{-1}.
How much faster was Concorde than Flyer I? [2]

  c There are approximately 4.336×10^9 stars in our galaxy and about 5.776×10^3 stars visible to the naked eye. What fraction of the galaxy can we see?
Write your fraction in the form $\frac{1}{x}$. [2]

  d A triathlon has 3 stages. The largest triathlon has a 3.8×10^0 km swim, a 1.8×10^2 km cycle ride and a $0.421\,95 \times 10^2$ km run.
How far is the race in total?
Give your answer as an ordinary number to 3 sf and in standard form. [3]

10 The highest point on the Earth's surface, Mount Everest, has a recorded height of 8.848×10^3 m.
The deepest point in the ocean, the Challenger Deep, has a depth of 1.1×10^4 m.
How far is it from the bottom of Challenger to the top of Everest?
Give your answer in standard form and as an ordinary number. [4]

11 a The Americas consist of Canada, area 3.852×10^6 miles^2; the USA, area 3.676×10^6 miles^2; Central America, area 2.022×10^5 miles^2 and South America, area 6.879×10^6 miles^2.

    i Calculate the total area of the Americas.
Give your answer in standard form and as an ordinary number. [2]

    ii Calculate the difference in area between the largest and smallest areas in the list. [1]

  b The total surface area of the Earth is 1.969×10^8 miles^2.
The surface area of the world's land masses is 5.749×10^7 miles^2.

    i Find the ratio (area of the Americas) : (total area of the land mass of the Earth).
Give your answer in the form $1:m$. [1]

    ii What area of the Earth's surface is covered by water? [1]

    iii What percentage of the Earth's surface is covered by water? [2]

Revision 3

1. **a** Mo says that the highest common factor of 4410 and 3300 is 2310, and that their lowest common multiple is 485 100. Correct his mistakes. [5]

    b Mo says that the smallest integers that 3300 and 4410 must be multiplied by to make them square numbers are 33 and 10. Explain why he is correct. [3]

2. **a** A rectangle with sides $(4 + \sqrt{3})$ and $(p - \sqrt{3})$ has an area q, where p and q are integers. Find p and q. [4]

    b The sides of an equilateral triangle are $(12 - a\sqrt{3})$, $(6 + a\sqrt{3})$ and $(b\sqrt{3} + 4\frac{1}{2})$. Find a and b. [5]

3. **a** Arrange these numbers from smallest to largest. [6]

    - **i** 4^2
    - **ii** $\sqrt{2.25}$
    - **iii** 1^7
    - **iv** $(-3)^3$
    - **v** 3.2^3
    - **vi** $\sqrt[3]{46.656}$
    - **vii** $\sqrt[3]{-592.704}$
    - **viii** 5^0

    b Which two adjacent numbers have the biggest difference? [1]

    c Which two adjacent numbers have the smallest difference? [1]

4. Two consecutive even numbers when squared and added together equal 100. What are the numbers? [2]

5. David has these solutions to the equations. He says that $a = 3$, $b = 1$ and $c = 2$. Is David correct? For the values that are not correct, give all of the possible correct values.

    a $(5^a)^2 = 625$ [2]

    b $6^{b-1} \times 36 = 1296$ [2]

    c $c^2 = 2^c$ [2]

6. A chord joins the points $(2, 1)$ and $(6, 5)$ on a circle.

    a Calculate the equation of the perpendicular bisector of this chord. [6]

    b Repeat part **a** for a second chord joining the points $(2, 9)$ and $(6, 5)$ on the same circle. [6]

    c Solve your equations simultaneously to find the centre of the circle. [3]

    d Sketch the circle. [3]

7. The graphs $y = (\frac{1}{2}x - 1)^2$ and $y = \frac{1}{2}x + 3$ intersect at points A and B.

    a Use a graphical method to estimate the x-coordinates of A and B. [6]

    b Use an algebraic method to find the exact values of the x-coordinates of A and B. [6]

8. Take the Earth to be a sphere of radius 6370 km. It has a mean density of $5.5\,g/cm^3$.

    a **i** Calculate the volume of the Earth. [2]

      ii Calculate the mass of the Earth. [2]

    b During a storm, water falls at a rate of 1460 litres per second. If none of the water is absorbed, how long would it take for the entire earth to be covered in water to a depth of 10 cm? [5]

9. The table shows the number of words per sentence in 50 sentences in a book.

Words per sentence	Number of sentences
$1 < W \leqslant 5$	3
$5 < W \leqslant 10$	15
$10 < W \leqslant 15$	21
$15 < W \leqslant 20$	9
$20 < W \leqslant 25$	2

    a Show that an estimate of the mean number of words per sentence is 11.73. [5]

    b Draw a cumulative frequency diagram and use it to estimate the median and interquartile range. [8]

Revision   Chapters 13 – 18

10 Ten students preparing for a music exam recorded how many hours per day, on average, they practised. The data was compared with the mark they scored in their Music exam.

Student	A	B	C	D	E	F	G	H	I	J
Hours of practice	4	7	9	7	8	6	10	10	9	5
Test mark	92	65	76	45	74	40	95	30	45	30

   a Draw a scatter graph and draw the line of best fit. [5]

   b Which two students do not fit the trend? Give possible reasons why. [2]

   c Another student missed the exam but had practised, on average, for 5.5 hours daily. What mark should he have achieved? [1]

11 Danish unemployment rates for the period October 2012 to September 2014 are shown.

Month	Oct 2012	Nov 2012	Dec 2012	Jan 2013	Feb 2013	Mar 2013
Rate (%)	7.8	7.7	7.8	7.8	7.9	7.8
Month	Apr 2013	May 2013	Jun 2013	Jul 2013	Aug 2013	Sep 2013
Rate (%)	7.8	7.8	7.8	7.7	7.7	7.6

Month	Oct 2013	Nov 2013	Dec 2013	Jan 2014	Feb 2014	Mar 2014
Rate (%)	7.6	7.4	7.1	7.2	7.2	6.9
Month	Apr 2014	May 2014	Jun 2014	Jul 2014	Aug 2014	Sep 2014
Rate (%)	6.8	6.6	6.5	6.4	6.2	6.0

   a Draw a time series diagram of these figures. Show the data for both years, 2012–13 and 2013–14, on the same axes labelled from October to September. [5]

   b Compare unemployment rates in these periods. [2]

12 a Evaluate these expressions without using a calculator.

     i 6^3 [1]
     ii 10^5 [1]
     iii $\sqrt[3]{125}$ [1]
     iv $\sqrt[3]{512}$ [1]
     v $\sqrt[4]{0}$ [1]
     vi $\sqrt[3]{-343}$ [1]

12 b Kia has attempted to write each of the following as a single power. In each case explain whether Kia is right and give the correct answer if she is wrong.

   i $12^4 \times 12^5 = 144^{20}$ [1]
   ii $(6^2)^3 = 6^5$ [1]
   iii $210^3 \div 210^2 = 210^{1.5}$ [1]
   iv $14^{12} \div 14^{13} = \frac{1}{14}$ [1]
   v $4 \div 4^5 = 4^{-5}$ [1]
   vi $\left(\dfrac{9^8 \times 9^8}{9^5 \times 9^7}\right)^2 = 9^{20}$ [3]

   c Solve each of these equations for x.

     i $5^2 \times 125 = 5^x$ [2]
     ii $16^5 = 4^x$ [2]
     iii $4^8 = 16^x$ [2]
     iv $(3^2)^3 = 9^x$ [2]

13 a A gym has 1.2×10^2 members, each of whom use, on average, 4.7×10^3 units of electricity per year. How many units does the gym use in a year? [2]

   b The mass of a nitrogen atom is 2.326×10^{-23} g. One litre of air contains 4.3602×10^{-1} g of nitrogen. How many nitrogen atoms are there in one litre of air? [2]

14 The fuel consumption in a car engine is modelled by the function $C = \dfrac{240}{v} + \dfrac{v}{8} + 10$, where C is the consumption in litres per hour and v is the speed in mph.

   a Taking values of v, in tens, from 10 to 100, draw up a table, calculate the values of C and draw the graph. [7]

   b Find the consumption when

     i $v = 17.5$ [1]
     ii $v = 67.5$ mph. [1]

   c Find the speed when

     i $C = 30$ l/h [1]
     ii $C = 23$ l/h. [2]

   d At what speed is the car running most efficiently? Give your reason. [2]

18 Pythagoras, trigonometry, vectors and matrices

Introduction

The highest mountain in the world is Mount Everest, located in the Himalayas. Its peak is now measured to be 8848 metres above sea level.

The mountain was first climbed in 1953, by Edmund Hilary and Sherpa Tenzing, almost 100 years after its height was first measured as part of the Great Trigonometrical Survey of India in 1856. The original surveyors, who included George Everest, obtained a height of 8840 metres by measuring the distance and angle of elevation between Mount Everest and a fixed location.

What's the point?

Once a right-angled triangle is seen in a particular problem then a mathematician only needs two pieces of information to be able to calculate all the other lengths and angles.

Objectives

By the end of this chapter, you will have learned how to …

- Use Pythagoras' theorem to find a missing side in a right-angled triangle or the length of a line segment on a coordinate grid.
- Use trigonometric ratios to find missing lengths and angles in triangles.
- Find the exact values of $\sin \theta$ and $\cos \theta$ for key angles.
- Use the sine and cosine rules to find missing lengths and angles.
- Use the sine formula for the area of a triangle.
- Calculate with vectors and use them in geometric proofs.
- Multiply matrices and use them to represent transformations.

Check in

1 Work out each of these.

   a 7^2 **b** $4^2 + 6^2$ **c** $3^2 + 5^2$ **d** $8^2 - 4^2$ **e** $7^2 - 2^2$ **f** $\sqrt{17^2 - 15^2}$

2 Rearrange these equations to make x the subject.

   a $y = \dfrac{x}{6}$ **b** $y = \dfrac{x}{5}$ **c** $y = \dfrac{x}{10}$ **d** $y = \dfrac{2}{x}$ **e** $y = \dfrac{5}{x}$ **f** $y = \dfrac{8}{x}$

Chapter investigation

An engineering company is building a ski lift.
The height of the lift is exactly 45 m.
An engineer suggests using a cable of length 200 m.

The maximum angle of incline is 12°.
Does the engineer's lift meet the criteria?
Design a lift that meets the criteria using the smallest possible length of cable.

SKILLS

18.1 Pythagoras' theorem

The longest side of a **right-angled triangle** is called the **hypotenuse**. The hypotenuse is always opposite the right angle.

In a right-angled triangle the square of the length of the hypotenuse is equal to the sum of squares of the lengths of the other two sides.

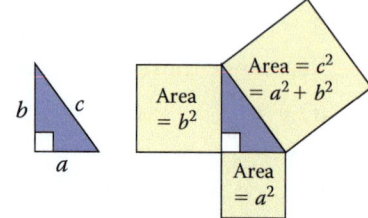

- In symbols, Pythagoras' theorem is
  $c^2 = a^2 + b^2$ where c is the length of the hypotenuse.

EXAMPLE

Calculate the length of the other side in each triangle.

a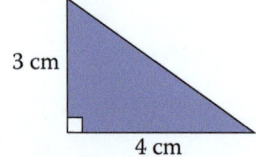

b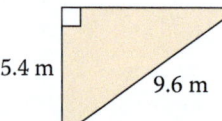

Round the answer when it is not exact. Two significant figures is sensible here.

a Add squares to find the hypotenuse.
$c^2 = 3^2 + 4^2$
$c^2 = 9 + 16$
$c^2 = 25$
$c = \sqrt{25} = 5\,\text{cm}$

This is sometimes called the 3, 4, 5 triangle.

b Subtract to find a shorter side.
$b^2 = 9.6^2 - 5.4^2$
$b^2 = 63$
$b = \sqrt{63} = 7.9372... = 7.9\,\text{m}$ (2 sf)

You can use Pythagoras to work out lengths in other shapes.

EXAMPLE

a A rectangle measures 6 cm by 3 cm. Find the *exact* length of its diagonal.

b A rhombus has sides of length 52 mm. The length of the shorter diagonal is 40 mm. Find the length of the longer diagonal.

Sketch the shape and find a right-angled triangle, then use Pythagoras.

a $c^2 = 3^2 + 6^2$
$= 45$
$c = \sqrt{45}$
$= \sqrt{9 \times 5}$
Length of diagonal $= 3\sqrt{5}\,\text{cm}$

b $d^2 = 52^2 - 20^2$
$= 2304$
$d = \sqrt{2304}$
$= 48$
Length of long diagonal
$= 2 \times 48 = 96\,\text{mm}$

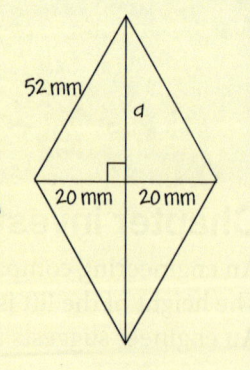

For the exact length, give the answer as a surd.

Geometry   Pythagoras, trigonometry and vectors

Fluency

Exercise 18.1S

1 List the first 20 square numbers.

2 In each diagram find the unknown area.

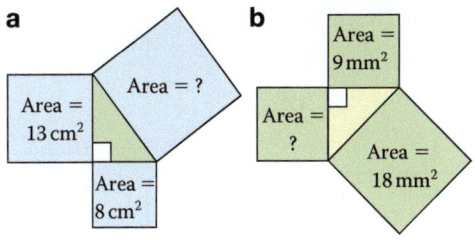

3 Calculate the length of the hypotenuse in these right-angled triangles. State the units.

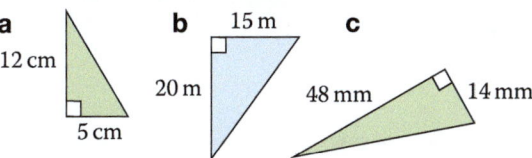

4 Calculate the lengths marked by letters.

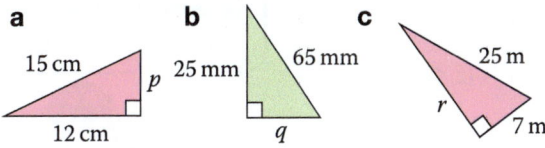

5 Calculate the lengths marked by letters.

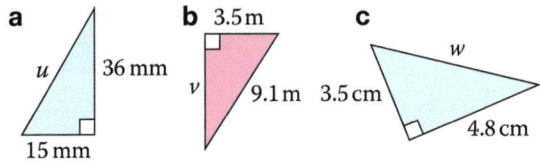

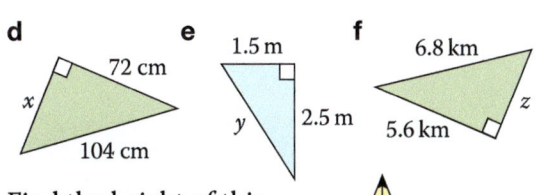

6 Find the height of this isosceles triangle.

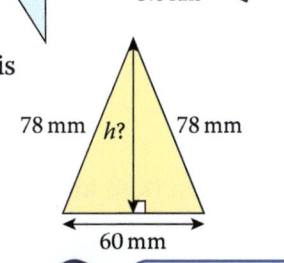

7 Find the length of the diagonals in each rectangle.

	Length of rectangle	Width of rectangle
a	21 mm	28 mm
b	5.8 cm	4.3 cm
c	24.6 m	15.7 m
d	156 km	89 km

8 Calculate the *exact* lengths marked by letters.

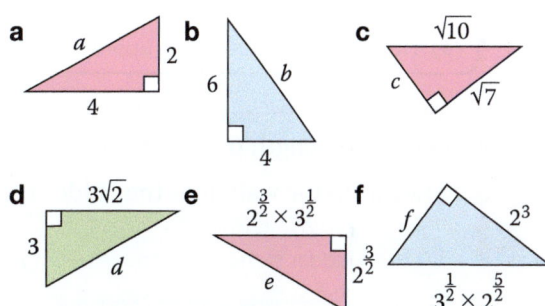

9 A rectangle is 18 cm long. The length of its diagonals is 24 cm. Find the *exact* width of the rectangle.

10 a A square has sides of length 10 cm. Find the *exact* length of its diagonals.

   b A square has diagonals of length 10 cm. Find the *exact* length of its sides.

11 a A rhombus has sides of length 60 mm. The length of the longer diagonal is 96 mm. Find the length of the shorter diagonal.

   b The lengths of the diagonals of a rhombus are 38.4 cm and 11.2 cm. Find the length of the sides of the rhombus.

*12 In kite PQRS, PQ = QR = 10 cm, RS = SP = 17 cm and PR = 16 cm. Calculate the length of QS.

13 An 8.6 metre tall post is supported by two 10 metre cables. Find the distance between the lower ends of the cables.

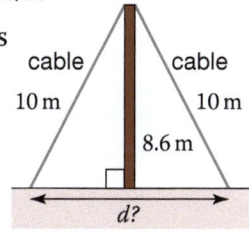

APPLICATIONS

18.1 Pythagoras' theorem

RECAP

- In a right-angled triangle the square of the length of the hypotenuse is equal to the sum of squares of the lengths of the other two sides.
  - To find the length of the hypotenuse use $c^2 = a^2 + b^2$
  - To find the length of a shorter side use $a^2 = c^2 - b^2$ or $b^2 = c^2 - a^2$

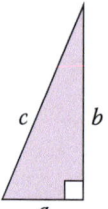

HOW TO

To solve a problem involving the sides of a right-angled triangle
1. Sketch a diagram (if needed) and decide which side is the hypotenuse.
2. Use Pythagoras to find the side you need.
3. Round your answer to an appropriate degree of accuracy and include any units.

EXAMPLE

A ladder of length 6 metres leans against a wall. Its base is 1.5 metres from the wall. How far up the wall does the ladder reach?

1. The ladder is the hypotenuse of a right-angled triangle.
2. For a short side subtract squares. $h^2 = 6^2 - 1.5^2 = 33.75$
3. Round sensibly. $h = \sqrt{33.75}$
   $= 5.8 \text{ m (1 dp)}$

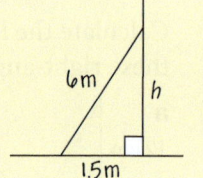

EXAMPLE

Calculate the *exact* distance between the points (2, 5) and (6, 3).

1. The line between (2, 5) and (6, 3) is the hypotenuse of a right-angled triangle.
2. Use $c^2 = a^2 + b^2$. $c^2 = 2^2 + 4^2$
   $= 20$
   $c = \sqrt{20} = \sqrt{4 \times 5}$
3. The exact distance is $2\sqrt{5}$ units

For an exact distance leave the answer as a surd.

EXAMPLE

A circle of radius 3.4 cm has a chord AB of length 5.2 cm.

Find the distance of the chord from the centre of the circle.

1. The radius is the hypotenuse.
2. Subtract squares for a short side.
   $a^2 = 3.4^2 - 2.6^2$
   $a^2 = 4.8$
   $a = \sqrt{4.8}$
   $= 2.190...$
3. Round sensibly.
   The chord is 2.2 cm (2 sf) from the centre.

Remember: a radius bisects a chord.

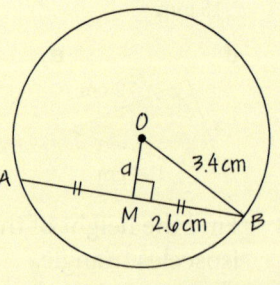

Geometry   Pythagoras, trigonometry and vectors

Reasoning and problem solving

Exercise 18.1A

1. The top of a 4 metre ladder leans against the top of a wall. The wall is 3.8 metre high. How far from the wall is the bottom of the ladder?

   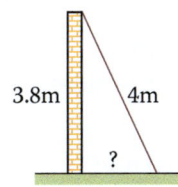

2. Find the *exact* distance between these points.

   a (1, 3) and (4, 7)   b (2, 1) and (6, 4)
   c (3, 2) and (4, 5)   d (0, 5) and (3, 2)
   e (3, 7) and (8, 0)   f (−2, 7) and (2, 4)
   g (2, 0) and (−3, 2) h (−5, 4) and (−1, −2)

3. A chord is 36 mm long. It is 24 mm from the centre of the circle. Find the radius of the circle.

4. A circle, radius 4.8 cm, has two parallel chords of length 5.2 cm. Find the distance between the chords.

5. The diagram shows a path across a rectangular field. How much further is it from *A* to *C* along the sides of the field than along the path?

   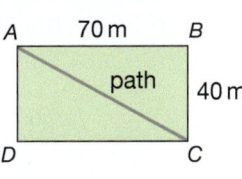

6. Jake draws a triangle in which $AB = 8$ cm, $BC = 4$ cm and $AC = 9$ cm.

   a Explain why angle *ABC* cannot be 90°.
   b Is angle *ABC* acute or obtuse? Explain your answer.

7. Safety advice says that the ratio of *d* to *h* should be 1 : 4.

   For a 5 metre ladder find *d* and *h* correct to the nearest cm.

   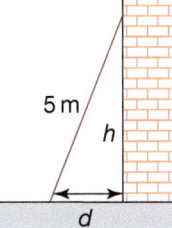

8. Use Pythagoras theorem to draw accurately lines of length

   a $\sqrt{13}$ cm   b $\sqrt{20}$ cm   c $\sqrt{34}$ cm

9. Find the *exact* area of this triangle.

   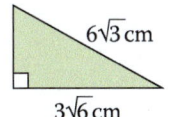

10. An isosceles triangle has sides 20 cm, 26 cm and 26 cm. Find the area of the triangle.

11. Find the area of this end of a building.

    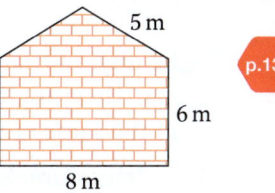

    p.136

12. Find the length *QP*. Give your answer as a surd.

    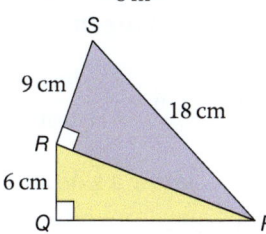

*13. a Use this diagram to prove Pythagoras' theorem.

    b i Prove that for any right-angled triangle the area of the semicircle on the hypotenuse is equal to the sum of the areas of the semicircles on the other two sides.

       ii State and prove a similar statement for equilateral triangles.

    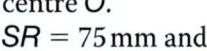

*14. *SR* is a tangent to the circle with centre *O*. $SR = 75$ mm and $OS = 85$ mm. Find the *exact* length of *PQ*.

    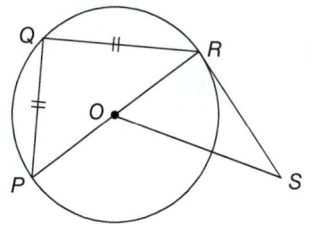

    p.226

15. When three whole numbers *a*, *b*, and *c* satisfy $a^2 + b^2 = c^2$ then *a*, *b* and *c* are called a *Pythagorean triple*.

    a How many Pythagorean triples can you find with $c < 100$?

    b Lola says that you can find other Pythagorean triples by multiplying those you have found by 2 or 3 or 4…. Is this true? Explain.

    *c Prove that, when *x* and *y* are positive integers with $x > y$ then $2xy$, $x^2 + y^2$ and $x^2 - y^2$ is a Pythagorean triple.

1064, 1112

SKILLS

18.2 Trigonometry 1

Right-angled triangles that all have an acute angle θ are similar. Their **angles** are equal and corresponding pairs of sides are in the same ratio.

These **trigonometric** ratios are called the **sine** (sin), **cosine** (cos) and **tangent** (tan) of θ.

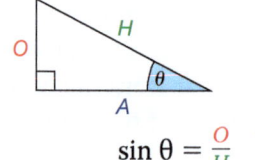

- $\sin \theta = \dfrac{\text{Opposite side}}{\text{Hypotenuse}}$ $\cos \theta = \dfrac{\text{Adjacent side}}{\text{Hypotenuse}}$ $\tan \theta = \dfrac{\text{Opposite side}}{\text{Adjacent side}}$

$\sin \theta = \dfrac{O}{H}$
$\cos \theta = \dfrac{A}{H}$
$\tan \theta = \dfrac{O}{A}$

If you know a side and another angle in a right-angled triangle, you can find the other sides.

EXAMPLE

Find the sides marked by letters.

a 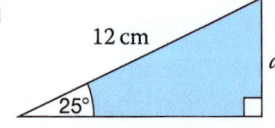   **b** 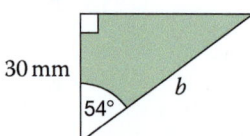   **c**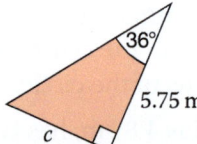

Identify which values you know and require.

a θ = 25°, O = a, H = 12   **b** θ = 54°, A = 30, H = b   **c** θ = 36°, O = c, A = 5.75

Choose the trigonometric ratio and substitute values.

$\sin 25° = \dfrac{a}{12}$ $\cos 54° = \dfrac{30}{b}$ $\tan 36° = \dfrac{c}{5.75}$

Rearrange to find the unknown value.

$12 \times \sin 25° = a$ $b = \dfrac{30}{\cos 54°}$ $5.75 \times \tan 36° = c$

$a = 5.1\,\text{cm}$ (2 sf) $b = 51\,\text{mm}$ (nearest 1 mm) $c = 4.18\,\text{m}$ (3 sf)

Round sensibly and check that the answers seem reasonable.

You could use a different trigonometric ratio or Pythagoras' theorem to find the third side.

If you know the lengths of two sides of a right-angled triangle, you can find the other angles.

EXAMPLE

Calculate the angles marked by letters.

a 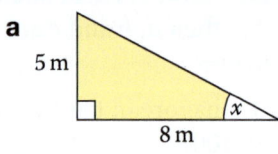   **b** 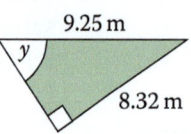   **c**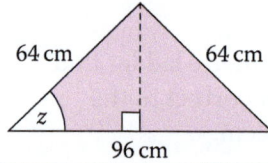

Identify the sides.

a O = 5, A = 8   **b** O = 8.32, H = 9.25   **c** A = 96 ÷ 2 = 48, H = 64

Choose the ratio and substitute values.

$\tan x = \dfrac{5}{8} = 0.625$ $\sin y = \dfrac{8.32}{9.25} = 0.8994...$ $\cos z = \dfrac{48}{64} = 0.75$

You can find the third angle in each triangle by subtracting from 180°.

Use an inverse trigonometric function to find the angle.

$x = \tan^{-1} 0.625$ $y = \sin^{-1} 0.8994...$ $z = \cos^{-1} 0.75$
$x = 32°$ (2 sf) $y = 64.1°$ (3 sf) $z = 41°$ (nearest 1°)

Geometry Pythagoras, trigonometry and vectors

Fluency

Exercise 18.2S

1 Find the sides marked by letters.

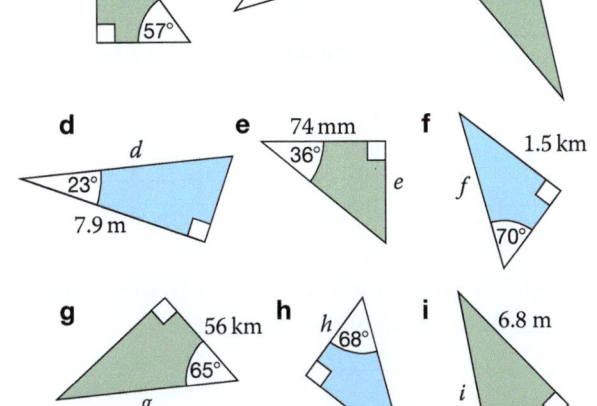

2 Find the angles marked by letters.

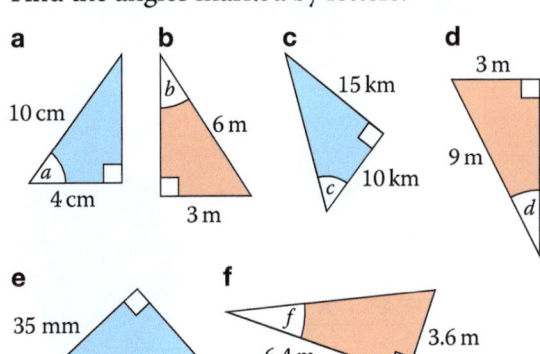

3 Calculate all the unknown sides and angles.

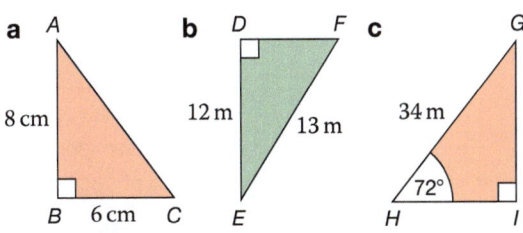

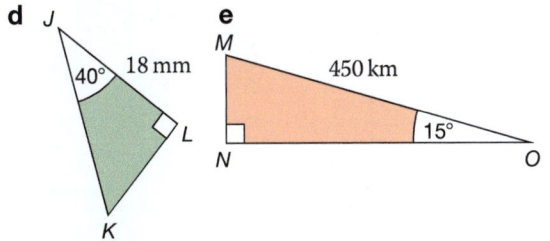

3 f

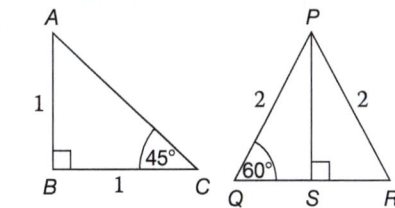

4 a In triangle ABC, $AB = AC = 25$ mm and angle $B = 51°$. Calculate BC.

 b Calculate all the angles of triangle PQR in which $PQ = QR = 7$ cm and $PR = 4$ cm.

5 $KLMN$ is a kite. $KX = 15$ cm, $XM = 27$ cm and $LN = 24$ cm. Calculate the sides and angles of kite $KLMN$.

6 A rhombus has sides of length 10 cm. The length of the longer diagonal is 17 cm.

 Find **a** the angles of the rhombus

 b the length of the shorter diagonal.

*7 a Copy the table.

Angle	sin	cos	tan
0°			
30°			
45°			
60°			
90°			∞

 b

 Use these triangles to find *exact* values for the 30°, 45° and 60° rows of the table. Use surds where necessary.

 c Complete the rest of your table.

8 Safety advice says that the angle between a ladder and the ground must be 75°. For this ladder, find

 a how far up the wall it reaches

 b how far the bottom of the ladder is from the wall.

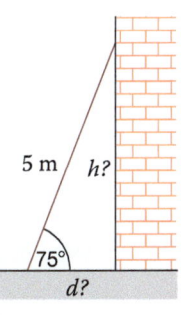

1131, 1133, 1943

371

APPLICATIONS

18.2 Trigonometry 1

RECAP

- In a right-angled triangle
  $\sin \theta = \dfrac{\text{Opposite side}}{\text{Hypotenuse}}$ $\cos \theta = \dfrac{\text{Adjacent side}}{\text{Hypotenuse}}$ $\tan \theta = \dfrac{\text{Opposite side}}{\text{Adjacent side}}$

$\sin \theta = \dfrac{O}{H}$
$\cos \theta = \dfrac{A}{H}$
$\tan \theta = \dfrac{O}{A}$

HOW TO

To solve a trigonometry problem
1. Draw a sketch and look for a right-angled triangle.
2. Decide which trigonometric ratio(s) to use and substitute values.
3. Work out the value(s) required. Check whether they look reasonable.

Angle	sin	cos	tan
30°	$\frac{1}{2}$	$\frac{\sqrt{3}}{2}$	$\frac{1}{\sqrt{3}}$
45°	$\frac{1}{\sqrt{2}}$	$\frac{1}{\sqrt{2}}$	1
60°	$\frac{\sqrt{3}}{2}$	$\frac{1}{2}$	$\sqrt{3}$

EXAMPLE

Haamid looks at a tree that is 30 metres from his bedroom window. The angle from horizontal to the top of the tree is 40°. The angle from horizontal to the base of the tree is 10°.

Haamid thinks the tree is over 30 metres tall.

Is Haamid correct? State any assumptions you make.

① The sketch does not need to be realistic – just use a line for the tree.

Assume the tree is vertical and 30 m is the horizontal distance.

② $\tan 40° = \dfrac{x}{30}$ Using $\tan \theta = \dfrac{O}{A}$.

③ $30 \times \tan 40° = x$ Rearrange to find x.

$x = 25.172...$ Store in your calculator.

② $\tan 10° = \dfrac{y}{30}$

③ $30 \times \tan 10° = y$ $y = 5.289...$

Carry on working on your calculator.

Height of tree = $y + x = 5.289... + 25.172...$
= 30.462... > 30 m

Haamid is correct, the tree is over 30 m.

Use letters like x and y to label lengths that you want to find.

EXAMPLE

Rhombus PQRS has sides of length 20 cm and angle P = 60°. Find the *exact* area of the rhombus.

① PQS is an equilateral triangle, so QS = 20 cm.
② You need the height PT to find the area of triangle PQS.
For an *exact* length, use an *exact value* from the table.

$\sin 60° = \dfrac{O}{H} = \dfrac{PT}{20}$

③ $PT = 20 \times \sin 60°$

$PT = 20 \times \dfrac{\sqrt{3}}{2} = 10\sqrt{3}$

Area of triangle PQS = $\dfrac{1}{2} \times 20 \times 10\sqrt{3} = 100\sqrt{3}$

Area of rhombus PQRS = $2 \times 100\sqrt{3} = 200\sqrt{3}$ cm²

The diagonals are perpendicular (and are also lines of symmetry).

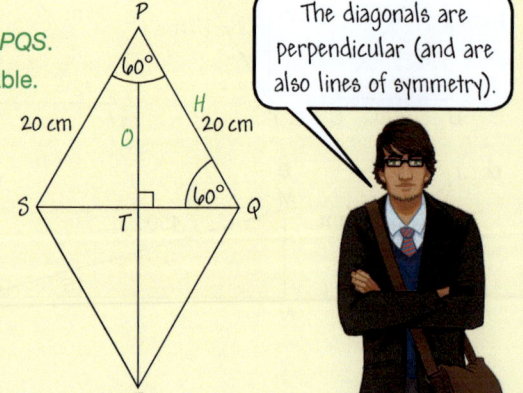

Geometry Pythagoras, trigonometry and vectors

Reasoning and problem solving

Exercise 18.2A

1. Carl looks at a tower from the beach. He estimates these measurements.

   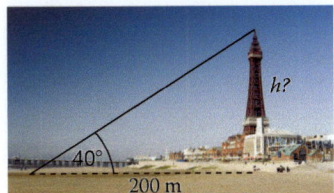

   a. Use Carl's estimates to find the height of the tower.
   b. Why might the answer to part **a** be inaccurate?

2. Thamra is 1.6 m tall. When she looks at a statue from 8 m away, the angle between horizontal and the top is 50°. Thamra says the statue is over 10 metres tall. Is Thamra correct?

3. The diagram shows a swimmer and a boat observed on opposite sides of a 30 m lighthouse.

   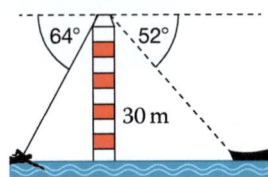

   Find the distance between the swimmer and the boat.

4. The diagram shows a flag on top of a building. How tall is the flag?

5. *PQRS* is a rectangle. The diagonal $PR = 24$ cm and angle $RPQ = 60°$. Find the *exact* area of *PQRS*.

6. A flowerbed is in the shape of a parallelogram. It has sides of length 4 m and 6 m and the smaller angle between them is 45°. Find the *exact* area of the flowerbed.

7. An equilateral triangle has sides of length 18 mm. Find the *exact* area of the triangle.

8. A ramp is needed to help wheelchairs go up three steps. Each step is 12 cm high and 80 cm wide.

   Safety advice says the angle between the ramp and horizontal must be less than 4°.
   Is this ramp safe? Explain your answer.

9. a. Find $\tan \theta$, where θ is the angle between the line $y = 2x$ and the positive x axis.
   b. Repeat part **a** for the line $y = 3x$.
   c. Repeat part **a** for the line $y = \frac{1}{2}x$.
   d. What do you notice?

10. The diagonals of a rectangle are twice as long as the shorter sides. Calculate the angle between a diagonal and a short side.

11. The diagram shows the end view of a house.

    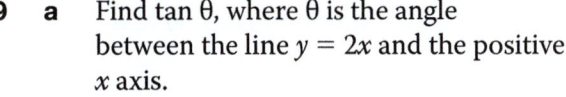

    a. Find the angle the roof makes with the horizontal.
    b. Show the exact area of this end is $24(21 + \sqrt{13})$ ft².

12. Find the *exact* perimeter and area of this isosceles trapezium.

    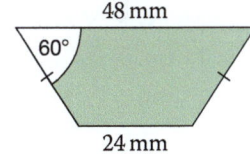

*13 Prove that, for any acute angle θ,
   a. $\sin \theta = \cos(90° - \theta)$
   b. $\sin^2 \theta + \cos^2 \theta = 1$
   c. $\tan \theta = \dfrac{\sin \theta}{\cos \theta}$
   d. $\tan \theta \times \tan(90° - \theta) = 1$

*14 a. The vertices of a regular pentagon lie on a circle of radius 5 cm. Calculate
      i. the perimeter
      ii. the area of the pentagon.
   b. Repeat part **a** for a regular octagon.

SKILLS

18.3 Trigonometry 2

There are two rules that you can use to work out sides and angles in triangles that are *not* right-angled and a formula you can use to find the area of any triangle.

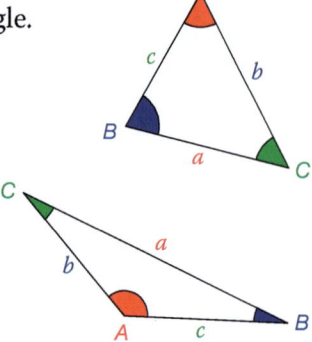

- If you know an angle and its opposite side together with any other side or angle you can use the **sine rule**.

  $$\frac{a}{\sin A} = \frac{b}{\sin B} = \frac{c}{\sin C} \quad \text{or} \quad \frac{\sin A}{a} = \frac{\sin B}{b} = \frac{\sin C}{c}$$

- If you know two sides and the angle between them or you know three sides you can use the **cosine rule**.

  $$a^2 = b^2 + c^2 - 2bc \cos A \quad \text{or} \quad \cos A = \frac{b^2 + c^2 - a^2}{2bc}$$

- If you know two sides and the angle between them you can use the area formula Area $= \frac{1}{2} ab \sin C$

EXAMPLE

Calculate a in this triangle.

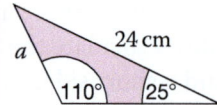

An angle and the opposite side are known: use the sine rule.

$$\frac{a}{\sin 25°} = \frac{24}{\sin 110°}$$ You only need to use two of the fractions.

$$a = \frac{24 \times \sin 25°}{\sin 110°}$$ Rearrange to find a.

$a = 10.793... = 11$ cm (2 sf)

EXAMPLE

In triangle ABC, $AB = 4.7$ m, $BC = 5.6$ m and angle $ABC = 68°$. Calculate all the other sides and angles.

Two sides and the angle between them are known: use the cosine rule.

$AC^2 = 4.7^2 + 5.6^2 - 2 \times 4.7 \times 5.6 \times \cos 68°$

$AC = \sqrt{33.730...} = 5.807... = 5.8$ m (1 dp)

Now that opposites are known, use the sine rule.

$\frac{\sin C}{4.7} = \frac{\sin 68°}{5.8078...}$ Write with the unknown value on top.

$\sin C = \frac{4.7 \times \sin 68°}{5.8078...} = 0.7503...$

$\angle C = \sin^{-1} 0.7503... = 48.618...° = 49°$ (nearest 1°)

$\angle B = 180° - 68° - 48.618... = 63.381... = 63°$ (nearest 1°)

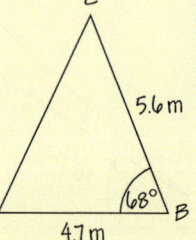

Check the largest angle is opposite the longest side and the smallest angle opposite the shortest side.

131° has the same sine as 49°, but is too big to be the answer here.

p.48

EXAMPLE

In parallelogram $PQRS$, $PQ = 10$ cm, $QR = 8$ cm and diagonal $PR = 15$ cm. Calculate

a   angle PQR

b   the area of $PQRS$.

a   In triangle PQR, you know three sides: use the cosine rule.

$$\cos Q = \frac{10^2 + 8^2 - 15^2}{2 \times 10 \times 8} = -0.38125$$

$\angle PQR = \cos^{-1}(-0.38125) = 112.411...°$

$= 112°$ (nearest degree)

b   In triangle PQR, you know two sides and the angle between them.

Area of triangle $PQR = \frac{1}{2} \times 10 \times 8 \times \sin 112.411...°$

$= 40 \times \sin 112.411...°$

Area of $PQRS = 80 \times \sin 112.411...° = 74$ cm² (2 sf)

Geometry   Pythagoras, trigonometry and vectors

Fluency

Exercise 18.3S

1 Calculate the sides marked by letters.

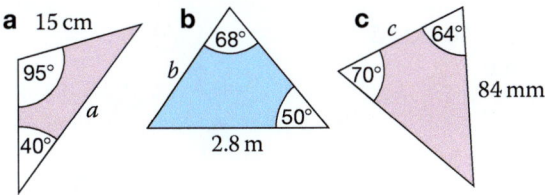

2 Calculate the unknown angles.

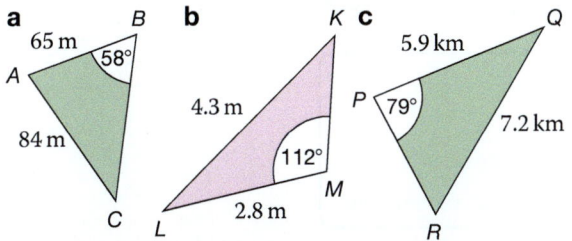

3 Calculate the sides marked by letters.

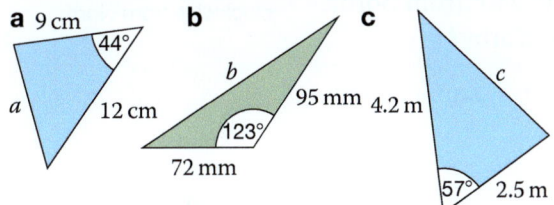

4 Calculate all the angles of each triangle.

Find the largest angle first.

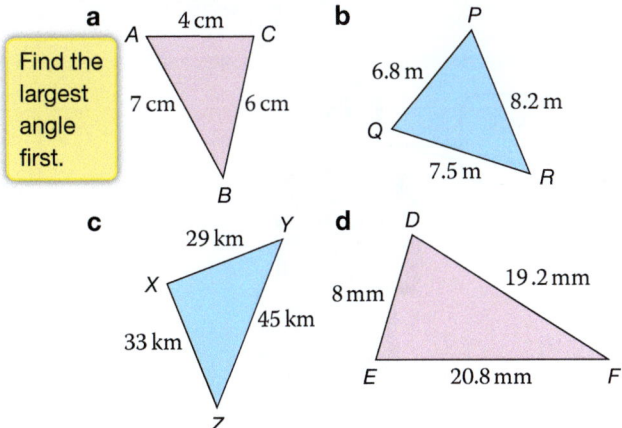

5 Work out the area of each triangle.

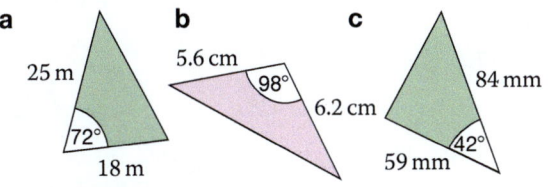

6 Work out all the other sides and angles.

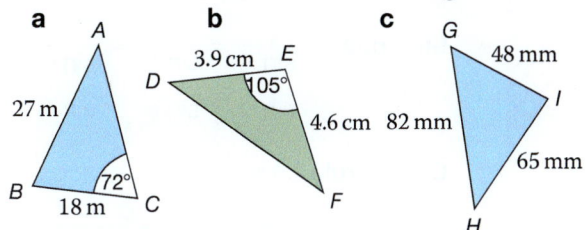

7 Calculate all the other sides and angles in each triangle.

a   Angle $A = 118°$, $AB = 2.6$ cm, $AC = 3.5$ cm

b   Angle $E = 92°$, $EF = 25$ mm, $DF = 48$ mm

c $GH = 15.2$ km, $HI = 13.5$ km, $GI = 9.4$ km

d   Angle $J = 43°$, angle $K = 61°$, $JK = 29$ m

8 Calculate the lengths of the diagonals of parallelogram PQRS.

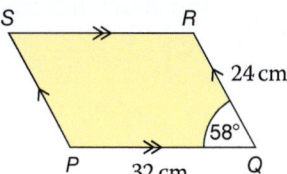

9 In triangle XYZ, $XY = 9$ cm, $YZ = 21$ cm and $ZX = 24$ cm.

Show that angle X is *exactly* 60° and find the other angles.

10 ABCD is an isosceles trapezium. Calculate the *exact* length of

a   AC

b   CD.

*11 In quadrilateral PQRS, the lengths of PQ, QR, RS, SP and PR are 6 cm, 8 cm, 7 cm, 9 cm and 10 cm respectively. Calculate

a   the angles of the quadrilateral

b   the length of QS.

12 The diagram shows a weight attached by ropes to points X and Y, 2.5 metres apart on the ceiling. Calculate the length of each rope.

375

APPLICATIONS

18.3 Trigonometry 2

RECAP

- **Sine rule** $\dfrac{a}{\sin A} = \dfrac{b}{\sin B} = \dfrac{c}{\sin C}$ At least one side & opposite ∠

  or $\dfrac{\sin A}{a} = \dfrac{\sin B}{b} = \dfrac{\sin C}{c}$

- **Cosine rule** $a^2 = b^2 + c^2 - 2bc \cos A$ Two sides & ∠ between

  or $\cos A = \dfrac{b^2 + c^2 - a^2}{2bc}$ Three sides

- **Area of any triangle** $= \dfrac{1}{2} ab \sin C$ Two sides & ∠ between

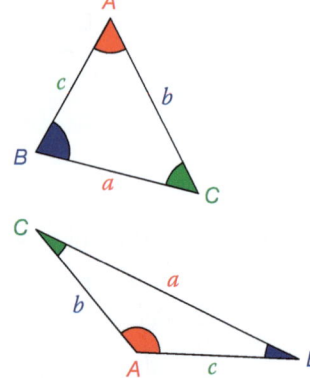

HOW TO

To solve for unknown sides and angles in a triangle
1. Draw a sketch and include all the known values.
2. Decide which trigonometric rule(s) to use and substitute values.
3. Work out the value(s) required. Check whether they look reasonable.

EXAMPLE (p.46)

Alton is 15 km due west of Borby.
The bearing of Calworth is 136° from Alton and 240° from Borby.
How much nearer is Calworth to Alton than to Borby?

Bearings are measured clockwise from North

(1) ∠BAC = 136° − 90° = 46°,   ∠ABC = 360 − 90° − 240° = 30°

∠ACB = 180 − 46° − 30° = 104°

Assume all distances are direct across a flat plane.

(p.190)

(2) $\dfrac{a}{\sin 46°} = \dfrac{15}{\sin 104°}$ A side and opposite ∠ are known.

(3) $a = \dfrac{15 \times \sin 46°}{\sin 104°}$ Rearrange to find a.

= 11.120… km

(2) $\dfrac{b}{\sin 30°} = \dfrac{15}{\sin 104°}$

(3) $b = \dfrac{15 \times \sin 30°}{\sin 104°} = 7.729…$ km

Difference = 11.120… − 7.729… = 3.390…

Calworth is 3.4 km (2 sf) nearer to Alton than to Borby.

Store 11.120… in your calculator to use later.

EXAMPLE

Tse says the area of this plot of land is more than 1000 m².

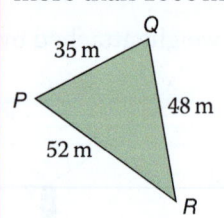

Is Tse correct?

(1)(2) The diagram gives three sides, so use the cosine rule to find an angle.

(3) $\cos R = \dfrac{52^2 + 48^2 - 35^2}{2 \times 52 \times 48} = 0.7578…$

∠R = cos⁻¹ 0.7578 = 40.728…°

Area of triangle PQR = $\dfrac{1}{2} \times 52 \times 48 \times \sin 40.728…°$

= 814.28…. m²

The area is less than 1000 m².

Tse is not correct.

Remember to answer the question!

Geometry   Pythagoras, trigonometry and vectors

Reasoning and problem solving

Exercise 18.3A

> State any assumptions you make.
> Check that each answer seems reasonable.

1. Aamina and Bishr stand 100 m apart on a beach as they watch their friend Sam swimming.

   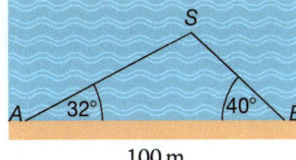

   How much nearer is Sam to Bishr than to Aamina?

2. A ship sailing due north sees a lighthouse on bearing 035°. After another 3 kilometres, the bearing of the lighthouse is 054°.

   How far is the ship now from the lighthouse?

3. A hole on a golf course is 300 m from the tee. A golfer hits the ball a distance of 190 m, but the direction is 15° off course.

   How far from the hole is the ball?

4. From a lighthouse, one boat is 6 km away on a bearing of 130° and another boat is 9 km away on a bearing of 195°.

   Calculate the distance between the boats.

5. A yacht sails due west for 12 km. It then sails on bearing 194° until it is south-west of its starting point.

   How far is the direct route back to the starting point?

6. A farmer says the area of this field is more than a hectare.

   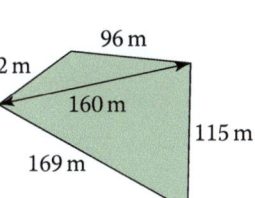

   Is the farmer correct?
   (1 hectare = 10 000 m²).

7. Packham is 57 km due south of Quinton. Rodley is south-east of Packham and 86 km from Quinton. Calculate
   a   the bearing of Rodley from Quinton
   b   how much nearer Rodley is to Packham than Quinton.

8. The hands of a clock are 3 cm and 4.5 cm long. Find the distance between the ends of the hands at
   a   8 am   b   1:15 pm.

9. The diagram shows the end of a building. Calculate

   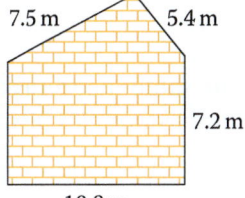

   a   the angle each side of the roof makes with the horizontal.
   b   the area of this end of the building.

10. An aircraft flies 86 km on bearing 294° followed by 120 km on bearing 198°.
    a   Use a scale drawing to find the distance and bearing of its final position from its starting position.
    b   Use trigonometry to check your answers.
    c   Which method is better and why?

11. Ships S and T leave a port at the same time. S sails at 9 km/h on bearing 164°. T sails on bearing 210°.

    After 40 minutes, the bearing of T from S is 259°. Work out the speed of T.

12. a   Use triangles ADC and BDC to write two different expressions for h.

    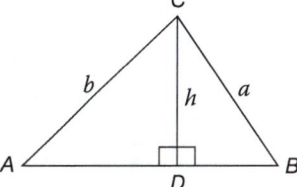

    b   Use your answer to part **a** to show that $\dfrac{a}{\sin A} = \dfrac{b}{\sin B}$

    c   Show that the area of triangle ABC can be found from $\frac{1}{2}bc \sin A$ or $\frac{1}{2}ca \sin B$.

*13 a   Show that $a^2 = b^2 + c^2 - 2bc \cos A$.

   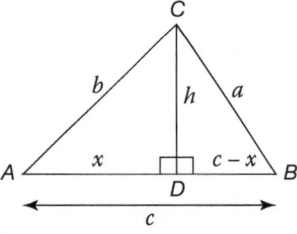

   b   Rearrange this result to show that
   $$\cos A = \dfrac{b^2 + c^2 - a^2}{2bc}$$
   c   Show that these results also apply when angle A is obtuse.

   Note that $\cos(180° - A) = -\cos A$.

SKILLS

18.4 Pythagoras and trigonometry problems

The same theorems and rules that you use to solve problems in two dimensions can be used to solve problems in three dimensions.

You can use these rules for right-angled triangles only.

- **Pythagoras' Theorem** $\quad O^2 + A^2 = H^2$
- **Trigonometry** $\qquad \sin\theta = \dfrac{O}{H} \quad \cos\theta = \dfrac{A}{H} \quad \tan\theta = \dfrac{O}{A}$

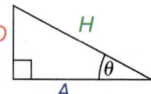

You can use these rules for any triangle.

- **Sine rule** $\qquad \dfrac{a}{\sin A} = \dfrac{b}{\sin B} = \dfrac{c}{\sin C} \quad$ or $\quad \dfrac{\sin A}{a} = \dfrac{\sin B}{b} = \dfrac{\sin C}{c}$
- **Cosine rule** $\qquad a^2 = b^2 + c^2 - 2bc\cos A \quad$ or $\quad \cos A = \dfrac{b^2 + c^2 - a^2}{2bc}$
- **Area of triangle** $= \dfrac{1}{2} ab \sin C$

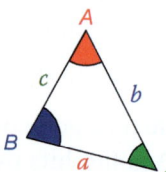

EXAMPLE

The base *BCDE* of this pyramid is a square with sides of length 8 cm. The length of each slant edge is 10 cm.

F is the centre of the base and *M* is the mid-point of *CD*.

Calculate   **a**   the height of the pyramid

            b $\angle ABF$, the angle between a slant edge and the base

            c $\angle AMF$, the angle between a triangular face and the base.

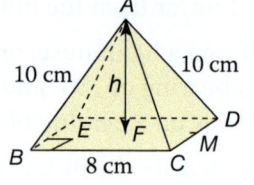

a   *h* is one of the short sides of the right-angle triangle *ABF* and *BF* extended is the hypotenuse of right-angle triangle *BCD*.

In triangle *BCD*

By Pythagoras $\quad BD^2 = 8^2 + 8^2 = 128$

$\qquad\qquad\qquad BD = \sqrt{128} = 8\sqrt{2} \quad$ or $\quad 11.3137... \text{ cm}$

$\qquad\qquad\qquad BF = BD \div 2 = 4\sqrt{2} \quad$ or $\quad 5.656... \text{ cm}$

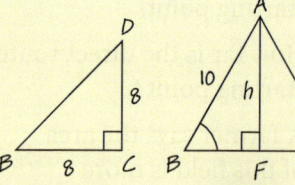

In triangle *ABF*

By Pythagoras $\quad h^2 = 10^2 - BF^2 = 100 - 32 = 68$

$\qquad\qquad\qquad h = \sqrt{68} = 2\sqrt{17} \quad$ or $\quad 8.2462... \text{ cm}$

The height of the pyramid $= 2\sqrt{17} \quad$ or $\quad 8.2$ cm (2 sf)

You could use the cosine rule in triangle ABD to find angle ABF but it is easier to use right-angled triangles.

b   In triangle *ABF*

$\cos \angle ABF = \dfrac{BF}{AB} = \dfrac{4\sqrt{2}}{10} = 0.5656...$

$\angle ABF = \cos^{-1} 0.5656... = 55.550...°$

The angle between a slant edge and the base is 56° (nearest 1°).

c   In triangle *AMF*

$\tan \angle AMF = \dfrac{2\sqrt{17}}{4} = 2.0615...$

$\angle AMF = \tan^{-1} 2.0615... = 64.123...°$

The angle between a triangular face and the base is 64° (nearest 1°).

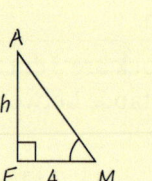
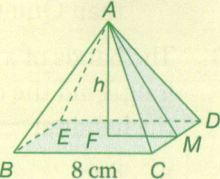

Geometry   Pythagoras, trigonometry and vectors

Fluency

Exercise 18.4S

1. Calculate
   a. AC
   b. AG
   c. ∠GAC, the angle between AG and the base
   d. ∠FAB, the angle between plane AFGD and the base.

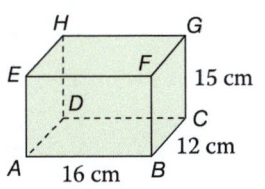

2. a. Calculate the height of this cone.
   b. i. Use triangle SPR to calculate angle SPR.
      ii. Use the cosine rule in triangle PQR to check your answer to part i.

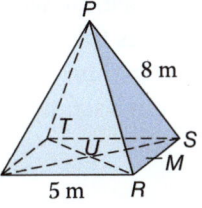

3. For this square-based pyramid, calculate
   a. the height
   b. ∠PSU, the angle between a slant edge and the base
   c. ∠PMU, the angle between a triangular face and the base.

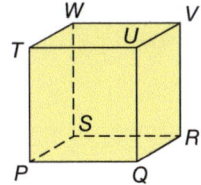

4. This cube has sides of length 20 cm.
   a. Calculate the following lengths, giving each answer as a surd.
      i. QV
      ii. PV
   b. Calculate
      i. ∠VQR (between VQ and the base)
      ii. ∠VPR (between VP and the base).

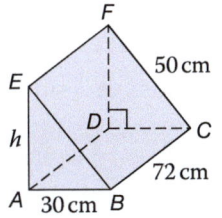

5. For this triangular prism, work out
   a. the height, h
   b. AC
   c. EC
   d. ∠ECA (between EC and the base)
   e. ∠EBA (between BCFE and the base).

6. In this prism, two faces are isosceles trapeziums and the other faces are rectangles.

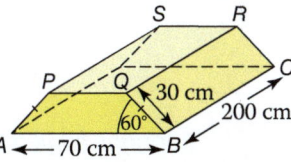

   a. Use triangle AQB to calculate
      i. AQ
      ii. ∠QAB
      iii. the area of AQB.
   b. Use different methods to check your answers.
   c. Calculate   i. AR   ii. ∠RAC.

7. This solid is made from two square-based pyramids. Calculate
   a. the lengths of the sloping edges
   b. the angle between a small triangular face and an adjoining large triangular face.

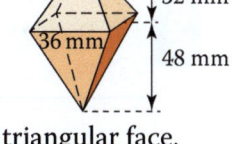

8. ∠PQR = ∠PQS = 90°
   Calculate
   a. ∠RPS
   b. the area of triangle RPS
   c. ∠RQS
   d. the area of triangle RQS.

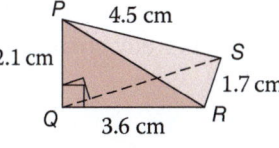

*9. The top and bottom of this frustum are squares. The other faces are congruent isosceles trapeziums.
    Calculate
    a. the length of a sloping edge
    b. the angle between a sloping edge and the base
    c. the angle between a trapezium and the base.

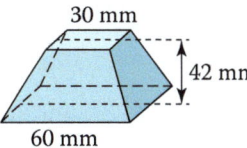

10. The diagram shows the dimensions of a waste skip.
    Calculate
    a. AE
    b. the angle the face ADHE makes with the horizontal.

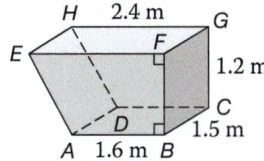

APPLICATIONS

18.4 Pythagoras and trigonometry problems

RECAP

For *right-angled* triangles only
- **Pythagoras' Theorem** $O^2 + A^2 = H^2$
- **Trigonometry** $\sin\theta = \dfrac{O}{H}$ $\cos\theta = \dfrac{A}{H}$ $\tan\theta = \dfrac{O}{A}$

For *any* triangle
- **Sine rule** $\dfrac{a}{\sin A} = \dfrac{b}{\sin B} = \dfrac{c}{\sin C}$
- **Cosine rule** $a^2 = b^2 + c^2 - 2bc\cos A$ or $\cos A = \dfrac{b^2 + c^2 - a^2}{2bc}$
- **Area of triangle** $\dfrac{1}{2}ab\sin C$

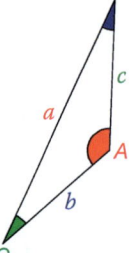

HOW TO

To solve problems involving triangles
1. Draw a sketch and include all the known values.
2. Decide which trigonometric rule(s) to use and substitute values.
3. Work out the value(s) required. Check whether they look reasonable.

Bearings are measured clockwise from North

EXAMPLE

Two observers, Kelly and Liam, are on a straight coastline with Kelly 16 km due north of Liam. Kelly observes a ship on bearing 224°. Liam observes the same ship on bearing 302°.
Kelly says the ship is more than 10 km from the coastline.
Is Kelly correct? Show how you decide and state any assumptions made.

① $\angle SKL = 224° - 180° = 44°$, $\angle SLK = 360° - 302° = 58°$ $\angle KSL = 180° - 44° - 58° = 78°$

Assume Kelly and Liam are at sea level.

② Using the sine rule in triangle KSL.
$\dfrac{SL}{\sin 44°} = \dfrac{16}{\sin 78°}$ A side and opposite ∠ are known.

③ $SL = \dfrac{16 \times \sin 44°}{\sin 78°} = 11.362...$ km

② In right-angled triangle STL
$\sin 58° = \dfrac{d}{11.362...}$

③ $d = 11.362... \times \sin 58° = 9.636...$ km < 10 km

This is less than 10 km, so Kelly is not correct.

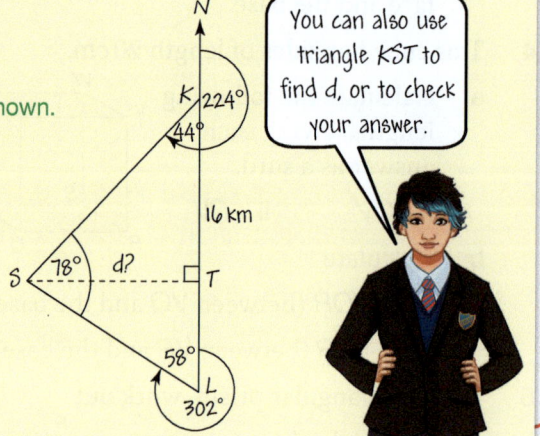

You can also use triangle KST to find d, or to check your answer.

EXAMPLE

Given that θ is acute and $\cos\theta = \dfrac{7}{25}$, find as fractions **a** $\sin\theta$ **b** $\tan\theta$

① Sketch a right-angled triangle in which $\cos\theta = \dfrac{7}{25}$.
② Use Pythagoras to find the unknown side x, then sin and tan can be found.
③ $x^2 = 25^2 - 7^2 = 576$
$x = \sqrt{576} = 24$

a $\sin\theta = \dfrac{O}{H} = \dfrac{24}{25}$ **b** $\tan\theta = \dfrac{O}{A} = \dfrac{24}{7}$

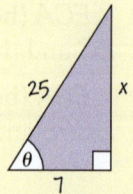

Geometry   Pythagoras, trigonometry and vectors

Reasoning and problem solving

Exercise 18.4A

1. A guidebook says this tower is over 50 m tall.
   Is it correct?

2. The diagram shows 2 sections AB and BC of a cable car route. Calculate angle ABC.

3. From a port P, a ship sails 46 km on a bearing of 104° followed by 32 km on a bearing of 310°.
   a. Calculate the distance and bearing of the ship from P after this journey.
   b. The ship travels west until it is due north of P. The captain says they are now less than 10 km from P.
      Is he correct?

4. In each part θ is an acute angle.
   a. Given that $\tan\theta = \frac{3}{4}$, find as fractions
      i. $\sin\theta$ ii. $\cos\theta$.
   b. Given that $\sin\theta = \frac{12}{13}$, find as fractions
      i. $\cos\theta$ ii. $\tan\theta$.
   c. Given that $\cos\theta = 0.8$, find as decimals
      i. $\sin\theta$ ii. $\tan\theta$.

5. Prove that, for any acute angle θ
   $(\sin\theta)^2 + (\cos\theta)^2 = 1$

6. A regular n-sided polygon has its vertices on a circle, radius r.
   Show that the area of the polygon is given by
   $A = \frac{1}{2}nr^2 \sin\left(\frac{360°}{n}\right)$

7. For this building, calculate
   a. the angle that the roof makes with the horizontal
   b. the cost of tiling the roof at $49 per m² plus 20% VAT.

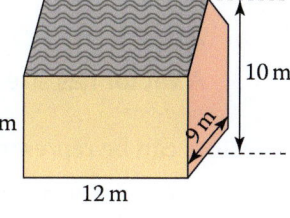

8. The diagram shows a rectangular flowerbed. The radius of the large circle is 50 cm. The radius of each small circle is 18 cm. Calculate the area of the flowerbed in square metres.

9. A rugby cross-bar, AB, is 5.6 m wide and 3 m above the ground.
   The rugby ball, R, is placed on the ground, 14 m from the bottom of one post, C, and 10 m from the bottom of the other post, D. Calculate angle ARB.

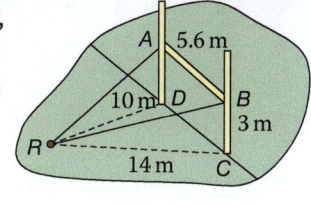

10. A hat is shaped as a cone with radius 8 cm and height 15 cm. Find, in terms of π, the area of card needed to make the hat.

*11. Find, in terms of π, the mass of this pedestal given that the density is 500 kg/m³.

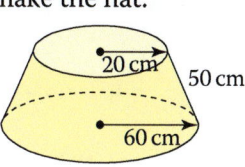

*12. The diagram shows the cross-section of a pipe with diameter 30 mm. The speed of the water in the pipe is 1.5 metres per second. How long will it take to fill a 40 litre tank?

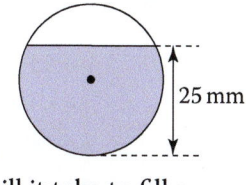

*13. The top of a table is made in the shape of a regular octagon by cutting congruent isosceles triangles from the corners of a 1 m square piece of wood. Show that the perimeter of the octagon is $8(\sqrt{2}-1)$ metres.

SKILLS

18.5 Vectors

p.140
- A **scalar** has size, but no direction.
- A **vector** has size and direction.

Scalars
Distance, mass, temperature

Vectors
Translations, velocities, forces

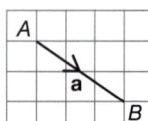

Vectors can be represented in several ways:

on a diagram;

using letters: **a**, a or $\overrightarrow{AB}$, the vector from point A to point B;

as a **column vector** $\binom{3}{-2}$, 3 right and 2 down.

You can add and subtract vectors or multiply them by a scalar.

The first number in a column vector is called the *x* **component** and the second is called the *y* **component**.

- **a + b = b + a**   is vector **a** followed by vector **b** which is equal to **b** followed by **a**.
- **a − b ≠ b − a**   is vector **a** followed by vector **−b** (or **−b** followed by **a**).
- **3a = a + a + a**   is a vector parallel to **a** but 3 times as long.
- **−b**   is a vector equal in size to **b** but in the opposite direction.
- If $\overrightarrow{AB}$ is a multiple of $\overrightarrow{BC}$, then points A, B and C are **collinear**.

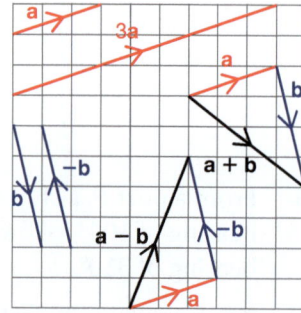

Collinear points lie on the same straight line.

EXAMPLE

$\mathbf{p} = \binom{2}{-1}$, $\mathbf{q} = \binom{0}{3}$, $\mathbf{r} = \binom{-3}{4}$ and $\mathbf{s} = \binom{-4}{5}$

a Calculate **i** **p + q + r** **ii** **2p − r**. **b** Write **s** in terms of **p** and **q**.

Work out the *x* component, then the *y* component or use a diagram.

a i $\underline{p} + \underline{q} + \underline{r} = \binom{2}{-1} + \binom{0}{3} + \binom{-3}{4} = \binom{-1}{6}$

ii $2\underline{p} - \underline{r} = 2\binom{2}{-1} - \binom{-3}{4} = \binom{7}{-6}$

b $\binom{-4}{5} = a\binom{2}{-1} + b\binom{0}{3}$

−4 = 2a   ⇒ a = −2
5 = −a + 3b ⇒ 5 = 2 + 3b ⇒ b = 1
$\underline{s} = -2\underline{p} + \underline{q}$

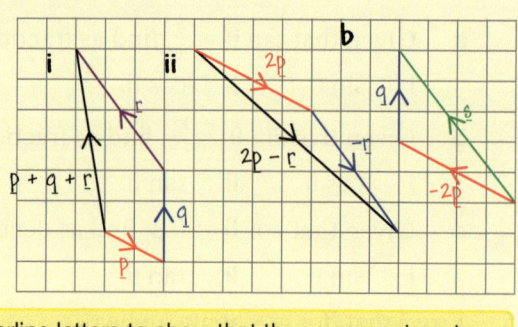

Underline letters to show that they represent vectors.

EXAMPLE

A is the point (1, 4), B is (3, 3) and C is (7, 1).

Use vectors to show that A, B and C are collinear.

Compare the coordinates. The vector from A to B is 2 right and 1 down. From B to C is 4 right and 2 down.

$\overrightarrow{AB} = \binom{2}{-1}$

$\overrightarrow{BC} = \binom{4}{-2} = 2\overrightarrow{AB}$

So A, B and C are collinear.

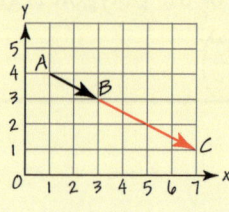

You can use a diagram to check.

Geometry   Pythagoras, trigonometry and vectors

Fluency

Exercise 18.5S

1. **a** Write each vector as a column vector.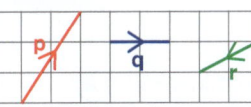
   b Calculate
   - i 2**p**
   - ii 3**r**
   - iii **p** + **q**
   - iv **p** + **r**
   - v **p** − **q**
   - vi **p** − **r**
   - vii **q** + **r**
   - viii **q** − **r**
   - ix **p** + **q** + **r**

2. Vectors $\mathbf{a} = \begin{pmatrix} -2 \\ 4 \end{pmatrix}$, $\mathbf{b} = \begin{pmatrix} 3 \\ -1 \end{pmatrix}$ and $\mathbf{c} = \begin{pmatrix} 0 \\ -2 \end{pmatrix}$.
   a Calculate
   - i **a** + **b** + **c**
   - ii 2**a** + **b**
   - iii **a** − **b** − **c**
   - iv **b** − 3**c**
   - v 2**a** + 3**b**
   - vi $3\mathbf{b} + \frac{1}{2}\mathbf{a} - \frac{1}{2}\mathbf{c}$

   b Use diagrams to check your answers.

3. *P* is the point (5, 3), *Q* is the point (8, 4), *R* is the point (5, 8) and *S* is the point (−1, 1).
   a Calculate these column vectors.
   - i $\overrightarrow{PQ}$
   - ii $\overrightarrow{QR}$
   - iii $\overrightarrow{PR}$
   - iv $\overrightarrow{RS}$
   - v $\overrightarrow{PS}$
   - vi $\overrightarrow{SP}$

   b Are *P*, *Q* and *S* are collinear?

   c Use a diagram to check your answers.

4. *K* is the point (−2, 3), *L* is the point (−1, −4), *M* is the point (4, −3) and *N* is the point (2, −1).
   a Write as column vectors
   - i $\overrightarrow{KL}$
   - ii $\overrightarrow{KM}$
   - iii $\overrightarrow{KN}$
   - iv $\overrightarrow{ML}$
   - v $\overrightarrow{LN}$
   - vi $\overrightarrow{NM}$.

   b Draw a diagram to show these vectors.

   c Which three of the points are collinear?

5. Write each of the vectors **c** to **j** in terms of **a** and/or **b**.

   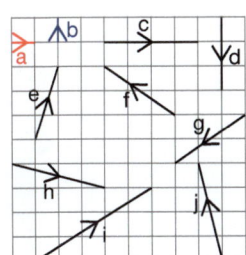

6. **a** On isometric paper, draw diagrams to show
   - i 4**p**
   - ii −2**r**
   - iii **p** + **q**
   - iv **p** − **r**
   - v **p** + **q** + **r**
   - vi 3**p** + **r**
   - vii 3**r** − **q**
   - viii **r** − **q** − **p**

   b Write **s**, **t** and **u** in terms of **q** and **r**.

   c Write **p** in terms of **q** and **r**.

7. Find *a*, *b* and *c* if each vector pair is parallel
   a $\begin{pmatrix} a \\ 3 \end{pmatrix} \begin{pmatrix} 20 \\ 15 \end{pmatrix}$
   b $\begin{pmatrix} -4 \\ b \end{pmatrix} \begin{pmatrix} -1 \\ 3 \end{pmatrix}$
   c $\begin{pmatrix} c \\ 2 \end{pmatrix} \begin{pmatrix} 12 \\ -6 \end{pmatrix}$

8. Show that $\begin{pmatrix} 3 \\ 2 \end{pmatrix}$, $\begin{pmatrix} 12 \\ 8 \end{pmatrix}$ and $\begin{pmatrix} -6 \\ -4 \end{pmatrix}$ are all parallel.

9. In each case find *x* and *y*
   a $\begin{pmatrix} x \\ y \end{pmatrix} + \begin{pmatrix} 1 \\ 3 \end{pmatrix} = \begin{pmatrix} 6 \\ 2 \end{pmatrix}$
   b $\begin{pmatrix} x \\ y \end{pmatrix} - \begin{pmatrix} 2 \\ 5 \end{pmatrix} = \begin{pmatrix} 8 \\ 4 \end{pmatrix}$
   ***c** $x\begin{pmatrix} 3 \\ 2 \end{pmatrix} + y\begin{pmatrix} 1 \\ 5 \end{pmatrix} = \begin{pmatrix} 11 \\ 3 \end{pmatrix}$
   ***d** $x\begin{pmatrix} 4 \\ 3 \end{pmatrix} - y\begin{pmatrix} 3 \\ -2 \end{pmatrix} = \begin{pmatrix} 6 \\ 13 \end{pmatrix}$

10. **p** = 3**a** + 4**b**, **q** = **a** − 2**b** and **r** = **b** − **a**.
    a Find in terms of **a** and **b**
    - i **p** + 2**q**
    - ii **q** + 2**r**
    - iii **q** + **r**
    - iv 3**q** − **p**.

    b What can you say about
    - i vectors **p** + 2**q** and **q** + 2**r**
    - ii vectors **q** + **r** and 3**q** − **p**?

*11. **s** = **p** + 3**q**, **t** = 5**p** + **q**, **u** = 5**q** − 3**p** and **v** = **p** − **q**.
    Find *k* given that **s** + **t** is parallel to **u** + *k***v**.

12. A boat is travelling east at 8 km per hour. It meets a current travelling at 2 km per hour from the north-east.
    a Draw a scale diagram to add these vectors.
    b Find
    - i the resulting speed
    - ii the direction in which the boat travels.

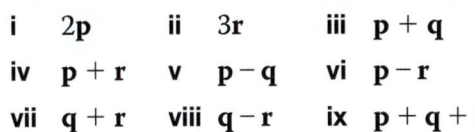

APPLICATIONS

18.5 Vectors

RECAP

- **Addition** $a + b$ is **a** followed by **b** (or **b** followed by **a**).
- **Subtraction** $a - b$ is **a** followed by $-b$ (or $-b$ followed by **a**).
- **Multiplication by a scalar** gives parallel vectors.
  $3a$ is parallel to **a**, but 3 times as long
  $-b$ is equal in size to **b** but in the opposite direction.
- If $\overrightarrow{AB}$ is a multiple of $\overrightarrow{BC}$, then points A, B and C are **collinear**.

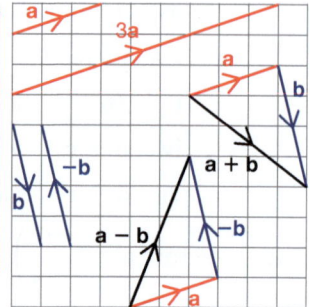

HOW TO

To solve problems involving vectors
1. Decide whether to use a diagram or vector algebra.
2. Find the vectors or components required to obtain the answer.

EXAMPLE

Show that
a $p + q = q + p$
b $p - q = -q + p$

1. Use diagrams to compare the vectors.
2.

This is sometimes called the parallelogram rule for addition.

Vectors **p** and **q** form two sides of a parallelogram; $p + q = q + p$ and $p - q = -q + p$ are the two diagonals.

The **resultant** vectors are equal since they have the same length and direction.

EXAMPLE   p.244

In triangle OAB, $\overrightarrow{OA} = a$ and $\overrightarrow{OB} = b$.

P is the point on OA with $OP:PA = 2:1$ and Q is the point on OB with $OQ:QB = 2:1$.

a   Write $\overrightarrow{PQ}$ in terms of **a** and **b**.
b   Describe the relationship between the line segments PQ and AB.

a $\overrightarrow{OP} = \frac{2}{3}a$ and $\overrightarrow{OQ} = \frac{2}{3}b$

$\overrightarrow{PQ} = \overrightarrow{PO} + \overrightarrow{OQ}$ Vector that starts at P and ends at Q.

$\overrightarrow{PQ} = -\frac{2}{3}a + \frac{2}{3}b$

b   ① Use the diagram to find vector $\overrightarrow{AB}$ then algebra to compare it with $\overrightarrow{PQ}$.

② $\overrightarrow{AB} = -a + b$ Vector that starts at A and ends at B.

$\overrightarrow{PQ} = \frac{2}{3}\overrightarrow{AB}$

PQ is parallel to AB and the length of PQ is $\frac{2}{3}$ of the length of AB.

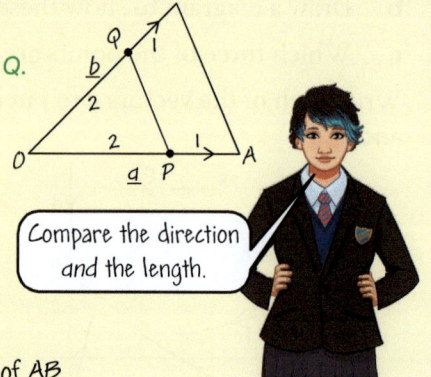

Compare the direction and the length.

Geometry   Pythagoras, trigonometry and vectors

Reasoning and problem solving

Exercise 18.5A

1 a Write down a series of vectors for a boat trip around this island. The trip must start and end at the harbour.

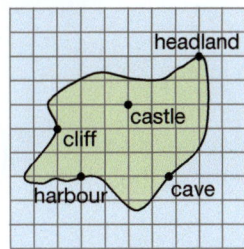

b Add your vectors and explain the result.

2 Use diagrams to show that
  a p + q = q + p
  b 3(p + q) = 3p + 3q
  c −(p + q) = −p − q
  d (p + q) + r = p + (q + r)

3 Find an expression for
  a the length of vector $\begin{pmatrix} x \\ y \end{pmatrix}$
  b the angle the vector makes with the horizontal.

4 In triangle OAB, $\overrightarrow{OA}$ = a and $\overrightarrow{OB}$ = b.
M is the mid-point of OA.
N is the mid-point of OB.
  a Write the following vectors in terms of a and b.
  $\overrightarrow{ON}$ $\overrightarrow{MO}$
  $\overrightarrow{MN}$ $\overrightarrow{AB}$
  b Write two facts about MN and AB.

5 $\overrightarrow{OP}$ = p and $\overrightarrow{OQ}$ = q.
  a Write $\overrightarrow{PQ}$ and $\overrightarrow{QP}$ in terms of p and q.
  b M is the mid-point of PQ.
  Write $\overrightarrow{OM}$ in terms of p and q.

6 a Work out $\overrightarrow{PQ}$ in terms of a and b.
  b Hence show that PQRS is a trapezium.

7 A point P (2, 3) is moved by a series of translations given by these vectors.

$\begin{pmatrix} 4 \\ 1 \end{pmatrix}$ $\begin{pmatrix} 5 \\ -2 \end{pmatrix}$ $\begin{pmatrix} -6 \\ 7 \end{pmatrix}$ $\begin{pmatrix} -3 \\ -5 \end{pmatrix}$

7 a Find the final position of P.
  b Describe the overall transformation.

8 A plane's journey should be 100 km due south, but it is blown 40 km off course by a wind from the north-west.
  a Use a scale diagram to find the resultant distance travelled and bearing.
  b Use trigonometry to check your answer.
  c Which method is better and why?

9 In triangle OAB, $\overrightarrow{OA}$ = a and $\overrightarrow{OB}$ = b.
P is the point on OA with OP:PA = 3:1.
M is the mid-point of OB and N is the mid-point of PB.
  a Write these vectors in terms of a and b.
    i $\overrightarrow{OP}$ **ii** $\overrightarrow{BP}$ **iii** $\overrightarrow{MN}$
  b Describe the relationship between the line segments MN and OA.

*10 In triangle OAB, $\overrightarrow{OA}$ = a and $\overrightarrow{OB}$ = b.
L is the mid-point of OA, M is the mid-point of OB and N is the mid-point of AB.
G is the point on ON with OG:GN = 2:1.
  a Use vectors to show that
    i AGM is a straight line and AG:GM = 2:1
    ii BGL is a straight line and BG:GL = 2:1.
  b What does this tell you about the lines joining each vertex of a triangle to the mid-point of the opposite side?

*11 ABCD is a quadrilateral. The mid-points of its sides are P, Q, R and S. Use vectors to prove that PQRS is a parallelogram.

*12 Use vector methods to prove that the diagonals of a parallelogram bisect each other.

18.6 Matrices

- A **matrix** is an array of numbers, usually written in brackets.
- The **order** of a matrix is its size, given in terms of the number of rows and the number of columns. Matrices can be any size, but you will only **need** to work with those that are 2×1 or 2×2.

Vectors such as $\begin{pmatrix} 3 \\ -1 \end{pmatrix}$ are 2×1 **matrices**.

$\begin{pmatrix} 3 & -1 \\ 0 & 4 \end{pmatrix}$ is a 2×2 matrix (read as '2 by 2')

The methods used in Section 18.5 for multiplying vectors by scalars can also be used with 2×2 matrices. A matrix can also be multiplied by another matrix when the number of columns in the first matrix is the same as the number of rows in the second matrix.

$4\begin{pmatrix} 1 & 2 \\ 3 & -1 \end{pmatrix} = \begin{pmatrix} 4 & 8 \\ 12 & -4 \end{pmatrix}$

To multiply matrices, multiply each row from the first matrix by each column from the second matrix. When multiplying a row by a column the elements are multiplied in pairs and the results added.

Examples:

$\begin{pmatrix} 4 & 5 \\ 2 & 6 \end{pmatrix}\begin{pmatrix} 3 \\ 2 \end{pmatrix} = \begin{pmatrix} 4 \times 3 + 5 \times 2 \\ 2 \times 3 + 6 \times 2 \end{pmatrix} = \begin{pmatrix} 12 + 10 \\ 6 + 12 \end{pmatrix} = \begin{pmatrix} 22 \\ 18 \end{pmatrix}$

$\begin{pmatrix} 2 & 3 \\ 0 & 6 \end{pmatrix}\begin{pmatrix} 7 & 1 \\ 5 & 4 \end{pmatrix} = \begin{pmatrix} 2 \times 7 + 3 \times 5 & 2 \times 1 + 3 \times 4 \\ 0 \times 7 + 6 \times 5 & 0 \times 1 + 6 \times 4 \end{pmatrix}$

$= \begin{pmatrix} 14 + 15 & 2 + 12 \\ 0 + 30 & 0 + 24 \end{pmatrix} = \begin{pmatrix} 29 & 14 \\ 30 & 24 \end{pmatrix}$

When two matrices are multiplied, the answer has the same number of rows as the first matrix and the same number of columns as the second matrix.

So a 2×2 matrix multiplied by a 2×1 matrix always gives a 2×1 matrix.

Any matrix is left unchanged when multiplied by the **identity matrix**, $I = \begin{pmatrix} 1 & 0 \\ 0 & 1 \end{pmatrix}$

EXAMPLE

Use these matrices.

$A = \begin{pmatrix} 1 & 3 \\ 2 & 4 \end{pmatrix} \quad B = \begin{pmatrix} 0 & 2 \\ 5 & -3 \end{pmatrix} \quad C = \begin{pmatrix} 4 \\ -1 \end{pmatrix} \quad D = \begin{pmatrix} x \\ y \end{pmatrix}$

a Calculate **i** AB **ii** BA **iii** AC

b Show that $AI = A$ and $IC = C$.

c Given that $BD = C$ find the values of x and y.

Order of multiplication matters. AB is not the same as BA.

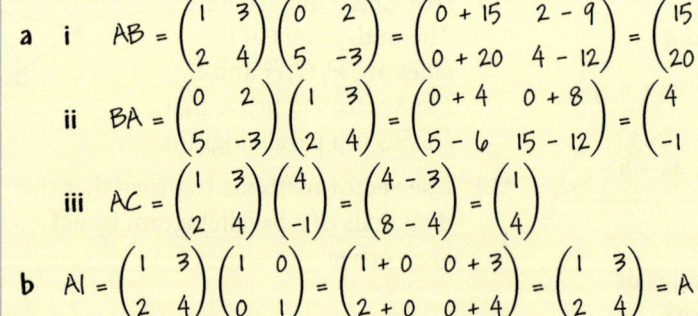

c $\begin{pmatrix} 0 & 2 \\ 5 & -3 \end{pmatrix}\begin{pmatrix} x \\ y \end{pmatrix} = \begin{pmatrix} 4 \\ -1 \end{pmatrix}$

$\begin{pmatrix} 2y \\ 5x - 3y \end{pmatrix} = \begin{pmatrix} 4 \\ -1 \end{pmatrix}$

$2y = 4$ so $y = 2$

$5x - 3y = -1$ gives $5x - 6 = -1$

so $5x = 5$ and $x = 1$

Geometry   Pythagoras, trigonometry and vectors

Fluency

Exercise 18.6S

1 Calculate

   a $\begin{pmatrix} 5 \\ 7 \end{pmatrix} + \begin{pmatrix} 3 \\ 4 \end{pmatrix}$ b $\begin{pmatrix} 5 \\ 7 \end{pmatrix} - \begin{pmatrix} 3 \\ 4 \end{pmatrix}$ c $4\begin{pmatrix} 5 \\ 7 \end{pmatrix}$

2 Calculate

   a $\begin{pmatrix} 5 & 1 \\ 3 & 2 \end{pmatrix}\begin{pmatrix} 6 \\ 4 \end{pmatrix}$ b $\begin{pmatrix} 2 & -1 \\ 0 & 3 \end{pmatrix}\begin{pmatrix} 5 \\ 6 \end{pmatrix}$

   c $\begin{pmatrix} -3 & 4 \\ 2 & -1 \end{pmatrix}\begin{pmatrix} -1 \\ 5 \end{pmatrix}$ d $\begin{pmatrix} 4 & -5 \\ -3 & 0 \end{pmatrix}\begin{pmatrix} -2 \\ -1 \end{pmatrix}$

3 Calculate

   a $\begin{pmatrix} 2 & 1 \\ 5 & 3 \end{pmatrix}\begin{pmatrix} 6 & 4 \\ 7 & 8 \end{pmatrix}$

   b $\begin{pmatrix} 3 & 2 \\ -4 & 0 \end{pmatrix}\begin{pmatrix} 1 & 5 \\ 9 & 7 \end{pmatrix}$

   c $\begin{pmatrix} 5 & -3 \\ 2 & 1 \end{pmatrix}\begin{pmatrix} 2 & -1 \\ 4 & 6 \end{pmatrix}$

   d $\begin{pmatrix} -1 & 3 \\ 4 & -2 \end{pmatrix}\begin{pmatrix} -6 & 0 \\ 1 & 2 \end{pmatrix}$

4 Calculate

   a $\begin{pmatrix} 9 & 3 \\ 0 & -1 \end{pmatrix}\begin{pmatrix} 5 \\ -6 \end{pmatrix}$

   b $\begin{pmatrix} 4 & -1 \\ 2 & 1 \end{pmatrix}\begin{pmatrix} 7 & -3 \\ 2 & 0 \end{pmatrix}$

   c $\begin{pmatrix} 0 & -3 \\ 5 & -4 \end{pmatrix}\begin{pmatrix} -4 & 6 \\ 0 & -8 \end{pmatrix}$

   d $\begin{pmatrix} -7 & 4 \\ 5 & 0 \end{pmatrix}\begin{pmatrix} -2 \\ -1 \end{pmatrix}$

5 Use these matrices.

   $K = \begin{pmatrix} 3 \\ 5 \end{pmatrix} L = \begin{pmatrix} 4 \\ 7 \end{pmatrix} M = \begin{pmatrix} 6 & 2 \\ 1 & 8 \end{pmatrix} N = \begin{pmatrix} 0 & 5 \\ 2 & -1 \end{pmatrix}$

   Calculate

   a K + L   b K − L   c 4K
   d 2M   e MK   f ML
   g NK   h NL   i MN
   j NM   k IK   l NI

6 $A = \begin{pmatrix} 1 & 0 \\ 3 & 5 \end{pmatrix} B = \begin{pmatrix} -4 \\ 2 \end{pmatrix} C = \begin{pmatrix} 6 & 5 \\ -1 & 3 \end{pmatrix}$

   a Write down the order of each matrix.

   b Calculate where possible
      i 3A   ii −5B   iii AB   iv BC
      v AC   vi CA   vii BA   viii CB

7 $A = \begin{pmatrix} 2 & 5 \\ 1 & 3 \end{pmatrix}$ and $B = \begin{pmatrix} 3 & -5 \\ -1 & 2 \end{pmatrix}$

   Show that AB = BA = I

8 Work out the values of x and y.

   a $\begin{pmatrix} 5 & -1 \\ 4 & 7 \end{pmatrix}\begin{pmatrix} 2 \\ -3 \end{pmatrix} = \begin{pmatrix} x \\ y \end{pmatrix}$

   b $\begin{pmatrix} 3 & 1 \\ 4 & 2 \end{pmatrix}\begin{pmatrix} x \\ 6 \end{pmatrix} = \begin{pmatrix} 0 \\ y \end{pmatrix}$

   c $\begin{pmatrix} 2 & -1 \\ 3 & 4 \end{pmatrix}\begin{pmatrix} x \\ y \end{pmatrix} = \begin{pmatrix} 9 \\ 19 \end{pmatrix}$

   d $\begin{pmatrix} x & 5 \\ 3 & 1 \end{pmatrix}\begin{pmatrix} x \\ y \end{pmatrix} = \begin{pmatrix} -15 \\ 7 \end{pmatrix}$

   e $\begin{pmatrix} 4 & -2 \\ 3 & y \end{pmatrix}\begin{pmatrix} x \\ y \end{pmatrix} = 2\begin{pmatrix} y \\ 5 \end{pmatrix}$

9 a Prove that, for any 2 × 1 matrix M, IM = M

   b Prove that, for any 2 × 2 matrix N, IN = N and NI = N

*10 $M = \begin{pmatrix} a & b \\ c & d \end{pmatrix}$ and $N = \frac{1}{ad - bc}\begin{pmatrix} d & -b \\ -c & a \end{pmatrix}$

   Show that NM = I and MN = I

11 A café sells cups of tea for £1.50 and cups of coffee for £2. The table shows the number of cups of tea and coffee sold during one day.

	Cups of tea	Cups of coffee
Morning	30	45
Afternoon	52	64

   Work out these matrix calculations.
   In each case describe the information given by the answer.

   a $\begin{pmatrix} 30 \\ 52 \end{pmatrix} + \begin{pmatrix} 45 \\ 64 \end{pmatrix}$

   b $\begin{pmatrix} 30 & 45 \\ 52 & 64 \end{pmatrix}\begin{pmatrix} 1.5 \\ 2 \end{pmatrix}$

APPLICATIONS

18.6 Matrices

RECAP

- To **multiply matrices**, multiply each row from the first matrix by each column from the second matrix.

$$\begin{pmatrix} a & b \\ c & d \end{pmatrix} \begin{pmatrix} x \\ y \end{pmatrix} = \begin{pmatrix} ax + by \\ cx + dy \end{pmatrix}$$

$$\begin{pmatrix} a & b \\ c & d \end{pmatrix} \begin{pmatrix} p & q \\ r & s \end{pmatrix} = \begin{pmatrix} ap + br & aq + bs \\ cp + dr & cq + ds \end{pmatrix}$$

The vector that moves a point from the origin (0, 0) to another point is called the **position vector** of that point. 2×2 matrices can be used to define transformations. The co-ordinates of an image point is found by **pre-multiplying** the position vector by the matrix representing the transformation.

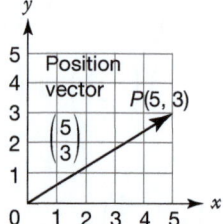

HOW TO

Here is a general strategy for solving problems

1. Multiply matrices to find the position vectors of image points.
2. Draw a diagram to show the object and its image.
3. Describe the transformation and answer any other questions.

Remember that order is important. $AB \neq BA$

EXAMPLE

The transformation matrix $\begin{pmatrix} 0 & -1 \\ 1 & 0 \end{pmatrix}$ maps triangle $P(2, 1)$ $Q(3, 1)$ $R(2, 4)$ to $P'Q'R'$.

a Draw PQR and $P'Q'R'$. **b** Give a full description of the transformation.

a ① Multiply the transformation matrix by the position vector of each vertex. ② Draw a diagram.

$P \begin{pmatrix} 0 & -1 \\ 1 & 0 \end{pmatrix} \begin{pmatrix} 2 \\ 1 \end{pmatrix} = \begin{pmatrix} -1 \\ 2 \end{pmatrix}$ $Q \begin{pmatrix} 0 & -1 \\ 1 & 0 \end{pmatrix} \begin{pmatrix} 3 \\ 1 \end{pmatrix} = \begin{pmatrix} -1 \\ 3 \end{pmatrix}$

$R \begin{pmatrix} 0 & -1 \\ 1 & 0 \end{pmatrix} \begin{pmatrix} 2 \\ 4 \end{pmatrix} = \begin{pmatrix} -4 \\ 2 \end{pmatrix}$

b ③ Describe the transformation.

The transformation is a rotation of 90° anticlockwise about (0, 0).

Remember to give all the information.

The unit vectors $\begin{pmatrix} 1 \\ 0 \end{pmatrix}$ and $\begin{pmatrix} 0 \\ 1 \end{pmatrix}$ are position vectors of vertices of the **unit square**.

The general result $\begin{pmatrix} a & b \\ c & d \end{pmatrix} \begin{pmatrix} 1 \\ 0 \end{pmatrix} = \begin{pmatrix} a \\ c \end{pmatrix}$ and $\begin{pmatrix} a & b \\ c & d \end{pmatrix} \begin{pmatrix} 0 \\ 1 \end{pmatrix} = \begin{pmatrix} b \\ d \end{pmatrix}$ is useful when no shape is given.

EXAMPLE

a Describe the single transformation represented by $\begin{pmatrix} -2 & 0 \\ 0 & -2 \end{pmatrix}$

b Find the transformation matrix that reflects shapes in the line $y = -x$.

a ① $\begin{pmatrix} 1 \\ 0 \end{pmatrix}$ maps to $\begin{pmatrix} -2 \\ 0 \end{pmatrix}$ and $\begin{pmatrix} 0 \\ 1 \end{pmatrix}$ maps to $\begin{pmatrix} 0 \\ -2 \end{pmatrix}$ **b** 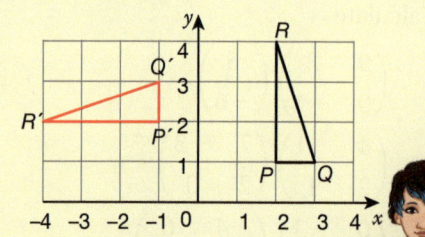 $\begin{pmatrix} 1 \\ 0 \end{pmatrix}$ maps to $\begin{pmatrix} 0 \\ -1 \end{pmatrix}$

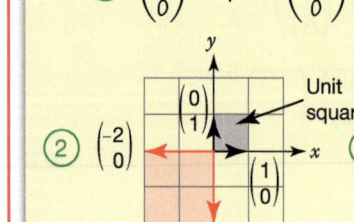

$\begin{pmatrix} 0 \\ 1 \end{pmatrix}$ maps to $\begin{pmatrix} -1 \\ 0 \end{pmatrix}$

③ This is enlargement centre (0, 0), scale factor −2

Matrix = $\begin{pmatrix} 0 & -1 \\ -1 & 0 \end{pmatrix}$

Geometry Pythagoras, trigonometry and vectors

Reasoning and problem solving

Exercise 18.6A

1. The transformation matrix $\begin{pmatrix} 1 & 0 \\ 0 & -1 \end{pmatrix}$ maps triangle $A(3, 0)$ $B(4, 3)$ $C(-1, 4)$ to $A'B'C'$.

   a Draw ABC and $A'B'C'$.

   b Give a full description of the transformation.

2. A trapezium has vertices $P(0, 4)$ $Q(6, 4)$ $R(6, 6)$ and $S(2, 6)$.

   a Draw $PQRS$ and its image under transformation matrix $\begin{pmatrix} \frac{1}{2} & 0 \\ 0 & \frac{1}{2} \end{pmatrix}$.

   b Describe the transformation fully.

3. By drawing a diagram of quadrilateral $K(2, -2)$ $L(-1, -2)$ $M(-2, -1)$ $N(0, 3)$ and its image, illustrate what happens to a shape under the transformation matrix $\begin{pmatrix} -3 & 0 \\ 0 & -3 \end{pmatrix}$.

   Give a full description of the transformation.

4.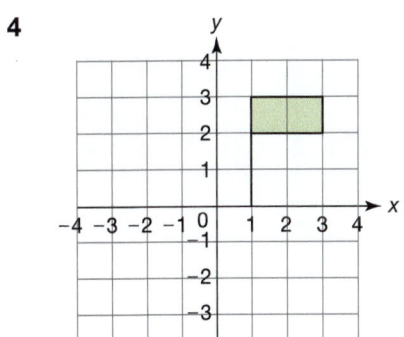

   Draw the image of the flag after each of these transformation matrices.

   Describe each transformation.

   a $\begin{pmatrix} 0 & 1 \\ -1 & 0 \end{pmatrix}$ b $\begin{pmatrix} -1 & 0 \\ 0 & 1 \end{pmatrix}$ c $\begin{pmatrix} -1 & 0 \\ 0 & -1 \end{pmatrix}$

5. a Describe the single transformation represented by $\begin{pmatrix} 5 & 0 \\ 0 & 5 \end{pmatrix}$.

   Use the unit square.

   b Check your answer by drawing another small shape and its image under this transformation matrix.

6. a Find the transformation matrix that reflects the unit square in the line $y = x$.

   b Check your answer by drawing a triangle and its image under your transformation matrix.

7. a The unit square $OABC$ is mapped to $OA'B'C'$ under the transformation matrix M.

   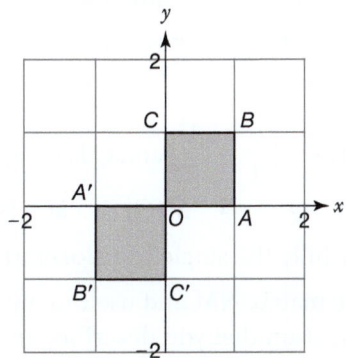

   i Work out the matrix M.

   ii Sally says that M represents more than one transformation.

   Is Sally correct? Explain your answer.

   b The unit square $OABC$ is mapped to $OA''B''C''$ under the transformation matrix $\begin{pmatrix} 0 & -1 \\ -1 & 0 \end{pmatrix}$.

   Draw a diagram showing $OABC$ and $OA''B''C''$ and give a full description of the transformation that is represented by this matrix.

8. Describe the transformation represented by each of these matrices.

   a $\begin{pmatrix} 3 & 0 \\ 0 & 3 \end{pmatrix}$ b $\begin{pmatrix} 0 & 1 \\ 1 & 0 \end{pmatrix}$ c $\begin{pmatrix} 0 & -1 \\ 1 & 0 \end{pmatrix}$

9. Find the matrix that represents each of the following transformations.

   a reflection in the line $y = 0$

   b reflection in the line $x = 0$

   c rotation of 270° anticlockwise about (0, 0)

   d enlargement, centre (0, 0), scale factor -4.

SKILLS

18.7 Combining transformation matrices

Points, shapes, transformations and their matrices can all be represented by letters. Such letters can be combined to represent the image of a point or shape, a combination of transformations or the matrix that represents this combination.

If **A** and **B** are transformation matrices and *P* is a point or shape

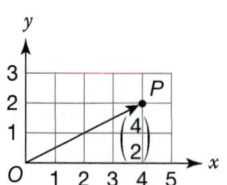

- **A**(*P*) is the image of *P* after transformation **A**.
- **BA**(*P*) is the image of **A**(*P*) after transformation **B**.
- **BA** is the matrix that represents **A** *followed by* **B**.

Order is important when transformation matrices are combined:

AB is **B** *followed by* **A**.

EXAMPLE

$M = \begin{pmatrix} 0 & 1 \\ 1 & 0 \end{pmatrix}$, $N = \begin{pmatrix} 0 & -1 \\ 1 & 0 \end{pmatrix}$ and *Q* is the quadrilateral with vertices at (1, 1), (3, 0), (4, 2) and (4, 3).

a Draw **i** *Q*   **ii** **M**(*Q*).   **iii** **NM**(*Q*).

b Describe fully the single transformation that has the same effect as **NM**.

c Calculate matrix **NM** and use the unit square to show that your answer represents the transformation that you described in part **b**.

a i Plot and join the given vertices. Label the shape *Q*.

ii Find the position vector of each vertex of *Q* after **M**.

$\begin{pmatrix} 0 & 1 \\ 1 & 0 \end{pmatrix}\begin{pmatrix} 1 \\ 1 \end{pmatrix} = \begin{pmatrix} 1 \\ 1 \end{pmatrix}$ $\begin{pmatrix} 0 & 1 \\ 1 & 0 \end{pmatrix}\begin{pmatrix} 3 \\ 0 \end{pmatrix} = \begin{pmatrix} 0 \\ 3 \end{pmatrix}$

$\begin{pmatrix} 0 & 1 \\ 1 & 0 \end{pmatrix}\begin{pmatrix} 4 \\ 2 \end{pmatrix} = \begin{pmatrix} 2 \\ 4 \end{pmatrix}$ $\begin{pmatrix} 0 & 1 \\ 1 & 0 \end{pmatrix}\begin{pmatrix} 4 \\ 3 \end{pmatrix} = \begin{pmatrix} 3 \\ 4 \end{pmatrix}$

Plot and join these vertices. Label the image **M**(*Q*).

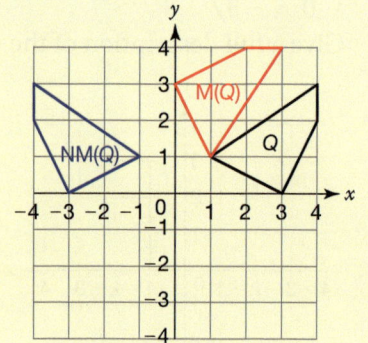

iii Find the position vector of each vertex of **M**(*Q*) after **N**.

$\begin{pmatrix} 0 & -1 \\ 1 & 0 \end{pmatrix}\begin{pmatrix} 1 \\ 1 \end{pmatrix} = \begin{pmatrix} -1 \\ 1 \end{pmatrix}$ $\begin{pmatrix} 0 & -1 \\ 1 & 0 \end{pmatrix}\begin{pmatrix} 0 \\ 3 \end{pmatrix} = \begin{pmatrix} -3 \\ 0 \end{pmatrix}$

$\begin{pmatrix} 0 & -1 \\ 1 & 0 \end{pmatrix}\begin{pmatrix} 2 \\ 4 \end{pmatrix} = \begin{pmatrix} -4 \\ 2 \end{pmatrix}$ $\begin{pmatrix} 0 & -1 \\ 1 & 0 \end{pmatrix}\begin{pmatrix} 3 \\ 4 \end{pmatrix} = \begin{pmatrix} -4 \\ 3 \end{pmatrix}$

Plot and join these vertices. Label the image **NM**(*Q*).

Remember to give all the information.

b Describe the transformation that maps *Q* onto **NM**(*Q*).

Reflection in the y axis (or line x = 0).

c Multiply the matrices – take care with the order.

$NM = \begin{pmatrix} 0 & -1 \\ 1 & 0 \end{pmatrix}\begin{pmatrix} 0 & 1 \\ 1 & 0 \end{pmatrix} = \begin{pmatrix} -1 & 0 \\ 0 & 1 \end{pmatrix}$

Use unit vectors to find the image of the unit square.

$\begin{pmatrix} 1 \\ 0 \end{pmatrix}$ maps to $\begin{pmatrix} -1 \\ 0 \end{pmatrix}$ and $\begin{pmatrix} 0 \\ 1 \end{pmatrix}$ maps to $\begin{pmatrix} 0 \\ 1 \end{pmatrix}$

The unit square has been reflected in the y axis.

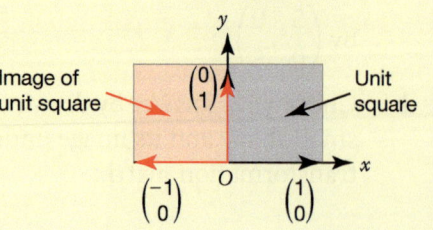

Geometry   Pythagoras, trigonometry and vectors

Fluency

Exercise 18.7S

1 Use **M**, **N** and **Q** as defined in the example.
  a Draw i Q   ii N(Q)   iii MN(Q).
  b Describe fully the single transformation that has the same effect as **MN**.
  c Use the unit square to show that the matrix **MN** represents the transformation that you described in part **b**.

2 $\mathbf{M} = \begin{pmatrix} 1 & 0 \\ 0 & -1 \end{pmatrix}$, $\mathbf{N} = \begin{pmatrix} 0 & 1 \\ -1 & 0 \end{pmatrix}$ and T is the triangle with vertices $(1, 2)$, $(4, 4)$ and $(1, 4)$.

  a Using x and y axes from -5 to 5, draw
     i T ii $\mathbf{M}(T)$ iii $\mathbf{NM}(T)$
  b Using new x and y axes from -5 to 5, draw
     i T ii $\mathbf{N}(T)$ iii $\mathbf{MN}(T)$.
  c i Calculate matrix **MN** and describe the single transformation represented by your answer.
     ii Repeat part **i** for **NM**.

3 Use transformation matrices
$\mathbf{A} = \begin{pmatrix} 2 & 0 \\ 0 & 2 \end{pmatrix}$ and $\mathbf{B} = \begin{pmatrix} 0 & 1 \\ -1 & 0 \end{pmatrix}$
and the pentagon P shown below.

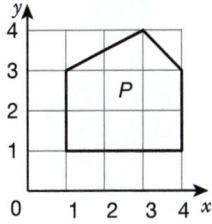

  a On axes of x and y from -8 to 8, copy P and draw the following images
     i $A(P)$ ii $BA(P)$
  b On a separate diagram, draw P and the following images
     i $B(P)$ ii $AB(P)$
  c Calculate the matrices **BA** and **AB**.
  d Describe what your results show about combining these transformation matrices.

4 $\mathbf{R} = \begin{pmatrix} -1 & 0 \\ 0 & -1 \end{pmatrix}$ and $\mathbf{S} = \begin{pmatrix} 0 & 1 \\ -1 & 0 \end{pmatrix}$
  a Calculate i **RS**   ii **SR**
  b Describe the transformation represented by
     i **R**   ii **S**   iii **RS**   iv **SR**

5

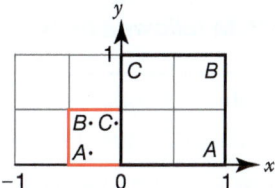

  a Write the transformation matrix that maps $OABC$ directly onto $OA'B'C'$.
  b Check by multiplying two transformation matrices.

6 Use matrices $\mathbf{M} = \begin{pmatrix} 5 & 0 \\ 0 & 5 \end{pmatrix}$ and $\mathbf{N} = \begin{pmatrix} -\frac{1}{5} & 0 \\ 0 & -\frac{1}{5} \end{pmatrix}$

  a Calculate i **MN**   ii **NM**
  b Describe the transformation represented by
     i **M**   ii **N**   iii **MN**

7

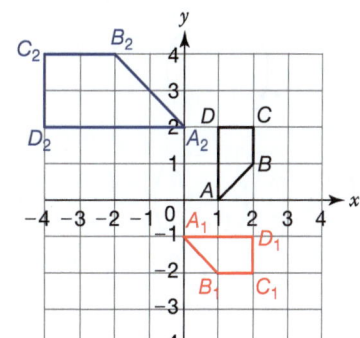

  a i Find the transformation matrix that maps trapezium $ABCD$ onto $A_1B_1C_1D_1$.
     ii Use the position vectors of A, B, C and D to check your answer.
  b i Find the transformation matrix that maps trapezium $A_1B_1C_1D_1$ onto $A_2B_2C_2D_2$.
     ii Use position vectors to check.
  c i Find the transformation matrix that maps trapezium $ABCD$ onto $A_2B_2C_2D_2$.
     ii Use position vectors to check.

APPLICATIONS

18.7 Combining transformation matrices

RECAP

When two transformations are represented by matrices, **M** and **N**,
- **MN** represents **N** *followed by* **M**
- **NM** represents **M** *followed by* **N**

Remember that **MN** is usually different from **NM**

HOW TO

Here is a general strategy for solving problems
1. Use the unit square to find transformation matrices or what they do.
2. Multiply matrices to combine transformations.
3. Describe transformations fully and answer any other questions.

EXAMPLE

a Find the matrix that represents **i** reflection in the x axis **ii** reflection in $y = -x$
   iii reflection in the x axis followed by reflection in $y = -x$.

b Describe the single transformation that is equivalent to the combined transformation.

a i ① $\begin{pmatrix}1\\0\end{pmatrix}$ maps to $\begin{pmatrix}1\\0\end{pmatrix}$ and $\begin{pmatrix}0\\1\end{pmatrix}$ to $\begin{pmatrix}0\\-1\end{pmatrix}$. Sketch what happens to the unit square.

Matrix = $\begin{pmatrix}1 & 0\\0 & -1\end{pmatrix}$

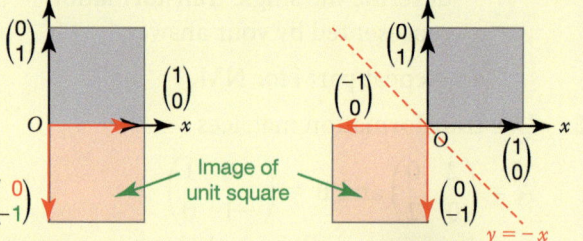

ii $\begin{pmatrix}1\\0\end{pmatrix}$ maps to $\begin{pmatrix}0\\-1\end{pmatrix}$ and $\begin{pmatrix}0\\1\end{pmatrix}$ to $\begin{pmatrix}-1\\0\end{pmatrix}$.

Matrix = $\begin{pmatrix}0 & -1\\-1 & 0\end{pmatrix}$

② **iii** Multiply the matrices, but take care with the order.

$\begin{pmatrix}0 & -1\\-1 & 0\end{pmatrix}\begin{pmatrix}1 & 0\\0 & -1\end{pmatrix} = \begin{pmatrix}0 & 1\\-1 & 0\end{pmatrix}$

Remember to give all the information.

b ① $\begin{pmatrix}1\\0\end{pmatrix}$ maps to $\begin{pmatrix}0\\-1\end{pmatrix}$ and $\begin{pmatrix}0\\1\end{pmatrix}$ to $\begin{pmatrix}1\\0\end{pmatrix}$

③ This is rotation through 90° clockwise about (0, 0).

EXAMPLE

Interpret $\begin{pmatrix}0 & 1\\-1 & 0\end{pmatrix}\begin{pmatrix}0 & -1\\1 & 0\end{pmatrix} = \begin{pmatrix}1 & 0\\0 & 1\end{pmatrix}$ in terms of transformations.

① ② ③ Find and describe the transformation represented by each matrix.

$\begin{pmatrix}0 & 1\\-1 & 0\end{pmatrix}$ rotates the unit square through 90° clockwise about (0, 0).

$\begin{pmatrix}0 & -1\\1 & 0\end{pmatrix}$ rotates the unit square 90° anticlockwise about (0, 0).

$\begin{pmatrix}1 & 0\\0 & 1\end{pmatrix}$ is the identity matrix – it maps the unit square onto itself.

A rotation of 90° anticlockwise about (0, 0) followed by a rotation of 90° clockwise about (0, 0) returns an object to its original position.

Geometry   Pythagoras, trigonometry and vectors

Reasoning and problem solving

Exercise 18.7A

1. **a** Find the matrix that represents
   - **i** rotation of 180° about (0, 0)
   - **ii** enlargement, centre (0, 0), scale factor 2
   - **iii** rotation of 180° about (0, 0) followed by enlargement, centre (0, 0), scale factor 2

   b Describe the single transformation that is equivalent to the combined transformation.

   c Are the answers to **a iii** and **b** the same if the order of the transformations is reversed?

2. **a** Find the matrix that represents
   - **i** reflection in $y = x$
   - **ii** rotation of 270° anticlockwise about (0, 0)
   - **iii** reflection in $y = x$ followed by rotation of 270° anticlockwise about (0, 0)

   b Describe the single transformation that is equivalent to the combined transformation.

   c Are the answers to **a iii** and **b** the same if the order of the transformations is reversed?

3.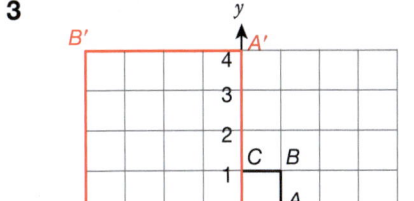

   a Write down the transformation matrix that maps $OABC$ to $OA'B'C'$.

   b Use a different method to check your answer.

4. **a** Find the matrix which represents reflection in $y = -x$ followed by enlargement, centre (0, 0), scale factor $\frac{1}{4}$.

   b Is the answer to **a** the same if the order of the transformations is reversed?

5. Interpret each part in terms of transformations.

   a $\begin{pmatrix} 0 & -1 \\ -1 & 0 \end{pmatrix}\begin{pmatrix} 0 & -1 \\ -1 & 0 \end{pmatrix} = \begin{pmatrix} 1 & 0 \\ 0 & 1 \end{pmatrix}$

   b $\begin{pmatrix} -1 & 0 \\ 0 & -1 \end{pmatrix}\begin{pmatrix} -1 & 0 \\ 0 & -1 \end{pmatrix} = \begin{pmatrix} 1 & 0 \\ 0 & 1 \end{pmatrix}$

6. Interpret each part in terms of transformations.

   a $\begin{pmatrix} \frac{1}{2} & 0 \\ 0 & \frac{1}{2} \end{pmatrix}\begin{pmatrix} 8 & 0 \\ 0 & 8 \end{pmatrix} = \begin{pmatrix} 4 & 0 \\ 0 & 4 \end{pmatrix}$

   b $\begin{pmatrix} -10 & 0 \\ 0 & -10 \end{pmatrix}\begin{pmatrix} 0.1 & 0 \\ 0 & 0.1 \end{pmatrix} = \begin{pmatrix} -1 & 0 \\ 0 & -1 \end{pmatrix}$

7. Interpret in terms of transformations:

   $\begin{pmatrix} -1 & 0 \\ 0 & 1 \end{pmatrix}\begin{pmatrix} 1 & 0 \\ 0 & -1 \end{pmatrix} = \begin{pmatrix} -1 & 0 \\ 0 & -1 \end{pmatrix}$

   $= \begin{pmatrix} 0 & -1 \\ -1 & 0 \end{pmatrix}\begin{pmatrix} 0 & 1 \\ 1 & 0 \end{pmatrix}$

8. **a** Find the transformation matrix that maps
   - **i** R onto S
   - **ii** S onto R
   - **iii** R onto T
   - **iv** T onto R

   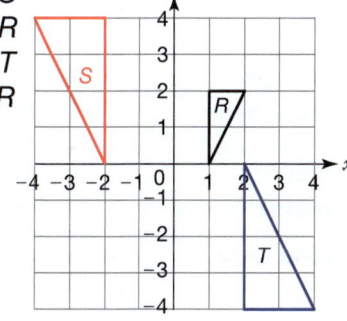

   b
   - **i** Multiply your answers to **a i** and **a ii**.
   - **ii** Multiply your answers to **a iii** and **a iv**.

*9. **a** Calculate $\begin{pmatrix} 0 & 1 \\ 1 & 0 \end{pmatrix}\begin{pmatrix} -1 & 0 \\ 0 & -1 \end{pmatrix}\begin{pmatrix} -1 & 0 \\ 0 & 1 \end{pmatrix}$

   b
   - **i** Interpret part **a** in terms of transformations.
   - **ii** Use a shape of your choice to illustrate your interpretation.

   c Investigate what happens if you change the order in which the matrices are multiplied.

10. Find a set of transformation matrices that move triangle A to the other positions in this pattern.

Summary

Checkout
You should now be able to...

	Test it Questions
✓ Use Pythagoras' theorem to find a missing side in a right-angled triangle or the length of a line segment on a coordinate grid.	1, 4, 5
✓ Use trigonometric ratios to find missing lengths and angles in triangles.	2 – 5
✓ Find the exact values of $\sin\theta$ and $\cos\theta$ for key angles.	6
✓ Use the sine and cosine rules to find missing lengths and angles.	7
✓ Use the sine formula for the area of a triangle.	8
✓ Calculate with vectors and use them in geometric proofs.	9, 10
✓ Multiply matrices and use them to represent transformations.	11

Language	Meaning	Example
Hypotenuse	The side opposite the right angle in a right-angled triangle.	(right triangle with Hypotenuse labelled)
Pythagoras' theorem	For a right-angled triangle, the area of the square drawn along the hypotenuse is equal to the sum of the areas of the squares drawn along the other two sides.	$c^2 = a^2 + b^2$
Multiply (matrices)	Multiply each row from the first matrix by each column from the second matrix.	$\begin{pmatrix} 2 & 3 \\ 4 & -1 \end{pmatrix}\begin{pmatrix} 5 & 1 \\ 7 & 0 \end{pmatrix} = \begin{pmatrix} 10+21 & 2+0 \\ 20-7 & 4-0 \end{pmatrix} = \begin{pmatrix} 31 & 5 \\ 13 & 3 \end{pmatrix}$
Identity matrix	$\begin{pmatrix} 1 & 0 \\ 0 & 1 \end{pmatrix}$ leaves any other matrix unchanged in multiplication	$\begin{pmatrix} 1 & 0 \\ 0 & 1 \end{pmatrix}\begin{pmatrix} 3 & -4 \\ -2 & 1 \end{pmatrix} = \begin{pmatrix} 3 & -4 \\ -2 & 1 \end{pmatrix}$
Sine ratio	The ratio of the length of the opposite side to the hypotenuse in a right-angled triangle.	$\sin\theta = \frac{3}{5} = 0.6$
Cosine ratio	The ratio of the length of the adjacent side to the hypotenuse in a right-angled triangle.	$\cos\theta = \frac{4}{5} = 0.8$
Tangent ratio	The ratio of the length of the opposite side to the adjacent side in a right-angled triangle.	$\tan\theta = \frac{3}{4} = 0.75$
Scalar	A quantity with size only.	Mass, temperature, speed
Vector	A quantity with both size and direction.	$\begin{pmatrix} -2 \\ 1 \end{pmatrix}$, $\begin{pmatrix} 1 \\ -2 \end{pmatrix}$
Resultant	The vector that is equivalent to adding or subtracting two or more vectors.	$a + b$, $a - b$
Multiple	The original vector multiplied by a scalar.	$3a = a + a + a$
Collinear	Points are collinear if they lie on the same straight line.	A B C

Geometry Pythagoras, trigonometry and vectors

Review

1. Calculate the lengths a and b. Give your answers to 1 decimal place.

   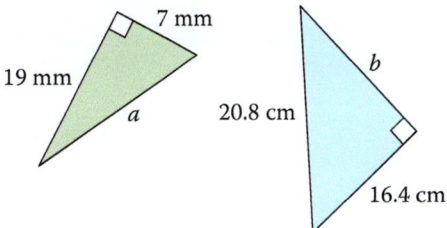

2. Calculate the lengths a, b and c to 3 s.f.

   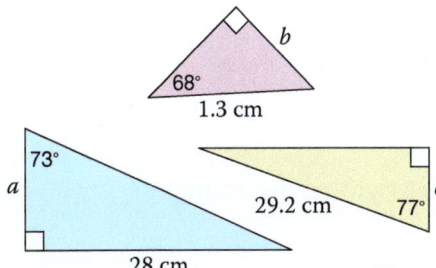

3. Calculate the size of angles A and B.

   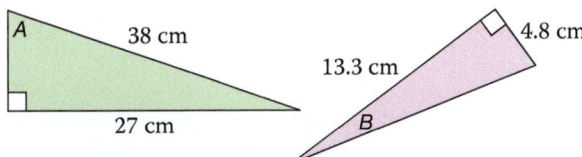

4. In this cuboid, calculate to 1 decimal place

   a EG b EC c $\angle FEG$ d $\angle HDF$

   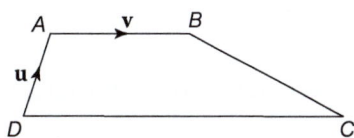

5. A ship starts at a port then sails 120 km due south then 75 km due west.

   a How far is the ship from port?

   b Give the bearing of the ship from the port.

6. Write the exact value of.

   a sin 90°   b cos 30°

7. Calculate the size of length x and angle θ.

   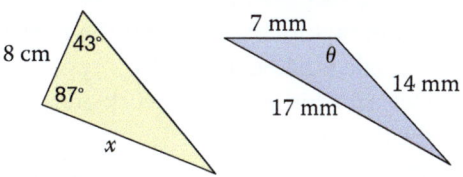

8. Calculate the area of the triangle ABC where $AB = 8$ cm, $BC = 11$ cm and angle $ABC = 35°$.

9. Work out these vector sums and draw the resultant vectors.

   $$\mathbf{u} = 5\begin{pmatrix}2\\-1\end{pmatrix} + 4\begin{pmatrix}-1\\0\end{pmatrix} \quad \mathbf{v} = 3\begin{pmatrix}3\\2\end{pmatrix} - 4\begin{pmatrix}3\\-1\end{pmatrix}$$

10. AB is parallel to DC and the ratio of the lengths DC to AB is $3 : 2$.

    a Write these vectors in terms of $\mathbf{u}$ and $\mathbf{v}$.

    i $\overrightarrow{DB}$ ii $\overrightarrow{DC}$ iii $\overrightarrow{BC}$

    M is the midpoint of AB and N is the midpoint of DC.

    b Write $\overrightarrow{MN}$ in terms of $\mathbf{u}$ and $\mathbf{v}$.

11. The transformation matrix $\begin{pmatrix}0 & -1\\-1 & 0\end{pmatrix}$ maps triangle $P(3, 0)$ $Q(2, 3)$ $R(-2, 2)$ to $P'Q'R'$.

    a Draw PQR and $P'Q'R'$.

    b Give a full description of the transformation.

What next?

Score		
	0 – 4	Your knowledge of this topic is still developing. To improve, look at MyiMaths: 1064, 1095, 1112, 1120, 1131, 1133, 1134, 1135
	5 – 8	You are gaining a secure knowledge of this topic. To improve your fluency look at InvisiPens: 19Sa – f
	9 – 11	You have mastered these skills. Well done you are ready to progress! To develop your problem solving skills look at InvisiPens: 19Aa – h

Assessment 18

1 a Nerea says incorrectly that the hypotenuse of this triangle is 181 m.
Describe her error and work out the correct hypotenuse. [3]

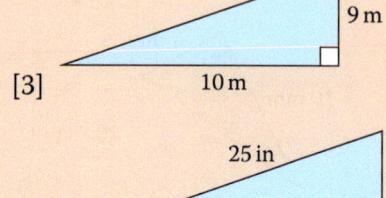

b Pawel says incorrectly that the missing side in this triangle is 34.66 in to 2 dp.
Describe his error and work out the correct value of the missing side. [3]

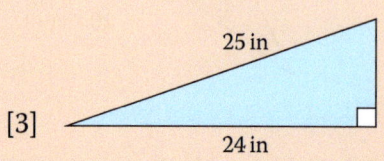

2 a Angelina says that 2, 3, 13 is a Pythagorean triple.
Say why she is wrong. [2]

b Darshna says that 20, 99 and 100 is a Pythagorean triple. Say why she is wrong and find a Pythagorean triple with highest value 100. [2]

3 Find the distance between the points $(-3, -5)$ and $(6, -1)$. [3]

4 Six triangles are drawn with these sides. Which of them are right-angled triangles?
Give reasons for your answers.

    a 5, 12, 13 km [2]   **b** 9, 12, 15 cm [2]
    c 9, 14, 17 m [2]   **d** 1.6, 3.0, 3.4 in [2]
    e 11, 19, 22 yds [2]   **f** 3.6, 7.7, 8.5 ft [2]

5 A tangent, PQ, is drawn to a circle of radius 5.875 cm.
Q is 11.6 cm from the centre of the circle.

Calculate

    a the distance PQ [3]
    b angle θ. [3]

6 a The base of a ladder is on horizontal ground and is leaning against a vertical wall.
The base is 1.75 m from the wall and the ladder makes an angle of 12.5° with the wall.
How long is the ladder? [3]

  b How far up the wall does the ladder reach? [2]

7 Carletta and Ivan are playing conkers.
Carletta's conker, C, is tied to the end of a string 59.5 cm long.
She pulls it back from the vertical until it is 28 cm horizontally from its original position. Calculate

    a the vertical distance, h, that the conker has risen [4]
    b angle θ. [3]

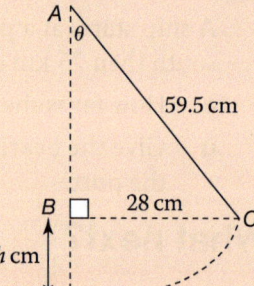

8 Quinton is 18 kilometres due east of Perville. Roxley is 32 kilometres from Perville on bearing 203°.
Calculate the distance and bearing of Roxley from Quinton. [8]

9 A paperweight is in the shape of a pyramid with a horizontal rectangular base *PQRS*.
The vertical height of the pyramid, *TX*, is 3.8 cm.
Calculate, to 3 significant figures,

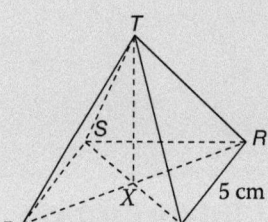

  a the length of the slant edge *TR* [4]
  b the angle that *TR* makes with *PQRS*. [3]

10 a $x = \begin{pmatrix} 4 \\ 6 \end{pmatrix}$ and $y = \begin{pmatrix} -2 \\ 7 \end{pmatrix}$. Draw these vectors on squared paper.

  i $x + y$ [2]   **ii** $y - x$ [2]   **iii** $2x$ [2]

  b Find the length of each vector in part **a**. Give your answer in surd form. [6]

11 $p = \begin{pmatrix} 4 \\ 1 \end{pmatrix}$, $q = \begin{pmatrix} 7 \\ -2 \end{pmatrix}$ and $r = \begin{pmatrix} -8 \\ 5 \end{pmatrix}$

  a Find these vectors.

   i $p + q + r$ [2]   **ii** $4q + 3p - 2r$ [3]   **iii** $3r - (p + q)$ [3]

  b Find the values of k and y for which $p - r = k\begin{pmatrix} -3 \\ y \end{pmatrix}$ [4]

12 a Describe the single transformation represented by $\begin{pmatrix} 2 & 0 \\ 0 & 2 \end{pmatrix}$. [3]

  b Find the matrix transformation that rotates shapes through 180° about the origin. [2]

  c i Find the result of the matrix multiplication $\begin{pmatrix} 0 & -1 \\ -1 & 0 \end{pmatrix}\begin{pmatrix} 0 & 1 \\ 1 & 0 \end{pmatrix}$. [2]

   ii Explain what part **c i** tells you in terms of transformations. [3]

13 *SR* is parallel to *PQ*.
$\overrightarrow{PQ} = \mathbf{a}$, $\overrightarrow{SR} = 4\mathbf{a}$ and $\overrightarrow{PS} = 2\mathbf{b} - \mathbf{a}$

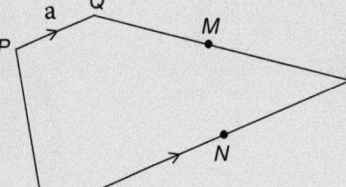

  a Show that $\overrightarrow{QR} = 2(\mathbf{b} + \mathbf{a})$ [2]
  b *M* is the mid-point of *QR*.
  $\overrightarrow{SN} : \overrightarrow{NR} = 5 : 3$
  Show that *MN* is parallel to *PS*. [4]

19 The probability of combined events

Introduction

There is an old myth, that says if it rains on St Swithin's Day (15th July) it will rain for 40 days afterwards. Weather statistics show that it is untrue. Unfortunately long-scale weather forecasting is fairly unreliable, particularly in the UK so people look for tell-tale signs to help their predictions. In probability language, if a particular event occurs (rain on St Swithin's Day), does this increase the probability of another event occurring (a wet summer)?

What's the point?

Quite often, the occurrence of one event will significantly increase the probability of another event occurring. For example, lung cancer appears unpredictably, but its occurrence is greatly increased if a particular person is a smoker. Understanding the probabilities attached to linked events helps us to evaluate everyday risks.

Objectives

By the end of this chapter, you will have learned how to …

- Use Venn diagrams to represent sets.
- Use a possibility space to represent the outcomes of two experiments and to calculate probabilities.
- Use a tree diagram to show the outcomes of two experiments.
- Calculate conditional probabilities.

Check in

1 Work out each of these expressions.

   a $1 - 0.45$
   b $1 - 0.96$
   c $1 - 0.28$
   d $1 - 0.375$
   e $0.2 + 0.4$
   f $0.3 + 0.04$
   g $0.65 + 0.25$
   h 0.5×0.36
   i 0.25×0.68
   j 0.64×0.3
   k $1 - 0.125 - 0.64$
   l $1 - 0.125 \times 0.64$

2 Work out each of these expressions.

   a $1 - \frac{5}{6}$
   b $1 - \frac{1}{5}$
   c $1 - \frac{7}{9}$
   d $\frac{1}{5} + \frac{2}{3}$
   e $\frac{3}{4} + \frac{1}{6}$
   f $\frac{2}{3} \times \frac{5}{6}$
   g $\frac{2}{9} \times \frac{4}{5}$
   h $1 - \frac{3}{4} \times \frac{4}{5}$

Chapter investigation

Two six-sided dice, labelled 1 to 6, are thrown. The score is given by adding the two numbers that appear uppermost.

Which result is most likely to occur or are all possible results equally likely?

What if, instead of adding the two numbers, you multiply them together?

SKILLS

19.1 Venn diagrams

- A **set** is a collection of numbers or objects.
- The objects in the set are called the **members** or **elements** of the set.
- $n(A)$ is the number of elements in the set A.

If the set X is 'the factors of 6', then you can write $X = \{1, 2, 3$ and $6\}$.

You can say that 3 is an element (member) of the set X.

If the set Y is 'the even numbers', then you can write $Y = \{2, 4, 6, 8, \ldots\}$.

- The universal set, which has the symbol ξ, is the set containing all the elements.
- The empty set, $\emptyset$, is the set with no elements.

You can use a Venn diagram to show the relationship between sets.

- The **intersection** of two sets, $A \cap B$, consists of the elements common to both sets A and B.
- The **union** of two sets, $A \cup B$, consists of the elements which appear in at least one of the sets A or B.
- The **complement** of a set, A', consists of the elements which are not in A.

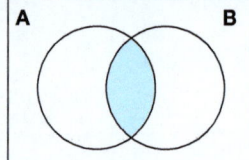
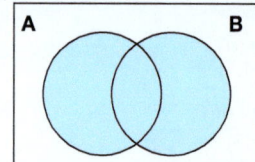
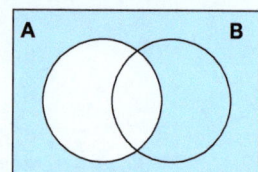

EXAMPLE

$\xi = \{1, 2, \ldots 11, 12\}$
$A = \{\text{Factors of 12}\}$ $B = \{2, 3, 5, 6, 11\}$.
Find **a** $A \cap B$ **b** $A \cup B$ **c** $(A \cup B)'$

a $A = \{1, 2, 3, 4, 6, 12\}$
 $A \cap B = \{2, 3, 6\}$
b $A \cup B = \{1, 2, 3, 4, 5, 6, 11, 12\}$
c $(A \cup B)' = \{7, 8, 9, 10\}$

- You list the elements of each set or show the number of elements in each region.

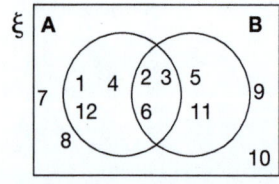
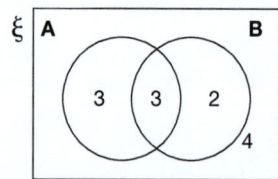

In the example, $P(A) = \dfrac{6}{12} = \dfrac{1}{2}$

- You can use Venn diagrams to work out probabilities.
- $P(A) = \dfrac{\text{Number of elements in set } A}{\text{Total number of elements in } \xi} = \dfrac{n(A)}{n(\xi)}$

You can write $P(A \text{ and } B)$ as $P(A \cap B)$, $P(A \text{ or } B)$ as $P(A \cup B)$ and $P(\text{not } A)$ as $P(A')$.

Probability The probability of combined events

Fluency

Exercise 19.1S

1. List the elements of these sets.
   a. P – the first ten square numbers
   b. R – countries in North America
   c. S – the first ten prime numbers
   d. T – factors of 36

2. a. Using the sets in question **1**, give
      i. P ∩ T
      ii. S ∩ T
      iii. P ∩ S
      iv. P ∪ S
   b. i. n(P ∩ T)
      ii. n(P ∩ S)

3. Give a precise description of each set.
   a. {1, 2, 5, 10}
   b. {2, 4, 6, 8, 10, 12, ...}
   c. {a, e, i, o, u}
   d. {HH, HT, TH, TT}
   e. {1p, 2p, 5p, 10p, 20p, 50p, £1, £2}
   f. {3, 6, 9, 12, 15, 18, 21, 24, 27, 30}

4. List the elements of these sets.
   a. A – the first ten positive integers
   b. B – single digit odd numbers
   c. C – single digit prime numbers
   d. D – single digit square numbers

5. Using the sets in question **4**, give the sets
   a. B ∩ C
   b. B ∩ D
   c. B ∪ D
   d. C ∪ D

6. Say why B, C and D must be subsets of A for the sets in question **4**.

7. A = {even numbers}
   B = {odd numbers}
   C = {multiples of 5}
   D = {prime numbers}
   E = {multiples of 3}
   F = {square numbers}
   G = {factors of 36}

   For these pairs of sets, state if they have any elements in common. If they do, then list them.

7. a. A, C   b. A, D   c. F, G
   d. E, F   e. A, B   f. C, G
   g. D, F   h. D, E   i. D, E and F

8. The Venn diagram shows information about the sport that 25 students are playing in PE this term.

   The results are shown on the Venn diagram.

   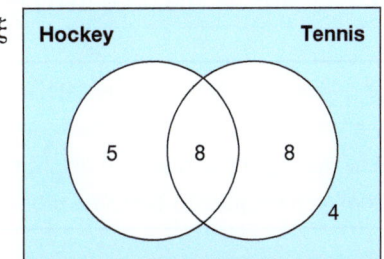

   a. How many students are
      i. in the intersection of hockey and tennis
      ii. playing hockey
      iii. not playing tennis
      iv. in the union of hockey and tennis?
   b. Describe the shaded region in words.
   c. What fraction of students are playing either hockey or tennis, but not both?

9. An insurance company surveys 50 customers. The customers are sorted into
   C = {Car insurance}
   H = {Home insurance}
   The results are shown on the Venn diagram.

   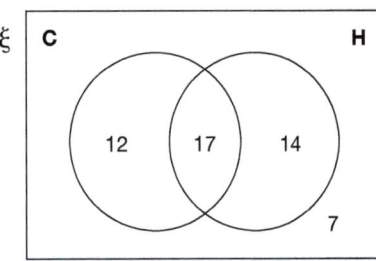

   a. A customer is chosen at random. Find
      i. P(C)   ii. P(H′)
      iii. P(C ∩ H)   iv. P(C ∪ H)
   b. How many customers had home insurance but not car insurance?

APPLICATIONS

19.1 Venn diagrams

RECAP
- The **intersection** of two sets, $A \cap B$, consists of the elements common to both sets.
- The **union** of two sets, $A \cup B$, consists of the elements which appear in at least one of the sets.
- The **complement** of a set consists of all the elements that are not in the set.
- The **universal set**, which has the symbol ξ, is the set containing all the elements.
- The **empty set**, $\varnothing$, is the set with no elements.
- **Venn diagrams** can be used to represent the relationships between sets.

HOW TO
To solve a problem involving several sets
1. Draw a Venn diagram to show the information.
2. Fill in all the regions of the Venn diagram.
3. Use the Venn diagram to calculate probabilities.

Don't forget to fill in the region outside the circles.

EXAMPLE

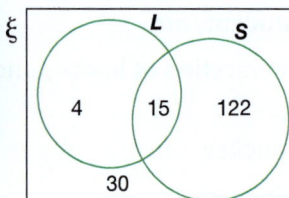

This Venn Diagram shows the *numbers* of pupils in a primary school classified by whether they are left handed (L) and whether they can swim (S).

a How many pupils are in the school?
b How many pupils can swim?
c What do you know about a pupil in $L \cap S$?
d How many pupils are in $L \cap S$?

a 171 = 4 + 15 + 122 + 30
b 137 = 15 + 122 add up just the numbers falling within S
c they are left handed and can swim
d 15 it is just the overlapping area in both L and S.

EXAMPLE

A teacher is organising a school trip to Rome for the students in year 11.

There are 120 students in the school year.

72 students study history. 90 students are visiting Rome.

12 students don't study history and aren't visiting Rome.

Find the probability that a student is studying history and visiting Rome.

Draw a Venn diagram to show the number of students in each region. Let the number of students studying history and visiting Rome be x. The total number of students is 120

$72 - x + x + 90 - x + 12 = 120$

$174 - x = 120$

$x = 174 - 120$

$= 54$

P (history and Rome) $= \dfrac{54}{120}$

$= 0.45$

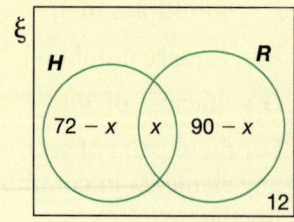

Probability The probability of combined events

Reasoning and problem solving

Exercise 19.1A

1. U = {All triangles}
   E = {Equilateral triangles}
   I = {Isosceles triangles}
   R = {Right-angled triangles}

   a. Sketch a member of $I \cap R$.
   b. Explain why $E \cap R = \varnothing$.

2. Thilá sorts a group of objects into the sets A and B. She draws a Venn diagram to show her results.

   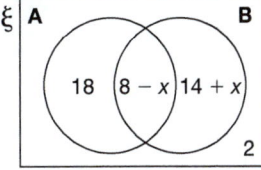

   a. If A and B are mutually exclusive, find the value of x.
   b. If $x = 8$, what can you say about the sets A and B?

3. At Newtown School there are 27 students in class 11B.
   17 students play tennis, 11 play basketball and 2 play neither game.
   How many students play both tennis and basketball?

4. A group of people took a literacy and a numeracy test. 85% of applicants passed the literacy test and 75% passed the numeracy test. What are the maximum and minimum proportions of people who did not pass either test?

5. David is organising a family reunion.
   They can take part in two activities, archery or paintball.
   David shows the results on a Venn diagram.

   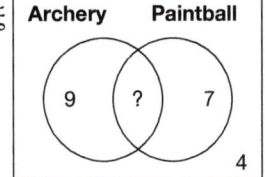

   Archery costs $22 and paintball costs $18. David collects $524 to pay for the activities. How many sign up for both activities?

6. The people in a tennis club (ξ) are classified by whether they have green eyes (G) or red hair (R).
   If a person is selcted at random from the club then
   $P(G) = 0.4$ $P(G \cap R') = 0.3$
   $P(G' \cap R') = 0.4$
   16 people have red hair but not green eyes.

   a. Draw a Venn diagram to show the numbers of people in each set.
   b. If a person is selcted at random from the club, calculate these probabilities.
      i. $P(G \cap R)$ ii. $P(G \cup R)$
      iii. $P(R')$ iv. $P(G' \cap R)$

7. The people in a football club (ξ) are classified by whether they have blue eyes (B) or grey hair (G).
   If a person is selected at random from the club then
   $P(G \cap B) = \frac{3}{16}$ $P(B) = \frac{1}{2}$
   15 people have grey hair.
   25 people have blue eyes but not grey hair.

   a. Draw a Venn diagram to show the numbers of people in each set.
   b. If a person is selcted at random from the club, calculate these probabilities.
      i. $P(G)$ ii. $P(G \cup B)$
      iii. $P(G')$ iv. $P(G' \cap B)$

*8. U = {All quadrilaterals}
    X = {Rhombuses}
    Y = {Rectangles}

    a. Sketch a member of $X \cap Y$.
    b. What sort of quadrilaterals are in the set $X \cap Y$?
    c. If X had been the set containing kites, how would your answers to parts **b** and **c** differ?

*9. At Highfield School there are 100 students in year 10.
    They all study one or more of these subjects.
    - 36 study French
    - 42 study Spanish
    - 47 study German
    - 10 study both French and Spanish
    - 12 study both Spanish and German
    - 9 study both French and German

    What is the probability that a student chosen at random studies all three subjects?

SKILLS

19.2 Sample spaces

- The list or table of all of the possible outcomes of a trial is called a **possibility space** or **sample space**.

	1	2	3	4	5	6
1	1,1	1,2	1,3	1,4	1,5	1,6
2	2,1	2,2	2,3	2,4	2,5	2,6
3	3,1	3,2	3,2	3,4	3,5	3,6
4	4,1	4,2	4,3	4,4	4,5	4,6
5	5,1	5,2	5,3	5,4	5,5	5,6
6	6,1	6,2	6,3	6,4	6,5	6,6

When two ordinary dice are thrown, there are 36 possible outcomes. If the dice are fair then the 36 outcomes are equally likely.

- You can use a sample space to calculate probabilities.

EXAMPLE

Jasper throws two dice and adds the results.
Loukia throws two dice and multiplies the results.

a Draw a sample space for Jasper's and Loukia's experiments.
b Find the probability that Jasper scores 8.
c Find the probability that Loukia scores 6.
d Find the probability that Loukia scores 5 or less.

a Jasper

+	1	2	3	4	5	6
1	2	3	4	5	6	7
2	3	4	5	6	7	8
3	4	5	6	7	8	9
4	5	6	7	8	9	10
5	6	7	8	9	10	11
6	7	8	9	10	11	12

Loukia

×	1	2	3	4	5	6
1	1	2	3	4	5	6
2	2	4	6	8	10	12
3	3	6	9	12	15	18
4	4	8	12	16	20	24
5	5	10	15	20	25	30
6	6	12	18	24	30	36

b The same sum appears on the diagonal.

$P(8) = \frac{5}{36}$

c $P(6) = \frac{4}{36} = \frac{1}{9}$

d $P(5 \text{ or less}) = \frac{10}{36} = \frac{5}{18}$

- You can use the sample space to write the set of all possible outcomes.

EXAMPLE

A fair coin is tossed and a fair die is thrown.
- If a head is seen then the score on the dice is doubled.
- If a tail is seen then the score is just the number on the dice.

Find   a P(6)   b P(>6)

Draw a sample space grid to help find the set of all outcomes.

	1	2	3	4	5	6
T	1	2	3	4	5	6
H	2	4	6	8	10	12

The set of all outcomes is {1, 2, 3, 4, 5, 6, 8, 10, 12}

a $P(6) = \frac{2}{12} = \frac{1}{6}$

b $P(>6) = \frac{3}{12} = \frac{1}{4}$

Probability   The probability of combined events

Fluency

Exercise 19.2S

1. Using Jasper's table shown opposite find the probability that the **sum of the scores** seen on two fair dice is
   a. exactly 10
   b. at least 10
   c. a square number
   d. less than 5.
   e. Write the set of all possible outcomes.

2. Using Loukia's table shown opposite find the probability that the *product of the scores* seen on two fair dice is
   a. exactly 10
   b. at least 10
   c. a square number
   d. less than 5.
   e. Write the set of all possible outcomes.

3. Two fair dice are thrown and the difference between the scores showing on the two dice is recorded.
   a. Make a table to show the sample space.
   b. Write the set of all possible outcomes.
   c. Find the probability that the difference is
      i. 0   ii. 3
      iii. 6   iv. a prime number.

4. A fair coin is tossed three times and the outcome recorded (for example HHT).
   a. Write the set of the 8 possible outcomes.
   b. In how many of these are exactly two heads seen?
   c. In how many do you see three of the same?

5. Two fair spinners are used. On one, the possible scores are 1, 2 and 4. On the other, the scores are 1, 3 and 5. The sum of the scores on the two spinners is recorded.
   a. Make a table to show the sample space.
   b. Write the set of all possible outcomes.
   c. Find the probability that the score is
      i. 2   ii. 3   iii. even

6. The two spinners in question **5** are used again but the score recorded is the product of the two scores.
   a. Make a table to show the sample space.
   b. Write the set of all possible outcomes.
   c. Find the probability that the score is
      i. 2   ii. 3   iii. even.

7. For the spinners used in questions **5** and **6**, if you wanted to get an even number, would you be better to use the sum or the product of the scores on the two spinners?

8. A pair of unbiased dice are thrown and the sum and product of the scores are recorded in two lists. The dice are thrown 100 times.
   a. Estimate the number of times a sum of exactly 10 will be seen.
   b. Estimate the number of times a product of exactly 10 will be seen.
   c. Would you expect to see 6 in the list of sums more often, less often or about the same number of times as in the list of products?
   d. Would you expect to see 3 in the list of sums more often, less often or about the same number of times as in the list of products?
   e. Marin says that 7 is certain to appear more often in the list of sums because you can't score 7 using the product. Is he correct?

9. Two fair dice are thrown together and the scores are added. One is an ordinary dice with the numbers 1 to 6, and the other has faces labelled 1, 2, 2, 3, 3, 3.
   a. Make a table to show the sample space of the total scores for a throw.
   b. Find the probability that the score is
      i. 6   ii. 7   iii. 9   iv. 3.
   c. What other scores are as likely to happen as 6?
   d. Why are some scores less likely to occur than 6?

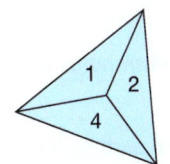 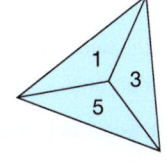

1199, 1263

APPLICATIONS

19.2 Sample spaces

RECAP
- A systematic list or table allows you to enumerate all the possible outcomes from a combination of events.
- For equally likely outcomes, P(Event)
$$= \frac{\text{Number of outcomes in event}}{\text{Total number of outcomes}}$$

Once you've drawn a table, the individual cells give you the outcomes.

HOW TO

To calculate probabilities for combined events
1. Create a list or table of all possible outcomes.
2. Count the number of entries which are in an event to calculate the probability of the event.
3. Answer the question

EXAMPLE

Rohini has 6 cards numbered from 1 to 6.
She takes two cards *without replacement*.
A = {Product is even} B = {Sum is even}
Show that P(A) = 2P(B).

> Could you work out what the probabilities of an even sum or product is without finding all the sums and products?

① Draw a sample space.

You cannot take the same card twice so there are 30 outcomes.

② Use the sample space to calculate probabilities.

$P(A) = \frac{24}{30} = \frac{4}{5}$ There are 24 cells with an even product.

×	1	2	3	4	5	6
1		2	3	4	5	6
2	2		6	8	10	12
3	3	6		12	15	18
4	4	8	12		20	24
5	5	10	15	20		30
6	6	12	18	24	30	

$P(B) = \frac{12}{30} = \frac{2}{5}$ There are 12 cells with an even sum.

③ P(A) = 2P(B)

+	1	2	3	4	5	6
1		3	4	5	6	7
2	3		5	6	7	8
3	4	5		7	8	9
4	5	6	7		9	10
5	6	7	8	9		11
6	7	8	9	10	11	

EXAMPLE

A fair coin is tossed and a fair die is thrown. If a head is seen then the score on the dice is squared. If a tail is seen then the score is just the number on the dice.

Explain why 1 and 4 are the most likely scores to be recorded.

① Draw a sample space.

	1	2	3	4	5	6
T	①	2	3	④	5	6
H	1	④	9	16	25	36

② 1 and 4 both occur twice in the list of outcomes whilst all other numbers only occur once.

③ 1 and 4 are each twice as probable as the other numbers.

Probability   The probability of combined events

Reasoning and problem solving

Exercise 19.2A

1. Two cards are taken from a set of cards numbered 1 to 6.
   Find the probability that the *difference* in the value of the two cards is

   a   3

   b   a factor of 6

   c   at least 1.

2. A fair coin is tossed until a head is seen. The number of times it has been tossed is recorded as the score on that trial.

   What is the probability that the score recorded is at least 3?

   Hint: consider the possible outcomes if a fair coin is tossed twice.

3. A fair dice in thrown and the spinner shown is spun. If the spinner lands on yellow you miss a turn (move 0 squares) otherwise you move the number of squares given by the product of the scores on the spinner and the dice.

   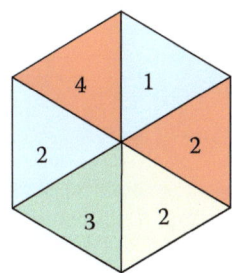

   a   Using labels R1, Y2 etc for the outcomes on the spinner construct a table to show the possible scores.

   b   What is the probability that the score recorded is

   i   4   ii   more than 6?

4. Cards numbered 1 to 100 are put in a box and Alessandra is asked to pick one at random. What is the probability that she chooses

   a   a single digit number

   b   a two digit number

   c   a number containing at least one 3?

   Don't write out the whole list, but imagine how many cards satisfy each condition.

5. Two fair spinners are used – one has sections showing the numbers 1, 1, 2, 3, 5 and the other has sections showing 3, 4, 5, 7, 8, 9.

   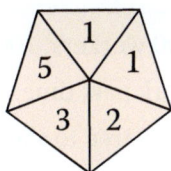 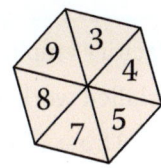

   a   What is the probability that the total score on the two spinners is

   i   6   ii   even?

   b   What is the probability that the score on one spinner is at least twice the score on the other spinner?

*6. Jack and Eija each roll a fair dice. Whoever gets the larger score wins the game.

   a   If a draw is allowed, what is the probability that Eija wins the game?

   b   A draw is not allowed and if the two dice show the same they roll again until one wins.
       What is the probability it will be Eija?

*7. A blue and a red dice are both fair and are thrown together. The following events are defined on the scores seen.

   A – The dice show the same score.

   B – The total score is at least 10.

   C – The total score is odd.

   D – The high score is a 4.

   E – The score on one dice is a proper factor of the score on the other. (A proper factor is a factor which is not 1 or the number itself.)

   a   Find these probabilities

   i   P(A and B)

   ii   P(E)

   iii   P(C and D).

   b   Find two pairs of mutually exclusive events.

   c   How many pairs of mutually exclusive events are there?

1199, 1263

407

SKILLS

19.3 Tree diagrams

- You can use a **frequency tree** to show the outcomes of two events.

EXAMPLE

A factory employs 300 workers. 100 workers are skilled and the rest are unskilled.
25 skilled workers work part time. 120 unskilled workers work full time.

a Draw a frequency tree to show this information.
b How many part-time workers are there altogether?
c If a worker is chosen at random what is the probability the worker is fulltime?

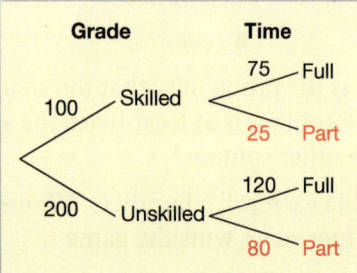

The frequency tree shows the number of workers along each branch.

The probability of each outcome is the relative frequency.

b 25 skilled and 80 unskilled workers are part time. $25 + 80 = 105$

c 75 skilled and 120 unskilled workers are full time, $\frac{75 + 120}{300} = \frac{13}{20}$

You can use a tree **diagram** to show the probabilities of two events.
- Write the outcomes at the end of each branch.
- Write the probability on each branch.
- The probabilities on each set of branches should add to 1.

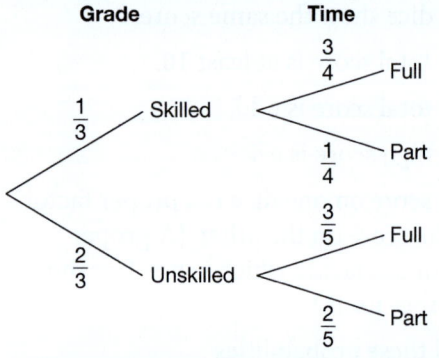

80 out of 200 unskilled workers work part time.
$\frac{80}{200} = \frac{2}{5}$

When you give a probability as a fraction, try to reduce it to its simplest form.

To find probabilities when an event can happen in different ways
- Multiply the probabilities along the branches.
- Add the probabilities for the different ways of getting the chosen event.

$P(\text{Full time}) = P(\text{Skilled and full time}) + P(\text{Unskilled and full time})$

$= \left(\frac{1}{3} \times \frac{3}{4}\right) + \left(\frac{2}{3} \times \frac{3}{5}\right)$

$= \frac{3}{12} + \frac{6}{15} = \frac{13}{20}$

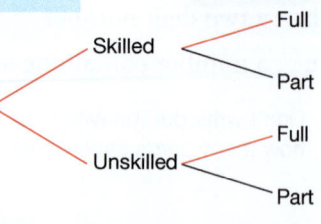

Probability   The probability of combined events

Fluency

Exercise 19.3S

1 A university surveys 3250 students. Approximately 2000 students live at home. A quarter of the students living at home are worried about debt.
Three in five of students living away from home are worried about debt.

  a Draw a frequency tree showing the number of students on each branch.

  b A student is chosen at random. What is the probability that the student is concerned about debt?

2 The MOT test examines whether cars are roadworthy.
10% of cars fail the MOT because there is something wrong with their brakes.
40% of the cars with faulty brakes also have faulty lights.
20% of cars whose brakes are satisfactory have faulty lights.

An MOT centre tested 3000 cars in March.

  a Draw a frequency tree showing the numbers of cars failing with brakes and lights in March.

  b How many cars fail on at least one of brakes and lights?

  c Tommy says that this means the rest of the 3000 cars passed the MOT test. Why is Tommy wrong?

3 A company has 360 employees.
120 employees are male.
80% of the male employees and 168 of the female employees are in the pension scheme.

  a Draw a frequency tree to show this information.

  b Find the probability that an employee chosen at random is in the pension scheme.

  c Find the probability that the employee is male.

  d A female employee is chosen at random. Find the probability that she is in the pension scheme.

  e Add the probabilities of each outcome onto each branch of your frequency diagram.

  f Find the probability that an employee is in the pension scheme. Does your answer agree with your answer in part **a**?

4 Jessica travels to work by car two days a week and by train on the other three. She is late for work 10% of the time when she travels by car, and late 20% of the time when she travels by train.

  a Draw a probability tree diagram showing the probability of each outcome.

  b Jessica works 150 days during the first 8 months of a year. Draw a frequency tree to show the numbers of days she travels to work by car and train, and on which she is late and on time.

  c Estimate how many days Jessica is late for work during this period.

  d What is the probability Jessica is late for work on a day chosen at random during this period?

5 Athletes are regularly tested for performance enhancing drugs.
If an athlete is taking a drug, the test will give a positive result 19 times out of 20. However, in athletes who are not taking the drug one in fifty tests is also positive.

It is thought that around 20% of athletes in a particular event are taking the drug.
The authorities do tests on 500 athletes.

  a Draw a probability tree showing the numbers of athletes expected to be taking or not taking the drug, and the results of the test.

  b Calculate the total number of positive tests expected to be seen.

  c How many of these came from athletes not taking the drug?

  d Find the probability that the test gives an accurate result.

APPLICATIONS

19.3 Tree diagrams

RECAP

- You can use tree **diagrams** to show the probabilities of two events.
  - Write the outcomes at the end of each branch
  - Write the probability on each branch
  - The probabilities on each set of branches should add to 1
- To find probabilities when an event can happen in different ways
  - Multiply the probabilities along the branches.
  - Add the probabilities for the different ways of getting the chosen event

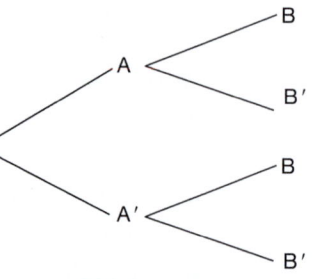

▲ The events A and not A and B and not B cover all the possible outcomes at each branching.

- If the outcome of one event does not affect what happens in another then the events are **independent**.
- If A and B are independent events then
  $P(A \text{ and } B) = P(B) \times P(A)$

If you can show that $P(A \text{ and } B) = P(B) \times P(A)$ then A and B are independent.

HOW TO

To prove that two events are independent
1. Draw a tree diagram showing the possible outcomes and probabilities of each outcome.
2. Use the tree diagram to find the probability of $P(A \text{ and } B)$, $P(A)$ and $P(B)$.
3. Test to see if $P(A \text{ and } B) = P(B) \times P(A)$ is satisfied.

EXAMPLE

A bag has 5 red and 3 blue balls in it.
A ball is taken at random and *not* replaced, and then a second ball is taken out.
Show that choosing a red ball on the second attempt is dependent on whether or not the first ball was red.

① Draw a tree diagram.

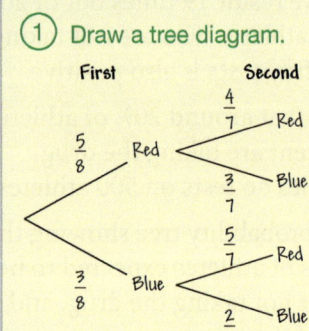

② Find the probabilities of each event.

$P(\text{first ball red}) = \dfrac{5}{8}$

$P(\text{second ball red}) = P(R, R) + P(B, R)$

$= \dfrac{5}{8} \times \dfrac{4}{7} + \dfrac{3}{8} \times \dfrac{5}{7}$

$= \dfrac{20}{56} + \dfrac{15}{56} = \dfrac{35}{56}$

$= \dfrac{5}{8}$

Don't cancel fractions to lowest form when you multiply the probabilities along the path – you are likely to have to add them!

$P(\text{both balls are red}) = \dfrac{5}{8} \times \dfrac{4}{7}$

$= \dfrac{20}{56} = \dfrac{5}{14}$

③ If the events are independent then

$P(\text{first ball red}) \times P(\text{second ball red}) = P(\text{both balls are red})$

$P(\text{first ball red}) \times P(\text{second ball red}) = \dfrac{5}{8} \times \dfrac{5}{8} = \dfrac{25}{64} \neq \dfrac{5}{14}$

So the events are dependent.

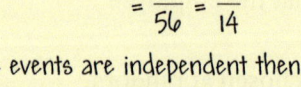

Probability The probability of combined events

Reasoning and problem solving

Exercise 19.3A

1. In a league, teams are awarded 3 points for a win, one for a draw and none for a loss. Amelie thinks that
   - her team has a probability of 0.6 of winning any match.
   - her team has a probability of 0.3 for a draw
   - the result of any game is independent of other results.

   a Find the probability that her team has at least three points after two games.

   b How have you used Amelie's assumption that the results of the game are independent in your answer to part **a**?

   c Do you think that Amelie's assumption that the results of the game are independent is reasonable? Give a reason for your answer.

2. A Year 13 pupil is taking her driving test. Records show that people taking the test at that age have a 70% chance of passing on the first attempt and 80% on any further attempts needed.

   a Show this information on a tree diagram showing up to three attempts.

   b Find the probability that
      i the pupil passes at the second attempt
      ii the pupil has still not passed after three attempts.

3. Denzel is going to the airport to catch a flight. He needs to travel on a bus and then catch a train to the airport.
   He catches a bus which has a probability of 0.8 of making a connection with a train which always gets to the airport on time. The next train has a probability of 0.7 of getting him to the airport on time.

   a Show this information on a tree diagram.

   b Find the probability that Denzel gets to the airport in time for his flight.

4. A bag has 4 red and 4 blue balls in it. A ball is taken from it at random and not replaced and then a second ball is taken out.

   X = the second ball is blue
   Y = the two balls are the same colour
   Z = both balls are blue

   a Amali says that $Z = X \cap Y$. Is she correct?

   b Show that X and Y are independent events.

5. A bag has 10 white and 5 black balls in it. A ball is taken from it at random, the colour noted and the ball is replaced and then a second ball is taken at random.

   A = the two balls are different colours
   B = at least one ball is black.

   Construct a tree diagram and decide if A and B are independent events.

*6 A spinner has a probability of $\frac{1}{6}$ of landing on blue.
   Green is three times as likely to occur as blue.
   Red, black and yellow are equally likely to occur.

   a Calculate the probability that the spinner lands on red.

   Sara is playing a game where she spins the spinner and if it lands on red, she takes double the score seen when she throws a fair dice.
   Sara needs a 4 to finish.

   b Draw a probability tree showing the outcomes of the spinner and the dice.

   c What is the probability Sara finishes on her go?

   d Is the use of the spinner a help or a hindrance to Sara getting a 4 to finish, or does it not matter? Give a reason.

   e How would Sara's experiment change if she needed
      i a 5 to finish?
      ii an 8 to finish?

1208, 1334, 1935, 1966

SKILLS

19.4 Conditional probability

In life, some events are dependent on other events occurring beforehand.

- In **conditional** probability, the probability of subsequent events depends on previous events occurring.

Your chances of exam success are conditional on whether you revise or not.

You can calculate probabilities based on the conditions you are given.

- For two events A and B. The probability of B happening, *given* that A has already happened, is written as P(B given A).
- To calculate P(B given A) you can
  - calculate the fraction of outcomes in event B that are also in event A
  - or use the formulae
    $$P(B \text{ given } A) = \frac{P(A \text{ and } B)}{P(A)} \text{ and } P(A \text{ and } B) = P(B \text{ given } A) \times P(A)$$

If A and B are independent, then P(B given A) = P(B).

EXAMPLE

An insurance company records information from 200 accident claims. The table shows the speed of the car and weather conditions during the accident.

	Wet	Dry	Total
Speeding	11	21	32
Not speeding	77	91	168
Total	88	112	200

A is the event 'it is wet' and
B is the event 'a car is speeding'.
Are A and B independent?
Give your reason.

$P(B) = \frac{32}{200} = 0.16$

$P(B \text{ given } A) = \frac{11}{88} = 0.125$ 11 out of the 88 cars were speeding when it was wet.

$P(B \text{ given } A) \neq P(B)$, so A and B are not independent.

You can use Venn diagrams to calculate conditional probabilities.

EXAMPLE

This Venn diagram shows the probabilities attached to two events A and B.

Find   **a** P(A given B)   and   **b** P(B given A)

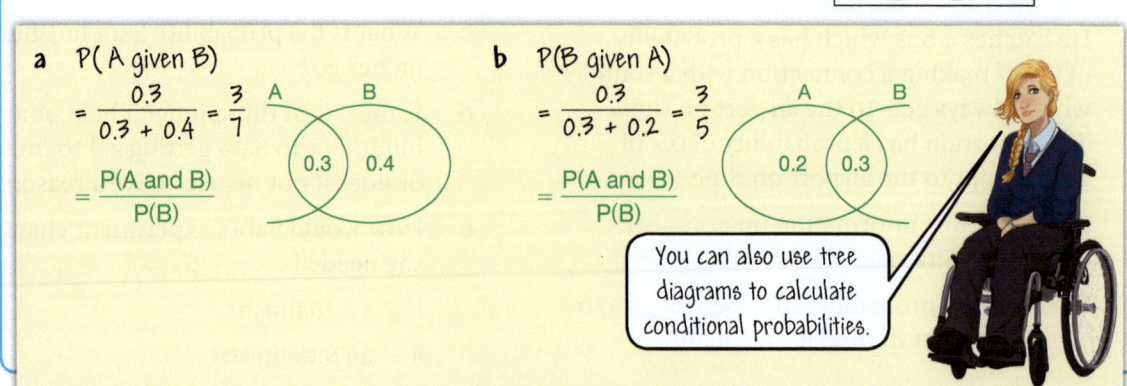

a P(A given B)
$= \frac{0.3}{0.3 + 0.4} = \frac{3}{7}$
$= \frac{P(A \text{ and } B)}{P(B)}$

b P(B given A)
$= \frac{0.3}{0.3 + 0.2} = \frac{3}{5}$
$= \frac{P(A \text{ and } B)}{P(B)}$

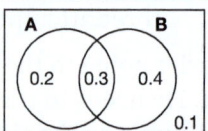

You can also use tree diagrams to calculate conditional probabilities.

Probability   The probability of combined events

Fluency

Exercise 19.4S

1. For the sets shown in the Venn diagram, find
   a P(S or R)
   b P(S given R)
   c P(R given S)
   d P(R and S)

   Are S and R independent?

   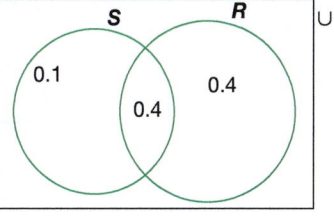

2. Three sets are defined

   U = {single digit integers}
   P = {prime numbers}
   F = {factors of 6}

   a Draw a Venn diagram showing the probability of a single digit number chosen at random lying in each of the regions created by the sets P and F.
   b Find i P(P) ii P(P and F) iii P(P given F)
   c Explain why P and F are not independent.

3.
	Walk	Other
Year 7	43	136
Year 11	98	83

   The table shows the way year 7 and year 11 pupils normally get to a particular school.
   a What is the probability a Year 7 pupil chosen at random walks to school?
   b What is the probability a Year 11 pupil chosen at random walks to school?
   c Is the way pupils travel to school independent of the year group they are in?

4. 95% of drivers wear seat belts.
   44% of car drivers involved in accidents die if they are not wearing a seat belt.
   92% of those that do wear a seat belt survive.
   a Draw a tree diagram to show this information.
   b What is the probability that a driver in an accident did not wear a seat belt and survived?

5.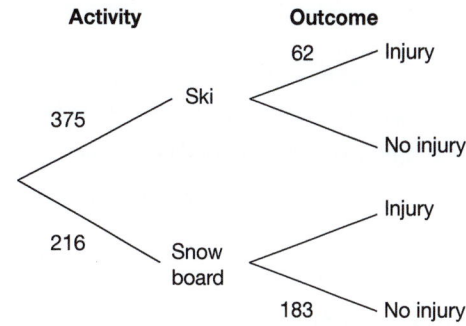

   a Copy and complete the tree diagram showing the number of people in a ski resort who completed a survey about their holiday.
   b If a person is chosen at random from the people *who were snowboarding*, what is the probability they suffered an injury?
   c If a person is chosen at random from those *who were injured*, what is the probability that they were snowboarding?

6. There are eight hundred thousand adults in a large city. A rare illness affects about one in 500 adults.
   A new screening test gives a positive result 98% of the time when a person has the illness.
   It also gives a positive result on 1% of people who do not have the illness.
   a Draw a tree diagram to show the numbers of adults in the city expected to have and not have the illness, and what the results of the screening test would be.
   b Find the probability that a person with a negative test result has the disease.
   c Find the probability that a person with a positive test result has the disease.

APPLICATIONS

19.4 Conditional probability

RECAP

- The **conditional** probability of B happening, *given* that A has already happened, is written as P(B given A).
- To calculate P(B given A) you can
  - calculate the fraction of outcomes in event B that are also in event A
  - or use the formulae

  $P(B \text{ given } A) = \dfrac{P(A \text{ and } B)}{P(A)}$ and $P(A \text{ and } B) = P(B \text{ given } A) \times P(A)$

If B happening does not depend on A happening then P(B given A) = P(B).

HOW TO

To solve problems involving conditional probability
1. As appropriate, use the information given to create a two-way table, Venn diagram or tree diagram.
2. Decide which events depend on the outcome of previous events and calculate the required probabilities.
3. Answer the question.

EXAMPLE

In a long jump competition, athletes have to jump at least 6.5 metres in the first three rounds to be eligible for three more jumps.
In the competition, 70% of the athletes qualify for the extra jumps and 50% record a best jump of at least 7 metres.

a Draw a Venn diagram to represent this information.

b Calculate the probability that an athlete makes a jump of at least 7 metres given that she qualified for extra jumps.

---

a Let Q = {athletes qualifying for extra jumps}
S = {athletes jumping over 7 metres}

S must be a **subset** of Q, because if you've jumped over 7 metres, then you've definitely jumped over 6.5 metres – so you've qualified.

① Jumping ≥ 7m ⇒ jumping ≥ 6.5m, so S is a subset of Q (S ⊂ Q).

b P(S given Q) = $\dfrac{0.5}{0.7} = \dfrac{5}{7}$ ② ③

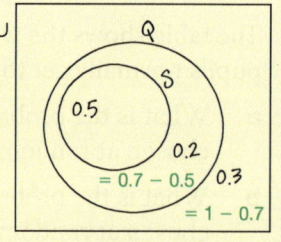

0.5
0.2
= 0.7 − 0.5   0.3
= 1 − 0.7

EXAMPLE

95% of drivers wear seat belts. 60% of car drivers involved in serious accidents die if they are not wearing a seat belt, whereas 80% of those that do wear a seat belt live.

a Draw a tree diagram to show this information.

b What is the probability that a driver in a serious accident did not wear a seat belt and lived?

c What is the probability that a driver who died in a serious accident was not wearing a seat belt?

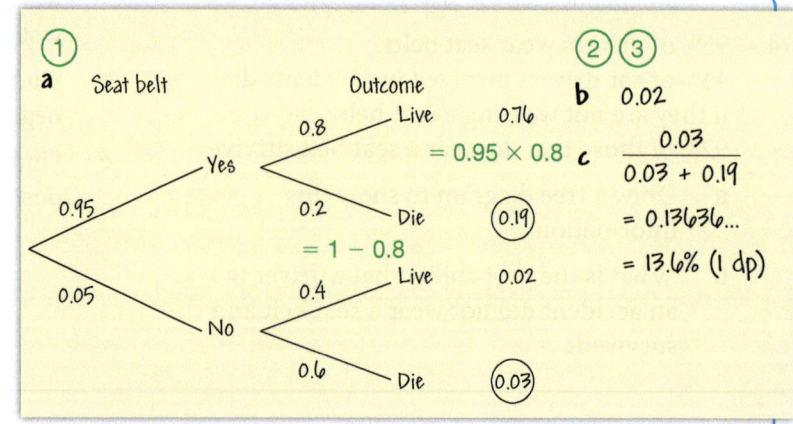

Probability The probability of combined events

Reasoning and problem solving

Exercise 19.4A

1. Of the employees in a large factory
   - $\frac{1}{3}$ travel to work by bus
   - $\frac{1}{4}$ by train
   - the rest by car.

   Those travelling by bus have a probability of $\frac{1}{4}$ of being late.
   Those by train will be late with probability $\frac{1}{5}$
   Those by car will be late with probability $\frac{1}{10}$.

   a Draw and complete a tree diagram, and calculate the probability that an employee chosen at random will be late.

   b Calculate the probability that an employee who is late travels by bus.

2. An insurance company classifies drivers in three categories.
   P is 'low risk', and they represent 25% of drivers who are insured.
   Q is 'moderate risk' and they represent 60% of the drivers.
   R is 'high risk'.

   The probability that a category P driver has one or more accidents in a twelve month period is 2%.
   The corresponding probabilities for Q and R are 6% and 10%.

   a Find the probability that a motorist, chosen at random, is assessed as a category Q risk and has one or more accidents in the year.

   b Find the probability that a motorist, chosen at random, has one or more accidents in the year.

   c If a customer has an accident in a twelve month period, what is the probability that they were a category Q driver?

3. A team plays 38 matches over a season in different conditions with these results.

	Wet	Dry
Win	6	17
Draw	2	5
Lose	4	4

   Find the probability that the team

   a won a match chosen at random

   b won a match that was played in wet conditions

   c played the match in wet conditions given that they won the match.

4. If a team wins a match the probability they win their next match is 0.6 and the probability of a draw is 0.2. If the match was a draw the probability of winning the next match is 0.5 and of drawing is 0.1. If they lose the match the probability of winning the next match is 0.3 and of drawing is 0.4.

   a Draw a tree diagram for two matches after the team has drawn a match.

   b A win is worth 3 points, a draw 1 point and a loss 0 points.
   What is the probability the team scores at least 3 points in the two matches?

   c Given that the team scores at least three points what is the probability they scored exactly 4 points?

5. A GP practice encourages elderly people to have a flu vaccination in the autumn. They claim it reduces the likelihood of having flu over the winter from 50% to 15%. The practice gives flu vaccination to 60% of the elderly people one year.
   What is the probability that an elderly person chosen at random from the practice

   a gets flu that winter

   b has been vaccinated given that they get flu

   c has been vaccinated given that they did not get flu?

Summary

Checkout
You should now be able to...

Test it
Questions

You should now be able to...	Questions
✓ Use tables and Venn diagrams to represent sets.	1
✓ Use a possibility space to represent the outcomes of two experiments and to calculate probabilities.	2, 3
✓ Use a tree diagram to show the outcomes of one or more experiments and to calculate probabilities.	4 – 6
✓ Calculate conditional probabilities.	1, 4, 5

Language	Meaning	Example
Set	A set is a collection of objects.	E = {Even numbers ⩽ 12} = {2, 4, 6, 8, 10, 12}
Element **Member**	An object in a set.	
Universal set, ξ	The set containing all the elements.	$ξ$ = {Positive numbers ⩽ 12}
Empty set, ∅	The empty set has no members.	{All odd numbers divisible by 2} = {} = ∅
Venn diagram	A way of showing sets and their elements.	T = {Multiples of three, less than 12} $ξ$ = {3, 6, 9, 12} (Venn diagram showing E and T with elements 2, 4, 8, 10 in E only; 6, 12 in intersection; 3, 9 in T only; 1 5 7 11 outside)
Intersection	For two or more sets, a new set containing the elements found in *all* of the sets.	Intersection, $E \cap T$ = {6, 12}
Union	For two or more sets, a new set containing the elements found in *any* of the sets.	Union, $E \cup T$ = {2, 3, 4, 6, 8, 9, 10, 12}
Complement	The complement of a set is all members which are not in that set but are in the universal set. The complement of A is A′.	O' = {odd numbers ⩽ 12} = {1, 3, 5, 7, 9, 11} O' = {2, 4, 6, 8, 10, 12}
Tree diagram	A diagram that shows all the outcomes of one or more consecutive events as successive branches. Probabilities are given on the branches.	(Tree diagram: 100 → H (60) → HH (35), HT (25); 100 → T (40) → TH (24), TT (16))
Frequency tree	A tree diagram in which each branch is labelled using the number of times that combination of outcomes occurred.	
Conditional probability	The probability of one event given the probability that another event has occurred. P(A given B) = P(A ∩ B) / P(B)	For a regular dice, P(2 given rolled an even number) = $\frac{1}{3}$
Independent	Two events are independent if the result of one event does *not* affect the result of the other event.	If A and B are independent then P(A given B) = P(A) P(A and B) = P(A) × P(B)

Probability   The probability of combined events

Review

1. Draw a Venn diagram to show the multiples of 3, even numbers and the factors of 20 less than 20.
   Simon selects an even number from the Venn diagram. What is the probability that it is a factor of 20?

2. Jakub has packs of sweets from 3 different brands. The relative frequency of each colour for each brand is shown in the table.

Colour	Brand 1	Brand 2	Brand 3
Red	0.25	0.2	0.15
Blue	0.25	0.3	0.35
Orange	0.25	0.4	0.45
Yellow	0.25	0.1	0.05

   Jakub selects one sweet from a pack of each brand.

   Estimate the probability he selects
   a 3 sweets of the same colour
   b at least 2 red sweets.

3. Jayden can choose fillings of egg, cheese or tuna for his sandwich, and can use white, brown or granary bread.
   a Draw a table to show all the possible types of sandwiches.

   Jayden is equally likely to pick each filling and equally likely to pick each bread type.

   b What is the probability that Jayden chooses
      i   granary bread
      ii   egg in brown bread
      iii not tuna and not white bread?

4. Madison must pass through two sets of traffic lights on her way to work. The probability the first set is red when she approaches is 0.15 and the probability the second set is red is 0.35.
   a Draw a tree diagram to show all the possible outcomes.
   b Calculate the probability that
      i   she doesn't have to stop at all
      ii she has to stop at least once.
   c On Monday the first light is green when Madison approaches it. What is the probability the second set is also green?

5. A bag contains 10 red and 6 black counters. A counter is removed at random and *not* put back. A second counter is then removed and not replaced.
   a Calculate the probability that
      i   both counters are red
      ii   both counters are black
      iii the counters are different colours.

   A third counter is now chosen.

   b Calculate the probability that all the counters are the same colour.
   c Two black counters are chosen without replacement. A third counter is then chosen. What is the probability that the third counter is also black?

6. The probability that Mia is late to work is $\frac{1}{20}$. If Mia is late to work then the probability that Sadjad is late to work is $\frac{1}{5}$, otherwise it is $\frac{1}{10}$.
   a Draw a tree diagram to show all the possible outcomes.
   b Calculate the probability that
      i   neither of them is late to work
      ii Sadjad is late to work.

What next?

Score		
	0 – 2	Your knowledge of this topic is still developing. To improve, look at MyiMaths: 1199, 1208, 1262, 1263, 1334, 1921, 1922, 1935
	3 – 5	You are gaining a secure knowledge of this topic. To improve your fluency look at InvisiPens: 20Sa – k
	6	You have mastered these skills. Well done you are ready to progress! To develop your problem solving skills look at InvisiPens: 20Aa – d

Assessment 19

1. The table shows the groupings of 80 people in a local Gymnastics club.

	Children under 12		Teenagers 13 to 19		Adults 20 to 30	
	Male	Female	Male	Female	Male	Female
Number of people	4	9	17	24	15	11

   a Use the information in the table to complete the Venn diagram. [4]

   b Calculate these probabilities
      i $P(T)$ ii $P(\text{not } M)$
      iii $P(T \text{ or } M)$ iv $P(T \text{ and } M)$ [4]

   c Are the events T and M independent? [3]

2. Gitanjali has two unbiased spinners.
   Each spinner is divided into eight equal sectors containing the numbers 1 to 8.
   The spinner are spun and their scores are added together.

   a Draw a sample space diagram for these spinners. [3]

   b Use your diagram to find
      i $P(3)$ [1]   ii $P(8)$ [1]   iii $P(12)$ [1]

   c What is the most likely score? [1]

   d How has the assumption that the spinners are unbiased affected your answers to parts b and c? [1]

3. $\xi = \{\text{Whole numbers from 1 to 20}\}$, $T = \{\text{factors of 18}\}$, $F = \{\text{factors of } n\}$ for some integer n.
   $P(F) = 0.25$ $P(T \text{ and } F) = 0.1$ and $P(T \text{ or } F) = 0.45$

   a Draw a Venn diagram to show the number of elements in each region. [4]

   b Find n. [3]

4. 32 editors in a publishing company went for a break in the staff room. The staff room has tea and coffee and a box containing chocolate biscuits and plain biscuits.
   17 editors had a cup of tea.
   12 of the editors who had tea chose a chocolate biscuit.
   22 editors ate chocolate biscuits altogether.
   Complete the frequency tree. [4]

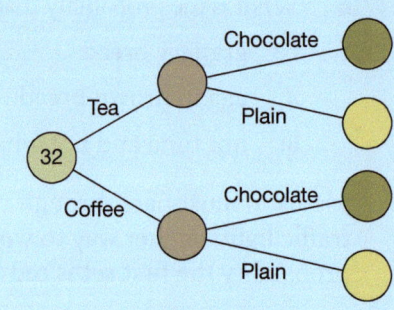

Probability   The probability of combined events

5 The probability that my postman delivers my mail between 10:00 and 10:30 in the morning is 0.15
   The probability it comes before 10 am is 0.1

   a Complete the tree diagram by writing the probabilities on each branch. [3]
   b Calculate the probability that the postman delivers my post before 10:30 on both days. [2]
   c Calculate the probability that the postman delivers my post before 10:30 on at least one day. [3]

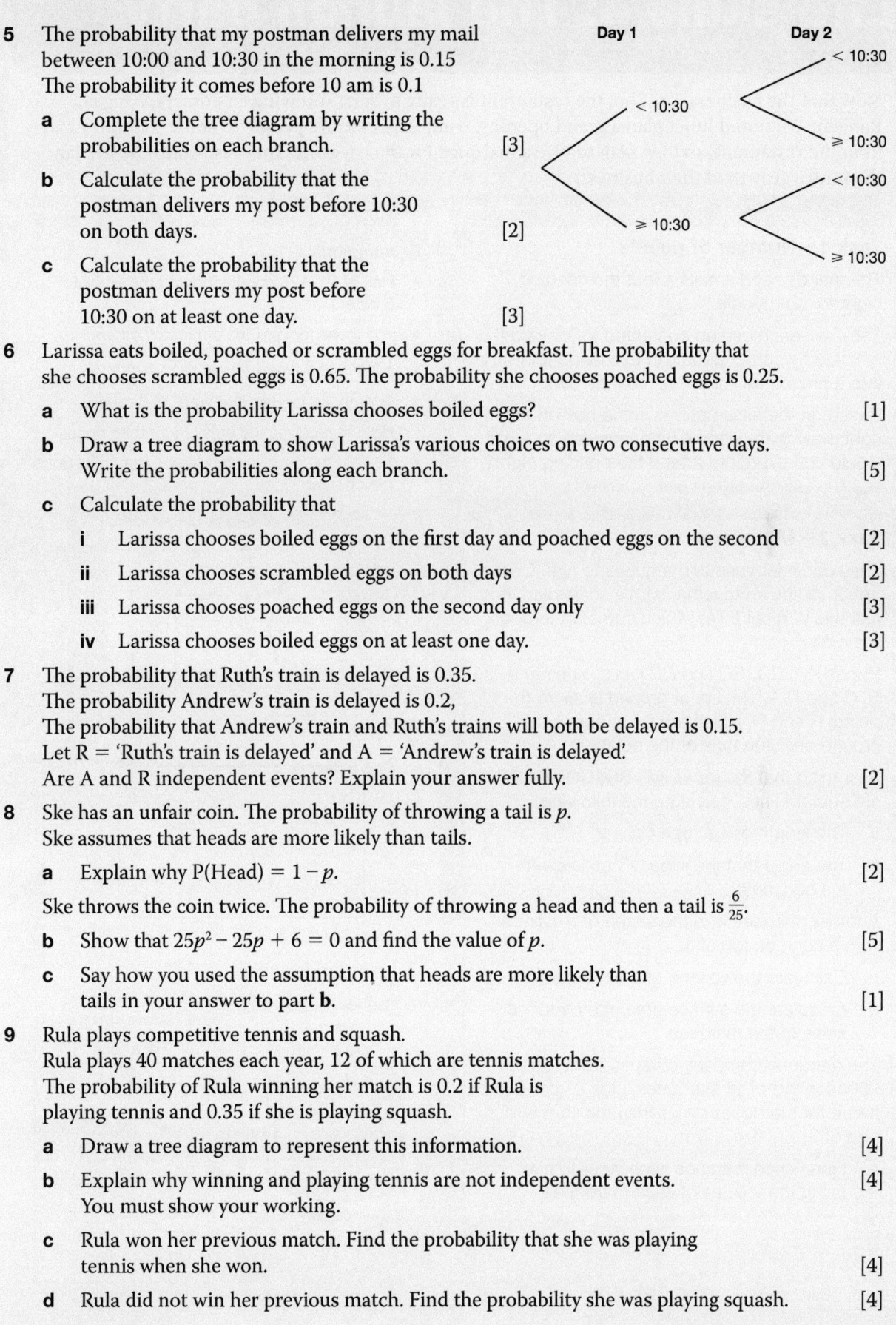

6 Larissa eats boiled, poached or scrambled eggs for breakfast. The probability that she chooses scrambled eggs is 0.65. The probability she chooses poached eggs is 0.25.

   a What is the probability Larissa chooses boiled eggs? [1]
   b Draw a tree diagram to show Larissa's various choices on two consecutive days. Write the probabilities along each branch. [5]
   c Calculate the probability that
      i   Larissa chooses boiled eggs on the first day and poached eggs on the second [2]
      ii Larissa chooses scrambled eggs on both days [2]
      iii Larissa chooses poached eggs on the second day only [3]
      iv Larissa chooses boiled eggs on at least one day. [3]

7 The probability that Ruth's train is delayed is 0.35.
   The probability Andrew's train is delayed is 0.2,
   The probability that Andrew's train and Ruth's trains will both be delayed is 0.15.
   Let R = 'Ruth's train is delayed' and A = 'Andrew's train is delayed'.
   Are A and R independent events? Explain your answer fully. [2]

8 Ske has an unfair coin. The probability of throwing a tail is p.
   Ske assumes that heads are more likely than tails.

   a Explain why P(Head) = $1 - p$. [2]

   Ske throws the coin twice. The probability of throwing a head and then a tail is $\frac{6}{25}$.

   b Show that $25p^2 - 25p + 6 = 0$ and find the value of p. [5]
   c Say how you used the assumption that heads are more likely than tails in your answer to part **b**. [1]

9 Rula plays competitive tennis and squash.
   Rula plays 40 matches each year, 12 of which are tennis matches.
   The probability of Rula winning her match is 0.2 if Rula is playing tennis and 0.35 if she is playing squash.

   a Draw a tree diagram to represent this information. [4]
   b Explain why winning and playing tennis are not independent events. You must show your working. [4]
   c Rula won her previous match. Find the probability that she was playing tennis when she won. [4]
   d Rula did not win her previous match. Find the probability she was playing squash. [4]

Life skills 4: The launch party

Now that the business is set up, the restaurant is ready to start receiving customers. Abigail, Raheem, Mike and Juliet plan a grand opening. They expect more people to come than they can fit in the restaurant, so they plan to hire a marquee for the car park. They also continue to plan the future growth of their business.

Task 1 – Number of guests

The friends send emails about the opening night to 128 people.

They ask each person contacted to forward the email to five other people in exchange for entry into a prize draw for a free meal for two.

Based on the assumptions in the box on the right, how many people who received an email would you expect to attend the opening night?

Assumptions
- Half of the 128 people forward the email to 5 others.
- $\frac{1}{4}$ of these forward the email to 5 others.
- $\frac{1}{8}$ of these forward the email to 5 others.
- $\frac{1}{16}$ of these forward the email to 5 others.
- No one receives the email more than once.
- 10% of the people who receive the email go to the opening night.

Task 2 – Marquees

They consider various marquees to hire. One option is shown together with a scale plan. It has two vertical poles which come up through the roof.

Ropes *AP*, *BQ*, *CQ* and *DP* join the points *A*, *B*, *C* and *D*, which are at ground level, to the points *P* and *Q*, which are 8.6 m above the ground near the tops of the poles.

Assuming that the ropes *AP*, *BQ*, *CQ* and *DP* are straight lines, calculate the following.

a   The length of the rope *QC*.

b   The angle that the rope *QC* makes with the horizontal.

Another marquee is in the shape of a cylinder with a cone on top of it.

c   Calculate the volume of the marquee.

d   Calculate the surface area of the roof and sides of the marquee.

The marquee company charges a flat fee of £500 for hire of all marquees, plus £1 per cubic metre for marquees larger than the standard size of 8 m × 6 m × 3 m.

e   Find which marquee is cheaper to hire using the volume of each marquee.

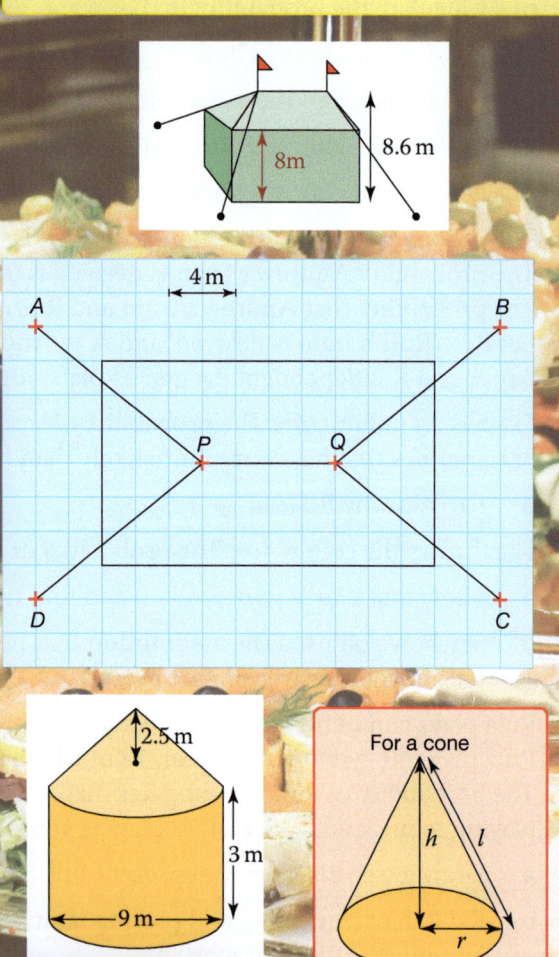

For a cone
$V = \frac{1}{3}\pi r^2 h$
Curved S.A. $= \pi r l$

Life skills

420

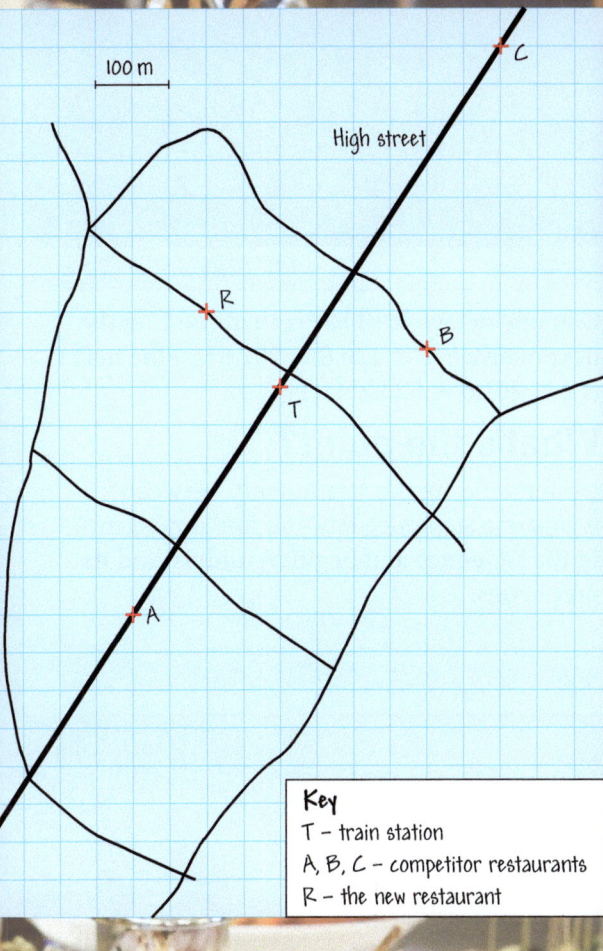

Key
T – train station
A, B, C – competitor restaurants
R – the new restaurant

Task 3 – Marketing slogan

The friends make the following claim in their marketing material for the opening night:

'Closest restaurant to the train station!'

Determine if their claim is correct by

a Finding the following as column vectors (using metres as units) $\vec{TB}$ $\vec{TA}$ $\vec{TR}$

b and so finding the distances TB, TA and TR.

c Use the cosine rule and your answers to part b to find the angle ARB.

Task 4 – Forecasting

The friends agree to price meals as follows.

All starters	£4
Main courses	£15 (expensive)
	£13 (medium)
	£11 (cheap)
All desserts	£3

Assuming all customers have a main course, but not all have a starter or dessert, use the market research and the prices to estimate the probability that a customer spends more than £17.50 on food.

Results of market research

- P(customer has a starter) = 0.7
- P(customer who has had a starter has an expensive main course) = 0.2
- P(customer who has had a starter has a medium main course) = 0.5
- P(customer who has had a starter has a cheap main course) = 0.3
- P(customer who has not had a starter has an expensive main course) = 0.4
- P(customer who has not had a starter has a medium main course) = 0.4
- P(customer who has not had a starter has a cheap main course) = 0.2
- P(Any customer has a dessert) = 0.4

Projections

Number of customers in first month = 700
Growth of 5% per month

Task 5 – Future growth

The friends have used the outcome of the opening night, and some additional market research, to make some projections about the growth of the business. They create a table as part of a report to the bank who gave them the business loan.

Based on their projections, copy and complete the following table for their report.

Number of customers in one years' time (start of 13th month)	
Total number of customers in the first year	
Month in which 2000 customers first reached	

20 Sequences

Introduction

Musical scales are typically written using eight notes: the C Major scale uses C D E F G A B C. The interval between the first and last C is called an octave.

The pitch of a musical note, measured in Hertz (Hz), corresponds to the number of vibrations per second.

The frequencies of the corresponding notes in each octave follow a geometric sequence. If the C in one octave is 130.8 Hz then the C in the next octave is $2 \times 130.8 = 261.6$ Hz, the next C is $2 \times 261.6 = 523.2$ Hz, etc.

What's the point?

Understanding the relationship between terms in a sequence lets you find any term in the sequence and begin to understand its properties.

Objectives

By the end of this chapter you will have learned how to …

- Find terms of a linear sequence using a term-to-term or position-to-term rule.
- Recognise special types of sequence and find terms using either a term-to-term or position-to-term rule.
- Find terms of a quadratic sequence using a term-to-term or position-to-term rule.

Check in

1 Complete the next two values in each pattern.

- **a** 2, 4, 6, 8, ...
- **b** 100, 94, 88, 82, 76, ...
- **c** 1, 2, 4, 7, 11, ...
- **d** 10, 7, 4, 1, ...
- **e** 3, 6, 12, 24, ...
- **f** $1, \frac{1}{2}, \frac{1}{3}, \frac{1}{4}, \frac{1}{5}, ...$

2 Given that $n = 3$, put these expressions in ascending order.

$2n + 7 \quad 4(n-1) \quad 2n^2 \quad \frac{9}{n} + 15 \quad 15 - n$

3 This pattern has been shown in three different ways.
It has a special name.
What is it called?
Describe how it got this name.

1	4	9	16
1 × 1	2 × 2	3 × 3	4 × 4
□	⊞	(3×3 grid)	(4×4 grid)

Chapter investigation

Alain, Bo and Cara are making patterns with numbers.

Alain makes a sequence by adding a fixed number onto a starting number.

1, 4, 7, 10, 13, 16, 19, 22, 25, 28, ...   Start with 1, add 3 each time

Bo makes a second sequence by adding Alain's sequence onto another starting number.

1, 2, 6, 13, 23, 36, 52, 71, 93, 118, ...   Start with 1, then add 1, then add 4, then add 7 then add 10, etc.

Cara takes the first two numbers in Bo's sequence and makes a third term by adding these together, then a fourth term by adding the second and third terms, etc.

1, 2, 3, 5, 8, 13, 21, 34, 55, 89, 144, ... $1 + 2 = 3, 2 + 3 = 5, 3 + 5 = 8$, etc.

How do sequences like these behave?

SKILLS

20.1 Linear sequences

Linear sequences can be generated and described using
- a **'term-to-term' rule**
- a **'position-to-term' rule**

For the sequence
5, 8, 11, 14, 17…
'add 3'
nth term = 3n + 2

EXAMPLE

Find the first five terms of the following sequences using the given term-to-term rules.
a First term 20   Rule Subtract 7
b First term 4   Rule Add 6

a

b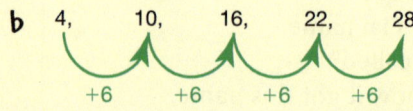

EXAMPLE

The *n*th term of a sequence is $5n - 2$.
a Find the first five terms of the sequence
b Find the 50th term.

> '5n' means '5 × n'

a 3, 8, 13, 18, 23

1st term (n = 1): (5 × 1) − 2 = 3
2nd term (n = 2): (5 × 2) − 2 = 8
3rd term (n = 3): (5 × 3) − 2 = 13

b 50th term = (5 × 50) − 2 = 248

4th term (n = 4): (5 × 4) − 2 = 18
5th term (n = 5): (5 × 5) − 2 = 23

EXAMPLE

The rule for a sequence is $T(n) = 2n + 3$, where $T(n)$ is the *n*th term of the sequence and *n* is the position of the term in the sequence.
Find the first three terms of the sequence.

> '2n' means '2 × n'

T(1) = 2 × 1 + 3 = 5
T(2) = 2 × 2 + 3 = 7
T(3) = 2 × 3 + 3 = 9
Sequence is 5, 7, 9, …

> T(1) is the first term of the sequence.
> T(2) is the second term of the sequence.

- **Linear sequences** are sequences in which the differences between terms are a constant. This is known as the 'first difference'.

EXAMPLE

Find the *n*th term for these sequences.
a 5, 9, 13, 17, 21, …
b 10, 4, −2, −8, −14, …

a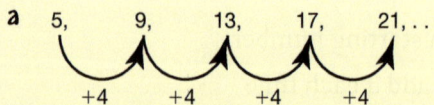

The difference between each term is +4.
*n*th term = 4n ± ☐
Compare the sequence with the first terms of the sequence 4n.

4n: 4, 8, 12, 16, 20, …
      5, 9, 13, 17, 21,
*n*th term = 4n + 1

b 10,   4,   −2,   −8,   −14, …
        −6   −6   −6   −6

The difference between each term is −6.
*n*th term = −6n ± ☐
Compare the sequence with the first terms of the sequence −6n.

−6n: −6, −12, −18, −24, …
      10,   4, −2, −8, …
*n*th term = −6n + 16
            = 16 − 6n

Algebra Sequences

Fluency

Exercise 20.1S

1 Find the first four terms of the following sequences using these term-to-term rules.

a	First term -10	Rule	Add 3
b	First term 2.5	Rule	Add 5
c	First term $1\frac{3}{4}$	Rule	Add $\frac{1}{2}$
d	First term 1.2	Rule	Add 0.15
e	First term 12.5	Rule	Subtract 3
f	First term 5	Rule	Subtract 4
g	First term 9	Rule	Subtract 1.5
h	First term $1\frac{3}{4}$	Rule	Subtract $\frac{1}{2}$

2 Find the missing terms in these linear sequences.

   a 3, 8, ☐, 18, 23
   b 4, 7, ☐, 13, 16
   c 20, ☐, 26, ☐, 32
   d 11, 15, ☐, 23, ☐
   e 30, 24, ☐, 12, 6
   f 10, 7, ☐, 1, ☐
   g $2\frac{1}{4}$, ☐, $1\frac{1}{4}$, ☐, $\frac{1}{4}$
   h 1.4, 1.25, ☐, ☐, 0.8

3 The rule for a sequence is $T(n) = 3n + 3$, where $T(n)$ is the nth term of the sequence and n is the position of the term in the sequence.

Find the first three terms of the sequence.

4 The rule for a sequence is $T(n) = 5n - 4$, where $T(n)$ is the nth term of the sequence.

Find the first three terms of the sequence.

5 The terms of a sequence can be generated using the rule $T(n) = 4n + 7$.

Calculate the

   a 10th term   b 15th term
   c 100th term   d 1000th term.

6 Find the first five terms of these sequences.

   a $3n$ b $4n + 2$ c $2n + 5$
   d $4n - 1$ e $5n - 3$ f $2n - 6$
   g $3n - 12$ h $0.5n + 5$
   i $2.5n - 0.5$ j $2 - \frac{1}{3}n$ k $\frac{1}{6} + \frac{1}{2}n$

7 Find the first five terms of these sequences.

   a $-2n$ b $15 - n$
   c $10 - 2n$ d $25 - 5n$
   e $7.5 - 6n$ f $4 - 0.5n$

8 Find the nth term of these sequences.

   a 3, 5, 7, 9, 13, ...
   b 4, 7, 10, 13, 16, ...
   c 5, 8, 11, 14, 17, ...
   d 4, 10, 16, 22, 28, ...
   e 7, 17, 27, 37, ...
   f 2, 10, 18, 26, 34, ...
   g 2, 3.5, 5, 6.5, 8, ...
   h 1.4, 2, 2.6, 3.2, 3.8, ...

9 Find the nth term of these sequences.

   a 15, 12, 9, 6, 3, ...
   b 10, 6, 2, -2, -6, ...
   c 5, 1, -3, -7, -11, ...
   d -2, -5, -8, -11, -14, ...
   e $\frac{1}{2}$, $-\frac{1}{4}$, -1, $-1\frac{3}{4}$, $-2\frac{1}{2}$, ...

10 For the sequence 2, 6, 10, 14, 18, ...

   a predict the 10th term
   b find the nth term.
   c Use your answer to part **b**, to evaluate the accuracy of your prediction.

11 Is 75 a term in the sequence described by the nth term $5n - 3$?

Give your reason.

APPLICATIONS

20.1 Linear sequences

RECAP

Know how to generate and describe linear sequences using
- the 'term-to-term' rule and
- 'position-to-term' rule using the nth term.

2, 6, 10, 14, 18, ...
differences all = +4
nth term rule = $4n \pm \square$
Compare $4n$: 4, 8, 12, 16, 20
           2, 6, 10, 14, 18 -2
Sequence = $4n - 2$

HOW TO

To describe a linear sequence using the nth term
1. Find the *constant* difference between terms.
2. This difference is the first part of the nth term.
3. Add or subtract a constant to adjust the expression for the nth term.

EXAMPLE

Kia says
'there will be 30 squares in the 12th diagram because there are 10 squares in the 4th diagram.'

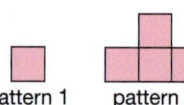 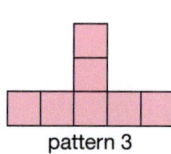 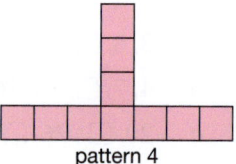

pattern 1   pattern 2   pattern 3   pattern 4

Do you agree with Kia? Explain your reasoning.

Pattern, n 1   2   3 $T(n) = 4$
Number of squares, $T(n) = $ 1   4   7   10
                         3   3   3
nth term, $T(n) = 3n \pm \square$

① Calculate the first differences.
② First difference is a constant, +3.
③ Compare the sequence with the first term of the sequence $3n$.

Compare $3n$: 3, 6, 9, 12, 15
              1, 4, 7, 10, 13 -2

Sequence = $3n - 2$

$T(12) = 3 \times 12 - 2 = 36 - 2 = 34$

No, I do not agree with Kia. There are 34 squares in the 12th diagram.

Just because $12 = 3 \times 4$ does not mean $S(12) = 3 \times S(4)$

EXAMPLE

Seung-Hee is generating a sequence using the rule $T(n) = 6n - 2$. She thinks that every term will be an even number.

Do you agree with Seung-Hee? Explain your reasoning.

$T(1) = 4$, $T(2) = 10$, $T(3) = 16$ .... $T(10) = 58$ Test the conjecture.
Yes, I agree with Seung-Hee.
When you multiply any number by 6 you produce an even number. Subtracting two from an even number always produces an even number.

You can also write $T(n) = 2(3n - 1)$ which is clearly a multiple of 2.

Algebra Sequences

Reasoning and problem solving

Exercise 20.1A

1. Kate thinks that the tenth term of the sequence 3, 7, 11, 15, 19, … will be 38.

   Do you agree with Kate?
   Give your reason.

2. Anika says that the nth term of the sequence 5, 9, 13, 17, 21, … is $n + 4$.

   Do you agree with Anika?
   Give your reason.

3. **a** Match each sequence with the correct 'term-to-term' rule and 'position-to-term' rule.

4, 1, −2, −5, …
6, 10, 14, 18, …
3, 10, 17, 24, …
6, 2, −2, −6, …

Subtract 4
Add 4
Subtract 3
Add 7

$T(n) = 4n + 2$
$T(n) = 7n - 4$
$T(n) = 7 - 3n$
$T(n) = 10 - 4n$

   b Create your own puzzle.

4. Sam is making patterns using matches.

    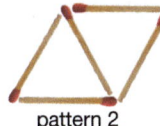 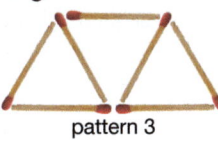
   pattern 1   pattern 2   pattern 3

   a Find a formula for the number of matchsticks, m, in the nth pattern.

   b How many matches will Sam need for the 50th pattern?

   c Which is the first pattern that will need more than 100 matches to make?

5. Draw a set of repeating patterns to represent the sequence given by the nth term $4n + 3$.

6. Matthew is building a fence.

   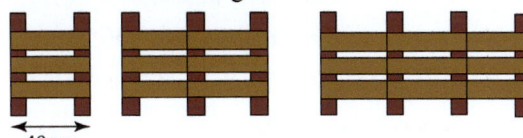
   40 cm

   The fence needs to be 11.2 m long. How many posts and planks does Matthew need?

7. How many squares are in the 15th pattern?

   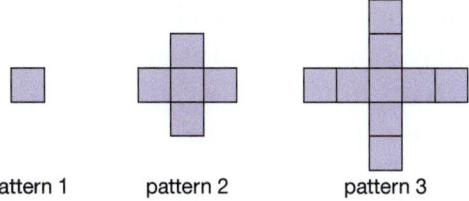
   pattern 1   pattern 2   pattern 3

8. Suha is generating a sequence using the rule $T(n) = 4n - 2$. She thinks that every term will be an even number.

   Do you agree with Suha?
   Give your reason.

9. The first part of this number has been blotted out. Jessica says the number will appear in the sequence $5n - 2$. How can she be sure?

10. Hannah is placing paving slabs around different sized ponds.

    She notices that the sequence can be described by the nth term $2n + 6$.

    a Justify that the nth term is $2n + 6$

    b Explain how the expression for the nth term relates to the structure of Hannah's patterns.

11. For each of these linear sequences find the formula for the nth term.

    a 5th term 63, decreases by 7 each time.

    b 7th term 108, 8th term 113.

    c 100th term 26, 101st term 31.

    d 22nd term 104, 122nd term 172.

*12. Reya says

   'The two sequences $T(n) = 7 + 5n$ and $S(n) = 54 - 6n$ don't have any terms in common because when I solve $T(n) = S(n)$ I get $n = 4\frac{3}{11}$ which isn't an integer'.

   Give a reason why Reya is wrong.

SKILLS

20.2 Quadratic sequences

Quadratic sequences can be generated and described using a
- 'term-to-term' rule
- 'position-to-term' rule.

For the sequence
- 2, 5, 10, 17, 26...
  'add the odd numbers starting with 3.'
- nth term = $n^2 + 1$

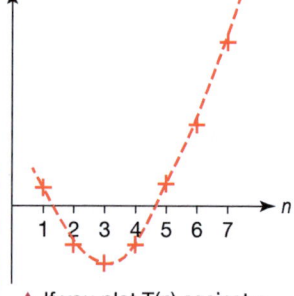

▲ If you plot T(n) against n eg 1, −2, −3, −2, 1, 6 for a quadratic sequence then you get points on a quadratic curve.

EXAMPLE

Find the first five terms of the sequence using this term-to-term rule.
First term   5   Rule   Add consecutive odd numbers

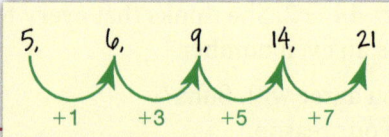

- Quadratic sequences are sequences in which
  - the second differences between terms are constant
  - the nth term rule contains an n^2 term and no higher powers of n.
  - The coefficient of n^2 is half the constant second difference.

EXAMPLE

The nth term of a sequence is $n^2 + n + 1$.
  a Find the first five terms of the sequence.
  b Check the sequence is a quadratic sequence.

a 3, 7, 13, 21, 31, ...

$T(1) = 1^2 + 1 + 1 = 3$
$T(2) = 2^2 + 2 + 1 = 7$ T(1) is the first term of the sequence.
$T(3) = 3^2 + 3 + 1 = 13$
$T(4) = 4^2 + 4 + 1 = 21$ T(2) is the second term of the sequence.
$T(5) = 5^2 + 5 + 1 = 31$

b 3, 7, 13, 21, 31, ...
       4   6   8   10   First difference
         2   2   2   Second difference

As the second difference is constant, the sequence is quadratic.

EXAMPLE

Find the nth term for these quadratic sequences.
  a 5, 11, 21, 35, 53, ...
  b 3, 8, 15, 24, 35, ...

a 5,   11,   21,   35,   53, ...
       +6 +10 +14 +18   First difference
         +4   +4   +4   Second difference

nth term = $2n^2 \pm \square$

Compare the sequence with the first terms of the sequence $2n^2$.

$2n^2$ 2, 8, 18, 32, 50
            5, 11, 21, 35, 53
Difference 3, 3, 3,   3,   3

Sequence = $2n^2 + 3$
Constant difference +3
Quadratic sequence, nth term = $2n^2 + 3$

b 3,   8,   15,   24,   35, ...
       +5 +7   +9   +11   First difference
         +2   +2   +2   Second difference

The coefficient of n^2 is half the value of the second difference.

nth term = $n^2 \pm \square$

Compare the sequence with the first term of the sequence n^2.

n^2 1, 4, 9, 16, 25
            3, 8, 15, 24, 35
Difference 2, 4, 6, 8,   10

Sequence = $n^2 + 2n$ By inspection
Linear sequence, nth term = $2n$
Quadratic sequence, nth term = $n^2 + 2n$

Algebra   Sequences

Fluency

Exercise 20.2S

1 Find the first four terms of these sequences using the term-to-term rules.

   a First term 2
   Rule Add consecutive odd numbers

   b First term 5
   Rule Add consecutive even numbers

   c First term 10
   Rule Add consecutive whole numbers

   d First term 5
   Rule Add consecutive multiples of 3

   e First term 10
   Rule Subtract consecutive odd numbers

   f First term 8
   Rule Subtract consecutive even numbers

2 Find the next two terms of these quadratic sequences.

   a 3, 7, 13, 21, □, □
   b 4, 6, 10, 16, □, □
   c 1, 5, 10, 16, □, □
   d 1, 4, 9, 16, □, □
   e 10, 7, 6, 7, □, □
   f 8, 18, 24, 26, □, □
   g 10, 7, −1, −14, □, □

3 Find the missing terms in these quadratic sequences.

   a 6, 12, 22, □, 54
   b □, 8, 18, 32, 50
   c 2, □, 10, 17, 26
   d 1, 5, □, 19, □
   e −4, 10, 28, □, □
   f 6, 12, 15, □, 12
   g 5, 8, 6, □, −13

4 **a** Find the first five terms for the sequences described by these rules.

   i $T(n) = n^2 + 3$
   ii $T(n) = n^2 + 2n$
   iii $T(n) = n^2 + n - 2$
   iv $T(n) = 2n^2$
   v $T(n) = 2n^2 + 4$

   b Calculate the 10th term of each sequence.

5 **a** Using the nth term for each sequence, calculate the first five terms.

   i $n^2 + 1$ **ii** $n^2 - 3$
   iii $n^2 + 2n + 1$ **iv** $n^2 + n - 4$
   v $2n^2 + n$ **vi** $0.5n^2 + n + 1$
   vii $2n^2 + 2n + 2$ **viii** $n(n - 2)$

   b Calculate the second difference in each case to check the sequences are quadratic.

6 Find the nth term of these quadratic sequences.

   a 4, 7, 12, 19, …   **b** −3, 0, 5, 12, …
   c 2, 8, 18, 32, …   **d** 2, 6, 12, 20, …
   e 5, 12, 21, 32, …   **f** 3, 8, 15, 24, …
   g 6, 14, 26, 42, …   ***h** 3, 4, 3, 0, …

7 Find the nth term of these sequences.

   a 9, 6, 1, −6, …   **b** 18, 12, 2, −12, …
   c 0, −2, −6, −12, …   **d** 10, 8, 4, −2, …

8 **a** For the sequence 2, 6, 12, 20, 30, … predict the position of the first term that will have a value greater than 1000.

   b Find the nth term for the sequence 2, 6, 12, 20, 30, …

   c Use your answer to part **b** to evaluate the accuracy of your prediction.

9 Is 150 a term in the quadratic sequence described by the nth term $n^2 + 3$?

   Give your reason.

10 Jarinda is generating a sequence using the rule $T(n) = 3n^2 - n$. She thinks that every term will be an even number. Do you agree with Jarinda? Give your reason.

> **Did you know…**
>
> The distance travelled by a falling object in successive seconds forms a quadratic sequence.

APPLICATIONS

20.2 Quadratic sequences

RECAP

Know how to generate and describe quadratic sequences using
- the 'term-to-term' rule and
- the 'position-to-term' rule using the nth term.

HOW TO

To describe a quadratic sequence using the nth term
1. Find the first difference between each term.
2. Find the difference between the first differences: the sequence is quadratic if the second difference is *constant*.
3. The coefficient of n^2 is half the value of the second difference.
4. Add a linear sequence to adjust the expression for the nth term.

3, 10, 21, 36, 55, ...
second differences all = +4
nth term rule = $2n^2 \pm \square$
$2n^2$: 2, 8, 18, 32, 50
         3, 10, 21, 36, 55
         1, 2, 3, 4, 5
nth term rule = $2n^2 + n$

EXAMPLE

One of these statements is correct.
Who is right, Ed or Asha? Give your reason.

Ed says,

'Since $12 = 3 \times 4$ and there are 13 squares in the 4th pattern then in the 12th pattern there will be $3 \times 13 = 39$ squares.'

Asha says,

'There is a pattern in this sequence with 420 squares.'

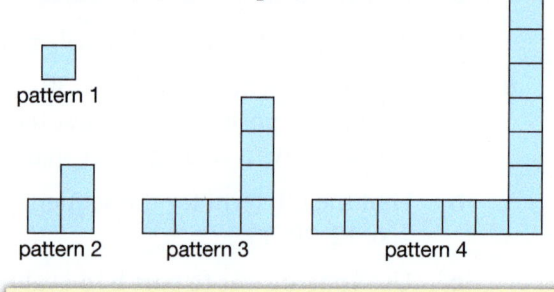

pattern 1, pattern 2, pattern 3, pattern 4

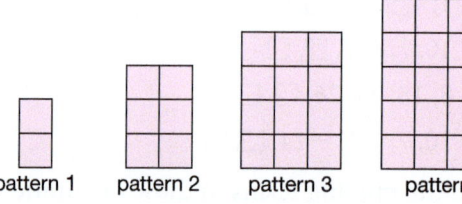

pattern 1, pattern 2, pattern 3, pattern 4

1, 3, 7, 13, ...
  2 4 6
   2 2

The coefficient of n^2 is 1.
nth term = $n^2 + \square$
n^2: 1 4 9 16
        1, 3, 7, 13, ...
        0 -1 -2 -3

① First difference
② Second difference
③ $1 = 2 \div 2$
④ By inspection

Linear sequence, nth term = $1 - n$
Quadratic sequence, nth term = $n^2 - n + 1$
The 12th pattern has $12^2 - 12 + 1 = 133$ squares.

2, 6, 12, 20, ...
  4 6 8
   2 2

The coefficient of n^2 is 1.
nth term = $n^2 + \square$
n^2: 1 4 9 16
        2 6 12 20, ...
        1 2 3 4

Linear sequence, nth term = n
Quadratic sequence, nth term = $n^2 + n = n(n + 1)$
$n^2 + n = 420$ Is there a positive integer solution?
$n^2 + n - 420 = (n + 21)(n - 20) = 0$
$n = -21$ or 20
The 20th pattern has 420 squares.

Ed is wrong and Asha is correct.   Remember to answer the question.

Algebra   Sequences

Reasoning and problem solving

Exercise 20.2A

1. Georgina thinks that the 10th term of the sequence 4, 7, 12, 19, 28, … will be 56. Do you agree with Georgina? Give your reason.

2. Hazeem thinks that the nth term of the sequence 5, 7, 11, 17, 25, … will start with '$2n^2$'.
   a. Do you agree with Hazeem? Give your reason.
   b. Find the full expression for the nth term of the sequence.

3. a. Match each sequence with the correct 'position-to-term' rule.
   b. Find the missing entries.

2, 6, 12, 20, …	$T(n) = n^2 - n$
3, 0, −5, −12, …	
0, 3, 8, 15, …	$T(n) = n^2 + 2$
	$T(n) = n(n + 1)$
3, 6, 11, 18, …	$T(n) = 4 - n^2$

   c. Create your own matching puzzle.

4. The first terms of a quadratic sequence are 3 and 7. Find five different quadratic sequences with this property.

5. Sam is making patterns using matches.

   a. Draw the next pattern.
   b. Find a formula for the number of matchsticks, m, in the nth pattern.
   c. Find the pattern that contains 139 matches.
   d. How many matches will Sam need for the 50th pattern?

6. Draw two sets of patterns to represent the sequence described by the nth term $n^2 + 2$.

7. Draw a set of rectangular patterns to describe sequences with these nth terms.
   a. $T(n) = (n + 2)(n + 3)$
   b. $T(n) = 2n^2 + 7n + 3$

8. How many squares are there in the 15th pattern?

9. Four boys and four girls sit in a row as shown.

   The boys want to swap places with the girls.

   However, they are only allowed to move by either
   – sliding into an empty chair or
   – jumping over one person into an empty chair.

   a. How many moves will it take?
   b. How many moves will it take if there are 3 boys and 3 girls, 2 boys and 2 girls, …?
   c. Copy and complete the table.

Number of pairs	1	2	3	4
Number of moves				

   d. Find a rule to predict the number of moves for 50 pairs of boys and girls.

10. Justify that the nth term of the triangular number sequence 1, 3, 6, 10, … is $\frac{1}{2}n(n + 1)$.
    a. Use an algebraic approach such as analysing the first and second differences to find the nth term.
    b. Use a geometric approach by drawing each term of the sequence.

*11. By looking at successive differences, or otherwise, find expressions for the nth term of these *cubic* sequences.
    a. 1, 8, 27, 64, 125, 216, …
    b. 2, 16, 54, 128, 250, 432, …
    c. 0, 6, 24, 60, 120, 210, …
    **d. −2, 2, 14, 40, 86, 158, …

SKILLS

20.3 Special sequences

- Square, cube and triangular numbers are associated with geometric patterns.

Square numbers
1, 4, 9, 16, 25, ...

Cube numbers
1, 8, 27, 64, 125, ...

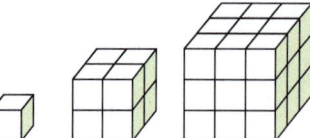

Triangular numbers
1, 3, 6, 10, 15, ...

EXAMPLE

a Find a square number that is also i a triangular number ii a cube number.
b Show that 33, 34 and 35 can each be written as the sum of no more than three triangular numbers.

List the numbers Square numbers 1, 4, 9, 16, 25, 36, 49, 64, ...
Triangular numbers 1, 3, 6, 10, 15, 21, 28, 36, ...
Cube numbers 1, 8, 27, 64, 125, 216, ...

a i 1 or 36   ii 1 or 64   Look for numbers appearing in both lists.
b 33 = 21 + 6 + 6   34 = 21 + 10 + 3   35 = 28 + 6 + 1
   or 15 + 15 + 3   or 28 + 6   or 15 + 10 + 10

- **Arithmetic** (linear) progressions have a constant difference between terms. $T(n+1) - T(n) = d$

EXAMPLE

Which of these sequences are arithmetic progressions?
a 5, 8, 11, 14, ...   b 1, −2, 4, −8, ...   c 3, $3\sqrt{3}$, 9, $9\sqrt{3}$, ...   d 7, 3, −1, −5, ...

a 8 − 5 = 11 − 8 = 14 − 11 = 3   b −2 ÷ 1 = 4 ÷ −2 = −8 ÷ 4 = −2
   Arithmetic   Constant difference, +3   not arithmetic   Constant ratio, −2
c $3\sqrt{3} ÷ 3 = 9 ÷ 3\sqrt{3} = 9\sqrt{3} ÷ 9 = \sqrt{3}$ d 3 − 7 = −1 − 3 = −5 − −1 = −4
   not arithmetic   Constant ratio, $\sqrt{3}$ Arithmetic   Constant difference, −4

- In a **quadratic** sequence the differences between terms form an arithmetic sequence; the second differences are constant.

EXAMPLE

Find the missing terms in this quadratic sequence.
2, 6, 12, 20, ☐, 42, ☐

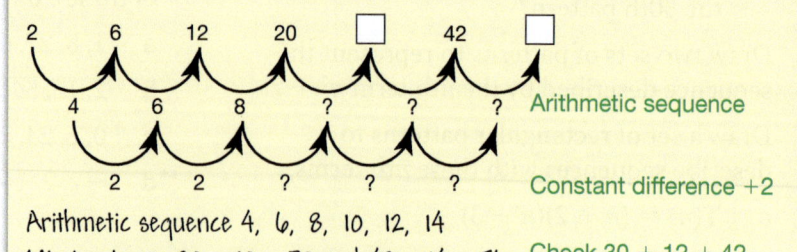

Arithmetic sequence
Constant difference +2

Arithmetic sequence 4, 6, 8, 10, 12, 14
Missing terms 20 + 10 = 30 and 42 + 14 = 56   Check 30 + 12 + 42

Algebra Sequences

Fluency

Exercise 20.3S

1. Write down the first ten terms of these number sequences.
   a. triangular b. square c. cube

2. Express these numbers as the sum of not more than three triangular numbers.
   a. 30   b. 31   c. 32

3. Describe these sequences using one of the words: Arithmetic, Quadratic, Another Type
   a. 2, 5, 8, 11, …
   b. 7, 11, 15, 19, …
   c. 2, 3, 5, 8, 13, …
   d. 2, 5, 10, 17, …
   e. 2, 6, 18, 54, …
   f. 18, 15, 12, 9, …
   g. 1, 4, 5, 9, …
   h. 1, 2, 4, 8, …
   i. 4, 7, 13, 22, …
   j. 0.5, 2, 3.5, 5, …
   k. $\frac{1}{4}, \frac{1}{2}, 1, 2, \ldots$
   l. $3, 1, \frac{1}{3}, \frac{1}{9}, \ldots$
   m. −2, 4, −8, 16, …
   n. 2, −2, 2, −2, …
   o. $2, 2+\sqrt{5}, 2+2\sqrt{5}, 2+3\sqrt{5}, \ldots$
   p. $\sqrt{5}, 5, 5\sqrt{5}, 25, \ldots$

4. Do the triangular numbers form a quadratic sequence? Give your reason.

5. Find the next three terms of the following sequences using the properties of the sequence.
   a. Arithmetic   2, 4, □, □, □
   b. Quadratic   2, 4, □, □, □

6. Generate the first four terms of an arithmetic progression using the following facts.

	First term	Constant difference
a	3	2
b	10	5
c	3	0.5
d	2	−3
e	$\frac{1}{2}$	$\frac{1}{2}$
f	−3	−2
g	4	$\sqrt{3}$
h	$\sqrt{3}$	$\sqrt{3}$
i	$2\sqrt{5}$	$\sqrt{5}$

7. Find the missing term in each sequence, giving your reason in each case.
   a. 5, 10, □, 20, 25, …
   b. 5, 10, □, 26, 37, …

8. By considering a number's prime decomposition, or otherwise, find three square numbers that are also cube numbers.

9. Iain thinks that the triangular number sequence can be generated using this rule.
   $$T(n) = \frac{1}{2}n(n+1)$$
   Do you agree with Iain? Give your reason.

10. Gulna thinks that the sequence 1, 4, 9, 16, 25, … is a quadratic sequence.
    Sami thinks that the sequence 1, 4, 9, 16, 25, … is a square sequence.
    Who is correct? Give your reason.

11. a. Find the first five terms of these sequences.
    i. $\dfrac{n}{n+1}$
    ii. $\dfrac{n}{n^2+1}$
    iii. $\dfrac{1}{n^2}$
    iv. $\dfrac{n^2}{n+1}$
    v. $(0.9)^n$
    vi. $(1.1)^n$

    b. As n increases, comment on the behaviour of $T(n)$.

APPLICATIONS

20.3 Special sequences

RECAP
- Square numbers 1, 4, 9, 16, ... n^2 ...
- Cube numbers 1, 8, 27, 64, ... n^3 ...
- Triangular numbers 1, 3, 6, 10, 15, ... – differences 2, 3, 4, 5, ...
- Arithmetic sequence – constant difference.
- Quadratic sequence – differences form an arithmetic sequence.

HOW TO
1. Memorise the square, triangular and cube number sequences.
2. Identify arithmetic and quadratic sequences by looking at the difference between terms.

EXAMPLE

Create two sequences with the following properties.
a Arithmetic sequence with starting term 5
b Geometric sequence with starting term 3

a 5, 8, 11, 14, 17, ... ② the rule is 'add 3'
5, 2, −1, −4, −7, ... the rule is 'subtract 3'
b 3, 6, 12, 24, 48, ... ③ the rule is '× 2'
$3\sqrt{5}$, 15, $15\sqrt{5}$, 75, $75\sqrt{5}$, ... the rule is '× $\sqrt{5}$'

EXAMPLE

Karl loves performing his amazing magic trick.

1. Pick any whole number
2. Write down 4 times the number
3. Add these two numbers together
4. Add the second and third numbers together
5. Add the third and fourth numbers together
6. Keep repeating until you get the 8th number
7. The 8th number is always 60 times the first number

a Try the trick out. Does it always work?
b Does it work for negative integers, fractions and decimals?
c Explain how the trick works.

a Start with 3 3, 4 × 3 = 12, 3 + 12 = 15, 12 + 15 = 27, 15 + 27 = 42, 27 + 42 = 69,
42 + 69 = 111, 69 + 111 = 180 Yes 180 = 60 × 3
b Start with −2 −2, −8, −10, −18, −28, −46, −74, −120 Yes −120 = 60 × −2
Start with $\frac{1}{2}$ $\frac{1}{2}$, 2, $2\frac{1}{2}$, $4\frac{1}{2}$, 7, $11\frac{1}{2}$, $18\frac{1}{2}$, 30 Yes 30 = 60 × $\frac{1}{2}$
Start with 1.25 1.25, 5, 6.25, 11.25, 17.5, 28.75, 46.25, 75 Yes 75 = 60 × 1.25
c Let the first number be N
N, 4N, 5N, 9N, 14N, 23N, 37N, 60N The 8th term is 60 × N

Algebra Sequences

Reasoning and problem solving

Exercise 20.3A

1. The first term of a sequence is 4. Create five more terms of
   a. an arithmetic sequence
   b. a quadratic sequence.

2. Complete the missing cells

	Type	1st term	5th term
a	Arithmetic	12	
b	Arithmetic		23
c	Arithmetic	0.5	
d	Arithmetic		
e	Quadratic	3	
f	Quadratic		256
g	Quadratic	$\sqrt{3}$	
h	Quadratic		$\frac{1}{16}$

3. Jenny thinks that the triangular number sequence can be created by starting with the number 1 and then adding on 2, adding on 3 and so on.
   a. Do you agree with Jenny? Give your reason.
   b. What are the pentagonal, hexagonal and tetrahedral numbers? How are they created?

4. Dara and Sam are given three options for a gift to celebrate their birthdays.

   Option 1 $500

   Option 2 $100 for the first month, $200 the next month, $300 the next month until the end of the year

   Option 3 During the month of their birthday 1 cent on Day 1, 2 cents on Day 2, 4 cents on Day 3, 8 cents on Day 4, …

   Dara's birthday is 20th February.
   Sam's birthday is 5th September.

   a. Which option should Dara choose? Give your reason.
   b. Which option should Sam choose? Give your reason.

5. The terms of a sequence are given by $T(n) = ar^{n-1}$ where a and r are constants.

   Describe how the terms $T(n)$ behave for different values of r. Use words like diverge, converge, constant and oscillate.

Summary

Checkout
You should now be able to...

	Test it Questions
✓ Generate a sequence using a term-to-term or position-to-term rule.	1, 2, 4
✓ Recognise a linear sequence and find a formula for its nth term.	3
✓ Recognise a quadratic sequence and find a formula for its nth term.	4, 5
✓ Recognise and use special sequences.	6 – 9

Language	Meaning	Example
Sequence	An ordered set of numbers or other objects.	Square numbers
Term	One of the separate items in a sequence.	1 4 9 16 …
Position	A number that counts where a term appears in a sequence.	$T(1) = 1$ First term, position 1 $T(2) = 4$ Second term, position 2
Term-to-term rule	A rule that links a term in a sequence with the previous term.	Sequence: 3, 5, 7, 9, 11, 13 …. Term-to-term rule: 'add 2' $T(n + 1) = T(n) + 2$
Position-to-term rule / General term / nth term	A rule that links a term in a sequence with its position in the sequence.	Position-to-term rule: $T(n) = 2n + 1$
Linear / Arithmetic sequence	A sequence with a constant difference between terms. A graph of $T(n)$ against n gives points on a straight line.	4, 9, 14, 19, 24, … Constant difference = 5 $T(n + 1) = T(n) + 5$, $T(n) = 5n - 1$
Triangular numbers	The sequence formed by summing the integers: $1, 1 + 2, 1 + 2 + 3, …, \frac{1}{2}n(n + 1), …$	1, 3, 6, 10, 15, …
Quadratic sequence	A sequence in which the differences between terms form an arithmetic sequence; the second differences are constant. A graph of $T(n)$ against n gives points on a quadratic curve.	4, 9, 16, 25, … +5, +7, +9, … First difference +2, +2, … Second difference $T(n + 1) = T(n) + 2n + 3$, $T(n) = n^2 + 2n + 1$

Review

1. **a** What are the next three terms of these sequences?
   - **i** 8, 17, 26, 35, …
   - **ii** 71, 58, 45, 32, …
   - **iii** 2.8, 4.4, 6, 7.6, …

   b Write the term-to-term rule for each of the sequences in part **a**.

2. Calculate the 13th term for the sequences with these position-to-term rules.
   - **a** $T(n) = 5n + 8$
   - **b** $T(n) = 12n - 15$

3. Write a formula for the nth term of these sequences.
   - **a** 1, 7, 13, 19, …
   - **b** 15, 22, 29, 36, …
   - **c** 51, 39, 27, 15, …
   - **d** $-6.5, -8, -9.5, -11, …$

4. The nth term of a sequence is given by $3n^2 + 4n$.

   Calculate
   - **a** the 7th term
   - **b** the 10th term.

5. Work out the rule for the nth term of these sequences.
   - **a** 4, 7, 12, 19, …
   - **b** 2, 10, 24, 44, …
   - **c** 4, 13, 26, 43, …

6. Classify each sequence using one of these words.

Linear	Quadratic
Arithmetic	Another type

   - **a** $2, -6, 18, -54, 162, …$
   - **b** 1, 3, 6, 10, 15, …
   - **c** 1, 3, 4, 7, 11, …
   - **d** 0.6, 0.45, 0.3, 0.15, 0, …
   - **e** 4, 6, 10, 16, 24, …
   - **f** 0.1, 0.01, 0.001, 0.0001, 0.00001, …

7. **a** Write down the next two terms of these sequences.
   - **i** 1, 3, 6, 10, …
   - **ii** $1, \frac{1}{4}, \frac{1}{9}, \frac{1}{16}$

   b Write down the nth term of the sequences in part **a**.

8. This sequence is formed by doubling the current term to get the next term.

   2, 4, 8, 16, …
   - **a** Write down the next three terms of the sequence.
   - **b** Write down the rule for the nth term of the sequence.

9. Write a rule for the nth term of these sequences and use it to work out the 10th term of each sequence.
   - **a** $\sqrt{2}, 3\sqrt{2}, 5\sqrt{2} …$
   - **b** $\frac{1}{3}, \frac{2}{5}, \frac{3}{7}, \frac{4}{9}, …$

What next?

Score		
0 – 4		Your knowledge of this topic is still developing. To improve, look at MyiMaths: 1054, 1165, 1166, 1173, 1920
5 – 8		You are gaining a secure knowledge of this topic. To improve your fluency look at InvisiPens: 21Sa – i
9		You have mastered these skills. Well done you are ready to progress! To develop your problem solving skills look at InvisiPens: 21Aa – e

Assessment 20

1 Chris is given the terms of some sequences and the descriptions of the sequences. He can't remember which description matches which group of terms. Can you match each group of terms to the correct description? [9]

- **a** The first five even numbers less than 5.
- **b** The first five multiples of 7.
- **c** The first five factors of 210.
- **d** The first five triangular numbers.
- **e** All the factors of 24.
- **f** Odd numbers from -3 to 5.
- **g** Prime numbers between 10 and 30.
- **h** $15 <$ square numbers $\leqslant 81$.
- **i** Multiples of 5 between 3 and 33.

- **A** $-3, -1, 1, 3, 5$
- **B** $5, 10, 15, 20, 25, 30$
- **C** $11, 13, 17, 19, 23, 29$
- **D** $7, 14, 21, 28, 35$
- **E** $4, 2, 0, -2, -4$
- **F** $1, 3, 6, 10, 15$
- **G** $1, 2, 3, 4, 6, 8, 12, 24$
- **H** $1, 2, 3, 5, 7$
- **I** $16, 25, 36, 49, 64, 81$

2 Nathan is given this sequence

$$1, \quad 11, \quad 21, \quad 31, \quad 41$$

He says that the common difference of this sequence is $+11$.

- **a** Is he correct? If not, work out the common difference. [1]
- **b** Complete this sentence. Show your working.
  'The nth term of this sequence is...' [2]

3 a For this sequence of patterns find the nth term for the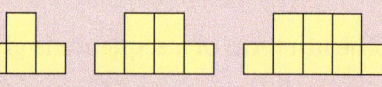
   - **i** perimeter [2]
   - **ii** area. [2]
- **b** A pattern in the sequence has an area of 48 cm². What is its perimeter? [1]
- **c** A pattern in the sequence has an perimeter of 48 cm. What is its area? [1]

4 Marguerite is given some nth terms, $T(n)$, for some sequences. She makes the following statements. Which of her statements are correct and which are not correct?

For Marguerite's statements that are not correct, rewrite the statement correctly.

- **a** $T(n) = 2n + 7$ — The 10th term is 27. [1]
- **b** $T(n) = 6n - 5$ — The first 3 terms are $-5, 1, 7$. [1]
- **c** $T(n) = 13 - 3n$ — The 100th term is 287. [1]
- **d** $T(n) = n^2 - 10$ — The 10th term is 100. [1]
- **e** $T(n) = 15 - 3n^2$ — The 100th term is $-29\,985$. [1]

Algebra Sequences

5 Here is a sequence of patterns.

Fabiane says that the ratio 'number of red squares':
'number of blue squares' stays the same as the shapes
get bigger. Is she correct? If you think she isn't correct
say whether the ratio is decreasing or increasing. Give your reason. [4]

6 a Jack and Dawn are looking at the start of this sequence
of patterns. Jack says the next two patterns will have
12 and then 18 dots. Dawn says that they will have 10
and then 15 dots. Who is correct? Draw the next two
of these patterns. [2]

b Find a formula connecting the pattern number, t, and the number of dots in
the pattern, $D(t)$. Show your working and explain your answer. [4]

c Use your formula to find the number of dots in pattern number

   i 10 [1]   **ii** 50 [1]   **iii** 100 [1]

7 These diagrams are of a quadrilateral and a
pentagon with all their diagonals.

a How many diagonals are there in the next
three polygons?
Draw the next three of these patterns. [3]

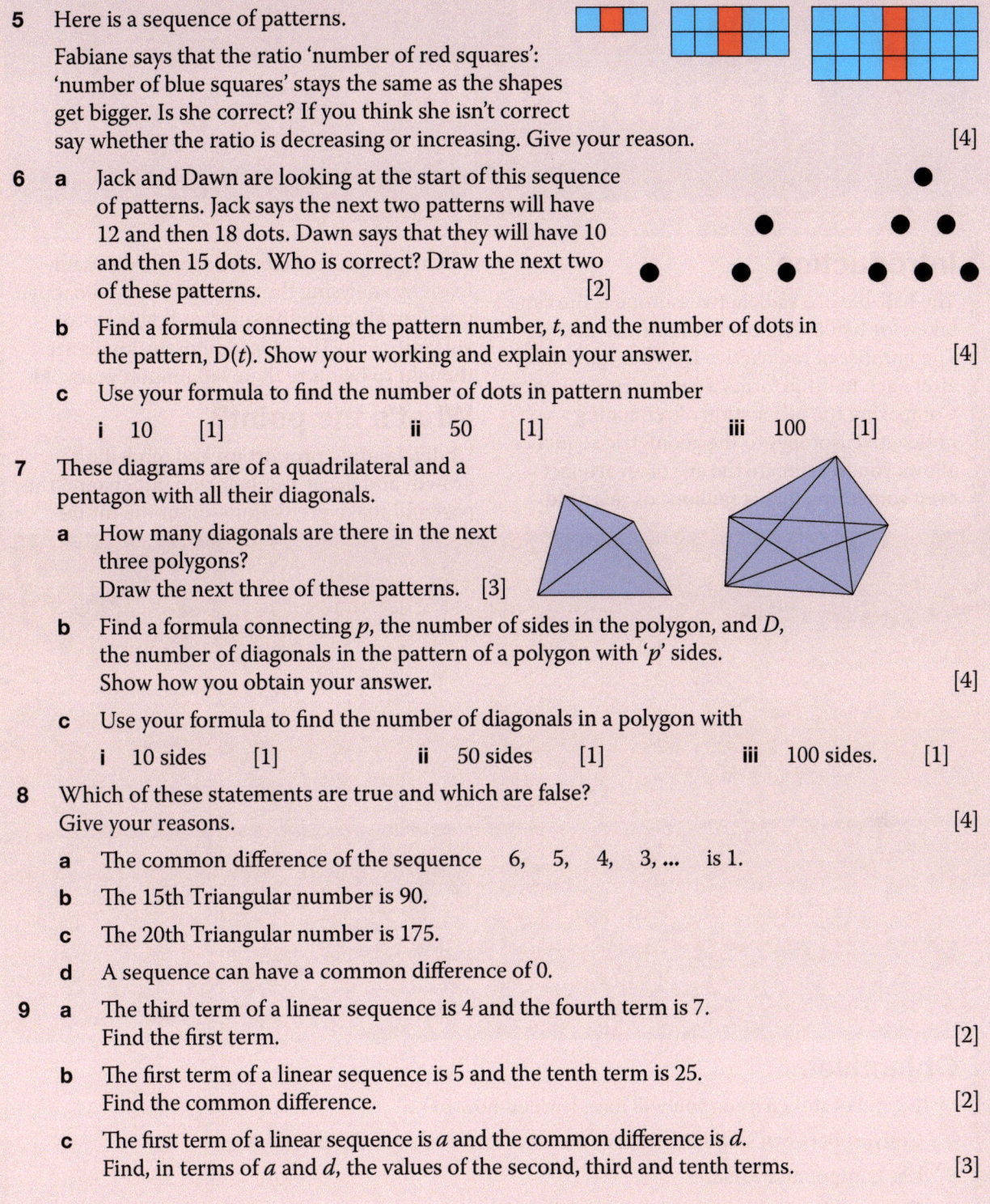

b Find a formula connecting p, the number of sides in the polygon, and D,
the number of diagonals in the pattern of a polygon with 'p' sides.
Show how you obtain your answer. [4]

c Use your formula to find the number of diagonals in a polygon with

   i 10 sides [1]   **ii** 50 sides [1]   **iii** 100 sides. [1]

8 Which of these statements are true and which are false?
Give your reasons. [4]

a The common difference of the sequence   6,   5,   4,   3, ...   is 1.

b The 15th Triangular number is 90.

c The 20th Triangular number is 175.

d A sequence can have a common difference of 0.

9 a The third term of a linear sequence is 4 and the fourth term is 7.
Find the first term. [2]

b The first term of a linear sequence is 5 and the tenth term is 25.
Find the common difference. [2]

c The first term of a linear sequence is a and the common difference is d.
Find, in terms of a and d, the values of the second, third and tenth terms. [3]

21 Units and proportionality

Introduction
The half-life of a radioactive isotope is the time taken for half its radioactive atoms to decay. The number of radioactive isotopes remaining after each half-life forms a geometric sequence. Comparing the proportion of remaining radioactive isotopes to the geometric sequence, allows you to estimate the age of an artefact – even something that is millions of years old.

Scientists can estimate the age of a dinosaur fossil by analysing the proportion of radioactive uranium atoms in the surrounding layers of volcanic rock. The oldest dinosaur fossils are thought to be more than 240 million years old.

What's the point?
Understanding proportion and modelling growth and decay can help you understand the past and make predictions about the future.

Objectives
By the end of this chapter, you will have learned how to …
- Convert between standard units of measure and compound units.
- Use compound measures.
- Compare lengths, areas and volumes of similar shapes.
- Solve direct and inverse proportion problems.
- Describe direct and inverse proportion relationships using an equation.
- Recognise graphs showing direct and inverse proportion and interpret the gradient of a straight line graph.
- Find the instantaneous and average rate of change from a graph.
- Solve repeated proportional change problems.

Check in

1. **a** Increase these amounts by 10%.

    i £45   **ii** 60 mm   **iii** 48 km   **iv** 4 hours

   b Decrease these amounts by a quarter.

    i 6 miles   **ii** 58 minutes   **iii** 14 kg   **iv** €62

2. Calculate the average speed of cars that made these journeys.

   a 480 miles in 12 hours   **b** 85 miles in two hours

   c 27 miles in 30 minutes   **d** 48 miles in 90 minutes

3. Find the volume of these prisms. State the units of your answers.

   a 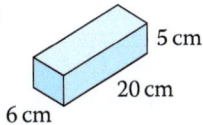   **b** 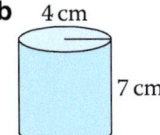   **c**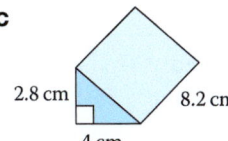

4. Work out the surface area of the prisms in question **3**.

Chapter investigation

In 1798 Thomas Malthus wrote that human populations typically grow by a fixed percentage each year but that agricultural yield grows by a fixed amount each year. Why does this lead to a 'Malthusian catastrophe'?

	Value in 1950	Growth per year
World population	2.5 billion	1.8%
Number of people who could be fed	4.0 billion	200 million

Using these numbers when would the catastrophe occur?

Could catastrophe be avoided if population growth was held fixed at its rate for the year 2000?

SKILLS

21.1 Compound units

Compound measures describe one quantity in relation to another.
These are examples of compound measures.

- **Speed** = $\dfrac{\text{Total distance travelled}}{\text{Total time taken}}$ Units such as m/s, km/h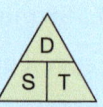

- **Density** = $\dfrac{\text{Mass}}{\text{Volume}}$ Units such as g/cm³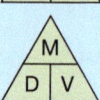

Use the triangle to work out which calculation to use.

Cover D (for distance)
You multiply
S (speed) × T (time)

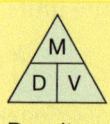

Density = $\dfrac{\text{Mass}}{\text{Volume}}$

EXAMPLE

Meena jogs at an average speed of 5 km/h for $1\tfrac{1}{2}$ hours.
What distance does she jog?

Distance = $5 \times 1\tfrac{1}{2}$
= 7.5 km

Force is measured in newtons (N).

EXAMPLE

Find the density of a piece of wood with cross-section area 42 cm², length 12 cm and mass 693 g.

Volume = 42 × 12 = 504 cm³
Density = 693 ÷ 504 = 1.375 g/cm³
Mass in grams. Volume in cm³.
So density is in g/cm³.

A rate is also a compound unit. It tells you how many units of one quantity there are compared with one unit of another quantity.

- **Rate of pay** = $\dfrac{\text{Amount of money}}{\text{Time}}$ Units such as £/h

Rate of flow is a compound measure. It is the volume of liquid that passes through a container in a unit of time.

- **Rate of flow** = $\dfrac{\text{Volume}}{\text{Time}}$ Units such as litres/s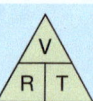

Best buy problems

When comparing best value between products finding *cost per unit* is a good strategy.

Roller ball pens are sold in bulk boxes of 20 or 35.

A box of 20 costs £9.20. A box of 35 costs £15.75. Which is the better value?

9.20 ÷ 20 = 46p per pen.
15.75 ÷ 35 = 45p per pen.
The box of 35 is better value.
In this example the compound unit is pence per pen.

EXAMPLE

Water empties from a tank at a rate of 1.5 litres per second.
It takes 10 minutes to empty the tank.
How much water was in the tank?

Use the triangle to work out which formula to use.
Volume = rate × time
Convert the time to seconds.
10 minutes = 10 × 60 s = 600 s
Volume in tank = 1.5 × 600 = 900 litres

Ratio and proportion   Units and proportionality

Fluency

Exercise 21.1S

1. The winners' times in some of the races at a sports day are
   a. 100 metres in 13 seconds
   b. 200 metres in 28 seconds
   c. 400 metres in 58.4 seconds
   d. 1500 metres in 4 minutes 52 seconds

   Calculate the speed of each winner in m/s, correct to 1 dp.

2. Work out the distance travelled in
   a. 2 hours at 80 km/h
   b. 7 hours at 23 mph
   c. 6 seconds, at 9 m/s
   d. 1 day at 12 mph.

3. Work out the time it takes to travel
   a. 180 kilometres at 60 km/h
   b. 280 miles at 70 mph
   c. 8 kilometres at 24 km/h
   d. 15 miles at 60 mph.

4. A cube with volume 640 cm³ has a mass of 912 g. Find the density of the cube in g/cm³.

5. An emulsion paint has a density of 1.95 kg/litre. Find
   a. the mass of 4.85 litres of the paint
   b. the number of litres of the paint that would have a mass of 12 kg.

6. The table shows the densities of different metals.

Metal	Density
Zinc	7130 kg/m³
Cast iron	6800 kg/m³
Gold	19320 kg/m³
Tin	7280 kg/m³
Nickel	8900 kg/m³
Brass	8500 kg/m³

   Use the information in the table to find
   a. the mass of 0.8 m³ of zinc
   b. the mass of 0.5 m³ of cast iron
   c. the mass of 3.2 m³ of gold
   d. the volume of 910 g of tin
   e. the volume of 220 g of nickel
   f. the volume of a brass statue that has mass 17 kg.

7. If 4 metres of fabric costs $8.40, find the price of the fabric in dollars per metre.

8. Jane is paid $478 a week. Each week she works 40 hours. What is her hourly rate of pay?

9. Find the rate of flow for pipes A and B in litres/s.
   a. Pipe A: 20 litres of water in 8 seconds.
   b. Pipe B: 48 litres of water in 30 seconds.

10. Water empties from a tank at a rate of 2 litres/s. It takes 10 minutes to empty the tank.
    How much water was in the tank?

11. An electric fire uses 18 units of electricity over a period of 7.5 hours.
    a. What is the hourly rate of consumption of electricity in units per hour?
    b. How many units of electricity are used in 24 hours?

12. A car has fuel efficiency of 8 litres per 100 km.
    a. How far can the car travel on 40 litres of petrol?
    b. How many litres of petrol would be needed for a journey of 250 km?
    c. What is the car's rate of fuel consumption in km per litre?

13. Sachina received 75 US dollars in exchange for £50.
    a. Calculate the rate of exchange in US dollars per £.
    b. How many US dollars would she get for £125?
    c. If Sachina received $120 US dollars, how many pounds did she exchange?

APPLICATIONS

21.1 Compound units

RECAP

Compound units describe one quantity in relation to another.
- The density of a material is its mass divided by its volume.
- Speed is the distance travelled divided by the time taken.
- A formula triangle is a useful way to remember the relationships between the different parts.
- A rate is also a compound unit.

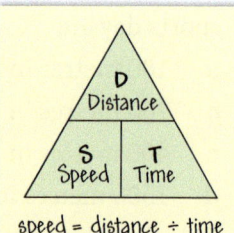

speed = distance ÷ time
Distance = speed × time
Time = distance ÷ speed

HOW TO

To solve problems involving compound units

1. Write the correct formula.
2. Convert units or work out quantities to apply the formula.
3. Work out the answer, making sure the units are correct.

EXAMPLE

A train leaves Paris at 13:40 and arrives in Martagny at 15:00. If the distance is 90 km find the average speed of the train.

1. Write the formula for speed.
2. Work out the time in hours.
3. Work out the answer using the correct units.

Speed = $\frac{\text{distance}}{\text{time}}$

Time = 1 hour 20 min = $1\frac{1}{3}$ h = 1.333... h

Speed = $\frac{90}{1.333}$ = 67.5 to 3 sf

The average speed of the train is 67.5 km/h.

EXAMPLE

Sand was falling from the back of a lorry at a rate of 0.4 kg/s. It took 20 minutes for all the sand to fall from the lorry.

How much sand was the lorry carrying?

1. Write the formula for mass.
2. Convert the time to seconds.
3. Work out the answer using the correct units.

The rate of flow is in kg/s, which is mass divided by time.
Mass = rate × time

20 minutes = 20 × 60 s = 1200 s

Mass = 0.4 × 1200 = 480

The lorry was carrying 480 kg of sand.

EXAMPLE

A metal cuboid has a length of 7 cm, a width of 5 cm and a height of 4 cm. It has a mass of 1.470 kg. Find its density in g/cm³.

1. Write the formula for density.
2. Work out the volume of the cuboid and convert mass to grams.
3. Work out the answer using the correct units.

Density = $\frac{\text{mass}}{\text{volume}}$

Volume of cuboid = length × width × height
= 7 × 5 × 4 cm³ = 140 cm³

Mass = 1.470 kg = 1470 g

Density = $\frac{1470}{140}$ = 10.5 g/cm³ to 3 sf

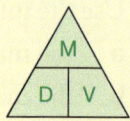

Ratio and proportion Units and proportionality

Reasoning and problem solving

Exercise 21.1A

1. A train leaves Euston at 8:57 a.m. and arrives at Preston at 11:37 a.m.
   If the distance is 238 miles find the average speed of the train.

2. Sand falls from the back of a lorry at a rate of 0.2 kg/s.
   It took 25 minutes for all the sand to fall from the lorry.
   How much sand was the lorry carrying?

3. A metal cuboid has a length of 9 cm, a width of 5 cm and a height of 4 cm.
   It has a mass of 1.53 kg.
   Find its density in g/cm³.

4. A car travels 24 miles in 45 minutes.
   Find the average speed of the car in miles per hour (mph).

5. Copy and complete the table to show speeds, distances and times for five different journeys.

Speed (kmph)	Distance (km)	Time
105		5 hours
48	106	
	84	2 hours 15 minutes
86		2 hours 30 minutes
	65	1 hour 45 minutes

6. A cube of side 2 cm has mass 40 grams.

   a. Find the density of the material from which the cube is made. Give your answer in g/cm³.

   Volume of cube = length³.

   b. A cube of side length 2.6 cm is made from the same material.
   Find the mass of this cube, in grams.

7. A solid block has a length and width of 22.50 mm, and a height of 3.15 mm. It has a mass of 9.50 g.

   a. Find the density of the metal from which the block is made.
   Give your answer in g/cm³.

   b. How many blocks can be made from 1 kg of the material?

8. In this question, give your answers in kg/m³.

   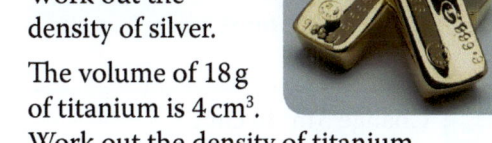

   a. The volume of 31.5 g of silver is 3 cm³. Work out the density of silver.

   b. The volume of 18 g of titanium is 4 cm³. Work out the density of titanium.

   c. A sheet of aluminium foil has volume 0.4 cm³ and mass 1.08 g. Work out the density of aluminium foil.

9. Grace earns £340 per week for 40 hours work.
   If Grace works overtime, she is paid 1.5 times her standard hourly rate.

   a. How much is Grace paid for 7 hours of overtime work?

   b. Grace earned £531.25 last week. How many hours of overtime did she work?

10. The toll charged for a car travelling on a motorway was $33.60 for a journey of 420 km.
    Cars with trailers are charged double.
    How much would it cost for a car with a trailer to travel 264 km?

11. A yacht race has three legs of 8 km, 6 km and 10 km.
    The average speed for the winning yacht was 6.2 km/h.
    The second yacht finished 8 minutes after the winner.
    How long did it take the second place yacht to finish the race?

SKILLS

21.2 Converting between units

The relationships between the metric units of area are

- $1\,cm^2 = 10\,mm \times 10\,mm$ $\quad$ • $1\,m^2 = 100\,cm \times 100\,cm$
  $ = 100\,mm^2$ $ = 10\,000\,cm^2$

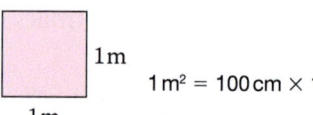

$1\,m^2 = 100\,cm \times 100\,cm$

EXAMPLE

Change $5\,m^2$ to cm^2.

$5\,m^2 = 5 \times 10\,000\,cm^2$
$ = 50\,000\,cm^2$

You expect a larger number, so multiply.

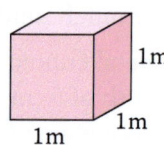

$1\,m^3 = 100\,cm \times 100\,cm \times 100\,cm$

The relationships between the metric units of **volume** are

- $1\,cm^3 = 10\,mm \times 10\,mm \times 10\,mm$ $\quad$ • $1\,m^3 = 100\,cm \times 100\,cm \times 100\,cm$
  $ = 1000\,mm^3$ $ = 1\,000\,000\,cm^3$

In **similar** shapes, corresponding sides are in the same **ratio**.

You can use the ratio to work out **areas** and **volumes** in similar shapes and solids.

p.52

- For similar shapes with length ratio $1:n$
  - the ratio of the areas is $1:n^2$
  - the ratio of the volumes is $1:n^3$

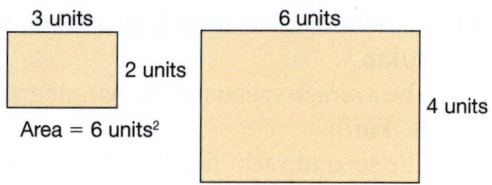

Length ratio $= 1:2$
Area ratio $= 6:24 = 1:4 = 1:2^2$

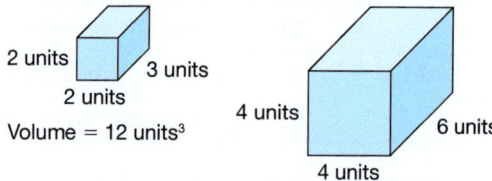

Length ratio $= 1:2$
Volume ratio $= 12:96 = 1:8 = 1:2^3$

EXAMPLE

Two similar Russian dolls are on display.

a   The surface area of the smaller doll is $7.2\,cm^2$.
Work out the surface area of the larger doll.

b   The volume of the larger doll is $145.8\,cm^3$.
Work out the volume of the smaller doll.

Ratio of lengths, smaller : larger $= 1.2:3.6 = 1:3$
a   Area ratio is $1:3^2 = 1:9$
Surface area of larger doll $= 9 \times 7.2 = 64.8\,cm^2$
b   Volume ratio is $1:3^3 = 1:27$
Volume of smaller doll $= 145.8 \div 27 = 5.4\,cm^3$

Ratio and proportion   Units and proportionality

Fluency

Exercise 21.2S

1 Convert these measurements to millimetres.
   a 5 cm   b 8 cm
   c 15 cm   d 6 cm 7 mm
   e 19 cm 3 mm   f 4.5 cm
   g 4.3 cm   h 10.6 cm
   i 80 cm   j 1 m

2 Convert these measurements to centimetres.
   a 60 mm   b 85 mm
   c 240 mm   d 63 mm
   e 4 mm   f 4 m
   g 10 m   h 3.5 m
   i 1.6 m   j 1.63 m

3 Convert these measurements to metres.
   a 400 cm   b 450 cm
   c 475 cm   d 470 cm
   e 50 cm   f 1 km
   g 4 km   h 0.5 km
   i 3.5 km   j 18 km

4 Here are two identical rectangles, A and B.

   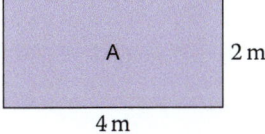

   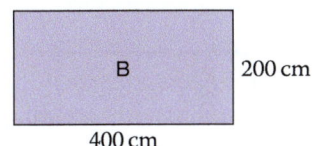

   a Calculate the area of rectangle A in m^2.
   b Calculate the area of rectangle B in cm^2.

5 a Calculate the area of this rectangle in m^2.
   b Convert your answer to cm^2.

   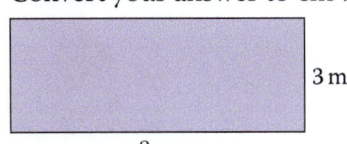

6 Convert these areas to mm^2.
   a $4 cm^2$
   b $7.3 cm^2$
   c $10.9 cm^2$
   d $2.5 cm^2$
   e $400 cm^2$

   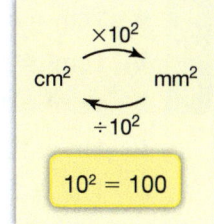

   $cm^2 \xrightarrow{\times 10^2} mm^2$
   $\xleftarrow{\div 10^2}$
   $10^2 = 100$

7 Convert these areas to cm^2.
   a $600 mm^2$ b $1200 mm^2$
   c $850 mm^2$ d $6500 mm^2$
   e $10000 mm^2$

8 Convert these areas to m^2.
   a $40000 cm^2$
   b $85000 cm^2$
   c $1000000 cm^2$
   d $125000 cm^2$
   e $5000 cm^2$

   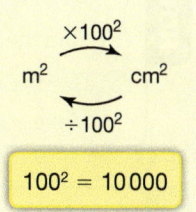

   $m^2 \xrightarrow{\times 100^2} cm^2$
   $\xleftarrow{\div 100^2}$
   $100^2 = 10000$

9 Convert these areas to cm^2.
   a $5 m^2$ b $10 m^2$ c $6.5 m^2$
   d $7.75 m^2$ e $0.6 m^2$

10 P and Q are two similar cuboids.

   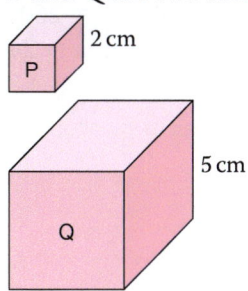

   a The surface area of cuboid P is $37.2 cm^2$.
      Work out the surface area of cuboid Q.
   b The volume of cuboid P is $12.4 cm^3$.
      Work out the volume of cuboid Q.

APPLICATIONS

21.2 Converting between units

RECAP

- 1 cm = 10 mm
- 1 cm² = 10 mm × 10 mm = 100 mm²
- 1 cm³ = 10 mm × 10 mm × 10 mm = 1000 mm³
- 1 m = 100 cm
- 1 m² = 100 cm × 100 cm = 10 000 cm²
- 1 m³ = 100 cm × 100 cm × 100 cm = 1 000 000 cm³
- If the scale factor for length is n then the scale factor for area is $n \times n = n^2$, and the scale factor for volume is $n \times n \times n = n^3$.

> In similar shapes, the corresponding sides are in the same ratio.

HOW TO

To solve problems involving similar shapes

① Compare corresponding lengths, areas or volumes.

② Find the ratio between lengths, areas or volumes.

③ Use the ratio to answer the question.
   Make sure that you include units in your answer.

EXAMPLE

P and Q are similar shapes.

Calculate the surface area of shape Q.

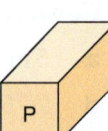
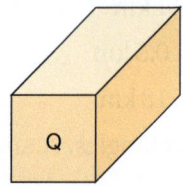

Volume = 260 cm³   Volume = 877.5 cm³
Surface area = 244 cm²   Surface area = ?

① Compare the volumes of the shapes.

Volume ratio
$$\frac{V_Q}{V_P} = \frac{877.5}{260} = \frac{27}{8}$$

② Find the length ratio first.

Length ratio
$$\frac{L_Q}{L_P} = \frac{\sqrt[3]{27}}{\sqrt[3]{8}} = \frac{3}{2}$$

Then work out the area ratio.

Area ratio
$$\frac{A_Q}{A_P} = \frac{3^2}{2^2} = \frac{9}{4}$$

③ Multiply the ratio by the surface area of shape P.

Surface area of Q
$$= 244 \times \frac{9}{4} = 549 \text{ cm}^2$$

> Always sense check your answer — are you expecting a larger or a smaller value?

Ratio and proportion   Units and proportionality

Reasoning and problem solving

Exercise 21.2A

1. A soft drink comes in a bottle.
   The bottle is 20 cm tall and contains 330 ml of juice.
   As an advertising stunt, a similar bottle is made the size of a man.
   The bottle is 1.8 m tall.

   a. The label has an area of 28 cm². What is the area of the label on the larger bottle?

   b. What is the capacity of the larger bottle?

2. J and K are two similar boxes.

   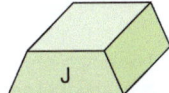 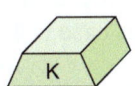

   The volume of J is 702 cm³.
   The volume of K is 208 cm³.
   The surface area of J is 549 cm².
   Calculate the surface area of K.

3. X and Y are two similar solids.

   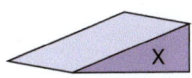 

   The total surface area of X is 150 cm².
   The total surface area of Y is 216 cm².
   The volume of Y is 216 cm³.
   Calculate the volume of X.

4. C and D are two cubes.

   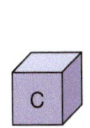 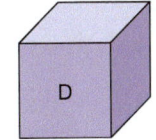

   a. Explain why any two cubes must be similar.

   b. Explain why any two cuboids are not necessarily similar.

   The ratio of the side lengths of cube C and cube D is 1 : 7.

   c. Write down the ratio of

      i. their surface areas

      ii. their volumes.

5. Two model cars made to different scales are mathematically similar.

   The overall widths of the cars are 3.2 cm and 4.8 cm respectively.

   a. What is the ratio of the radii of the cars' wheels?

   The cars are packed in mathematically similar boxes so that they just fit inside the box.

   b. The surface area of the larger box is 76.5 cm².
      Work out the surface area of the smaller box.

   c. The volume of the smaller box is 24 cm³.
      Work out the volume of the larger box.

6. At the local pizzeria Alina is offered two deals for €9.99.

   > Deal A: One large round pizza with radius 18 cm.
   > Deal B: Two smaller round pizzas each with radius 9 cm.

   Which deal gives the most pizza?

7. A frustum is created by removing a cone of height 10 cm from a cone of height 15 cm.

   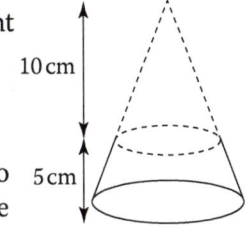

   Show that the volume of the frustum is in the ratio 19 : 8 to the volume of the cone removed.

8. Cone A has surface area 260.3 cm² and volume 188.5 cm³.
   Cone B has surface area 1017.9 cm² and volume 1352.2 cm³.
   Are the cones similar?
   Explain your answer.

SKILLS

21.3 Direct and inverse proportion

- Numbers or quantities are in **direct proportion** when the *ratio* of each pair of corresponding values is the same.

One litre of a shade of purple paint is made by mixing 200 ml of red paint and 800 ml of blue paint.
The ratio of red paint to blue paint is 1 : 4.

- When you multiply (or divide) one of the variables by a certain number, you have to multiply or divide the other variable by the same number.

250 ml of red paint has to be mixed with 4 × 250 = 1000 ml of blue paint.

You can always find the amount of blue paint by multiplying the amount of red paint by a fixed number, the constant of proportionality, in this case is 4.

- You write 'y is proportional to x' as $y \propto x$. This can also be written as $y = kx$, where k is the **constant of proportionality**.

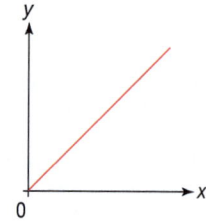

▲ The graph of $y = kx$ is a straight line that passes through the origin.

EXAMPLE

If 35 metres of steel cable has mass 87.5 kilograms, what is the mass of 25 metres of the same cable?

If x = length of the cable (metres) and w = mass (kilograms), then $w = kx$
The constant of proportionality k represents the mass of one metre of cable.
$87.5 = k \times 35 \Rightarrow k = 87.5 \div 35 = 2.5$ so $w = 2.5x$
Substitute $x = 25$ into the formula: $w = 2.5 \times 25 = 62.5$ kg

- When variables are in **inverse proportion**, one of the variables increases as the other one decreases, and vice-versa.
- 'y is inversely proportional to x' can be written as $y \propto \frac{1}{x}$, or $y = \frac{k}{x}$, where k is the constant of proportionality.
- If two variables are in inverse proportion, the product of their values will stay the same.

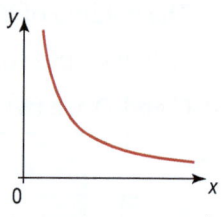

▲ The graph of $y = \frac{k}{x}$ is a reciprocal graph.

EXAMPLE

A pot of paint, which will cover an area of 12 m², is used to paint a garage floor. Find the width of the floor when the length is

    a 1 m   **b** 3 m   **c** 6.5 m

Use l for the length of the floor and w for the width.
$lw = 12 \Rightarrow w = \frac{12}{l}$

Substitute the given values into the formula.
a $w = \frac{12}{1} = 12$ m   **b** $w = \frac{12}{3} = 4$ m   **c** $w = \frac{12}{6.5} = 1.85$ m

Ratio and proportion   Units and proportionality

Fluency

Exercise 21.3S

1 The table shows corresponding values of the variables w, x, y and z.

w	3	6	9	15
x	8	14	20	32
y	5	10	15	25
z	4	7	10	16

Which of these statements could be true?

  a $w \propto x$
  b $z \propto x$
  c $z \propto w$
  d $y \propto w$
  e $w = ky$, where k is a constant

2 Using the values from the table in question **1**, plot graphs of

  a w against x
  b w against y
  c x against y
  d x against z.

Use your results to describe the key features of a graph showing the relationship between two variables that are in direct proportion.

3 The mass, w, of a piece of wooden shelving is directly proportional to its length, l.

  a Write a formula that connects w and l including a constant of proportionality, k.
  b Given that 2.5 m of the shelving has mass 6.2 kg, find the value of the constant k.
  c Use your previous answers to calculate the mass of a 2.9 m length of the shelving.

4 You are told that the variable y is directly proportional to the variable x.
Explain what will happen to the value of y when the value of x is

  a doubled
  b halved
  c multiplied by 6
  d divided by 10
  e multiplied by a factor of 0.7

5 You are told that the variable w is inversely proportional to the variable z.
Explain what will happen to the value of w when the value of z is

  a doubled
  b halved
  c multiplied by 6
  d divided by 10
  e multiplied by a factor of 0.7.

6 It takes 20 hours for 5 people to decorate 100 cakes.

  a How many people are needed to decorate 100 cakes in 10 hours?
  b How many hours will it take 4 people to decorate 100 cakes?
  c How long will it take 5 people to decorate 200 cakes?

> Take care! Hours and people are inversely proportional.
> People and cakes are directly proportional.

7 You are told that y is inversely proportional to x, and that when $x = 4$, $y = 4$. Find the value of y when x is equal to

  a 8
  b 2
  c 40
  d 1
  e 100
  f 0.5

8 Given that $y \propto \frac{1}{w}$, and that $y = 10$ when $w = 50$, write an equation connecting y and w.

9 You are told that $y = \frac{k}{x}$, and that $y = 20$ when $x = 40$. Find the value of the constant k.

10 The variables u and v are in inverse proportion to one another. When $u = 6$, $v = 8$. Find the value of u when $v = 12$.

11 The distance, d, that Freya travels while driving at a constant speed is directly proportional to the time spend driving, t.

  a Write the statement 'd is directly proportional to t' as a formula including a constant of proportionality, k.
  b Given that Freya travels 72 miles in $1\frac{1}{2}$ hours, find the value of the constant k.
  c Use your previous answers to calculate the distance travelled in $2\frac{1}{2}$ hours.
  d How long would it take Freya to drive 160 miles? Give your answer in hours and minutes.

1048, 1948, 1949   SEARCH

APPLICATIONS

21.3 Direct and inverse proportion

RECAP

- Numbers or quantities are in **direct proportion** when the ratio of each pair of corresponding values is the same.
- You write 'y is proportional to x' as $y \propto x$ or $y = kx$, where k is the **constant of proportionality**.
- 'y is inversely proportional to x' can be written as 'y is proportional to $\frac{1}{x}$' or $y = \frac{k}{x}$, where k is the constant of proportionality.
- When two quantities are in inverse proportion, then their graph is a reciprocal curve.

- One quantity can be directly proportional to the square, cube, square root, cube root ... of another quantity.
- One quantity can be inversely proportional to the square, cube, square root, cube root ... of another quantity.

HOW TO

To solve proportion problems
1. Write the proportional relationship as an equation.
2. Substitute values into the equation to find the constant, k.
3. Use the formula to answer the question.

EXAMPLE

A wedding cake is going to be made from two round tiers. The baking time, T minutes, is directly proportional to the square of the individual cake's radius, R mm.
When $R = 150$, $T = 50$.
Find T when $R = 180$.

1. T is directly proportional to R^2.
   $T \propto R^2$ so $T = kR^2$
2. Substitute values into the equation to find k.
   When $R = 150$, $T = 50$, so $50 = k \times 150^2$
   $k = 0.0022222...$ or $\frac{1}{450}$
3. Use the formula to find T when $R = 180$.
   $T = 0.00222... \times 180^2$ $T = 72$

You could leave your value for k as an exact fraction instead of a decimal.

EXAMPLE

The force of attraction, F, between two magnets is inversely proportional to the square of the distance, d, between them. Two magnets are 0.5 cm apart and the force of attraction between them is 50 newtons.

When the magnets are 3 cm apart what will be the force of attraction between them?

1. F is inversely proportional to d^2.
   $F \propto \frac{1}{d^2}$, so $F = \frac{k}{d^2}$
2. Substitute values into the formula to find k.
   When F is 50, d is 0.5 $50 = \frac{k}{0.5^2}$
   $k = 50 \times 0.5^2 = 12.5$
3. Use the formula to find d.
   So $F = \frac{12.5}{d^2}$
   When $d = 3$ $F = \frac{12.5}{3^2} = 1.4$ newtons

Ratio and proportion   Units and proportionality

452

Reasoning and problem solving

Exercise 21.3A

1. A shop sells drawing pins in two different packs.

   Pack A contains 120 drawing pins and costs £1.45.

   Pack B contains 200 of the same drawing pins, and costs £2.30.

   Calculate the cost of one drawing pin from each pack, and explain which pack is better value.

2. A store sells packs of paper in two sizes.

Regular	Super
150 sheets	500 sheets
Cost $1.05	Cost $3.85

   Which of these two packs gives better value for money? You must show all of your working.

3. R varies with the square of s. If R is 144 when s is 1.2, find

   a. a formula for R in terms of s

   b. the value of R when s is 0.8

   c. the value of s when R is 200.

4. For a circle, explain how you know that the area is directly proportional to the square of its radius.

   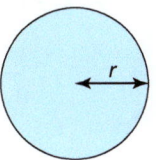

   State the value of the constant of proportionality in this case.

   What other formulae do you know that show direct proportion?

5. The surface area of a sphere varies with the square of its radius. If the surface area of a sphere with radius 3 cm is 113 cm², find the surface area of a sphere with radius 7 cm.

6. Imogen carries out a scientific experiment. She sets two magnets at varying distances apart and notices that the force between them decreases as the square of the distance increases. She performs the experiment three times. These are her results.

Distance d (cm)	1	2	3	4	5
d^2	1	4	9	16	25
Force F (newtons)	100	25	?	?	4

   a. Study Imogen's results. Write a formula for F in terms of d and use it to complete the rest of the table.

   b. Plot a graph of F against d and join the points to form a smooth curve.

7. Match the four sketch graphs with the four proportion statements.

   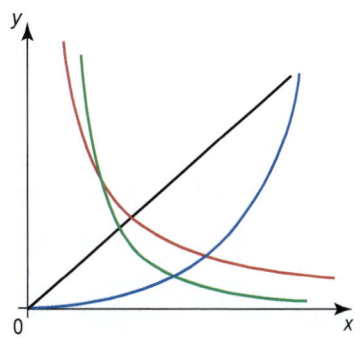

   A $y \propto \dfrac{1}{x^2}$ B $y \propto \dfrac{1}{x}$

   C $y \propto x$ D $y \propto x^2$

8. A carpet layer is laying square floor tiles in two identical apartments. The time t (mins) that it takes to lay the tiles is inversely proportional to the square of the length of each tile. Given that tiles with a side length of 20 cm take 4 hours and 10 minutes to lay, how long will the second apartment take using tiles with side length 40 cm?

9. Given that y is inversely proportional to the square root of x and that y is 5 when x is 4, find

   a. a formula for y in terms of x

   b. the value of

   i. y when x is 100   ii. x when y is 12.

10. Given that P varies inversely with the cube of t and that P is 4 when t is 2, find

    a. a formula for P in terms of t

    b. without a calculator

    i. P when t is 3   ii. t when P is $\frac{1}{2}$.

SKILLS

21.4 Growth and decay

- To increase something by $r\%$, multiply it by $\dfrac{100 + r}{100} = 1 + \dfrac{r}{100}$
- To decrease something by $r\%$, multiply it by $\dfrac{100 - r}{100} = 1 - \dfrac{r}{100}$

▲ To increase something by 20%, multiply by 1.2
To reduce something by 20%, multiply by 0.8

1.2 and 0.8 are called **multipliers**.

Sometimes percentage increases or decreases are repeated.

If interest is calculated on the interest it is **compound interest** otherwise it is **simple interest**.

EXAMPLE

The number of grey squirrels in a forest is 540. The population increases by 10% each year.
a Draw a graph to illustrate the growth over the next 6 years.
b After how long will the population double?

This is called **exponential growth**.

a $100\% + 10\% = 110\%$, multiplier = 1.1

Years	Population
0	540
1	540 × 1.1 = 594
2	594 × 1.1 = 653
3	653 × 1.1 = 719
4	719 × 1.1 = 791
5	791 × 1.1 = 870
6	870 × 1.1 = 957

Find values to plot on a graph.

Try using the 'Ans' key on your calculator to repeat calculations.

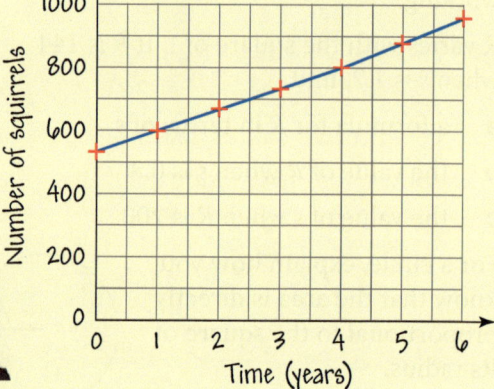

b After 7 years, population = 957 × 1.1 = 1053   Not quite double, 540 × 2 = 1080.
After 8 years, population = 1053 × 1.1 = 1158
It will take just over 7 years for the population to double.

EXAMPLE

The value of a car is $25 000. The value decreases by 22% each year.
a Write a formula for the value of the car after n years.
b Use your formula to find the value of the car after 8 years.
c After how long will the car be worth less than $1000?

This is called **exponential decay**.

a $100\% - 22\% = 78\%$, multiplier = 0.78
Value after n years = $25\,000 × 0.78^n$

b Value after 8 years = $25\,000 × 0.78^8$ = $3425   (nearest $)

You could check this by multiplying by 0.78 eight times.

c Continuing to multiply $3425 by 0.78 gives
2672, 2084, 1625, 1268, 989
After 13 years the car will be worth less than $1000.

Ratio and proportion   Units and proportionality

Fluency

Exercise 21.4S

1. Find the decimal multiplier for each percentage change.
   a. Increase of 25%
   b. Decrease of 25%
   c. Increase of 2.5%
   d. Decrease of 2.5%

2. The number of trout in a lake is 800. The number decreases by 15% each year.
   a. Draw a graph to illustrate the fall in the population over the next 8 years.
   b.
      i. Find a formula for the number of trout after n years.
      ii. Use your formula to check three values on your graph.

3. A bacteria population doubles every 20 minutes.
   a. Starting with 1 bacterium, draw a graph of the population growth over 3 hours.
   b. Estimate the number of bacteria after
      i. 150 minutes
      ii. 170 minutes
   c. What has happened to the population between 150 minutes and 170 minutes?

4. The table gives information about the population growth of two bacteria colonies.

Colony	Population now	Increase per hour
A	200	50%
B	400	35%

   a. Show that the population of Colony A after n hours is 200×1.5^n
   b. Find an expression for the population of Colony B after n hours.
   c. When does the population of Colony A become bigger than that of Colony B?

5. The value of a new car is £16000. The car loses 15% of its value at the start of each year.
   a. Find a formula for the value of the car after n years.
   b. Find the value of the car after 4 years.
   c. After how many complete years will the car's value drop below £4000?

6. The population of a town is 52000. The population increases by 1.5% each year.
   a. Find the population after 6 years.
   b. When will the population reach 60000?

7. Sadie invests $2000 in a savings account. The bank adds 4% compound interest at the end of each year. Sadie does not add or take any money from the account for 10 years.
   a. Copy and extend this table to show how Sadie's investment grows.

End of year	Amount in the account ($)
1	2000 × 1.04 = 2080
2	

   b. Work out the percentage interest that Sadie's investment earns in 10 years.
   c. $P is invested with compound interest r% added at the end of each time period,
      i. Show that the total amount at the end of n time periods is
      $$A = P\left(1 + \frac{r}{100}\right)^n$$

      A time period is usually a year or a number of months.

      ii. Use this formula to check the last amount in your table in part **a**.

8. The half-life of a radioactive substance is the time it takes for the amount to go down to half of the original amount.
   After how many half-lives will there be less than 1% of the radioactive substance left?

*9. The table shows how the population of the world has grown since 1900.
   a. Draw a graph of this data.
   b. A growth function $P = 1.65 \times 1.0125^{(y-1900)}$ where P is the population in billions in year y has been suggested as a model.
      i. Show this function on your graph.
      ii. What annual percentage increase in world population does the model assume?

Year	Population (billions)
1900	1.65
1910	1.75
1920	1.86
1930	2.07
1940	2.30
1950	2.56
1960	3.04
1970	3.71
1980	4.45
1990	5.29
2000	6.09
2010	6.87

1070, 1238

APPLICATIONS

21.4 Growth and decay

RECAP (p.244)
- To increase/decrease something by $r\%$, multiply by $1 \pm \dfrac{r}{100}$
- When a principal amount £P is invested with compound interest $r\%$ added at the end of each time period, the total accrued at the end of n time periods is
$$A = P\left(1 + \dfrac{r}{100}\right)^n$$

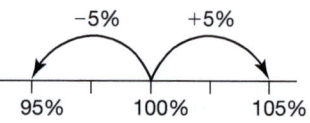

To increase something by 5%, multiply by 1.05
To decrease something by 5%, multiply by 0.95

HOW TO

To solve a repeated percentage change problem
1. Find the multiplier.
2. Decide whether to use a single calculation or a step-by-step approach.
3. Calculate the value required and answer the question.

Repeating steps is using an iterative process

EXAMPLE

Adira invests $4000 in an account. Interest of 1.5% is added at the end of every 6 months.
a How much compound interest is earned in 4 years?
b How much longer will it take for the amount in the account to reach $5000?
c What assumptions have you made in answering these questions?

① The multiplier = $1 + \dfrac{1.5}{100} = 1.015$

a Amount after 4 years = $4000 \times 1.015^8 = \$4505.97$ ② 4 years = 8×6 months. Round to the nearest pence.

Interest = $\$4505.97 - \$4000 = \$505.97$ ③ Take away the original amount.

b 4573.56, 4642.16, 4711.80, 4782.47, 4854.21, 4927.02, 5000.93   ② Multiplying by 1.015.
This takes 7×6 months = $3\tfrac{1}{2}$ years.

c The interest rate does not change and Harry does not add or take out any money.   ③

EXAMPLE (p.198)

An angling club says that there are 1500 trout in a lake and each year the trout population falls by 20%. The club decides to add 500 trout to the lake at the end of each year.
a Show that $T_{n+1} = 0.8T_n + 500$ where T_n is the number of trout in the lake after n years.
b Estimate the number of fish in the lake after 4 years.
c Sketch a graph to show how the trout population varies in this time. State any assumptions.

(p.364)

a 100% − 20% = 80%   ①

Multiplying T_n by 0.8, then adding 500 gives $T_{n+1} = 0.8T_n + 500$

b $T_1 = 0.8 \times 1500 + 500 = 1700$ ②
$T_2 = 0.8 \times 1700 + 500 = 1860$
$T_3 = 0.8 \times 1860 + 500 = 1988$
$T_4 = 0.8 \times 1988 + 500 = 2090$ ③

Round to the nearest whole number.

c Assumes the same conditions each year.   ③

Decay curves fall more quickly at first.

Ratio and proportion   Units and proportionality

Reasoning and problem solving

Exercise 21.4A

1 For each account in the table below, find the compound interest earned.

	Account Original amount	Compound interest rate	Number of years
a	£250	4% per year	6
b	£840	2.5% per 6 months	5
c	£4500	1.25% per 3 months	3

2 A building society offers two accounts: Vishna says that they would give the same interest on an investment. Is Vishna correct? Explain your answer.

> **Easy Saver**
> 4% interest added at the end of each year
> **Half-yearly saver**
> 2% interest added at the end of every 6 months

3 A road planner uses the formula 2400×1.08^n to estimate the number of vehicles per day that will travel on a new road n months after it opens.

  a Describe two assumptions the planner has made.

  b Sketch a graph to show what the planner expects to happen.

  c Give a reason why the planner's assumptions may not be appropriate.

4 There are 250 rare trees in a forest, but each year the number of trees falls by 30%. A woodland trust aims to plant 60 more trees in the forest at the end of each year.

  a Show that $T_{n+1} = 0.7\, T_n + 60$ where T_n denotes the number of trees in the forest after n years.

  b Work out the number of trees after 5 years.

  c Sketch a graph to show how the number of trees varies in this time. State any assumptions you make.

5 Igor takes out a loan for $500. Interest of 2% is added to the amount owing at the end of each month, then Igor pays off $90 or all the amount owing when it is less than $90.

  a How long will it take Igor to pay off the loan? Show your working.

  b Work out the percentage interest that Igor will pay on the loan of $500.

6 Sally invests €8000 in an account that pays 3.5% interest at the end of each year. Sally has to pay 20% tax on this interest. Calculate how much Sally will have in her account at the end of 4 years.

7 Liam finds a formula for the compound interest earned by $P invested for 6 years at a rate of 4.5%. Here is Liam's method.

> Interest in 1 year = 0.045 × $P
> Interest for 6 years = 6 × 0.045 × $P = $0.27P

  a Why is Liam's method incorrect?

  b Find a correct formula.

  c After 6 years the interest earned is $1934.46. Find, to the nearest one pound, the original amount $P.

8 Find the minimum rate of interest for an investment of $500 to grow to $600 in 6 years.

9 Tanya measures the temperature of a cup of coffee as it cools.

Time t(min)	0	10	20	30	40	50	60
Temperature T(°C)	85	68	55	45	39	34	31

  a **i** Use Tanya's data to draw a graph.

  ii Find the rate at which the coffee is cooling after half an hour.

  b Tanya says $T = 20 + 65 \times 0.97^t$ is a good model of the data.

  i Is Tanya correct? Show how you decide.

  *__ii__ Explain each term in Tanya's model.

*__10__ The half-life of caesium-137 is 30 years.

  a Show that when 1 kilogram of caesium-137 decays, the amount left after t years is
  $$f(t) = 2^{-\frac{t}{30}} \text{ kg}.$$

  b Sketch a graph of amount against time.

  c Describe how the function and graph would change if $f(t)$ was given in terms of grams instead of kilograms.

  1070, 1238 SEARCH

Summary

Checkout
You should now be able to...

Test it
Questions

Checkout	Test it Questions
✓ Use compound measures.	1 – 3
✓ Convert between standard units of measure and compound units.	4
✓ Compare lengths, areas and volumes of similar shapes.	5
✓ Solve direct and inverse proportion problems.	6 – 8
✓ Describe direct and inverse proportion relationships using an equation.	6 – 8
✓ Recognise graphs showing direct and inverse proportion and interpret the gradient of a straight line graph.	9
✓ Find the instantaneous and average rate of change from a graph.	10
✓ Solve repeated proportional change problems.	11

Language	Meaning	Example
Speed	A measure of the distance travelled by an object in a certain time.	The speed limit on the motorway is 70 mph.
Density	A measure of the amount of matter in a certain volume.	Metal / Mass of 1 cm³ / Density — Gold: 19.3 g, 19.3 g/cm³ — Iron: 7.87 g, 7.87 g/cm³
Rate of change	One quantity measured per unit of time or per unit of another quantity.	Total pay for 3 hours work = £16.50. Rate of pay = $\frac{16.50}{3}$ = £5.50 per hour
Similar	The same shape but different in size.	(triangles: 4 cm, 5 cm and 8 cm, 10 cm)
Directly proportional	Two variables are proportional if one is always the same multiple of the other.	$y = kx$ where k is a constant.
Inversely proportional	Two variables are inversely proportional if one is proportional to the reciprocal of the other.	$y = \frac{k}{x}$ where k is a constant.

Review

1. A box of cereal costs £1.55 for 750 g. What is the cost per 100 g?

2. The nutritional information from a pack of chips is shown in the box.

per 100 g when baked	
Energy	270 kcal
Protein	4.2 g
Carbohydrate	39 g
Fat	10.5 g
Salt	0.9 g

   A suggested serving is 135 g.

   a For a serving, calculate the amount of
      i energy   ii fat   iii salt.
   b What size serving contains 513 kcal?

3. Convert
   a 6.7 m to mm
   b 2651 s to minutes and seconds
   c 450 cm^3 to litres
   d 7 m/s to km/h.
   e 67 253 mm^2 into m^2.

4. A 3D shape is enlarged by scale factor 3.
   a The volume of the smaller shape is 19.5 cm^3, what is the volume of the larger shape?
   b The surface area of the larger shape is 360 cm^2, what is the surface area of the smaller shape?

5. y is proportional to x. When $x = 3.5$, $y = 4.2$
   a Write a formula linking x and y.
   b What is the value of
      i y when $x = 7.9$ ii x when $y = 11.6$?

6. y is proportional to the square of x. When $x = 8$, $y = 128$.
   a Write a formula to link x and y.
   b What is the value of
      i y when $x = 11$ ii x when $y = 50$?

7. X is inversely proportional to Y. When $X = 12$, $Y = 5.5$
   a Write a formula to link X and Y.
   b What is the value of
      i Y when $X = 22$ ii X when $Y = 132$?

8. Sketch graphs to show the relationship
   a between x and y in question 6
   b between X and Y in question 7.

9. The graph shows the change in the depth of liquid in a container over time.

   Use the graph to estimate
   a the rate of change of the depth at 3 hours
   b the average rate of change over the whole 5 hours.

   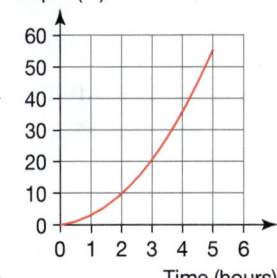

10. $1500 is invested in a bank account that pays 2.5% interest per year.
    a How much is in the account after
       i one year   ii five years?
    b Write a formula to find the value of the account, V, after t years.

What next?

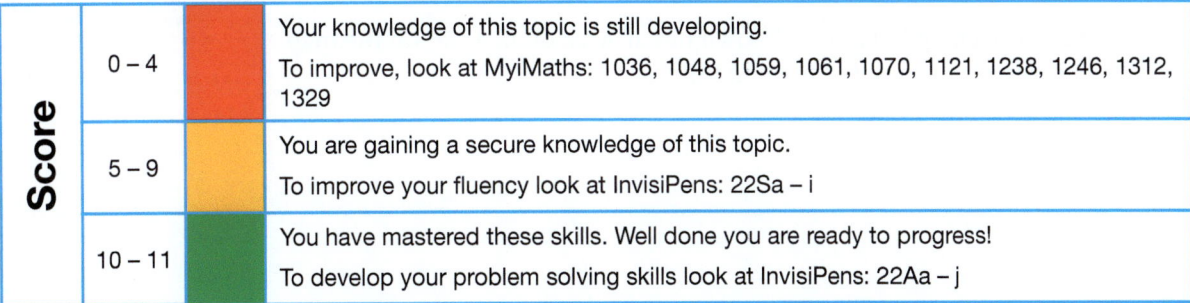

Assessment 21

1 Light travels at 186 000 miles per second. The Sun is 93 000 000 miles from Earth. Calculate how long, it takes for light from the Sun to reach the Earth. Give your answer in minutes and seconds. [3]

2 a 1 inch is equivalent to 2.54 cm. How many inches are there in 1m? [1]

b A pack of 8 batteries cost $6.50. How much do 12 batteries cost? [2]

c It takes 15 minutes to cook 8 omelettes. How long would it take to cook 28 omelettes? [2]

3 The Olympic sprinter Usain Bolt can run 100 m in 9.8 seconds.
A wombat can maintain a speed of 40 km/h for 150 m.
Who would win a 100 m race, Usain Bolt or a wombat?
Show your working. [4]

4 a A statue has a mass of 3850 grams and volume of 529 cm^3? What is its density? [2]

b A silver bar has a mass of 250 g and a density of 10.5 g/cm^3. What is its volume? [2]

c A litre of milk has a density of 1.03 g/cm^3. What is the mass of the milk? [2]

5 A dog eats three tins of dog food in two days.

a How many tins does it eat in 30 days? Give your answer to the nearest tin. [2]

b Will 100 tins feed the dog for 67 days? Give your reason. [2]

c A second dog eats twice as much as the first. Will 30 tins be enough to feed the two dogs for two weeks?
Show your working. [4]

6 A cylindrical container of water emptied in $17\frac{1}{2}$ minutes. The container was full.
The water flowed out at an average rate of 150 ml/second.

a How many litres of water did the container hold? [3]

b The base of the cylinder is 23 cm in diameter. How high is it? [3]

7 A cuboid has volume 9600 cm^3.
The density of the cuboid is 7.2 g/cm^3.

a Find the mass of the cuboid in kg. [2]

The pressure that the base of the cuboid exerts on the floor is 0.7 N/cm^2.
The force of the cuboid on the floor can be calculated using this formula.
Force = 9.8m, where m is the mass in kg.
The cuboid has length 40 cm, width x cm and height y cm.

b Find the value of x to the nearest cm. [3]

c Find the value of y to the nearest cm. [2]

8 The scale on a map is 1 : 7500. A town has an area of 37.5 cm^2 on the map.
What is the real area of the town in km^2? [3]

9 a The diagram shows two similar cylinders.
The smaller cylinder has radius r cm.
The larger cylinder has radius $2r$ cm.
The cost of making the cylinders depends on the surface area.
The cost of making the small cylinder is £2.50.
How much does it cost to make the larger cylinder? [2]

b The larger cylinder has volume $1.2\,m^3$.
Find the volume of the smaller cylinder. Give your answer in cm^3. [2]

10 The length of the extension, e cm, when a spring is stretched is directly proportional to T, the tension in newtons. $T = 10$, when $e = 2.5$

a Write a formula connecting e and T. [2]

b Find the value of e when $T = 17$ N. [1]

c Find the value of T when $e = 4.5$ cm. [1]

11 The area, A, of a shape is directly proportional to x^2. When $x = 4$, $A = 50.272$

a Write a formula connecting A and x. [2]

b Elsa says that the shape is a circle. Is Elsa correct? Give your reason. [2]

c Find A when $x = 10$ [1]

d Find x when $A = 254.502$ [1]

12 It is claimed that monthly rent for a one bedroom flat is inversely proportional to the distance in km to the city centre. This is the data from eight available flats.

	Flat 1	Flat 2	Flat 3	Flat 4	Flat 5	Flat 6	Flat 7	Flat 8
Distance to city centre (km)	2.1	2.8	3.7	4.1	4.7	5.6	6.2	6.5
Monthly rent (€)	900	625	475	450	525	325	300	275

a Draw a graph to show the data. [4]

b Which flat does not support the claim that the rent and distance are inversely proportional? Give your reason. [1]

c Nina has a maximum budget of €525 per month to spend on rent.
How close to the city centre can Nina afford to live? [1]

13 P is inversely proportional to the square root of t. When P is 5, t is 4.

a Find an expression for P in terms of t. [2]

b Calculate P when $t = 100$. [2]

14 The height, h metres, of a ball when it is thrown is given by $h = 15t - 5t^2$, where t is the time in seconds.

a Plot a graph of the height by taking readings at 0.5 second intervals. [4]

b Estimate the speed of the ball 1 second after being thrown. [3]

15 The value, V, of a car can be calculated using the formula $V = 27\,000 \times 0.95^t$, where t is the number of years since the car was bought.

a Complete this sentence.
The car was bought for $\$_____$
and the value of the car decreases by $\_\_\_$% each year. [2]

b After 10 years the value of the car decreases by 10% each year.
How much will the car be worth after 15 years? [4]

22 Differentiation

Introduction
Very little stays the same. In most situations there is change. For example, populations vary over time; the heights of waves vary with position; the number of goods sold vary with their price; and the speeds of chemical reactions vary with temperature.

What's the point?
If you want to describe the real world then you need a language for describing how one variable changes as a function of another variable. In mathematics this is called differential calculus.

Objectives
By the end of this chapter, you will have learned how to ...
- Find the rate of change of a function using a chord.
- Find the rate of change of a simple function using differentiation.
- Find the tangent to a curve at a point.
- Use the derivative of a function to identify any turning points and their type.

Check in

1. Work out
   - **a** the gradient of this line.
   - **b** the equation of this line.

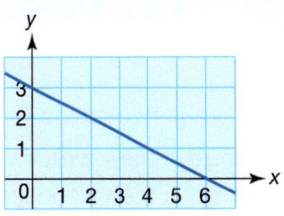

2. Expand the brackets and simplify these expressions
   - **a** $(2x + 5)(x - 3)$
   - **b** $5x^3(3x^2 - 4)$

3. Simplify these algebraic fractions.
   - **a** $\dfrac{3x^7 + 6x^5}{3x^4}$
   - **b** $\dfrac{x^2 - 4x - 21}{x + 3}$

4. Solve these equations.
   - **a** $4 - 1.6t = 0$
   - **b** $3x^2 - 6x = 0$

Chapter investigation

A piece of wire is bent into the shape of the function $y = x^5 - 2x^4 - 35x^3$

A bead is threaded on to the wire and is free to slide along it.

There are three places where you can balance the bead.

Find the three places and decide if the bead will be stable if you give it a small nudge.

SKILLS

22.1 Rates of change

p.370

- The gradient of a line segment is
  $$\frac{\text{Change in the } y \text{ direction}}{\text{Change in the } x \text{ direction}}.$$

The chord from (0, 0) to (2, 8) has gradient 4. The graph of $y = x^3$ also passes through the points (0, 0) and (2, 8).
However, the gradient of the curve changes from point to point.

- To estimate the gradient of a curve at a point P, find the gradient of a chord between points on either side of P.

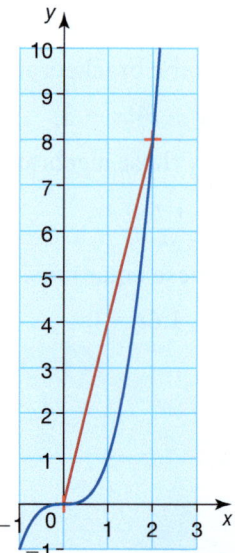

EXAMPLE

Estimate the gradient of the graph of $y = x^3$ at (1, 1) by finding the gradient of the chord joining (0.999, 0.999³) to (1.001, 1.001³).

Use the formula gradient = $\frac{\text{change in the } y \text{ direction}}{\text{change in the } x \text{ direction}}$

$\frac{1.001^3 - 0.999^3}{1.001 - 0.999} = 3.000001$

Zoom in on the graph of $y = x^3$ at (1, 1) and you will see that the curve starts to look more and more like a straight line.

Advanced mathematics can be used to prove that the gradient of $y = x^3$ at (1, 1) is precisely 3.

The straight line through (1, 1) with gradient 3 is called a tangent to the curve.

- The average gradient of a curve **between two points** is the gradient of the chord joining the two points.
- The gradient of a curve **at a point** is the gradient of the tangent at that point.

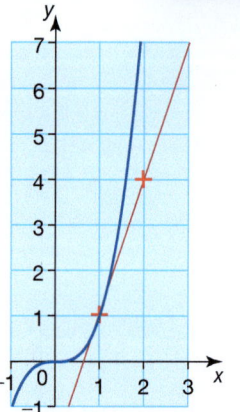

p.274

EXAMPLE

Find the equation of the tangent to the curve $y = x^3$ at the point (1, 1).

The tangent is a straight line through (1, 1), with gradient 3.
$y = mx + c$ $1 = 3 + c$ so $c = -2$
The equation is $y = 3x - 2$.

Algebra   Differentiation

Fluency

Exercise 22.1S

1. **a** Estimate the gradient of the curve $y = x^3$ at the point $(2, 8)$ by finding the gradient of the chord joining $(1, 1)$ to $(3, 27)$.
    b Improve the estimate of part **a** by using a chord closer to $(2, 8)$.
    c Use your best estimate to find the equation of the tangent at $(2, 8)$.

2. **a** Estimate the gradient of the curve $y = x^3 - 9x$ at the point $(2, -10)$ by finding the gradient of the chord joining $(1, -8)$ to $(3, 0)$.
    b Improve the estimate of part **a** by using a chord closer to $(2, -10)$.
    c Use your best estimate to find the equation of the tangent at $(2, -10)$.

3. **a** Plot the graph of $y = x^2$ for $-3 \le x \le 3$.
    b Draw, as accurately as possible, tangents at the points $(\pm 1, 1)$ and $(\pm 2, 4)$.
    c Hence complete the table of gradients.

x	-2	-1	0	1	2
Gradient					

    d Suggest a possible formula for the gradient at any point (p, p^2) on the graph of $y = x^2$.

4. A graph consists of two straight line segments; from O to A and from A to B as shown.

    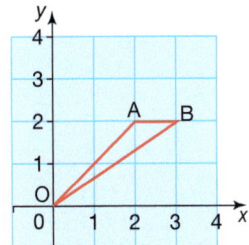

    a Find the gradients of
       i OA **ii** AB.
    b What is the gradient of the graph from O to B?
    c Explain why the gradient of OB cannot be found by simply taking the average of your answers to **a i** and **a ii**.

5. **a** Sketch the graph of $y = x^2 + x - 2$.
    b Find the gradients of the chords through
       i $(0, -2)$ and $(6, 40)$
       ii $(1, 0)$ and $(5, 28)$
       iii $(2, 4)$ and $(4, 18)$
    c What is the gradient of the curve at $(3, 10)$?
    d Use your sketch of the curve to illustrate the answers to **b** and **c**.

*6. The sketch shows the graph of $y = x^2$ with the tangent drawn at point $P(p, p^2)$.

The tangent crosses the y-axis at point Q.

Use your answer to **3 d** to find the gradient and hence the equation of the tangent at P. Finally, show that point Q has coordinates $(0, -p^2)$.

*7. **a** For the graph of $y = x^2 - 10$, find the gradient of the chord joining the point where $x = 4 - h$ to the point where $x = 4 + h$.
    b Use your answer to **a** to explain why the gradient of $y = x^2 - 10$ at the point $(4, 6)$ is precisely 8.

APPLICATIONS

22.1 Rates of change

RECAP
- The gradient of a graph at a point can be estimated geometrically by drawing a tangent at that point.
- The gradient of a graph at a point can be estimated numerically by using a chord near that point.

HOW TO

To solve problems involving rates of change
1. Use the fact that the gradient at a point can represent instantaneous rates of change, especially speed.
2. Interpret rates of change in different contexts.

EXAMPLE

A ball is thrown vertically upwards with an initial speed of $20\,\text{ms}^{-1}$. Its height – time graph is as shown. The graph has equation $h = 20t - 5t^2$.

a How is the initial speed represented on this graph?

b Use a chord to estimate the vertical speed of the ball 1 second after being thrown.

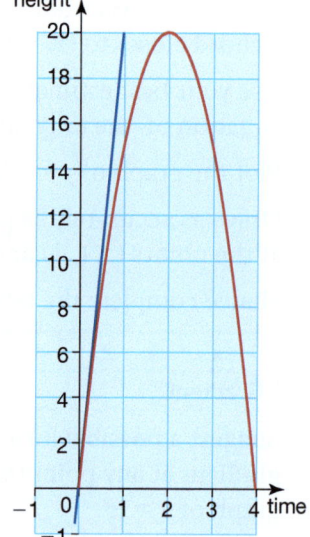

① ② Each gradient at a point represents vertical speed.

a The gradient of the tangent at the origin.
The gradient of the tangent is 20, so initial speed = $20\,\text{ms}^{-1}$.

b The gradient of the chord from $t = 0.9$ to $t = 1.1$ is given by $\dfrac{15.95 - 13.95}{1.1 - 0.9} = 10\,\text{ms}^{-1}$.

② Interpret this in the context of throwing the ball.
After 1 second, the ball's vertical speed is $10\,\text{ms}^{-1}$.

EXAMPLE

Water is poured at a constant rate into each of these containers.

The containers are initially empty.

Draw graphs of depth of water against time for each container.

a  b  c

a ② Cylinder has constant width, so depth increases at a steady rate and gradient is constant.

b ② As conical flask gets narrower depth increases more quickly and gradient gets steeper.

c ② As bowl gets wider, depth increases more slowly and gradient gets less steep.

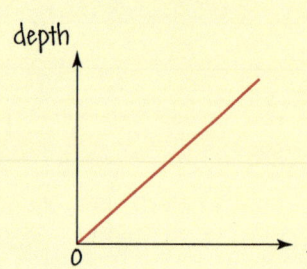

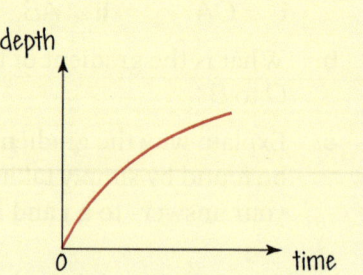

Algebra Differentiation

Reasoning and problem solving

Exercise 22.1A

1. A ball is thrown vertically upwards with an initial velocity of $30\,\text{ms}^{-1}$. Its height t seconds later is given by $h = 30t - 5t^2$.

   a. Use a chord to estimate its velocity 4 seconds after being thrown.

   b. Interpret your answer.

2.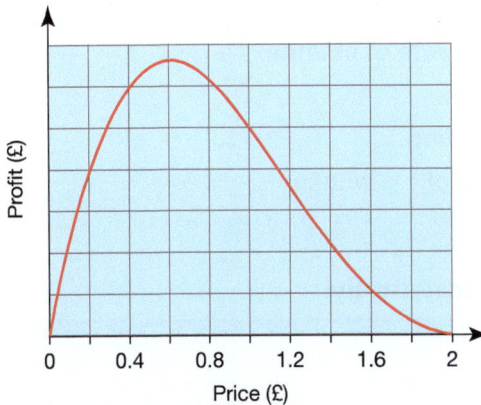

   The graph shows how the expected weekly profit from the manufacture of a *Choco* bar depends upon its selling price.

   a. Explain the meaning of the two points where the profit is zero.

   b. What is the importance of the point where the gradient of the graph is zero?

   c. What selling price would you recommend and why?

3.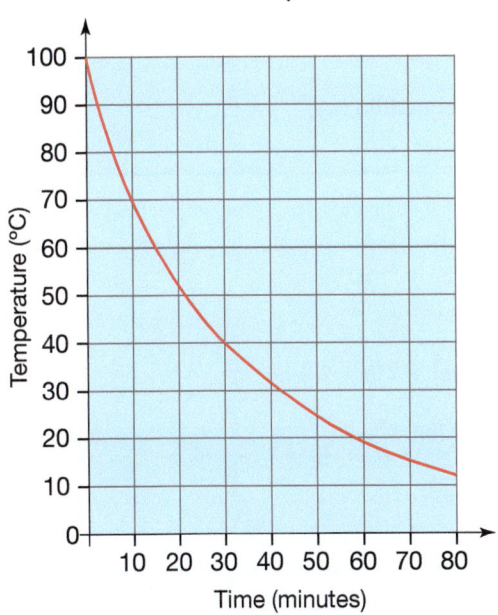

   The graph shows a cooling curve for a container of water.

   a. Use tangent lines to obtain rough estimates of the gradients of the curve at times 0, 30 and 60 minutes.

   b. What are the units of these gradients?

   c. Use your answers to **a** and **b** to write a brief description of the cooling of the water.

4. The velocity, $v\,\text{ms}^{-1}$, of a projectile at time t seconds is given by
   $$v = 20 - 10t.$$

   a. Sketch a graph of v against t.

   b. What is measured by the gradient of this graph?

   c. Find the gradient of the graph and state its units.

5. The space shuttle used two solid fuel rockets which burned for just two minutes and were then parachuted back to Earth from a height of 45 kilometres.

   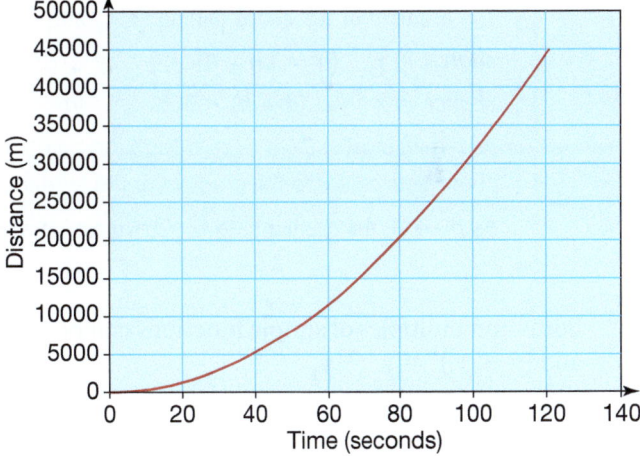

   a. Calculate the average speed of the space shuttle during the first two minutes of flight.

   b. Use a tangent line to estimate the speed of the space shuttle when the solid fuel rockets detached.

SKILLS

22.2 Differentiating polynomials

You have already seen that

- The gradient of a straight line is the rate of change. It is found using $\frac{\text{change in } y}{\text{change in } x}$
- The gradient of a curve **at a point** is the gradient of the tangent at that point.
- The gradient of a graph at a point can be estimated geometrically by drawing a tangent to the curve at that point. It can be estimated numerically by using a chord near that point.

> - The gradient of a graph at a point is an **instantaneous rate of change**.
> - This instantaneous rate of change is called a **derivative**.
> - The process of finding a derivative is called **differentiation**.
> - If y is a function given in terms of x, then the derivative of y is written either as y', or as $\frac{dy}{dx}$.

The derivative of a function at given points can be found **analytically**.

EXAMPLE

If $y = x^2$, find the value of y' at $x = 3$

When $x = 3$, $y = 3^2 = 9$

Take another point close to $(3, 9)$. Say $x = 3 + h$, where h is a positive number. Then $y = (3 + h)^2 = h^2 + 6h + 9$

The gradient of the chord joining $(3, 9)$ and $(3 + h, h^2 + 6h + 9)$ is

$$\frac{\text{change in } y}{\text{change in } x} = \frac{(h^2 + 6h + 9) - 9}{(3 + h) - 3} = \frac{h^2 + 6h}{h} = h + 6$$

As h gets smaller ($h \to 0$, or 'h **tends to** zero'), the gradient of the chord gets closer to the gradient of the tangent.

As $h \to 0$, the gradient $\to 6$. Therefore, at $x = 3$, $y' = 6$.

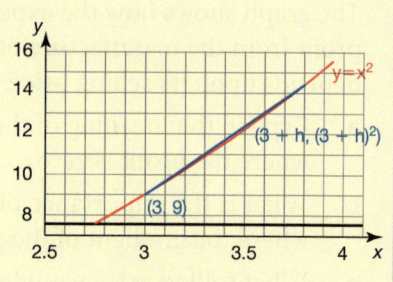

Sums and multiples of simple functions can be found using the results $(ky)' = ky'$ and $(y_1 + y_2)' = y_1' + y_2'$

EXAMPLE

If $y = x^3 + 4x^2$, find the value of y' when $x = 3$

Consider $y_1 = x^3$. When $x = 3$, $y = 3^3 = 27$

When $x = 3 + h$, $y = (3 + h)^3 = h^3 + 9h^2 + 27h + 27$

The gradient of the chord joining $(3, 27)$ and $(3 + h, h^3 + 9h^2 + 27h + 27)$ is

$$\frac{\text{change in } y}{\text{change in } x} = \frac{(h^3 + 9h^2 + 27h + 27) - 27}{(3 + h) - 3} = \frac{h^3 + 9h^2 + 27h}{h} = h^2 + 9h + 27$$

As $h \to 0$, the gradient $\to 27$. Therefore, at $x = 3$, $y_1' = 27$.

Now consider $y_2 = 4x^2$. Using $(ky)' = ky'$, and the result in the previous example, we can see that $y_2' = 4 \times 6 = 24$ when $x = 3$.

Combining these results we see that $y' = 27 + 24 = 51$ when $x = 3$.

Algebra   Differentiation

Fluency

Exercise 22.2S

1 Here is the graph of $y = x^2$.

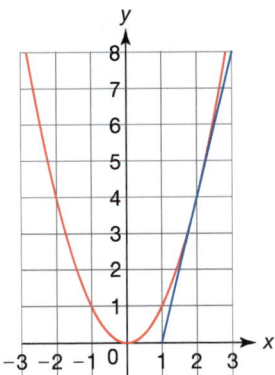

A tangent is drawn at the point where $x = 2$.

a Use this tangent to state the instantaneous rate of change when $x = 2$.

b Verify your result in **a** analytically.

c Find the derivative of $y = x^2$ when $x = 4$.

d Use the results in the first example on page 468, and your solutions in **a** and **c**, to suggest a general rule for the **gradient function** when $y = x^2$:

$$y' = \frac{dy}{dx} = \ldots$$

2 Here is the graph of $y = x^3$. A tangent is drawn at the point where $x = 1$.

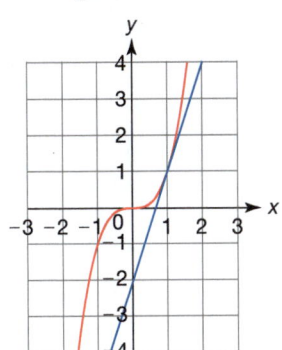

a State the derivative when $x = 1$.

b State the derivative when $x = 0$.

c Find the derivative of $y = x^3$ when $x = 2$.

d Choose two other values for x. Find the derivative at each of these points.

e Suggest a general rule for the gradient function when $y = x^3$.

3 $y = c$, where c is a constant.

Explain why $y' = 0$.

4 For each of these functions, work out the value of the derivative when

    i $x = 2$ **ii** $x = -1$

a $y = x^2 + 1$ **b** $y = x^2 - 6$

c $y = x^3 - 8$ **d** $y = x^3 + 3$

5 Consider the function $y = x$.

Explain why $y' = 1$ for any value of x.

6 For each of these functions, work out the value of y' when

    i $x = 1$ **ii** $x = -3$

a $y = x^2 + x$ **b** $y = x^2 - x$

c $y = x^2 + x - 2$ **d** $y = x^3 - x + 3$

7 For each of these functions, work out the value of $\frac{dy}{dx}$ when $x = 2$

a $y = 2x^3 + x^2$

b $y = 2x^3 + 3x^2 + 1$

c $y = 4x^3 + 2x^2 + x$

d $y = x^3 + 4x^2 - 5x + 1$

*8 Here is the graph of $y = x^4$.

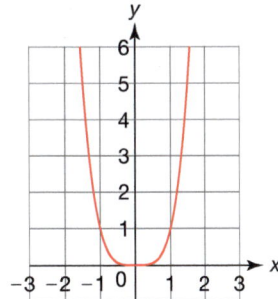

a Find the derivative of $y = x^4$ when $x = 1$.

b Choose two other non-zero values for x. Find y' at each of these points.

c Suggest a general rule for the gradient function when $y = x^4$.

APPLICATIONS

22.2 Differentiating polynomials

RECAP

For the function, y, in terms of x, it is often necessary to find the gradient function, $\frac{dy}{dx}$. Here are some commonly used results (a is a constant).

$y = x$ $\frac{dy}{dx} = 1$ $y = ax$ $\frac{dy}{dx} = a$

$y = x^2$ $\frac{dy}{dx} = 2x$ $y = ax^2$ $\frac{dy}{dx} = 2ax$

$y = x^3$ $\frac{dy}{dx} = 3x^2$ $y = ax^3$ $\frac{dy}{dx} = 3ax^2$

$y = x^4$ $\frac{dy}{dx} = 4x^3$ $y = ax^4$ $\frac{dy}{dx} = 4ax^3$

It is also true that if $y = a$, than $\frac{dy}{dx} = 0$

HOW TO

To find the derivative of a function that can be written as a polynomial

① If necessary, manipulate the function into polynomial form.

② Use the generalisation that $y = x^n$ $\frac{dy}{dx} = nx^{n-1}$

③ Differentiate each term in turn.

④ Simplify where possible.

EXAMPLE

Find the derivative of each of these functions:

a $y = 8x^5$ **b** $y = x^3(x - 5)$ **c** $y = \dfrac{5x^6 + 7x^3}{2x^2}$

a ① No simplification is required

② ③ $y = 8x^5 \Rightarrow y' = 8 \times 5x^4$

④ $y' = 40x^4$

b ① $y = x^3(x - 5) = x^4 - 5x^3$

② ③ $y = x^4 - 5x^3 \Rightarrow y' = 4x^3 - 5 \times 3x^2$

④ $y' = 4x^3 - 15x^2$

c ① $y = \dfrac{5x^6 + 7x^3}{2x^2} = \dfrac{5x^6}{2x^2} + \dfrac{7x^3}{2x^2} = \dfrac{5}{2}x^4 + \dfrac{7}{2}x$

② ③ $y = \dfrac{5}{2}x^4 + \dfrac{7}{2}x \Rightarrow y' = \dfrac{5}{2} \times 4 \times x^3 + \dfrac{7}{2}$

④ $y' = 10x^3 + \dfrac{7}{2}$

HINT: 'Find the derivative' means the same as 'find the gradient function'.

There are several different ways of referring to the derivative. You need to know two of them:

- $\dfrac{dy}{dx}$ was introduced by the German mathematician Gottfried Leibniz.
- y' was introduced by the French mathematician Joseph-Louis Lagrange.

Algebra   Differentiation

Reasoning and problem solving

Exercise 22.2A

1 For each of these functions, work out $\dfrac{dy}{dx}$.

  a $y = 12x^3$
  b $y = 2x^6$
  c $y = 3x^4 + 7x$
  d $y = 15x^3 - 8x$
  e $y = 5x^6 - 3x^4$
  f $y = 12x^3 + 5x^2 + 1$
  g $y = x^3 + x^2 + x + 1$
  h $y = 5x^3 + 2x^2 - 9x - 3$

2 For each of these functions, work out y'.

  a $y = 8x(x^3 + 1)$
  b $y = x^2(3x^4 - 4)$
  c $y = x^5(3x + 7)$
  d $y = 6x^2(5x^2 - 3)$
  e $y = (x + 2)(x + 5)$
  f $y = (x - 2)(x + 3)$
  g $y = (x - 1)(x - 7)$
  h $y = (2x - 1)(3x + 4)$

3 Find the derivative of each of these functions.

  a $y = \dfrac{7x^6}{3}$
  b $y = \dfrac{5x^4}{12}$
  c $y = \dfrac{2x^3 + 3x}{6}$
  d $y = \dfrac{20x^4 - 6x^3}{2x^2}$
  e $y = \dfrac{7x^5 - 2x^3}{x}$
  f $y = \dfrac{8x^4 + 7x^3}{3x^2}$
  g $y = \dfrac{x^2 - 3x - 4}{x + 1}$
  h $y = \dfrac{x^2 - 5x + 6}{x - 2}$

4 For each of these functions, find the value of the gradient when

  i $x = 0$ **ii** $x = 2$ **iii** $x = -1$

  a $y = 2x^5$
  b $y = 2x^5 + 3x - 1$
  c $y = 4x^2(2x + 3)$
  d $y = (x + 4)(x - 3)$
  e $y = \dfrac{3x^5 - x^4}{x}$
  f $y = \dfrac{x^2 - x - 2}{x - 2}$

5 A company uses a profit function of $P = -2s^2 + 900s - 100\,000$ where P = profit (\$) and s = selling price (\$). Here is a graph of this function.

  a Find the gradient function $\dfrac{dP}{ds}$.
  b What is the instantaneous rate of change in profit when the selling price is
   i \$205 **ii** \$224 **iii** \$230
  c Interpret these solutions.

6 The height, in metres, above ground of some lead in a shot tower is given by the formula $h = 44.1 - 4.9t^2$, where t is time in seconds.

  a Find the gradient function $\dfrac{dh}{dt}$.
  b What is the instantaneous rate of change in the height of the lead after
   i 1 second **ii** 2 seconds?
  c Explain the significance of the sign of your answer.

Use graph plotting software for question 7

7 Plot the graph of $y = x(x + a)(x - b)$. Use the constant controller to set $a = b = 2$.

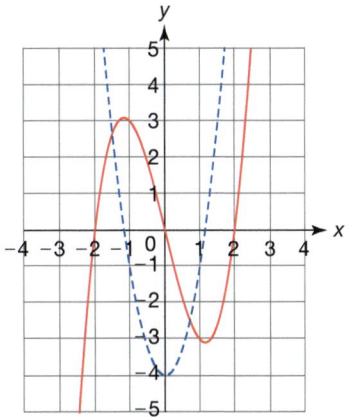

  a Use the software to construct the gradient function.
  b Use the constant controller to vary the values of a and b. In particular, study the graph when
   i $b = 0$ **ii** $b = -1$ **iii** $b = -2$
  c Write about what you notice.

SKILLS

22.3 Tangents to curves

A function $y = f(x)$ can be differentiated to find the gradient function $\dfrac{dy}{dx} = f'(x)$.

The gradient function is used to find the gradient of the tangent to the curve at any chosen point. The equation of the tangent will always be of the form $y = mx + c$

- The equation of a tangent will always be of the form $y = mx + c$.
- It is found by calculating the derivative at the given point to find m.
- c is then found by analysing the result.

EXAMPLE n.22.2

Find the equation of the tangent to the curve $y = x^3 - 2x^2$ at the point $(2, 0)$.

The problem can be solved analytically:

When $x = 2 + h$, $y = (2 + h)^3 - 2(2 + h)^2 = h^3 + 6h^2 + 12h + 8 - 2(h^2 + 4h + 4) = h^3 + 4h^2 + 4h$

The gradient of the chord joining $(2, 0)$ and $(2 + h, h^3 + 4h^2 + 4h)$ is

$$\dfrac{\text{change in } y}{\text{change in } x} = \dfrac{(h^3 + 4h^2 + 4h) - 0}{(2 + h) - 2} = \dfrac{h^3 + 4h^2 + 4h}{h} = h^2 + 4h + 4$$

As $h \to 0$, the gradient $\to 4$. Therefore, at $x = 2$, $\dfrac{dy}{dx} = 4$.

Using $y = mx + c$, we now know that the equation of the tangent is
$y = 4x + c$

$(2, 0)$ lies on the line, therefore $0 = 8 + c \Rightarrow c = -8$
The equation of the tangent is $y = 4x - 8$
The problem could also be solved using differentiation:

$y = x^3 - 2x^2 \Rightarrow \dfrac{dy}{dx} = 3x^2 - 4x$

$x = 2 \Rightarrow \dfrac{dy}{dx} = 12 - 8 = 4$

And then as before, the equation of the tangent is $y = 4x - 8$

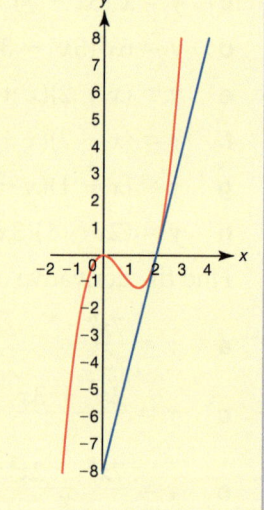

EXAMPLE

The quadratic function $y = ax^2 + bx$ has a tangent drawn at the point $(3, 3)$. The equation of the tangent is $y = 4x - 9$. Find the values of a and b.

Sketch a graph of the situation.

Find the gradient function: $\dfrac{dy}{dx} = 2ax + b$

At $x = 3$, the gradient is 4. Substitute into $\dfrac{dy}{dx} = 2ax + b$
to get $6a + b = 4$ ①

The point $(3, 3)$ lies on the line. Substitute into $y = ax^2 + bx$ to get $9a + 3b = 3$

$9a + 3b = 3 \Rightarrow 3a + b = 1$ ②

Solve the simultaneous equations:

① − ② $\Rightarrow 3a = 3 \Rightarrow a = 1 \Rightarrow b = -2$

The quadratic function is $y = x^2 - 2x$

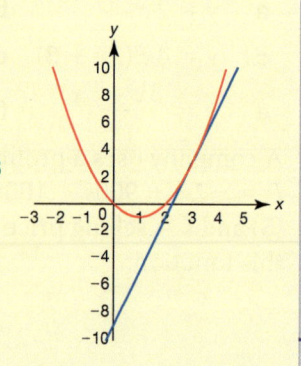

Algebra Differentiation

Fluency

Exercise 22.3S

1 a Find, analytically, the equation of the tangent to the curve $y = x^3 - 2x^2$ at the point

  i $(1, -1)$

  ii $(-1, -3)$

  iii $\left(\dfrac{3}{2}, -\dfrac{9}{8}\right)$

  b Verify your results in **a** by using the gradient function $\dfrac{dy}{dx}$.

2 Here is a graph of the function $y = 2x^3 - x^2$. The tangent is drawn at the point $(1, 1)$.

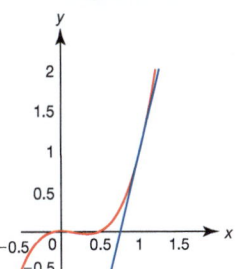

  a Find the equation of the tangent analytically.

  b Find $\dfrac{dy}{dx}$.

  c Use the gradient function to verify your result in **a**.

  d Find the equation of the tangent at the point $\left(\dfrac{1}{2}, 0\right)$.

3 Here is a graph of the function $y = x^2 - 7x - 8$. The tangent is drawn at the point $(2, -18)$.

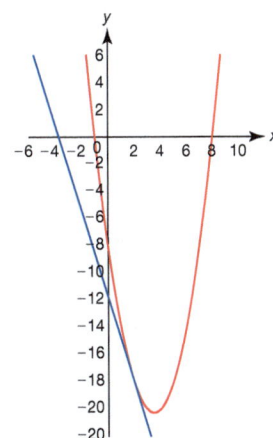

  a Find the equation of the tangent analytically.

  b Find $\dfrac{dy}{dx}$.

  c Use the gradient function to verify your result in **a**.

  d Find the equation of the tangent at the point $(8, 0)$.

4 The quadratic function $y = ax^2 + bx$ has a tangent $y = 7x - 1$ at the point $(1, 6)$. Find the values of a and b.

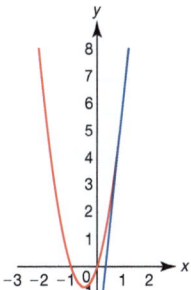

5 The quadratic function $y = ax^2 + bx$ has a tangent $y = -2(x + 1)$ at the point $(1, -4)$. Find the values of a and b.

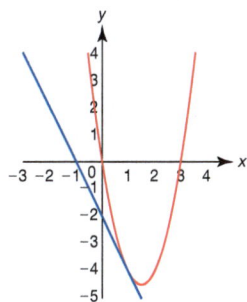

*6 The quadratic function $y = ax^2 + bx + c$ has a tangent $y = 4x + 6$ at the point $(0, 6)$. Find the values of b and c.

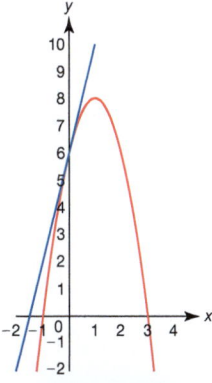

APPLICATIONS

22.3 Tangents to curves

RECAP

A normal to a curve is perpendicular to the tangent at that point.

If two lines are perpendicular then their gradients have a product of -1.

Here is the graph of $y = x^2 - 5x$

At the point $(2, -6)$ the tangent has a gradient of -1.
The normal has a gradient of 1.
$1 \times -1 = -1$

At the point $(4, -4)$ the tangent has a gradient of 3.
The normal has a gradient of $-\frac{1}{3}$ $3 \times -\frac{1}{3} = -1$

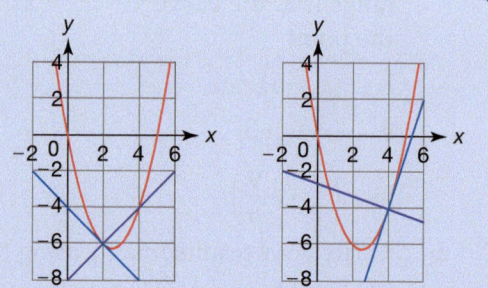

HOW TO

To find the equation of a normal to a curve at a given point

① Differentiate to find $\frac{dy}{dx}$ and find the gradient of the tangent at the given point.

② Find the gradient of the normal at that point.

③ Substitute into $y = mx + c$

④ Use the given point to establish the value of c.

EXAMPLE

Find the normal to the curve $y = x^2 + 2x - 3$ at

a $(1, 0)$ **b** the point where $x = -2$

a ① $\frac{dy}{dx} = 2x + 2$. $x = 1 \Rightarrow \frac{dy}{dx} = 4$.

② The gradient of the normal is $-\frac{1}{4}$

③ The equation of the normal is $y = -\frac{1}{4}x + c$

④ The point $(1, 0)$ lies on the line so $0 = -\frac{1}{4} + c \Rightarrow c = \frac{1}{4}$

The equation of the normal is therefore $y = -\frac{1}{4}x + \frac{1}{4}$

HINT: This function factorises so it can be sketched easily.
$y = (x + 3)(x - 1)$

b ① $x = -2 \Rightarrow \frac{dy}{dx} = -2$.

② The gradient of the normal is $\frac{1}{2}$

③ The equation of the normal is $y = \frac{1}{2}x + c$

④ When $x = -2$, $y = (-2)^2 + 2 \times -2 - 3 = -3$,
so $-3 = \frac{1}{2}x - 2 + c \Rightarrow c = -2$

The equation of the normal is therefore $y = \frac{1}{2}x - 2$

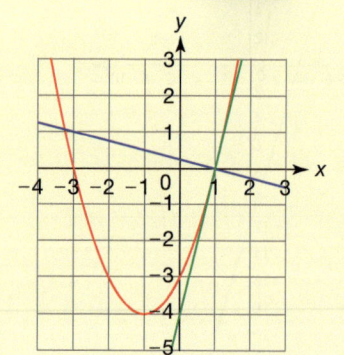

Algebra   Differentiation

Reasoning and problem solving

Exercise 22.3A

1. Find the equation of the normal to the curve $y = x^2 - 5x - 6$ at
   a. $(3, -12)$
   b. $(4, -10)$
   c. $(-1, 0)$
   d. $(5, -6)$
   e. the point where $x = 6$
   f. the point where $x = 0$

2. Find the equation of the normal to the curve $y = x^3 - 2x - 6$ at
   a. $(0, -6)$
   b. $(1, -7)$
   c. $(2, -2)$
   d. $(-1, -5)$
   e. the point where $x = -\frac{1}{2}$
   f*. the point where $y = -10$

3. Find the equation of the normal to the curve $y = x^4 - 3x^2$ at
   a. $(1, -2)$
   b. $(2, 4)$
   c. $\left(-\frac{1}{2}, -\frac{11}{16}\right)$
   d. $(-2, 4)$
   e. the point where $x = 0$
   f. the point where $x = 1.5$

4. Here is the graph of the function $y = x^4 - 3x^3 + 5x + 1$.

   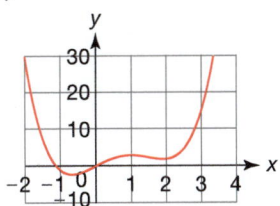

   a. Find the equation of the tangent to the curve at the point $(3, 16)$
   b. Find the equation of the normal to the curve when $x = 2$
   c. Find the equation of the normal to the curve when $x = 0$
   d. State the equation of the tangent to the curve $y = x^4 - 3x^3 + 5x - 2$ when $x = 3$
   e. State the equation of the normal to the curve $y = x^4 - 3x^3 + 5x + 5$ when $x = 0$

5. The graphs of $y = x^3 - x^2 + 3$ and $y = x^2 - 6$ are shown below.

   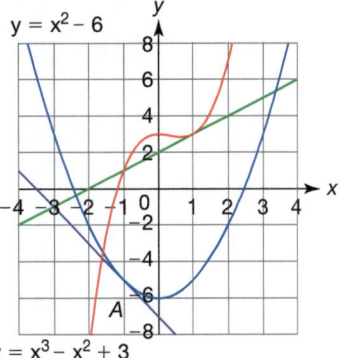

   The tangent to $y = x^3 - x^2 + 3$ at $x = 1$ is drawn.

   The tangent to $y = x^2 - 6$ at $x = -1$ is drawn.

   The two tangents intersect at a point labelled A.

   Find the coordinates of A.

6. The curve $y = x^3 + ax^2 + b$ has the tangent $x + y = 8$ at the point $(1, 7)$. Find the values of a and b.

7. The curve $y = x^3 + ax^2 + bx$ has the tangent $y + x + 2 = 0$ at the point $(-1, -1)$. Find the values of a and b.

*8. The graphs of $y = 2x^3 + 4x^2$ and $y = \frac{1}{10}x^2 + \frac{1}{10}x - \frac{6}{5}$ are shown below.

   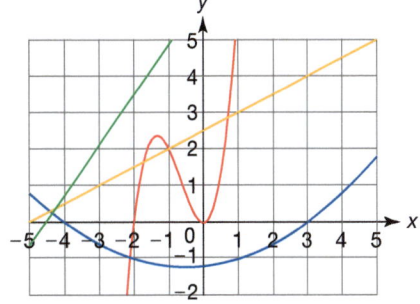

   The normal to $y = 2x^3 + 4x^2$ at $x = -1$ is drawn.

   The normal to $y = \frac{1}{10}x^2 + \frac{1}{10}x - \frac{6}{5}$ at $x = -4.5$ is drawn.

   Find the coordinates of the point of intersection of the two normals.

SKILLS

22.4 Turning points

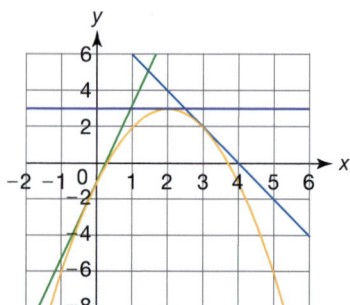

If the gradient of a curve is positive then the function is increasing at that point.

If the gradient of a curve is negative then the function is decreasing at that point.

A turning point occurs when the function is neither increasing nor decreasing.

Turning points are also known as stationary points.

- Points on a curve with a gradient of 0 are called **turning points**.
- A turning point can be a **maximum**, a **minimum**, or a **point of inflection**.
- Differentiation can be used when identifying the position of turning points.

Not all points of inflection are turning points. At a point of inflection $\frac{dy}{dx}$ may not be zero.

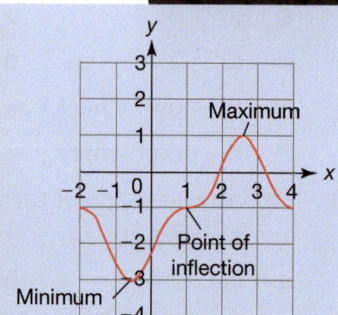

EXAMPLE

a Find the coordinates of the turning point of $y = x^2 + 3x$.
b State the nature of the turning point.
c Sketch the graph of the function.

a Differentiate to find $\frac{dy}{dx}$. $\frac{dy}{dx} = 2x + 3$

Create and solve the equation $\frac{dy}{dx} = 0$. $2x + 3 = 0 \Rightarrow -\frac{3}{2}$

Substitute into the original function to find y. $y = \left(-\frac{3}{2}\right)^2 + 3 \times -\frac{3}{2} = \frac{9}{4} - \frac{9}{2} = -\frac{9}{4}$

There is a turning point at $\left(-\frac{3}{2}, -\frac{9}{4}\right)$.

b The function is a positive quadratic. Therefore the turning point is a minimum.

c The function can be factorised. $y = x^2 + 3x \Rightarrow y = x(x + 3)$

Find the x-intercepts by solving for $y = 0$. $x(x + 3) = 0 \Rightarrow x = 0$ and $x = -3$

The graph is a smooth curve through $(0, 0)$ and $(-3, 0)$ with a minimum at $\left(-\frac{3}{2}, -\frac{9}{4}\right)$.

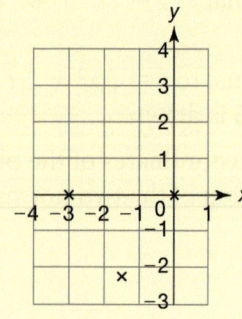

There is only one way to create this curve using the known facts. Sometimes more information will be needed to sketch a graph.

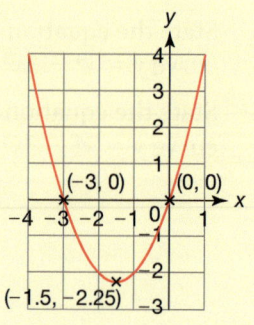

Algebra   Differentiation

Fluency

Exercise 22.4S

1. For each of these quadratic functions
   i. Find the coordinates of any turning point
   ii. State the nature of the turning points
   iii. Sketch a graph, clearly labelling all turning points and points of intersection with the axes.

   a. $y = x^2 - 4x$
   b. $y = x^2 - 3x - 4$
   c. $y = 7 + 3x - x^2$
   d. $y = 7 + 3x - 4x^2$

2. For each of these cubic functions
   i. Find the coordinates of any turning point
   ii. State the nature of the turning points
   iii. Sketch a graph, clearly labelling all turning points and points of intersection with the axes.

   a. $y = x^3 + 3x^2$
   b. $y = 2x^3 + 3x^2 + 4$
   c. $y = 8 - x^3$
   d. $y = x^3 + x^2 + 4$

3. Here is the graph of the function $y = x^3 - 9x$

   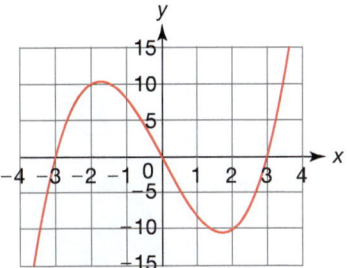

   a. Find the coordinates of the points of intersection with the x-axis.
   b. Find the exact coordinates of the two turning points.

4. The graph shows the function $y = 1.5x^4 - 10x^3 - 18x^2 + 6$.

   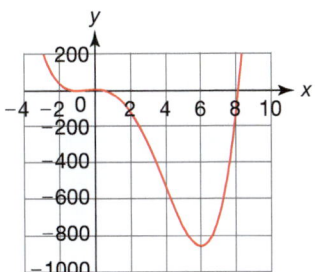

   a. Find the coordinates of the turning points.
   b. State the nature of each turning point.

5. Here is the graph of the function $y = x^4 + x^3 - 3x^2 - 5x + 1$

   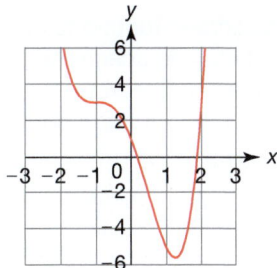

   a. Given that $\dfrac{dy}{dx} = (x + 1)^2 (ax - b)$, find the values of a and b.
   b. Find the coordinates of each turning point.
   c. State the nature of the turning points.

6. The graph shows the function $y = x^3 + 3x + 9$

   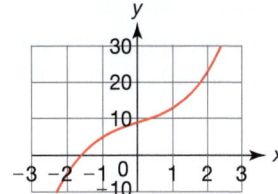

   a. Find $\dfrac{dy}{dx}$
   b. Explain how your result in **a** proves that there are no turning points for the function.

   NOTE: This is an example of a function with a point of inflection that is not a turning point.

7. Here is the graph of $y = x^3 + 3x^2 - 9x - 16$.

   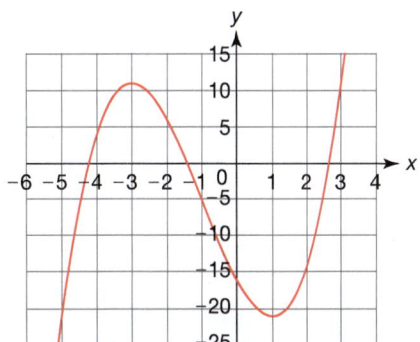

   a. Sketch this graph on a coordinate grid.
   b. On the same diagram, sketch the graph of $\dfrac{dy}{dx}$.
   c. Explain the links between the two sketches.

APPLICATIONS

22.4 Turning points

RECAP

- Many problems can be solved using the turning point of a function.
- In particular, quadratic functions can be used to model a wide range of real-life scenarios. For example, there are applications in engineering, business, science, manufacturing, statistics and sport.
- Optimisation problems often involve differentiating and finding turning points in these situations.

HOW TO

To solve an optimisation problem

1. Establish the relevant function, $y = f(x)$
2. Differentiate to find $\dfrac{dy}{dx}$
3. Solve the equation $\dfrac{dy}{dx} = 0$ to find where turning points occur.
4. Interpret the result in the context of the problem.

EXAMPLE

A ball is thrown straight up into the air at a velocity of 15 m/s. The height above ground (h metres) at time t seconds is given by the formula $h = 1.5 + 15t - 4.9t^2$

a At what time does the ball reach its maximum height?

b What is the maximum height reached by the ball?

a ① In this case h and t replace y and x: $h = 1.5 + 15t - 4.9t^2$

② $\dfrac{dh}{dt} = 15 - 9.8t$

③ $15 - 9.8t = 0 \Rightarrow t = \dfrac{15}{9.8} = 1.5306...$

④ 1.5306... seconds = 1.53 seconds to 3 significant figures

> This 'negative parabola' will have a turning point which is a maximum.

b Avoid using a rounded value as part of a multi-step calculation: Maximum height $= 1.5 + 15 \times 1.5306... - 4.9 \times 1.5306...^2 = 12.9795... = 13.0$ metres to 3 significant figures

EXAMPLE

Another ball is thrown straight up into the air at a velocity of 12 m/s from a starting height of 2 metres. The height at time t is given by the formula $h = 4 + 12t - 4.9t^2$. Which of the two balls is thrown higher?

① $h = 4 + 12t - 4.9t^2$

②③ $\dfrac{dh}{dt} = 12 - 9.8t = 0 \Rightarrow t = \dfrac{12}{9.8} = 1.2244...$

④ Maximum height $= 4 + 12 \times 1.2244... - 4.9 \times 1.2244...^2 = 11.3469... = 11.3$ metres to 3 significant figures. The first ball is thrown higher.

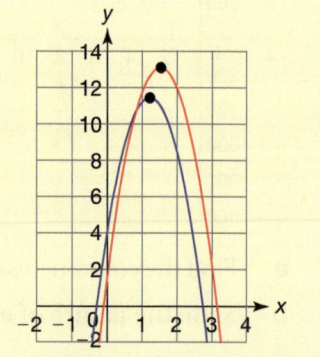

Algebra   Differentiation

Reasoning and problem solving

Exercise 22.4A

1. $y = 4x^3 + px + q$ has a turning point at $(-2, 73)$.

   a Work out the values of p and q.

   b Establish whether the turning point is a minimum or a maximum. Explain your reasoning.

   c Find the coordinates of the second turning point.

2. $y = x^3 + bx + c$ has a turning point at $\left(\frac{1}{2}, \frac{15}{8}\right)$.

   a Work out the values of b and c.

   b Find the coordinates of the second turning point.

   c State the nature of the turning points.

3. The curve $y = x^3 + ax^2 + bx + c$ passes through $(0, 4)$ and has a minimum point at $(2, 8)$.

   a Find the values of a, b and c.

   b Find the coordinates of the maximum point.

   c Sketch the curve.

4. The curve $y = -x^3 + ax^2 + bx + c$ has a y-intercept of 2 and a maximum point at $(3, 20)$.

   a Find the values of a, b and c.

   b Find the coordinates of the minimum point.

   c Sketch the curve.

5. Look at the graph of the function $y = 2x^2 + 6x + 8$

   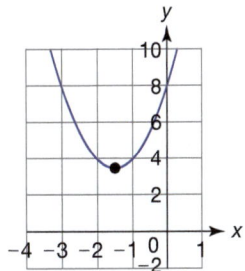

   a Find the coordinates of the minimum point by completing the square.

   b Verify your solution to **a** using differentiation.

6. In economics, a loss function can help model financial scenarios. An example of a quadratic loss function is
   $y = 5(120 - x)^2 - 17$

   a Find $\dfrac{dy}{dx}$

   b How could you find the minimum point of this function?

7. A company establishes a profit function $P = -3s^2 + 800s - 25\,000$ to show how profit (£P) varies against selling price per unit (£s).

   a Find $\dfrac{dp}{ds}$

   b Find the price per unit which generates maximum profit.

   c What is the profit at this price?

8. A tennis ball is thrown into the air from a height of 2 metres at a velocity of 8 m/s. The height at time t seconds is given by the formula
   $h = 2 + 8t - 4.9t^2$

   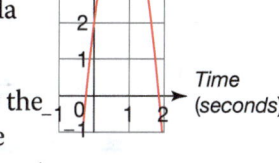

   a Calculate exactly the time at which the ball starts its descent.

   b Calculate exactly the maximum height of the tennis ball.

   c The ball is hit by a racquet 1.6 seconds after it is thrown. What distance has the ball travelled in this time?

*9 A piece of card measures 30 cm by 20 cm. You can make an open-top box by marking a square in each corner of the card. Cut one side of each square and fold another side of each square, to create four tabs for construction.

   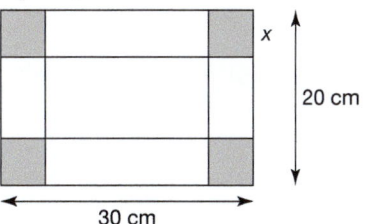

   Let x be the side length of the square. What value of x gives the greatest possible volume?

Summary

Checkout
You should now be able to...

	Test it Questions
✓ Find the rate of change of a function using a chord.	1 – 2
✓ Find the rate of change of a simple function using differentiation.	3 – 7
✓ Find the tangent to a curve at a point.	6, 8
✓ Use the derivative of a function to identify any turning points and their type.	9 – 10

Language | Meaning | Example

Language	Meaning	Example
Derivative	The rate of change of a function. For the curve $y = f(x)$ the gradient, written $\frac{dy}{dx}$ or y', is the gradient of the tangent line at the point $(x, f(x))$.	The derivative of x^n is nx^{n-1}
Turning point	A point at which a function is (instantaneously) not changing: the derivative is zero.	A ball thrown in the air has a turning point at the top of its path – it's height as a function of time. This turning point is a maximum.
Maximum / Minimum	A turning point at which the function on both sides is lower / higher than at the turning point.	
Point of inflection	A turning point at which the function is lower on one side and higher on the other side of the turning point.	The function $y = x^3$ has a point of inflexion at the origin (0, 0).
Tangent	A straight line which (locally) touches a curve at a single point.	If the point (a, b) is a maximum or minimum of a function then the line $y = b$ is a tangent at that point.
Chord	A straight line joining two points on a curve.	As the two points get closer and closer the gradient of a chord gets closer and closer to the gradient of the tangent line at the point where they meet.

Algebra   Differentiation

Review

1. Estimate the gradient of the curve $y = x^3 - 2x$ at the point (2, 4) by finding the gradient of the chord joining
   a. (1, −1) to (3, 21)
   b. (1.5, 0.375) to (2.5, 10.625)

2. Work out the value of the derivative of $y = x^3 + 9$ when
   a. $x = 1$
   b. $x = 2$

3. For each of these functions, work out the value of y' when $x = -1$
   a. $y = x^3 + x^2 + 9x$
   b. $y = 3x^2 + 6x - 4$

4. For each of these functions, work out the value of $\dfrac{dy}{dx}$ when $x = -2$
   a. $y = 3x^3 - x^2$
   b. $y = 5x^3 - 7x + 1$

5. Work out $\dfrac{dy}{dx}$ for each of these functions
   a. $3x(x^2 + 1)$
   b. $(x - 3)(x + 9)$
   c. $\dfrac{3x^2 - 6x^4}{3x}$
   d. $\dfrac{x^2 - 4x + 3}{x - 1}$

6. Find the equation of the tangent to the curve $y = x^3 + 3x^2 - 2$ at the point
   a. (1, 2)
   b. (−1, 0)

7. The distance, in metres, of a car from a point after t seconds is given by the formula $y = 2t^2 + 3t$.
   a. Find the gradient function $\dfrac{dy}{dt}$.
   b. What is the instantaneous rate of change in the distance after
      i. 1 second
      ii. 5 seconds?
   c. What do these values describe?

8. For each of the following functions
   i. Find the coordinates of any turning point
   ii. State the nature of the turning points.
   a. $y = 3x^2 - 5$
   b. $y = x^2 - 10x + 24$
   c. $y = 10 - 3x - x^2$
   d. $y = x^3 - 3x^2 - 24x$

9. A ball is projected into the air at a velocity of 16 m/s. The height, in metres, at time t seconds is given by the formula $h = 16t - 4.9t^2$
   a. At what time does the ball reach its maximum height and start dropping?
   b. Calculate the maximum height of the ball.
   c. How fast is the ball travelling after 1.5 seconds?

   Give your answers to 3 significant figures where appropriate.

What next?

Score		
0 – 3		Your knowledge of this topic is still developing. To improve, look at MyiMaths.
4 – 7		You are gaining a secure knowledge of this topic. To improve your fluency look at InvisiPens: 23Sa
8 – 10		You have mastered these skills. Well done, you are ready to progress! To improve your problem-solving skills look at InvisiPens: 23Aa

Revision 4

1 Three children are standing at points *A*, *B* and *C* as shown and throwing a ball to each other around the triangle. *B* has the ball and throws it to *C*. How far does she throw it? [3]

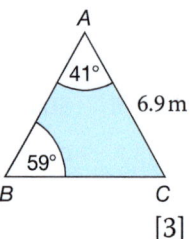

2 A man walks 7.6 km through the jungle on a bearing of 200°. How far south is he from his starting point? [3]

3 The sun casts a shadow on Dan 230 cm long. The shadow makes an angle with the ground of 35°. How tall is Dan, to the nearest cm? [2]

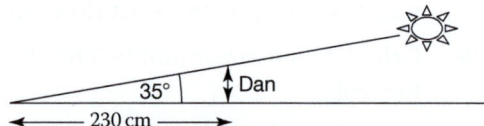

4 A plane coming in to land has 10 miles to go. It is currently 1.5 miles above the ground. Calculate its angle of descent, θ. [2]

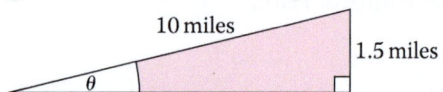

5 *D* is the midpoint of *AB* in the triangle *ABC*.

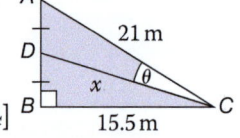

i Calculate the length, *x*, of *CD*. [4]

ii Find the value of θ. [3]

6 A flagpole *OP* stands at one corner of a rectangular parade ground *ABCO*. The dimensions of the parade ground are 150 m by 100 m where *OC* and *AB* are the shorter sides and the distance of *P* from the corner *C* is 125 m.

a Find the height of the flagpole. [3]
b Find angle *PCO*. [2]
c Find lengths *BO*, and *PB*. [4]
d Find angle *PBO*. [2]

7 *M* and *N* are the midpoints of *PQ* and *PR*. Helen says that the vectors $\overrightarrow{RQ}$ and $\overrightarrow{NM}$ are related by $\overrightarrow{RQ} = -\overrightarrow{NM}$. Express the vectors $\overrightarrow{RQ}$ and $\overrightarrow{NM}$ in terms of $\mathbf{a} = \overrightarrow{PQ}$ and $\mathbf{b} = \overrightarrow{PN}$ and give the correct relationship between $\overrightarrow{RQ}$ and $\overrightarrow{NM}$. [3]

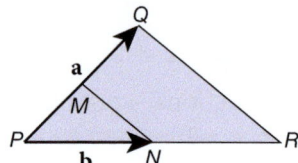

8 Abdullah writes a sequence of numbers that begins 2 6 12 20 ....

a He says that the *n*th term of this sequence is $n^2 + n$. Is he correct? Show your working. [2]

b Work out the 50th and 100th terms. [2]

A second sequence is formed of the differences between each term and the next in the first sequence.

c Work out the *n*th term of this new sequence. Hence find the 50th and 100th terms. [4]

9 The diagram shows a pattern of black and white triangles.

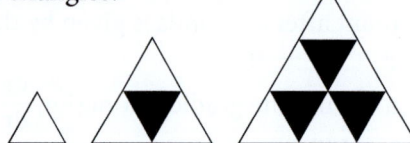

a Copy and complete the table. [5]

Rows (*r*)	1	2	3	4	5
White triangles △	1	3			
Black triangles ▼	0	1			
Total number of triangles	1	4			

b The formula for the number of white triangles is $n = \dfrac{r(r+1)}{2}$.

Find the number of white triangles in a pattern with

i 50 rows [1]

ii 100 row patterns. [1]

9 **c** Find the formula for the total number of triangles in a pattern with r rows. [1]

d Use this formula and the one given in part **c** to derive a formula for the number of black squares in any row. Give your answer in its simplest form. [3]

e Test your formula by working out the number of black triangles in row 2. [1]

f Use your formula to find the number of black triangles in the 60th pattern. [1]

10 **a** Describe the transformation defined by matrix $\mathbf{M} = \begin{pmatrix} 0 & -1 \\ 1 & 0 \end{pmatrix}$ [1]

b Find the transformation matrix, $\mathbf{N}$, that reflects shapes in the x axis. [1]

c **i** Calculate the matrix that transforms shapes by $\mathbf{M}$ followed by $\mathbf{N}$.

ii Describe the single transformation that is equivalent to $\mathbf{M}$ followed by $\mathbf{N}$. [2]

11 **a** A recipe for 6 people uses 4 eggs. How many eggs do you need to make pancakes for 9 people? [2]

b The supermarkets 'Liddi' and 'Addle' both sell teacakes. 'Liddi' sells a pack of 5 for £1.35 and 'Addle' sells a pack of 4 for £1.10. Which supermarket gives better value for money? [2]

12 Philadelphia to Ventnor is 72 miles.

a Richard drives from Philadelphia to Ventnor in 108 minutes. What is his average speed? Give your answer in mph. [2]

b Jamahl drives from Ventnor to Philadelphia at an average speed of 50 mph. How long, to the nearest minute, did his journey take? [2]

c Jamahl drove from Ventnor to Philadelphia at an average speed 4 mph faster than his return journey. How many minutes did he save on the outward journey? [3]

d Richard and Jamahl both leave on their outward journeys at 10:00. They meet in a village café for coffee. How far from Ventnor is the village and at what time do they meet? [4]

13 **a** An object has a mass of 1.26 kg and volume of 180 cm³. What is its density? [2]

b A cylindrical metal rod with radius 3.5 cm and length 12.5 cm has a density of 11.4 g/cm³. What is the mass of the rod? [3]

c A silver bar has a mass of 30 g and a density of 10.5 g/cm³. What is its volume? [2]

14 **a** If $a = kb^2$, where k is a constant, we say that a is proportional to b^2 and write $a \propto b^2$. Write the following formulae in the same way.

i $m = \dfrac{45}{d^3}$ [1]

ii $T = \dfrac{2\pi\sqrt{l}}{10}$ [1]

iii $S = 4(r^2 - 3)5$ [1]

b $d \propto \sqrt{h}$. When $d = 80$, $h = 64$. Calculate

i d when $h = 56.25$ [2]

ii h when $d = 22$. [1]

15 Boyle's law states that, under certain conditions, the pressure P exerted by a particular mass of gas is inversely proportional to the volume V it occupies. A volume of 150 cm³ exerts a pressure of 6×10^4 Nm^{-2}. The volume is reduced to 80 cm³. What is the new pressure? [3]

16 Ally plays tennis, netball or squash every morning. The probability of choosing each sport is:

Tennis 0.28   Netball 0.35   Squash P.

a Show that $P = 0.37$ [2]

b Draw a tree diagram showing her possible choices over two consecutive days. [3]

c Find the probability that Ally

i plays tennis on day 1 and squash on day 2 [2]

ii plays netball at least once [3]

iii doesn't play tennis on either day. [4]

Formulae

Make sure that you know these formulae.

Quadratic formula

The solutions of $ax^2 + bx + c = 0$ $a \neq 0$ are $x = \dfrac{-b \pm \sqrt{b^2 - 4ac}}{2a}$

Circles

Circumference of a circle $= 2\pi r = \pi d$

Area of a circle $= \pi r^2$

Pythagoras' theorem

In any right-angled triangle
$$a^2 + b^2 = c^2$$

Trigonometry formulae

In any right-angled triangle $\quad \sin A = \dfrac{a}{c} \quad \cos A = \dfrac{b}{c} \quad \tan A = \dfrac{a}{b}$

In *any* triangle

sine rule $\quad \dfrac{a}{\sin A} = \dfrac{b}{\sin B} = \dfrac{c}{\sin C} \quad$ cosine rule $\quad a^2 = b^2 + c^2 - 2bc \cos A \quad$ Area $= \tfrac{1}{2}ab \sin C$

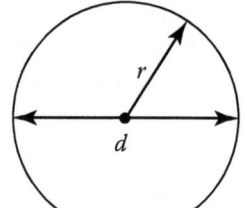
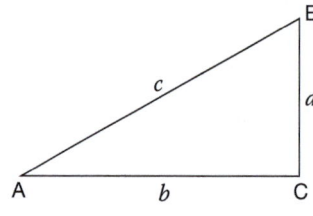
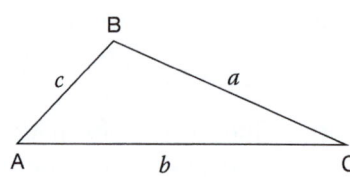

Make sure that you know and can derive these formulae.

Perimeter, area, surface area and volume formulae

Area of a trapezium $= \tfrac{1}{2}(a + b)h$

Volume of a prism $=$ area of cross section $\times$ length

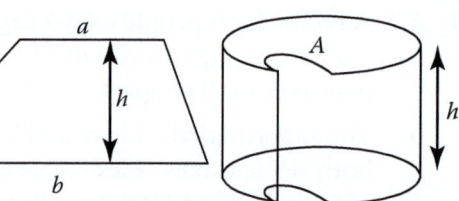

Compound interest

If P is the principal amount, r is the interest rate over a given period and n is number of times that the interest is compounded then

Total accrued $= P\left(1 + \dfrac{r}{100}\right)^n$

Probability

If P(A) is the probability of outcome A and P(B) the probability of outcome B then

P(A or B) = P(A) + P(B) − P(A and B)   P(A and B) = P(A given B)P(B)

Make sure that you can use these formulae if you are given them.

Perimeter, area, surface area and volume formulae

Curved surface area of a cone $= \pi r l$

Surface area of a sphere $= 4\pi r^2$

Volume of a sphere $= \tfrac{4}{3}\pi r^3$

Volume of a cone $= \tfrac{1}{3}\pi r^2 h$

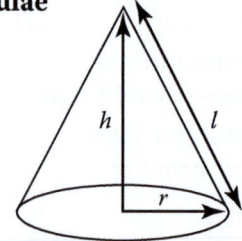
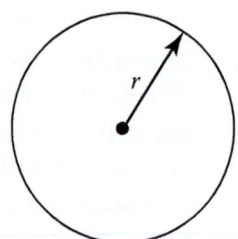

Kinematics formulae

If a is constant acceleration, u is initial velocity, v is final velocity, s is displacement from the position when $t = 0$ and t is time taken then

$v = u + at \qquad s = ut + \tfrac{1}{2}at^2 \qquad v^2 = u^2 + 2as$

Key phrases and terms

Use our handy grid to understand what you need to do with a maths question.

Circle	Draw a circle around the correct answer from a list.
Comment on	Give a judgement on a result. This could highlight any assumptions or limitations or say whether a result is sensible.
Complete	Fill in any missing information in a table or diagram.
Construct	Draw a shape accurately.
Construct, using ruler and compasses	Compasses must be used to create angles. Do *not* erase construction lines.
Criticise	Make negative judgement. This could highlight any assumptions or limitations or say why a result is not sensible.
Deduce	Use the information supplied or the preceding result(s) to answer the question.
Describe	Usually used for describing a graph- one mark per descriptive sentence. When describing transformations give all the relevant information.
Diagram not accurately drawn	Calculate any angles or sides, do *not* measure them on the diagram.
Draw	Accurately plot a straight line, bar chart or transformations.
Draw and label	Draw and mark on values.
Estimate	Simplify a calculation, by rounding, to find an approximate value.
Evaluate	Calculate the value of a numerical or algebraic expression.
Expand	Multiply out brackets.
Factorise fully	Put in brackets with the highest common factor outside the brackets.
Give a reason	Write an explanation of your argument, including the information you use and your reasons or rules used.
Give the exact value	Do not use rounding or approximations in your calculations. You should give your answers as fractions, surds and multiples of π.
Give your answer in terms of π.	Do not use a numerical approximation to π, instead treat it as an algebraic variable (letter).
Give your answer in its simplest form/ simplify	Collect all like terms. Any fractions or ratios should be in lowest terms.
Give your answer to an appropriate degree of accuracy	For example, if the numbers in the question are given to 2 decimal place, give your answer to 2 decimal place.
Measure	Use a ruler or protractor to accurately measure lengths or angles.
Prove that	Formally obtain a required result using a sequence of logical steps. Each line of working should follow from the previous line and any general results used should be quoted.
Rearrange	Change the subject of a formula.
Shade	Use hatching to indicate an area on a diagram.
Show	Present information by drawing the required diagram.
Show that	Obtain a required result showing each stage of your working.
Show working	Usually asked for to support a decision.
Sketch	Represent using a diagram or graph. This should show the general shape and any important features, such as the position of intercepts, using labels as necessary. It does not require an accurate drawing.
Solve	Find an answer using algebra or arithmetic. This often means find the value of *x* in an equation.
State	Write one sentence answering the question.
Tick a box	Choose the correct answer from a list.
Work out	Work out the answer and show your working.
Write	Write your answer; written working out is not usually required.
Use an approximation	Estimate.
Use a line of best fit	Draw a line of best fit and use it.
Use	Use supplied information or the preceding result(s) to answer the question.
You must show your working.	Marks will be lost for not writing down how you found the answer to the question.

Answers

Chapter 1

Check in

1. **a** Four thousand **b** Four hundred
   c Four tenths **d** Four thousandths
2. **a** [number line from −5 to 5 with arrows at −2.5, −3, 0, +4]
   b −3, −2.4, −1.8, 0, +1.5, +5, +6

1.1S

1. **a** One thousand, three hundred and seven
   b Twenty-nine thousand and six
   c Three-hundred thousand
   d Six-hundred-and-five thousand and thirty
2. **a** 8043 **b** 70 000 000 **c** 200 051 **d** 2010
3. **a** 0.1, 0.3, 1, 1.3, 2, 3.1 **b** 6.07, 7.06, 27.6, 77.2, 607
   c 7.03, 7.08, 7.3, 7.38, 7.8, 7.83
   d 2.18, 2.4, 4.18, 4.2, 8.24, 8.4
4. **a** 8000.6, 6008, 6000.8, 862.6, 682.8
   b 97.4, 94.7, 79.4, 74.9, 49.7, 47.9
   c 18.7, 18.16, 17.6, 17.16, 16.7, 16.18
   d 13.2, 13.145, 2.5, 2.38, 1.1, 1.06
5. **a** 250 > 205 **b** 1.377 < 1.73 **c** $\frac{3}{8}$ < 0.4
   d −17 > −71 **e** −0.09 < −0.089 **f** $\frac{5}{8}$ = 0.625
6. **a** 4.52, 5 tenths > 0 tenths. **b** 5.5, 5 tenths > 0 tenths.
   c 16.8, 8 tenths > 7 tenths. **d** 16.8, 8 tenths > 1 tenth.
7. **a** F **b** T **c** T **d** T
   e F **f** T
8. **a** i 3050 ii 3000 iii 3000
   b i 1760 ii 1800 iii 2000
   c i 290 ii 300 iii 0
   d i 50 ii 0 iii 0
   e i 40 ii 0 iii 0
   f i 740 ii 700 iii 1000
   g i 2960 ii 3000 iii 3000
   h i 1450 ii 1500 iii 1000
   i i 20 ii 0 iii 0
   j i 24 600 ii 24 600 iii 25 000
   k i 16 340 ii 16 300 iii 16 000
   l i 167 730 ii 167 700 iii 168 000
9. **a** i 39.1 ii 39.11 **b** i 7.1 ii 7.07
   c i 5.9 ii 5.92 **d** i 512.7 ii 512.72
   e i 4.3 ii 4.26 **f** i 12.0 ii 12.01
   g i 0.8 ii 0.83 **h** i 26.9 ii 26.88
   i i 0.1 ii 0.08
10. **a** i 0.1 ii 0.07 iii 0.070
    b i 15.9 ii 15.92 iii 15.918
    c i 128.0 ii 128.00 iii 127.998
    d i 887.2 ii 887.17 iii 887.172
    e i 55.1 ii 55.14 iii 55.145
    f i 0.0 ii 0.01 iii 0.007
11. **a** 200 **b** 2000 **c** 5 **d** 10
    e 0.0005 **f** 100 000
12. **a** 480 **b** 1200 **c** 490 **d** 14 000
    e 530 **f** 15 000 **g** 0.36 **h** 0.42
    i 0.057 **j** 0.0047 **k** 1.4 **l** 0.000 004 2
13. **a** 9.73 **b** 0.36 **c** 147.5 **d** 29
    e 0.53 **f** 4.20 **g** 1245.400 **h** 0.004
    i 270 **j** 460.0
14. **a** 167 **b** 248 **c** 7.16 **d** 10.95
    e 2430 **f** 2813
15. **a** 21.4 **b** 6.73 **c** 410.6 **d** 20.07
    e 0.6025 **f** 8.6
16. **a** 1306 **b** 2.085 **c** 1085 **d** 2.487
    e 0.0008 **f** 6.19 **g** 0.045 13 **h** 0.0045
17. **a** 1760 **b** 93

1.1A

1. $3.60
2. **a** 5 mm
   b i 5.05 cm ii 3.55 cm iii 12.3 cm
   c i 5.1 cm ii 3.6 cm iii 12.35 cm
3. **a** A 26 000, B 3000, C 26 000, D 6000
   b A 26 000, B 2700, C 26 100, D 5700
4. **a** No. Difference = 0.24 s < 0.25 s
   b 4th place.
   c Adams and Bolyai.
5. **a** 86 × 57 **b** 44 × 29 **c** 73 × 55 **d** 121 × 90
   e 2%
*6. $\sqrt{3}$ = 1.732 051... 25^3 = 15 625, 1.73 × 15 600 = 26 988
   Ian has rounded values at each stage, not just at the end.
7. 3.118 > 3.112, 3.118 ⩾ 3.112, 4.5 ⩽ $\frac{9}{2}$, 4.5 = $\frac{9}{2}$,
   4.5 ⩾ $\frac{9}{2}$, 3.004 > 2.9961, 3.004 ⩾ 2.9961
8. 1.45 ⩽ x < 1.495 or 1.505 ⩽ x < 1.55
9. Yes. The upper bounds (in kg) are 85, 95.5, 96.55, 72.5 so the upper bound for the total weight is 349.55 kg < 350 kg.

1.2S

1. **a** 22 **b** −12 **c** −2 **d** −14
   e 12 **f** 12 **g** 19 **h** −15
   i 11 **j** −61 **k** 344 **l** 49
2. **a** −10.8 **b** 0.7 **c** 13.5 **d** −0.7
   e −38.2 **f** 112.7
3. **a** 10 **b** 10.1 **c** 10.3 **d** 11.1
   e 16.4 **f** 7.5
4. **a** 7.77 **b** 5.25 **c** 3.6 **d** 3.9
   e 2.13 **f** 13.04
5. **a** 2.5 **b** 2.1 **c** 2.2 **d** 3.9
   e 1.4 **f** 0.7
6. **a** 0.74 **b** 1.05 **c** 1.51 **d** 7.14
   e 1.08 **f** 0.64
7. **a** 35.96 **b** 34.61 **c** 96.35 **d** 104.42
   e 14.931 **f** 10.401 **g** 26.211 **h** 63.836
8. **a** 10.68 **b** 0.43 **c** 4.097 **d** 0.76
   e 2.12 **f** 4.19 **g** 2.04 **h** 5.7
9. **a** 2.63 **b** 0.45 **c** 24.32 **d** 14.72
10. **a** 3.97 **b** 0.095 **c** 12.44 **d** 58.34
11. **a** 2.6 **b** 7.86 **c** 5.9 **d** 92.54
    e 0.97 **f** 24.27
12. **a** 13.896 **b** 19.45 **c** 359.79 **d** 7.683
    e 0.326 **f** 11.42
13. See question 12.
14. **a** 4.1 **b** 40.2 **c** 11.288 **d** 5.968
    e 0.892 **f** 0.469
15. **a** 99.88 **b** 28.986 **c** 14.54 **d** 8.6757
    e 457.75 **f** 93.79 **g** 19.802 **h** −42.0705
16. **a** When borrowing from the column to the left, instead of subtracting 1 from this number add 1 to the number below it. This will have the same effect when the numbers are subtracted.
    b i 6 7 .¹3 6 ¹2
         − 2 ⁴3̶ . 4 ⁶5̶ 4
         ─────────────
           4 3 . 9 0 8
       ii 6 ¹0 .¹0 ¹0 9
         − ²1̶ 2 .⁴3̶ 4 7
         ─────────────
           4 7 . 6 6 2

1.2A

1. 73
2. **a** £150.01 **b** £120.01 **c** £55 **d** £24.50
   e £79

486

3 a 'Ants', 'Cows', and 'Bees'.   b 'Ants and 'Cows'.
   c Two copies of 'Cows' and one copy of 'Bees'.
4 119.43; 57.97; 26.09 and 35.37; 19.15 and 3.45
5 a 11 688   b 18 340
   c 292 (291 would only *equal* Lockton)
6 a, b, c

Name	Score	Ahead of next person by
Bryony	12 653	5303
Callum	7350	350
Edward	7000	477
Asha	6523	5559
Dora	964	
TOTAL	34 490	

   d 11 690 points (11 689 to be in equal first place).
7 a 342 and 1026   b 724 and 1448
   c 342 and 432, 522 and 1026.
8 a 3249 + 3073 = 6322   b 36.58 − 27.79 = 8.79

1.3S

1 a −25   b −32   c −72   d −20
   e 30   f 49   g 16   h −20
   i −18   j 26   k −42   l −48
2 a −2   b −5   c 5   d 4
   e −22   f −1   g 40   h 4
   i 5   j −17   k −3   l 27
3 a 98   b 152   c 273   d 323
   e 308   f 609   g 2499   h 296
   i 1136   j 3599
4 a 56   b 259   c 716   d 53
   e 22   f 430   g 300   h 56
   i 51   j 123
5 a 9.3   b 700   c 12.2   d 10.6
   e 463.33   f 130   g 570   h 0.416
   i 0.000125   j 0.0000121
6 See questions 3−5.
7 a 24.91   b 4.284   c 105.84   d 130.8985
   e 42.9442   f 369.5328   g 22.1778   h 0.035 978 4
8 a 3.87   b 0.775   c 0.916   d 7.53
   e 18.13   f 3.45   g 4.15   h 7.74
   i 4.08   j 2.35
9 a 5.26   b 28.88   c 1384.29   d 175.56
   e 28.65   f 111.51   g 0.04   h 861.30
10 See questions 7−9.
11 a 28.81   b 28.81   c 4.3   d 0.067
    e 10   f 100
12 a 196   b 4   c 18   d 289
13 a 121   b 39   c 18   d 5.29 (3 sf)
    e 10   f 18   g 8   h 1533

1.3A

1 a 136   b 56
2 a £24   b £312
   c Other months have 30 or 31 days (more than 4 weeks).
3 a 24 bags   b $113
   c 4 separate bags cost 50 cents more than 5 bags.
4 €1087.30
5 a $(8 − 4) \times (3 − 1) = 8$
   b $16 \div (2 \times 3 − 2) = 4$
   c $(5 \times 3 − 3 \times 9) \div (3 − 6) = 4$
   d $((4 − 2) \times 2)^2 \div 2^2 = 4$
   e $2 + 3^2 \times (4 + 3) = 65$
6 a 7 small, 8 medium, 8 large, 8 rectangles.
   b £503.05
   c £632.05
*7 $32.14
8 yes

Review

1 a 24.3 < 24.5   b −0.5 > −0.9
   c 0.5 > 0.06   d 1.456 < 1.46
2 a False, 0.85 ÷ 10 = 85 ÷ 1000.   b True
   c True   d False, $3^2 > 3 \times 2$.   e True
3 a 45.9   b 0.08   c 0.085   d 2000
   e 78 000   f 3.5
4 a 8300   b 25.9   c 310   d 5
   e 76.4   f 5.49   g 0.085   h 0.008
5 a 3842   b 192   c 309.32   d 53.053
   e 7726   f 34.6   g 917.4   h 9.91
6 a 540   b 63   c 0.56
   d 500   e 51.522   f 2.653 35
   g 1.9   h 290
7 a −11   b −241   c 12.6   d −0.43
   e −42   f 5.5   g −17   h 5
8 a 1   b 40   c 112   d 112
   e 10.2   f 30
9 a 56.8112   b −1

Assessment 1

1 a 57.6 < 303; 0.8, 1.9, 3.3, 44, 57.6, 303
   b −2.19 < −0.07; −2.19, −0.07, 30, 43.56, 188.0, 194.7
2 Yes, 0.42, 3, 4.236, 51.6, 4200, 216 000.
3 a No, LB = 271.75 cm < 271.78 cm.
   b No, Yao Defen UB = 233.345 cm > 233.341 cm.
4 a Nadya or Jane?
   b No, Dave 40, Nadya 50 (1 sf).
   c Dave 45, Nadya 53
5 a Abena 13 000, Edward 8100 (2 sf)
   b 15 395.8 km
   c The estimate would be smaller, the approximation for Abena's *x* value would decrease more than Edward's value.
6 2.125 km
7 a 25 sweets   b 25 000 (nearest 1000)
8 a 30, −20   b 7
   c Students' answers, for example, 2 correct, 8 wrong; 1 correct, 4 wrong, 5 unanswered; 10 unanswered.
9 Sue 19.9, Clive 27.8, Kivi 61.7, Henry 1.83
10 Yes
11 a −2.2, −4.5, −7.8   b 7.49, 3.67, −3.82
12

−2	3	−8	5
−7	4	−1	2
7	−6	1	−4
0	−3	6	−5

13 Students' answer, for example, Lana (more accurate, not enough money), Gardener (less accurate, enough money).
14 a 1
    b 0, 1 ÷ 0 is undefined.
    c Students' answers < 1, for example, $\frac{1}{2} < 2$.
15 a $(3 + 4) \times 5 = 35$
    b $(4 \div 3) + 5 = 6\frac{1}{3}$
    c $5(2^3 + 0.4) \div (4 − 3 \times −1) = 6$

Chapter 2

Check in

1 a 45   b 52   c −26   d 196
   e 7   f 13   g −50   h 30
2 a 9 + 6 = 15   b 8 + 7 = 15
   c 5 × 3 = 15   d 27 − 12 = 15
3 a 3   b 4   c 10   d 6
   e 4   f 25   g 33   h 7

2.1S

1. $5x + 8y$
2. $2b - 3a$
3. No – the correct answer is $7d + 7e$
4.
   a. $2a + 2b$ b. $6t + 3s$
   c. $d + 6e$ d. $6y + 8w$
   e. $5p + 4f$ f. $e + 2f + 3g + 4h$
   g. $7k + 4 + 2d$ h. $10y + 16$
   i. $5p + 2q + 4$ j. $8b + 5 - 2a + 6ab$
5.
   a. $4d$ b. $3q$ c. $7a$ d. $4y$
   e. t f. 0 g. $3k + 4$
   h. $6y + 6$ i. $7 - 4p$ j. $8b + 4d - 12 + 4bd$
6. Eric is partly correct – the correct answer is $3k$. Ian is incorrect.
7. No – the correct answer is $3a^2$
8.
   a. $2d^2$ b. $4b^2$ c. $9a^3$ d. $3y^2 + 4y$
   e. $3t^3$ f. $2m^2 + 5m$ g. $3k^3 + 4$ h. $8y^2 - 2y + 6$
9.
   a. $2ad^2$ b. $13ab^2$ c. $9xy^3$
   d. $3xy^2 + 4xy$ e. $5s^3t - 2st^3$ f. $6m^2n + mn^2$
   g. $5gh^2 + 4g^2h - 2gh$ h. $8y^2 - 5y - 2x + 11x^2$
10. Multiple answers possible, for example $2(a^2 + b^2) + b^2$, $5a^2 + b^2 - 3a^2 + 2b^2$. No
11.
    a. $3ab$ b. $3ab$ c. $3ab$ d. $6b$
    e. $20a$ f. $4pq$ g. $2\sqrt{3}p$ h. $4a^2$
    i. a^2b j. a^2b^2 k. $2ab^2$ l. $3\sqrt{2}a^2b$
12.
    a. $\frac{3}{a}$ b. $\frac{a}{3}$ c. $\frac{5}{p}$ d. $\frac{2m}{5}$
    e. $\frac{y}{\sqrt{3}}$ f. $\frac{5b}{\sqrt{3}}$
13. a. $\frac{4b}{a}$ b. $4\sqrt{3}b$
14. a. 24 b. 32 c. 144
    d. 432 e. 4 f. 2
15. a. -4 b. 4 c. 4
    d. 12 e. $\frac{-1}{5}$ or -0.2 f. -5

*16. $7b - 26$

17. $4p + 7q$, $4p + 7q$, $4p + 2q$, $6mn$, $10mn$, $6mn$, $2d$, 2, $2d$, $2n - 8$, $2n$, $2n - 8$

2.1A

1. a i. $8(p + 2)$,   ii. $32p$ b. $3x$ by $2y$
2. $20p + 22q + 10$
3. $21st$ cm^2
4. a i. £4   ii. £7.25   iii. £3.10   b. $10p$
5. £175
6. Abdul
7. Paul, $2x^2 = 2 \times 36 = 72$
8.
   a. $3m + 10$; $2m + 5$, $m + 5$; $2m$, 5, m; $m - 5$, $m + 5$, $-m$, $2m$
   b. $4a + 7b$; $a - 2b$, $3a + 9b$; $a + 2b$, $-4b$, $3a + 13b$; $3b$, $a - b$, $-a - 3b$, $4a + 16b$
9.
   a. $9a = 4a + 3a - -2a$
   b. $(5b + c) = (3b + 7c) + (3b - 2c) - (b + 4c)$
   c. $16d + 3e = 3d \times 4 + 4d + 8e \div 2 - e$
10. $x + 2y - 2$, $y - 4$, $2x + 6$; $2x - y + 8$, $x + y$, $3y - 8$; $2y - 6$, $2x + y + 4$, $x + 2$

2.2S

1. 2^6
2. a. 16 b. $5^{11} = 48\,828\,125$
3. a. 125 b. 1024 c. 1
   d. 36 e. -1 f. $\frac{1}{4}$
4. a. 3^6 b. 2^9 c. 5^7 d. 6^7
   e. 7^9 f. 11^{10} g. 5^4 h. 3^6
   i. 8^5 j. 2^7 k. 6^{-4} l. 2^{-1}
5. a. 2^6 b. 5^{24} c. 3^{21} d. 8^{16}
   e. 7^{-8} f. 2^{-6}
6. No, the answer is a^7, Firas should add the indices.
7. a. a^6 b. y^{10} c. b^4 d. p^{-7}
   e. h^{12} f. s^3t^3 g. $x^{15}y^5$ h. x^5y^{-1}
   i. p^4q^7 j. p^2q^2r
8. No, the answer is $8y^7$. Rula should multiply the coefficients.
9. a. $3x^7$ b. $5y^7$ c. $12b^8$ d. $2p^{11}$
   e. $30h^{11}$ f. $4\sqrt{3}s^3t^4$
10. No – the answer is $4p^8$. Diego should divide the coefficients.
11. a. $2y^4$ b. $2a^6$ c. $5k^4$ d. $3p^5$
    e. $0.5x^6$ f. $\sqrt{3}y^4$
12. a. a^6 b. y^{12} c. k^{15} d. p^{56}
    e. a^8 f. q^8 g. $16a^6$ h. $729y^{12}$
    i. $32k^{15}$ j. $4p^{28}$ k. $3\sqrt{3}a^9$ l. $25\sqrt{5}a^{20}$
13. a. a^{-8} b. b^6 c. $27c^{-9}$ d. $60d^{-1}$
    e. $3e^6$ f. $2f^{-9}$ g. g^{40} h. 2
14. a. 1 b. 5 c. $\frac{1}{16}p^{-2}$ d. $\frac{1}{16}p^{-4}$
    e. $5\sqrt{a}$ f. $2\sqrt[3]{d}$

*15. a. $16\sqrt{2}x^{-\frac{9}{4}}$ b. $\left(\frac{1}{16}\right)^{\frac{4}{5}}y^{\frac{16}{15}}$

16. a. $x = -5$ b. $x = 3$ c. $x = -2$ d. $x = -2$

2.2A

1. a. $52x^3y^3$ b. $10a^4b^2$
2. a. $40p^3q^4$ b. $21s^2t^{-1}$ c. $2v^2(3u^2 + 4u^{-1})$
3. $52a^2b^6$
4. $49a^6b^{-2}$
5. $30w^{-9}$
6. a. $a^6b^2 = (a^3)^2 \times b^2$ b. $c^3d(c^7d^3 \div c^5d^3)^{\frac{1}{2}} = c^4d$
   c. $e^3f^2 = e^4f^6 \times e^{-2}f^6 \div e^{-1}f^2$
7. $-(xy^2)^{101}$
8. $27x^6$
9. a. False, $3^x \times 3^y = 3^{x+y}$.
   b. Yes, if $x + y = 0$ then $3^x \times 3^y = 9^{x+y} = 1$.
10. $5t^2 \times 10t^{-4} \div (5t^{-3})^2 = 2t^4$ instead of $2t^{-6}$.
11. a. 1 b. $\frac{1}{2}$ c. $\frac{7}{2}$ d. $\frac{1}{2}$
12. b i. $3^{3x} + 3^{x+2}$ ii. $3^{2x} - 3^{-x}$ c. $u^4 - \frac{u^2}{9}$
13. $1 = x^1 \div x^1 = x^{1-1} = x^0$, $x^0 \div x^1 = x^{0-1} = x^{-1}$, $\sqrt{x} \times \sqrt{x} = x$ and $x^{\frac{1}{2}} \times x^{\frac{1}{2}} = x^{\frac{1}{2}+\frac{1}{2}} = x$
14. a. 4 b. 16 c. $9x^4$ d. $8x^6$
15. $X = x$, $Y = 8x^2y^2$, $Z = 2x^{-1}y^{-2}$

2.3S

1. a. $4y + 8$ b. $6b + 42$ c. $7y + 21$
   d. $36d + 60$ e. $15t + 24$ f. $\frac{w}{2} + 5$
2. No, $5(x + 4) = 5x + 20$. Sandra should multiply both terms in the bracket by 5.
3. a. $-4x - 20$ b. $-6b - 18$ c. $-t - 2$
   d. $-3d - 24$ e. $-30t - 80$ f. $-32w - 72$
4. a. $-3x + 15$ b. $-2b + 16$ c. $-t + 8$
   d. $7d - 70$ e. $81 - 18t$ f. $-48 + 40w$
5. a. $y^2 + 2y$ b. $b^2 - 7b$ c. $-y^2 - 3y$
   d. $2d^2 + 5d$ e. $3t^2 - 8t$ f. $2sw + 5w$
6. No, $x(x + 4) = x^2 + 4x$. Karl should multiply both terms in the bracket by x.
7. a. $y^3 - 2y$ b. $b^3 - 6b^2$ c. $3y^2 + 9y$ d. $2d^2 - 10d$
   e. $7t^2 - 56t$ f. $63w - 9w^2$ g. $a^2b + 5ab^2$ h. $\frac{5}{2}st + \frac{15}{4}t^2$
8. a. $9x + 42$ b. $13y + 9$ c. $13t + 6$
   d. $13p - 43$ e. $8b - b^2$ f. $9m^3 - 34m^2$
   g. $x^2 + 22x + 27$ h. 0
   i. $2rs + 2rt + 2st$ j. 0
9. b. 8 c. 7 d. 6 e. q^2
   f. $3r$ g. $9s^2t^2$
10. a. $4(p + 2)$ b. $5(y + 2)$ c. $3(d + 7)$ d. $9(k + 8)$
    e. $6(b + 4)$ f. $6(w + 9)$
11. a. $6b(1 + 4c)$ b. $6w(1 + 9y)$
    c. $8b(2a - 5)$ d. $15(q - 3p)$
    e. $p(p + 8)$ f. $y(1 + 6y)$
    g. $w(1 + 4w^2)$ h. $ab(1 - 4b)$
    i. $6b(b + 4c)$ j. $12xyz(x + 3y)$
    k. $5m(3n - 1 + 2m^2)$ l. $fc(mu + r)$
    m. $me(icky + ous)$

Answers

12 No. The highest common factor is $4p$ so the complete factorisation is $4p(3 + 5q)$.

13 a $(p + 1)(p + 3)$ **b** $(2q + 3)^2(2q + 4)$
   c $32(2r + 1)(r + 1)$ **d** $9s^2(u + v)^3(3tu + 3tv + 2s)$

2.3A

1 Clare $5x \times 0 = 0 \not\equiv 5x$, Amal $3 \times 1 = 3 \not\equiv 3p$, Vicky $7p \times 14 = 98p \not\equiv 21p$, $7p \times 7q = 49pq \not\equiv 14pq$.

2 a $(x + y)(a + b)$ **b** $(d + m)(c + b)$ **c** $(a + 2)(a + b)$
   d $(c - m)(d + e)$ **e** $(a^2 + c^3)(b + d^4)$ **f** $(x + 2)(x^2 + y)$

3 a $p = -4, q = 21$ **b** $p = 3, q = 6$ **c** $p = -2, q = 2$
   d $p = 3, q = 11$ **e** $p = 1, q = -10, r = 2$
   f $p = 4, q = -3, r = 9$

4 a $4(x - 1)$ **b** $25(b + 2)$

5 $3(2x - 1) = 15 \Rightarrow 6x - 3 - 15 = 0 \Rightarrow 6(x - 3) = 0$

6 a $5(4x + 7)$ **b** 1
   c Students' answers, for example, $(5x + 5) + (15x + 30)$.

7 a $\frac{2y(y + y + 2)}{2}$ **b** $2y(y + 1)$

8 a 6 **b** 16.5 **c** 58.6 **d** 33.2

9 Area $= 4(3x + 2) - 2(2x - 1) = 8x + 10 = 2(4x + 5)$

10 a Students' results.
   b i The result is always 2.
      ii The result is always the starting number.
   c i $x \to 2x \to 2x + 4 \to x + 2 \to 2$
      ii $x \to 2x \to 2x + 1 \to 10x + 5 \to 10x \to x$

11 a $xz + yz$ **b** $x(x + y) + y(x + y) = x^2 + 2xy + y^2$
   c $x^2 + 2xy + y^2$
   d i $4x^2 + 12x + 9$ **ii** $2x^2 + 13x + 20$

12 $(x + 3)(x + 4)$

13 Perimeter A $= 18x + 24y = 3(6x + 8y)$, Perimeter B $= 6x + 8y$

2.4S

1 a $\frac{1}{3}$ **b** $\frac{11}{19}$ **c** $\frac{1}{6}$
   d $\frac{3}{10}$ **e** $\frac{1}{4}$ **f** $\frac{1}{11}$

2 a $\frac{t^2}{3}$ **b** $\frac{p^7 q^3}{r^3}$ **c** $\frac{20xy}{3}$

3 a $2x + 3$ **b** $x + 5$ **c** $\frac{1}{x}$ **d** 3

4 No, he cannot cancel only part of the numerator. This expression cannot be factorised.

5 a x **b** p **c** $\frac{1}{y}$
   d $\frac{6}{y-1}$ **e** 3 **f** This cannot be simplified

6 a $y + 2$ **b** $x - 4$ **c** $\frac{1}{x + 3}$
   d $(p-1)^2$ **e** $y + 4$ **f** $\frac{1}{(b - 2)^3}$

7 a $\frac{5}{8}$ **b** $\frac{3}{7}$ **c** $\frac{7}{6} = 1\frac{1}{6}$
   d $-\frac{1}{18}$ **e** $\frac{29}{54}$ **f** $\frac{31}{42}$

8 a $\frac{5p}{7}$ **b** $\frac{10q}{9}$ **c** $\frac{2}{3r}$
   d $\frac{6s-5}{s^2}$ **e** $\frac{48t + 35}{20t^2}$ **f** $\frac{6v - 25u}{5uv}$

9 Paloma is incorrect $\frac{2}{x} + \frac{4}{2x + 3} = \frac{2(2x + 3) + 4x}{x(2x + 3)} = \frac{8x + 6}{x(2x + 3)}$

10 a $\frac{7x + 4}{x(x + 2)}$ **b** $\frac{8x - 3}{x(x - 1)}$ **c** $\frac{8x + 13}{(x + 2)(2x + 3)}$
   d $\frac{x^2 - x + 4}{(x + 2)(x - 3)}$ **e** $\frac{2(x + 8)}{(x - 2)(x + 3)}$ **f** $\frac{(3x - x^2 - 6)}{(x - 1)(x - 3)}$

11 a $\frac{8}{9}$ **b** $\frac{5}{3} = 1\frac{2}{3}$ **c** $\frac{4}{11}$
   d $\frac{11}{13}$ **e** $\frac{9}{8} = 1\frac{1}{8}$ **f** $\frac{133}{9} = 14\frac{7}{9}$

12 a $\frac{21x^2}{10}$ **b** 6 **c** $\frac{15}{x^2}$
   d $\frac{6x}{yz}$ **e** $6y^2z^2$ **f** $\frac{189y^3}{4x^3z^3}$

13 a $\frac{10}{x(x + 2)}$ **b** $\frac{3}{x - 1}$ **c** $\frac{9}{2x + 3}$
   d 1 **e** $\frac{2}{x + 2}$ **f** y **g** $\frac{10(x + 3)}{(2 - 3x)}$

2.4A

1 a $-3(x - 5) = -3x + 15$, $\frac{3(3x + 7)}{(x - 5)(2x + 1)}$
   b The second fraction should have been turned upside-down, $4x - 10 = 2(2x - 5)$. $\frac{3x(x + 2)^2}{2(2x - 5)^2}$

2 A and E, C and G, D and F, B and $\frac{5x}{12}$.

3 a $\frac{x + 7}{6}$

4 $\frac{3x}{(x + 3)(2x - 3)}$

5 Common difference $= \frac{(1 - x)}{(x + 1)^2}$

6 a $\frac{x}{2}$ **b** $\frac{4(x^2 + xy + 2y^2)}{xy^2}$

7 a $\frac{8(2p - 3)}{(p + 1)(p - 3)}$
   b Cannot divide by 0, the width must be positive.

8 a $12y$ **b** $3y(3x + 8)$

9 a $\frac{6x(x - 3)(x + 1)}{x - 2}$ **b** Cannot divide by 0,

10 $25, \frac{6}{25(x + 1)}, 4(x + 1)^2, \frac{5(x + 1)^2}{12}, \frac{1}{15}$

11 1

12 $\frac{4(2x + 1)}{(2x - 3)(5 - 2x)}$

Review

1 a 3 **b** 7 **c** 1

2 a $13y$ **b** $7xy$ **c** $3x$ **d** y^3
   e $\frac{x}{2}$ **f** $5xy$

3 a $6a - 3b$ **b** $2 - 5a + 7b$ **c** $4a + 4a^2$
   d $5a^2b - 5ab^2$ **e** $6ab + 3b - 9$ **f** $2a - a^2$

4 a 10 **b** -21 **c** -4 **d** 28
   e 12 **f** -14 **g** $1\frac{1}{3}$ **h** 3

5 a 10 **b** $3\frac{1}{2}$

6 a a^7 **b** b^5 **c** c^6 **d** $20d^9$
   e $4e^6$ **f** 6

7 a x^5 **b** y^2 **c** $8z^6$ **d** 1
   e $28u^4$ **f** $5p^{-8}$ **g** $27r^{-6}$ **h** $\frac{1}{4}s^{-4}t^6$

8 a $10a + 15$ **b** $18b - 9c$ **c** $-32d^2 + 8cd$
   d $y - y^2$

9 a $5x(x + 2)$ **b** $7a(3b^2 - 2)$ **c** $15p(2p + q^2 - 3q^3)$

10 a $\frac{x + 2}{4}$ **b** $\frac{x}{x + 1}$

11 a $\frac{3}{a}$ **b** $\frac{b}{40}$ **c** $\frac{3}{2c}$ **d** $\frac{3d - 2}{4d^2}$
   e $\frac{3f + 5d}{df}$ **f** $\frac{5}{a}$ **g** $\frac{ab}{3}$
   h $\frac{a}{(a + 1)(a + 2)}$

Assessment 2

1 a $157\,\text{cm}^2$ **b** $3.58\,\text{m}$ **c** $3.27\,\text{in}$

2 $10z^2$

3 a The sum of the areas of the rectangles should equal the area of the square.
   $4y + 28 + 7y + y^2 = y^2 + 11y + 28 > y^2$ as $y > 0$.
   This sum is greater than the area of the square so Sasha must be incorrect.
   b $y^2 - 11y + 28$

4 a i $80\,\text{m}$ **ii** $245\,\text{m}$
   b 8 seconds.

5 a Correct.
   b Not complete. $abc^2(ac - b^3 + a^4b^2c^2)$
   c Incorrect. $6p^3q^2r^4(2r^5 - 3q^3r - 5p^2q^5)$
   d Correct.
   e Incorrect. $g(g^2 + g - 1)$
   f Incorrect, this cannot be factorised.
   g Not complete. $2(p - q)[2 - 3(p - q)]$
   h Incorrect. $3(y + 2z)^2[1 + 3(y + 2z)]$
   i Not complete. $-2(x - 2)(x - 1)$

6 a $2y^2 + 80y$
   b No. The volume is $y \times y \times 20 = 20y^2$ and $20y^2 = 200\,\text{cm}^3$.
      Thus $y^2 = 10$ so $y = \sqrt{10}$.
7 a i $5W - 10$ ii $5(W - 2)$
   b i $8W + 4$ ii $4(2W + 1)$
8 $w = 5$, $x = 8\tfrac{1}{3}$
9 a $32p^2 - 22p$ b $2p(16p - 11)$
10 a $\tfrac{5}{9}$ b $\tfrac{11}{20}$ c $\tfrac{13}{18}$ d $\tfrac{2q}{5}$
   e $5x$ f $\tfrac{p}{2q}$ g $\tfrac{27a^2}{2}$ h $\tfrac{z-4}{6}$
11 a $\tfrac{10z-9}{24}$ b $\tfrac{59-44y}{35}$
   c $\tfrac{7x+11}{(x+1)(x+2)}$ d $\tfrac{13(p-2)}{(2p+3)(3p-2)}$
   e $\tfrac{a(7a+12)}{(a-2)(2a+9)}$ f $\tfrac{6x-1}{(2x+1)^3}$
12 a $\tfrac{4(w+3)}{5}$ b $w = 12$ c 4 and 3

Chapter 3

Angle rules will be abbreviated in the following way:
VO vertically opposite angles are equal;
CA corresponding angles are equal;
AA alternate angles are equal;
IA interior angles sum to 180°;
ASL angles on a straight line sum to 180°;
AP angles at a point sum to 360°;
AST the angle sum of a triangle is 180°;
ASQ the angle sum of a quadrilateral is 360°.

Check in

1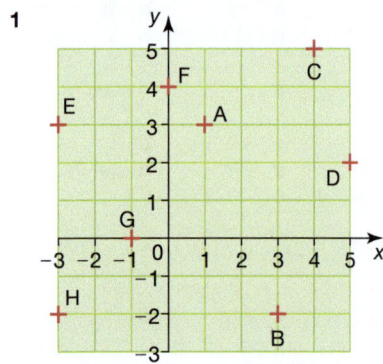

2 $a = 56°$, $b = 112°$, $c = 235°$, $d = 160°$
3 $e = 72°$, $f = 63°$, $g = 118°$, $h = 75°$

3.1S

1 a $a = 360° - 20° - 90° = 250°$ (AP) reflex,
   b $b = 180° - 90° - 54° = 36°$ (ASL) acute,
      $c = 54°$ (VO) acute,
      $d = 180° - 54° = 126°$ (ASL) obtuse
2 a $3p = 180° - 63° = 117°$ (ASL), $p = 39°$
   b $6q = 360° - 90°$ (AP), $q = 45°$
   c $r + 90° = 180° - 36°$ (ASL), $r = 54°$
      $s - 20° = 144°$ (VO), $s = 164°$
      $4t = 36°$ (VO), $t = 9°$
3 a i $a = 68°$ (AA), $b = 180° - 68° = 112°$ (IA), $c = 68°$ (CA),
         $d = b = 112°$ (VO),
      ii $e = 180° - 81° = 99°$ (ASL),
         $f = 81°$ (CA), $g = 360° - 81° = 279°$ (AP)
   b i a, c and f ii b, d and e iii g
      iv a and b, b and c, c and d, d and a,
         e and f are supplementary.
4 a $a = 75°$ (CA)
      $3b = 180° - a = 105°$ (ASL), $b = 35°$
   b $2c = 114°$ (AA), $c = 57°$
      $d + 30° = 114°$ (VO), $d = 84°$

   c $e = 180° - 120° = 60°$ (IA)
      $6f = 360° - e = 300°$ (AP), $f = 50°$
      $4g = 120°$ (VO), $g = 30°$
5 a $s = 98°$ (VO), $t = 98°$ (CA),
      $u = 180° - 98° = 82°$ (ASL),
      $p = 73°$ (CA), $q = 73°$ (VO),
      $r = 180° - p = 107°$ (ASL)
   b $a = 57°$ (CA), $b = a = 57°$ (AA),
      $c = 180° - 57° = 123°$ (ASL), $d = c = 123°$ (CA)
6 66°
7 70°
*8 $\angle MNR = 74°$ (AA), $\angle MNS = 180° - 74° = 106°$ (IA)
    $\angle YNR = \angle MNS = 106°$ (VO)
*9 a $\angle EFB = 96°$ (CA)
   b $\angle GFB = 180° - 96° = 84°$ (IA)
   c $\angle AFG = \angle EFB = 96°$ (VO)
   d $\angle CGH = 96°$ (VO)
   e $\angle DGH = 180° - 96° = 84°$ (ASL)

3.1A

1 a Amadi 027°, Kate 117°, Neil 222°, Sonja 310°
   b 132° anti-clockwise (to his left).
2 a i 150°   ii 135°   iii 52.5°   iv 102.5°
   b i 2964°   ii 247°
3 a 234°   b 116°   c 347°   d 015°
4 a L 031°, P 282°, Y 240°.   b L 211°, P 102°, Y 060°.
5 025° to S, then 270° to R
6 a 180°   b 118° to B, then 225° to P
7 $x = a$ (AA)
   $y = b$ (CA)
   $a + b + c = x + y + c = 180°$ (ASL)
8 a Using interior angles gives: angle
      $Q =$ angle $S = 116°$ and angle $R = 64°$
   b Any two of: Opposite angles are equal.
      The angles add up to 360°. Adjacent angles sum to 180°.
*9 a $000° \leqslant x < 180°$
   b For bearings $\geqslant 180°$, subtracting 180° from the bearing of A
      from B gives the bearing of B from A.
   c Let α be the bearing of A from B and β be the bearing of B
      from A.
      $\beta = \alpha + 180°$ or $\alpha - 180°$
      so if α becomes $\alpha + x$, $\beta = \alpha + 180° + x$ or $\alpha - 180° + x$.
10 Let the angles of a parallelogram be (in anticlockwise order) a,
   b, c, and d.
   $a + b = 180°$ (IA), $a + d = 180°$ (IA)
   so $a = 180° - b = 180° - d$, $b = d$.
   $c + d = 180°$ (IA), $c = 180° - d = a$

3.2S

1 a i $a = 39°$ ii $b = 24°$ iii $c = 96°$ iv $d = 214°$
   b i c ii d
2 a $p = 128°$ b $q = 69°$
   c $r = 124°$, $s = 56°$.   d $t = 102°$, $u = 66°$.
   e $v = 90°$, $w = 82°$.   f $x = 117°$, $y = 63°$.
3 $x = 35°$, $y = 90°$, $z = 55°$.
4 a $x = 30°$ b $y = 40°$ c $z = 50°$.
5 a i $\angle ABC = \tfrac{180° - 84°}{2} = 48°$ (AST, equal base angles)
      ii $\angle BCD = 180° - 84° - 23° - 48° = 25°$ (AST ACD)
   b Triangle BCD is not isosceles because it does not have 2
      equal base angles.
6 a i 58°   ii 64°   iii 32°
   b Isosceles.
7 a $\angle BCD = 81°$, $\angle BAD = 121°$ b Trapezium

Answers

490

3.2A

1. $c = a = 50°, d = b = 70°, e = 60°$
2. **a** $a = 44°, b = 224°$ **b** $p = 75°, q = 42°, r = 117°$
3. $\angle RST = 148°$
4. $\angle A = 85°, \angle B = 60°, \angle C = 35°$
5. $\angle P = 75°, \angle Q = 85°, \angle R = 95°, \angle S = 105°$
6. An obtuse angle is greater than 90°, so two obtuse angles would have a total of more than 180°. This is impossible, so a triangle cannot have more than one obtuse angle.
7. All the statements are correct as a square satisfies the requirements for a rhombus, rectangle and parallelogram.
8. **a** (1, 2)   **b** (4, 8)   **c** (6, 4)
9. A kite contains *one* pair of equal angles.
   If 50° is one of a pair, the fourth angle $= 360° - 50° - 50° - 130° = 130°$.
   If 130° is one of a pair, the fourth angle $= 360° - 130° - 130° - 50° = 50°$.
   In each case the shape has *two* angle pairs.
   The unknown angles must form a pair: $2x = 360° - 130° - 50° = 180°, x = 90°$
10. $\angle A = \angle D = 60°, \angle C = \angle E = 120°$
11. $\angle QTR = 150°$
12. **a** One cut gives 2 identical scalene triangles or 2 different isosceles triangles.
    Two cuts give 2 pairs of identical right-angled triangles.
    b One cut gives 2 identical obtuse-angled isosceles triangles or 2 identical acute-angled isosceles triangles.
    Two cuts give 4 identical right-angled triangles.
    c One cut gives 2 identical right-angled triangles.
    Two cuts give 2 pairs of triangles – one pair are identical isosceles acute-angled triangles and the other pair are identical isosceles obtuse-angled triangles.
    d One cut gives 2 identical right-angled isosceles triangles.
    Two cuts give 4 identical right-angled isosceles triangles.
13. **a** a square   **b** a rhombus   **c** a rectangle
    d a rectangle   **e** a parallelogram   **f** a parallelogram

3.3S

1. **a** i 4   ii 4   **b** i 5   ii 5
   c i 1   ii 1   **d** i 2   ii 0
   e i 2   ii 2   **f** i 8   ii 8
   g i 1   ii 1   **h** i 4   ii 0
2. **a** No lines of symmetry, rotation symmetry order 3
   b 1 line of symmetry, no rotation symmetry (or order 1)
   c no lines of symmetry, no rotation symmetry (or order 1)
   d no lines of symmetry, no rotation symmetry (or order 1)
   e 2 lines of symmetry, rotation symmetry order 2
   f no lines of symmetry, rotation symmetry, order 2
3. **a**

Shape	Number of lines of symmetry	Order of rotation symmetry
Equilateral triangle	3	3
Isosceles triangle	1	1
Scalene triangle	0	1
Rectangle	2	2
Square	4	4
Parallelogram	0	2
Rhombus	2	2
Kite	1	1
Trapezium	0	1

   b i rectangle, square, parallelogram, rhombus
   ii rectangle, square, parallelogram, rhombus
   iii square, rhombus
   iv rectangle, square
   v square, rhombus, kite
   vi rectangle, square, parallelogram, rhombus
4. Each has one line of symmetry through the centre, but no rotation symmetry (order 1)

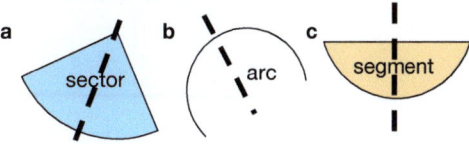

5. **a** rotation symmetry order 2, no lines of symmetry
   b no rotation symmetry(order 1), 1 line of symmetry
   c rotation symmetry order 3, no lines of symmetry
   d rotation symmetry order 3, 3 lines of symmetry

3.3A

1. **No stopping** - 4 lines of symmetry, rotation symmetry of order 4
   Hospital - 2 lines of symmetry, rotation symmetry of order 2
   Ice or snow - 3 lines of symmetry, rotation symmetry of order 3
2. **Suriname** - 1 line of symmetry, no rotation symmetry (or order 1)
   UK - no lines of symmetry, rotation symmetry of order 2.
3.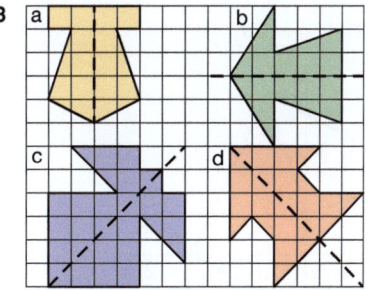
4. **a** isosceles triangle   **b** parallelogram
5.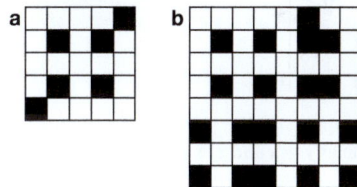
6. **a** Shaded copy of diagram that has rotation symmetry 2, but no lines of symmetry.   Example:
   b Shaded copy of diagram that has one line of symmetry, but no rotation symmetry.   Example: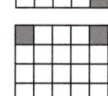
7. **a** Shaded copy of diagram that has one line of symmetry, but no rotation symmetry.   Example:
   b Altered copy of diagram that has rotation symmetry order 2, but no lines of symmetry.   Example:
8. **a** Diagram with 4 trapeziums shaded, with result having rotation symmetry of order 2 and 2 lines of symmetry.   Example:
   b Diagram with 4 trapeziums shaded, with result having rotation symmetry of order 4, but no lines of symmetry.   Example: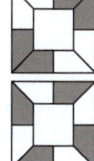

9   a (4, 3) and (0, −1)
    b isosceles trapezium.

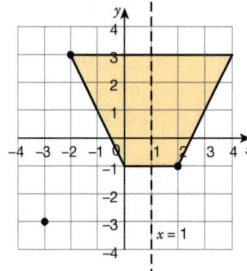

10 a (2, 4), (4, 2), (4, −2),
       (2, −4), (−2, −4), (−4, 2)
    b 4
    c $y = x$ and $y = -x$

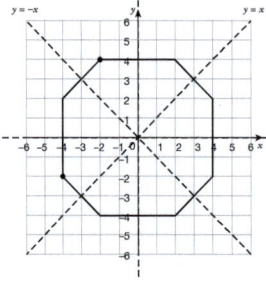

11 The other corners are (1, 4) and (−1, −2).

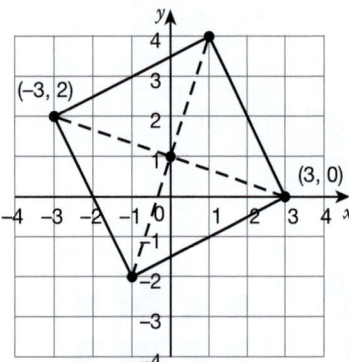

12 Shape for which the lines $y = x$ and $x + y = 3$ are both lines of symmetry.

Example:

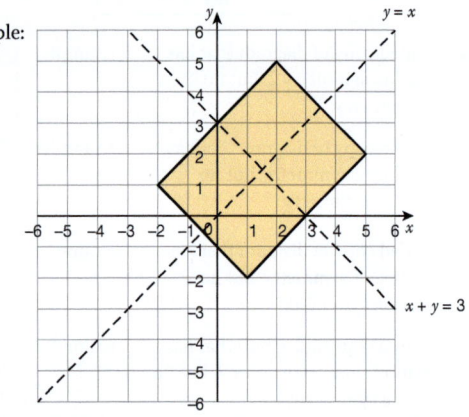

3.4S

1   a B and F, A and E
    b C and D, F and H, B and H, A and G, E and G
2   a A is congruent to B (RHS).
    b K is not congruent to L because the marked sides are not corresponding.
    c P is not congruent to Q because the equal angles are not between the equal sides.
    d S is congruent to T (ASA).

3   B is congruent to C.
    The angles of each triangle are 90°, 39° and 51° and both have hypotenuse 8 cm (RHS). A is not congruent to either B or C because the 8 cm is not on the hypotenuse.
4   a 4   b 16   c 64
5   a 6.25   b 15.625
6   a Angle $ACB = 37°$ (VO), Angle $DEC = 53°$ (AST).
       Triangles ABC and DEC have equal angles so are similar.
    b $DE = 12$ cm, $AC = 8$ cm.
7   a i $CD = 15$ cm   ii $BD = 20$ cm
    b 216 cm²
8   a 8.75 cm   b 78 cm² (2sf)   c 690 cm³ (2sf)
*9 a $PS = 4$ cm   b $PQTS = 21$ cm²

3.4A

1   a $PQ = RQ$ (equal sides of isosceles triangle) and $PS = RS$
       (S is the mid-point of PR).
       Triangles PQS and RQS are congruent (SSS)
    b $\angle PSQ = \angle QSR$
       $2 \times$ angle $PSQ = 180°$ (angles on a straight line add to 180°), so angle $PSQ = 90°$
2   a $AD = BC, DC = AB, DB = BD$ (side common to both triangles). Triangles ABD and BCD are congruent by (SSS).
    b i $\angle CBD$ ii $\angle DBA$
3   In triangles ABC and ADC
    $AB = AD$ and $BC = DC$
    Triangles ABC and ADC are congruent (SSS), so, angle B = angle D.
4   a 9.6 cm   b $6\frac{2}{3}$ cm
5   32 cm long, 16 cm wide and 60 cm tall.
6   The Mini size is similar to the Medium size scale factor $\frac{3}{5}$ or $\frac{5}{3}$
    (or equal width : height ratio 4 : 5).
    The Small size is similar to the Extra Large size scale factor $\frac{1}{2}$ or 2
    (or equal width : height ratio 3 : 4).
7   a Angle RSQ = angle RQS
       (base angles of isosceles triangle RQS)
       Angle RSQ = angle PQS (AA)
       Angle PQS = angle PSQ (base angles of isosceles triangle PQS)
       So all the angles marked in red are equal.
       Using similar reasoning proves that all the angles marked in yellow are equal.
       Also the sides PQ, QR, RS and SP are all equal.
       The 4 triangles are all congruent (ASA)
       The 4 angles at T are equal and total 360°, so each angle is 90°.
    b i square, kite
       ii For the square the proof is the same as the proof in part a as a square is a special kind of rhombus.
           For the kite:
           Triangle ADC is congruent to triangle ABC by SSS, so $\angle DAE = \angle BAE$. Triangle AED is congruent to AEB by SAS. Therefore $\angle AED = \angle AEB = 180° \div 2 = 90°$ (ASL). Triangle DCE is congruent to triangle BCE by SAS so $\angle CED = \angle CEB = 180° \div 2 = 90°$ (ASL).

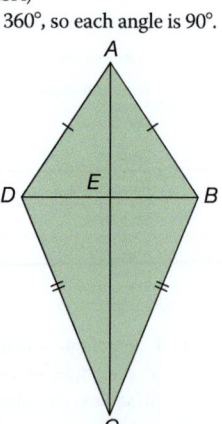

8   Height 40 cm, base diameter 33 cm and top diameter 70 cm.
9 a 50 cm   b 704 cm²
*10 a i EF = 28.8 cm
       ii BC = 48 cm
     b 1300 cm²
*11 PU = PS (equal sides in triangle PUS)
   PQ = PR (equal sides in triangle PQR)
   ∠QPU = ∠RPS (both equal 60° − ∠TPS)
   Triangles PUQ and PSR are congruent (SAS)

3.5S

1 81°
2 150°
3 Angle sum of pentagon = 3 × 180° = 540°
   5x = 540° − 90° − 95° − 110° = 245°
   x = 245° ÷ 5 = 49°
   Angle sum of hexagon = 4 × 180° = 720°
   6y = 720° − 149° − 156° − 121° = 294°
   y = 294° ÷ 6 = 49° = x
4 Triangle 3, 120°, 60°; quadrilateral 4, 90°, 90°; pentagon 5, 72°, 108°; hexagon 6, 60°, 120°; heptagon 7, $51\frac{3}{7}°$, $128\frac{4}{7}°$; octagon 8, 45°, 135°; nonagon, 9, 40°, 140°; decagon 10, 36°, 144°.
5 106°
6 a 156°.
  b Exterior angle = 360° ÷ 15 = 24°, Interior angle = 180° − 24° = 156°.
7 a Rotational symmetry of order 5 and 5 lines of symmetry.
  b Rotational symmetry of order 8 and 8 lines of symmetry.
  c Rotational symmetry of order 10 and 10 lines of symmetry.
8 85°
9 a x = 40°   b y = 70°
10 y = 55°
11 ∠APB = 36°
*12 ∠BXC = 100°
*13 ∠RWS = 22.5°
*14 Angle sum of hexagon = 4 × 180° = 720°
   ∠BCD = interior angle = 720° ÷ 6 = 120°
   ∠BCA = base angle of isosceles triangle
   ∠BCA = $\frac{180° − 120°}{2}$ = 30°
   ∠ACD = ∠BCD − ∠BCA
        = 120° − 30° = 90°
15 30 sides

3.5A

1 Equilateral triangle
2 20 sides.
3 120 sides.
4 12 sides.
5 20 sides.
6 a z = 150°
  b A square and equilateral triangle will fit the space.
     A dodecagon (12-sided polygon) would fit in the space.
7 a 10   b Decagon
8 a i   Yes, 36 sides.
     ii Yes, 40 sides.
     iii Yes, 45 sides.
     iv No, 360° ÷ 7° is not a whole number, so it is not possible to have a regular polygon with exterior angles of 7°.
  b It is only possible to have a regular polygon with a given exterior angle when that angle is a factor of 360°

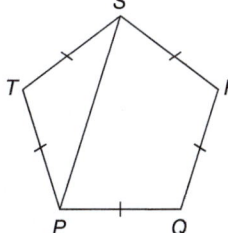

*9 Interior angle of a pentagon = 108°.
   ∠TPS = $\frac{180° − 108°}{2}$ = 36° (base angle of isosceles triangle PST)
   ∠SPQ = ∠TPQ − ∠TPS = 108° − 36° = 72°
   ∠SPQ + ∠PQR = 72° + 108° = 180°
   PS is parallel to QR (interior angles add to 180°)
*10 Triangles ABC, CDE, EFG and GHA all have two equal sides (equal sides of regular octagon). They also have the included angle equal (equal angles of regular octagon)
   These triangles are congruent (SAS)
   So AC = CE = EG = GA
   ∠FEG = ∠FGE = (180 − 135) ÷ 2 = 22.5° (isosceles triangles have equal base angles). As the isosceles triangles are congruent ∠DEC = ∠FEG = 22.5 so ∠GEC = 135 − 22.5 − 22.5 = 90°.
   The same argument can be used to show that ∠ECA = ∠CAG = ∠AGE = 90°.
   ACEG has equal sides and equal angles of 90° so is a square.
11 a E.g. 1 hexagon and 4 triangles, 2 hexagons and 2 triangles, 2 squares and 3 triangles, 2 octagons and 1 square.
   b Schlafli tessellations are described using the number of sides of each of the regular shapes. For example a tessellation may be given by 3, 4², 6 (one 3-sided polygon, 2 four-sided polygons and one 6-sided polygon at each vertex where they meet)
12 a Yes.
   b When a side is extended at the point where the shape is concave, the angle formed is inside the shape, rather than outside it. If angles turned through at such points are taken to be negative, the sum of these and the other exterior angles is still 360°.
*13 ∠ACD = $\frac{270°(n − 2)}{n}$ − 90°
*14 ∠HGC = 168°

Review

1 a = 105°, b = 60°, c = 120°
2 a = 105° (CA)
   b = 75° (ASL)
   c = 65° (AA)
3 a 243°   b 321°   c 025°
4 40°
5 a One line of symmetry but no rotation symmetry (or rotation symmetry, order 1)
  b a = 70°
6 A(1, 1), B(3, 3), C(5, 1) should be plotted and labeled. The triangle is isosceles and right-angled.
7 a Trapezium   b Rhombus
8 No, in triangle ABC the hypotenuse is 13 cm but in triangle EFD it is one of the shorter sides which is 13 cm. Since the triangle is right-angled these sides cannot both be 13 cm so they are not congruent.
9 x = 2.25 cm, y = 18 cm².
10 a 5 × 180° = 900°
   b Geometrical proof that the interior angles sum to 540°
      Interior ∠s = 540 ÷ 5 = 108°.
      Exterior ∠s = 180 − 108 = 72°.
      Sum of exterior ∠s = 72° × 5 = 360°

Assessment 3

1 a 182.5° or 177.5°
  b At 1:05 the hour hand is on a bearing of 032.5° and the minute hand is on a bearing of 030°.
     At 1:06 the hour hand is on a bearing of 033° and the minute hand is on a bearing of 036°.
     The minute hand is on a lesser bearing than the hour hand at 1:05 but a greater bearing than the hour hand at 1:06 so the two must be on the same bearing at some point between the two times.

493

2 103° anti-clockwise.
3 a 130° **b** 255° **c** 215° **d** 310°
4 Rafa was south of Sunita.
5 a Incorrect. $a = 62°$ (IA).
  b Correct (CA).
  c Correct (CA, ASL).
  d Incorrect. $d = 87°$ (ASL, CA).
  e Incorrect. $e = 88°$ (exterior angles of a polygon).
  f Incorrect. $f = 92°$ (ASL).
  g Correct (isosceles triangle has base angles 52°, ASL).
  h Correct (ASQ).
6 a False. Two obtuse angles add to more than 180° which is the sum of the three angles in a triangle.
  b True.

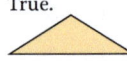

  c False. One right angle = 90° and one obtuse angle is more than 90°. This comes to more than 180° before the acute angle has been added.
  d True.

7 a 93° **b** 170° **c** 73° **d** 174°
8 a 108° **b** 25 slabs. **c** 72°, 54°, 54°
9 120 cm
10 a ∠PTS = ∠PQR (CA). ∠PST = ∠PRQ (CA). ∠SPT = ∠RPQ (common angle).
  b 15 cm
  c The ratio of shorter to longer parallel sides is $\frac{AB}{10}$ in trapezium ABFE and $\frac{AB}{15}$ in trapezium ABCD. $\frac{AB}{10} > \frac{AB}{15}$. The ratios between sides are different so the shapes are not similar.
11 a 2 cm **b** 18 cm² **c** 2 cm²
  d FD = AB = 6 cm and FD is parallel to AB so ABDF is a parallelogram and DB is parallel to FA.
  ∠DBC and ∠FAB are corresponding angles so are equal.
12 a Any shape with rotation symmetry of order 3 but no lines of symmetry.
  b Any shape with 3 lines of symmetry, but no rotation symmetry.

Chapter 4

Check in

1 a $22\frac{1}{2}$ **b** 21 **c** 6 **d** 21
2 a 29, 45, 55, 65, 66, 89, 90, 98, 101, 112
  b 7.88, 8.78, 8.87, 8.9, 8.95, 9.0, 9.8, 10.9, 11.25, 11.3

4.1S

1 8
2 Yes, $\frac{55}{60} > \frac{30}{40}$.
3

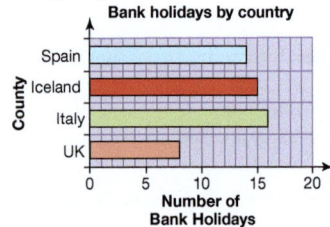

4

5 a 30° **b** Boys 210°, Girls 150°
  c Pie chart with angles given in part **b**.
6 a 6°
  b Sunny 90°, Cloudy 108°, Rainy 84°, Snowy 18°, Windy 60°
  c Pie chart with angles given in part **b**.
7 Pie chart with the following angles. 1: 30°, 2: 25°, 3: 35°, 4: 80°, 5: 70°, 6: 60°, 7: 60°
8 a 20°
  b i 7 **ii** 10 **iii** 1

4.1A

1 The vertical scale does not start at 0.
2 A pie chart shows the proportion not the number of jars sold. There is no information about what sauces are in the 'other' category. The angles for BBQ, Chili and Pesto are similar so it is difficult to compare the number of jars sold.
3 a There are too many categories. A pie chart does not show how many drinks were sold.
  b It is easy to compare the frequencies of two or more drinks. A bar chart shows how many drinks were sold.
4 a No, the angle for German is the same as the angle for Spanish.
  b No, Lydia doesn't know how many students are represented in each pie chart.
5 a Pie chart with angles. Monday 60°, Tuesday 100°, Wednesday 60°, Thursday 40°, Friday 80°, Saturday 20°.
  b Pie chart with angles. Monday 80°, Tuesday 40°, Wednesday 0°, Thursday 40°, Friday 80°, Saturday 120°.
  c Pie chart, the angles represent the proportion.
  d Bar chart, the heights of the bars show the frequencies.

4.2S

1 a 8 **b** 9 **c** 2 **d** 6
  e 2 **f** 24 **g** 18 **h** 104
  i 15 **j** 5
2 a i 5 **ii** 10 **iii** 35 **iv** 98
  v 2
  b Set **ii**, the outlier has not affected the median.
3 a Mode = 1, Range = 10 **b** Mode = 8, Range = 3
  c Mode = 11, Range = 4 **d** Mode = 25, Range = 15
  e Mode = 8, Range = 3 **f** Mode = 5, Range = 3
  Part **a** contains the outlier −6 which increases the range but does not affect the mode. Part **d** contains the outlier 36 which increases the range but does not affect the mode.
4 a 3, 3, 3, 3, 4, 4, 5, 5. Mean = 3.75, mode = 3, median = 3.5, range = 2.
  b See part **a**.
5 a 1, 1, 1, 1, 2, 2, 2, 2, 2, 3, 3, 4, 4, 4, 4, 4, 5, 5, 5, 5, 5, 5. Mean = 3.28, mode = 5, median = 4, range = 4.
  b Not required Mode = 5, Median = 4, Mean = 3.28
6 a i 7 **ii** 6 **iii** 5.82 **iv** 6 **v** 2
  b i 75 **ii** 63 **iii** 60.1 **iv** 63 **v** 27
  c i 8 **ii** 96 **iii** 95.6 **iv** 96 **v** 2
  d i 71 **ii** 22, 37 **iii** 40.4 **iv** 37 **v** 38
  e i 26 **ii** 88, 89 **iii** 84.2 **iv** 87 **v** 7
  f i 72 **ii** 27 **iii** 46.9 **iv** 34 **v** 37
  g i 8 **ii** 105 **iii** 105.2 **iv** 105 **v** 3

Answers

7 a 1, 6, 8, 2, 8, 5, 6, 9, 3, 5, 7, 4, 4, 5, 5
  b i 8 **ii** 5 **iii** 5.2 **iv** 5 **v** 3
  c Range and IQR stay the same, other values decreases by 100.

4.2A

1 a 2, 4
  b The amount of bottles delivered to number 47 varies more.
2 a 1, 3, 4, 1, 1
  b Mean = 1.8, mode = 2, median = 2
  c On average, houses in Ullswater Drive have more cars than those in Ambleside Close.
3 23 or 42
4 −3.1, −2.6, 3.5, 4.1, 4.1
5 Students' answers, for example, 0, 1, 2, 6, 7, 6, 14 or 6, 6, 6, 6, 7, 11, 20 or 3, 4, 5, 6, 9, 9, 17. In general, $a, b, c, 6, d, b + 5, a + 14$ with $a, 1 \leq b \leq c \leq 6 \leq d \leq b + 5$.
6 Four of the numbers are 6, 6, 8, and 16 and the fifth lies between 8 and 16. At its lower bound this gives a mean of 8.8, at its upper bound this gives a mean of 10.4
7 a Mean = 5.85, mode = 4, median = 6, range = 7, interquartile range = 4
  b Mean = 6.05, Mode = 4, Median = 6, Range = 8, interquartile range = 4
8 41 years old.
9 14 raisins.
*10 78.6%

4.3S

1 a 4, 9, 16, 11, 10   **b** 50
2 a 8, 6, 7, 10, 3, 6   **b** 40
3 a 7, 9, 13, 6, 5
  b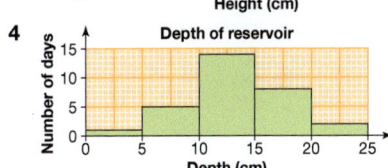
4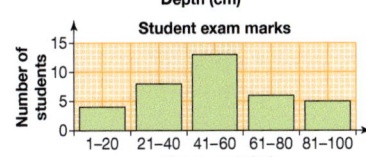
5
6 a i 2 **ii** 1
  b 85–90 m **c** 70–75 m **d** 10 athletes

4.3A

1 a Class width: 2, 1, 1, 1, 3
  Frequency density: 6, 17, 19, 11, 6
  b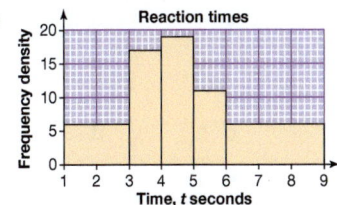
2 a Class width: 5, 5, 10, 20, 20, 40;
  Frequency density: 1.2, 2, 2.3, 1.45, 1.2, 0.2

b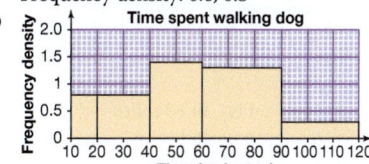
3 a Time, t minutes: $90 \leq t < 120$
  Class width: 10, 20, 20, 30
  Frequency: 16, 28, 39
  Frequency density: 0.8, 0.3
  b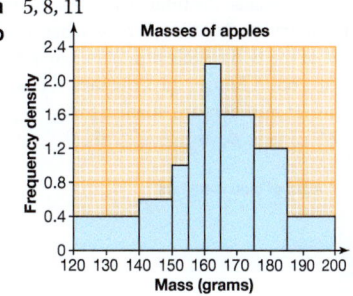
4 a 10 **b** 4, 18, 53, 20, 10 **c** 105
5 a 5, 8, 11
  b

Review

1 11
2 a Pie chart with labeled sectors.
    Warehouses 80°, Corporate/Finance 40°,
    Staff Training 140°, Transport 30°, IT Support 70°.
  b $648 000
  c Heights of bars ($1000s):
    144, 72, 252, 54, 126
3 a i 36.1 s **ii** 55 s
  b i 37 s **ii** 40 s **iii** 29 s
  c 11 s
4 a i 47 **ii** 22
  b i 37 **ii** 25
  c The women are younger in general and their ages vary more.
5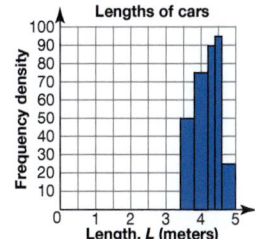

Assessment 4

1 a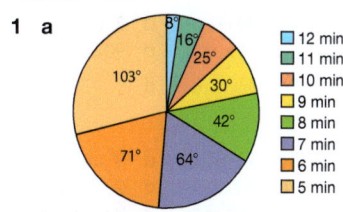
  b i 5 min **ii** 7 min
  c 7 min 1 second

2

	Glasses	No glasses
Left-handed	1	4
Right-handed	14	16

3 a

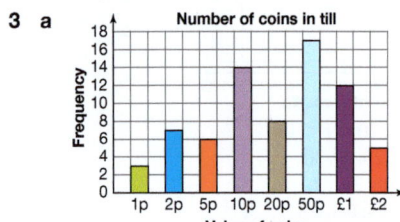

  b 50p   **c** 20p   **d** 47p
4 a Walk   **b** Car   **c** 20%   **d** 120
  e 300
5 a i 103 miles   **ii** 99.5 miles   **iii** 88 miles
  b Yes, if she completes approximately one journey per day she travels approximately 103 miles per day.
  c i 107 miles   **ii** 52.25 miles
  d Mean increases by 1, because the total has increased but not the number of journeys. Median increases, because 98 replaces 90 as one of the two middle numbers. No mode as each number occurs once. Range, stays the same because the highest and lowest numbers are unaffected.
6 a 9   **b** 2   **c** 8   **d** 0
  e 1.2
7 a

Ages of club members (histogram, Age Y years on x-axis, Frequency density on y-axis)

  b 32   **c** 31   **d** 59
8 a 70   **b** 60   **c** 360

Chapter 5

Check in

1 a 6 squares shaded   **b** 8 squares shaded
  c 9 squares shaded   **d** 10 squares shaded
  e 7 squares shaded
2 a $\frac{1}{2}$ **b** $\frac{3}{4}$ **c** $4\frac{4}{5}$ **d** $\frac{19}{20}$
  e $\frac{3}{4}$
3 a $20 \times 30 = 600$ **b** $360 \div 60 = 6$
  c $1200 - 800 = 400$ **d** $7000 + 6000 = 13\,000$

5.1S

1 a $3\frac{3}{4}$ **b** $1\frac{3}{5}$ **c** $3\frac{1}{3}$ **d** $\frac{13}{15}$
  e $2\frac{2}{5}$ **f** $1\frac{17}{25}$
2 a 10   **b** 21   **c** 12   **d** 13
  e 22   **f** 8
3 a 24 m   **b** 36 m
4 a 15 litres   **b** 180 cl   **c** 48 cl   **d** 250 ml
5 a 30 m   **b** 8 km   **c** 16 mm   **d** 90 m
6 a $1\frac{2}{3}$ **b** $5\frac{1}{2}$ **c** $4\frac{1}{2}$ **d** $\frac{2}{3}$
  e $9\frac{1}{3}$ **f** $28\frac{4}{5}$
7 a $2\frac{2}{3}$ miles   **b** $33\frac{1}{2}$ miles   **c** $12\frac{1}{2}$ miles   **d** $3\frac{3}{4}$ miles
8 a 6.7 m   **b** 5.3 mm   **c** 8.7 cm   **d** 22.3 km
9 a 10.5   **b** 126   **c** 240   **d** 99
  e 300   **f** 132   **g** $225   **h** 990 m

10 a 135   **b** 91   **c** 744   **d** 18
  e 387   **f** 192   **g** 1104 mm   **h** 1952 kg
  i €1443   **j** £493
11 a 0.5   **b** 0.6   **c** 0.25   **d** 0.51
  e 0.64   **f** 0.22   **g** 0.15   **h** 0.7
  i 0.07   **j** 0.085   **k** 0.0015   **l** 0.0001
12 a 5.7   **b** 200   **c** 15.93   **d** 99.84
  e 16.81   **f** 20
13 a €3.84   **b** $53.55   **c** $13.95   **d** £170.10
  e £28.16   **f** €15.90
14 a €9   **b** £2.70   **c** $4.20

5.1A

1 $82.90
2 1.8 pints
3 These times add up to 25 hours 24 min so Phil cannot be correct.
4 €6656
5 5.225 g
6 24°
7 £425
8 $\frac{2}{5}$
9 $\frac{3}{10}$
10 120
11 59%
12 Oldest gets 9, middle gets 6, youngest gets 2. 9+6+2 = 17. Strictly speaking they do not get the exact fractions of their father's wishes, but at least no horses were injured in the share out!

5.2S

1 a $\frac{3}{5}$ **b** 1   **c** 1   **d** $\frac{3}{5}$
  e $\frac{2}{3}$ **f** $\frac{3}{5}$
2 a $\frac{20}{30}$ **b** $\frac{18}{42}$ **c** $\frac{35}{45}$ **d** $\frac{25}{40}$
  e $\frac{3}{9}$ **f** $\frac{35}{42}$
3 a $\frac{3}{10}$ **b** $\frac{5}{6}$ **c** $\frac{11}{20}$ **d** $\frac{3}{8}$
  e $\frac{17}{20}$ **f** $\frac{37}{56}$
4 a $1\frac{1}{4}$ **b** $1\frac{4}{5}$ **c** $1\frac{5}{8}$ **d** $4\frac{1}{4}$
5 a $\frac{7}{4}$ **b** $\frac{23}{16}$ **c** $\frac{14}{9}$ **d** $\frac{18}{7}$
6 a $\frac{33}{10}$ **b** $\frac{25}{8}$ **c** $\frac{111}{14}$ **d** $\frac{496}{53}$
  e $\frac{27}{20}$ **f** $\frac{3}{4}$ **g** $\frac{39}{20}$ **h** $\frac{69}{14}$
7 a $\frac{3}{20}$ **b** $\frac{4}{27}$ **c** $\frac{1}{14}$ **d** $\frac{1}{4}$
  e $\frac{20}{63}$ **f** $\frac{1}{12}$ **g** $\frac{12}{65}$ **h** $\frac{1}{15}$
  i $\frac{5}{21}$ **j** $\frac{1}{12}$
8 a $\frac{3}{4}$ **b** $\frac{3}{4}$ **c** $\frac{4}{5}$ **d** $1\frac{4}{5}$
  e $3\frac{1}{7}$ **f** $2\frac{3}{4}$
9 a $\frac{5}{32}$ **b** $\frac{1}{8}$ **c** $\frac{2}{21}$ **d** $\frac{1}{48}$
  e 20   **f** 9   **g** $\frac{25}{2}$ **h** $\frac{77}{3}$
  i 9
10 a $\frac{1}{3}$ **b** $\frac{1}{4}$ **c** $\frac{1}{6}$ **d** $\frac{1}{4}$
  e $\frac{8}{9}$ **f** $\frac{45}{56}$ **g** $\frac{2}{9}$ **h** $\frac{8}{21}$
  i $\frac{27}{28}$
11 a $\frac{9}{8}$ **b** $\frac{11}{10}$ **c** $2\frac{16}{25}$ **d** 3
  e $\frac{115}{24}$ **f** $\frac{17}{14}$ **g** $\frac{43}{22}$ **h** $\frac{112}{27}$
12 a $\frac{49}{8}$ **b** $\frac{5}{8}$ **c** $\frac{165}{16}$ **d** $\frac{2}{3}$
13 a $\frac{10}{3}$ **b** $\frac{35}{64}$ **c** $\frac{15}{38}$ **d** $\frac{697}{160}$

Answers

5.2A

1 $\frac{5}{8}$ inch
2 a $3\frac{1}{8}$
   b Asif has added the numerators (1, 1, 1, and −1) and added the denominators.
3 €59.38
4 a $\frac{1}{2}+\frac{5}{2}, 2\frac{1}{2}+\frac{1}{2}, \frac{5}{2}+2\frac{1}{2}$ b $\frac{3}{4}\times\frac{4}{3}$ c $\frac{3}{5}\times\frac{5}{2}$ or $\frac{3}{5}\times 2\frac{1}{2}$
   d $\frac{3}{4}=\frac{3}{8}\times 2$ e $2\frac{3}{8}$
   f $\frac{1}{8}$ goes in to $\frac{5}{2}$ 20 times.   g $\frac{3}{4}$ and $\frac{4}{3}$
5 a i $\frac{5}{14},\frac{3}{7},\frac{7}{8}$ ii $\frac{2}{7},\frac{2}{3},\frac{5}{6}$
   b i $\frac{3}{4},\frac{17}{40},\frac{2}{5},\frac{3}{8}$ ii $\frac{5}{6},\frac{5}{8},\frac{7}{12},\frac{11}{24}$
6 $\frac{6}{11}\cdot\frac{6}{11}-\frac{1}{2}=\frac{1}{22}$ and $\frac{1}{2}-\frac{4}{9}=\frac{1}{18},\frac{1}{22}<\frac{1}{18}$.
7 $\frac{6}{35}$
8 $\frac{3}{8},\frac{1}{10}$. These are the smallest fractions in the list.
9 No. For example, $4>2$ but $\frac{1}{4}<\frac{2}{4}=\frac{1}{2}$
10 First row $\frac{2}{3}$; second row $\frac{3}{10},\frac{1}{2},\frac{3}{20}$; third row $\frac{2}{3}$. ✶ is ×.
11 28
12 a Zosia: $\frac{28}{35}\div\frac{15}{35}=\frac{28}{15}$ Ahmad: $\frac{28}{5}\div 3=28\div 15=\frac{28}{15}$, Mara $\frac{4}{5}=\text{Ans}\times\frac{3}{7}\Rightarrow\frac{4}{5}\times\frac{7}{3}=\text{Ans}\times\frac{3}{7}\times\frac{7}{3}\Rightarrow\text{Ans}=\frac{28}{15}$
   b Zosia $\frac{ad}{bd}\div\frac{bc}{bd}=\frac{ad}{bc}$ Ahmad $\frac{ad}{b}\div c = ad\div bc = \frac{ad}{bc}$, Mara $\frac{a}{b}=\text{Ans}\times\frac{c}{d}\Rightarrow\frac{a}{b}\times\frac{d}{c}=\text{Ans}\times\frac{c}{d}\times\frac{d}{c}\Rightarrow\text{Ans}=\frac{a}{b}\times\frac{d}{c}=\frac{ad}{bc}$

5.3S

1 a i 0.5   ii 50%
   b i 0.75   ii 75%
   c i 0.4   ii 40%
   d i 0.1   ii 10%
   e i 0.2   ii 20%
   f i 0.25   ii 25%
   g i 0.125   ii 12.5%
   h i 0.625   ii 62.5%
2 a 62.5%   b 80%   c 87.5%   d 60%
   e 37.5%   f 12.5%   g 18.75%   h 21.875%
4 a i 0.0625   ii 6.25%
   b i 0.28   ii 28%
   c i 0.056   ii 5.6%
   d i 0.075   ii 7.5%
   e i 0.4375   ii 43.75%
   f i 0.03125   ii 3.125%
   g i 0.5125   ii 51.25%
   h i 0.20625   ii 20.625%
5 a No, the denominator has prime factors 2 and 5 only.
   b No, the denominator has prime factors 2 and 5 only.
   c Yes, the denominator has prime factor 11.
   d No, the denominator has prime factors 2 and 5 only.
6 a $0.\dot{1}$ b $0.\dot{5}$ c $0.7\dot{5}$ d $0.3\dot{4}\dot{6}$
   e $0.76\dot{5}$ f $0.1\dot{6}$
7 a i $0.\dot{3}$ ii 33.$\dot{3}$%
   b i $0.1\dot{6}$ ii 16.$\dot{6}$%
   c i $0.\dot{6}$ ii 66.$\dot{6}$%
   d i $0.\dot{1}4285\dot{7}$ ii 14.$\dot{2}$8571$\dot{4}$%
   e i $0.\dot{1}$ ii 11.$\dot{1}$%
   f i $0.8\dot{3}$ ii 83.$\dot{3}$%
   g i $0.\dot{0}\dot{9}$ ii 9.$\dot{0}\dot{9}$%
   h i $0.08\dot{3}$ ii 8.$\dot{3}$%
8 a $0.\dot{4}2857\dot{1}$ b 0.1875   c 0.2125   d $0.\dot{5}$
   e 0.16   f $0.\dot{7}1428\dot{5}$ g 0.328   h $0.91\dot{6}$
10 a 0.43   b 0.86   c 0.94   d 0.455
    e 0.0375   f 1.05

11 a $\frac{1}{2}$ b $\frac{1}{4}$ c $\frac{1}{5}$ d $\frac{1}{8}$
    e $\frac{3}{4}$ f $\frac{9}{10}$
12 a $\frac{51}{100}$ b $\frac{43}{100}$ c $\frac{413}{1000}$ d $\frac{719}{1000}$
    e $\frac{91}{100}$ f $\frac{871}{1000}$
13 a $\frac{8}{25}$ b $\frac{11}{20}$ c $\frac{1}{25}$ d $\frac{31}{200}$
    e $\frac{16}{25}$ f $\frac{53}{200}$
14 a $\frac{49}{100}$ b $\frac{53}{100}$ c $\frac{73}{100}$ d $\frac{81}{100}$
    e $\frac{37}{100}$ f $\frac{19}{100}$
15 a $\frac{11}{20}$ b $\frac{31}{50}$ c $\frac{21}{25}$ d $\frac{13}{20}$
    e $\frac{18}{25}$ f $\frac{37}{200}$
16 a $\frac{7}{9}$ b $\frac{2}{3}$ c $\frac{5}{33}$ d $\frac{17}{33}$
    e $\frac{41}{333}$ f $\frac{4}{27}$ g $\frac{1}{30}$ h $\frac{7}{45}$
    i $\frac{1}{36}$ j $\frac{7}{110}$ k $\frac{27}{110}$ l $\frac{19}{135}$
17 a $\frac{1}{3}$ b $\frac{1}{90}$ c $\frac{1}{2250}$ d $\frac{111}{550}$
    e $\frac{1}{540}$ f $\frac{1107}{11000}$
18 $\frac{1}{81}$
19 a 45%   b $\frac{1}{6}$ c $\frac{10}{81}$ d $\frac{7}{40}$

5.3A

1 a 10D   b 10B
2 a 0.33, 33.3%, $33\frac{1}{3}$%, 33
   b $0.\dot{4}$, 44.5%, 0.45, 0.454
   c 22.3%, 0.232, 23.22%, 0.233, $0.2\dot{3}$
   d $0.6\dot{5}$, 0.66, 66.6%, 0.6666, $\frac{2}{3}$
   e 14%, 14.$\dot{1}$%, 0.142, $\frac{1}{7}$, $\frac{51}{350}$
   f $\frac{5}{6}, \frac{6}{7}$, 0.866, $0.8\dot{6}$, 89%
3 $\frac{4}{29}$
4 No. $\frac{800-46}{800}=0.9425=94.25\%$ of customers wait longer than 5 minutes $<95\%$.
5 a The two shops charge the same price.
   b Shop A is cheaper.
   If first book costs x and second book costs y with $x>y$ then cost in Shop B − cost in Shop A $=\frac{1}{4}(x-y)>0$ as $x>y$.
   So Shop B is more expensive.
6 $\frac{9}{6400}$
7 It is equidistant from both. $0.6\dot{3}-0.5=\frac{3}{22}$, $0.5-0.3\dot{6}=\frac{3}{22}$
8 $\frac{59}{165}$
9 a 1, this is $3\times\frac{1}{3}$.   b $\frac{1}{2}$, this is $0.4+\frac{1}{10}\times$ (part **a**).
10 a 91.4 g   b 70 g fat, 19.3 g saturates, 6.7 g salt
    c 66.5 g
11 a $x=\frac{1}{9}$ b $x=\frac{49}{110}$

Review

1 a 4   b 15   c 12   d 24
2 a 21   b 10.2   c 1   d 5
3 a $\frac{20}{7}$ b $\frac{16}{11}$
4 a $2\frac{2}{9}$ b $7\frac{1}{7}$
5 a $\frac{1}{5}$ b 3   c $\frac{7}{2}$ d $\frac{5}{7}$
6 a $\frac{1}{10}$ b $\frac{27}{8}$ or $3\frac{3}{8}$ c $\frac{2}{3}$ d 18
   e $\frac{1}{2}$ f $\frac{1}{2}$ g $\frac{17}{33}$ h $\frac{41}{28}$ or $1\frac{13}{28}$
7 a 0.8   b 0.85   c 0.03   d 0.375
   e 0.22   f 0.004   g 2   h 0.055
8 a $\frac{7}{10}$ b $\frac{107}{500}$ c $\frac{9}{25}$ d $\frac{1}{100}$
   e $\frac{3}{25}$ f $\frac{3}{2}$ g $\frac{111}{250}$ h $\frac{1}{500}$
9 a 60%   b 65%   c 0.7%   d 45%
   e 35%   f 80%   g 9%   h 180%

10 a $0.\dot{2}$ b $0.\dot{8}5714\dot{2}$
11 a $\frac{8}{9}$ b $\frac{7}{30}$
12 a $\frac{9}{16}, \frac{5}{8}, \frac{2}{3}$ b $\frac{1}{5}, 22.2\%, 0.\dot{2}$

Assessment 5

1 a $\frac{11}{50}$
   b Labour $5.60, materials $2, advertising $8, profit $4.40.
   c $20.09
2 a Blue sky   b Space   c Chat-chat
   d No, 8 is 32% of 25 but 40% of 20.
3 a 36.4% (3 sf)   b 14%
4 a 0.12   b 4th   c 3rd
5 $1\frac{1}{2}$ hectares
6 8.5 yards
7 a 540 m².   b 252 m²
8 $\frac{10}{33}$
9 Yes, % increase = 60%.
10 $\frac{1}{7}$ is not exactly 14%. Doubling 14% also doubles the error.
   $\frac{1}{7} = 14.2857....\%$ so $\frac{2}{7} = 28.5714......\% = 29\%$ (nearest %).
11 If 102 people represent *exactly* 34% of the total, total = 102 ÷ 0.34 = 300, and 66% of 300 = 198 not 200.
12 33.3%, $33\frac{1}{3}\%$, 0.334, 0.34, $\frac{5}{14}, \frac{3}{8}$
13 $\frac{1}{3} = 0.\dot{3} = 0.3333...$
   $1 = \frac{1}{3} + \frac{1}{3} + \frac{1}{3}$
   0.333333...+
   0.333333...+
   0.333333...
   ———————
   $0.999999... = 0.\dot{9}$
14 a i $0.\dot{2}$ ii 0.52   iii $0.91\dot{6}$ iv 0.35
    v $0.6\dot{3}$ vi 0.6875
   b If the denominator of a fraction has a prime factor other than 1, 2 and 5 the decimal expansion will be recurring.
   c $\frac{14}{30}, \frac{19}{99}, \frac{425}{612}$
15 a $\frac{2}{11}$ b $\frac{4}{15}$ c $\frac{34}{333}$ d $\frac{4165}{9999}$
   e $\frac{5}{22}$ f $\frac{3479}{9900}$

Life skills 1

1 a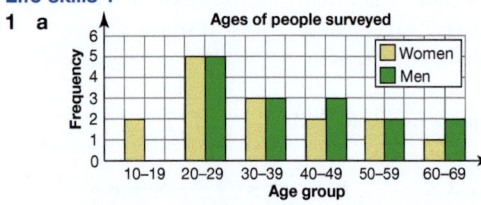

Overall, the men interviewed were older than the women.
   b Yes, but the difference is small. Women's mean = £28.60, men's mean = £30.
2 a £222 222   b $P = R - G - S - C$
   c 52 222   d $S = R - G - C - P$
                   $S = £97\,778$
3 (16–24) 38, (25–49) 93, (50–64) 36, (65+) 33.
4 a $\frac{7}{40}$
   b Pie chart angles: A 144°, R 90°, J 63°, M 63°.
   c Abigail £20 000, Raheem £12 500, Juliet £8750, Mike £8750.
5 a i £21 061.82   ii £19 963.55
   b i £7121.89   ii £7513.69
   c 10.4%
   d $C = \frac{A}{1+i}\left(\frac{1-\left(\frac{1}{1+i}\right)^5}{1-\left(\frac{1}{1+i}\right)}\right) = \frac{A}{1+i}\left(\frac{1-\left(\frac{1}{1+i}\right)^5}{\frac{1+i-1}{1+i}}\right) = \frac{A}{1+i} \times \frac{1-\left(\frac{1}{1+i}\right)^5}{\frac{i}{1+i}}$
      $= \frac{A}{i}\left(1-\left(\frac{1}{1+i}\right)^5\right)$
   e $A = \dfrac{Ci}{1-\left(\frac{1}{1+i}\right)^5}$

Chapter 6

Check in

1 a i 12   ii 15   iii 8   iv 45   v 1
   b x
2 a $4n + 7$ b $3(n - 6)$ c $10 - n^2$ d $\frac{3n-6}{2}$
   e n^5

6.1S

1 a 23   b 30   c 28   d 81
   e 162   f 3
2 a 18   b 4   c 6   d 2
   e 5
3 77°F
4 a $A = p + 6$ b $A = m - 5$
   c $A = \frac{2k}{7}$ d $A = 6 + 5t$
   e $A = 3d + 7$ f $A = 3(t + 7)$
   g $A = \sqrt{y} + 8$
5 a $T = 2\pi\sqrt{\frac{25}{9.8}} = 10$ s (3 sf)   b 52.2 cm (3 sf)
6 a $x = y - 3$ b $x = y + c$ c $x = \frac{y}{5}$
   d $x = \frac{y-P}{3}$ e $x = 5y$ f $x = c(y - d)$
   g $x = \pm\sqrt{y}$ h $x = y^2$ i $x = \sqrt[3]{y}$
   j $x = y^3$ k $x = (y - 2)^2$
   l $x = \pm\sqrt{5T - 2}$
7 a $p = \frac{A-5}{2}$ b $p = \frac{A-d}{2}$ c $p = \frac{A+6}{3}$
   d $p = \frac{H}{4d}$ e $p = 4(H - t)$ f $p = 4H - t$
   g $p = 5W^2$ h $p = \frac{w^2 - D}{w}$
8 $x = 10 - y$
9 a $x = \frac{p-n}{m-o}$ b $x = \frac{b+d}{a-c}$
   c $x = \frac{g-f}{h+e}$ d $x = \frac{t-r}{u-s}$
*10 $x = \frac{3y+2}{y-1}$
11 a $x = \frac{qz+py}{p+q}$ b $x = \frac{t+5k}{k-1}$
   c $x = \frac{3q-4p}{7}$ d $x = -\frac{(a+16b)}{15}$
   e $x = \sqrt{\frac{4n^2+3m}{11}}$
*12 $R = \frac{30(p+5)}{15-2p}$

6.1A

1 a 511 185 932.5 km²   b $r = \sqrt{\frac{S}{4\pi}}$
   c 1737.3 km (1 dp)
2 a $r = \sqrt{\frac{V}{\pi h}}$ b 4.2 m (1 dp)
3 a $h = \frac{S-2\pi r^2}{2\pi r} = \frac{S}{2\pi r} - r$
   b $h = \sqrt{\left(\frac{S-\pi r^2}{\pi r}\right)^2 - r^2} = \sqrt{\left(\frac{S}{\pi r}\right)^2 - \frac{2S}{\pi}}$
4 a $f = \frac{uv}{u+v}$ b $v = \frac{uf}{u-f}$
5 a $V = \frac{P}{I}$ b $R = \frac{V}{I}$ c $E = VIt$ d $E = IRQ$
   e $E = I^2Rt$
6 a $c = 0.64d + 6.99$ b $d = \frac{c-6.99}{0.64}$ c $65
7 a i 11.6 cm²   ii 29.4 cm² (1 dp)
    iii 7.9 cm² (1 dp)   iv 63.7 cm² (1 dp)
8 a 1243.6 cm³   b $R = 15$ mm, $r = 7$ mm

6.2S

1 a $f(x) = 3x$ b $f(x) = x - 2$
   c $f(x) = 2x + 1$ d $f(x) = x^2$
   e $f(x) = \frac{1}{x}$ f $f(x) = x^3$
2 a 5   b 11   c −5   d 1
3 a 2   b 7   c −1   d 0
4 −2, 0, 3, 4.5 $\xrightarrow{f(x)}$ −12, −2, 13, 20.5

5 a $\frac{1}{4}$ b $\frac{1}{2}$ c $\frac{2}{5}$
  d Not possible/undefined
6 a $f^{-1}(x) = x - 4$ b $g^{-1}(x) = x + 3$
  c $h^{-1}(x) = 2x$ d $f^{-1}(x) = \frac{x}{5}$
  e $g^{-1}(x) = \frac{x+3}{2}$ f $h^{-1}(x) = \frac{x-2}{7}$
  g $f^{-1}(x) = 3(x + 4)$ h $f^{-1}(x) = 2x - 5$
7 a $f^{-1}(x) = \frac{1}{x}$ b $g^{-1}(x) = 2 - x$
  The inverse is the same as the function.
8 a 17 b 80 c 27 d 36
  e 43 f 44 g $8x + 3$ h $8x + 12$
9 a $(3x + 1)^2$ b $3x^2 + 1$ c $x = \frac{1}{3}$
10 $f^{-1}(x) = \frac{x-5}{4}$
  $ff^{-1}(x) = 4\left(\frac{x-5}{4}\right) + 5 = x - 5 + 5 = x$
*11 For example: *As a function maps an input 'x' to an output 'y' then the inverse function maps 'y' back onto 'x'.*
12 For example: *for the function $f(x) = x^2$, 3 will map to 9 but so will -3, Therefore, the inverse function might map 9 to 3 or -3. The domain has to be restricted to ensure a one-to-one mapping.*
13 a $\frac{x-1}{2}$ b $\frac{x+2}{3}$ c $6x - 3$ d $\frac{x+3}{6}$
  e $\frac{x+3}{6}$ f $(fg)^{-1}(x) = g^{-1}f^{-1}(x)$

6.2A

1 a i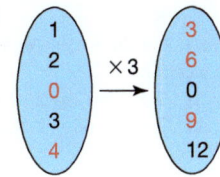
    ii $y = 3x$
    iii $f(x) = 3x$
    iv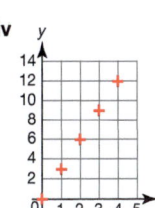
  b i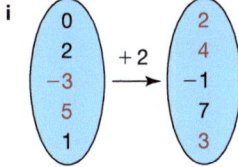
    ii $y = x + 2$
    iii $f(x) = x + 2$
    iv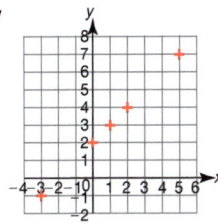
  c i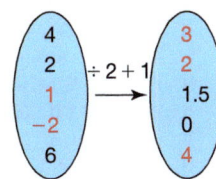
    ii $y = \frac{1}{2}x + 1$
    iii $f(x) = \frac{1}{2}x + 1$
    iv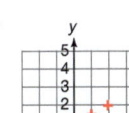
  d i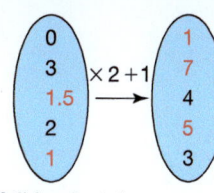
    ii $y = 2x + 1$
    iii $f(x) = 2x + 1$
    iv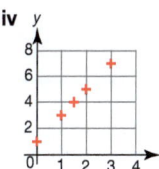

2 a i $f(x) = \frac{1}{3}x$
    ii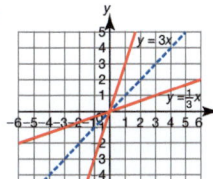
  b i $f(x) = x - 2$
    ii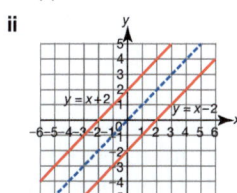

2 c i $f(x) = 2(x - 1)$
    ii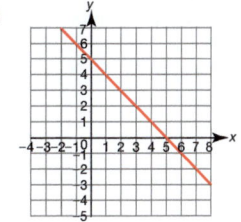
  d i $f(x) = \frac{x-1}{2}$
    ii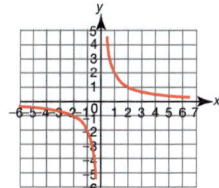
    iii The inverse function is always a reflection in the line $y = x$

3 a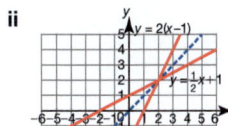
  b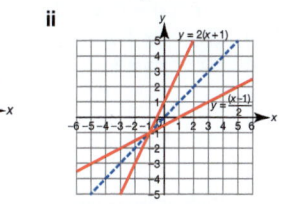
  c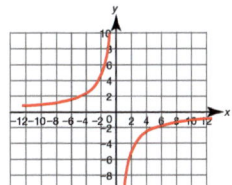
  d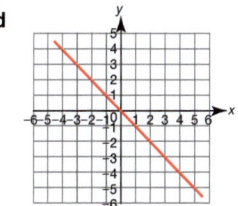
  Each function and its inverse have the same graph

4 a is a many-to-one function, every value in the domain is mapped to one value in the range.
  b is not a function as 0 maps to more than one value in the range.

5 a

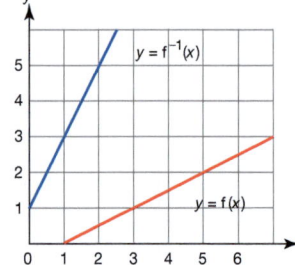

b Many answers possible, for example $f(x) = \frac{x-1}{2}$

6

$f(x)$ is one-to-many so is not a function.

7 a $f(x) = x$ or $f(x) = c - x$
b $f(x) = x$ or $f(x) = c - x$ where c is a number.

8 a $f_n f_m(x) = f_n(x^m) = (x^m)^n = x^{nm} = x^{mn} = f_{mn}(x)$
b i $m \times n = n \times m$, $f_n f_m(x) = x^{mn} = x^{nm} = f_m f_n(x)$
ii Let $f_n^{-1}(x) = f_m(x)$.
$f_n f_m(x) = f_m f_n(x) = f_{mn}(x) = x^{mn} = x^1$, $mn = 1$ so $m = \frac{1}{n}$.
$f_n^{-1}(x) = f_m(x) = f_{\frac{1}{n}}(x)$

9 a $g(x) = 3x + 2$ **b** $g(x) = \frac{1}{2}x - 2$
c $g(x) = 3x - 4$ ***d** $g(x) = 3x - 5$

6.3S

1 $x + 3 < 10$ inequality, $x(x+3) \equiv x^2 + 3x$ identity, $2x + 1 = 6$ equation, $x^2 + 3x$ expression, $v = u + at$ formula.

2 Multiple answers possible, for example
a $3x + 2$ **b** $4x - 5 = 7$
c $F = ma$ **d** $3x < x + 2$
e $x(x + 2) \equiv x^2 + 2x$

3 a No, $4(a + 2) \equiv 4a + 8$ **b** Yes, $3(x + 2) \equiv 3x + 3 \times 2$
c Yes, $5(y - 2) \equiv 5y - 5 \times 2$ **d** No, $y(y + 3) \equiv y^2 + 3 \times y$
e Yes, $x(x - 4) \equiv x^2 - 4 \times x$

4 a $5(a + 2) \equiv 5 \times a + 5 \times 2 \equiv 5a + 10$
b $3(x + 4) \equiv 3 \times x + 3 \times 4 \equiv 3x + 12$
c $5(y - 3) \equiv 5 \times y - 5 \times 3 \equiv 5y - 15$
d $y(y + 3) \equiv y \times y + y \times 3 \equiv y^2 + 3y$
e $x^2(x - 4) \equiv x^2 \times x^1 - x^2 \times 4 \equiv x^3 - 4x^2$

5 a $4a + 8 + 2a + 2 \equiv 6a + 10$
b $3x + 6 + 4x - 4 \equiv 7x + 2$
c $5y - 10 + 3y - 9 \equiv 8y - 9$
d $y^2 + 3y + 2y + 6 \equiv y^2 + 5y + 6$
e $x^2 - 4x + x^2 + 2x \equiv 2x^2 - 2x$

6 a $a = 7, b = 9$ **b** $a = 7, b = 6$
c Multiple solutions possible including $a = 1, b = 8$
d $a = 1$ and $b = 2$ **e** $a = 3$ and $b = 10$
f $a = 6$ and $b = -22$

7 a No, LHS $\equiv a^2 + 7a + 10$
b No, LHS $\equiv x^2 + 7x + 12$
c Yes, $b^2 + 2b + 6b + 12 \equiv b^2 + 8b + 12$
d Yes, $y^2 + 3y - 2y - 6 \equiv y^2 + y - 6$
e No, LHS $\equiv 2y^2 - 4y - 6$
f No, LHS $\equiv 3p^2 - 9p + 6$

8 a $8(1 + 3p)$ **b** $9(2q - 3)$ **c** $5(5r + 11)$ **d** $4s(s + -3)$
e $7t(3 + 4t)$ **f** $18u^2(3v + u)$

9 a Formula **b** Identity **c** Equation **d** Equation
e Equation **f** Identity **g** Identity **h** Formula
i Formula **j** Equation **k** Identity

10 Student's own answer.

6.3A

1 a i $5(a + 2) \equiv 5a + 10$ **ii** $5a + 10 = 80$
b i $7(b + 5) \equiv 7b + 35$ **ii** $7b + 35 = 105$
c i $2(c + 1) \equiv 2c + 2$ **ii** $2c + 2 = 24$
d i $12b + 120 \equiv 12(b + 10)$ **ii** $12(b + 10) = 240$

2 a i $x < 10$ **ii** $190 - 7x \equiv 120 + 7(10 - x)$
iii Student's suggestion, e.g. $x = 5$
iv Student's suggestion, e.g. $190 - 7x = 155$
b i $x < 11$
ii $132 - 4x \equiv 88 + 4(11 - x)$
iii Student's suggestion, e.g. $x = 4$
iv Student's suggestion, e.g. $132 - 4x = 116$

3 a $2(g + 7) \equiv 2g + 14$
b $2(m + n) \equiv 2m + 2n$
c $a(b + c) \equiv ab + ac$
d $f(3 + b + d) \equiv 3f + bf + df$
e $(2a)^2 \equiv 4a^2$
f $(a+b)^2 \equiv a^2 + 2ab + b^2$

4 a Sometimes true **b** Always true
c Always true **d** Sometimes true
e Never true **f** Never true
g Sometimes true **h** Sometimes true

5 a Student's own examples e.g.
$3 + 4 + 5 + 6 + 7 = 25 = 5 \times 5$
b $n + (n + 1) + (n + 2) + (n + 3) + (n + 4) = 5n + 10$
$= 5(n + 2)$ which has 5 as a factor.

6 a $7 + 3 = 10 = 2 \times 5$ **b** $5 + 7 = 12 = 2 \times 6$
c $0.5^2 = 0.25 < 0.5$ **d** $7 \times 8 = 56 = 2 \times 28$

7 a $(2n)^2 \equiv 2^2 \times n^2 = 4n^2$
b $n(n + 1) - n \equiv n^2 + n - n \equiv n^2$
c $2n(2m + 1) \equiv 4mn + 2n \equiv 2(2mn + n)$
d $(2n + 1) + (2m + 1) = 2n + 2m + 2 = 2(n + m + 1)$

6.4S

1 a $x^2 + 8x + 12$ **b** $x^2 + 8x + 15$
c $x^2 + 6x - 16$ **d** $x^2 - 4x - 12$
e $x^2 - 2x - 15$ **f** $x^2 - 6x - 16$

2 a $x^2 + 7x + 10$ **b** $x^2 + 7x + 12$
c $y^2 - 5y - 14$ **d** $y^2 - 36$
e $b^2 + 5b - 24$ **f** $b^2 - 81$
g $a^2 + 6a + 9$ **h** $a^2 - 14a + 49$

3 a $2x^2 + 9x + 10$ **b** $3x^2 - 5x - 12$
c $6y^2 + 20y + 14$ **d** $9y^2 - 36$
e $8b^2 + 26b - 24$ **f** $25b^2 - 81$
g $4a^2 + 12a + 9$ **h** $9a^2 - 42a + 49$

4 a $x^2 - 8x + 12$ **b** $x^2 - 8x + 15$
c $x^2 - 10x + 16$

5 a $x^2 - 7x + 10$ **b** $x^2 - 7x + 12$
c $y^2 - 4y + 4$ **d** $y^2 - 12y + 36$

6 a $3x^2 - 11x + 10$ **b** $2x^2 - 10x + 12$
c $6x^2 - 16x + 10$ **d** $8x^2 - 22x + 12$
e $25x^2 - 20x + 4$ **f** $4x^2 - 20x + 25$

7 a $x^3 + 8x^2 + 31x + 30$ **b** $x^3 + 8x^2 + 19x + 12$
c $y^3 - 2y^2 - 29y - 42$ **d** $y^3 + 4y^2 - 36y - 144$
e $b^3 + 3b^2 - 34b + 48$ **f** $b^3 - 2b^2 - 81b + 162$
g $a^3 + 9a^2 + 27a + 27$ **h** $y^3 - 21y^2 + 147y - 343$

8 a $x(x + 5)$ **b** $x(x + 7)$ **c** $2x(x + 6)$ **d** $6x(2x + 1)$

9 a $(x + 2)(x + 3)$ **b** $(x + 2)(x + 5)$
c $(a + 3)(a + 8)$ **d** $(a + 12)(a + 1)$
e $(x + 7)(x + 13)$ **f** $(x + 19)(x + 1)$
g $(x + 3)(x - 2)$ **h** $(x + 5)(x - 2)$
i $(y + 4)(y - 3)$ **j** $(y + 5)(y - 3)$
k $(b - 12)(b + 1)$ **l** $(a - 8)(a + 3)$
m $(p - 6)(p - 3)$ **n** $(x - 20)(x + 1)$

10 a $(2x + 1)(x + 3)$ **b** $(2y + 3)(y + 1)$
c $3(b + 2)(b + 1)$ **d** $(3b + 2)(b + 3)$

e $(2x - 3)(x + 2)$
f $2(x + 3)(x - 1)$
g $(4x + 3)(x + 4)$
h $2(x - 2)(2x + 1)$
i $(3x - 2)(5x - 7)$
j $(2x + 5)(9x - 4)$
11 a $(x + 3)(x - 3)$
b $(y + 5)(y - 5)$
c $(b + 10)(b - 10)$
d $(h + 9)(h - 9)$
e $(y + 8)(y - 8)$
f $(a + 15)(a - 15)$
g $4(x - 5)(x + 5)$
h $(3x + 8)(3x - 8)$
i $5(b + 5)(b - 5)$
j Cannot be factorised
12 a $\left(x + \frac{1}{2}\right)\left(x - \frac{1}{2}\right)$
b $\left(x + \frac{7}{9}\right)\left(x - \frac{7}{9}\right)$
c $(20x + 13)(20x - 13)$
d $(x + \sqrt{2})(x - \sqrt{2})$
e $(3x + 2y)(3x - 2y)$
f $\left(5x + \frac{y}{3}\right)\left(5x - \frac{y}{3}\right)$
g $\left(\frac{4}{7}x + \frac{8}{9}y\right)\left(\frac{4}{7}x - \frac{8}{9}y\right)$
h $x^2(x - 16)$
13 a $(x + 2y)^2$
b Cannot be factorised
c $5xy(1 + 2xy)$
d $-(x + 5)(x - 2)$
e $-(2x - 3)(4x + 1)$
f $-y(3x^2 - 15x - 2)$
g $(x^2 - 3)(x + 2)(x - 2)$
h $(2x^2y^2 - 3)(x^2y^2 + 2)$

6.4A

1 a $3x(2x + 3) \equiv 3x \times 2x + 3x \times 3$
b $(2x + 6)^2 \equiv 2x \times 2x + 6 \times 2x + 2x \times 6 + 6 \times 6$
2 $(2n + 5)(2n - 1) + 3 \times 3 \equiv 4n^2 + 8n + 4 \equiv (2n + 2)^2$
Yes. She can make a square with side $2n + 2$.
3 a 3400 b 480 c 240 d 1000
e 3596 f 1551 g 2491 h 2484
4 a $(a + b)^2 \equiv a^2 + 2ab + b^2$
b i $(1.32 + 2.68)^2 = 16$ ii $(21 - 11)^2 = 100$
5 Let n^2 and $(n + 1)^2$ be two consecutive square numbers
Then $(n + 1)^2 - n^2 \equiv n^2 + 2n + 1 - n^2 \equiv 2n + 1 \equiv n + (n + 1)$
6 Let $(2n)^2$ and $(2n + 2)^2$ be two consecutive even square numbers (an even square number must have an even square root).
Then $(2n + 2)^2 - (2n)^2 \equiv 4n^2 + 8n + 4 - 4n^2 \equiv 8n + 4$
$\equiv 4(2n + 1)$
7 Let $(n - 1), n, (n + 1)$ be three consecutive numbers.
$(n - 1)(n + 1) \equiv n^2 + n - n - 1 \equiv n^2 - 1 < n^2$
8 The nth term formula is $(n + 1)(n + 2) - n(n + 3)$
$\equiv n^2 + 3n + 2 - n^2 - 3n \equiv 2$
*9 a 14 b 1 c 19
d 13 e 14 f 31
*10 $(a + b)(a - b)(a^2 + b^2)$
*11 a i $a^2 + 2ab + b^2$
ii $a^3 + 3a^2b + 3ab^2 + b^3$
iii $a^4 + 4a^3b + 6a^2b^2 + 4ab^3 + b^4$
b When written in the standard way (descending powers of a or descending powers of b) the numerical coefficients of the terms in $(a + b)^n$ are the numbers from the $(n + 1)$th row of Pascal's triangle.
12 a $(x^2 - 1)(x + 1) \equiv x^3 + x^2 - x - 1$
b $(x^2 + x - 2)(x + 2) \equiv x^3 + 2x^2 + x^2 + 2x - 2x - 4$
$\equiv x^3 + 3x^2 - 4$
c $x^3 + x^2 + x - x^2 - x - 1 \equiv x^3 - 1$
13 a $a = 1, b = 3, c = 11$ b $a = 1, b = 5, c = 22$
c $a = -1, b = -3, c = 23$ d $a = 2, b = \frac{3}{2}, c = \frac{11}{2}$
e $a = -6, b = -3, c = 78$
14 a i Smallest value ii -3 iii 11
b i Smallest value ii -5 iii 22
c i Largest value ii 3 iii 23
d i Smallest value ii $-\frac{3}{2}$ iii $\frac{11}{2}$
e i Largest value ii 3 iii 78
*15 a $\frac{x + 5}{4}$ b $\frac{2y - 3}{y - 10}$ c $\frac{x - 3}{5}$ d $\frac{2(x + 4)}{x + 3}$
e $\frac{x + 4}{x - 2}$ f $\frac{x + 2}{3x - 2}$
16 $\frac{1}{p - 4}$

Review

1 a -11 b 20 c 9 d 5

2 a $X = \frac{A - 3}{2}$ b $X = \frac{3C + B}{A}$
c $X = \frac{20Z - Y}{3}$ d $X = 4Y^2 - 4$
e $X = \pm\sqrt{L^2 + 2K}$
3 a i 35 ii -2
b i $\frac{x}{5}$ ii $x - 3$
c i $5x + 3$ ii $5(x + 3)$
4 a $v^2 = u^2 + 2as$ b $3(x + 2) \equiv 3x + 6$
c $5x^2 + 3$ d $7a + 5 = 19$
5 a $x > -2$ b $y \leq 0$
6 $5(2x + 3) + 2(4 - 5x) \equiv 10x + 15 + 8 - 10x \equiv 23$
7 Let n be an integer.
$n + (n + 1) + (n + 2) \equiv 3n + 3 \equiv 3(n + 1)$
8 a $x^2 + 11x + 18$ b $x^3 - 13x^2 + 42x$
c $3x^2 + 31x - 22$ d $12x^2 + 2x - 2$
9 a $(x + 9)(x - 9)$ b $(4x + 7)(4x - 7)$
c $x(x - 7)$ d $7x(3x + 4)$
10 a $(x - 4)(x + 1)$ b $(x - 2)(x - 5)$
c $(x - 1)(x + 9)$ d $(2x + 1)(x + 3)$
e $(2x + 3)(3x - 1)$ f $2(2x - 3)(2x - 1)$
11 a $x + 2$ b $\frac{x - 4}{x + 4}$ c $\frac{x - 1}{x + 1}$

Assessment 6

1 a Stina, $s = \frac{1}{2}(-4 + 12)8 = 32$.
b Stina, $v^2 = 0^2 + 2 \times 2 \times 16, v = \sqrt{64} = 8$.
2 a $\frac{a}{n^2} = -x + \frac{h}{n^2}, x = \frac{h - a}{n^2}, n^2 = \frac{h - a}{x}, n = \sqrt{\frac{h - a}{x}}$
b $\frac{t}{x} = \sqrt{\frac{1 - e}{1 + e}}, \frac{t^2}{x^2} = \frac{1 - e}{1 + e}, t^2(1 + e) = x^2(1 - e),$
$t^2 + et^2 = x^2 - ex^2, e(x^2 + t^2) = x^2 - t^2, e = \frac{x^2 - t^2}{x^2 + t^2}$
c $\frac{1}{S} = \frac{r}{Rr} + \frac{R}{Rr} = \frac{(r + R)}{(Rr)}, (r + R)S = Rr, S = Rr/(r + R)\frac{Rr}{(r + R)}$
3 a $C = 15p + 12q + 14t$ b £5.82 c £13.01
4 a 58.8 m (1 dp) b 44.7 m c 101 km/h
5 a 10.4 km b 5.4 m
c No. $3.57\sqrt{(85)} = 32.9$
6 9
7 a i No, $p^2 - 11p + 28$. ii No, $26v^2 - 38v - 47$.
b i Yes ii No, $(v + 10)(v - 10)$.
iii No, $(6x - y)(5x + 3y)$.
8 a i $(2.4 - 3.6)(2.4 - 3.6)$ ii 1.44
b i $= (89 - 11)(89 + 11) = 78 \times 100 = 7800$
ii $= (6.89 - 3.11)(6.89 + 3.11) = 3.78 \times 10 = 37.8$
9 $(x - 4)^2 + 14$
10 a F $5 \times 4 = 20$ b F 5 has 2 factors, 1 and 5.
c T Let the two numbers be $2x$ and $2y$. $2x + 2y = 2(x + y)$ so the sum is even.
d F $0^2 = 0$ e F $2 \times 7 = 14$
f T Let the numbers be $2x, 2x + 2$ and $2x + 4$.
$2x + 2x + 2 + 2x + 4 = 6x + 6 = 6(x + 1)$, is divisible by 6.
g F x^2 has 3 factors, 1, x and x^2.
11 a $y = 4x + 1$ b $y = \frac{x}{3} - 2$
12 a No, $g^{-1}(x) = \frac{x + 1}{2}$
b i 17 ii 107

Revision 1

1 Jenni would save $17.50 - $15 = $2.50
2 a £4928.40 b 30 096 miles
3 a i $18x + 9$ ii $9x - 18$
iii $2x^2 + 51x - 20$ iv $18x^2 - 27x - 18$
b 76.6 cm³ (3 sf)
4 a i QRS = 28° (**Isosceles** △) so QSR = 180 − 90 − 28 = 62° (Angles in a △); PSQ = 62° (Angles in a △) so RST = 180 − 2 × 62 = 56° (Angles on a straight line); RSP = 2 × 62 = 124°
ii SP = SR, ∠PSQ = ∠RSQ and QS is common to both triangles so the triangles are congruent by SAS, PQ and QR are corresponding sides. (Other reasons possible)

b   i   Isosceles
      ii ∠ABD = 72°, ∠DBC = 36°
      iii No. It has different angles.
5 a   i   Correct. HBD is an isosceles right-angled triangle.
      ii Incorrect. HBDF is a square.
      iii Incorrect. HBGC is an isosceles trapezium.
  b   i   One of the following: OPF, OBQ, ODQ
      ii One of the following: HOF, HOB, BOD, FOD, BHF, BDF, HBD, HFD, ACO, CEO, EGO, GAO.
  c   i   ∠ABC = 135°
      ii ∠POF = 45°
      iii ∠HOC = 135°
      iv ∠ABH = 22.5°
6

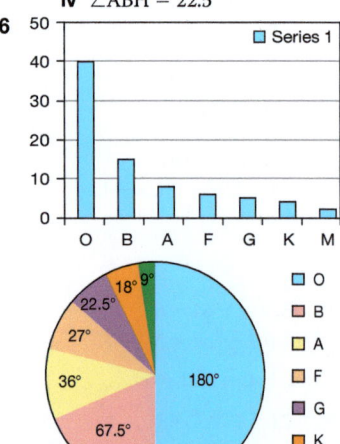

7 a   i   6.5   ii 7 nights.
  b   i   11   ii 5 nights
8 a   22 450   b   31.2%
  c   Maximum = 35 271, Minimum = 35 216
9 a   55.7   b   69
10 a $\frac{1}{5}$
  b   i   16 ml   ii 4 ml
  c   i   Acid: 64 ml, Water: 16 ml
      ii Acid: 16 ml, Water: 84 ml
11 No, for example, if $x = -1$, $10x = -10 < -1 = x$
12 a 248.5 ft   b   88.2 ft (3 sf)   c   75.1 mph (3 sf)
   d Yes. He will pass the train in 2326 ft < 5280 ft.

Chapter 7

Check in

1 a   i   16 cm   ii 160 mm   iii 12 cm²
  b   i   18 cm   ii 180 mm   iii 20.25 cm²
  c   i   19 cm (2 sf) ii 190 mm (2 sf) iii 14 cm²
2 A(3, 2), B(5, 5), C(4, −1), D(−2, −3), E(1, 6)
3 a $x = 3$ b $x = 5$ c $y = x$
  d $y = 2$ e $y = -1$

7.1S

1 a   Distance = 205 km ± 10 km, Bearing = 062° ± 4°
  b   Distance = 120 km ± 10 km, Bearing = 258° ± 4°
  c   Distance = 85 km ± 10 km, Bearing = 116° ± 4°
  d   Distance = 85 km ± 10 km, Bearing = 296° ± 4°
  e   Distance = 60 km ± 10 km, Bearing = 210° ± 4°
2 a   75 km, 010°   b   200 km, 160°
  c   125 km, 200°   d   160 km, 309°
  e   105 km, 240°
3 a   250 m   b   Length = 225 m, Width = 100 m
  c   i   20 cm   ii 5 km

4 Different scales may be used. If the scale 1 : 50 is used the lengths will be as follows. 5.6 m (11.2 cm), 4.2 m (8.4 cm), 4.1 m (8.2 cm), 2.7 m (5.4 cm), 2.9 m (5.8 cm), 1.9 m (3.8 cm), 3 m (6 cm), 2.1 m (4.2 cm).

7.1A

1 a   486 miles, 184°
  b   The distance would decrease.
2 Yes (actually a little less than 5 km nearer to Liam)
3 a   i   Estimated cost of fencing $2250
      ii Area ≈ 140 000 m²
  b   Make the scale diagram 10 times larger.
4 a   ≈ 850 km   b   5 hours   c   ≈ 24 063 km²
5 a   Check students' drawings.   b   i 048° ± 2°   ii 329° ± 2°

7.2S

1 a   i   Area = $\frac{1}{2}$ (25 + 43) × 24 = 816 cm²
      ii Area = 25 × 24 + $\frac{1}{2}$ × 18 × 24 = 816 cm²
  b   i   Area = $\frac{1}{2}$(60 + 24) × 30 = 1260 mm²
      ii Area = 30 × 24 + $\frac{1}{2}$ × 30 × 8 + $\frac{1}{2}$ × 30 × 28 = 1260 mm²
2 a   112.5 cm²   b   5.1 m²
3 Rectangle height = 15 mm, parallelogram base = 40 cm, triangle height = 2.5 m.
4 a   90%   b   87.8% (to 3 sf)
5 7 cm
6 Height of rectangle = 9 cm
7 Length = 18 cm
*8 75.3% (to 3 sf)
9 70.2 m²
10 a 300   b   2400

7.2A

1 Yes, 2540 m² > 2500 m²
  b   The assumption that the measurements are accurate.
2 a   15 cm   b   The minimum width would reduce.
3 a $\frac{1}{2}$ b $\frac{1}{2}$ c $\frac{2}{3}$ d $\frac{1}{3}$
4 a, b Several methods are possible, three are given below.

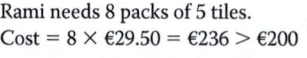

Total area   Total area = 6 + 6 +   Total area = 9 + 2 +
= 4 + 5.5 + 5.5 + 4   1 + 6 = 19 cm²   2 + 6 = 19 cm²
= 19 cm²
5 125 cm²
6 No number of tiles = 4 × 2 + 8 × 4 = 40
  Rami needs 8 packs of 5 tiles.
  Cost = 8 × €29.50 = €236 > €200

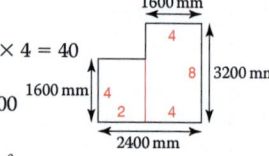

7 Area of map = 30 × 20 = 600 cm²
  Area more than 4 cm from edge = 22 × 12 = 264 cm². This is less than half of 600 cm², so Amy is correct – more than half of the map is within 4 cm of the edge.
*8 a   44 badges (using a tessellating pattern)   b   15.5% (3 sf)
*9

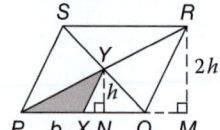

Let the base of triangle PYX be b and the height h, then area of triangle $PYX = \frac{1}{2}bh$

Base of parallelogram = $2b$ (X is mid-point of PQ)
Height of parallelogram = $2h$ (similar triangles)
Area of parallelogram $PQRS = 2b \times 2h = 4bh = 8 \times \frac{1}{2}bh$
So area triangle $PXY = \frac{1}{8}$ area of parallelogram $PQRS$

10 a 20 cm² **b** 17.5 cm²

7.3S

1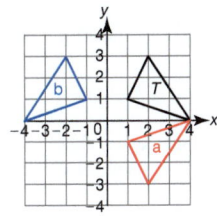
2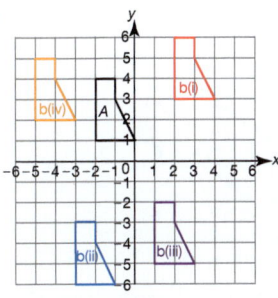

3 pentagon
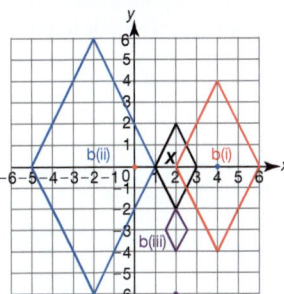

4 rhombus
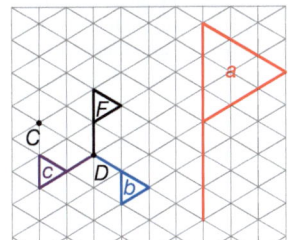

5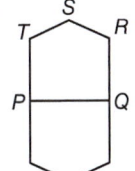

6 a i 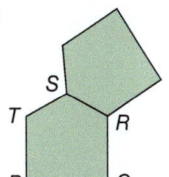 **ii**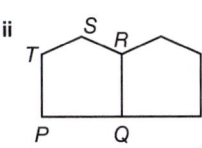

iii 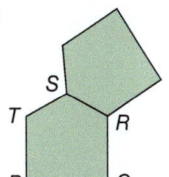 **iv**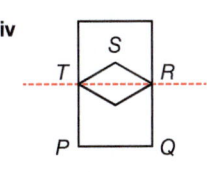

b i The mirror line is the line through S, perpendicular to PQ (the line of symmetry of the pentagon).
ii Q

7 a kite
b
c
d
e
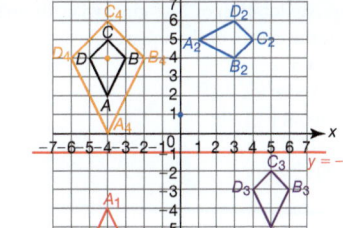

8 a trapezium
b
c
d
e
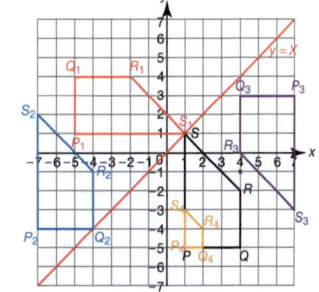

7.3A

1 a Translation by vector $\binom{2}{-4}$
b Rotation 90° clockwise (or 270° anti-clockwise) about $(-4, 3)$
c Rotation 90° anti-clockwise (or 270° clockwise) about $(-1, 2)$
d Reflection in $y = 3$
e Reflection in $x = 4$
f Rotation 180° clockwise (or anti-clockwise) about $(4, 3)$
g Enlargement, centre $(0, -2)$, scale factor 2
h Enlargement, centre $(-3, -6)$, scale factor $\frac{3}{2}$

2 a All co-ordinates are multiplied by 3
b Enlargement, centre $(0, 0)$, scale factor 3

3 a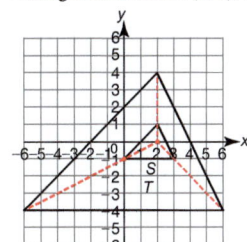

b Enlargement, centre $(2, 0)$, scale factor $\frac{1}{4}$

4 Yes, $(1, 4)$ is equidistant from each vertex of triangle A and the corresponding vertex in triangle B and each edge has been rotated 90° clockwise.

5 a Translation by vector $\binom{4}{0}$, Reflection in $x = 3$ Rotation 180° clockwise (or anti-clockwise) about $(3, 3)$
b All points move under the translation, the points on the line $x = 3$ do not move under the reflection, the point $(3, 3)$ does not move under the rotation.

*6 a Reflection in $y = x$
b Rotation 90° clockwise (or 270° anti-clockwise) about $(3, 3)$
c Reflection in $x + y = 6$

*7 a Reflection in BD **b** Reflection in AC
c Rotation of 180° clockwise (or anti-clockwise) about the centre of the square.

503

d Rotation of 90° clockwise (or anti-clockwise) about the centre of the square. Rotation of 270° clockwise (or anti-clockwise) about the centre of the square.

*8 **a** Reflection in $y = x + 2$
b Rotation 90° clockwise (or 270° anti-clockwise) about $(-1, 0)$

9 **a** **i** $(x, y) \Rightarrow (x, -y)$
   ii $(x, y) \Rightarrow (-x, y)$
   iii $(x, y) \Rightarrow (y, x)$
   iv $(x, y) \Rightarrow (-y, -x)$
   *v Reflection in $x = a$: $(x, y) \Rightarrow (2a - x, y)$
       Reflection in $y = b$: $(x, y) \Rightarrow (x, 2b - y)$
b **i** The point (x, y) is transformed to $(y, -x)$
   ii The point (x, y) is transformed to $(-x, -y)$
   iii The point (x, y) is transformed to $(-y, x)$

7.4S

1 a, b Reflection in $y = -x$ followed by reflection in $y = x - 1$

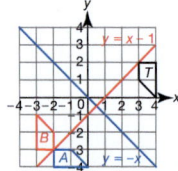

c The order does not affect the final result. T has again rotated 180° about the point $(\frac{1}{2}, -\frac{1}{2})$
d $(\frac{1}{2}, -\frac{1}{2})$

2 a Rotation of 180° clockwise (or anti-clockwise) about $(0, 0)$
b

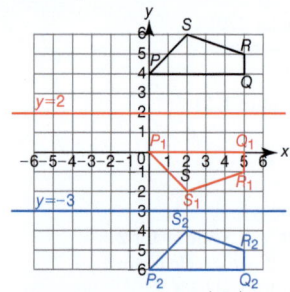

Translation by vector $\binom{0}{-10}$

3 a **b**

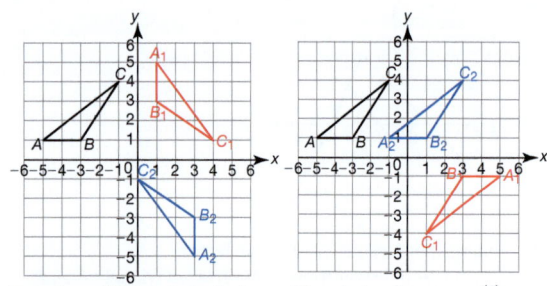

Rotation of 90° anti-clockwise   Translation by vector $\binom{4}{0}$
(or 270° clockwise) about $(2, 2)$

4

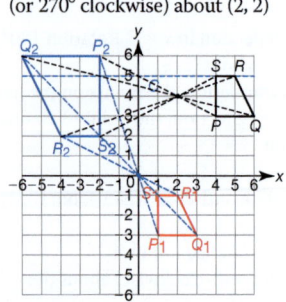

Enlargement, centre $(2, 4)$, scale factor -2

5 a

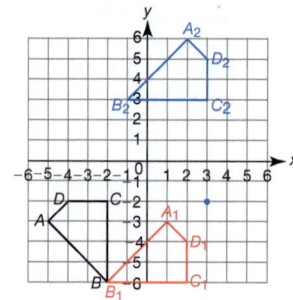

Rotation of 90° clockwise (or 270° anti-clockwise) about $(3, -2)$
b Rotation of 90° anti-clockwise (or 270° clockwise) about $(3, -2)$

6 a, b

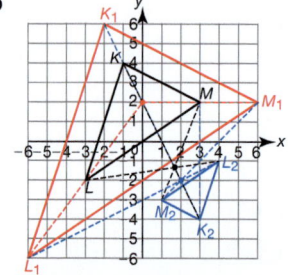

c Enlargement, centre $(1\frac{2}{3}, -1\frac{1}{3})$ scale factor $-\frac{1}{2}$
d **i** Enlargement, centre $(1\frac{2}{3}, -1\frac{1}{3})$ scale factor -2
    ii $(1\frac{2}{3}, -1\frac{1}{3})$

*7 **a** **i**

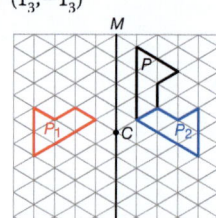

ii Reflection in a mirror line through C that is rotated 60° clockwise about C from M's position.

b **i**

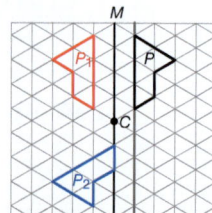

ii Reflection in a mirror line through C that is rotated 60° anti-clockwise about C from M's position.

*8 **a** **i**

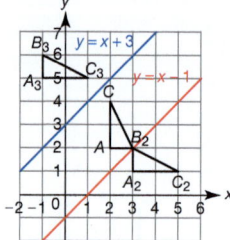

ii Reflection in the line $y = x + 3$
b **i**

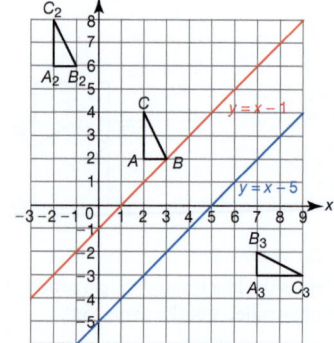

ii Reflection in $y = x - 5$

Answers

504

7.4A

1 a i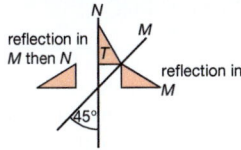

  ii Translation right by six squares.

  b The translation is to the left, rather than the right.

2 a i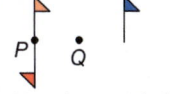

  ii Rotation of 90° anti-clockwise about the point of intersection of the mirror lines.
  The angle of rotation is twice the angle between the mirrors.

  b The rotation changes direction to clockwise 90° about the same centre of rotation.

  c 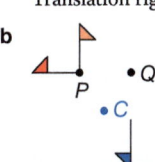 4 lines of symmetry and rotational symmetry, order 4

3 Enlargement, scale factor 2, centre at the centre of the hexagon (starting with the smallest hexagon)
  Enlargement, scale factor $\frac{1}{2}$, centre at the centre of the hexagon (starting with the largest hexagon)

4 a Translation right by distance 2PQ

  b Rotation of 180° about C

5 a Multiple answers possible, for example a translation by the vector $\binom{4}{3}$ followed by a translation by the vector $\binom{1}{-4}$.

  b There are an infinite number of possible pairs of translations. In all pairs the x components of vectors total 5 and the y components total -1.

6 Rotation of 180° clockwise (or anti-clockwise) about (0, 0).

*7 There are many possible pairs. Each pair must include a reflection; the other transformation could be a translation or a rotation. For example, translation by the vector $\binom{0}{-8}$ followed by reflection in the line $y = -x - 7$ or reflection in the y axis followed by rotation 90° anti-clockwise about (4, −3)

*8 PQT is mapped onto RQT by reflection in TQ or by rotation of 90° anti-clockwise about T.
  PQT is mapped onto PST by reflection in TP or by rotation 90° clockwise about T.
  PQT is mapped onto SRT by reflection in a line through T parallel to PQ or by rotation of 180° clockwise (or anti-clockwise) about T or enlargement, centre T, scale factor −1.

9 a square
  b rhombus.

10 Students, own patterns

11 a Always true   **b** Never true
  c Sometimes true – usually, but not in the case of 2 half turns of 180° (see Skills Q3b) or full turns of 360°
  d Sometimes true – usually not true, but true when the translation is perpendicular to the mirror line

  e Sometimes true – always except when the rotation is a full turn of 360°.

Review

1 a 063°   **b** 243°   **c** 17.5 km
2 a 60 m²   **b** 30 cm² **c** 2.25 mm² **d** 49.5 cm²
  e 32.5 m²

3 a, b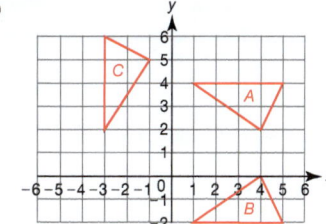

  c Reflection in the line $y = -x$

4 a 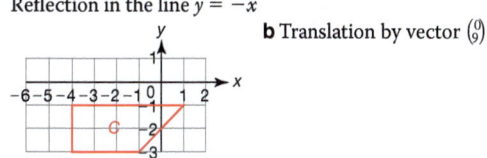   **b** Translation by vector $\binom{9}{9}$

5 a, b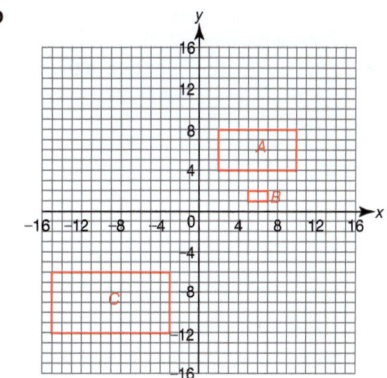

  c Enlargement of scale factor 0.5 centre of enlargement $(-10, 6)$.

6 a reflection in the x axis
  b No – the combined transformation is then equivalent to reflection in the y axis.

Assessment 7

1 a ÷100 000   **b** ×1000   **c** ÷1 000 000
2 a Check student's drawings.
  b i 6.5 cm (± 3 mm)   **ii** 60° (± 3°)
    iii 135° (± 3°)
  c Kite   **d** Rectangle
3 a Yes   **b** No, right angle.
  c Yes   **d** Yes
  e Yes   **f** No, acute.
  g No, right angle.   **h** No, acute.
  i Yes   **j** No, acute.
  k No, reflex.
4 a Check students' drawings. **b** 15–16 cm²
5 a i 195°   **ii** 110°   **iii** 015°   **iv** 290°
    v 245°
  b The bearings differ by 180°.
6 a 49
  b i No, 2 × 2 = 4 and 4 is not a factor of 49.
    ii No, 3 × 3 = 9 and 9 is not a factor of 49.

c No. Each side has length 7 m so must include two 2 × 2 slabs and a 3 × 3 slab. There are exactly three ways of arranging slabs around the edges, all of which include gaps or overlap.

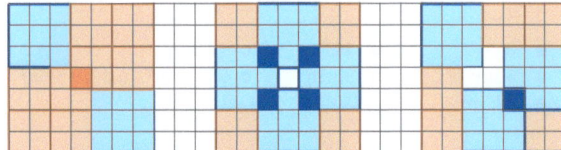

(Other lines of reasoning possible.)

7 a $\frac{1}{4}$ b $\frac{2}{3}$

8 Yes Area of lawn = 40.56 m² Fertiliser needed = 1419.6 g < 1500 g

9 Check students' drawings.

10 a, c

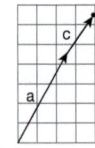

b $\binom{25}{45}$ d $\binom{15}{20}$

e $\binom{40}{65}$ f $\binom{25}{45} + \binom{15}{20} = \binom{40}{65}$

11 a Rotation 180° about the point (2, 2).

b, c

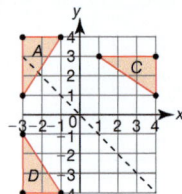

d Reflection in the x axis or the line $y = 0$.

Chapter 8

Check in

1 a $\frac{3}{5}$ b $\frac{3}{7}$ c $\frac{5}{8}$ d $\frac{7}{11}$
  e $\frac{5}{6}$ f $\frac{9}{20}$ g $\frac{23}{24}$ h $\frac{5}{8}$

2 a 40 b 40 c 21 d 56

3 a 0.7 b 0.75 c 0.375 d 0.4
  e 0.$\dot{3}$ f 0.0625 g 0.$\dot{1}$ e 0.625

8.1S

1 a 0.5 b 0.3
2 a i $\frac{1}{5}$ ii 0.2 iii 20%
  b i $\frac{7}{40}$ ii 0.175 iii 17.5%
  c i $\frac{13}{40}$ ii 0.325 iii 32.5%
  d i $\frac{3}{10}$ ii 0.3 iii 30%
3 a 15 times. b 5 times.
4 a Bag A. In bag A the probability of getting a red is $\frac{1}{3}$ compared with $\frac{1}{5}$ in bag B.
  b Bag B. Bag A only contains 1 red ball so it is impossible to take two reds without replacement from this bag.
5 a 50
  b i $\frac{9}{50}$ ii $\frac{7}{25}$ iii $\frac{27}{50}$
  c 2 red, 3 green, 5 blue.
  d Increase the number of trials.
6 9 days.
7 14 shares.
8 a Students' results
  b There are infinite possible outcomes with decreasing probabilities.
9 e.g. Roll the dice 600 times and record the number of times each side comes up, compare this with the expected frequencies of 100 for each side. A large difference between expected and obtained frequencies suggests bias.

8.1A

1 No. The results should start to reflect these proportions after a higher number of trials.
2 a Xavier's statement is consistent with his experiment but his sample size is too small to be reliable.
  b No. This is an estimate, not a fact.
  c i This estimate combines all the results (115 reds) and divides by the total number of trials (125). A higher number of trials gives a more reliable probability.
    ii Not necessarily, this is an estimate only.
3 a

Colour	Red	White	Blue
Frequency	10	5	10
Relative Frequency	0.4	0.2	0.4

The relative frequencies sum to 1.
  b A or C. B could not be Alik's as his table shows that white was an observed result and this spinner does not include white.
  c Multiple answers possible provided they include the colours red, white and blue.
4 Jareer may need a larger and more diverse sample before a difference in numbers becomes apparent. The difference between weekend and weekday births may be too small to be observable in this sample.
5 a 10
  b More often; white balls make up a higher proportion of the total in the blue bag than they do in the red bag.
6 a 6 times.
  b 25p per go.

8.2S

1 a $\frac{1}{6}$ b $\frac{2}{6} = \frac{1}{3}$ c $\frac{3}{6} = \frac{1}{2}$ d $\frac{2}{6} = \frac{1}{3}$
2 a Yes, each outcome has probability $\frac{1}{6}$.
  b No. It is more likely that the pin will land point down due to the instability of the object.
  c Yes, each outcome has probability $\frac{1}{7}$.
  d No. Getting 999 tails in a row is less likely than getting 4 tails in a row.
  e No, some letters are more common than others. For example, 'e' and 't' are more common than 'x' and 'q'.
3 6 black counters.
4 a Highest relative frequency $\frac{9}{20} = 0.45$.
    Lowest relative frequency $\frac{3}{20} = 0.15$.
  b Total 6's = 161, total throws = 24 × 20 = 480; relative frequency = 161 ÷ 480 = 0.34
5 a $\frac{1}{6}$ b $\frac{4}{9}$ c 1
6 a 0.5, 0.45, 0.47, 0.43, 0.38, 0.38, 0.41, 0.4, 0.39, 0.38, 0.39, 0.39, 0.38, 0.38
  b 0.38
7 Dominika's method is best.

8.2A

1 a $\frac{14}{30} = \frac{7}{15}$ b $\frac{9}{30} = \frac{3}{10}$
2 1, 2, 3, 4 and 6 are factors of 12, 5 is not a factor of 12. The theoretical probability of getting a 5 is $\frac{1}{6} = 0.17$ and the experimental probability of this is $(100 - 71) \div 100 = 0.29$. The experimental probability is a lot higher than the theoretical probability so there is reason to believe the dice is biased towards 5.
3 $\frac{1}{8}$
4 0.785 (3 sf)
5 a $\frac{3}{5}$ b $\frac{13}{32}$
6 $\frac{1}{6}$

7 The experimental probability suggests the following:
  $(17 \div 80) \times 19 = 4.0$ 4 white sectors
  $(32 \div 80) \times 19 = 7.6$ 7.6 black sectors
  $(31 \div 80) \times 19 = 7.4$ 7.4 red sectors

Given that there are equal numbers of black sectors and red sectors, these results suggest two possibilities. There could be 7 black sectors, 7 red sectors and $19 - 2 \times 7 = 5$ white sectors, in which case the experiment results suggest the spinner is biased **against** white. There could be 8 black sectors, 8 red sectors and $19 - 2 \times 8 = 3$ white sectors, in which case the experiment results suggest the spinner is biased **in favour** of white.

8.3S
1 a Not mutually exclusive. **b** Mutually exclusive.
  c Not mutually exclusive. **d** Not mutually exclusive.
2 a Not mutually exclusive **b** Mutually exclusive
  c Mutually exclusive **d** Mutually exclusive
3 a 0   **b** $1 - \frac{1}{3} = \frac{2}{3}$
4 a $\frac{1}{6}$ **b** $\frac{1}{2}$
5 a $\frac{1}{3}$ **b** $\frac{1}{2}$
6 a $\frac{1}{7}$ **b** $\frac{3}{7}$ **c** $\frac{11}{21}$
7 a 0.2   **b** 0.6   **c** 0.5
8 $P(\text{green}) = \frac{1}{7}$ $P(\text{blue}) = \frac{2}{7}$ $P(\text{white}) = \frac{4}{7}$
  Green, white and blue are mutually exclusive options so
  $P(\text{green or white or blue}) = \frac{1}{7} + \frac{2}{7} + \frac{4}{7} = 1$
  Green, white and blue are exhaustive, there can be no other colours in the bag.
9 a $P(\text{green}) = 0.1$ **b** $P(\text{white}) = 0.6$

8.3A
1 a

	1	2	3	4	5	6
1	0	1	2	3	4	5
2	1	0	1	2	3	4
3	2	1	0	1	2	3
4	3	2	1	0	1	2
5	4	3	2	1	0	1
6	5	4	3	2	1	0

  b i $\frac{10}{36} = \frac{5}{18}$ **ii** $\frac{6}{36} = \frac{1}{6}$
  iii $1 - \frac{6}{36} - \frac{10}{36} = \frac{20}{36} = \frac{5}{9}$

2 No, taking a green, taking a blue, and taking a black ball are mutually exclusive events $P(\text{green or blue or black}) = \frac{1}{4} + \frac{1}{3} + \frac{1}{2}$
  $= \frac{3}{12} + \frac{4}{12} + \frac{6}{12} = \frac{13}{12} > 1$
  A probability cannot be greater than 1 so Serena must be wrong.
3 a $\frac{1}{6} + \frac{1}{3} + \frac{1}{8} + \frac{3}{8} = 1$
  Arinda **may** be correct as her probabilities do not exceed 1.
  b If Arinda is correct there are no other colours. Her probabilities indicate $P(\text{green or blue or white or black}) = 1$ so the probability of another colour is 0.
4 a $\frac{1}{2}$ **b** $\frac{3}{4}$ **c** $\frac{5}{12}$
5 a C and D   **b** C and D
  c A and C are not mutually exclusive because the number 2 is both even and prime.
6 a True, a student cannot be both male and female.
  b True, a student must be either male or female.
  c False, there are 9 students in the class who are both male and dark-haired.
  d True, there are no students in the class that are both female and have red hair.
7 a

	1	2	3	4	5	6
1	0	−1	−2	−3	−4	−5
2	1	0	−1	−2	−3	−4
3	2	1	0	−1	−2	−3
4	3	2	1	0	−1	−2
5	4	3	2	1	0	−1
6	5	4	3	2	1	0

  b i $\frac{4}{36} = \frac{1}{9}$ **ii** $\frac{1}{36}$ **iii** $\frac{35}{36}$
8 a 0.1   **b** 0.3   **c** 0.5
9 a 0.001   **b** 0.500

Review
1 a 0.15   **b** 10
2 a 0.15   **b** Red 0.15; Green 0.15; Yellow 0.45
  c 60
3 a $\frac{6}{19}$ **b** 0   **c** 1
4 a $\frac{1}{9}$ **b** $\frac{1}{18}$
5 $\frac{9}{20}$
6 a $\frac{7}{25}$ **b** $\frac{6}{25}$ **c** $\frac{12}{25}$
7 a $0.2\dot{3}$ **b** $\frac{1}{6}$ **c** it would be close to $0.1\dot{6} = \frac{1}{6}$

Assessment 8
1 No, the probability of this event is greater than zero.
2 $P(H) = 0.5$ because each throw is an independent event.
3 $\frac{51}{100}$
4 a 200   **b** 2   **c** 4
5 a If $x = 0.2$ then the total probability $= 1.02 > 1$, $x = 0.19$
  b 52   **c** 128   **d** 118
6 a i 0.18   **ii** 0.17
  b Ben's, because he sampled more packets.
7 a $\frac{1}{5}$ **b** $\frac{1}{2}$
8 9 packets
9 a Yes, $P(\text{Late}) = 0.51 > 0.49$ **b** 70
10 a i $\frac{3}{25}$ **ii** $\frac{8}{25}$ **iii** $\frac{7}{25}$ **iv** $\frac{11}{25}$
  b $P(\text{Blue 4}) = 0$
11 a $\frac{2}{5}$ **b** $\frac{1}{5}$ **c** $\frac{1}{2}$ **d** $\frac{3}{5}$
12 a i No, 102 is a multiple of 3.   **ii** No, 5 and 37 are prime.
  iii Yes, no white numbers are multiples of 7
  b i No, 81 is a multiple of 3.   **ii** No, 29 is prime.
  iii No, 84 is a multiple of 7.
13 No, sunny and snowing are not mutually exclusive events.
14 a $P(\text{blue}) = \frac{72}{360} = \frac{1}{5}$
  b All angles are equally likely to be chosen.

Chapter 9
Your estimates may differ from the estimates given here. In some questions exact values are given in brackets.

Check in
1 a i 38.5   **ii** 39
  b i 16.1   **ii** 16
  c i 103.9   **ii** 100
  d i 0.1   **ii** 0.082
  e i 0.4   **ii** 0.38
2 a 954   **b** 337.415
  c 48.99 (2 dp)   **d** 22.37 (2 dp)
  e 105.59 (2 dp)   **f** −45.19 (2 dp)

9.1S
1 a i 8   **ii** 8.4   **iii** 8
  b i 20   **ii** 18.8   **iii** 19
  c i 40   **ii** 35.8   **iii** 36
  d i 300   **ii** 278.7   **iii** 279
  e i 1   **ii** 1.4   **iii** 1
  f i 4000   **ii** 3894.8   **iii** 3895
  g i 0.008   **ii** 0.0   **iii** 0
  h i 2000   **ii** 2399.9   **iii** 2400
  i i 9   **ii** 9.0   **iii** 9
  j i 10   **ii** 14.0   **iii** 14
  k i 1000   **ii** 1403.1   **iii** 1403
  l i 100 000   **ii** 140 306.0   **iii** 140 306

2	a	10	b	5	c	7			
3	a	8	b	560	c	0.008	d	10	
	e	4	f	45	g	16	h	400	
	i	2	j	5	k	30	l	40	
	m	2	n	25	o	50	p	230	
	q	12 100							
4	a	1.4	b	2.8	c	3.2	d	3.9	
	e	4.5	f	5.1	g	5.7	h	6.7	
	i	8.4							
5	a	100	b	1	c	900	d	60	
	e	7.1	f	7.1					

6 a The denominator would become zero, and you can't divide by zero.
   b 1.52 and 1.49 are both close to 1.5 so their difference is close to 0, when rounded their difference is approximated by 1.

7	a	6000	b	3500	c	2.5	d	44
	e	250	f	10	g	2.5	h	10
	i	100	j	4	k	5	l	15
8	a	55, larger.	b	3.4, larger.	c	4200, larger.		
	d	4900, larger.	e	15, smaller.	f	2, smaller.		

10 $47.3 \approx 50$, $18.9 \approx 20$, $8.72 \approx 10$

9.1A

1 a 108
   b No, $2.54 \times 36 = 91.44$.
      This estimate is 16.56 cm too large. This is a large difference relative to 91.44.
2 a 4 pots.
      Fence area $\approx 100\,m^2$. One pot covers $30\,m^2$, so 3 pots covers $90\,m^2$, not enough. 4 pots cover $120\,m^2$.
3 a, b
      i   700 (overestimate as values rounded up, exact value 620)
      ii 2 (underestimate as numerator rounded down and denominator rounded up, exact value $2.3\dot{7}\dot{0}$)
      iii 12 000 (underestimate as values rounded down, exact value 13 279.86)
      iv 30 (underestimate as positive value rounded down and negative value rounded up, exact answer 31.9)
4 He is wrong – out by a factor of 10.
   $20 \times 200 = 4000$, $4000 \div 4 = 1000$
5 a 4 jars   b 50 ml (16.99 ml)
6 a 70 (notice 490 is a multiple of 7, so work out $490 \div 7$) (76.00 (2 dp))
   b 210
7 £3500 (£3525.10)
8 i $2 \times 5 = 10$, so 10.10 is a reasonable answer.
   ii $20 \times 550 = 11\,000$, so 90 000 is not a reasonable answer.
   iii $10 \div 20 = 0.5$, 1.7 is not a reasonable answer. The value must be less than 1 as $12.15 < 17.55$.
9 a 30 litres (32.33 litres)
   b Less   c 3.30
10 240

9.2S

1	a	21, 20.1	b	19, 18.6	c	13, 12.5	d	9, 9.8
	e	10, 11.1	f	3, 3.1	g	400, 412.4		
2	a	71.1	b	29.624	c	2.078 853 046 59		
	d	186.408	e	0.150 885 603 9		f	19.05	
3	a	5.75, 5.8	b	1.3148, 1.3	c	1.73, 1.7	d	2.16421, 2.2
4	a	13 hours, 53 minutes, 20 seconds						
	b	1 day, 3 hours, 46 minutes, 40 seconds						
	c	5 days, 18 hours, 53 minutes, 20 seconds						
	d	1 week, 4 days, 13 hours, 46 minutes, 40 seconds						
	e	16 weeks, 3 days, 17 hours, 46 minutes, 40 seconds						
	f	49 weeks, 4 days, 5 hours, 20 minutes						

5	a	5.6, 5.6	b	73.045 28, 73	c	0.085 437 225, 0.085
	d	−35.052, −35				
	e	109.665, 110		f	38.235 294 117 647, 38	
6	a	464.592 396 7		b	0.453 608 414 2	
7	a	178.412 383 5		b	0.196 708 950 5	
	c	3.210 178 253		d	3.350 190 476	
	e	1.157 007 415		f	0.135 604 500 7	
8	a	i $T(2) = 2.25$, $T(3) = 2.370\,37$, $T(10) = 2.593\,742$, $T(100) = 2.704\,816$, $T(1000) = 2.716\,924$				
		ii $T(2) = 4$, $T(3) = 4.629\,63$, $T(10) = 6.191\,736$, $T(100) = 7.244\,646$, $T(1000) = 7.374\,312$				
	b	$T(n)$ in part ii is the square of $T(n)$ in part i.				

9.2A

1 a Correct
   b i Not correct   ii $(36 \div 2.5) + 5.5 = 19.9$
   c i Not correct   ii $36 \div (2.5 + 5.5) = 4.5$
   d Correct.
2 a €23   b €13.20
3 a €120   b €128.28   c €71.37
   d No. He would need another box of style A.
4 a D   b C (56)   c B (119.02)   d A (−2.63)
5 a £21.13   b £16.37
   c Yes, she would pay less under the new deal in both February and March.
*6 a 35.3 ft/s (3 sf)   b 757 mph (3 sf)

9.3S

1	a	Tonnes	b	Milligrams (mg)				
	c	Centimetres (cm) or millimetres (mm)			d	Litres (l)		
	e	Millilitres (ml)						
2	a	200 cm	b	4 m	c	4.5 m	d	4 km
	e	5 mm	f	4500 g	g	6 kg	h	6.5 kg
	i	2.5 tonnes	j	3000 ml				
3	a	2.5 m/h	b	32 km/h	c	80 km/h		
4	a	2 hours 30 minutes						
5	37.5 miles							
6	a	70 km/hr			b	14 km/litre		
7	a	480 kg/m³	b	41.7 g/cm³	c	200 kg/m³	d	3.9 kg/cm³
8	a	333.3 m/min			b	5.56 m/s		
9	a	5 g/cm³			b	87.88 g		
10	a	9.4575 kg			b	6.15 l		
11	a	5.75, 5.85 m			b	16.45, 16.55 l		
	c	0.85, 0.95 kg			d	6.25, 6.35 N		
	e	10.05, 10.15 s			f	104.65, 104.75 cm		
	g	15.95, 16.05 km			h	9.25, 9.35 m/s		
12	a	6.65, 6.75 m			b	7.735, 7.745 l		
	c	0.8125, 0.8135 kg			d	5.5, 6.5 N		
	e	0.0005, 0.0015 s			f	2.535, 2.545 cm		
	g	1.1615, 1.1625 km			h	14.5, 15.5 m/s		
13	a	32.5 − 37.5 mm			b	37.5 − 42.5 mm		
	c	107.5 − 112.5 mm			d	42.5 mm, 47.5 mm		
	e	21.75 cm, 22.25 cm			f	0.9975 m, 1.0025 m		
	g	0.4975 m, 0.5025 m			h	0.0029975 km, 0.0030025 km		
14	a	Max: 174 kg, min: 162 kg			b	Max: 28.4 kg, min: 27.6 kg		

9.3A

1 a 11 cm   b 4.9 m
2 a 1:45 pm   b £1.35
3 a No, mass = 449 kg.
   b No, each person could be 0.5 kg heavier, so the overall load could be 452 kg, which is not safe.
4 Jayne could be correct, maximum volume $= 3.45^3 > 40\,cm^3$, minimum volume $= 3.35^3 < 40\,cm^3$.
5 a 14 hours 54 minutes.   b 25 mph
6 Yes, Ben's maximum speed $= \frac{80}{10.25} = 7.80$ (2 dp) > Andreas' minimum speed $= \frac{50}{6.45} = 7.75$ (2 dp).

7 **a** 6 m³ **b** 14400 kg (or 14.4 tonnes)
 c $792 **d** 1.25 m³ (2 dp) **e** 1.1 m³ (1 dp)
*8 **a** 42 mph
 b 50 litres = 11.1 gallons, 11.1 × 45 = 500 miles.
 No. The car should have done 185 miles more than it did if the advert was correct.
9 Yes, assuming the rod contains only one type of metal. Using the implied degree of accuracy the density lies between 7497 kg/m³ and 8226 kg/m³. Steel is the only metal in the table with a density in this range.
*10 **a** 5.4 cm **b** 1.7 cm

Review

1 a i 93021.00 **ii** 93000
 b i 27.94 **ii** 28
 c i 0.01 **ii** 0.0063
 d i 0.90 **ii** 0.90
2 a i 64 **ii** 106 **iii** −1.1 **iv** 2
 b i 61.13 **ii** 108.26 **iii** −3.67 **iv** 2.22
3 a e.g. centimetres, millimetres
 b e.g. grams, kilograms
 c cm³, m³ **d** millilitres, litres
4 a 6.72 litres **b** 0.205 litres **c** 3.5 litres
5 20:15
6 16:25
7 300 cm³
8 a i 12.55 cm **ii** 12.45 cm
 b i 11.55 kg **ii** 11.45 kg
 c i 1.005 m **ii** 0.995 m
 d i 0.0255 km **ii** 0.0245 km
9 84.95 ⩽ x < 85.05
10 a 11.55, 5.95 **b** 0.37 (2 dp), 0.23
 c 225, 90.25 **d** 4.31 (2 dp), 3.60 (2 dp)
11 a 20.625 cm² **b** 27.625 cm²

Assessment 9

1 a 3.23 (3 sf) **b** 29 (2 sf)
 c 0.2 (1 sf) **d** 310 (nearest 10)
 e 5700 (nearest 100) **f** 256000 (nearest 1000)
2 a Bart **b** Bart **c** Christian **d** Ahmed
 e Christian
3 6 × 60 × 25 = 9000, Esther used 25 as an estimate for the number of hours in a day.
4 14
5 a 2 × 2² = 8, 2
 b 0.191109 0054, the difference is larger than his estimate.
6 40
7 63%
8 a 4.6 + (4.1 + 1.2) × 2.6 = 18.38
 b 14.9 − 6.8 ÷ (3.7 − 1.2) = 12.18
 c (3.4 × 1.6) + (5.9 − 2.8) = 8.54
 d 2.6 + 7.56 ÷ (1.8 − 0.72) = 9.6
 e (12.3 − 5.2 × 1.6 + 3.4) × 2 = 14.76
9 a ≈ 100 − 1000 litres **b** ≈ 20 kilograms
 c ≈ 2 grams **d** Students' answers in metres
 e ≈ 100 metres **f** ≈ 50 − 300 tonnes
 g ≈ 30 millilitres **h** ≈ 15 millimetres
 i ≈ 75 litres **j** ≈ 40 kilometres
 k ≈ 15 centimetres **l** ≈ 300 millilitres
10 a 66 km/h **b** 150 km/h **c** 481.25 km **d** 3410 miles
 e 17 min 52 s
11 a Yes, 0.81 g/cm³ < 1. **b** 2.28 cm³ (3 sf)
 c 5400 tonnes
12 a LB 221455 miles, UB 221465 miles
 b LB 85 cm, UB 95 cm
 c LB 452.5 g, UB 457.5 g
 d LB 3 min 28.75 s, UB 3 min 28.85 s
 e LB 238 bags, UB 242 bags
 f LB 27.5 tonnes, UB 28.5 tonnes
 g LB 585.5 mm, UB 586.5 mm
13 a 390.1625 cm³, 261.4375 cm³
 b 0.293 g/cm³ (3 sf), 0.194 g/cm³ (3 sf)

Chapter 10

Check in

1 a $\frac{2}{5}$ **b** $\frac{19}{45}$ **c** $\frac{1}{2}$ **d** $5\frac{17}{30}$ **e** $\frac{7}{12}$ **f** $3\frac{1}{8}$
2 a 9 **b** 4 **c** 30 **d** 10
3 a $x + 16y$ **b** $9x^2 + 3x$
 c $7p − 9$ **d** $24x − 27$
 e $5y + 11$ **f** $3x^2 − 2xy$
 g $x^2 + 2x − 63$ **h** $6w^2 − 32w + 32$
 i $p^2 + q^2 − 2pq$
4 a $(x + 2)(x + 3)$ **b** $(x − 6)(x + 4)$
 c $(x − 3)^2$ **d** $(x − 10)(x + 10)$

10.1S

1 a 16 **b** 8 **c** 5 **d** 5
 e 7 **f** 8
2 a 4 **b** 3 **c** 9 **d** 24
 e 4 **f** 16 **g** −11 **h** 4
3 a 20 **b** 45 **c** 60 **d** 16
4 a −8.5 (±0.1) **b** −0.5 (±0.1)
 c −4.5 (±0.1) **d** −8 (±0.1)
5 a

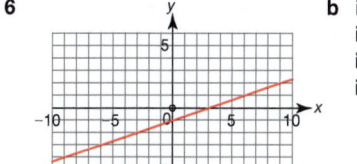

■ Equation 1: $y = 2x + 3$

 b i 5
 ii 4.5
 iii 8
 iv 2

6

■ Equation 1: $y = \frac{x}{3} − 1$

 b i 9
 ii −9
 iii 0
 iv 3

7 a 7 **b** 5 **c** −2 **d** −3
 e 5 **f** −7 **g** −2 **h** $-\frac{1}{2}$
8 a 3 **b** 1 **c** −2 **d** 2
 e 3 **f** 5 **g** $\frac{3}{2}$ **h** 5
9 a −7 **b** 2 **c** 47 **d** $1\frac{13}{17}$
 e 3 **f** $9\frac{1}{3}$ **g** $9\frac{9}{14}$ **h** $\frac{4}{23}$
 i −4 **j** 2
*10 **a** −13 **b** 1
*11 **a** −2.25 **b** 9 **c** 5 **d** 3

10.1A

1 Lengths: 1 m, 14 m
2 Angles 40°, 70°, 70°
3 Angles 40°, 60°, 80°
4 Square side length 8 m, rectangle 10 m by 6 m
5 22 m, 22 m and 16 m or 13 m, 13 m and 7 m, or 14.5 m, 14.5 m and 17.5 m.
6 a 160 **b** 80 **c** −8
7 $x = 37\frac{2}{3}$ Set 1: $74\frac{1}{3}$, 115, $192\frac{1}{3}$, $263\frac{2}{3}$, 222, $-65\frac{1}{3}$
 Set 2: 106, $196\frac{1}{3}$, $-24\frac{2}{3}$, 228, $162\frac{2}{3}$
8 Rectangle 64 m by 49 m, square side length 56 m.
9 5 m
10 32

11 7.2 m
12 $\frac{15}{11}$
13 35
14 Lengths: 3 m and 4 m, 1 m and 4 m

10.2S

1
a	$-3, -2$	**b**	-4, or $x = -3$	**c**	$x = -5, -3$
d	-4	**e**	$-17, -1$	**f**	$-13, -2$
g	$-3, 2$	**h**	$-4, 3$	**i**	$-6, 3$
j	$-11, 2$	**k**	$3, -2$	**l**	$4, -3$
m	$5, -3$	**n**	$-2, 8$	**o**	$15, -2$
p	$7, -4$	**q**	$6, 4$	**r**	$7, 5$
s	$-1, -5$	**t**	$-\frac{2}{3}, -4$	**u**	3
v	$-\frac{1}{4}, 3$				

2
a	-5	**b**	-3	**c**	4
d	8	**e**	$-17.31, -0.69$	**f**	$-7.6, -2.4$
g	$-4.4, 1.4$	**h**	$-8.4, 1.4$	**i**	$-13.3, 1.3$
j	$-3.3, 1.3$	**k**	$-0.5, 5.5$	**l**	$-1.2, 3.2$

3 Li needs to factorise first, $2(x^2 + 3x - \frac{3}{2})$

4
a	$-4, 5$	**b**	$1, -13$	**c**	$-4, 2$	**d**	$-7, 3$
e	$5, 9$	**f**	$3, 10$	**g**	$-1.5, 2$	**h**	$3, -\frac{2}{3}$
i	$-3, -7$	**j**	$1, 4$				

5
a	$-11.2, 2.2$	**b**	$-5.1, 2.1$	**c**	$0.8, 7.2$
d	$0.5, 6.5$	**e**	$-22.0, 2.0$	**f**	$-13.2, 0.2$
g	$-1.5, 0.5$	**h**	$-2.4, 1.0$	**i**	$-2.9, 0.9$
j	$-3.9, 1.2$					

6 Using the formula, the discriminant ($b^2 - 4ac$) is negative and so the square root cannot be calculated.
Or the graph of $y = x^2 + 4x + 20$ does not cross the x axis.

7
a	$-3, 0$	**b**	$-3.6, 0.6\ (\pm 0.1)$
c	$-4, 1$	**d**	$-2, -1$

8
a	$-0.4, 3.1\ (\pm 0.1)$	**b**	$-0.3\ (\pm 0.1), 3$
c	The graph does not intersect the line $y = -9$.		

*9 $x = -4.4\ (\pm 0.1)$ or $x = 1.4\ (\pm 0.1)$ or $x = 0$

10 a Applying Pythagoras' Theorem:
$x^2 + (x - 14)^2 = (x + 1)^2$
$x^2 + x^2 - 28x + 196 = x^2 + 2x + 1$
$x^2 - 30x + 195 = 0$
b $x = 9.5$ or $x = 20.5\ (\pm 0.1)$
$x = 9.5$ is not possible as $(x - 14)$ would be negative.
Longest side $= 20.5 + 1 = 21.5$

10.2A

1 a $w(w + 7) = 60$
$w^2 + 7w = 60$
$w^2 + 7w - 60 = 0$
b 5 cm, 12 cm

2 $p = -3$ – graph has just one solution so touches axis but doesn't go below.

3 No solution since root of a negative number cannot be calculated.

4
a	$9, 13$	**b**	$17, 21$	**c**	$6.16, -0.16$
d	$0.6, -0.5$	**e**	$8, 15$	**f**	10

5 a Surface area $= 5 \times 2\pi r + \pi r^2 = 100$, $\pi r^2 + 5\pi r = 50$,
$\pi r^2 + 5\pi r - 50 = 0$
b 4.42 cm

6
a	8 cm, 15 cm	**b**	34 cm
c	25 cm	**d**	Cylinder has larger volume

7
a	-1 or 0.3	**b**	-0.5 or 0.6
c	-2.36 or 0.76	**d**	-0.32 or 2.32
e	$-3, -2, 2$ or 3		

8 $ax^2 + bx + c = 0$
$x^2 + \frac{b}{a}x + \frac{c}{a} = 0$
$\left(x + \frac{b}{2a}\right)^2 - \frac{b^2}{4a^2} + \frac{c}{a} = 0$
$\left(x + \frac{b}{2a}\right)^2 = \frac{b^2}{4a^2} - \frac{4ac}{4a^2}$
$x + \frac{b}{2a} = \pm \frac{\sqrt{b^2 - 4ac}}{\sqrt{4a^2}}$
$x = \frac{-b}{2a} \pm \frac{\sqrt{b^2 - 4ac}}{2a}$

10.3S

1 a i Multiple solutions possible, for example $x = 5, y = 1$; $x = -3, y = 17$; $x = 4.5, y = 2$.
ii Multiple solutions possible, for example $x = 4, y = 3$; $x = -1, y = 33$; $x = 3.5, y = 6$.
b $x = 4, y = 3$

2
a	$(1, 3)$	**b**	$(5, 2)$	**c**	$m = 4, n = 2$
d	$(2, 5)$	**e**	$(4, -1)$	**f**	$e = 1, f = 3$
g	$m = 4, n = 2$	**h**	$(3, -2)$		

3
a	$x = 6, y = 1$	**b**	$x = 7, y = -2$
c	$p = 6, q = 1$	**d**	$a = 7, b = 1$

4
a	$x = 6, y = 1$	**b**	$x = 7, y = 2$
c	$p = \frac{40}{11}, q = -\frac{41}{11}$	**d**	$a = 5, b = -3$

5 $x = -0.3, y = 1.4$

6
a	$(1.4, 0.9)$	**b**	$(0.6, 2.7)$
c	$(-0.5, -2.5)$	**d**	$(2.5, 0.5)$

7
a	$x = 6, y = 2$	**b**	$a = 8, b = -1$
c	$p = 8\frac{8}{19}, q = -\frac{7}{19}$	**d**	$s = 12, t = -8$

8
a $(-3.37, 11.37)$ and $(2.37, 5.63)$
b $(-3.85, 12.85)$ and $(2.85, 6.15)$
c $(-1.45, 4.23)$ and $(1.20, 2.90)$
d $(-1.84, 6.76)$ and $(1.09, 2.37)$
e $(-0.38, 1.69)$ and $(1.58, 0.71)$
f $(0, 2)$ and $(1.6, 1.2)$
g $(3.53, 3.23)$ and $(-10.20, 10.10)$
h $(0.84, 3.58)$ and $(7.16, 0.42)$
i $(3, 7)$ and $(5, 17)$ **j** $(-2, -1)$ and $(2, 1)$

9 Graphical: The graphs do not intersect. Algebraically: The quadratic $x^2 - 6 = x - 8$ has a negative discriminant so cannot be solved.

10 a $x = -2, y = 1$
b $y = 2x + 5$ is a tangent to the circle $x^2 + y^2 = 5$

*11 a One – the only point of intersection of $x - y = 2$ and $y = x^3$.
b Two – the graphs of $x + y = 5$ and $y = \frac{1}{x}$ intersect twice.

12 62 and 73.

10.3A

1 a 17 cents   **b** 6.4 cm
2 a 17, 24   **b** 17, 23
c 4 large coaches, 1 small coach.
d 37.5°, 37.5°, 105°
e Paperback €5, hardback €10.

3 a 14, 9   **b** 4, -2

4 ☀ $= 29$ ☾ $= 17$

5 a 0 and 2 or $-\frac{2}{5}$ and $1\frac{3}{5}$
b James 1 year, Isla 3 years.

6 a The lines $y = 2x - 1$ and $y = 2x + 4$ do not intersect.
b Yes, if at least one of the equations is non-linear (e.g. quadratic) two graphs can cut in more than one place.

7 $(1, 7)$

8
a	$2.56, -1.56$	**b**	$1.62, -0.62$	**c**	$3, -1$
d	$2, -1$	**e**	$2, -2$		

9 a $x = -3, y = 2$ or $x = 3, y = -2$
b $p = -1, q = 6$

10.4S

1
a	1.4142	**b**	1.6180	**c**	2.2056
d	4.5698	**e**	7.9344	**f**	3.3219

2 a i 1, 1.5, 1.4167, 1.4142, 1.4142, ...
  ii 1, 3.25, 2.5103, 2.0407, 1.8247, 1.7800, 1.7783, 1.7783, ...
  b i $\sqrt{2}$ **ii** $\sqrt[4]{10}$
3 a -2.627
  b $-2.627^2 - 5 \times -2.627 + 5 = 0.00573 > 0$. This solution is not exact but is very close to the true value.
4 a $x_3 = -2.1044754$, $x_4 = -2.1117932$, $x_5 = -2.1139788$, $x_6 = -2.1146306$, $x_7 = -2.1148250$
  b These values appear to converge to -2.115.
  c 1.8608
5 a $x_{n+1} = \sqrt[3]{6x_n - 3}$
  $x = \sqrt[3]{6x - 3}$
  $x^3 = 6x - 3$
  b 2.1451
6 a $x_{n+1} = \sqrt[3]{5 - 2x_n}$ **b** 1.328
7 a $x^4 = 2x + 8$
  $x = \sqrt[4]{2x + 8}$
  $x_{n+1} = \sqrt[4]{2x_n + 8}$
  b 2 (or -1 or -2)   **c** 1.849
8 $x = 1.0000762$
9 $(7*A1 + 1)\wedge(\frac{1}{3})$

10.4A

1 $12000(1 + A)^3 = 6000(3 + 3A + A^2)$
  $2(1 + 3A + 3A^2 + A^3) = 3 + 3A + A^2$
  $2A^3 + 5A^2 + 3A - 1 = 0$
  $2(0.225)^3 + 5(0.225)^2 + 3(0.225) - 1 < 0$ and $2(0.235)^3 + 5(0.235)^2 + 3(0.235) - 1 > 0$ so the answer lies in the interval $0.225 < A < 0.235$, $A = 0.23$ (1 dp).
  b No, the formula should be $A_{n+1} = \sqrt{\frac{1 - 2A_n^3 - 3A_n}{5}}$
  c 23.4%
2 a $1000(1 + 0.04n) = 1000(1.03)^n$
  $1 + 0.04n = 1.03^n$
  $1 = 1.03^n - 0.04n$
  b 19.5 years
  c

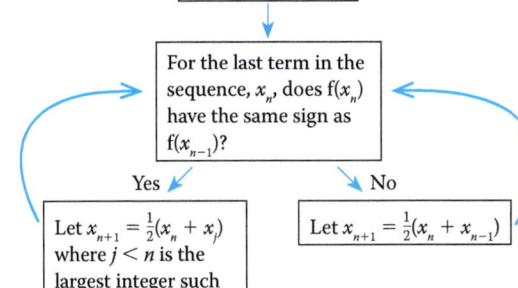

3 a 1.414215686   **b** 2.00001
4 a 1.75   **b** The sequence oscillates between 1 and 3.
5 a 240, 278, 312, 339, 360
  b Rapid growth over the first few years, then the rate of growth decreases.
  c 400

10.5S

1 $n = -1, 0, 1, 2, 3, 4$
2 a $x > 6$ **b** $x < 8$ **c** $x < 7$ **d** $x \leq 6$
  e $x < 14$ **f** $x \leq 6$ **g** $5 \leq x$ **h** $-\frac{2}{3} < x$
  i $5 > x > 1$ **j** $0 < x < 8$ **k** $3 \leq x < 7$ **l** $1 < x \leq 6$
  m $-1 \leq x \leq 4$ **n** $-1 < x < 2$
  o $-5 < x < 8$ **p** $-5 < x < -4$
3 a 9   **b** 18   **c** 7   **d** 4
  e 4   **f** 2
4 No, it is not possible for $x \leq 0$ and $x > 2$.
5 a i 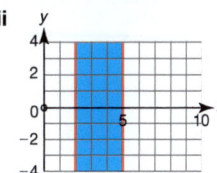   **b i**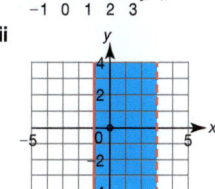
  ii 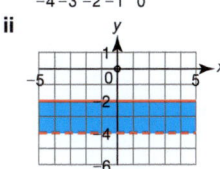   **ii**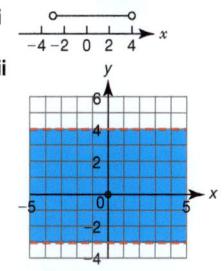
  c i 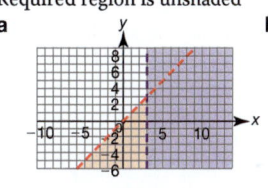   **d i**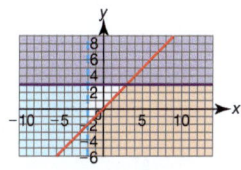
  ii 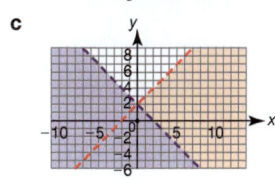   **ii**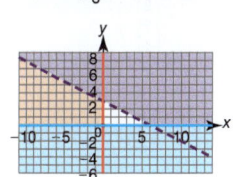

6 a $-2 \leq x < 3$, $3 < y \leq 6$ **b** $x \geq -2$, $y \leq 6$, $y > x$
  c $y \leq 7$, $y > x$, $x + y \geq 5$
  d $x + y < 5$, $y > x + 3$, $y \leq 4x + 10$
  e $y > x^2$, $x + y < 5$ **f** $y < 4 - x^2$, $x + y > 2$
7 Required region is unshaded
  a   **b**
  c   **d**

8 a $-8 < x < 8$ **b** $x < -1$, $x > 1$
  c $x < -2$, $x > 0$ **d** $0 \leq x \leq 6$
  e $-4 < x < -2$ **f** $-4 < x < 3$
  g $-0.5 \leq x \leq 3$ **h** No solutions
9 Several solutions possible, for example $x^2 \leq 1$
10 No – the inequality simplifies to $x \geq 7$ which is an incomplete solution. The correct answer is $x \geq 7$ and $x \leq -7$.
11 Graphical: The graph of $y = 3x^2 + 2$ does not touch or cross the x-axis.
  Algebraic: It is not possible to find a real square root of the negative number $-\frac{2}{3}$.

*12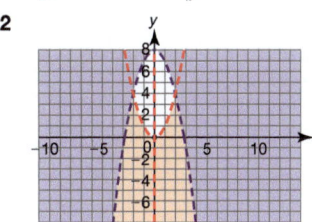

13 34 years old.

10.5A

1 a $x > 7\frac{2}{3}$ **b** 8

2 These simplify to $y \leq 6$ and $y > 6$, a number cannot be simultaneously smaller than or equal to six and greater than six.

3 a 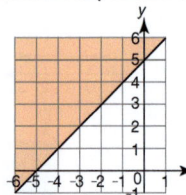 **b**  **c** (figure)

d 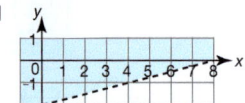 **e** (figure) **f**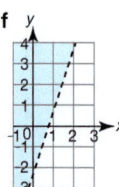

g 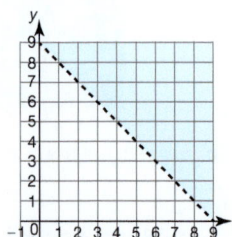 **h**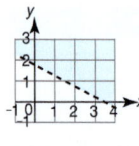

4 a $y < 2x + 4$ **b** $y \geq \frac{8}{3}x - 1$

5 a $x^2 \leq y \leq 9$ **b** $(x+2)(x-3) \leq y < 12$

6 a $(-1,-4), (-1,-3), (-1,-2),$ **b** $(3,0), (3,1), (4,0)$
$(-1,-1), (-1,0), (-1,1),$
$(0,-1), (0,0), (0,1)$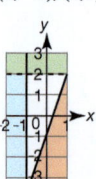

7 a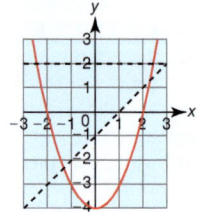

8 a $x \geq 0, y \geq 0$ as you cannot hire a negative number of coaches. The number of people accommodated by the coaches is given by $20x + 48y$, this must be at least equal to the 316 students and adults. $20x + 48y \geq 316$ which simplifies to $5x + 12y \geq 79$.
16 adults can supervise at most $16 \div 2 = 8$ coaches so total number of coaches $x + y \leq 8$

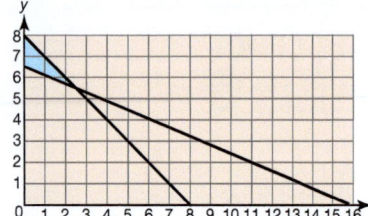

9 a For example, $x \geq 0, y \geq 0, x + y \leq 2$

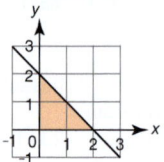

b For example $1 \geq y \geq 2, y \geq x, 2y + x > 7$

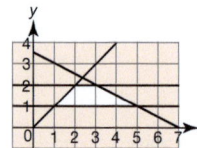

Review

1 a 14 **b** 100 **c** -3
  d 6 **e** 12 **f** 3.5

2 a $2x + 0.6 = 7$
  b Shakeel = 3 hr, 12 min; Andy = 3 hr, 48 min

3 a $(x+3)(x-4) = 0, x = -3, 4$
  b $(x-5)(x-7) = 0, x = 5, 7$
  c $(x+6)^2 = 0, x = -6$
  d $(4x+11)(4x-11) = 0, x = -\frac{11}{4}, \frac{11}{4}$
  e $3x(2x-3) = 0, x = 0, \frac{3}{2}$
  f $(3x+4)(x+1) = 0, x = -\frac{4}{3}, -1$
  g $(2x-3)(5x-2) = 0, x = \frac{3}{2}, \frac{2}{5}$

4 $(2x+3)(x-5) = 0, x = -\frac{3}{2}, 5$

5 a i $(x-2)^2 - 5$ **ii** $(x+\frac{3}{2})^2 - \frac{1}{4}$
  b i $x = 2, 8$ **ii** $x = -6, 2$

6 a $x = -1.14, 6.14$ **b** $x = -2.26, -0.736$

7 a $x = 3, y = -1$ **b** $v = 0.5, w = 1.5$
  c $(-1, 1)$ and $(2, 4)$ **d** $(-5, 0)$ and $(1, 6)$

8 a 2.888 939
  b $x = \sqrt[3]{27 - x}$
  $x^3 = 27 - x$
  $x^3 + x - 27 = 0$

9 a $x > 5$ **b** $x \leq -3$
  c $x < 4$
  d $x > 6$
  e $-7 < x < -1$
  f $x \leq -6, x \geq 3$

10 a (figure) **b**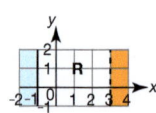

c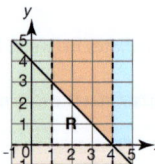

Assessment 10

1 a 8.75 kg **b** Albert 3, Oliver 9 **c** 154 g
  d Brendan €710, Arsene €430, José €860
  e Fizz 4, Milo 3   **f** 147 miles

2 a $5k - 2(30-k) = 115, k = 25$ **b** 9 km

Answers

3 a $1 + 3\sqrt{11}$ b 53, 59   c 100
4 a 1 s, the ball is going up or 3 s, the ball is coming down.
   b 2 s, the ball has reached its greatest height.
   c $25 = 20t - 5t^2, t = 2 \pm \sqrt{-1}, \sqrt{-1}$ is not real and 25 m > 20 m.
   d $t = 4$. $t = 0$ is the start, $t = 4$ the ball hits the ground.
5 12
6 No, 0.47 or -8.53.
7 a $x = 4, y = 6$ and $x = -6, y = -4$
   b $x = 4, y = -3$ and $x = -3, y = 4$
8 a i £6.50   ii £4.50   b 50 g
   c i 72p   ii 63p   d i $4.99   ii $11.95
   e i £250   ii £500   f i 4p   ii 3p
9
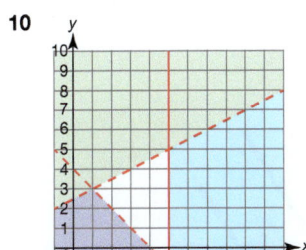
10

11 a $22g \leqslant 345, g \leqslant 15.7$ (3 sf)   b 101 or 102
   c i $f = 3, p = 5; f = 4, p = 6; f = 5, p = 7$
      ii $f = 5, p = 7$ iii $f = 3, p = 5$ gives 4
      empty platforms

Chapter 11

Circle rules are abbreviated as follows.
ACC Angle at centre is twice angle at circumference;
ASC Angle in a semicircle is 90°;
ASE Angles in same segment are equal;
OAS Opposite angles in a cyclic quadrilateral sum to 180°;
TPR Tangent and radius at a point are perpendicular;
TEL Tangents from an external point are equal;
PBC The perpendicular from the centre bisects the chord;
AS The alternate segment theorem.
Angle rules will be abbreviated in the following way:
VO Vertically opposite angles are equal;
CA corresponding angles are equal;
AA alternate angles are equal;
IA interior angles sum to 180°;
ASL angles on a straight line sum to 180°;
AP angles at a point sum to 360°;
AST the angle sum of a triangle is 180°;
ASQ the angle sum of a quadrilateral is 360°.

Check in

1 $a = 43°, b = 107°, c = 233°, d = 139°$
2 $e = 102°, f = 29°, g = 123°, h = 73°$
3 a 17.5 cm²   b 15 m²   c 50 m²

11.1S

1 a i <900   ii >4900   iii >90   iv >200
   b i 729   ii 5110   iii 94.3   iv 209
2 a Circumference = 126 cm (3 sf)
      Area = 1260 cm² (3 sf)
   b Circumference = 201 mm (3 sf)
      Area = 3220 mm² (3 sf)
   c Circumference = 11.3 m (3 sf)
      Area = 10.2 m² (3 sf)
   d Circumference = 1.32 km (3 sf)
      Area = 0.139 km² (3 sf)

3 28.0 cm (3 sf)
4 0.740 m (3 sf)
5 a 12 cm, 37.7 cm, 113 cm² (3 sf)
   b 2.3 m, 14.5 m, 16.6 m² (3 sf)
   c 15.6 mm, 31.2 mm, 764 mm² (3 sf)
   d 2.20 m, 4.40 m, 13.8 m² (3 sf)
6 a i $(5\pi + 24)$ cm   ii $(\frac{25}{2}\pi + 70)$ cm²
   b i $(9\pi + 90)$ m   ii $(\frac{81}{2}\pi + 648)$ m²
7 1640 mm² (3 sf)
8 a 8.00 m (3 sf)   b 3.85 m² (3 sf)
9 a i 28.6 cm (3 sf)   ii 49.1 cm² (3 sf)
   b i 6.71 m (3 sf)   ii 3.93 m² (3 sf)
*10 $(408 + 60\pi)$ mm²

11.1A

1 a Yes, $\frac{1}{2} \times \pi \times 3.2 + 2.6 + 1.6 + 2.6 = 11.8265...$ m < 12
   b No. Area = 9.7812... m²
      Weight of fertiliser = 9.7812... × 35 × 6 = 2.054 kg
2 a Total perimeter = $\frac{1}{2}\pi D + \frac{1}{2}\pi d + \frac{1}{2}\pi(D - d) = \pi D$
   b Total area = $\frac{1}{2}\pi \left(\frac{D}{2}\right)^2 + \frac{1}{2}\pi \left(\frac{d}{2}\right)^2 - \frac{1}{2}\pi \left(\frac{D - d}{2}\right)^2 = \frac{1}{4}\pi Dd$
3 a 123 m (3 sf)   b 56.5 m (3 sf)
4 No, number of revolutions = $\frac{1760 \times 36}{26\pi} = 775.696... < 1000$
5 $\frac{1}{2}\pi(2s) + 2s = 2\pi r$
   $(\pi + 2)s = 2\pi r$
   $s = \frac{2\pi}{\pi + 2}r$
6 $\frac{3}{\pi}$
7 a i 600
      ii Yes, % waste = $\frac{120 \times 80 - 600(4\pi)}{120 \times 80} \times 100 = 21.46\% < 25\%$
   b 3.95 cm, would still give 600 badges % wasted = $\frac{2247.49...}{9600} \times 100 = 23.41...\% < 25\%$. Sally is correct.
      4.05 cm, would only give 551 badges % wasted = $\frac{2501.746...}{9600} \times 100 = 26.06...\% > 25\%$. Sally is incorrect.
*8 13.0 inches (3 sf)
*9 a Area of circle = πr^2 Area of square = $2r \times 2r = 4r^2$
       Shaded area of circle = $4r^2 - \pi r^2 = r^2(4 - \pi)$
    b Percentage = $\frac{r^2(4 - \pi)}{4r^2} \times 100\% = \frac{(4 - \pi)}{4} \times 100\%$
       $\pi > 3$, so $\frac{(4 - \pi)}{4} < \frac{1}{4}$ and the percentage is less than 25%.
10 Let d_1, d_2, d_3, d_4, d_5 and d_6 be the diameters of the smaller semicircles, the diameter of A is $(d_1 + d_2 + d_3 + d_4 + d_5 + d_6)$.
   Length A = $\frac{1}{2} \times \pi \times (d_1 + d_2 + d_3 + d_4 + d_5 + d_6)$
   Length B = $\frac{1}{2} \times \pi \times d_1 + \frac{1}{2} \times \pi \times d_2 + ... + \frac{1}{2} \times \pi \times d_6$
   = $\frac{1}{2} \times \pi \times (d_1 + d_2 + d_3 + d_4 + d_5 + d_6)$ = Length A

11.2S

1 a 8π b 10π c 8π d 54π
2 a 6.28 mm, 28.3 mm² (3 sf)   b 1.88 m, 0.754 m² (3 sf)
   c 7.70 cm, 6.73 cm² (3 sf)   d 1.25 km, 0.144 km² (3 sf)
3 a i $(\pi + 6)$ m   ii $\frac{3\pi}{2}$ m²
   b i $(14\pi + 24)$ cm   ii 84π cm²
4 $(900 - 225\pi)$ cm²
5 a $\frac{1573\pi}{4}$ or 1240 mm² (3 sf)
   b $\frac{1573\pi}{4} - 420$ or 815 mm² (3 sf)
6 a $\frac{3\pi}{4}$ m²   b $(\frac{3\pi}{2} + 2)$ m
7 a $(4\pi + 24)$ cm   b $(16\pi + 32)$ cm²
8 50 cm²
9 $93\frac{1}{3}$ cm
*10 80%
11 a 7.09 m (3 sf)   b 2.62 m² (3 sf)

11.2A

1 $(24\pi + 96)$ inches
2 $15.6 - \frac{3\pi}{2} = 10.9$ cm² (to 3 sf)

3 a Length of arc = $\frac{90}{360} \times 2\pi \times 20 = 10\pi$ cm
       Perimeter of leaf = 20π cm
    b Area of one pink section = $400 - \frac{90}{360}\pi \times 20^2$
       Area of leaf = $400 - 2(400 - 100\pi) =$
       $200\pi - 400$ or $200(\pi - 2)$ cm^2
4 Pipe A: Area of water = $64\pi - 128$ mm^2
   Pipe B: Area of water = $192\pi + 128$ mm^2
            Difference = $192\pi + 128 - (64\pi - 128)$
                      = $128(\pi + 2)$ mm^2
5 a Area of large sector = $\frac{30}{360} \times \pi \times R^2 = \frac{1}{12}\pi R^2$
       Area of small sector = $\frac{1}{12}\pi r^2$
       $A = \frac{1}{12}\pi(R^2 - r^2) = \frac{1}{12}\pi(R + r)(R - r)$
    b $\frac{3}{2}\pi$
6 $\frac{45\pi}{2}$ cm^2
7 a 144°
    b Yes. Let a be the angle turned through by the small cog and b be the angle turned through by the large cog.
       $b = \frac{(\frac{a}{360} \times 16\pi)}{40\pi} \times 360 = \frac{2a}{5}$ so $2.5b = a$.
*8 a 19.1 m (3 sf)
    b Yes, area of entrance = $\frac{113}{360} \times \pi \times 3.6^2 + 2 \times \frac{1}{2} \times (1 + 3) \times 3 = 24.8$ m^2 (3 sf) > 24 m^2
*9 Area = $(1600 - 400\pi)$ mm^2

11.3S

1 $a = 2 \times 55° = 110°$ (ACC) $b = 240° \div 2 = 120°$ (ACC)
   $c = 90°$ (ASC) $d = 2 \times 50° = 100°$ (ACC)
2 $a = 52°$ (ASE) $b = 180° - 84° = 96°$ (OAS)
   $c = 180° - 98° = 82°$ (OAS) $d = 61°$ (ASE)
   $e = 180° - 61° - 87° = 32°$ (AST) $f = e = 32°$ (ASE)
   $g = 180° - 113° = 67°$ (OAS) $h = 90°$ (ASC)
   $i = 180° - 90° - 67° = 23°$ (AST)
3 $x = 90°$ (TPR) $y = 180° - 90° - 18° = 72°$ (AST)
4 $p = \frac{180° - 64°}{2} = 58°$ (base angle of isosceles triangle formed by equal tangents)
   $q = 180° - 58° = 122°$ (ASL)
   $r = 360° - 90° - 90° - 140° = 40°$ (ASQ, TPR)
5 $a = 48°, b = 72°$
6 $\angle QOS = 150°$ $\angle QRS = 75°$ $\angle OQR = 33°$
7 $\angle ACD = 64°$ $\angle CAD = 26°$ $\angle ADF = 64°$ $\angle CDE = 26°$
*8 $\angle PTS = 78°$ $\angle QRS = 127°$ $\angle PQR = 131°$
   $\angle TPQ = \angle RST = 102°$
   Check: Angle sum of pentagon = $3 \times 180° = 540°$ and $102° + 131° + 127° + 102° + 78° = 540°$

11.3A

1 a i $\angle AOC = 180° - 2x$ (AST, isosceles)
      ii $\angle AOD = 2x$ (ASL)
    b $\angle BOD = 2y$
    c The angle subtended by an arc at the centre of a circle is equal to twice the angle subtended at the circumference.
    d i Each angle at the circumference is half of the angle at the centre, so they must equal each other.
      ii When the angle at the centre forms a diameter, the arc is a semi-circle, the angle at the centre is 180° and so the angle subtended at the circumference is half of this, 90°.
2 Let $\angle PQR = x$, then $\angle PSR = 180° - x$ (OAS)
   $\angle PST = 180° - (180° - x) = x$ (ASL)
3 a $\angle PQO = \angle PRO = 90°$ (TPR) $OQ = OR$ as both are radii
       OP is in both triangles.
       Triangle PQO is congruent to triangle PRO (RHS)
    b The line joining the centre to the point of intersection of the tangents bisects the angle between the tangents.

4 In triangles OMA and OMB
   $MA = MB$ since M is the mid-point of AB
   $OA = OB$ as both are radii
   OM is in both triangles
   Triangle OMA is congruent to triangle OMB (SSS)
   $\angle OMA = \angle OMB = 90°$ (equal angles, ASL)
   OM is perpendicular to AB.
5 a $\angle BAX = 90° - x$ (TPR)
       $\angle BXA = 180° - 90° - (90° - x) = x$ (AST)
       $\angle BCA = \angle BXA = x$ (ASE)
    b $\angle PAC = \angle ABC$
6 $AP = AS, BP = BQ, CR = CQ$ and
   $DR = DS$ (TEL)
   Adding gives $AP + BP + CR + DR = AS + BQ + CQ + DS$
   $(AP + BP) + (CR + DR) = (BQ + CQ) + (DS + AS)$
   $AB + CD = BC + DA$

Life skills 2

1 a 10 000 : 1
   b, c

2 a 7.24 cm^2 b 45.25 cm^2 c 640
3 a $x + y \leq 32$ $x \leq 8y$
    b

    c 28
4 a Style B. b Style A 1.54 m^2, Style B 1.62 m^2.
    c Multiple answers possible.
5 a 11
    b Upper bound 97.0025 m^2, lower bound 95.0025 m^2.
    c No. If the lower bound is accurate the tables will be less than 1.2 m apart which does not meet the requirements.
6 a Juliet
    b 20
    c Multiple answers possible e.g. they call at similar times to the previous week, they don't repeat calls to people who are rarely in.

Answers

*7 ∠ABC = ∠ACB (base angles of isosceles triangle ABC)
∠ABC = ∠CAQ (alternate segment theorem)
So ∠ACB = ∠CAQ
AQ must be parallel to BC (alternate angles are equal)
The tangent at A is parallel to BC.

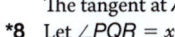

*8 Let ∠PQR = x
∠APQ = ∠PQR = x
(alternate angles equal, BA parallel to RQ)
∠BPR = ∠PQR = x (AS)
∠PRQ = ∠APQ = x (AS)
∠DQR = ∠PRQ = x
(alternate angles equal, CD parallel to PR)
∠RPQ = ∠DQR = x (AS)
So the angles of triangle PQR are all equal to x and PQR is an equilateral triangle.

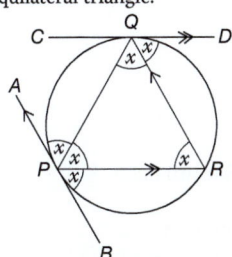

9 ∠OPT = 90° (TPR)
∠ACB = 90° (ASC)
∠CBA = 180° − 90° − 48° = 42° (AST)
∠OBT = 180° − 42° = 138° (ASL)
∠BTP = 360° − 112° − 138° − 90° = 20°

10 AB = AC, OA = OB = OC (radial lines).
Triangles AOB and AOC are congruent (SSS).
∠OAB = ∠OAC, ∠BAC = ∠OAB + ∠OAC = 2 × ∠OAC
∠BOC = 2 × ∠BAC (ACC)
∠BOC = 2 × 2 × ∠OAC = 4 × ∠OAC

11.4S

1 a, b

2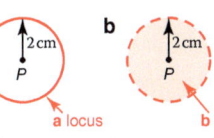

3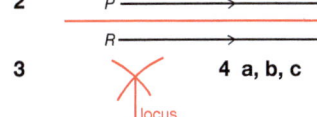

4 a, b, c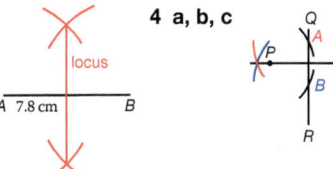

5 a, b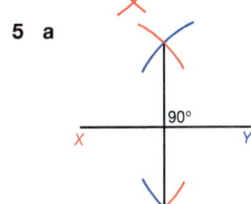

6 a, b

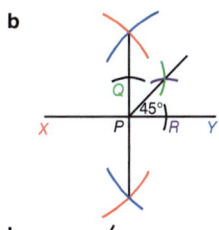
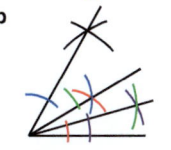

7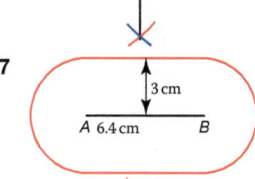

8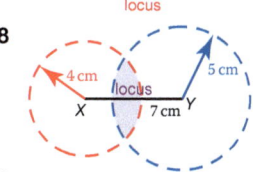

9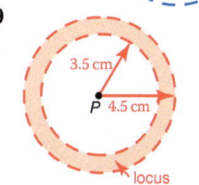

10 a 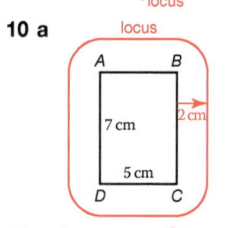 b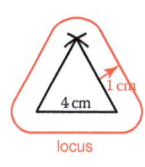

11 a, b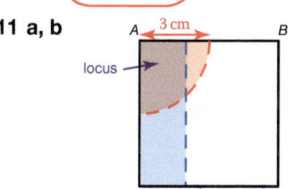

12 a, b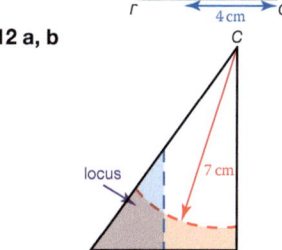

*13 a, b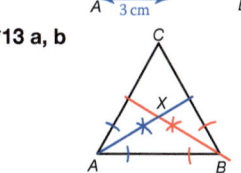

*14 a, b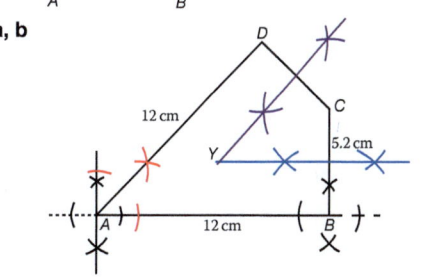

15 Two solutions are possible, the first triangle has angles 30°, 108° and 42°. The second has angles 30°, 12° and 138°.

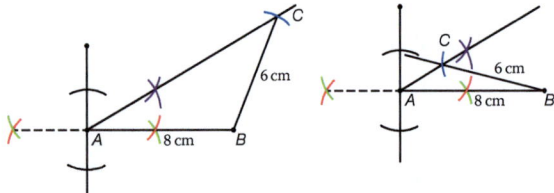

16 a   **b**

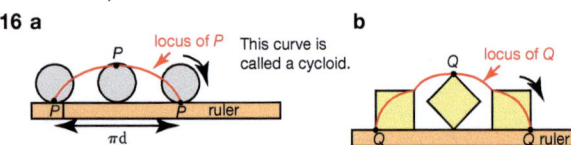

11.4A

1

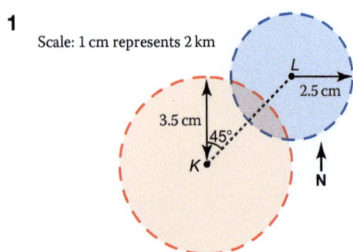

2

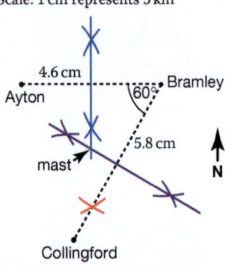

Distance to each village = 5 × 3 = 15 km

3 a

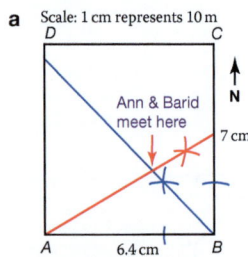

b Length need to reach $Q = PQ$
$PQ = 7.9 \times 2 = 15.8$ m
Length need to reach $R = PS + SR = 10 + 3.5 \times 2 = 17$ m
Cable needs to be at least 17 m long

4 a

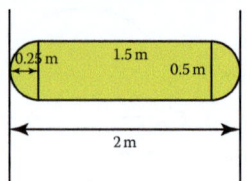

b No. Barid has walked $3.3 \times 10 = 33$ m and Ann has walked $4.7 \times 10 = 47$ m. Ann has walked further in the same space of time so is faster.

5

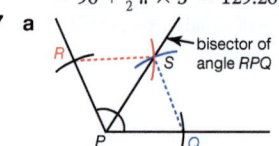

Total area = 0.946 m² (to 3 sf)

6 a 110 ft²
b The goat will reach most grass if the rail is in the centre of PS.
This gives an extra quarter circle which cannot be given by any other position along the perimeter.
Calculation for this area:
$= 18 \times 5 + \frac{1}{4}\pi \times 5^2 + \frac{1}{4}\pi \times 5^2$
$= 90 + \frac{1}{2}\pi \times 5^2 = 129.269... = 129$ ft²

***7 a**

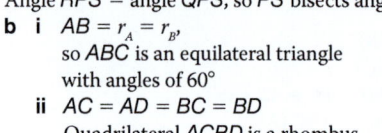

Jamie is correct.
In triangles PRS and PQS
$PR = PQ$ (equal radii), $RS = QS$ (equal radii).
PS is a side in both triangles.
Triangles PRS and PQS are congruent (SSS).
Angle RPS = angle QPS, so PS bisects angle RPQ.

b i $AB = r_A = r_B$,
so ABC is an equilateral triangle with angles of 60°
ii $AC = AD = BC = BD$
Quadrilateral $ACBD$ is a rhombus.
The diagonals of a rhombus bisect each other at 90°.
So CD is the perpendicular bisector of AB.

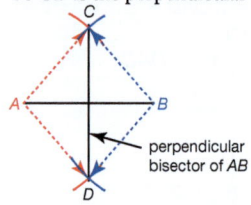

Review

1 a 133 cm² (3 sf)   **b** 40.8 cm (3 sf)
2 14.3 cm
3 a 124 m² (3 sf)   **b** 48.0 m (3 sf)
4 a 45.8 cm² (3 sf)   **b** 11.4 cm (3 sf)   **c** 27.4 cm (3 sf)
5 $a = 65°$, angles in same segment are equal
$b = 33°$, angles in same segment are equal
$c = 90°$, angles in semi-circles are 90°
$d = 36°$, angles in triangle add up to 180°
6 $a = 70.5°$, angles in an isosceles triangle
$b = 39°$, alternate segment theorem
7 a, b check students' drawing.
8 Check circle of radius 4 cm.
9 Check students' drawing.

Assessment 11

1 a i A, $61.58 \div 2\pi = 9.80$ cm (3 sf) $= x$
ii C, $314.16 \div 2\pi = 50.0$ cm (3 sf)
iii B, $2.01 \div 2\pi = 0.320$ cm (3 sf)
b i B, $\sqrt{\frac{176.7}{\pi}} = 7.50$ m (3 sf)   **ii** A, $\sqrt{\frac{81.1}{\pi}} = 5.08$ m (3 sf)
iii C, $\sqrt{\frac{0.407}{\pi}} = 0.360$ m (3 sf) $= y$
2 63.7 times

3 a 233 mm² (3 sf)
  b Outer = 59.7 mm (3 sf), inner = 25.1 mm (3 sf).
4 36 589 times
5 24 minutes
6 a 44.0 cm (3 sf)   b 117 cm² (3 sf)
7 $r = \frac{\theta}{360} \times 2 \times \pi \times r$, $\theta = \frac{360}{2\pi} = 57.3°$
8 $p = 55.5°$ (isosceles triangle), $q = 90 - 55.5 = 34.5°$ (angle in a semicircle is 90°), $r = q = 34.5°$ (isosceles triangle).
9 ∠BOC = 90 − 34 = 56°
  ∠BCO = $\frac{1}{2}$(180 − 56) = 62° (Isosceles triangle)
  ∠ABC = 90° (Angle in a semicircle)
  ∠CAB = 180 − 62 − 90 = 28° (Angle sum of a triangle)
10 $x = 90°$ (angle in semicircle), $w = 49°$ (Alternate Segment Theorem), $v = 41°$ (angle sum of a triangle).
11 ∠ABO = 90 − x (tangent perpendicular to radius), ∠OAB = 90 − x (isosceles triangle), ∠AOB = 180 − 2(90 − x) = $2x$ (angle sum of a triangle), $y = x$ (angle at centre) = 2 × angle at circumference

12 a 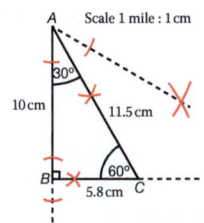   b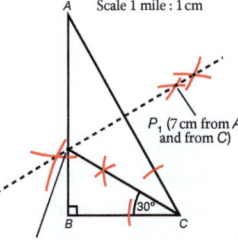

  c 7.3 miles (±0.5 miles).   d approx 3 miles/second
13 a Angles in triangle: ∠LPF = 107°, ∠PFL = 29.5°, ∠FLP = 43.5°.
    LP = 21.3 cm, PF = 29.7 cm (different scales possible)
  b 413 (± 10) miles on a bearing of 285° (± 10°).
  c Parts of LP and FL inside circle centre L and radius 50 miles.

Chapter 12

Check in

1 a 80   b 21   c 19.6   d 168
2 a 60%   b $\frac{7}{20}$ c $\frac{13}{20}$ d 0.35
3 a 1:3   b 1:6   c 2:3   d 5:2

12.1S

1 a 93.75 g   b 5.625 g   c 125 g   d 156.25 g
2 a 90 cm³   b 228 cm³   c 1.2 m³   d 168 cm³
3 a 60 g   b 36 mm   c 380 g   d 31.2 km
4 a i $\frac{1}{3}$ ii 33.3%
  b i $\frac{1}{2}$ ii 50%
  c i $\frac{3}{10}$ ii 30%
  d i $\frac{1}{4}$ ii 25%
  e i $\frac{2}{5}$ ii 40%
  f i 2   ii 200%
  g i $\frac{1}{8}$ ii 12.5%
  h i $\frac{1}{16}$ ii 6.3%
  i i $\frac{1}{12}$ ii 8.3%
  j i $\frac{1}{4}$ ii 25%
5 a i $\frac{3}{10}$ ii 30%   b i $\frac{5}{8}$ ii 62.5%
  c i $\frac{4}{9}$ ii 44.4%   d i $\frac{5}{12}$ ii 41.7%
6 a i $\frac{7}{20}$ ii 35%
  b i $\frac{2}{25}$ ii 8%
  c i $\frac{3}{10}$ ii 30%
  d i $\frac{3}{40}$ ii 7.5%
  e i $\frac{9}{40}$ ii 22.5%
  f i $\frac{3}{20}$ ii 15%
  g i $6\frac{2}{3}$ ii 666.7%
  h i $2\frac{1}{2}$ ii 250%
  i i $\frac{103}{25}$ ii 412%
  j i $6\frac{2}{5}$ ii 640%
7 a 135 kg   b 32.085 m   c $105   d 22.5 cm³
  e 510.15 g   f €16.50
8 A 30%, B 25.9%, C 23.3%, D 20.8%

12.1A

1 a $\frac{3}{25}$ b 12%
2 a £2100   b $\frac{8}{15}$, or 53.3%
3 a 64 cheeses
  b 32%
4 a i Maths   ii English
  b Overall score = (48 + 39 + 55) ÷ (60 + 50 + 70) = $\frac{71}{90}$
5 50 cars
6 Yes, $\frac{(45 \times 1 + 60 \times 2)}{220} = \frac{3}{4}$
7 Yes. Tropical ~ 0.124 g/ml > Pineapple ~ 0.115 g/ml, Blackcurrant ~ 0.117 g/ml.
8 a 40, 50   b 50 employees.
9 No, the shapes are both half shaded.
10 $\frac{5}{16}$

12.2S

1 a 1:3   b 3:1   c 1:3   d 7:2
  e 1:100   f 6:125   g 2:13   h 4:5
  i 5:8   j 37:3   k 7:19   l 16:19
2 a 2:5   b 11:16   c 5:2   d 50:3
  e 5:3   f 8:5
3 a B   b D   c A   d C
4 a $27:$63   b 287 kg:82 kg
  c 64.5 tonnes:38.7 tonnes
  d 19.5 litres:15.6 litres
  e €6:€12:€18
5 a €40, €35   b $350, $650   c 260 days, 104 days
  d 142.86 g, 357.14 g   e 214.29 m, 385.71 m
6 a $100:$250:$150   b 180°:45°:135°
  c 900 m:400 m:700 m
7 a 1:3   b 1:4   c 1:2   d 1:5
  e 1:2   f 1:3   g 1:5   h 1:3
  i 1:2   j 1:5   k 1:2   l 1:6
8 a 1000 m   b 4000 m   c 5000 m
  d 250 m   e 7250 m
9 a 325 m   b 0.6 cm
10 a 50 m   b 4 cm
11 a 200 m   b 12 cm
12 a 116 m   b 180 cm

12.2A

1 a 66.7%   b 360 cm
2 a 120%   b 102 kg
3 a 72 g   b 115 g Copper, 69 g Aluminium
4 $\frac{11}{50}$
5 a 300 m   b 120 m
6 75 cm by 200 cm
7 a 4 m, 6 m, 12 m   b 72 m²
8 0.7
9 a £280   b £80, £200
10 a 3:6:2   b $20 ($60 compared to $40)
11 8 sides
12 P = 55°, Q = 125°

12.3S

1. **a** 0.5 **b** 0.6 **c** 0.25 **d** 0.085
   e 0.0015 **f** 0.0001
2. **a** $40 **b** 260 cm **c** 3.2 kg **d** 20 m
   e 190 p **f** $35 **g** 3 kg **h** €6.20
3. **a** 325.35 kg **b** $120 **c** 10.35 kg **d** 5.88 kg
   e €21.70 **f** 78.2 m
4. **a** $224 **b** £385.20 **c** €1458 **d** $77
   e €13.35 **f** £465.83
5. 10% of £350 (£35 > £30)
6. 8% of $28 ($2 < $2.24)
7. **a** $549.60 **b** $2519 **c** $842.72 **d** $1167.90
8. **a** $790 **b** $1109.25 **c** €54.60 **d** €132.43
9. **a** 1.2 **b** 1.3 **c** 1.45
10. **a** 0.6 **b** 0.4 **c** 0.65
11. **a** £495 **b** 672 kg **c** $756 **d** 392 km
    e €658 **f** 256 m
12. **a** £275 **b** $2264 **c** €18 060 **d** €2520
    e $4.23 **f** €2000
13. **a** £385 **b** 70.3 kg **c** $550.20 **d** 491.4 km
    e 1128 kg **f** $216
14. **a** €397.80 **b** 524.9 kg **c** £1758.96
    d 599.56 km **e** $3423.55 **f** 2154.75 m

12.3A

1. **a** 60% **b** 48% **c** 60%
2. **a** 51.35% **b** 28.57% **c** 3.83%
3. **a** 20% **b** 20%
4. **a** 12.5% **b** 20%
5. 10.8% per year
6. £56 is 80% of the original price, the original price is not 120% of £56. The original price is £70.
7. $5
8. **a** 50 **b** 40 **c** 80 **d** 110
9. **a** €6 **b** €80
10. **a** Francesca £13.13, Maazin £12.80. Francesca has the bigger pay increase.
    b $78.03
11. **a** $50 **b** €30.94
12. $8450
13. 13 600
14. **a** €321.63 **b** €1749.60

Review

1. **a** $\frac{16}{25}$ **b** 36% **c** 45
2. **a** $\frac{1}{3}$ **b** 1:2
3. **a** 2:3 **b** 5:3 **c** 1:250
4. **a** £35, £7 **b** $22, $44, $33
5. **a** 2:1 **b** $\frac{1}{2}$
6. **a** 400 g milk, 600 g flour **b** 300 g **c** $\frac{2}{5}$
7. **a** 35 m **b** 962 m²
8. **a** 8 km **b** 3.125 cm
9. **a** 80.4 **b** 80.64
10. £6114
11. **a** 6.6% **b** $464 000 **c** 232%
12. **a** €1.24 **b** €9060

Assessment 12

1. **a** $N = 1$, $W = 4$
   b Brazil nuts 60%, walnuts 25%, hazelnuts 15%
   c Hazelnuts **d** Brazil nuts
   e 60% **f** $\frac{3}{4}$ **g** $\frac{1}{4}$
2. **a i** 0.52 = 52% **ii** £117 000 **b** $13 110
3. Chocolate 27, plain 36.
4. **a** 8:2:3 **b** Butter 40 g, cheese 60 g
5. **a** Small 150 cm³, large 600 cm³.
   b Small 250 ml, medium 625 ml.

6. **a** 6:5 **b** $\frac{5}{3}$:1
7. **a** Girls 32, boys 56. **b** Girls 44, boys 77, 44:77 = 4:7
8. **a** 3:4 **b** 9:16
9. **a** 9.1 s **b** 21.1 s
10. **a** 40.5 miles **b** 25.6 inches **c** 236 miles
11. **a** 200 m **b** 1.17 km **c** 4.8 cm
    d Distance = 0.2 × 20.75 = 4.15 km. Time = (4.15 ÷ 80) hours = 187 seconds (3 sf) < 200 seconds.
12. **a** Felix
    b Multiply by (100% − 45.4%). 550 × (1 − 0.454) = 300.3 ml
13. **a** 42.22% **b** 31%
14. **a** $2 712 500 **b** 90.8%
    c Yes, (35 − 24) ÷ 24 = 45.83% > 45%.
15. **a** 230 **b** 217
16. **a** $4236.31 **b** £2480 **c** €237.11

Revision 2

1. 9
2. 72 cm²
3. **a** 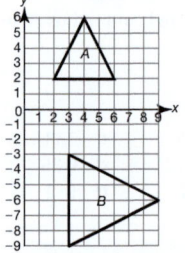 **b** 90° **c** 30 km²

4. **a** 39 609 m² **b** 3.96 hectares (3 sf)
5. **a** Enlargement, scale factor 3, centre (0, 0)
   b

6.

Rotation 90° clockwise about (0, 0) and enlargement of scale factor 1.5, centre (0, 0) (either order).

7. **a** 120 cm **b** 40 000 g or 40 kg
   c Length = 110.2 cm, mass = 41 800 g or 41.8 kg.
   d Percentage error for length = 8.89%, percentage error for mass = −4.31%
8. **a** LB 76.637 m³, UB 75.880 m³
   b Upper bound = 214 583.6 kg = 215 tonnes (3 sf)
   Lower bound = 212 464 kg = 212 tonnes (3 sf)
9. **a, b** 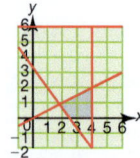 **c** 1 Lupin, 2 Stocks.

10. 4 cm × 25 cm
11. 26 10p coins and 15 50p coins.
12. $r < 100$
13. **a** 21.5%
    b i 9 **ii** 11

14 a ∠ABD = 120° (ASL), AB = AD (TEL), thus ∠BAD = ∠BDA = 30° (Isosceles △). Thus ∠DEA = 30° (AST), ∠EDA = 90° (ASC) so ∠EDF = 180 − 90 − 30 = 60° (ASL).
  b ∠DOA = 2 × ∠DEA = 60° (ACC). Thus ∠OAD = ODA = 60° (Isosceles triangle), so △ODA is equilateral and AD = AO = OD = r.
  c Arc AD = $\frac{\pi r}{3}$, area of sector AOD = $\frac{\pi r^2}{6}$

15 a, b 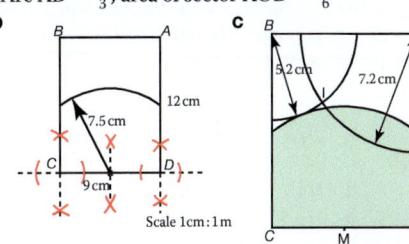   **c**

No. The point of intersection of the two arcs is not in the shaded region.

d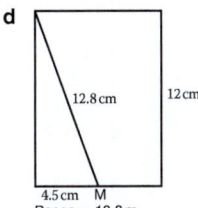

Range = 12.8 m

16 a 56 men and 48 women.
  b The numbers become 70 men and 60 women. 70 : 60 = 7 : 6 so the ratio does not change.
  c 70 men and 60 women.   **d** 65 men and 65 women.
17 7.3 inches
18 a 2.97% (3 sf)   **b** £10 669 (nearest £)
19 a 1 can pair with any of the five remaining cards, 3 can pair with any of the four remaining cards (as the pair 1 and 3 has already been counted), and so on. Number of pairings = 5 + 4 + 3 + 2 + 1 = 15
  b i $\frac{2}{15}$ **ii** $\frac{4}{15}$

Chapter 13

Check in

1 a 1, 2, 3, 4, 6, 12
  b 1, 2, 3, 5, 6, 10, 15, 30
  c 1, 2, 3, 4, 5, 6, 8, 10, 12, 15, 20, 24, 30, 40, 60, 120
  d 1, 2, 3, 4, 5, 6, 8, 9, 10, 12, 15, 18, 20, 24, 30, 36, 40, 45, 60, 72, 90, 120, 180, 360
2 2, 3, 5, 7, 11, 13, 17, 19, 23, 29, 31, 37, 41, 43, 47, 53, 59, 61, 67, 71, 73, 79, 83, 89, 97
3 a 2^7 **b** 3^3 **c** 5^7 **d** 6^2
  e 7^3 **f** 4^4
4 a 1   **b** 1   **c** 5   **d** 0.5

13.1S

1 a 12   **b** 8   **c** 4   **d** 6
2 a 77 = 7 × 11   **b** 51 = 3 × 17
  c 65 = 5 × 13   **d** 91 = 7 × 13
  e 119 = 7 × 17   **f** 221 = 13 × 17
3 9, 3, 3; 18 = 2 × 3 × 3
4 a 225   **b** 40   **c** 441   **d** 300
5 a 36 = $2^2 × 3^2$ **b** 120 = $2^3 × 3 × 5$
  c 34 = 2 × 17   **d** 48 = $2^4 × 3$
  e 27 = 3^3 **f** 105 = 3 × 5 × 7
  g 99 = $3^2 × 11$ **h** 37 = 37   **i** 91 = 7 × 13

6 a $2^2 × 263$ **b** $2^9 × 5$
  c $3 × 5^2 × 11$ **d** $5 × 11 × 13$
  e 7 × 11 × 13   **f** 3 × 73
  g 17^2 **h** $2^3 × 5 × 71$
  i $5 × 7^2 × 11$ **j** 7 × 13 × 19
  k $2 × 3^2 × 11 × 17$ **l** $2^2 × 11 × 13 × 17$
  m $2 × 5 × 7 × 13^2$ **n** $2^2 × 23 × 31$
  o $3^3 × 13 × 29$
7 36
8 a 20   **b** 36   **c** 30   **d** 60
  e 70   **f** 40
9 a 5   **b** 16   **c** 3   **d** 5
  e 14   **f** 15
10 a 48   **b** 800   **c** 66   **d** 416
  e 280   **f** 5040
11 a 1260, 60   **b** 8085, 7   **c** 1680, 48   **d** 9216, 2
  e 314 706, 2   **f** 82 944, 16
12 a HCF = 6, LCM = 1890   **b** HCF = 2, LCM = 34 650
  c HCF = 1, LCM = 510 510   **d** HCF = 1, LCM = 27 588
13 a The digits sum to a multiple of 3 so it is divisible by 3 and the last digit is 5 so it is divisible by 5.
  b The three-digit sequence 262 is repeated in this six-digit number. 262 262 = 262(1000 + 1)
14 Abundant: 72, 40, 30   Perfect: 6, 28   Deficient: 86, 50, 64, 27

13.1A

1 Let p be prime. The only factorisations possible are $p^3 = 1 × p^3 = p × p^2$ so the only factors of p^3 are p^3, p^2, p and 1.
2 The factors of the square of a prime p^2 are 1, p and p^2. Sum of proper factors = 1 + p.
  $1 + p < p^2$ if $0 < p^2 − p − 1$ for any prime p.
  From the graph, $0 \geqslant p^2 − p − 1$ for $\frac{1 − \sqrt{5}}{2} \leqslant p \leqslant \frac{1 + \sqrt{5}}{2}$ only, and there are no prime numbers in this region so we have the result.

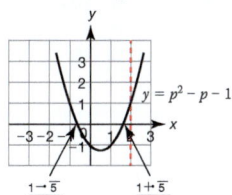

3 Other possibilities are 450 and 60 or 90 and 300.
4 45
5 a Any pair of numbers will support the statement.
  b

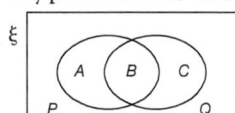

Let A, B and C be the products of the numbers in each section.
$P = A × B$, $Q = B × C$, $P × Q = A × B^2 × C$
HCF = B, LCM = $A × B × C$, HCF × LCM = $A × B^2 × C$ as required.
6 a 9, 15, ... (lots more)   **b** 2 (the only one that isn't odd)
  c i 6 − 1 = 5, 12 − 1 = 11, 18 − 1 = 17 (there are lots more)
    ii 36 − 1 = 35 (there are more)
7 Both online at 9:40.
8 120 seconds
9 18 cm
10 a e.g. 8, 12, 105   **b** e.g. 16, 24, 36, 54, 81
  c e.g. 108, 112, 120, 162   **d** e.g. 64
11 15 × 11 × 11, 33 × 5 × 11, 55 × 3 × 11, 3 × 5 × 121
12 HCF = $x − 1$
  LCM = $(x + 1)(x − 1)(2x + 1) = 2x^3 + x^2 − 2x − 1$

13 a $11x$ **b** $22x$
14 a 4 **b** 12 **c** 12 **d** 5
 e 6 **f** 9
15 a [triangle diagram: 13, 182, 585, 14, 45, 630] **b** [triangle diagram: $x+2$, $2(x^2+2x)$, $3x^2-5x-2$, $2x$, $6x^2-2x$, $3x-1$]

13.2S

1 a 4.5 (1 dp)
 b i 6.3 (1 dp) **ii** 7.7 (1 dp) **iii** 9.7 (1 dp)
2 a 2.7 (1 dp) **b** 3.7 (1 dp) **c** 4.3 (1 dp) **d** 5.3 (1 dp)
3 a ±36.67 (2 dp) **b** ±6.21 (2 dp)
 c ±84.22 (2 dp) **d** ±15.32 (2 dp)
4 a 23 **b** -6 **c** -4.12 (2 dp) **d** 0.25
5 a 6^5 **b** 4^9 **c** 11^7 **d** 1
 e 3^{12} **f** 9^{10}
6 a 7^2 **b** 8^4 **c** 3 **d** 4^6
 e 1 **f** 12^2
7 a 3^4 **b** 3^{12} **c** 3^{12} **d** 1
 e 3^{18} **f** 3^{18}
8 a 8^5 **b** 5^5 **c** 2^6 **d** 9^2
 e 8^{13} **f** 7^{29} **g** 4^{18} **h** 6^4
9 a 3 **b** 5^{10} **c** 4^2 **d** 7^2
 e 8^8 **f** 9^{19}
10 a 4^2 **b** 9^2 **c** 8 **d** 5^6
 e 6^5 **f** 8^{16}
11 a $3^3 \times 4^4$ **b** $7^2 \times 8^3$ **c** $5^6 \times 6^4$ **d** $2^3 \times 5^2$
 e $7^8 \times 9^{11}$ **f** $\frac{8^{14}}{5^6}$ **g** $2^4 \times 9^{10}$ **h** $3^8 \times 8^7$
12 a $8^3 \times 5^2$ **b** $6^2 \times 7^2$ **c** $5^2 \times 6^2$ **d** $5^{15} \times 7^{22}$
 e $\frac{8^7}{3^2}$ **f** $4^7 \times 5^4$ **g** $6^4 \times 7^2$ **h** $4^4 \times 7^5$
13 a 15.21 **b** 4.41 **c** 0.49 **d** 175.56 (2 dp)
 e -157.46 (2 dp) **f** 970.30 (2 dp)
 g -0.001 **h** 4784.09 (2 dp) **i** 31.01 (2 dp)
14 a 16 **b** 64 **c** 32 **d** 100
 e 1000 **f** 27 **g** 8 **h** 9
15 a 9^2 **b** 5^3 **c** 2^7 **d** 10^5
 e 3^4 **f** 7^3
16 a 4 **b** 3 **c** 3
17 2.1 (1 dp)

13.2A

1 Many answers possible including $a = 2, b = 3$ or $a = 3, b = 5$
2 Many answers possible including $p = 3, q = 6$ or $p = 4, q = 21$
3 a 20 **b** 4^5
 c Kiera. Her probability of getting full marks is $\frac{1}{1024}$ compared with $\frac{1}{2187}$ for Haneef.
4 $(-3, 2), (-3, 6), (-1, 0), (-1, 8), (3, 0), (3, 8), (5, 2), (5, 6)$
5 a 2.645751 is a rounded answer, the exact value has a decimal expansion that goes on forever without falling into a pattern. As 2.645751 is not exactly equal to $\sqrt{7}$ its square will not equal 7.
 b 56 and 57
6 $2 < \sqrt[3]{27} < 4 < \sqrt{20}, \sqrt[3]{125}, \sqrt[3]{210} < 6 < \sqrt{50} < 8 < \sqrt{75}, \sqrt{81}, \sqrt[3]{999} < 10$
7 a 27
 b No. Taking one condition away results in a non-unique answer.
 Multiple of 9 and less than 40: 9, 18, 27, 36.
 Cube number and less than 40: 1, 8, 27.
 Multiple of 9 and cube number: 27, 729, ...
8 a $64 = 8^2 = 4^3$ **b** $3^6 = (3^3)^2 = (3^2)^3$ **c** $4^6 = (4^2)^3 = (4^3)^2$
9 a $36 \times 216 = 6^2 \times 6^3 = 6^{(2+3)} = 6^5 = 7776$
 b $46656 \div 36 = 6^6 \div 6^2 = 6^{(6-2)} = 6^4 = 1296$
10 a kB 2.34%, MB 4.63%, GB 6.87%, TB 9.05%
 b Yes, $(2^{10})^2 = 2^{10 \times 2} = 2^{20}$ which is a megabyte.
11 a $a = 16$ **b** $b = 2$ **c** $c = -\frac{2}{3}$ **d** $d = 2$

12 $\frac{2 \times 4^{x+3}}{2^{2x+4}} = \frac{2^1 \times (2^2)^{x+3}}{2^{2x+4}} = \frac{2^1 \times 2^{2x+6}}{2^{2x+4}} = \frac{2^{2x+7}}{2^{2x+4}} = 2^{(2x+7)-(2x+4)}$
 $= 2^3 = 8$
13 a 320 cm^3 **b** 3.4 cm
 c $\sqrt[3]{320} = \sqrt[3]{8} \times \sqrt[3]{40} \approx 2 \times 3.4 = 6.8$
14 The equations can be written $y = x^{3-a}$ and $y = x^{5-a}$. At the point(s) of intersection $y = x^{3-a} = x^{5-a}$. As $3 - a \neq 5 - a$ for any value of a, the base number x must be one which is not altered by a change in power. The only numbers which are the same to any power are 0 and 1, so $x = 0$ or 1.

13.3S

1 a $2\sqrt{3}$ **b** $2\sqrt{5}$ **c** 5 **d** $3 + \sqrt{2}$
 e $7 + 2\sqrt{7}$ **f** $14 + \sqrt{17}$
2 a $\sqrt{6}$ **b** $\sqrt{15}$ **c** $\sqrt{143}$ **d** $\sqrt{231}$
3 a $\sqrt{2} \times \sqrt{7}$ **b** $\sqrt{3} \times \sqrt{11}$ **c** $\sqrt{3} \times \sqrt{7}$ **d** $\sqrt{5} \times \sqrt{7}$
 e $\sqrt{2} \times \sqrt{23}$ **f** $\sqrt{3} \times \sqrt{17}$
4 a $2\sqrt{5}$ **b** $3\sqrt{3}$ **c** $7\sqrt{2}$ **d** $4\sqrt{3}$
 e $2\sqrt{7}$ **f** $3\sqrt{5}$ **g** $3\sqrt{7}$ **h** $11\sqrt{3}$
 i $16\sqrt{2}$ **j** $8\sqrt{3}$ **k** $10\sqrt{5}$ **l** $13\sqrt{5}$
5 a $\sqrt{48}$ **b** $\sqrt{50}$ **c** $\sqrt{80}$ **d** $\sqrt{200}$
 e $\sqrt{405}$ **f** $\sqrt{343}$
6 a 6 **b** 12 **c** 66 **d** 42
 e $11\sqrt{21}$ **f** $14\sqrt{10}$ **g** $30\sqrt{5}$ **h** 32
 i $42\sqrt{5}$ **j** $156\sqrt{2}$
7 a $5\sqrt{5}$ **b** $7\sqrt{7}$ **c** $8\sqrt{3}$ **d** $-\sqrt{7}$
 e $\sqrt{2}$ **f** $-5\sqrt{14}$
8 a 2 **b** 5 **c** $3 + 3\sqrt{3}$ **d** $2\sqrt{3} + 8$
 e $3\sqrt{15} + 5\sqrt{3}$ **f** $8\pi - 2\pi\sqrt{5}$
 g $7 + 3\sqrt{5}$ **h** $38 - 14\sqrt{7}$
 i $7 - 4\sqrt{3}$ **j** $73 + 12\sqrt{35}$
9 a 3 **b** $\frac{7}{16}$ **c** $\frac{\sqrt{2}}{5}$ **d** $\frac{4 + 3\sqrt{2}}{10}$
 e $\frac{13\sqrt{7}}{20}$ **f** $\frac{5\sqrt{3}}{3}$
10 a 5.66 **b** 3.24 **c** 4.24 **d** 106.10
11 a $\frac{\sqrt{2}}{2}$ **b** $\frac{\sqrt{3}}{3}$ **c** $\frac{3\sqrt{7}}{7}$ **d** $\frac{5\sqrt{6}}{6}$
 e $\frac{\sqrt{10}}{2}$ **f** $\frac{3\sqrt{15}}{5}$
12 a $\frac{\sqrt{2}}{2}$ **b** $\frac{\sqrt{10}}{5}$ **c** $\frac{\sqrt{3}}{2}$ **d** $\frac{\sqrt{30}}{6}$
 e $\frac{2\sqrt{10}}{5}$ **f** $\frac{\sqrt{30}}{20}$

13.3A

1 a Perimeter $= 4\sqrt{5} + 2$, area $= \sqrt{5} + 5$
 b Perimeter $= 4$, area $= 2\sqrt{3} - 3$
2 a $\frac{3\sqrt{2} + 6}{5}$ **b** $\frac{54 + 36\sqrt{2}}{25}$
3 a $\frac{4\sqrt{7}}{7}$ cm **b** $\left(\frac{\sqrt{9023}}{7} + 5\sqrt{7}\right)$ cm
4 a $1.414 + 1.732 = 3.146$
 b $\sqrt{2} \times \sqrt{5} \approx 1.414 \times 2.236 = 3.162$
 c $5\sqrt{5} \approx 5 \times 2.236 = 11.18$
 d $2\sqrt{2} \times \sqrt{5} \times \sqrt{3} = 10.95$
5 e.g. $\sqrt{6} \approx 1.414 \times 1.732 = 2.449, \sqrt{12} \approx 1.414^2 \times 1.732 = 3.463, \sqrt{5} - \sqrt{3} \approx 2.236 - 1.732 = 0.504, \sqrt{20} \approx 2 \times 2.236 = 4.472$
6 B, A, C. The expressions can be written A $\frac{54 + 27\sqrt{3}}{36}$, B $\frac{60 - 36\sqrt{7}}{36}$, C $\frac{80 + 44\sqrt{5}}{36}$. $80 > 54$, 60 and $44\sqrt{5} > 27\sqrt{3}$, $-36\sqrt{7}$. $60 - 36\sqrt{7} = 54 + 6 - 36\sqrt{7}$, $6 - 36\sqrt{7} < 0$ and $27\sqrt{3} > 0$.
7 $20 + 22\sqrt{3}$
8 a $3\sqrt{2}$ cm **b** $18\sqrt{2}$ cm
***9 a** $\sqrt{64}$ and $\sqrt{100}$ **b** $2 - \sqrt{5}$ and $2 + \sqrt{5}$
 c $\sqrt{5}$ and $\sqrt{20}$ **d** $\sqrt{10}$ and $\sqrt{100}$
 e $3 + \sqrt{5}$ and $3 - \sqrt{5}$ **f** $8\sqrt{2}$ and $3\sqrt{2}$
 g $8\sqrt{2}$ **h** $2 - \sqrt{5}$

10 a $-2 + \frac{4}{3}\sqrt{3}$
  b $Y^2 = XZ = \sqrt{5}(1 - \sqrt{5}) = \sqrt{5} - 5 < 0$ as $\sqrt{5} < 5$ but Y^2 cannot be negative as it is a square number.

Review

1 a 2, 3, 37, 101   **b** 1, 3, 15, 105   **c** 63, 105
2 a $3 \times 5 \times 7$ **b** 37   **c** $2^2 \times 3 \times 5^2$ **d** $2 \times 3^2 \times 7$
3 a i 35   **ii** 1
  b i 39   **ii** 13
  c i 180   **ii** 12
  d i 540   **ii** 6
4 a 5.48   **b** 3.56
5 a 4   **b** 5   **c** 64   **d** 81
6 a 7^4 **b** 3^9 **c** 3^3 **d** 7^{21}
  e 5
7 a $6\sqrt{3}$ **b** $2\sqrt{3} + 2\sqrt{2}$ **c** 10   **d** $2\sqrt{3}$
  e 6   **f** $\frac{\sqrt{2}}{2}$ **g** $13 - 7\sqrt{3}$ **h** 3
8 a $\frac{\sqrt{7}}{7}$ **b** $\frac{\sqrt{6}}{2}$ **c** $\frac{\sqrt{5}}{2}$ **d** $\frac{2\sqrt{6}+3}{3}$
  e $\frac{6 - 5\sqrt{2}}{8}$ **f** $\frac{12 + 5\sqrt{6}}{27}$

Assessment 13

1 Isa, $2^3 \times 3^2 \times 5^2 \times 11$.
2 a $16 = 3 + 13 = 5 + 11$
  b $64 = 3 + 61 = 5 + 59 = 11 + 53 = 17 + 47 = 23 + 41$
  c Odd square number = even + odd, 2 is the only even prime, there is only one answer.
3 123
4 a $41 - 0 + 0^2 = 41$, $41 - 3 + 3^2 = 47$, $41 - 6 + 6^2 = 71$ are all prime.
  b $41, 41 - 41 + 41^2 = 41^2$.
5 a No, HCF = $2 \times 3^3 = 54$.
  b $96 = 2^5 \times 3$, $270 = 2 \times 3^3 \times 5$, LCM = $2^5 \times 3^3 \times 5 = 4320$.
6 6
7 a i HCF = 6   **ii** HCF = 84
  b Yes. Students' answers, for example HCF (84, 228, 504) = 12
8 9 pm
9 a Zainab, 3^5.   **b** Gino, 14^6.
10 a $p = 5$ **b** $q = 2$
11 a i 45   **ii** $15\sqrt{3} + 6\sqrt{21}$ **iii** $25\sqrt{3} + 10\sqrt{21}$ 2
  b $90 + 80\sqrt{3} + 32\sqrt{21}$ **c** $10 + 22\sqrt{3} + 4\sqrt{7}$
  d $45(5 + 2\sqrt{7})$
12 a $2\sqrt{3}, \sqrt{6} = \sqrt{2}\sqrt{3}$.   **b** 120, $\sqrt{36} = 6$.
  c $8\sqrt{3} + 19$, there are two $4\sqrt{3}$ terms.
  d $10 - 4\sqrt{6}$, there are two $-2\sqrt{6}$ terms.
  e $18 + 4\sqrt{3}$, the $\sqrt{3}$ terms don't cancel.
  f $2, \sqrt{5} \times -\sqrt{5} = -5$
  g $-14 - 28\sqrt{22}$, $12\sqrt{64} = 12 \times 8 = 96$, and the $\sqrt{11}$ terms don't cancel.
13 a $\frac{4\sqrt{5}}{5}$ **b** $\sqrt{8}$ **c** $\frac{\sqrt{2} - 2}{6}$ **d** $\frac{4\sqrt{5} - 3\sqrt{10}}{5}$
  e $\frac{15\sqrt{7} - 7\sqrt{3}}{35}$ **f** $\frac{1}{2}$
14 a $h^2 = (4 + \sqrt{2})^2 + (4 - \sqrt{2})^2 = 16 + 8\sqrt{2} + 2 + 16 - 8\sqrt{2} + 2 = 36$, $h = 6$.
  b $P = 6 + 4 + \sqrt{2} + 4 - \sqrt{2} = 14$, $A = \frac{1}{2}(4 + \sqrt{2})(4 - \sqrt{2}) = \frac{1}{2}(16 - 4\sqrt{2} + 4\sqrt{2} - 2) = 7$

Chapter 14

Check in

1 a i 9   **ii** 4
  b i 27   **ii** -8
  c i 18   **ii** 8
  d i 30   **ii** -10
  e i 24   **ii** 14
  f i 60   **ii** -20
  g i 63   **ii** 8
  h i 15   **ii** -30

2

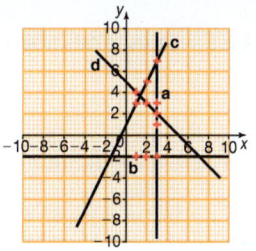

3 a $n = 2$ **b** $m = 1$ **c** $p = \frac{1}{2}$
4 a i 3   **ii** 4   **iii** Positive slant
  b i -4 **ii** 10   **iii** Negative slant
  c i 4   **ii** 5   **iii** Positive slant
  d i 2   **ii** 7.5   **iii** Positive slant
  e i 0   **ii** 7   **iii** Horizontal
  f i 0.5   **ii** 2   **iii** Positive slope

14.1S

1 a

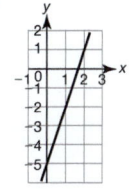

2 a

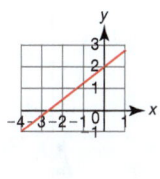

  b $3x - 5$ **b** $y = \frac{3}{4}x + 2$
3 a B   **b** C   **c** D   **d** A   **e** F   **f** E
4 a i gradient = 1, y-intercept = 2
    ii equation: $y = x + 2$
  b i gradient = 2, y-intercept = 3
    ii equation: $y = 2x + 3$
  c i gradient = 3, y-intercept = 0
    ii equation: $y = 3x$
  d i gradient = -2, y-intercept = 3
    ii equation: $y = -2x + 3$
  e i gradient = 0, y-intercept = 2
    ii equation: $y = 2$
  f i gradient undefined, no y-intercept
    ii equation: $x = 5$
5 $y = 3x + 5, y = 5x - 2, y = -2x + 7, y = \frac{1}{2}x + 9, y = -\frac{1}{4}x - 3, y = 4, y = x$
6 $y = -4, y = 6, x = -1, x = 3$
7 a $y = 6x + 2$ **b** $y = -2x + 5$
  c $y = -x + \frac{1}{2}$ **d** $y = -3x - 4$
8 Any value of the constant c is allowed.
  a $y = 2x + c$ **b** $y = -5x + c$
  c $y = -\frac{1}{4}x + c$ **d** $y = -4x + c$
  e $y = \frac{3}{4}x + c$ **f** $y = \frac{9}{2}x + c$
9 a $y = -4x - 2$ **b** $2y - 3x = -4$
10 Any value of the constant c is allowed.
  a $y = -\frac{1}{2}x + c$ **b** $y = \frac{1}{5}x + c$
  c $y = 4x + c$ **d** $y = \frac{1}{4}x + c$
  e $y = -\frac{4}{3}x + c$ **f** $y = -\frac{2}{9}x + c$
11 a $y = -\frac{1}{2}x + 8\frac{1}{2}$ **b** $y = \frac{3}{2}x + 2\frac{2}{3}$
12 a $y = 4x - 7$ **b** $y = 2x - 4$
  c $y = \frac{1}{3}x + \frac{1}{3}$ **d** $y = -2x + 1$
13 A and E, B and F, D and G. C is the odd one out.

14.1A

1 a A, B, E   **b** C, F   **c** D   **d** A   **e** C, F

2 a $y = 6x + 50$
$m = 6$, a child grows 6 cm each year. $c = 50$, the child's height is 50 cm at birth.
b $y = \frac{1}{2}x + 10$
$m = \frac{1}{2}$, for every extra year of age a person has on passing their driving test they will need $\frac{1}{2}$ a lesson more. It is not sensible to interpret c.

3 The line of best fit has equation $y = x - 10$. On average students scored 10% less on paper 2.

4 a $y = -2x + 21$ **b** $y = \frac{3}{2}x + \frac{67}{4}$
c $y = \frac{6}{17}x - \frac{63}{34}$

5 $\frac{t}{3}$

6 FALSE.

7 $y = -\frac{1}{2}x + 4$

*8 $y = -\frac{3}{4}x + \frac{25}{4}$

14.2S

1 a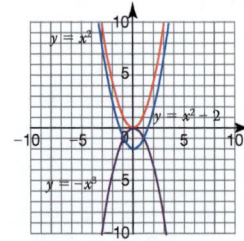
b $y = -x^2$ **c** $y = x^2 - 2$

2 a i $x = 4$ **ii** 0   **iii** (4, 4)
  b i $x = 3.5$ **ii** 2   **iii** (3.5, 12.25)
  c i $x = 0$ **ii** 2   **iii** (0, -4)
  d i $x = -5$ **ii** 0   **iii** (-5, -2)
  e i $x = 1.5$ **ii** 2   **iii** (1.5, 6.25)
  f i $x = 4$ **ii** 1   **iii** (4, 0)

3 a A   **b** F   **c** D   **d** C
  e E   **f** B

4 a $x = 0, x = -7$ **b** $x = 0, x = 8$
  c $x = 0, x = -5$ **d** $x = 0, x = -2$
  e $x = 0, x = -4$ **f** $x = 0, x = -2.5$
  g $x = 0, x = -3$ **h** $x = 0, x = -0.6$

5 a $x = -1, x = -6$ **b** $x = 1, x = 2$
  c $x = -2, x = -3$ **d** $x = -4, x = 3$
  e $x = 3, x = -2$ **f** $x = 1, x = 4$
  g $x = 5, x = -3$ **h** $x = -2, x = -7$

6 a $x = -2$ **b** $x = 4$ and $x = -4$
  c $x = 2$ and $x = -2$ **d** $x = -5$

7 a $y = (x + 3)^2 - 6$ **b** (0, 3)   **c** (-3, -6)
  d
  e 2

8 a i $y = (x + 1)^2 - 4$ **ii** (-1, -4)   **iv** 2
  b i $y = (x + 1)^2 + 2$ **ii** (-1, 2)   **iv** 0
  c i $y = (x + 4)^2 - 4$ **ii** (-4, -4)   **iv** 2
  d i $y = (x - 3)^2 + 3$ **ii** (3, 3)   **iv** 0
  e i $y = (x + \frac{5}{2})^2 - \frac{37}{4}$ **ii** $(-\frac{5}{2}, -\frac{37}{4})$ **iv** 2
  f i $y = (x - 3)^2$ **ii** (3, 0)   **iv** 1

g i $y = -(x - 5)^2 + 24$ **ii** (5, 24)   **iv** 2
h i $y = -(x - \frac{3}{2})^2 + \frac{37}{4}$ **ii** $(\frac{3}{2}, \frac{37}{4})$ **iv** 2
iii

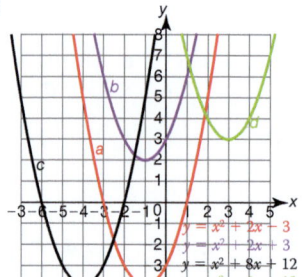

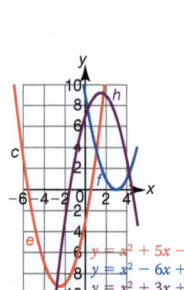

*9 a $y = 3(x + 1)^2 - 5$ **b** (0, -2)   **c** (-1, -5)
d   **e** 2

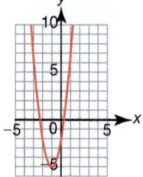

*10 $b^2 - 4ac = -20 < 0$, this appears under a square root in the formula for solving quadratic equations. You cannot find the square root of a negative number so there are no solutions to the equation $x^2 + 4x + 9 = 0$.

*11 $b^2 - 4ac = 100 > 0$, this appears within a square root in the quadratic formula. There are two square roots for every positive number so there are two solutions to the equation $3x^2 + 2x - 8 = 0$.

*12 17 metres

14.2A

1 a 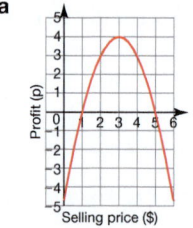   **b** $1 and $5   **c** $3, 4 cents

2 a €225, €1250   **b** €200, €250

3 a 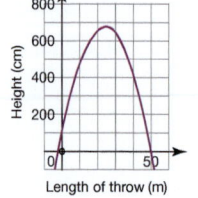   **b** 676 cm   **c** 50 m

4 a 50 cm   **b** 62.5 cm

Answers

5 a 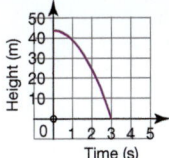　　**b** 44.1 metres　　**c** 3 seconds

6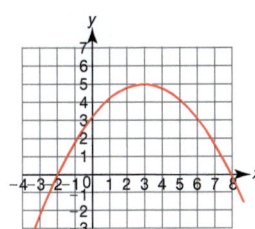

7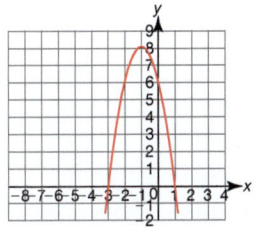

8 1

9 a 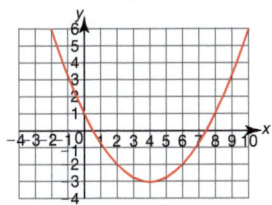　　**b** $x = 0.5, x = 7.5$

*10 a $\frac{1}{20}(10)^2 + 5 = 5 + 5 = 10$
　b Multiple solutions possible, e.g. (20, 25), (30, 50)
　c, d Students' research.

*11 a $(x - 4)^2 + 4$　　**b** $-(x - 3.5)^2 + 12.25$
　c $x^2 - 4$　　**d** $-(x + 5)^2 - 2$
　e $-(x - 1.5)^2 + 6.25$　　**f** $(x - 4)^2$
　i The equation of the line of symmetry is $x = q$.
　ii Number of roots is 0 if p and r have the same sign, 1 if $p = 0$ or $r = 0$, 2 if p and r have different signs.
　iii The turning point is (q, r).

14.3S

1 a 13.3 km/h (1 dp)　　**b** 40 km/h
2 a 5.5 km/h (1 dp)　　**b** 8 km/h
3 a 2 pm to 2:30 pm. This section of line is the steepest.
　b 46.7 km/h to (1 dp)　　**c** 35 km/h
　d Mark was travelling in the opposite direction.
4 a 6 m/s　　**b** 1 second and 12 seconds
　c 10 m/s²　　**d** 120 metres
5 a 2 m/s²　　**b** 18 metres　　**c** 143.5 metres
*6 The object accelerates during the first two seconds, at first slowly and then more quickly. It then accelerates steadily for a further three seconds. At five seconds the object begins to slow down steadily, coming to a halt at 15 seconds.
7 a 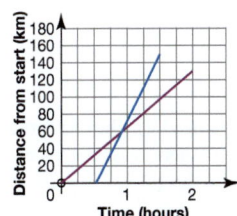　　**b** Approximately 11:53

14.3A

1 a i 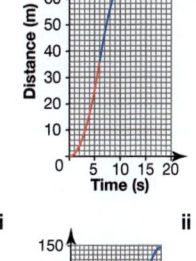　　**ii**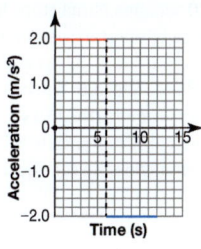

　b i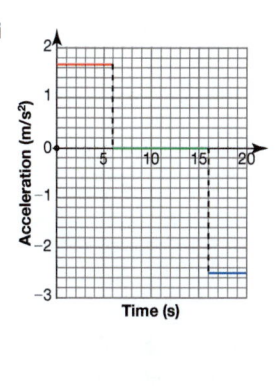

　ii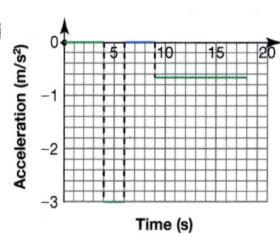

　c i 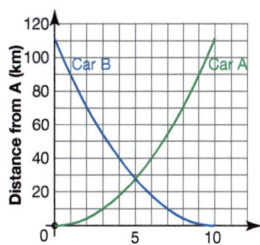　　**ii**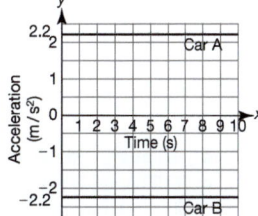

2 a The two cyclists are the same distance from a given point.
　b Chris: 15 km/h, Joanna: 8 km/h
3 a Car A and Car B both travel 111 metres (to 3 s.f.)
　b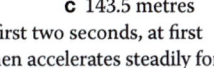

　c

4 a The second section suggests that time is going backwards while distance is increasing.
　b It is not possible to have a negative distance in the first section.
　c The second section suggests that velocity is changing instantaneously.
　d The second and fourth sections suggests that time is going backwards while velocity is increasing (or decreasing).

5 Initially Sara runs faster than Nima (8 m/s against 5 m/s). After 20 seconds Nima stops for a 5 second rest. 25 seconds in to the race Nima starts running again at 13.3 m/s and Sara slows her pace to 5 m/s. 37 seconds into the race Nima overtakes Sara. 40 seconds in to the race Nima slows her pace to 6.7 m/s and runs for a further 15 seconds. 45 seconds into the race Sara increases her pace to 6.7 m/s and runs for a further 15 seconds. Nima wins the race by 5 seconds.

14.4S

1 a 2   b $-\frac{2}{3}$ c 2   d -1
    e 6   f -4 g $-\frac{1}{3}$ h $\frac{1}{2}$
    i 1   j $-\frac{3}{8}$

2 a 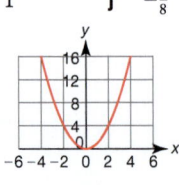   b i 2
                              ii 6
                              iii -2
                              iv -4
                              v 0
                              vi -6

3 a 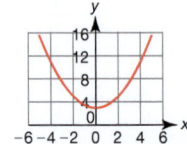   b i 2
                              ii -3
                              iii -2
                              iv 0
                              v 1
                              vi 3
   c The gradient will be 5 since it is always equal to the value of x in this case.

4 a 10.5
   b An overestimate since the errors in the smaller trapezium cancel each other out.

5 a 20   b 19.625

6 31 m

7 a i 1   ii 4   iii 9   iv 16
   b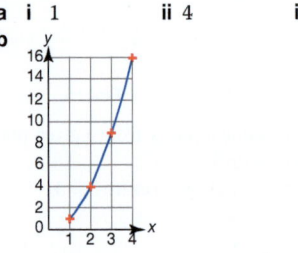
     The curve is a section of $y = x^2$.

8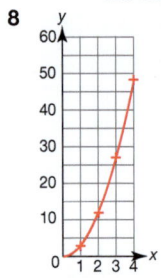
     This curve is an approximation to $y = x^3$.

14.4A

1 a 3 m/s²   b 0 m/s²   c 4.5 m/s²   d 20 metres
2 a £5560   b 153   c £153 pounds per year
   d i £247 pounds per year   ii £399 pounds per year
3 a $15 081.69
   b $23,000 per year (2 sf)
4 -6.25 °C per minute
5 $T = 6$ seconds
6 a $9\frac{1}{3}$ b 6.7   c 12.95

14.5S

1 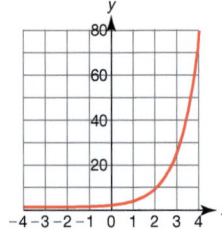   **2**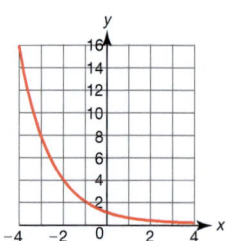

3 a

x	$-45°$	$0°$	$45°$	$90°$	$135°$
cos x	0.71	1	0.71	0	-0.71

x	$180°$	$225°$	$270°$	$315°$	$360°$
cos x	-1	-0.71	0	0.71	1

x	$405°$	$450°$
cos x	0.71	0

   b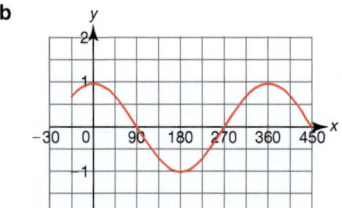

   c $360°$ d Max = 1, Min = -1

4 a

x	$-30°$	$0°$	$30°$	$60°$	$90°$
tan x	-0.58	0	0.58	1.73	–

x	$120°$	$150°$	$180°$	$210°$	$240°$
tan x	-1.73	-0.58	0	0.58	1.73

x	$270°$	$300°$	$330°$	$360°$	$390°$
tan x	–	-1.73	-0.58	0	0.58

   b, c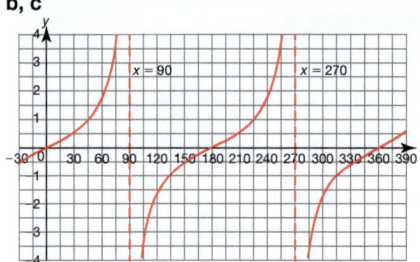

   d $180°$

5 a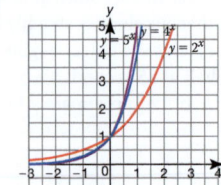

   b They all pass through the point (0, 1). They have the same characteristic shape. None of them cross the x-axis. The larger the 'base' number, the quicker the rate of increase shown in the graph

6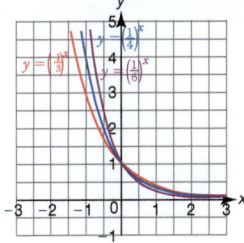

They all pass through the point (0, 1). They have the same characteristic shape. None of them cross the x-axis.
The larger the 'base' number, the quicker the rate of decrease shown in the graph.

7 It is the line $y = 1$
8 a C   b D   c B   d A
9 a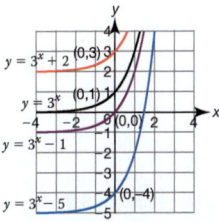

b The graphs all have the same shape, but are moved up or down the y-axis

10 a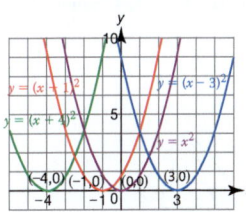

b The graphs all have the same shape and minimum y value. They are all displaced along the x-axis.

11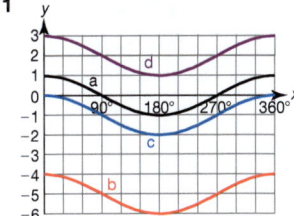

12

13 a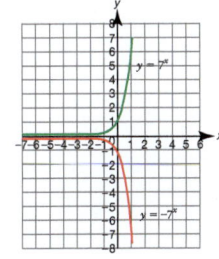

b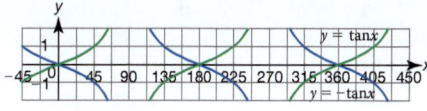

c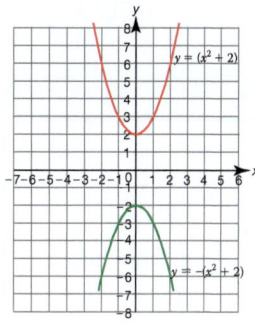

d The negative lines are reflections of the positive lines.
14 No. $\cos(x + 90) = -\sin x$

14.5A

1 a Any three of: 45°, 225°, 405°, 585°, 765°,...
   b Keep adding (or subtracting) 180°.
2 a 30°, 150°, 390°
   b Using the graph, continue the sine curve and measure when $y = 0.5$, or calculate 360k with the starting integer of 30°
3 a −30-40°, 30−40°, 330−340°, 390−400°   b 5
4 a 75 000s   b 3.3 days (accept 3.25 to 3.5)
5 a (135, 1)   b $p = 45$ (or 405°, 765°, ..)
   c $q = 135$ (or 495°, 855°, 1215°, ...)
6 a True. The graph is symmetric about the y-axis.
   b False. The period of sine is 360° > 180° so a horizontal translation of 180° will not map it onto itself.
   c True. The period of tangent is 180° so a horizontal translation of 180° will map it onto itself.
   d True. If the sine curve is translated 90° to the left it maps onto the cosine curve.
   e False. If the cosine curve is reflected in the y-axis and translated 90° to the left the result will be $y = \sin(x)$.
   f True. If the tangent curve is reflected in the x-axis and the y-axis it maps onto itself.
7 In the first function an increase in the size of a corresponds with an increase in amplitude. In the second function an increase in the size of a corresponds with a decrease in period.

Review

1 a 2.5   b $y = 2.5x + 2$
2 $y = -3x + 5$
3 a Gradient −2, y-intercept 7   b Gradient 1, y-intercept 9
   c Gradient −1, y-intercept 3   d Gradient $-\frac{2}{3}$, y-intercept $\frac{5}{3}$
4 a $y = 5x + 6$ b $y = -\frac{1}{5}x + 1$
5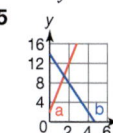

6 a 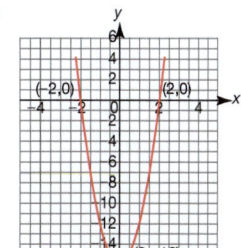   b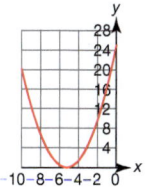

c 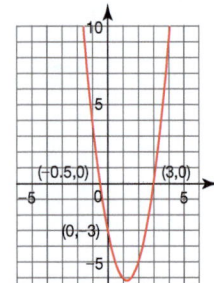   d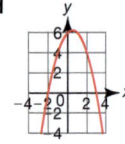

b (graph: Speed km/h vs Distance m)

7 a $(0, -6)$ **b** $(-2, 0), (1.5, 0)$
   c $x = -2, 1.5$ **d** -0.35
8 a i $(x + 1)^2 - 6$ **ii** $2(x - 3)^2 - 13$
   iii $-\left(x - \frac{5}{2}\right)^2 + \frac{25}{4}$
   b i $(-1, -6)$, min   **ii** $(3, -13)$, min
   iii $(2.5, 6.25)$, max
9 a 50 m   **b** 3.33 m/s   **c** 21.2 km/h
   d Increasing/accelerating.   **e** 2 m/s²

Assessment 14

1 a A: 2   B: $\frac{3}{2}$ C: $-\frac{5}{3}$ D: $-\frac{2}{3}$
   b i C   **ii** B   **iii** A   **iv** D
   c i $\frac{20}{3}$ **ii** $\frac{9}{2}$ **iii** 0   **iv** 0
2 a 2, 7   **b** 4, 9   **c** 6, -11 **d** $-4, 12$
   e $\frac{4}{7}, -2$ **f** $-\frac{15}{14}, \frac{35}{14}$
3 a $y = -5x - 61$ **b** $y = 3x - \frac{1}{2}$
   c $y = \frac{1}{3}x - 2$ **d** $y = \frac{1}{4}x - \frac{3}{4}$
4 a 1   **b** -8 **c** -1 **d** $-\frac{1}{2}$
5 a $y = 7x + 5$ **b** $y = -3x - \frac{29}{4}$
   c $y = -\frac{1}{3}x + \frac{8}{3}$
6 a A, H, J   **b** I and B or G. D and A, H, or J
   c I   **d** B, G   **e** D, H   **f** C, E
7 a $(3, 4), -\frac{1}{4}$ **b** $y = 4x - 8$
8 a $7y = 780 - 25x$ **b** 'decreases', '3.57'
   c Yes, it lies far from the line of best fit.
9 a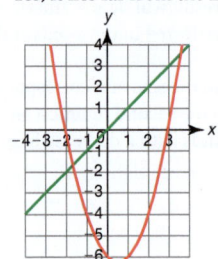
   b $(-1.6, -1.6)$ and $(3.6, 3.6)$ **c** $-2, 3$
   d $-2, 3$. The estimate was accurate.
10 a 8.7, 9.3, 10, 10.7, 11.3, 12, 12.7, 13.3, 14, 14.7
   b (scatter graph: Distance cm vs Mass g)
   c i 13 cm   **ii** 15 g
   d 8 cm
11 a 0, 6, 17, 33, 53, 79, 109, 144, 183, 228, 277

   c i 24 m   **ii** 58 m   **iii** 171 m

Chapter 15

Check in

1 a Several nets possible.   **b** 54 cm²   **c** 27 cm³
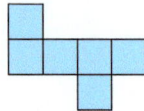

2 a i 78.5 cm² (3 sf)   **ii** 31.4 cm (3 sf)
   b i 172 cm² (3 sf)   **ii** 46.5 cm (3 sf)

15.1S

1 a Cube: 12, 8, 6. Triangular prism: 9, 6, 5. Hexagonal prism: 18, 12, 8. Triangle-based pyramid: 6, 4, 4. Square-based pyramid: 8, 5, 5.
   b $E = V + F - 2$
2 a, c, d, e
3 a triangular prism   **b** pentagon–based pyramid

4 Examples are given here, but other arrangements of the faces are possible.
   a 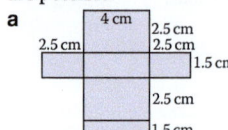   **b**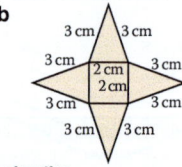

5 a Plan   Front elevation   Side elevation
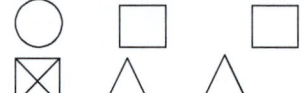
   b
   c
   d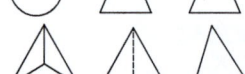
   e i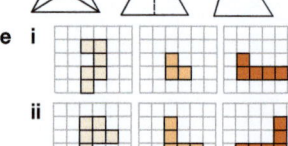
   ii
   *f i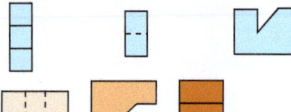
   ii

6 a   b   c
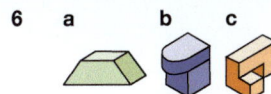

*7 Students' own 3D models.

15.1A

1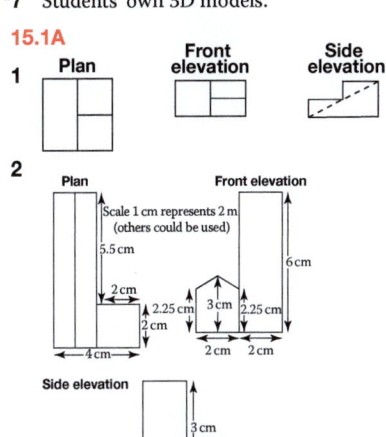

2

3 a Note that other arrangements of the faces are possible.
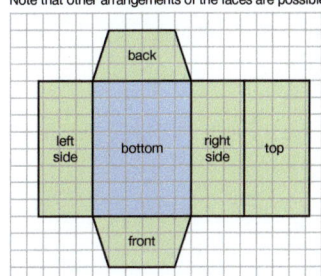

b Total length of edges = 64.6 cm > 0.5 m, not enough lace.
Number of vertices = 8 < 10, Tina has enough flowers.

4 365 cm² (3 sf)

5 a Other arrangements of the faces are possible.
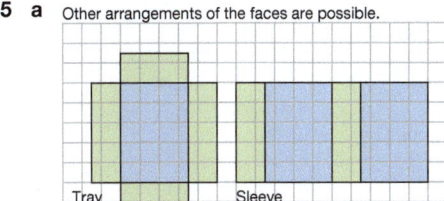

b The tray and sleeve have the same width and depth so the tray may not go inside the sleeve. Make the tray slightly smaller or the sleeve larger by at least the thickness of the card.

6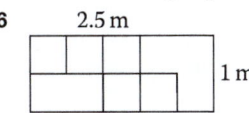
50 cm long.

*7 Net B. The three nets have the same area, the least wasteful net fits on the smallest piece of card.
Area of card needed:
A Rectangle 18 by 18 = 324 cm²
B Rectangle 24 by 13 = 312 cm²
C Rectangle 26 by 14 = 364 cm²
Net B wastes the smallest area of card.

This assumes that the card used is the smallest rectangle possible, which may not be the case if card comes in fixed sizes, and that no extra is needed eg. for tabs to be used to join sides.

8 Sketches of possible nets are given below.
a   b   c
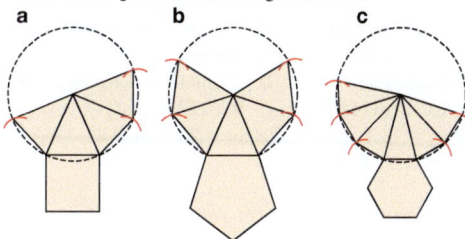

15.2S

1 a 3500 cm³   b 94.5 m³
  c 75.4 m³ (to 3 sf) d 1570 mm³ (to 3 sf)
  e 4800 mm³   f 270 cm³
2 Volume = 756 mm³, Height = 3 cm,
  Length = 8 m, Width = 7 mm
3 a 1782 cm³ (4 sf)   b 1072 mm³ (to 4 sf)
4 4410 g or 4.41 kg
5 73 500 g or 73.5 kg
6 19 200 kg or 19.2 tonnes
7 a 126 litres (to 3 sf) b 8.02 cm (to 3 sf)
  c 7.48 cm (to 3 sf)
8 7.5 cm
*9 532 g (to 3 sf)
10 66.24 litres

15.2A

1 a 2000 ÷ ($\pi \times 4^2 \times 7$) = 5.684..., 5 cups
  b 5.684 $\times \frac{7}{6}$ = 6.631..., 6 cups
2 a 160   b 160 × 0.95 = 152
3 6 tins
4 45.3 cm (3 sf)
5 a 1200 ÷ [$\pi(0.04^2 - 0.03^2) \times 600$] = 909 m (3 sf)
  b 909.4568 ... ÷ 4 = 227 m (3 sf)
6 a 17 hours 48 mins, assuming filled to the top.
  b The real depth to which the pool is filled.
7 1440
8 4.5 cm
9 a 1.64 kg (to 3 sf)
  b Multiple answers possible, for example:
    i Length = 20 cm, width = 16 cm, height = 15 cm
    ii 78.5% (3 sf)
10 9.67 g (3 sf)

15.3S

1 a Surface area = 144π cm², volume = 288π cm³
  b Surface area = 2304π mm², volume = 18432π mm³
  c Surface area = 216π cm², volume = 324π cm³
  d Surface area = 8.1π m², volume = 2.7π m³
2 a Volume = $618\frac{2}{3}$ cm³, surface area = 422.4 cm²
  b Volume = 112π = 352 m³ (3 sf), surface area = 84.8π m² or 266 m² (3 sf)
3 a 42π or 132 cm³ (3 sf)   b 2.08π or 6.53 m³ (3 sf)
4 a 98.8 cm³ (3 sf)   b 133 cm² (3 sf)
5 8 cm, 12 mm
6 a 343 cm³.   b r = 14 cm   c Diameter = 30 cm
7 x = 15 mm
8 a 493 cm³ (to 3 sf)   b 405 cm² (to 3 sf)
9 a Surface area = 528π cm², total volume = 1920π cm³
  b Surface area = 33π cm², volume = 30π cm²
10 Capacity in litres = 2.16 litres (3 sf)
11 Diameter = 9.85 cm (3 sf)

15.3A

1. **a** Steel, density = 7.325... g/cm³, this is nearer to 7.5 g/cm³ than 2.6 g/cm³.
   b 405 g
2. Total surface area of cylinder = $2\pi r \times 2r + 2\pi r^2 = 6\pi r^2$
   $\frac{2}{3} \times 6\pi r^2 = 4\pi r^2$ = surface area of sphere
3. **a** $h = 7.46$ cm (3 sf)   **b** Diameter = 7.82 cm (3 sf)
4. **a i** $h = 15.72$ cm (4 sf)   **ii** 572 cm² (3 sf)
   b Diameter = 15.4 cm (3 sf), height = 26.9 cm (3 sf), surface area = 1670 cm² (3 sf)
5. **a** 6 cm
   b Radius of sector = slant edge of cone, l
   Length of arc of sector = $\frac{\theta}{360} \times 2\pi \times l = 2\pi r$
   Dividing both sides by 2π gives $r = \frac{\theta}{360} \times l$
   Curved surface area of cone = Area of sector = $\frac{\theta}{360} \times \pi l^2$
   $= \frac{\theta}{360} \times l \times \pi l = r \times \pi l = \pi r l$
6. **a** 161.84π or 508 cm² (3 sf)   **b** $66\frac{2}{3}\%$
7. **a** One of several possible solutions is given below.

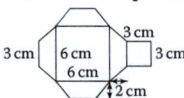

   Total area = 81 cm²
   b Length 13 cm and width 10 cm.
   c 37.7% (3 sf)
8. 2.09 mm (3 sf)
9. $h = 2.41$ m (to 3 sf).
10. No, mass = 2145 kg > 2000 kg.
11. There are many possible answers – one example is given below.
    a Radii of 4 cm and 8 cm, height 8.53.
    b Multiply answers to part **a** by $\sqrt[3]{2}$
    e.g. Radius of top = 10.1 cm, radius of bottom = 5.04 cm, height = 10.7 cm (all to 3 sf)
12. **a** $\frac{H-h}{H} = \frac{r}{R}$
    $R(H-h) = rH$
    $RH - Rh = rH$
    $H(R-r) = Rh$
    $H = \frac{Rh}{R-r}$

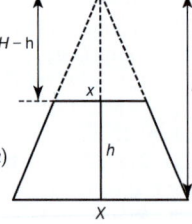

   b Volume of frustum = $\frac{1}{3}\pi R^2 H - \frac{1}{3}\pi r^2(H-h)$
   $V = \frac{1}{3}\pi \frac{R^2 hR}{R-r} - \frac{1}{3}\pi r^2\left(\frac{hR}{R-r} - h\right)$
   $= \frac{1}{3}\pi\left(\frac{hR^3 - r^2hR + r^2hR - r^3h}{R-r}\right)$
   $= \frac{1}{3}\pi h\left(\frac{R^3 - r^3}{R-r}\right)$
   $= \frac{1}{3}\pi h \frac{(R-r)(R^2 + Rr + r^2)}{(R-r)}$
   $= \frac{1}{3}\pi h(R^2 + Rr + r^2)$

13. $\frac{H-h}{H} = \frac{x}{X}$
    $X(H-h) = xH$
    $H(X-x) = Xh$
    $H = \frac{hX}{X-x}$
    Volume of frustum = $\frac{1}{3}X^2H - \frac{1}{3}x^2(H-h)$
    $V = \frac{1}{3}\frac{X^2hX}{X-x} - \frac{1}{3}x^2\left(\frac{hX}{X-x} - h\right)$
    $= \frac{1}{3}\left(\frac{hX^3 - x^2hX + x^2hX - x^3h}{X-x}\right)$
    $= \frac{1}{3}\left(\frac{hX^3 - x^3h}{X-x}\right)$

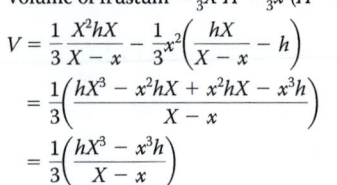

   $= \frac{1}{3}h\left(\frac{X^3 - x^3}{X-x}\right)$
   $= \frac{1}{3}h\frac{(X-x)(X^2 + Xx + x^2)}{(X-x)}$
   $V = \frac{1}{3}h(X^2 + Xx + x^2)$

Review

1. **a**   **b**

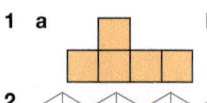   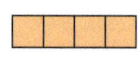

2.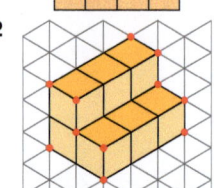

3. **a** 245 m³   **b** 273 m²   **c** 367.5 kg
4. **a** 9557 cm³ (4 sf)   **b** 2532 cm² (4 sf)
5. **a** 1767 cm³ (4 sf)   **b** 66.7 cm³ (3 sf)
6. Volume = 67.9 cm³ (3 sf), Surface area = 102 cm² (3 sf)
7. **a** 6 cm   **b** 10 cm²

Assessment 15

1. **a** Triangle-based pyramid or tetrahedron
   b

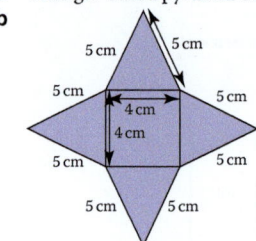

2. **a** A   **b** B
3. **a** 512   **b** 64   **c** 8
   d i 18   **ii** 26
4. **a** 432
   b No, ratio of volumes = 432 : 1,
   ratio of surface areas = 666 : 13.
5. **a** Yes. $9.6 \times 1 \times 1 = 9.6$ cm³ and $6.7 \times 1.2 \times 1.2 = 9.648$ cm³
   b Fat chips, 35.04 cm² < 40.4 cm²
6. **a i** 21.2 cm³ (3 sf)   **ii** 70.4 cm³ (3 sf)
   b 170 cm² (3 sf)
7. **a i** 50.4 in²   **ii** 13.5 in³
   b i 47 666.7 cm³ (1 dp)   **ii** 91
   c i 4.71 cm³ (3 sf)   **ii** 12.3 g (3 sf)
8. **a** 94.2 cm³ (3 sf)   **b** 104 cm² (3 sf)
9. **a** 41.6 mm² (3 sf)   **b** 2700 mm³ (3 sf)
   c 2630 mm³ (3 sf)
10. **a** 10 900 in³ (3 sf)   **b** 1020 in³ (3 sf)   **c** 3540 in² (3 sf)
11. **a** 654 ml   **b** 3 hr 38 min   **c** 3 cm
    d 141 ml (3 sf)   **e** 513 ml (3 sf)   **f** 02:51
    g As the cross section of the cone decreases, the water level decreases at an increasingly faster rate. A cylinder.

Life skills 3

1. 40
2. **a**

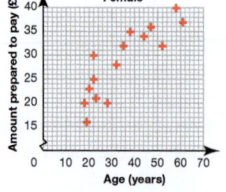

b Males – no correlation. There is no observable connection between age and amount prepared to pay. Females – positive correlation. Older females are prepared to pay more than younger females.

c No. Several reasons possible e.g. data may be unreliable due to a small sample size, sample may not give an indication of the relative populations of different age groups.

3 a

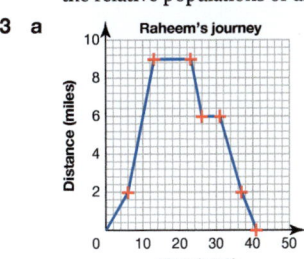

b 1:41 pm **c** 41 minutes.

4 a

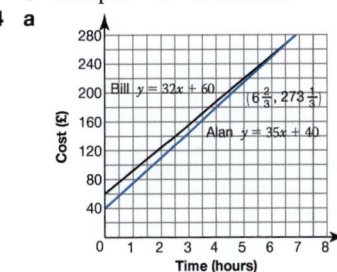

b $x > 6$ hours 40 minutes

c Multiple answers possible. If the quote is accurate Alan is cheaper, if it takes more than 10 minutes extra Bill is cheaper.

5 a 17 241 cm³
b 1 449 000 cm³ = 1.449 m³
$1\,m^3 = 1\,000\,000\,cm^3$
c $30 \times 17\,240.7312 = 517\,220$
Volume of cupboard = 1 449 000
$\frac{517\,220}{1\,449\,000} \times 100 = 35.7\%$
d 396.5 cm²   **e** 3.72 cm   **f** 18
g Cartons, they allow more efficient use of space.

6 a £854
b

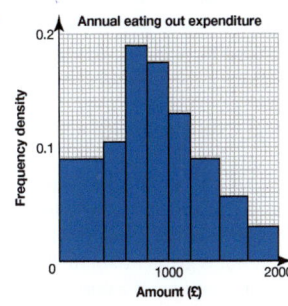

c

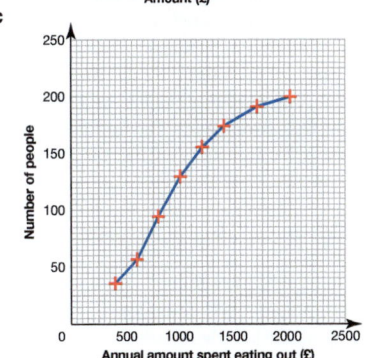

d Median ≈ 830, IQR ≈ 600.
e

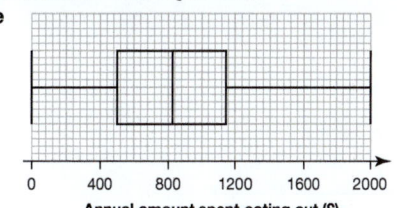

f The median is close to the estimate for the mean so supports this estimate.
g £213.50

Chapter 16

Check in

1 a 62   **b** 60   **c** 18.5   **d** 15
  e 7   **f** 30   **g** 108   **h** 150
2 a i £50   **ii** £70   **b** 2 days

16.1S

1 a 1, 6, 8, 3, 4, 3   **b** 70 kg to 74 kg   **c** 70 kg to 74 kg
2 a 3   **b** 13, 18, 23, 28
  c Mid-value × Number of cars: 13, 108, 46, 28
    Mean (estimate) = 195 ÷ 10 = 19.5 mph
3 a i $10 < t \leq 15$ **ii** $15 < t \leq 20$ **iii** 16.1
  b i $10 < t \leq 20$ **ii** $20 < t \leq 30$ **iii** 23.5
  c i $5 < t \leq 10$ **ii** $10 < t \leq 15$ **iii** 14.7
  d i $15 < t \leq 25$ **ii** $25 < t \leq 35$ **iii** 30
4 a $165 \leq h < 170$ **b** 164.3 cm   **c** $165 \leq h < 170$

16.1A

1 a December 50, January 59.7
  b Dec: modal class $40 < m \leq 60$, median class $40 < m \leq 60$.
    Jan: modal class $40 < m \leq 60$, median class $40 < m \leq 60$
  c Less variation in miles travelled in January, fewer short journeys.
    The most common journey length does not change.
    The mean length is greater for January than December.
2 a Teachers 23.1 minutes, office workers 38.3 minutes
  b Modal class: teachers $20 < t \leq 30$, office workers $30 < t \leq 40$. Median class: teachers $20 < t \leq 30$, office workers $30 < t \leq 40$.
  c On average, office workers take longer to travel home.
3 No, the maximum range of the sample is 40 g and an estimate for the mean is 53.125. The range is similar to that for Granny Smith apples but the mean is much higher than that for Granny Smith apples.
4 Yes, modal and median class is $24.5 \leq w < 25.5$. Estimated mean = 25.28 g which is close to 25 g. However as data is grouped the weight of the packets could be less than 25 g.
5 Machine A, $0.3007 - 0.3 < 0.3 - 0.2907$

16.2S

1

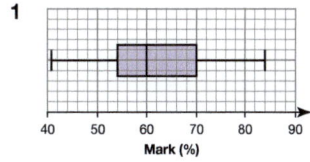

2

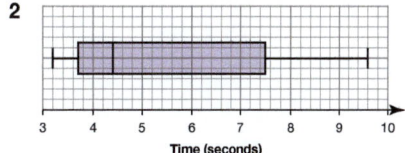

529

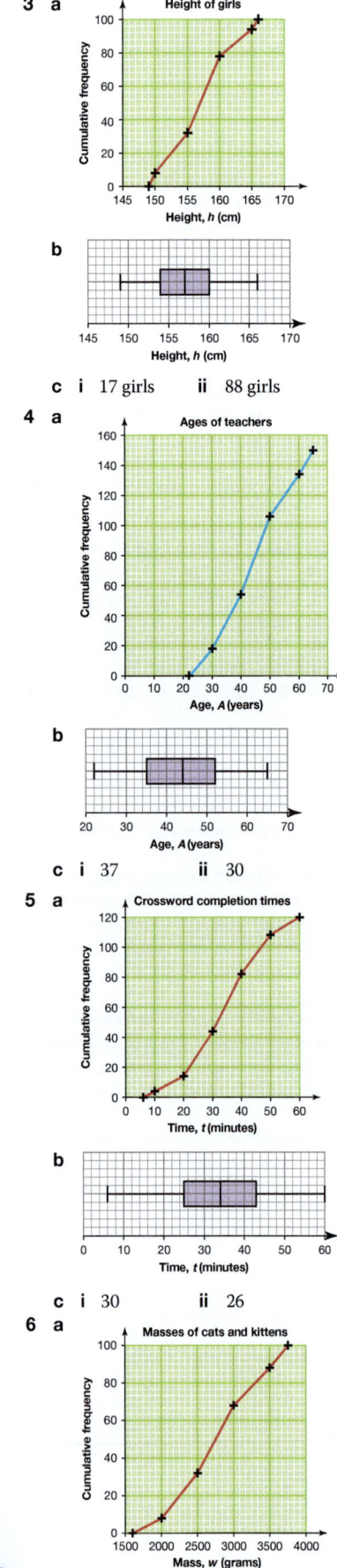

3 a

b

c i 17 girls **ii** 88 girls

4 a

b

c i 37 **ii** 30

5 a

b

c i 30 **ii** 26

6 a

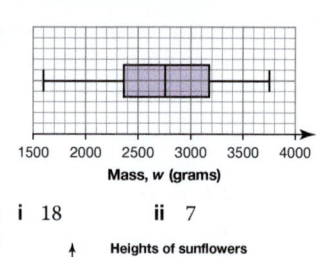

b

c i 18 **ii** 7

7 a

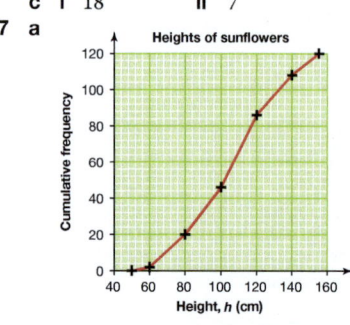

b

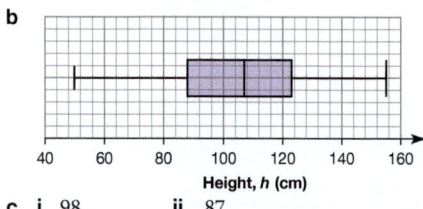

c i 98 **ii** 87

16.2A

1. The boys' results are higher than the girls', in general. The middle 50% of the girls' results is less varied than that of the boys. The range is the same for the girls and the boys.
2. Farmer Jenkins' sunflowers are shorter than Farmer Giles', in general. The middle 50% of Farmer Jenkins' sunflowers vary in height more than those of Farmer Giles. The range of heights of Farmer Jenkins' sunflowers is greater than that of Farmer Giles'.
3. Boys have higher mobile phone bills, on average. The middle 50% of the mobile phone bills has the same variation for girls and boys. The range of mobile phone bills is the same for girls and boys.
4. On average, waiting times are higher at the dentist. The highest waiting time at the doctor is greater than that for the dentist. The range of waiting times is greater at the doctor. The middle 50% of waiting times varies more at the doctor than at the dentist.
5. The average reaction time is the same for boys and girls. The quickest reaction time for the boys is lower than that for the girls. The range of reaction times is greater for girls. The middle 50% of reaction times varies less for girls than for boys.
6. On average, results are the same in the English and French tests. The highest French result is greater than the highest English result. The ranges of results are the same. The middle 50% of results varies more in the English test.
7. On average, 17-year-old girls make longer phone calls than 13-year-olds. The range of the length of calls made is greater for 17-year-olds. The longest phone call for the 17-year-olds is greater than that for the 13-year-olds. The middle 50% of the calls made varies more for 13-year-olds than for 17-year-olds.

16.3S

1. **a** Negative correlation **b** No correlation
   c Positive correlation
2. **a** A: Poor exam mark, lots of revision;
   B: Very good exam mark, lots of revision;
   C: Very good exam mark, little revision;
   D: Poor exam mark, little revision;
   E: Average exam mark, average amount of revision

b A: Not much pocket money, equal eldest;
   B: Lots of pocket money, equal eldest;
   C: Lots of pocket money, equal youngest;
   D: Not much pocket money, equal youngest;
   E: Average pocket money, medium age.
c A: Low fitness level, lots of hours in gym,
   B: Good fitness level, lots of hours in gym,
   C: Good fitness level, few hours in gym,
   D: Low fitness level, few hours in gym,
   E: Medium fitness level, medium hours in gym

3 a No correlation   b Positive correlation
  c Negative correlation

4 a
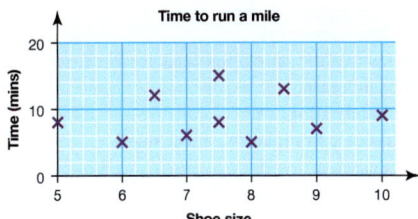
  b Positive correlation
  c i Increases   ii Decreases

5 a
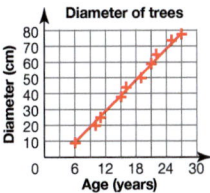
  b No correlation   c No relationship

16.3A

1 a Positive correlation.
  b Paper 1 = 20, paper 2 = 80. Student performed poorly in paper 1 and well in paper 2.
  c 30
  d No, the line predicts that the student scores 110% which is impossible.

2 a 16   b 24
  c One student scored 20 marks on paper 1, but only 5 marks on paper 2.

3 a, c
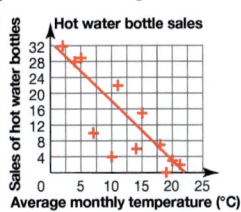
  b Positive correlation.
  d 20 years   e −9 cm
  f The estimate in part d uses interpolation, the estimate in part e uses extrapolation which can be unreliable.

4 a, c
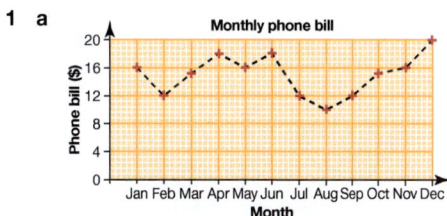

  b Negative correlation, as the temperature increases sales of hot water bottles decrease.
  d 9°C   e No, this is extrapolation and is unreliable.

16.4S

1
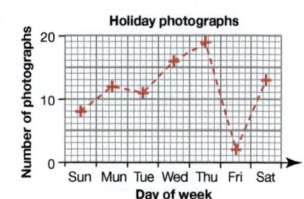

2
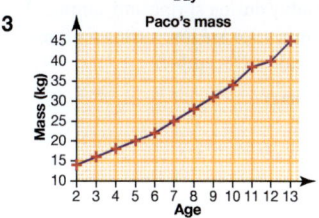

3
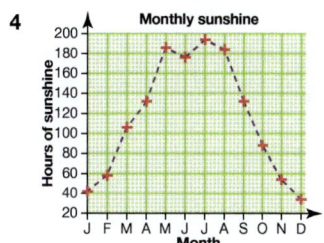

4
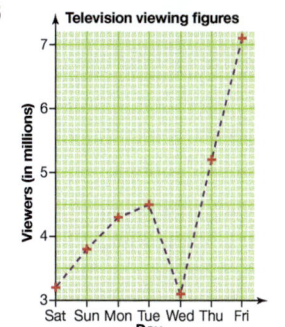

5

16.4A

1 a
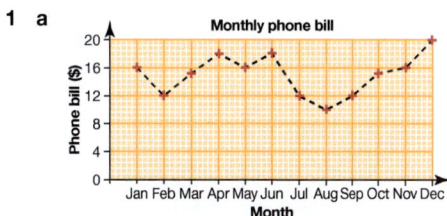

  b Typical phone bills are about $15. These fall during the summer months and peak in December

531

2 a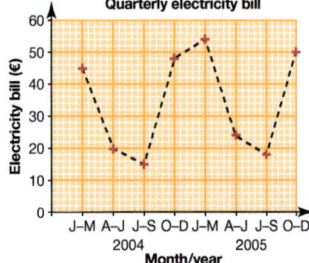

b Electricity bills are highest in the winter months and lowest in the summer months. This annual pattern repeats itself; there is a slight trend for bills to rise from year-to-year.

3 a

b Ice cream sales grow steadily during spring and summer but drop sharply in the autumn. sales are low during autumn and winter except for a peak in December.

4 a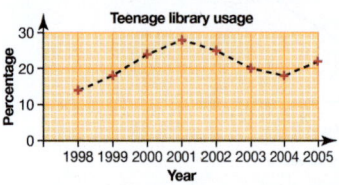

b The percentage of students using the library grows steadily from about 15% in 1998 to 28% in 2001. It has since fallen back to around 20%.

5 a

b Christabel's earnings have grown steadily during the three years. Her earnings peak in the winter months.

6 a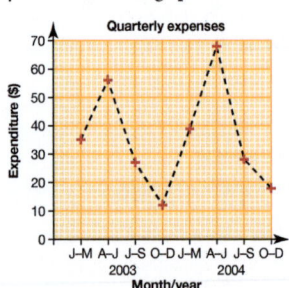

b Steve's expenses grow to a peak in spring and fall back in winter. The level of expenses appears to be growing from year-to-year.

Review

1 a $30 < s \leq 40$ **b** $30 < s \leq 40$ **c** $34\,750$

2 a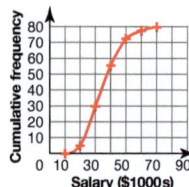

b i $26\,000   **ii** $42\,000   **iii** $33\,000
    iv $16\,000   **v** 80%

c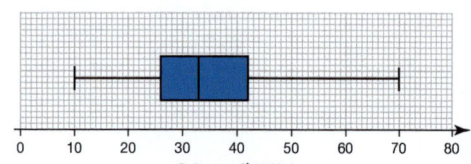

3 a, c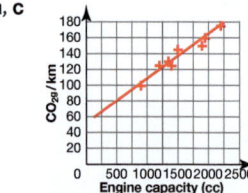

b Positive correlation, as engine capacity increases CO_2 emissions increase.

d Approximately 165 g/km

e This would be extrapolation, unreliable.

4 a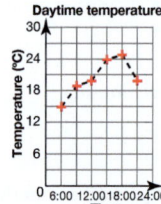

b The temperature increases to a peak at 18:00 and then begins to decrease.

Assessment 16

1 a i $160 \leq h < 165$ **ii** $160 \leq h < 165$
    iii $5875 \div 36 = 163.2$ (1 dp) cm

b i $160 \leq h < 165$ **ii** $155 \leq h < 160$
    iii $6235 \div 39 = 159.9$ (1 dp) cm

c i $160 \leq h < 165$ **ii** $160 \leq h < 165$
    iii $12\,110 \div 75 = 161.5$ (1 dp) cm

2 a $30 - 35$ **b** $25 - 30$

c $1435 \div 48 = 30$ hours (2sf)

3 a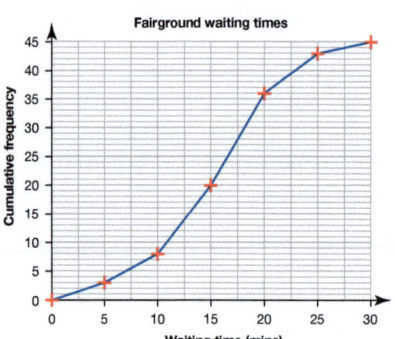

b i 8   **ii** 25

c i 16 **ii** $19 - 11 = 8$
d

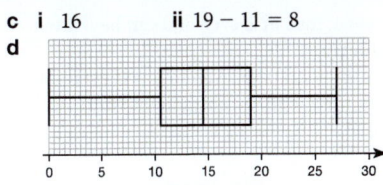

4 a, b

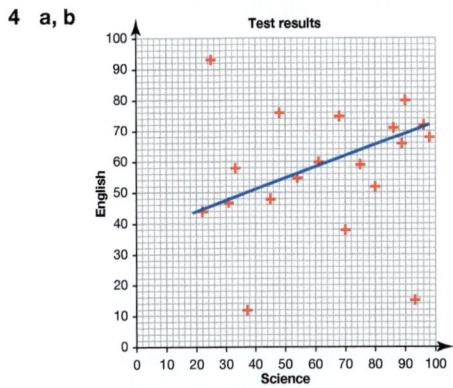

c Weak positive correlation
d (93, 15) shows a student who is very good at Science but poor at English, (25, 93) shows a student who is very good at English but poor at Science.
e i Approximately 44 **ii** Approximately 57
5 a 22 **b** 18 **c** 24
d The graph just shows a trend and there are no actual figures recorded at this time.
e 41
f Students' answers, for example, lunch break or end of working day.
6 a

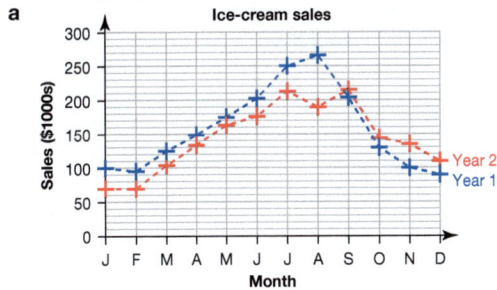

b Winter sales in Year 1 were lower than in Year 2. Summer sales in Year 1 were higher than in Year 2.

Chapter 17

Check in

1 a 7π **b** 13π **c** 36π **d** 4
e $3\sqrt{2}$ **f** $\sqrt{3}$
2 a i 1, 2, 3, 6 **ii** 1, 2, 3, 4, 6, 12
iii 1, 2, 4, 7, 14, 28 **iv** 1, 2, 3, 4, 6, 9, 12, 18, 36
b 2, 3, 5, 7, 11, 13, 17, 19, 23, 29, 31, 37, 41, 43, 47
3 a 3^2 **b** 4^5 **c** 6^3 **d** 5^4
4 a 16 **b** 27 **c** 16 **d** 125
e 128 **f** 7

17.1S

1 a 10 **b** 4 **c** 7 **d** 2
e 11 **f** 12 **g** 2 **h** 3
i 10 **j** 9 **k** 3 **l** 5
m 0 **n** 10 **o** 4

2 a 2^{-1} **b** 5^{-1} **c** -7^{-1} **d** 2^{-1}
e 10^{-1} **f** 3^{-1}
3 a 7^{-2} **b** 9^{-2} **c** 2^{-2} **d** 2^{-5}
e 3^{-4} **f** 6^{-4}
4 a $\frac{1}{64}$ **b** $\frac{1}{343}$ **c** $\frac{1}{25}$ **d** $\frac{1}{6561}$
e $\frac{1}{9}$ **f** $\frac{1}{729}$
5 a $\frac{1}{9}, 0.\dot{1}$ **b** $\frac{1}{8}, 0.125$ **c** $\frac{1}{10\,000}, 0.000\,01$
d 1, 1 **e** $\frac{1}{8}, 0.125$ **f** $\frac{1}{16}, 0.0625$
6 a $\frac{1}{16}$ **b** $\frac{1}{4}$ **c** 1 **d** 2
e 4 **f** 16 **g** 64 **h** $\frac{1}{2}$ **i** $\frac{1}{64}$
7 a $3^{-\frac{1}{2}}$ **b** $5^{-\frac{1}{2}}$ **c** $11^{-\frac{1}{2}}$
8 a 5 **b** $\frac{1}{5}$ **c** 1 **d** 125
e $\frac{1}{25}$ **f** $\frac{1}{125}$ **g** 64 **h** 9
i $\frac{1}{8}$ **j** $\frac{1}{3}$ **k** $\frac{1}{25}$ **l** $\frac{1}{10\,000\,000}$
m 8 **n** 4 **o** 243 **p** $\frac{1}{10}$
q $\frac{1}{64}$ **r** 100 **s** $\frac{1}{20}$ **t** 2197
9 a $2^{\frac{1}{2}}$ **b** $2^{\frac{1}{6}}$ **c** $2^{\frac{9}{2}}$ **d** $2^{-\frac{2}{5}}$
e $2^{\frac{5}{2}}$ **f** 1
10 a 3 **b** 3^{-6} **c** 3^{-2} **d** 3^{-6}
e 3^{-1} **f** 3^4
11 a 5 **b** $5^{\frac{2}{3}}$ **c** $5^{\frac{2}{3}}$ **d** $5^{\frac{1}{6}}$
e 5^{-4} **f** 5^{-6} **g** 5^3 **h** $5^{\frac{5}{2}}$
i $5^{-\frac{3}{2}}$ **j** $5^{-\frac{3}{5}}$ **k** $5^{-\frac{2}{5}}$ **l** $5^{-\frac{5}{6}}$
12 a $2^{-\frac{5}{4}}$ **b** $2^{\frac{1}{3}}$ **c** $2^{\frac{9}{8}}$ **d** $2^{-\frac{3}{4}}$
e $2^{-\frac{3}{8}}$ **f** $2^{-\frac{9}{2}}$
13 a $4^{-1}, 16^{-\frac{1}{2}}$ **b** $4^2, 16^1$ **c** $4^{-2}, 16^{-1}$ **d** $4^{-\frac{1}{2}}, 16^{-\frac{1}{4}}$
14 a 10^{-1} **b** $10^{\frac{1}{2}}$ **c** $10^{\frac{3}{2}}$ **d** $10^{-\frac{5}{2}}$
15 a 5^{-2} **b** 5^{-3} **c** $5^{-\frac{1}{3}}$ **d** $5^{\frac{2}{3}}$
e $5^{-\frac{3}{2}}$ **f** $5^{-\frac{4}{3}}$
16 a i $\frac{3}{2}$ **ii** $\frac{2}{3}$ **b i** $\frac{5}{2}$ **ii** $\frac{2}{5}$
c i $\frac{4}{3}$ **ii** $\frac{3}{4}$ **d i** $-\frac{3}{4}$ **ii** $-\frac{4}{3}$
17 $\frac{2}{3}$

17.1A

1 a i 32 cm **ii** 256 cm
b i Wednesday **ii** Sunday
c i 2^5 **ii** 2^8
2 a i 10^3 times, or 1000 times more powerful.
ii 10^4 times, or 10 000 times more powerful.
b Level 8
3 a $6x^{\frac{2}{3}}$ **b** 27 cm²
c No. Let side length be x. Surface area $= 6x^2$, volume $= x^3$.
$6x^2 = x^3 \Rightarrow x^2(x - 6) = 0 \Rightarrow x = 0$ or 6
Side length must be > 0 so a cube with side length 6 units is the only solution.
4 $27^{-\frac{2}{3}}, 25^0, 16^{\frac{3}{4}}, \left(\frac{1}{2}\right)^{-4}, 9^{\frac{3}{2}}, \left(\frac{1}{8}\right)^{-\frac{5}{3}}$
5 a 4 **b** 3 **c** -2 **d** 0
e 3 **f** $\frac{2}{3}$
6 a $\frac{1}{2}$ **b** $\frac{1}{3}$ **c** $\frac{1}{3}$ **d** $\frac{1}{4}$
7 a $x = -1$ **b** $x = 3$ **c** $x = -1$ **d** $x = \frac{1}{6}$
e $x = -\frac{5}{2}$
8 a Second row: $2^{\frac{2}{3}}, 2^{-\frac{2}{3}}$ bottom row: $2^{-\frac{5}{6}}$
b Second row: $2^{-\frac{5}{4}}, 2^{-\frac{1}{4}}$ bottom row: $2^{-\frac{3}{4}}$
9 a $\frac{1}{9}, \frac{1}{3}, 1, 3, 9$

b, c

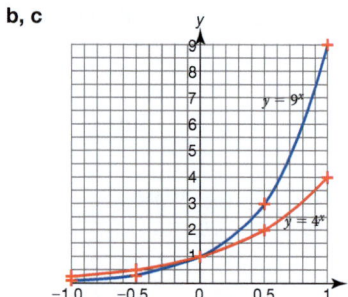

Both graphs pass through (0, 1). $y = 9^x$ is steeper. Both graphs tend to 0 as x gets more negative. Both graphs get steeper as x increases. Both graphs lie entirely above the x-axis.

17.2S

1 a $\pi + 11$ b $3(\pi + 1)$ c $\pi^2 + 3\pi - 6$
   d $2\pi^2 + \pi - 3$ e $9\pi^2 - 24\pi + 16$ f $3\pi^2 + 2\pi - 2$
2 a $5\sqrt{5}$ b $5\sqrt{3} - 2\sqrt{6}$
   c $3 - \sqrt{3}$ d $8\sqrt{3} + 6\sqrt{7} + \sqrt{14}$
   e $3\sqrt{6} - 12$ f $6 - 6\sqrt{2} + 4\sqrt{3}$
   g $9 + 5\sqrt{3}$ h $11 - 6\sqrt{2}$
   i 18 j $9 + 14\sqrt{5}$
3 a $\frac{1}{2}$ b $\frac{5}{3}$ c $1\frac{9}{16}$ d $\frac{8}{15}$
   e $1\frac{5}{9}$ f $\frac{143}{10}$
4 a $\frac{8}{15}$ b $\frac{29}{35}$ c $\frac{5}{56}$ d $\frac{44}{45}$
   e $4\frac{7}{8}$ f $\frac{59}{60}$
5 a $\frac{17}{30}$ b $1\frac{67}{87}$ c $\frac{1}{3}$ d $\frac{207}{1715}$
   e $1\frac{133}{324}$
6 a 294π b $126\frac{9}{25}\pi$ c 404π d $8\pi(2 + \sqrt{13})$
   e $2\frac{7}{20}\pi$ f $9\frac{1}{4}\pi$ g $1\frac{11}{18}\pi + 2(\sqrt{7} - \sqrt{2})$
   h $2\pi\frac{\sqrt{15}}{3}$
7 a $\frac{(6 + 3\sqrt{3})}{4}$ b $\frac{(5 - 3\sqrt{7})}{3}$ c $\frac{(30 + 10\sqrt{2})}{9}$ d $\frac{(20 + 7\sqrt{5})}{9}$
8 a $\frac{(5 + \sqrt{5})}{5}$ b $\frac{(7 + 2\sqrt{7})}{14}$ c $\frac{(5\sqrt{7} - \sqrt{77})}{21}$ d $\frac{(6 + \sqrt{3})}{12}$
   e $\frac{(5\sqrt{6} + 6\sqrt{2})}{12}$ f $\frac{(7\sqrt{2} - 2\sqrt{10})}{10}$
9 a 5.66   b 3.24   c 4.24   d 106.10
10 No, a–e $\frac{5}{4}$ f $\frac{7}{4}$

17.2A

1 Area $= 800\pi$ cm², perimeter $= 80 + 40\pi$ cm
2 $16 - 4\pi$ cm²
3 a $x = \pi + \sqrt{5}, y = -\pi + 2\sqrt{5}$
   b $x = 2\sqrt{5} + \pi, y = 3\sqrt{5} - 2\pi$
   c $x = 2\pi - \frac{12\sqrt{5}}{5}, y = \frac{13\sqrt{5}}{5} - 2\pi$
4 $10\sqrt{\pi}$ cm
5 Volume of cube $= 5^3 = 125$ cm³. Volume of cone $= \frac{1}{3}\pi r^2(3) = \pi r^2$.
   $125 = \pi r^2, r^2 = \frac{125}{\pi}, r = \frac{\sqrt{125}}{\sqrt{\pi}} = \frac{5\sqrt{5}}{\sqrt{\pi}}$ cm
6 $\frac{25}{6} + 40\sqrt{3}$ cm²
7 1 cm
8 $\frac{80\sqrt{5}}{3}\pi$ mm³
9 $4\sqrt{3}$ cm
10 a Using Pythagoras' theorem $h^2 = 2^2 - 1^2 = 3, h = \sqrt{3}$
    b 3 cm   c $2\sqrt{3}$ cm
11 250π cm³
*12 Neiva. First note that the square of an even number is even and the square of an odd number is odd as $(2a)^2 = 4a^2 = 2 \times 2a^2$ and $(2a + 1)^2 = 4a^2 + 4a + 1 = 2(2a^2 + 2a) + 1$.
Assume for a contradiction that $\sqrt{2} = \frac{m}{n}$ where m and n are non-zero integers with **no common factors**. This gives $2 = \frac{m^2}{n^2}$

so $2n^2 = m^2$. m^2 is even, thus m is even and can be written $m = 2x$ for some integer x.
$2n^2 = (2x)^2 = 4x^2$ so $n^2 = 2x^2$. n^2 is even thus n is even and shares a common factor 2 with m contradicting the initial statement as required.

17.3S

1 a 1000   b 1 000 000   c 100 000
2 a 1   b 0.01   c 0.00001
3 a 9×10^3 b 6.5×10^2 c 6.5×10^3
   d 9.52×10^2 e 2.358×10^1 f 2.5585×10^2
4 a 3.4×10^{-4} b 1.067×10^{-1} c 9.1×10^{-6} d 3.15×10^{-1}
   e 5.05×10^{-5} f 1.82×10^{-2} g 8.45×10^{-3} h 3.06×10^{-10}
5 $9 \times 10^3, 1.08 \times 10^4, 3.898 \times 10^4, 4.05 \times 10^4, 4.55 \times 10^4, 5 \times 10^4$
6 a 63 500   b 910 000 000 000 000 000
   c 111   d 299 800 000
7 a 0.0045   b 0.000 031 7   c 0.000 001 09
   d 0.000 000 979
8 a 6×10^2 b 4.5×10^4 c 6.5×10^0 d 5×10^6
   e 2.15×10^4 f 7×10^{13} g $1.225 16 \times 10^{20}$
   h 1.5×10^7 i 2.8×10^{-1} j 4×10^{-2} k 1.35×10^{-3}
   l 1.2×10^{-7}
9 a 2.5×10^8 b 2.4×10^{13} c 5×10^{-1} d 9.2×10^{-8}
10 a 5×10^1 b 7.5×10^2 c 2×10^{-2} d 4.2×10^{-8}
11 a 2×10^{-2} b 1×10^{-1} c 1.5×10^{-8}
    d 2×10^3 e 2.5×10^{-6} f 3.1×10^4
12 a 5.2×10^{-1} b 4.6×10^{-2} c 2.09×10^{-2}
    d 1.3×10^{-2} e -4.3×10^5 f 5.993×10^2
13 a 7.74×10^{-3} b 9.63×10^5 c 4.38×10^{-5}
    d 2.55×10^2 e 3.4×10^5 f 4.47×10^{-2}
14 See questions **9–13**.
15 a b i $3 \times 10^3 \times 2.1 \times 10^2 = 6.3 \times 10^5$
        ii $4.2 \times 10^{-3} \div 6 \times 10^{-1} = 7 \times 10^{-3}$
        iii $1.2 \times 10^3 \times 9 \times 10^{-2} = 1.08 \times 10^2$
        iv $4.9 \times 10^{-2} \div 7 \times 10^1 = 7 \times 10^{-4}$

17.3A

1 3.35×10^5 times heavier.
2 10^{45}
3 500 times bigger.
4 1.80×10^6
5 108 times smaller
6 Minimum $= 7.83 \times 10^7$ km, Maximum $= 3.775 \times 10^8$ km
7 a 3.3×10^{-9} s   b 9.46×10^{15} m
8 1.64×10^{-27}
9 a Jupiter, $2.668 612 \times 10^{27}$ kg   b 0.22%
10 a Correct   b 5 times more.   c HD 1080 p
11 Yes, $6.878 \times 10^{11} \times 2 = 1.3756 \times 10^{12} \approx 1.328 \times 10^{12}$
*12 Yes. Distance of one orbit $= 2 \times \pi \times 1.08 \times 10^8 = 2.16\pi \times 10^8$ km. Venutian year = time for one orbit $= \frac{2.16\pi \times 10^8}{1.26 \times 10^5} = 5385.587...$ hours $= (5385.587... \div 24)$ Earth days $= 224$ Earth days < 243 Earth days $=$ Venutian day.

Review

1 a 9^{-1} b 5^{-7} c $12^{\frac{1}{3}}$ d $8^{-\frac{1}{2}}$
2 a 10 000 000   b 125   c 1   d 729
   e 10   f 8   g $\frac{1}{5}$ h $\frac{1}{36}$
   i 8   j 13   k 9   l $\frac{1}{11}$
3 a $2^{\frac{5}{7}}$ b $2^{\frac{1}{7}}$ c $2^{\frac{6}{7}}$ d 2^{-7}
   e 2   f 2^{12} g 1   h $2^{-\frac{5}{3}}$
4 a 3   b $\frac{1}{2}$ c $\frac{2}{3}$ d -3
5 a $2\frac{6}{7}$ b $\frac{1}{12}$ c $1\frac{2}{15}$ d $\frac{27}{56}$
   e $3\frac{13}{15}$ f 5   g $\frac{2}{33}$ h $\frac{1}{6}$
   i $\frac{27}{40}$

Answers

6 a $4 - 3\pi$ b $1 - \sqrt{3}$ c 33π d $\frac{34 + 19\sqrt{3}}{6}$
  e $-5 - 7\sqrt{7}$ f $\frac{\sqrt{5} - 1}{3}$
7 a $\frac{\sqrt{6}}{2}$ b $2\sqrt{10}$
8 a i $2\sqrt{2}$ ii 12
  b i $2\sqrt{2}\pi$ ii $6\pi + 12$
9 a 6.3×10^7 b 1.496×10^8 km
  c 2.2×10^{-5} mm   d 3×10^{-2} kg
10 a $2\,180\,000\,000$ b $310\,000$
   c $0.000\,000\,5$ d $0.000\,099\,2$
11 a 8.4×10^{11} b 4.8×10^{10}
   c 4.53×10^7 d -1.916×10^{-4}

Assessment 17

1 a Soraya   b Leon   c Soraya
2 a No, 15^{12} b Yes, 3^{20}
  c i No   ii Yes   iii No
  d No, 7^3
3 a Yes   b No, 36 c No, $\frac{1}{7}$ d No, 2744
  e No, $\frac{1}{243}$ f No, 1
4 a $\frac{1}{2}$ b $\frac{1}{5}$ c 1 d $\frac{1}{4}$
5 a i 3^1 ii $9^{\frac{1}{2}}$
  b i 3^2 ii 9^1
  c i 3^4 ii 9^2
  d i 3^{-1} ii $9^{-\frac{1}{2}}$
  e i 3^{-2} ii 9^{-1}
  f i 3^{-4} ii 9^{-2}
  g i 3^3 ii $9^{\frac{3}{2}}$
  h i 3^{-3} ii $9^{-\frac{3}{2}}$
  i i $3^{\frac{1}{2}}$ ii $9^{\frac{1}{4}}$
6 a $V = \frac{\pi}{3}r^2 h$, $r = h$ so $V = \frac{\pi}{3}x^3$, $k = \frac{\pi}{3}$
  b $p = \pi\sqrt{2}$, $q = 2$ c $\frac{64\pi}{3}$ m^3
7 a $\frac{9\sqrt{3}}{2}$ in^2 b $9\pi\sqrt{3}$ in^3 c $0.04\,\pi^2$ in^3
  d $(9\pi\sqrt{3} - 0.04\,\pi^2)$ in^3
8 a English Channel $29\,000$ miles^2
    Baltic Sea $146\,000$ miles^2
    Bering Sea $876\,000$ miles^2
    Caribbean Sea $1\,060\,000$ miles^2
    Malay Sea $3\,140\,000$ miles^2
    Indian Ocean $28\,400\,000$ miles^2
  b 1×10^9 by 1
  c i 4 ii 2 iii 6 iv -3
  d i $10\,763$ km   ii 0.0008 joules/second
  e i 4×10^{-6} km   ii 4 mm
9 a 8706 b 199 times   c $\frac{1}{751\,000}$
  d 226 km $(3\,\text{sf}) = 2.26 \times 10^2$ km
10 1.9848×10^4 m $= 19\,848$ m
11 a i 1.46092×10^7 miles$^2 = 14\,609\,200$ miles^2
     ii 6.6768×10^6 miles^2
   b i $1 : 3.94\,(3\,\text{sf})$ ii 1.39×10^8 miles^2 iii 70.8%

Revision 3

1 a HCF $= 30$, LCM $= 485\,100$
  b $3300 \times 33 = (2^2 \times 3 \times 5^2 \times 11) \times 3 \times 11$
    $= 2^2 \times 3^2 \times 5^2 \times 11^2$;
    $4410 \times 10 = (2 \times 3^2 \times 5 \times 7^2) \times 2 \times 10$
    $= 2^2 \times 3^2 \times 5^2 \times 7^2$
2 a $p = 4$, $q = 13$ b $a = \frac{9 - 7\sqrt{3}}{2}$, $b = \frac{9}{2}$
3 a iv, vii, iii, viii, ii, vi, i, v   b iv and vii   c iii and viii
4 6 and 8
5 a No. $a = 2$ b No. $b = 3$ c No. $c = 2$ or 4
6 a $y = -x + 7$ b $y = x + 3$
  c $(2, 5)$
  d Check centre $(2, 5)$, radius $= 4$

7 a $x = -1.1, 7.1\,(1\,\text{dp})$ b $x = 3 \pm \sqrt{17}$

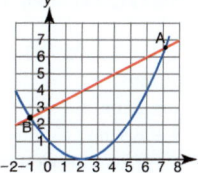

8 a i 1.0827×10^{12} km^3 ii 5.95×10^{24} kg
  b $11\,074\,631$ years (nearest year)
9 a $\frac{(3 \times 3 + 15 \times 7.5 + 21 \times 12.5 + 9 \times 17.5 + 2 \times 22.5)}{50} = \frac{586.5}{50} = 11.73$
  b Median $= 11.5$, IQR $= 6$

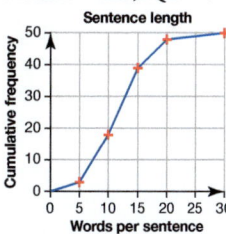

10 a

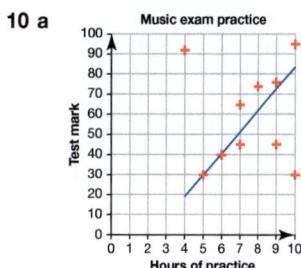

   b A, H, students' answers.   c 36
11 a

UK unemployment rate chart (Oct 2012 – Sep 2013; Oct 2013 – Sep 2014)

   b The trend over the two years is one of consistently falling unemployment rates. Year 2 is lower overall than Year 1
12 a i 216 ii $100\,000$ iii 5 iv 8
     v 0 vi -7
   b i No. 12^9 ii No. 6^6 iii No. 210^1 iv Yes
     v No. 9^{16}
   c i 5 ii 10 iii 4 iv 3
13 a $564\,000$ units   b 1.95×10^{22}
14 a

Graph of C vs v

   b i 26 l/h   ii 22 l/h
   c i 13 mph   ii 24 mph, 80 mph
   d 44 mph as this is where fuel consumption is lowest.

Chapter 18

Check in

1 a 49 b 52 c 34
  d 48 e 45 f 8
2 a $x = 6y$ b $x = 5y$ c $x = 10y$
  d $x = \frac{2}{y}$ e $x = \frac{5}{y}$ f $x = \frac{8}{y}$

18.1S

1. 1, 4, 9, 16, 25, 36, 49, 64, 81, 100, 121, 144, 169, 196, 225, 256, 289, 324, 361, 400
2. **a** $21\,\text{cm}^2$ **b** $9\,\text{mm}^2$ **c** $14\,\text{m}^2$ **d** $40\,\text{m}^2, 40\,\text{m}^2$
3. **a** $13\,\text{cm}$ **b** $25\,\text{m}$ **c** $50\,\text{mm}$
4. **a** $p = 9\,\text{cm}$ **b** $q = 60\,\text{mm}$ **c** $r = 24\,\text{m}$
5. **a** $u = 39\,\text{mm}$ **b** $v = 8.4\,\text{m}$
   c $w = 5.9\,\text{cm}$ (2sf) **d** $x = 75\,\text{cm}$ (nearest cm)
   e $y = 2.9\,\text{m}$ (2sf) **f** $z = 3.9\,\text{km}$ (2sf)
6. $h = 72\,\text{mm}$
7. **a** $35\,\text{mm}$ **b** $7.2\,\text{cm}$ (2sf) **c** $29.2\,\text{m}$ (3sf)
   d $180\,\text{km}$ (nearest km)
8. **a** $a = 2\sqrt{5}$ **b** $b = 2\sqrt{13}$ **c** $c = \sqrt{3}$ **d** $d = 3\sqrt{3}$
   e $e = 4\sqrt{2}$ **f** $f = 4\sqrt{2}$
9. $w = 6\sqrt{7}\,\text{cm}$
10. **a** $10\sqrt{2}\,\text{cm}$ **b** $5\sqrt{2}\,\text{cm}$
11. **a** $72\,\text{mm}$ **b** $20\,\text{cm}$
*12. $QS = 21\,\text{cm}$
13. $d = 10.2\,\text{m}$ (1 dp)

18.1A

1. $1.2\,\text{m}$ (1dp)
2. **a** 5 **b** 5 **c** $\sqrt{10}$ **d** $3\sqrt{2}$
   e $\sqrt{74}$ **f** 5 **g** $\sqrt{29}$ **h** $2\sqrt{13}$
3. $r = 30\,\text{mm}$
4. $8.1\,\text{cm}$ (1 dp)
5. $29\,\text{m}$ (to nearest m)
6. **a** $4^2 + 8^2 = 80 \neq 9^2$ (Pythagoras' theorem)
   b Obtuse, 9 cm is longer than the hypotenuse of a right-angled triangle with shorter sides 8 cm and 4 cm.
7. $d = 1.21\,\text{m}$ and $h = 4.85\,\text{m}$ (nearest cm)
8. **a** The diagonal of a rectangle with sides 2 cm and 3 cm.
   b The diagonal of a rectangle with sides 2 cm and 4 cm.
   c The diagonal of a rectangle with sides 3 cm and 5 cm.
   Note
   Either rectangles or right-angled triangles could be used.
9. $27\,\text{cm}^2$
10. $240\,\text{cm}^2$
11. $60\,\text{m}^2$
12. $3\sqrt{23}\,\text{cm}$
*13. **a** Area of large square $= (a + b)^2$
    $= c^2 + 4 \times \frac{1}{2}ab$ = (area of small square + 4 triangles)
    $a^2 + 2ab + b^2 = c^2 + 2ab$
    $a^2 + b^2 = c^2$
    b **i** Area of semi-circle $= \frac{1}{2}\pi r^2 = \frac{1}{2}\pi \left(\frac{d}{2}\right)^2 = \frac{1}{8} \times \pi d^2$
    Using a and b for the diameters of the smaller semi-circles and c for that of the largest circle.
    Sum of areas of smaller semi-circles $= \frac{1}{8}\pi a^2 + \frac{1}{8}\pi b^2 = \frac{1}{8}\pi(a^2 + b^2) = \frac{1}{8}\pi c^2$ (Pythagoras' theorem)
    = area of largest semi-circle
    ii Area of equilateral triangle $= \frac{1}{2}bh$ where $h^2 = x^2 - \left(\frac{x}{2}\right)^2 = \frac{3}{4}x^2$
    Area of equilateral triangle
    $= \frac{1}{2} \times x \times \frac{\sqrt{3}}{2}x = \frac{\sqrt{3}}{4}x^2$
    Sum of equilateral triangles on two shorter sides of right-angled triangle
    $= \frac{\sqrt{3}}{4}a^2 + \frac{\sqrt{3}}{4}b^2 = \frac{\sqrt{3}}{4}(a^2 + b^2)$
    $= \frac{\sqrt{3}}{4}c^2$ (Pythagoras' theorem)
    = area of equilateral triangle on the hypotenuse of the right-angled triangle.
*14. $40\sqrt{2}\,\text{mm}$
15. **a** There are many – some examples are given below:
    3, 4, 5   5, 12, 13   7, 24, 25
    b Yes, multiplying all the sides of a right-angled triangle by the same number gives a similar right-angled triangle.

c $(2xy)^2 = 4x^2y^2$
$(x^2 + y^2)^2 = x^4 + 2x^2y^2 + y^4$
$(x^2 - y^2)^2 = x^4 - 2x^2y^2 + y^4$
$(2xy)^2 + (x^2 - y^2)^2 = 4x^2y^2 + x^4 - 2x^2y^2 + y^4 = (x^2 + y^2)^2$

18.2S

All given to nearest whole number when not exact.

1. **a** $a = 7.5\,\text{cm}$ (2 sf)
   b $b = 29\,\text{cm}$ (nearest cm) **c** $c = 4.1\,\text{m}$ (2 sf)
   d $d = 8.6\,\text{m}$ (2 sf) **e** $e = 54\,\text{mm}$ (nearest mm)
   f $f = 1.6\,\text{m}$ (2 sf) **g** $g = 133\,\text{km}$ (nearest km)
   h $h = 3.8\,\text{m}$ (2 sf) **i** $i = 7.1\,\text{m}$ (2 sf)
2. **a** $a = 66.4°$ (1 dp) **b** $b = 30°$
   c $c = 56.3°$ (1 dp) **d** $d = 19.5°$ (1 dp)
   e $e = 51.3°$ (1 dp) **f** $f = 29.4°$ (1 dp)
3. **a** $A = 37°$ **b** $E = 23°$
   $C = 53°$ $F = 67°$
   $AC = 10\,\text{cm}$ $DF = 5\,\text{m}$
   c $GI = 32\,\text{m}$ **d** $KL = 15\,\text{mm}$
   $HI = 11\,\text{m}$ $JK = 23\,\text{mm}$
   $G = 18°$ $K = 50°$
   e $MN = 116\,\text{km}$ **f** $P = 54°$
   $NO = 435\,\text{km}$ $R = 36°$
   $M = 75°$ $PR = 72\,\text{cm}$
4. **a** $BC = 31\,\text{mm}$ (nearest mm)
   b $\angle PRQ = \angle RPQ = 73°$ (nearest °)
   $\angle PQR = 33°$ (nearest °)
5. $KN = KL = 19\,\text{cm}$ (nearest cm)
   $NM = LM = 30\,\text{cm}$ (nearest cm)
   $\angle LKN = 77°$ (nearest °)
   $\angle LMN = 48°$ (nearest °)
   $\angle KNM = \angle KLM = 117°$ (nearest °)
6. **a** $\angle DCB = \angle DAB = 116°$ (nearest °)
   $\angle ABC = \angle ADC = 64$ (nearest °)
   b $AC = 11\,\text{cm}$ (nearest cm)
*7. **a, b, c**

Angle	sin	cos	tan
0°	0	1	0
30°	$\frac{1}{2}$	$\frac{\sqrt{3}}{2}$	$\frac{1}{\sqrt{3}} = \frac{\sqrt{3}}{3}$
45°	$\frac{1}{\sqrt{2}} = \frac{\sqrt{2}}{2}$	$\frac{1}{\sqrt{2}} = \frac{\sqrt{2}}{2}$	1
60°	$\frac{\sqrt{3}}{2}$	$\frac{1}{2}$	$\sqrt{3}$
90°	1	0	∞

8. **a** $h = 4.8\,\text{m}$ (1 dp) **b** $d = 1.3\,\text{m}$ (1 dp)

18.2A

1. **a** $h = 168\,\text{m}$
   b It is difficult to estimate the distance and angle accurately. The horizontal distance may be inaccurate because of the buildings that surround the base of the tower. The ground may not be horizontal. (The real height is 158 m.)
2. Yes, height $= 9.534... + 1.6 = 11.134...\,\text{m} > 10\,\text{m}$
3. $38\,\text{m}$ (nearest m)
4. $9.0\,\text{m}$ (2 sf)
5. Area $= 144\sqrt{3}\,\text{cm}^2$
6. $12\sqrt{2}\,\text{m}^2$
7. $81\sqrt{3}\,\text{mm}^2$
8. No, $\tan \theta = \frac{36}{240} = 0.15$, $\theta = 8.53...° > 4°$
9. **a** $\tan \theta = 2$ **b** $\tan \theta = 3$ **c** $\tan \theta = \frac{1}{2}$
   d $\tan \theta$ = gradient m in $y = mx$
10. $60°$
11. **a** $\theta = 31°$ (nearest °)
    b Let x be the vertical height of the roof. $x^2 = 14^2 - 12^2 = 52$,
    $x = \sqrt{52} = 2\sqrt{13}$
    Total area $= 24 \times 21 + \frac{1}{2} \times 24 \times 2\sqrt{13} = 24(21 + \sqrt{13})\,\text{ft}^2$

Answers

12 Perimeter = 120 mm
Area = $432\sqrt{3}$ mm²

*13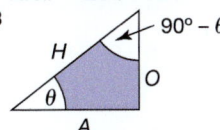

a $\sin\theta = \frac{O}{H} = \cos(90° - \theta)$
b $(\sin\theta)^2 + (\cos\theta)^2 = \left(\frac{O}{H}\right)^2 + \left(\frac{A}{H}\right)^2 = \frac{O^2 + A^2}{H^2} = \frac{H^2}{H^2} = 1$
c $\tan\theta = \frac{O}{A} = \frac{\frac{O}{H}}{\frac{A}{H}} = \frac{\sin\theta}{\cos\theta}$
d $\tan\theta \times \tan(90° - \theta) = \frac{O}{A} \times \frac{A}{O} = 1$

*14 a i 29 cm (nearest cm)   ii 59 cm² (nearest cm²)
    b i 31 cm (nearest cm)   ii 71 cm² (nearest cm²)

18.3S

1 a $a = 23$ cm (2 sf)   b $b = 2.3$ m (2 sf)
    c $c = 64$ mm (2 sf)

2 a $\angle C = 41°$ (nearest °)   b $\angle K = 37°$ (nearest °)
    $\angle A = 81°$ (nearest °) $\angle L = 31°$ (nearest °)
    c $\angle R = 54°$ (nearest °)
    $\angle Q = 47°$ (nearest °)

3 a $a = 8.3$ cm (1 dp)   b $b = 147$ mm (nearest mm)
    c $c = 3.5$ m (1 dp)

4 a $\angle A = 59°$ (nearest °)   b $\angle P = 59°$ (nearest °)
    $\angle B = 35°$ (nearest °) $\angle Q = 70°$ (nearest °)
    $\angle C = 86°$ (nearest °) $\angle R = 51°$ (nearest °)
    c $\angle X = 93°$ (nearest °)   d $\angle D = 90°$ (nearest °)
    $\angle Y = 47°$ (nearest °) $\angle E = 67°$ (nearest °)
    $\angle Z = 40°$ (nearest °) $\angle F = 23°$ (nearest °)

5 a Area = 214 m² (to nearest m²)
    b Area = 17 cm² (to nearest cm²)
    c Area = 1658 mm² (to nearest mm²)

6 a $\angle A = 39°$ (nearest °)   b $DF = 6.8$ cm (1 dp)
    $\angle B = 69°$ (nearest °) $\angle D = 41°$ (nearest °)
    $AC = 26$ m (2sf) $\angle F = 34°$ (nearest °)
    c $\angle I = 92°$ (nearest °)
    $\angle G = 52°$ (nearest °)
    $\angle H = 36°$ (nearest °)

7 a $BC = 5.2$ cm (1 dp)   b $\angle D = 31°$ (nearest °)
    $\angle B = 36°$ (nearest °) $\angle F = 57°$ (nearest °)
    $\angle C = 26°$ (nearest °) $DE = 40$ mm (2sf)
    c $\angle I = 81°$ (nearest °)   d $\angle L = 76°$
    $\angle G = 61°$ (nearest °) $KL = 20$ m (2 sf)
    $\angle H = 38°$ (nearest °) $JL = 26$ m (2 sf)

8 $PR = 28$ cm (2 sf)
    $QS = 49$ cm (2 sf)

9 $\cos X = \frac{9^2 + 24^2 - 21^2}{2 \times 9 \times 24} = 0.5$
    $\angle X = \cos^{-1} 0.5 = 60°$, $\angle Z = 22°$ (nearest °), $\angle Y = 98°$ (nearest °)

10 a $AC = 70$ cm   b $CD = 30$ cm

*11 a $\angle PQR = 90°$, $\angle QRS = 98°$, $\angle RSP = 76°$, $\angle SPQ = 96°$
    b 11 cm (2 sf)

12 $YZ = 1.3$ m (1dp)
    $XZ = 1.7$ m (1dp)

18.3A

1 Sam is 12 m nearer to Bishra than to Aamina (2 sf).
2 Ship is now 5.3 km from the lighthouse (2 sf).
3 126 m (nearest m)
4 8.4 km (2 sf)
5 Distance back is 23 km (nearest km)
6 Yes, area = 10 861 m² (to nearest m²) > 10 000 m²
7 a Bearing of Rodley from Quinton = 163° (nearest °)
    b Rodley is 50 km nearer to Packham than Quinton (2 sf).
8 a 6.5 cm (1dp)   b 3.6 cm (1dp)

9 a 27°, 40° (nearest °)   b Total area = 96 m² (2sf)

10 a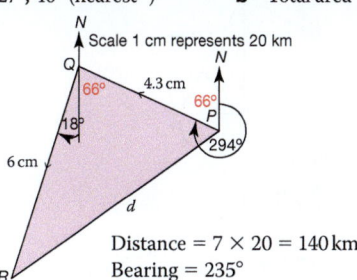

Distance = $7 \times 20 = 140$ km
Bearing = 235°

    b $d^2 = 86^2 + 120^2 - 2 \times 86 \times 120 \times \cos 84°$
    $d = 140$ km (nearest km)
    $\frac{\sin P}{120} = \frac{\sin 84°}{140.137...}$ and $\angle P = 58.387...°$
    Bearing from $P = 294° - 58.387...° = 236°$ (nearest °)
    c The second method is better. Trigonometry is more accurate and not subject to errors in measurements.

11 Speed = 12 km/h (2 sf)

12 a In triangle ADC $\sin A = \frac{h}{b}$, $h = b \sin A$
     In triangle BDC $\sin B = \frac{h}{a}$, $h = a \sin B$
    b $a \sin B = b \sin A$
     Dividing by $\sin A \sin B$ gives $\frac{a}{\sin A} = \frac{b}{\sin B}$
    c Area = $\frac{1}{2} \times AB \times h = \frac{1}{2} \times c \times b \sin A$ or $\frac{1}{2} \times c \times a \sin B$
     Area = $\frac{1}{2} bc \sin A$ or $\frac{1}{2} ca \sin B$

*13 a $b^2 = h^2 + x^2$ (Pythagoras' theorem)
     $a^2 = h^2 + (c - x)^2 = h^2 + c^2 - 2cx + x^2$ (Pythagoras' theorem)
     Subtracting gives $a^2 - b^2 = c^2 - 2cx$
     so $a^2 = b^2 + c^2 - 2cx$ (1)
     In triangle ADC $\cos A = \frac{x}{b}$ so $x = b \cos A$
     Substituting in (1) gives $a^2 = b^2 + c^2 - 2bc \cos A$
    b $2bc \cos A = b^2 + c^2 - a^2$
     Dividing by $2bc$ gives
     $\cos A = \frac{b^2 + c^2 - a^2}{2bc}$
    c $a^2 = h^2 + (x + c)^2 = h^2 + x^2 + 2cx + c^2$ (Pythagoras' theorem)
     $b^2 = h^2 + x^2$
     Subtracting gives $a^2 - b^2 = c^2 + 2cx$
     so $a^2 = b^2 + c^2 + 2cx$ (1)
     In triangle ADC $\cos(180° - \angle A) = \frac{x}{b}$
     so $x = b \cos(180° - \angle A)$
     Substituting in (1) gives
     $a^2 = b^2 + c^2 + 2bc \cos(180° - \angle A)$
     But $\cos(180° - \angle A) = -\cos A$
     Therefore $a^2 = b^2 + c^2 - 2bc \cos A$

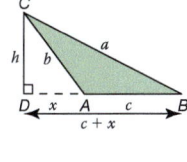

18.4S

1 a $AC = 20$ cm   b $AG = 25$ cm
    c $\angle GAC = 37°$ (nearest °)   d $\angle FAB = 43°$ (nearest °)

2 a $PS = 15$ cm
    b i $\angle SPR = 28°$ (nearest °)
       ii $\cos \angle QPR = \frac{17^2 + 17^2 - 16^2}{2 \times 17 \times 17} = 0.5570...$
       $\angle QPR = \cos^{-1} 0.5570... = 56.144...°$
       $\angle SPR = 56.144...° \div 2 = 28°$ (nearest °)

3 a 7.2 cm (2sf)   b $\angle PSU = 64°$ (nearest °)
    c $\angle PMU = 71°$ (nearest °)

4 a i $QV = 20\sqrt{2}$ cm   ii $PV = 20\sqrt{3}$ cm
    b i $\angle VQR = 45°$ ii $\angle VPR = 35°$ (nearest °)

5 a $h = 40$ cm   b $AC = 78$ cm
    c $EC = 88$ cm (nearest cm)   d $\angle ECA = 27°$ (nearest °)
    e $\angle EBA = 53°$ (nearest °)

6 a i $AQ = 61$ cm (nearest cm)   ii 25° (nearest °)
    iii 909 cm² (3 sf)

b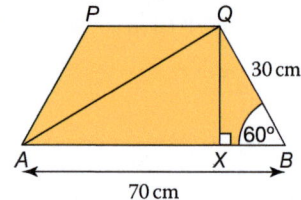

i $\cos 60° = \frac{XB}{30}, XB = 15$

$\sin 60° = \frac{QX}{30}, QX = 15\sqrt{3}$

In triangle AQX, $AX = 70 - 15 = 55$

$AQ^2 = 55^2 + (15\sqrt{3})^2 = 3700, AQ = 61$ cm (nearest cm)

ii Using triangle AQX,

$\tan \angle QAB = \frac{QX}{AX} = \frac{15\sqrt{3}}{55}$, $\angle QAB = 25°$ (nearest °)

iii Area of $AQB = \frac{1}{2} \times 70 \times 15\sqrt{3} = 909$ cm² (3 sf)

c i $AR = 209$ cm (nearest cm)

ii $\angle RAC = 8.1°$ (to 1dp)

7 a Shorter edges = 41 mm (nearest mm)
Longer edges = 54 mm (nearest mm)

b 130° (nearest °)

8 a $\angle RPS = 22°$ (nearest °)

b Area of triangle $RPS = 3.5$ cm² (1dp)

c $\angle RQS = 25°$ (nearest °)   d 3.1 cm² (1dp)

*9 a 47 mm (nearest mm)   b 63° (nearest °)

c 70° (nearest °)

10 a $AE = 1.4$ m (2 sf)   b 56° (nearest °)

18.4A

1 Yes, $h = 50.154... > 50$ m.

2 $\angle ABC = 161°$ (nearest °)

3 a $d = 22$ km (nearest km)
Bearing = 065° (nearest °)

b Yes, distance from $P = 9.440...$ km < 10 km.

4 a i $\sin \theta = \frac{3}{5}$ ii $\cos \theta = \frac{4}{5}$

b i $\cos \theta = \frac{5}{13}$ ii $\tan \theta = \frac{12}{5}$

c i $\sin \theta = 0.6$ ii $\tan \theta = 0.75$

5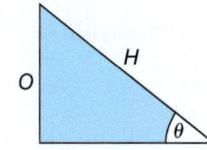

$(\sin \theta)^2 + (\cos \theta)^2 = \left(\frac{O}{H}\right)^2 + \left(\frac{A}{H}\right)^2$

$= \frac{O^2 + A^2}{H^2} = \frac{H^2}{H^2} = 1$

6 Angle subtended at centre of circle by one side $= \frac{360°}{n}$

Let r be the radius of the circle.

Area of each triangle $= \frac{1}{2}r^2 \sin\left(\frac{360°}{n}\right)$

Area of polygon, $A = \frac{1}{2}nr^2 \sin\left(\frac{360°}{n}\right)$

7 a 34° (nearest °)

b $7632.23

8 Area of flowerbed = 1.58 m²

9 $\angle ARB = 19°$ (nearest °)

10 136π cm²

*11 Mass = 26π kg

*12 42 seconds (nearest second)

*13 Using Pythagoras

$a^2 = 2x^2, a = \sqrt{2}x$

$a + 2x = 1$, so $a + \frac{2}{\sqrt{2}}a = 1$

$a(1 + \sqrt{2}) = 1$

$a = \frac{1}{\sqrt{2}+1} = \sqrt{2} - 1$

Perimeter $= 8a = 8(\sqrt{2} - 1)$ metres

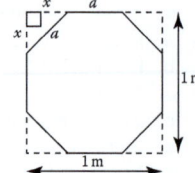

18.5S

1 a $\underline{p} = \binom{2}{3}, \underline{q} = \binom{2}{0}, \underline{r} = \binom{-2}{-1}$

b i $\binom{4}{6}$ ii $\binom{-6}{-3}$ iii $\binom{4}{3}$

iv $\binom{0}{2}$ v $\binom{0}{3}$ vi $\binom{4}{4}$

vii $\binom{0}{-1}$ viii $\binom{4}{1}$ ix $\binom{2}{2}$

2 a i $\binom{1}{1}$ ii $\binom{-1}{7}$ iii $\binom{-5}{7}$

iv $\binom{3}{5}$ v $\binom{5}{5}$ vi $\binom{8}{0}$

b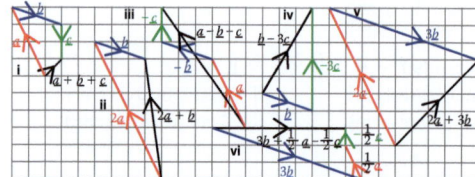

3 a i $\overrightarrow{PQ} = \binom{3}{1}$ ii $\overrightarrow{QR} = \binom{-3}{4}$ iii $\overrightarrow{PR} = \binom{0}{5}$

iv $\overrightarrow{RS} = \binom{-6}{-7}$ v $\overrightarrow{PS} = \binom{-6}{-2}$ vi $\overrightarrow{SP} = \binom{6}{2}$

b Yes, $\overrightarrow{SP} = 2\overrightarrow{PQ}$

c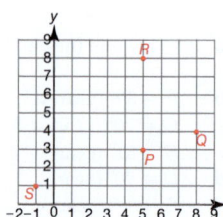

4 a i $\overrightarrow{KL} = \binom{1}{-7}$ ii $\overrightarrow{KM} = \binom{6}{-6}$ iii $\overrightarrow{KN} = \binom{4}{-4}$

iv $\overrightarrow{ML} = \binom{-5}{-1}$ v $\overrightarrow{LN} = \binom{3}{3}$ vi $\overrightarrow{NM} = \binom{2}{-2}$

b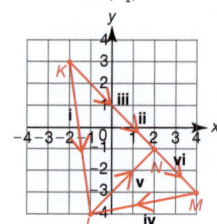

c K, M and N

5 $\underline{c} = 4\underline{a}$ $\underline{d} = -3\underline{b}$ $\underline{e} = \underline{a} + 3\underline{b}$

$\underline{f} = -3\underline{a} + 2\underline{b}$ $\underline{g} = -3\underline{a} - 2\underline{b}$ $\underline{h} = 4\underline{a} - \underline{b}$

$\underline{i} = 5\underline{a} + 3\underline{b}$ $\underline{j} = -\underline{a} + 4\underline{b}$

6 a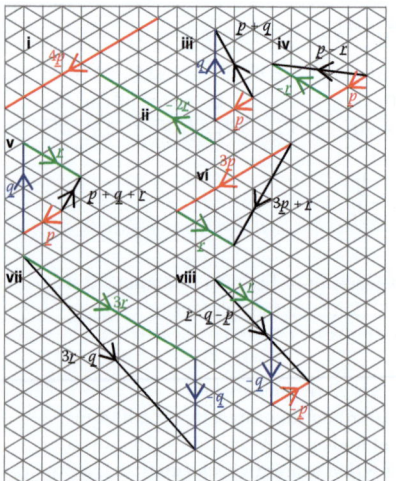

b $\underline{s} = \underline{q} + \underline{r}$ $\underline{t} = 2\underline{q} + \underline{r}$ $\underline{u} = \underline{r} - \frac{1}{2}\underline{q}$
c $\underline{p} = -\frac{1}{2}\underline{q} - \frac{2}{3}\underline{r}$

7 a $a = 4$ **b** $b = 12$ **c** $c = -4$

8 $\binom{12}{8} = 4\binom{3}{2}$ and $\binom{-6}{-4} = -2\binom{3}{2}$

9 a $x = 5, y = -1$ **b** $x = 10, y = 9$
  c $x = 4$ **d** $x = 3$
     $y = -1$ $y = 2$

10 a i $5\underline{a}$ **ii** $-\underline{a}$ **iii** $-\underline{b}$ **iv** $-10\underline{b}$
  b i Vectors $\underline{p} + 2\underline{q}$ and $\underline{q} + 2\underline{r}$ are both parallel to $\underline{a}$.
        $\underline{q} + 2\underline{r}$ is also equal in length to $\underline{a}$, but $\underline{p} + 2\underline{q}$ is 5 times as long.
    ii Vectors $\underline{q} + \underline{r}$ and $3\underline{q} - \underline{p}$ are both parallel to $\underline{b}$, but in the opposite direction.
        $\underline{q} + \underline{r}$ is equal in length to $\underline{b}$, but $3\underline{q} - \underline{p}$ is 10 times as long.

*__11__ 4.2

12

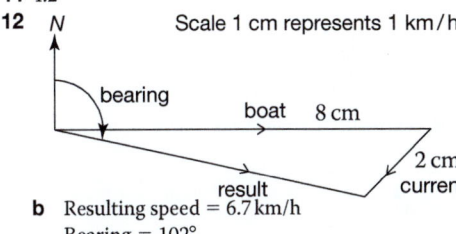

  b Resulting speed = 6.7 km/h
     Bearing = 102°

18.5A

1 a There are many possibilities. One of these is given below.
   $\binom{2}{-2}\binom{4}{3}\binom{0}{5}\binom{-3}{-1}\binom{-3}{0}\binom{-2}{-2}\binom{0}{-2}\binom{-1}{-2}\binom{3}{1}$
  b $\binom{0}{0}$ because the boat starts and ends at the same place.

2 a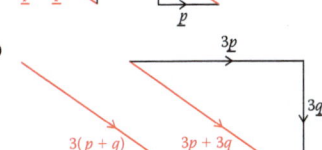

  b

  c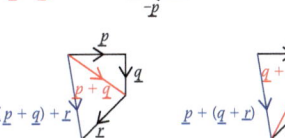

  d

3 a Using Pythagoras, length = $\sqrt{x^2 + y^2}$
  b Using trigonometry, angle = $\tan^{-1}\frac{y}{x}$

4 a $\overrightarrow{ON} = \frac{1}{2}\underline{b}$ $\overrightarrow{MO} = -\frac{1}{2}\underline{a}$
     $\overrightarrow{MN} = -\frac{1}{2}\underline{a} + \frac{1}{2}\underline{b}$ $\overrightarrow{AB} = -\underline{a} + \underline{b}$
  b MN is parallel to AB and half as long as AB.

5 a $\overrightarrow{PQ} = -\underline{p} + \underline{q}$ (or $\underline{q} - \underline{p}$)
     $\overrightarrow{QP} = -\underline{q} + \underline{p}$ (or $\underline{p} - \underline{q}$)
  b $\overrightarrow{OM} = \frac{1}{2}(\underline{p} + \underline{q})$

6 a $\overrightarrow{PQ} = 6\underline{a} + 2\underline{b}$
  b $\overrightarrow{PQ} = 2\overrightarrow{SR}$ This means that PQ is parallel to SR and so $PQRS$ is a trapezium.

7 a $(2, 4)$ **b** translation by vector $\binom{0}{1}$

8 a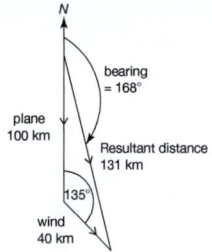

  b Using the cosine rule
     $(\text{resultant distance})^2 = 100^2 + 40^2 - 2 \times 100 \times 40 \times \cos 135°$
     resultant distance = 131 km (3sf)
     Using the sine rule $\frac{\sin \theta}{40} = \frac{\sin 135°}{131.365...}$, $\theta = 12.433...°$
     Bearing = $180° - \theta = 167.566...° = 168°$ (nearest °)
  c Trigonometry is more accurate because lengths on scale diagrams cannot be drawn or measured precisely.

9 a i $\overrightarrow{OP} = \frac{3}{4}\underline{a}$ **ii** $\overrightarrow{BP} = -\underline{b} + \frac{3}{4}\underline{a}$ **iii** $\overrightarrow{MN} = \frac{3}{8}\underline{a}$
  b MN is parallel to OA and the length of MN is $\frac{3}{8}$ of the length of OA.

*__10__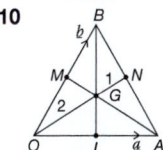

  a $\overrightarrow{AB} = -\underline{a} + \underline{b}$
     $\overrightarrow{ON} = \overrightarrow{OA} + \overrightarrow{AN} = \underline{a} + \frac{1}{2}(-\underline{a} + \underline{b}) = \frac{1}{2}(\underline{a} + \underline{b})$
     $\overrightarrow{OG} = \frac{2}{3}\overrightarrow{ON} = \frac{1}{3}(\underline{a} + \underline{b})$
    i $\overrightarrow{AG} = \overrightarrow{AO} + \overrightarrow{OG} = -\underline{a} + \frac{1}{3}(\underline{a} + \underline{b})$
        $= -\frac{2}{3}\underline{a} + \frac{1}{3}\underline{b}$
        $\overrightarrow{GM} = \overrightarrow{GO} + \overrightarrow{OM}$
        $= -\frac{1}{3}(\underline{a} + \underline{b}) + \frac{1}{2}\underline{b} = -\frac{1}{3}\underline{a} + \frac{1}{6}\underline{b}$
        $\overrightarrow{GM} = \frac{1}{2}\overrightarrow{AG}$
        AGM is a straight line and $AG:GM = 2:1$
    ii $\overrightarrow{BG} = \overrightarrow{BO} + \overrightarrow{OG} = -\underline{b} + \frac{1}{3}(\underline{a} + \underline{b}) = \frac{1}{3}\underline{a} - \frac{2}{3}\underline{b}$
        $\overrightarrow{GL} = \overrightarrow{GO} + \overrightarrow{OL} = -\frac{1}{3}(\underline{a} + \underline{b}) + \frac{1}{2}\underline{a} = \frac{1}{6}\underline{a} - \frac{1}{3}\underline{b}$
        $\overrightarrow{GL} = \frac{1}{2}\overrightarrow{BG}$
        BGL is a straight line and $BG:GL = 2:1$
  b The lines joining each vertex of a triangle to the mid-point of the opposite side all meet at a point G which is $\frac{2}{3}$ of the way along each line.

*__11__ Let $\overrightarrow{AB} = \underline{x}$, $\overrightarrow{BC} = \underline{y}$ and $\overrightarrow{CD} = \underline{z}$
     $\overrightarrow{PQ} = \frac{1}{2}\underline{x} + \frac{1}{2}\underline{y}$
     $\overrightarrow{AD} = \underline{x} + \underline{y} + \underline{z}$
     $\overrightarrow{SR} = \overrightarrow{SD} + \overrightarrow{DR} = \frac{1}{2}\underline{x} + \frac{1}{2}\underline{y}$
     PQ and SR are parallel and equal in length.
     This is sufficient to prove that $PQRS$ is a parallelogram.

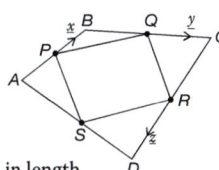

*__12__ The statement can be proved by showing that the mid-points of the diagonals coincide.
     In parallelogram $PQRS$, let $\overrightarrow{PQ} = \underline{a}$ and $\overrightarrow{PS} = \underline{b}$
     $\overrightarrow{PR} = \underline{a} + \underline{b}$ and its mid-point M is such that $\overrightarrow{PM} = \frac{1}{2}(\underline{a} + \underline{b})$
     $\overrightarrow{QS} = -\underline{a} + \underline{b}$ and its mid-point N is such that $\overrightarrow{QN} = \frac{1}{2}(-\underline{a} + \underline{b})$
     Therefore $\overrightarrow{PN} = \overrightarrow{PQ} + \overrightarrow{QN} =$
     $\underline{a} + \frac{1}{2}(-\underline{a} + \underline{b}) = \frac{1}{2}(\underline{a} + \underline{b})$
     Therefore $\overrightarrow{PN} = \overrightarrow{PM}$ and the mid-points of the diagonals are the same point.

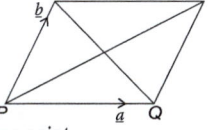

18.6S

1   a $\binom{8}{11}$ b $\binom{2}{3}$ c $\binom{20}{28}$

2   a $\binom{34}{26}$ b $\binom{4}{18}$ c $\binom{23}{-7}$ d $\binom{-3}{6}$

3   a $\begin{pmatrix}19 & 16\\51 & 44\end{pmatrix}$ b $\begin{pmatrix}21 & 29\\-4 & -20\end{pmatrix}$ c $\begin{pmatrix}-2 & -23\\8 & 4\end{pmatrix}$ d $\begin{pmatrix}9 & 6\\-26 & -4\end{pmatrix}$

4   a $\binom{27}{6}$ b $\begin{pmatrix}26 & -12\\16 & -6\end{pmatrix}$ c $\begin{pmatrix}0 & 24\\-20 & 62\end{pmatrix}$ d $\binom{10}{-10}$

5   a $\binom{7}{12}$ b $\binom{-1}{-2}$ c $\binom{12}{20}$ d $\begin{pmatrix}12 & 4\\2 & 16\end{pmatrix}$

   e $\binom{28}{43}$ f $\binom{38}{60}$ g $\binom{25}{1}$ h $\binom{35}{1}$

   i $\begin{pmatrix}4 & 28\\16 & -3\end{pmatrix}$ j $\begin{pmatrix}5 & 40\\11 & -4\end{pmatrix}$ k $\binom{3}{5}$ or K   l $\begin{pmatrix}0 & 5\\2 & -1\end{pmatrix}$ or N

6   a   A is 2 × 2, B is 2 × 1 and C is 2 × 2

   b   i $\begin{pmatrix}3 & 0\\9 & 15\end{pmatrix}$ ii $\binom{20}{-10}$ iii $\binom{-4}{-2}$
   iv   not possible   v $\begin{pmatrix}6 & 5\\13 & 30\end{pmatrix}$ vi $\begin{pmatrix}21 & 25\\8 & 15\end{pmatrix}$
   vii   not possible   viii $\binom{-14}{10}$

7   AB $= \begin{pmatrix}2 & 5\\1 & 3\end{pmatrix}\begin{pmatrix}3 & -5\\-1 & 2\end{pmatrix} = \begin{pmatrix}6-5 & -10+10\\3-3 & -5+6\end{pmatrix} = \begin{pmatrix}1 & 0\\0 & 1\end{pmatrix} = I$
   and BA $= \begin{pmatrix}3 & -5\\-1 & 2\end{pmatrix}\begin{pmatrix}2 & 5\\1 & 3\end{pmatrix} = \begin{pmatrix}6-5 & 15-15\\-2+2 & -5+6\end{pmatrix} = \begin{pmatrix}1 & 0\\0 & 1\end{pmatrix} = I$

8   a $x = 13, y = -13$ b $x = -2, y = 4$
   c $x = 5, y = 1$ d $x = 5, y = -8$ or $x = 10, y = -23$
   e $x = y = -5$ or $x = y = 2$

9   a   For any 2 × 1 matrix M $= \binom{a}{b}$
       IM $= \begin{pmatrix}1 & 0\\0 & 1\end{pmatrix}\binom{a}{b} = \binom{a+0}{0+b} = \binom{a}{b} = M$
   b   For any 2 × 2 matrix N $= \begin{pmatrix}a & b\\c & d\end{pmatrix}$
       IN $= \begin{pmatrix}1 & 0\\0 & 1\end{pmatrix}\begin{pmatrix}a & b\\c & d\end{pmatrix} = \begin{pmatrix}a+0 & b+0\\0+c & 0+d\end{pmatrix} = \begin{pmatrix}a & b\\c & d\end{pmatrix} = N$
       and NI $= \begin{pmatrix}a & b\\c & d\end{pmatrix}\begin{pmatrix}1 & 0\\0 & 1\end{pmatrix} = \begin{pmatrix}a+0 & 0+b\\c+0 & 0+d\end{pmatrix} = \begin{pmatrix}a & b\\c & d\end{pmatrix} = N$

10   NM $= \frac{1}{ad-bc}\begin{pmatrix}d & -b\\-c & a\end{pmatrix}\begin{pmatrix}a & b\\c & d\end{pmatrix} = \frac{1}{ad-bc}\begin{pmatrix}da-bc & db-bd\\-ca+ac & -cb+ad\end{pmatrix}$
   $= \frac{1}{ad-bc}\begin{pmatrix}ad-bc & 0\\0 & ad-bc\end{pmatrix} = \begin{pmatrix}1 & 0\\0 & 1\end{pmatrix} = I$
   and MN $= \begin{pmatrix}a & b\\c & d\end{pmatrix}\frac{1}{ad-bc}\begin{pmatrix}d & -b\\-c & a\end{pmatrix} = \begin{pmatrix}a & b\\c & d\end{pmatrix}\begin{pmatrix}\frac{d}{ad-bc} & \frac{-b}{ad-bc}\\\frac{-c}{ad-bc} & \frac{a}{ad-bc}\end{pmatrix}$
   $= \begin{pmatrix}\frac{ad}{ad-bc}-\frac{bc}{ad-bc} & \frac{-ab}{ad-bc}+\frac{ba}{ad-bc}\\\frac{cd}{ad-bc}-\frac{dc}{ad-bc} & \frac{-cb}{ad-bc}+\frac{da}{ad-bc}\end{pmatrix}$
   $= \begin{pmatrix}\frac{ad-bc}{ad-bc} & \frac{-ab+ab}{ad-bc}\\\frac{cd-dc}{ad-bc} & \frac{-cb+da}{ad-bc}\end{pmatrix} = \begin{pmatrix}1 & 0\\0 & 1\end{pmatrix} = I$

11   a $\binom{75}{116}$ This gives the total number of coffees and teas sold in the morning and the afternoon.
   b $\binom{135}{206}$ This gives the money (in £) taken on coffees and teas in the morning and the afternoon.

18.6A

1   a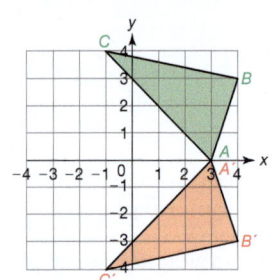
   b   reflection in the x axis (or $y = 0$)

2   a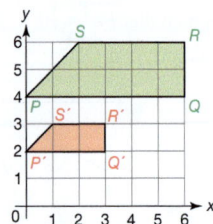
   b   enlargement, centre (0, 0) scale factor $\frac{1}{2}$

3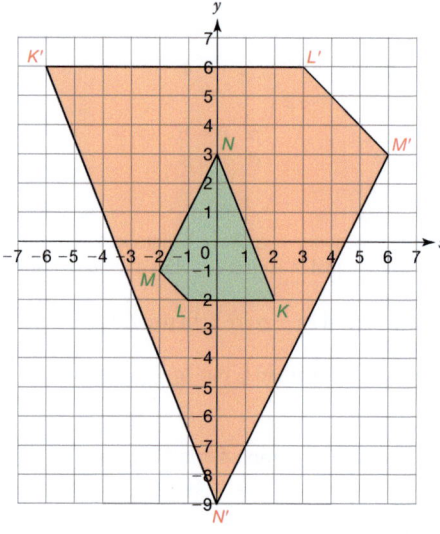
   enlargement, centre (0, 0), scale factor -3

4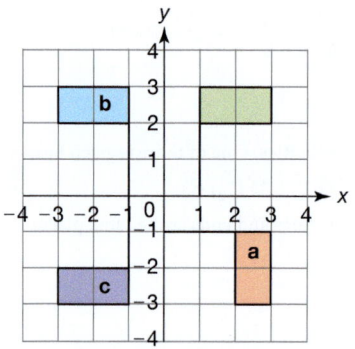
   a   rotation 90° clockwise (or 270° anticlockwise) about (0, 0)
   b   reflection in the y axis (or $x = 0$)
   c   rotation 180° clockwise (or anticlockwise) about (0, 0)

5   a   enlargement, centre (0, 0), scale factor 5
   b   Check students' diagrams.

6   a $\begin{pmatrix}0 & 1\\1 & 0\end{pmatrix}$ b   Check students' diagrams.

7   a   i $\begin{pmatrix}-1 & 0\\0 & -1\end{pmatrix}$ b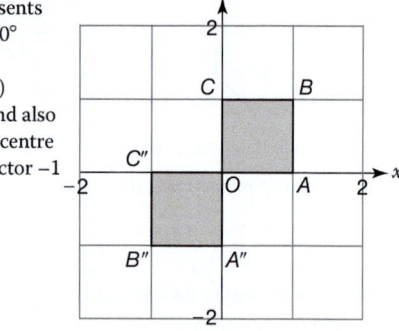
   ii   Yes - M represents rotation of 180° clockwise (or anticlockwise) about (0, 0) and also enlargement, centre (0, 0), scale factor -1

   reflection in $y = -x$

8   a   enlargement, centre (0, 0) scale factor 3
   b   reflection in $y = x$
   c   rotation 90° anticlockwise (or 270° clockwise) about (0, 0)

9   a $\begin{pmatrix}1 & 0\\0 & -1\end{pmatrix}$ b $\begin{pmatrix}-1 & 0\\0 & 1\end{pmatrix}$
   c $\begin{pmatrix}0 & 1\\-1 & 0\end{pmatrix}$ d $\begin{pmatrix}-4 & 0\\0 & -4\end{pmatrix}$

18.7S

1 a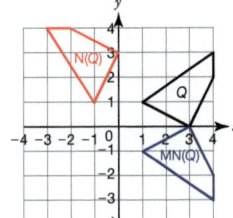
b reflection in the x axis (or $y = 0$ line).
c $MN = \begin{pmatrix} 1 & 0 \\ 0 & -1 \end{pmatrix}$

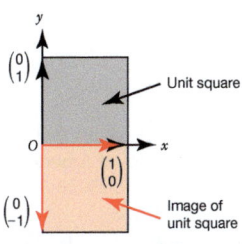

2 a i ii / **b i ii**

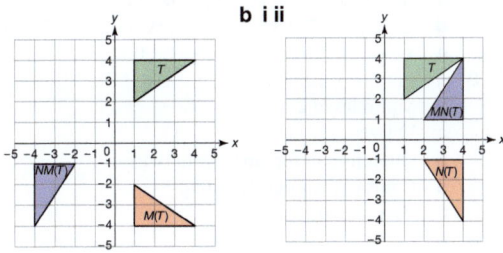

c i $MN = \begin{pmatrix} 0 & 1 \\ 1 & 0 \end{pmatrix}$ Reflection in $y = x$
   ii $NM = \begin{pmatrix} 0 & -1 \\ -1 & 0 \end{pmatrix}$ Reflection in $y = -x$

3 a

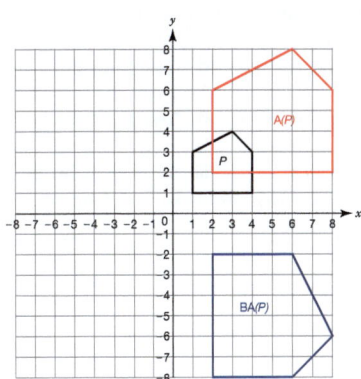

b i ii

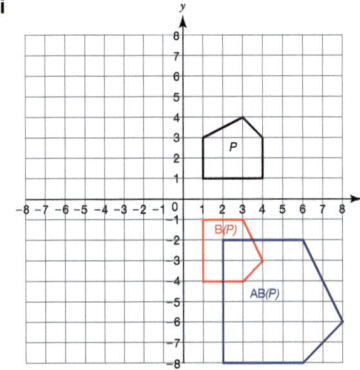

c $BA = \begin{pmatrix} 0 & 2 \\ -2 & 0 \end{pmatrix} = AB$
d The order in which the matrices are combined does not make any difference to the final image.

4 a i $RS = \begin{pmatrix} 0 & -1 \\ 1 & 0 \end{pmatrix}$ **ii** $SR = \begin{pmatrix} 0 & -1 \\ 1 & 0 \end{pmatrix}$
b i rotation through 180° clockwise (or anti-clockwise) about (0, 0) (or enlargement, centre (0, 0), scale factor -1)
   ii rotation through 90° clockwise (or 270° anti-clockwise) about (0, 0)
   iii rotation through 90° anti-clockwise (or 270° clockwise) about (0, 0)
   iv rotation through 90° anti-clockwise (or 270° clockwise) about (0, 0)

5 a $\begin{pmatrix} -\frac{1}{2} & 0 \\ 0 & \frac{1}{2} \end{pmatrix}$ **b** $\begin{pmatrix} -1 & 0 \\ 0 & 1 \end{pmatrix}\begin{pmatrix} \frac{1}{2} & 0 \\ 0 & \frac{1}{2} \end{pmatrix}$ or $\begin{pmatrix} \frac{1}{2} & 0 \\ 0 & \frac{1}{2} \end{pmatrix}\begin{pmatrix} -1 & 0 \\ 0 & 1 \end{pmatrix} = \begin{pmatrix} -\frac{1}{2} & 0 \\ 0 & \frac{1}{2} \end{pmatrix}$

6 a i $\begin{pmatrix} -1 & 0 \\ 0 & -1 \end{pmatrix}$ **ii** $\begin{pmatrix} -1 & 0 \\ 0 & -1 \end{pmatrix}$
b i enlargement, centre (0, 0) scale factor 5
   ii enlargement, centre (0, 0) scale factor $-\frac{1}{5}$
   iii enlargement, centre (0, 0), scale factor -1
   or rotation 180° clockwise (or anticlockwise) about (0, 0)

7 a $\begin{pmatrix} 0 & 1 \\ -1 & 0 \end{pmatrix}$ **b** $\begin{pmatrix} -2 & 0 \\ 0 & -2 \end{pmatrix}$ **c** $\begin{pmatrix} 0 & -2 \\ 2 & 0 \end{pmatrix}$

18.7A

1 a i $\begin{pmatrix} -1 & 0 \\ 0 & -1 \end{pmatrix}$ **ii** $\begin{pmatrix} 2 & 0 \\ 0 & 2 \end{pmatrix}$ **iii** $\begin{pmatrix} -2 & 0 \\ 0 & -2 \end{pmatrix}$
b enlargement, centre (0, 0), scale factor -2
c yes

2 a i $\begin{pmatrix} 0 & 1 \\ 1 & 0 \end{pmatrix}$ **ii** $\begin{pmatrix} 0 & 1 \\ -1 & 0 \end{pmatrix}$ **iii** $\begin{pmatrix} 1 & 0 \\ 0 & -1 \end{pmatrix}$
b reflection in the x axis (or $y = 0$ line)
c No The answers would be $\begin{pmatrix} -1 & 0 \\ 0 & 1 \end{pmatrix}$ and reflection in the y axis (or $x = 0$ line)

3 $\begin{pmatrix} 0 & -4 \\ 4 & 0 \end{pmatrix}$

4 a $\begin{pmatrix} 0 & -\frac{1}{4} \\ -\frac{1}{4} & 0 \end{pmatrix}$ **b** yes

5 a When an object is reflected twice in $y = -x$ it returns to its original position.
b A rotation of 180° (clockwise or anticlockwise) about (0, 0) followed by another rotation of 180° (clockwise or anticlockwise) about (0, 0) returns an object to its original position. (Accept repeated enlargements, centre (0, 0), scale factor -1 instead of rotations.)

6 a Enlargement, centre (0, 0), scale factor 8 followed by enlargement, centre (0, 0), scale factor $\frac{1}{2}$ is equivalent to enlargement, centre (0, 0), scale factor 4.
b Enlargement, centre (0, 0), scale factor 0.1 followed by enlargement, centre (0, 0), scale factor -10 is equivalent to rotation by 180° anticlockwise (or clockwise) about (0, 0) or enlargement, centre (0, 0), scale factor -1.

7 Reflection in the x axis followed by reflection in the y axis is equivalent to rotation through 180° (clockwise or anticlockwise) about (0, 0). Reflection in $y = x$ followed by reflection in $y = -x$ is also equivalent to rotation through 180° (clockwise or anticlockwise) about (0, 0).
(Accept enlargement, centre (0, 0), scale factor -1, instead of each rotation.)

8 a i $\begin{pmatrix} -2 & 0 \\ 0 & 2 \end{pmatrix}$ **ii** $\begin{pmatrix} -\frac{1}{2} & 0 \\ 0 & \frac{1}{2} \end{pmatrix}$
   iii $\begin{pmatrix} 2 & 0 \\ 0 & -2 \end{pmatrix}$ **iv** $\begin{pmatrix} \frac{1}{2} & 0 \\ 0 & -\frac{1}{2} \end{pmatrix}$
b i $\begin{pmatrix} 1 & 0 \\ 0 & 1 \end{pmatrix}$ **ii** $\begin{pmatrix} 1 & 0 \\ 0 & 1 \end{pmatrix}$

9 a $\begin{pmatrix} 0 & 1 \\ 1 & 0 \end{pmatrix}\begin{pmatrix} -1 & 0 \\ 0 & -1 \end{pmatrix}\begin{pmatrix} -1 & 0 \\ 0 & 1 \end{pmatrix} = \begin{pmatrix} 0 & -1 \\ 1 & 0 \end{pmatrix}$
b i Reflection in the y axis, followed by rotation of 180° (clockwise or anticlockwise) about (0, 0) followed by reflection in $y = x$ is equivalent to rotation through 90° anticlockwise (or 270° clockwise) about (0, 0).
(Accept enlargement, centre (0, 0), scale factor -1, instead of the 180° rotation.)
   ii Check students' diagrams.

c The direction of rotation is reversed in cases where the order of the reflections is reversed for example,
$\begin{pmatrix}-1 & 0 \\ 0 & 1\end{pmatrix}\begin{pmatrix}-1 & 0 \\ 0 & -1\end{pmatrix}\begin{pmatrix}0 & 1 \\ 1 & 0\end{pmatrix} = \begin{pmatrix}0 & 1 \\ -1 & 0\end{pmatrix}$

10 $A \to B \begin{pmatrix}0 & 1 \\ 1 & 0\end{pmatrix}$ $A \to C \begin{pmatrix}0 & -1 \\ 1 & 0\end{pmatrix}$ $A \to D \begin{pmatrix}-1 & 0 \\ 0 & 1\end{pmatrix}$ $A \to E \begin{pmatrix}-1 & 0 \\ 0 & -1\end{pmatrix}$
$A \to F \begin{pmatrix}0 & -1 \\ -1 & 0\end{pmatrix}$ $A \to G \begin{pmatrix}0 & 1 \\ -1 & 0\end{pmatrix}$ $A \to H \begin{pmatrix}1 & 0 \\ 0 & -1\end{pmatrix}$

Review

1 $a = 20.2$ mm, $b = 12.8$ cm
2 $a = 8.56$ cm, $b = 1.21$ cm, $c = 6.57$ cm
3 $\angle A = 45.3°$ (1 dp), $\angle B = 19.8°$ (1 dp)
4 a 15.3 cm   b 15.8 cm
  c 11.3°   d 75.3°
5 a 142 km (3 sf)   b 212° (3 sf)
6 a 1   b $\frac{\sqrt{3}}{2}$
7 $x = 7.12$ cm (3 sf), $\theta = 103°$ (nearest °)
8 25.2 cm² (3 sf)
9 $\mathbf{u} = \begin{pmatrix}6 \\ -5\end{pmatrix}$ $\mathbf{v} = \begin{pmatrix}-3 \\ 10\end{pmatrix}$

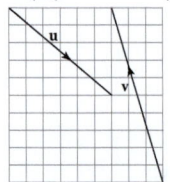

10 a i $\mathbf{u} + \mathbf{v}$ ii $1.5\mathbf{v}$ iii $0.5\mathbf{v} - \mathbf{u}$
   b $0.25\mathbf{v} - \mathbf{u}$
11 a

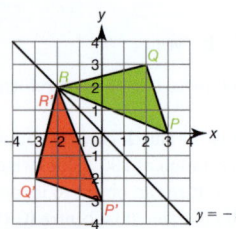

   b reflection in $y = -x$

Assessment 18

1 a Nerea has not taken the square root of 181.
    $a = 13.45$ m (2 dp)
  b Pawel calculated the square root of $24^2 + 25^2$ instead of the square root of $25^2 - 24^2$. Side = 7 in.
2 a $2^2 + 3^2 = 13$, not 13^2 as in a Pythagorean triple.
  b $20^2 + 99^2 = 10201 \neq 100^2$. 60, 80, 100 or 28, 96, 100.
3 9.85 (3 sf)
4 a Yes, $5^2 + 12^2 = 169 = 13^2$ b Yes, $9^2 + 12^2 = 225 = 15^2$
  c No, $9^2 + 14^2 = 277 \neq 17^2$
  d Yes, $1.6^2 + 3.0^2 = 11.56 = 3.4^2$
  e No, $11^2 + 19^2 = 482 \neq 22^2$
  f Yes, $3.6^2 + 7.7^2 = 72.25 = 8.5^2$
5 a 10.0 cm (3 sf)   b 59.6° (3 sf)
6 a 8.09 m (3 sf)   b 7.89 m (3 sf)
7 a 7 cm.   b 28.1° (1 dp)
8 distance = 42.4 km (3 sf), bearing = 226° (nearest °)
9 a 4.97 cm (3 sf)   b 49.9° (3 sf)
10 a Check students' drawings
   b i $\sqrt{173}$ ii $\sqrt{37}$ iii $4\sqrt{13}$
11 a i $\begin{pmatrix}3 \\ 4\end{pmatrix}$ ii $\begin{pmatrix}56 \\ -15\end{pmatrix}$ iii $\begin{pmatrix}-35 \\ 16\end{pmatrix}$
   b $k = -4, y = 1$
12 a enlargement, centre (0, 0), scale factor 2
   b $\begin{pmatrix}-1 & 0 \\ 0 & -1\end{pmatrix}$
   c i $\begin{pmatrix}-1 & 0 \\ 0 & -1\end{pmatrix}$ ii Reflection in $y = x$ followed by reflection in $y = -x$ is equivalent to rotation through 180° about the origin

13 a $\overrightarrow{QR} = \overrightarrow{QP} + \overrightarrow{PS} + \overrightarrow{SR} = -\mathbf{a} + 2\mathbf{b} - \mathbf{a} + 4\mathbf{a} = 2\mathbf{b} + 2\mathbf{a} = 2(\mathbf{b} + \mathbf{a})$
   b $\overrightarrow{MN} = \overrightarrow{MR} + \overrightarrow{RN} = \mathbf{b} + \mathbf{a} + \frac{3}{8}(-4\mathbf{a}) = \mathbf{b} + \mathbf{a} - 1.5\mathbf{a} = \mathbf{b} - 0.5\mathbf{a} = \frac{1}{2}\overrightarrow{PS}$ MN is parallel to PS

Chapter 19

Check in

1 a 0.55   b 0.04   c 0.72   d 0.625
  e 0.6   f 0.34   g 0.9   h 0.18
  i 0.17   j 0.192 k 0.235 l 0.92
2 a $\frac{1}{6}$ b $\frac{4}{5}$ c $\frac{2}{9}$ d $\frac{13}{15}$
  e $\frac{11}{12}$ f $\frac{5}{9}$ g $\frac{8}{45}$ h $\frac{2}{5}$

19.1S

1 a 1, 4, 9, 16, 25, 36, 49, 64, 81, 100
  b Canada, United States of America, Mexico
  c 2, 3, 5, 7, 11, 13, 17, 19, 23, 29
  d 1, 2, 3, 4, 6, 9, 12, 18, 36
2 a {1, 4, 9, 36}   b {2, 3}   c ∅
  d {1, 2, 3, 4, 5, 7, 9, 11, 13, 16, 17, 19, 23, 25, 29, 36, 49, 64, 81, 100}
  e 4   f 0
3 a Factors of 10
  b Multiples of 2
  c Vowels
  d Outcomes of flipping a coin twice
  e Coins of the pound sterling
  f First ten multiples of 3
4 a 1, 2, 3, 4, 5, 6, 7, 8, 9, 10   b 1, 3, 5, 7, 9
  c 2, 3, 5, 7   d 1, 4, 9
5 a {3,5,7}   b {1, 9}
  c {1, 3, 4, 5, 7, 9}   d {1, 2, 3, 4, 5, 7, 9}
6 Single digit numbers are less than ten and are integers.
7 a Yes, 10 and 20.   b Yes, 2.
  c Yes, 4, 9, and 36.   d Yes, 9 and 36.
  e No   f No
  g No   h Yes, 3.
  i No
8 a i 8   ii 13   iii 9   iv 21
  b Students who play **neither** hockey **nor** football.   c $\frac{13}{25}$
9 a i $\frac{29}{50}$ ii $\frac{19}{50}$ iii $\frac{17}{50}$ iv $\frac{43}{50}$
  b 14

19.1A

1 a

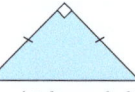

  (right-angled isosceles triangle)
  b An equilateral triangle has three 60° angles and cannot contain a right angle.
2 a $x = 8$ b A and B are disjoint.
3 3
4 Maximum 0.15, Minimum 0
5 5
6 a

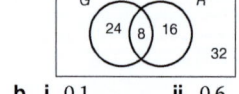

  b i 0.1   ii 0.6   iii 0.7   iv 0.2
7 a

  B ⎯⎯⎯⎯⎯⎯⎯⎯⎯⎯ G
  25 15 0
              40

  b i 0.1875   ii 0.5   iii 0.8125   iv 0.3125

*8 a (square) **b** Squares. **c** The answers would not differ.

*9 0.06

19.2S

1 a $\frac{1}{12}$ **b** $\frac{1}{6}$ **c** $\frac{7}{36}$ **d** $\frac{6}{36} = \frac{1}{6}$
  e {2, 3, 4, 5, 6, 7, 8, 9, 10, 11, 12}

2 a $\frac{2}{36} = \frac{1}{18}$ **b** $\frac{19}{36}$ **c** $\frac{8}{36} = \frac{2}{9}$ **d** $\frac{8}{36} = \frac{2}{9}$
  e {1, 2, 3, 4, 5, 6, 8, 9, 10, 12, 15, 16, 18, 20, 24, 25, 30, 36}

3 a

	1	2	3	4	5	6
1	0	1	2	3	4	5
2	1	0	1	2	3	4
3	2	1	0	1	2	3
4	3	2	1	0	1	2
5	4	3	2	1	0	1
6	5	4	3	2	1	0

  b {0, 1, 2, 3, 4, 5}
  c i $\frac{1}{6}$ **ii** $\frac{1}{6}$ **iii** 0 **iv** $\frac{4}{9}$

4 a {HHH, HHT, HTH, THH, TTH, THT, HTT, TTT}
  b 3 **c** 2

5 a

		Spinner 2	
	1	3	5
Spinner 1 1	2	4	6
2	3	5	7
4	5	7	9

  b {2, 3, 4, 5, 6, 7, 9}
  c i $\frac{1}{9}$ **ii** $\frac{1}{9}$ **iii** $\frac{1}{3}$

6 a

		Spinner 2	
	1	3	5
Spinner 1 1	1	3	5
2	2	6	10
4	4	12	20

  b {1, 2, 3, 4, 5, 6, 10, 12, 20}
  c i $\frac{1}{9}$ **ii** $\frac{1}{9}$ **iii** $\frac{2}{3}$

7 The product.

8 a $\frac{1}{12} \times 100 = 8\,(1\,\text{sf})$ **b** $\frac{1}{18} \times 100 = 6\,(1\,\text{sf})$
  c You should expect to see a 6 more often in the list of sums than in the list of products.
  d You should expect to see a 3 about the same number of times in the list of sums as in the list of products.
  e No, it is not certain that 7 will appear in the list of sums.

9 a

	1	2	2	3	3	3
1	2	3	3	4	4	4
2	3	4	4	5	5	5
3	4	5	5	6	6	6
4	5	6	6	7	7	7
5	6	7	7	8	8	8
6	7	8	8	9	9	9

  b i P(6) = $\frac{1}{6}$ **ii** P(7) = $\frac{1}{6}$ **iii** P(9) = $\frac{1}{12}$ **iv** P(3) = $\frac{1}{12}$
  c 4, 5 and 7.
  d 6 can be obtained from any of the outcomes on the unusual dice. This is not so for 2, 3, 8, and 9 which all have lower probabilities than 6.

19.2A

1 a $\frac{1}{6}$ **b** $\frac{2}{3}$ **c** $\frac{5}{6}$

2 $\frac{1}{4}$

3 a

	B1	R2	Y2	G3	B2	R4
1	1	2	0	3	2	4
2	2	4	0	6	4	8
3	3	6	0	9	6	12
4	4	8	0	12	8	16
5	5	10	0	15	10	20
6	6	12	0	18	12	24

  b i $\frac{1}{9}$ **ii** $\frac{5}{12}$

4 a $\frac{9}{100}$ **b** $\frac{9}{10}$ **c** $\frac{19}{100}$

5 a i $\frac{2}{15}$ **ii** $\frac{3}{5}$ **b** $\frac{2}{3}$

*6 a $\frac{5}{12}$ **b** $\frac{1}{2}$

*7 a i $\frac{1}{18}$ **ii** $\frac{1}{6}$ **iii** $\frac{1}{9}$
  b One of the following pairs: A and C, A and E, B and D, B and E.
  c Four pairs

19.3S

1 a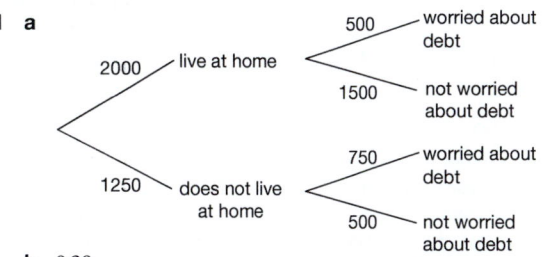
  b 0.38

2 a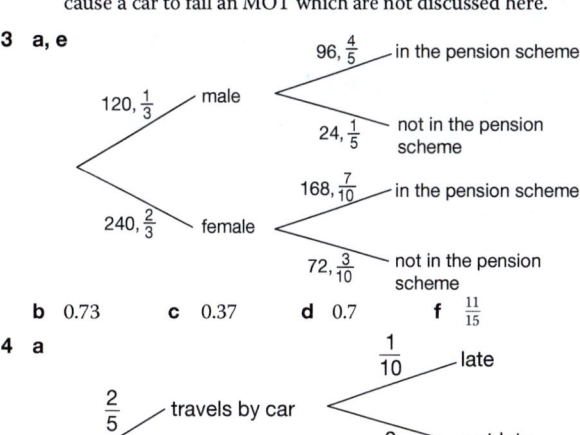
  b 840
  c Tommy is wrong because there are other features that can cause a car to fail an MOT which are not discussed here.

3 a, e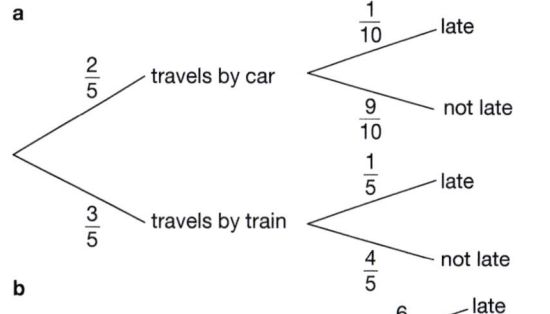
  b 0.73 **c** 0.37 **d** 0.7 **f** $\frac{11}{15}$

4 a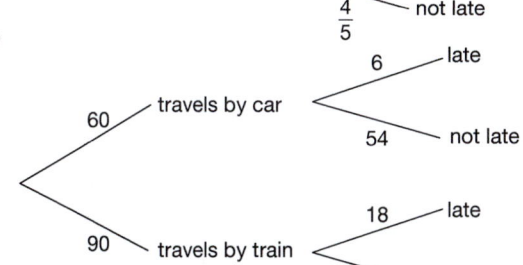
  b
  c 24 days **d** 0.16

5 a

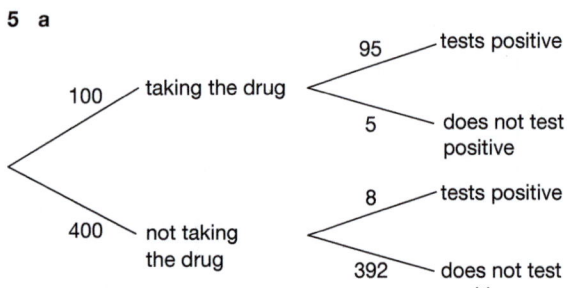

b 103 **c** 8 **d** 0.974

19.3A

1 a 0.84
b P(Draw then win) = P(Draw) × P(Win),
P(Lose then win) = P(Lose) × P(Win)
c Students' own responses e.g. No, losing the first match results in a drop in confidence, reducing the chance of winning the next match.

2 a

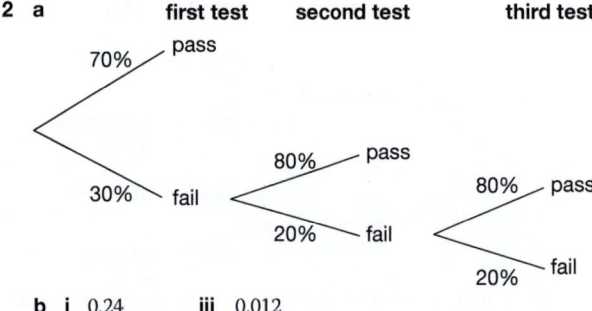

b i 0.24 **iii** 0.012

3 a

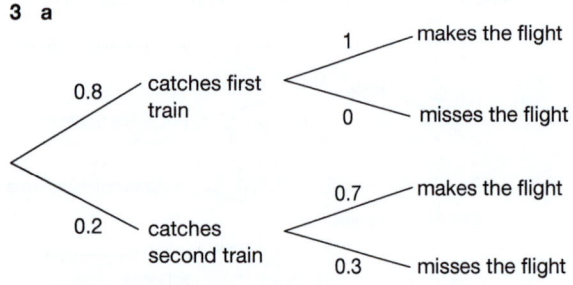

b 0.94

4 a Yes, $(X \cap Y)$ contains all outcomes where both balls are the same colour **and** that colour is blue.
b $P(X) = \frac{1}{2}$, $P(Y) = \frac{3}{7}$
$P(X \cap Y) = P(Z) = \frac{1}{2} \times \frac{3}{7} = P(X) \times P(Y)$

5

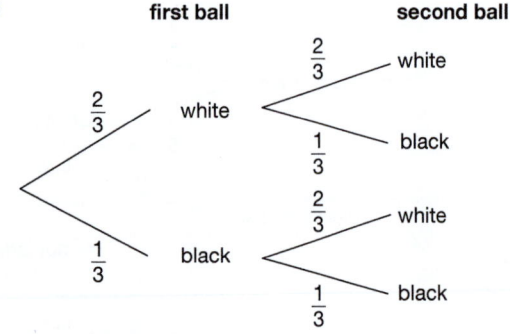

$P(A) = \frac{4}{9}$, $P(B) = \frac{5}{9}$
$P(A \cap B) = \frac{4}{9} > P(A) \times P(B) = \frac{20}{81}$
A and B are **not** independent.

Answers

*6 a $\frac{1}{9}$
b

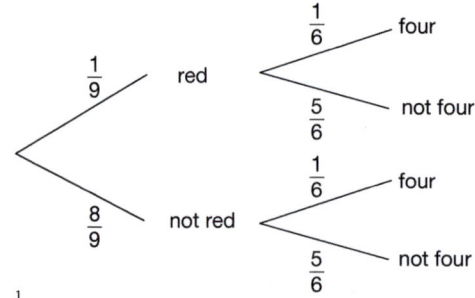

c $\frac{1}{6}$
d Using the spinner does not affect Sara's likelihood of getting a four. If the spinner lands on red, the possible scores are 2, 4, 6, 8, 10, and 12 and if the spinner does not land on red the possible scores are 1, 2, 3, 4, 5, and 6. Either way the possible scores are equally likely and $P(4) = \frac{1}{6}$.
e i P(finish) = P(not red **and** 5) = $\frac{4}{27} < \frac{1}{6}$.
ii P(finish) = P(red **and** 4) = $\frac{1}{54} < \frac{1}{6}$.

19.4S

1 a 0.9 **b** 0.5 **c** 0.8
d $P(S \cap R) = 0.4$
$P(S) \times P(R) = 0.5 \times 0.8 = 0.4$ S and R are independent.

2 ξ

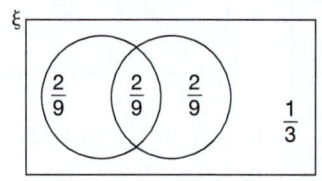

b i $\frac{4}{9}$ **ii** $\frac{2}{9}$ **iii** $\frac{1}{2}$
c P and F are not independent because $P(P \mid F) \neq P(P)$.
3 a $\frac{43}{179}$ **b** $\frac{98}{181}$
c No. P(Walks to school | in year 11) = $\frac{98}{181}$ is higher than P(Walks to school) = $\frac{141}{360}$.
4 a The following tree diagram describes drivers who are involved in a serious accident.

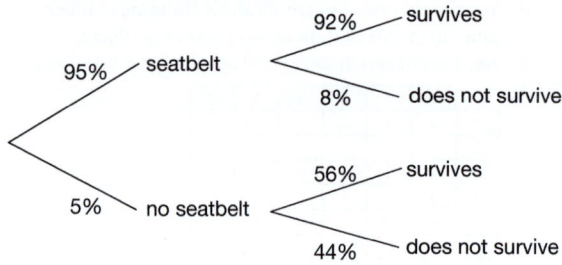

b P(No seatbelt and survives) = $0.05 \times 0.56 = 0.028$

5 a

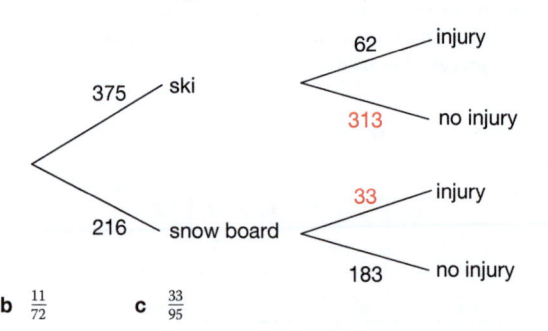

b $\frac{11}{72}$ **c** $\frac{33}{95}$

6 a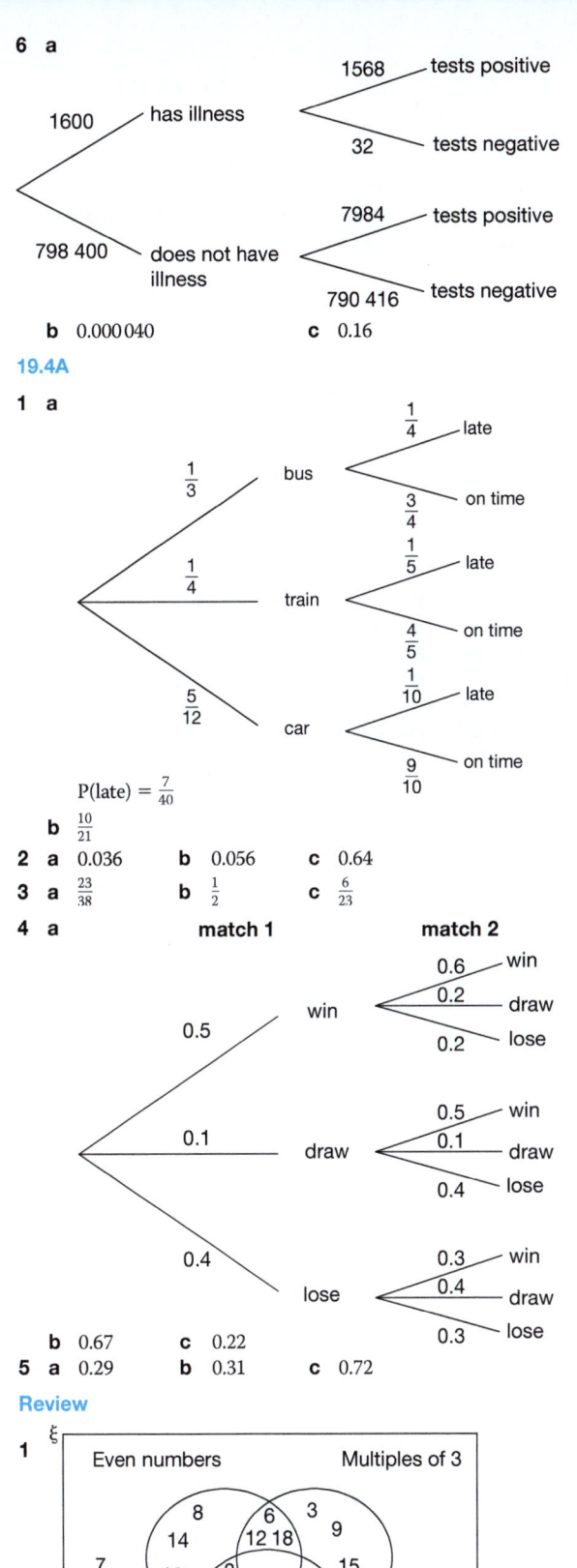

b 0.000040 c 0.16

19.4A

1 a

P(late) = $\frac{7}{40}$

b $\frac{10}{21}$
2 a 0.036 b 0.056 c 0.64
3 a $\frac{23}{38}$ b $\frac{1}{2}$ c $\frac{6}{23}$
4 a

b 0.67 c 0.22
5 a 0.29 b 0.31 c 0.72

Review

1

$\frac{4}{10} = \frac{2}{5}$

2 a 0.08 ii 0.1025
3 a

	Egg	Cheese	Tuna
White	W, E	W, C	W, T
Brown	B, E	B, C	B, T
Granary	G, E	G, C	G, T

b i $\frac{1}{3}$ ii $\frac{1}{9}$ iii $\frac{4}{9}$

4 a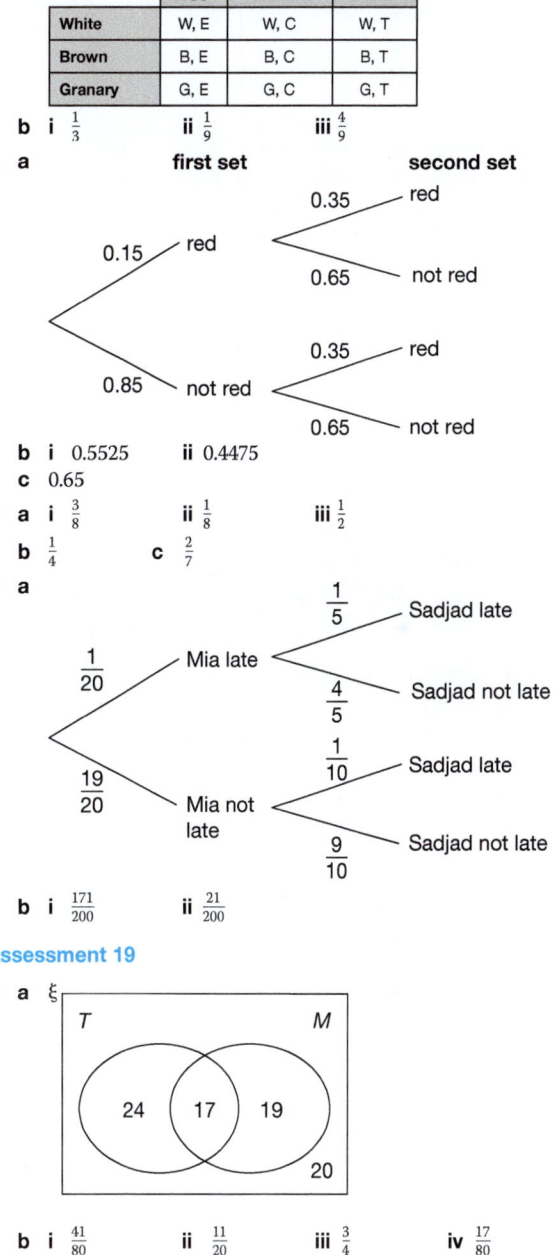

b i 0.5525 ii 0.4475
c 0.65
5 a i $\frac{3}{8}$ ii $\frac{1}{8}$ iii $\frac{1}{2}$
b $\frac{1}{4}$ c $\frac{2}{7}$
6 a

b i $\frac{171}{200}$ ii $\frac{21}{200}$

Assessment 19

1 a

b i $\frac{41}{80}$ ii $\frac{11}{20}$ iii $\frac{3}{4}$ iv $\frac{17}{80}$
c No, P(T) × P(M) = $\frac{41}{80} \times \frac{9}{20} = \frac{369}{1600} \neq \frac{17}{80}$ = P(T∩M).

2 a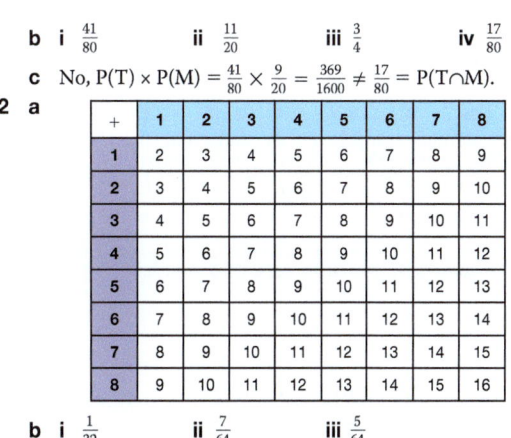

b i $\frac{1}{32}$ ii $\frac{7}{64}$ iii $\frac{5}{64}$
c 9
d All outcomes were assumed to be equally likely.

3 a 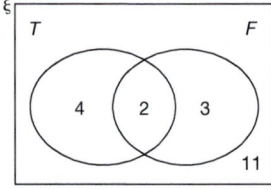   b $n = 16$

4 Tea 17, Coffee 15; Tea-Chocolate 12, Tea-Plain 5; Coffee-Chocolate 10, Coffee-Plain 5

5 a B $\frac{1}{4}$, A $\frac{3}{4}$; BB $\frac{1}{16}$, BA $\frac{3}{16}$, AB $\frac{3}{16}$, AA $\frac{9}{16}$
  b $\frac{1}{16}$ c $\frac{7}{16}$

6 a 0.1
  b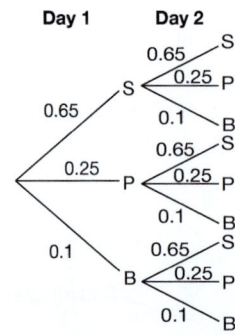

  c i 0.025   ii 0.4225
     iii 1875 (0.65 × .25 + .1 × .25)   iv 0.19

7 No, P(A) × P(R) = 0.07 ≠ P(A and R) = 0.15.
8 a P(H) + P(T) = 1, so P(H) = 1 − P(T) = 1 − p
  b $p(1-p) = \frac{6}{25} \Rightarrow 25p(1-p) = 6 \Rightarrow 25p - 25p^2 = 6, p = \frac{2}{5}$
  c P(T) < P(H), so p must be the smaller solution.
9 a

  b P(W|T) = $\frac{2}{10}$ ≠ P(W) = $\frac{61}{200}$
  c P(T|W) = $\frac{12}{61}$ d P(S|L) = $\frac{91}{139}$

Life skills 4
1 118
2 a 15.4 m   b 33.9°   c 244 m³   d 158 m²
   e The marquee comprised of a cylinder and cone is cheaper.
3 a $\vec{TB} = \begin{pmatrix} 200 \\ 50 \end{pmatrix}$ $\vec{TA} = \begin{pmatrix} -200 \\ -300 \end{pmatrix}$ $\vec{TR} = \begin{pmatrix} -100 \\ 100 \end{pmatrix}$
   b TB = 206 m, TA = 361 m, TR = 141 m   c 94.6°
4 0.412
5 1257 customers, 11 142 customers, 23rd month.

Chapter 20
Check in
1 a 10, 12   b 70, 64   c 16, 22
  d −2, −5   e 48, 96   f $\frac{1}{6}, \frac{1}{7}$
2 $4(n-1), 15-n, 2n+7, 2n^2, \frac{9}{n}+15$
3 Square numbers, the numbers give the areas of squares with integer sides.

20.1S
1 a −10, −7, −4, −1   b 2.5, 7.5, 12.5, 17.5
  c $1\frac{3}{4}, 2\frac{1}{4}, 2\frac{3}{4}, 3\frac{1}{4}$ d 1.2, 1.35, 1.5, 1.65
  e 12.5, 9.5, 6.5, 3.5   f 5, 1, −3, −7
  g 9, 7.5, 6, 4.5   h $1\frac{3}{4}, 1\frac{1}{4}, \frac{3}{4}, \frac{1}{4}$
2 a 13   b 10   c 23, 29   d 19, 27
  e 18   f 4, −2   g $1\frac{3}{4}, \frac{3}{4}$ h 1.1, 0.95
3 6, 9, 12
4 1, 6, 11
5 a 47   b 67   c 407   d 4007
6 a 3, 6, 9, 12, 15   b 6, 10, 14, 18, 22
  c 7, 9, 11, 13, 15   d 3, 7, 11, 15, 19
  e 2, 7, 12, 17, 22   f −4, −2, 0, 2, 4
  g −9, −6, −3, 0, 3   h 5.5, 6, 6.5, 7, 7.5
  i 2, 4.5, 7, 9.5, 12   j $1\frac{2}{3}, 1\frac{1}{3}, 1, \frac{2}{3}, \frac{1}{3}$
  k $\frac{2}{3}, 1\frac{1}{6}, 1\frac{2}{3}, 2\frac{1}{6}, 2\frac{2}{3}$
7 a −2, −4, −6, −8, −10   b 14, 13, 12, 11, 10
  c 8, 6, 4, 2, 0   d 20, 15, 10, 5, 0
  e 1.5, −4.5, −10.5, −16.5, −22.5
  f 3.5, 3, 2.5, 2, 1.5
8 a $2n+1$ b $3n+1$ c $3n+2$ d $6n-2$
  e $10n-3$ f $8n-6$ g $1.5n+0.5$ h $0.6n+0.8$
9 a $18-3n$ b $14-4n$ c $9-4n$
  d $1-3n$ e $1.25-0.75n$
10 a 38   b $4n-2$
   c 10th term = $4 \times 10 - 2 = 38$
11 No. Multiple explanations possible, for example
   If $5n − 3 = 75, 5n = 78, n = 15.6$
   The sequence is 2, 7, 12, 17, ... the units digit is always 2 or 7 so 75 will not be term.

20.1A
1 No – the 10th term will be 39 (*it is not double the 5th term*)
2 No – it's $4n + 1$
3 a Match each sequence with the correct 'term-to-term' rule and 'position-to-term' rule.

4, 1, −2, −5, …	Subtract 3	T(n) = 7 − 3n
6, 10, 14, 18, …	Add 4	T(n) = 4n + 2
3, 10, 17, 24, …	Add 7	T(n) = 7n − 4
6, 2, −2, −6, …	Subtract 4	T(n) = 10 − 4n

  b Students' own response
4 a $m = 2n + 1$ b 101   c 50th
5 Students' own response
6 29 post and 84 planks
7 57 squares
8 Yes – multiplying by 4 will always give an even number. Subtracting two from an even number will always give an even number.
9 Every number ending in 3 is 2 less than a number ending in 5. Any number ending in 5 is a multiple of 5. Therefore, any number ending in 3 is 2 less than a multiple of 5 and so can be represented by the formula $5n − 2$ for positive integer values of n.
10 a

Pattern	1	2	3	4
Squares	8	10	12	14

  The term-to-term rule is 'add 2' so the formula is $2n + a$ for some constant a. The first term is $2(1) + a = 8$ so $a = 6$.
  b There are always six squares in each pattern at the ends (three at each end). This relates to the '+6' part of the nth term. The pattern needs double the number of blue squares – top and bottom – which relates to the '$2n$' part of the nth term.

11 a $98 - 7n$ **b** $5n + 73$ **c** $5n - 474$ **d** $0.68n + 89.04$

*12 The two sequences might contain the same term in different positions.

20.2S

1 a 2, 3, 6, 11 **b** 5, 7, 11, 17 **c** 10, 11, 13, 16
  d 5, 8, 14, 23 **e** 10, 9, 6, 1 **f** 8, 6, 2, −4
2 a 31, 43 **b** 24, 34 **c** 23, 31
  d 25, 36 **e** 10, 15 **f** 24, 18
  g −32, −55
3 a 36 **b** 2 **c** 5 **d** 11, 29
  e 50, 76 **f** 15 **g** −1
4 a i 4, 7, 12, 19, 28 **ii** 3, 8, 15, 24, 35
    iii 0, 4, 10, 18, 28 **iv** 2, 8, 18, 32, 50
    v 6, 12, 22, 36, 54
  b i 103 **ii** 120 **iii** 108
    iv 200 **v** 204
5 a i 2, 5, 10, 17, 26 **ii** −2, 1, 6, 13, 22
    iii 4, 9, 16, 25, 36 **iv** −2, 2, 8, 16, 26
    v 3, 10, 21, 36, 55 **vi** 2.5, 5, 8.5, 13, 18.5
    vii 6, 14, 26, 42, 62 **viii** −1, 0, 3, 8, 15
  b i 2 **ii** 2 **iii** 2 **iv** 2
    v 4 **vi** 1 **vii** 4 **viii** 2
6 a $n^2 + 3$ **b** $n^2 - 4$ **c** $2n^2$ **d** $n^2 + n$
  e $n^2 + 4n$ **f** $n^2 + 2n$ **g** $2n^2 + 2n + 2$
  h $4n - n^2$
7 a $10 - n^2$ **b** $20 - 2n^2$
  c $n - n^2$ **d** $n + 10 - n^2$
8 a Estimates between 30 and 40
  b $n^2 + n$
  c If $n^2 + n \geq 1000$ then $n^2 + n - 1000 \geq 0$
    Smallest value of $n = 32$.
9 No – if $n^2 + 3 = 150$ then $n^2 = 147$. 147 is not a square number so 150 is not in the sequence.
10 Yes: $3n^2 - n = n(3n - 1)$
  If n is even: $3n - 1$ is odd so and even × odd = even
  If n is odd: $3n - 1$ is even and odd × even = even

20.2A

1 No – the 10th term is not usually double the 5th term (or the 10th term = 103)
2 a No – as the second difference = 2, the nth term will start with 'n^2'
  b $n^2 - n + 5$
3 $T(n) = n^2 - n$: 0, 2, 6, 12; $T(n) = n^2 - 1$: 0, 3, 8, 15; $T(n) = n^2 + 2$: 3, 6, 11, 18; $T(n) = n(n + 1)$: 2, 6, 12, 20; $T(n) = 4 - n^2$: 3, 0, −5, −12
4 Many answers possible including $n^2 + n + 1$, $2n^2 - 2n + 3$, $\frac{1}{2}n^2 + 2\frac{1}{2}n$, $3n^2 - 5n + 5$, $4n^2 - 8n + 7$
5 a

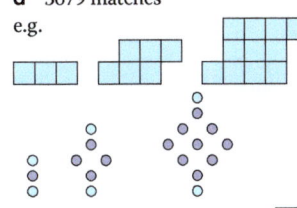

  b $m = \frac{3}{2}n^2 - \frac{3}{2}n + 4$ **c** 10th pattern
  d 3679 matches
6 e.g.

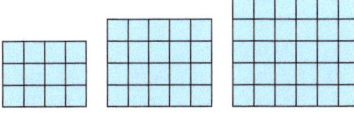

7 a

b

8 316
9 a 24 moves **b** 15 moves, 8 moves.
  c

Number of pairs	1	2	3	4
Number of moves	3	8	15	24

  d Number of moves = $n^2 + 2n$ where n = number of pairs.
    number of moves for 50 pairs = 2600
10 a 1 3 6 10 15
    1st Diff 2 3 4 5
    2nd Diff 1 1 1
  The second difference is 1 so the first part of the nth term is $\frac{1}{2}n^2$

$\frac{1}{2}n^2$	$\frac{1}{2}$	2	$4\frac{1}{2}$	8
Sequence	1	3	6	10
Difference	$\frac{1}{2}$	1	$1\frac{1}{2}$	2

  The nth term for the difference is $\frac{1}{2}n$
  So the overall nth term is: $\frac{1}{2}n^2 + \frac{1}{2}n = \frac{1}{2}n(n + 1)$
  b Consider each term of the triangular sequence being doubled and rearranged as rectangle:

  x   x xx   xxxx   xxxxx
           xx x   xxxx   xxxxx
                     xxxx   xxxxx
                              xxxxx

  The nth term of the sequence can be arranged as a rectangle with dimensions $n \times (n + 1)$.
  But this is twice the value of the triangular number and so:
  nth triangular number = $\frac{1}{2}n(n + 1)$
*11 a n^3 **b** $2n^3$ **c** $n^3 - n$ **d** $n^3 - 2n^2 + 3n - 4$

20.3S

1 a 1, 3, 6, 10, 15, 21, 28, 36, 45, 55
  b 1, 4, 9, 16, 25, 36, 49, 64, 81, 100
  c 1, 8, 27, 64, 125, 216, 343, 512, 729, 1000
2 a 21 + 3 + 6 **b** 28 + 3 **c** 28 + 3 + 1
3 a arithmetic **b** arithmetic
  c another type **d** quadratic
  e another type **f** arithmetic
  g another type **h** another type
  i quadratic **j** arithmetic
  k another type **l** another type
  m another type **n** another type
  o arithmetic **p** another type
4 Yes, as they have a constant second difference of 1. They can be represented by the quadratic expression $\frac{1}{2}n^2 + \frac{1}{2}n$.
5 a 6, 8, 10 **b** 8, 16, 32
6 a 3, 6, 12, 24 **b** 10, 50, 250, 1250
  c 3, 1.5, 0.75, 0.375 **d** 2, −6, 18, −54
  e $\frac{1}{2}, \frac{1}{4}, \frac{1}{8}, \frac{1}{16}$ **f** −3, 6, −12, 24
  g 4, $4\sqrt{3}$, 12, $12\sqrt{3}$ **h** $\sqrt{3}$, 3, $3\sqrt{3}$, 9
  i $2\sqrt{5}$, 10, $10\sqrt{5}$, 50
7 a 15, arithmetic sequence with difference 5.
  b 17, quadratic sequence with second difference 2.
8 Several answers possible including: $1^2 = 1^3 = 1$, $4^3 = 8^2 = 64$, $9^3 = 27^2 = 729$.
9 Yes – using this rule, T(1) = 1, T(2) = 3), T(3) = 6,... which is the triangular number sequence.
10 Both are correct, the square numbers form a quadratic sequence.

11 a i $\frac{1}{2}, \frac{2}{3}, \frac{3}{4}, \frac{4}{5}, \frac{5}{6}$ **ii** $\frac{1}{2}, \frac{2}{5}, \frac{3}{10}, \frac{4}{17}, \frac{5}{26}$
iii $1, \frac{1}{4}, \frac{1}{9}, \frac{1}{16}, \frac{1}{25}$ **iv** $\frac{1}{2}, \frac{4}{3}, \frac{9}{4}, \frac{16}{5}, \frac{25}{6}$
v 0.9, 0.81, 0.729, 0.6561, 0.59049
vi 1.1, 1.21, 1.331, 1.4641, 1.61051
b i The sequence approaches 1.
ii The terms get smaller.
iii The terms get smaller.
iv The terms get larger.
v The terms get smaller.
vi The terms get larger.

20.3A
1 Many answers possible including:
a 4, 7, 10, 13, 16, 19 **b** 4, 6, 10, 16, 24, 34
2 a 32 **b** 7 **c** −19.5 **d** −1, 7
e 768 **f** 1 **g** $9\sqrt{3}$ **h** 16
3 a Yes. This method creates the pattern shown below which represents the triangle numbers.
```
x x   x
  xx  xx
      xxx
```
b Pentagonal numbers: 1, 5, 12, 22 …

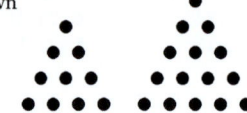

Hexagonal numbers: 1, 6, 15, 28 …

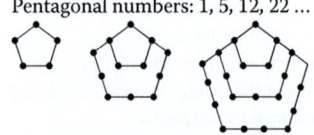

Tetrahedral numbers: 1, 4, 10, 20, …

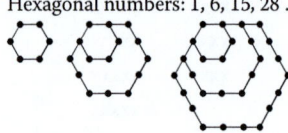

4 a Option 3: By the end of the February: $2.7 million
b Option 3: By the end of the September: $10.7 million
5 Assuming $a \neq 0$
For $r > 1$ or $r < -1$ terms diverge.
For $r = 1$ terms remain constant at a.
For $-1 < r < 1$ terms converge to 0.
For $r = -1$ terms oscillate between a and $-a$.

Review
1 a i 44, 53, 62 **ii** 19, 6, −7 **iii** 9.2, 10.8, 12.4
b i +9 **ii** −13 **iii** +1.6
2 a 73 **b** 141
3 a $6n - 5$ **b** $7n + 8$
c $63 - 12n$ **d** $-1.5n - 5$
4 a 175 **b** 340
5 a $n^2 + 3$ **b** $3n^2 - n$ **c** $2n^2 + 3n - 1$
6 a Another type **b** Quadratic **c** Another type
d Arithmetic **e** Quadratic **f** Another type
7 a i 15, 21 **ii** $\frac{1}{25}, \frac{1}{36}$
b i $\frac{1}{2}n(n+1)$ **ii** $\frac{1}{n^2}$
8 a 32, 64, 128 **b** 2^n
9 a $2\sqrt{2}n - \sqrt{2}; 19\sqrt{2}$ **b** $\frac{n}{2n+1}; \frac{10}{21}$

Assessment 20
1 a E **b** D **c** H **d** F
e G **f** A **g** C **h** I
i B
2 a +10 **b** $10n - 9$
3 a i $2n + 8$ **ii** $2n + 2$
b 54 **c** 42
4 a Each term is the sum of the two previous terms.
b 55, 89 **c** $144 = 12^2$ **d** Fibonacci

5 a Correct. $2 \times 10 + 7 = 27$
b Incorrect. $6 \times 1 - 5 = 1, 6 \times 2 - 5 = 7, 6 \times 3 - 5 = 13$
c Incorrect. $13 - 3 \times 100 = -287$
d Incorrect. $10^2 - 10 = 100 - 10 = 90$
e Correct. $15 - 3 \times 100^2 = 15 - 30000 = -29985$
6 No. The ratio is decreasing, 1 : 2, 1 : 4, 1 : 6
7 a Dawn

b $D = \frac{t(t+1)}{2}$
c i 55 dots **ii** 1275 dots **iii** 5050 dots
8 a 9, 14, 20 **b** $D = \frac{p(p-3)}{2}$
c i 35 **ii** 1175 **iii** 4850
9 a False. −1 **b** False. $15 \times 16 \div 2 = 120$.
c False 210 **d** True. All terms would be the same.
10 a −2 **b** $\frac{20}{9}$ **c** $a + d, a + 2d, a + 9d$

Chapter 21
Check in
1 a i £49.50 **ii** 66 mm
iii 52.8 km **iv** 4 hours 24 minutes
b i 4.5 miles **ii** 43.5 minutes
iii 10.5 kg **iv** €46.50
2 a 40 mph **b** 42.5 mph **c** 54 mph **d** 32 mph
3 a 600 cm³ **b** 351.9 cm³ **c** 45.92 cm³
4 a 500 cm² **b** 276.5 cm² **c** 107 cm²

21.1S
1 a 7.7 m/s **b** 7.1 m/s **c** 6.8 m/s **d** 5.1 m/s
2 a 160 km **b** 161 miles **c** 54 m **d** 288 miles
3 a 3 hours **b** 4 hours **c** 20 minutes **d** 15 minutes
4 1.425 g/cm³
5 a 9.46 kg **b** 6.15 litres
6 a 5704 kg **b** 3400 kg **c** 61 824 kg **d** 125 cm³
e 24.7 cm³ **f** 2000 cm³
7 $2.10/m
8 $11.95/hour
9 a 2.5 litres/s **b** 1.6 litres/s
10 1200 litres
11 a 2.4 units/hour **b** 57.6 units
12 a 500 km **b** 20 litres **c** 12.5 km/litre
13 a $1.5 per £ **b** $187.5 **c** £80

21.1A
1 89.25 mph
2 300 kg
3 8.5 g/cm³
4 32 mph
5 525 km, 2 hours 12 minutes 30 seconds, 37.3 kmph, 215 km, 37.14 kmph.
6 a 5 g/cm³ **b** 87.88 g
7 a 5.96 g/cm³ **b** 105 blocks
8 a 10 500 g/cm³ **b** 4500 g/cm³ **c** 2700 kg/m³
9 a £89.25 **b** 15 hours
10 $42.24
11 4 hours

21.2S
1 a 50 mm **b** 80 mm **c** 150 mm **d** 67 mm
e 193 mm **f** 45 mm **g** 43 mm **h** 106 mm
i 800 mm **j** 1000 mm
2 a 6 cm **b** 8.5 cm **c** 24 cm **d** 6.3 cm
e 0.4 cm **f** 400 cm **g** 1000 cm **h** 350 cm
i 160 cm **j** 163 cm

3 a 4 m   b 4.5 m   c 4.75 m   d 4.7 m
   e 0.5 m   f 1000 m   g 4000 m   h 500 m
   i 3500 m   j 18 000 m
4 a 8 m²   b 80 000 cm²
5 a 24 m²   b 240 000 m²
6 a 400 mm²   b 730 mm²   c 1090 mm²   d 250 mm²
   e 40 000 mm²
7 a 6 cm²   b 12 cm²   c 8.5 cm²   d 65 cm²
   e 100 cm²
8 a 4 m²   b 8.5 m²   c 100 m²   d 12.5 m²
   e 0.5 m²
9 a 50 000 cm²   b 100 000 cm²   c 65 000 cm²
   d 77 500 cm²   e 6000 cm²
10 a 232.5 cm²   b 193.75 cm³

21.2A

1 a 2268 cm³   b 240.57 litres
2 244 cm²
3 125 cm³
4 a Their sides are all the same length, so any scale factor will be constant on all dimensions.
   b The scale factor on one dimension could be different from that on another.
   c i 1 : 49   ii 1 : 343
5 a 2 : 3   b 34 cm²   c 81 cm³
6 Deal A
7 Volume of original cone : volume of removed cone = $3^3 : 2^3$ so volume removed cone : volume frustum = 27 − 8 : 8.
8 No, $\left(\frac{260.3}{1017.9}\right)^{\frac{1}{2}} \neq \left(\frac{188.5}{1352.2}\right)^{\frac{1}{3}}$.

21.3S

1 a false   b true   c false   d true   e true
2 a 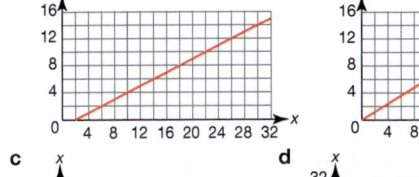 b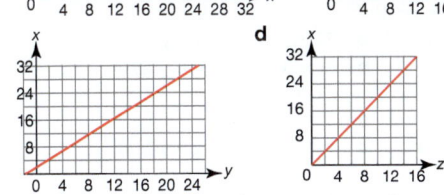

   Two variables that are in direct proportion have a straight-line graph passing through (0, 0).
3 a $w = kl$ b 2.48 kg/m   c 7.192 kg
4 a doubled   b halved
   c multiplied by 6   d divided by 10
   e multiplied by 0.7
5 a halved   b doubled
   c divided by 6   d multiplied by 10   e divided by 0.7
6 a 10 people   b 25 hours   c 40 hours
7 a 2   b 8   c 0.4
   d 16   e 0.16   f 32
8 $y = \frac{500}{w}$
9 800
10 4
11 a $d = kt$ b 48   c 120 miles
   d 3 hours 20 minutes

21.3A

1 Pack A: 1.2p per pin > Pack B: 1.15p per pin. Pack B is better value.
2 Regular, 0.7 cents/sheet < 0.77 cents/sheet.
3 a $R = 100s^2$ b 64   c 1.41

4 $A = \pi r^2$ so the area is directly proportional to the square of the radius, with constant of proportionality π.
   Students' answers e.g. $C = 2\pi r$, the circumference of a circle is directly proportional to the radius, with constant of proportionality 2π.
5 615 cm² (3 sf)
6 a $F = \frac{100}{d^2}$; 11.1 N, 6.25 N
   b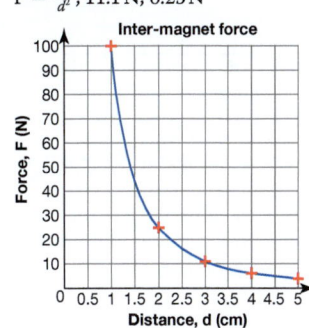

7 a green   b red   c black   d blue
8 1 hour 2 minutes 30 seconds
9 a $y = \frac{10}{\sqrt{x}}$ b i 1   ii $\frac{25}{36}$
10 a $P = \frac{32}{t^3}$ b i $1\frac{5}{27}$ ii 4

21.4S

1 a 1.25   b 0.75   c 1.025   d 0.975
2 a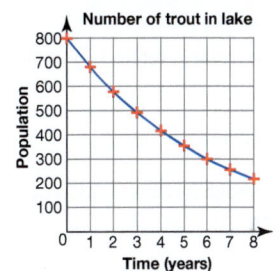
   b i Trout = 800×0.85^n
3 a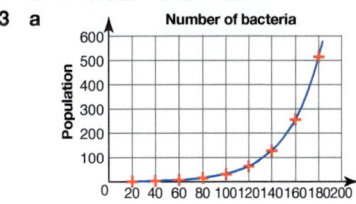
   b i 181   ii 362
   c The population has doubled.
4 a A number is increased by 50% by multiplying by 1.5, this increase occurs n times in n hours from a starting value of 200.
   b 400×1.35^n c 7 hours
5 a $16\,000 \times 0.85^n$ b £8352.10   c 9 years
6 a 56 859   b 10 years
7 a

End of year	Amount in the account ($)
2	2163.20
3	2249.73
4	2339.72
5	2433.31
6	2530.64
7	2631.86
8	2737.14
9	2846.62
10	2960.49

   b 48%
   c i Multiplier for percentage increase = $1 + \frac{r}{100}$. Number of times increase occurs = n. Initial amount = P.

8 7
9 a, b i 1.25%

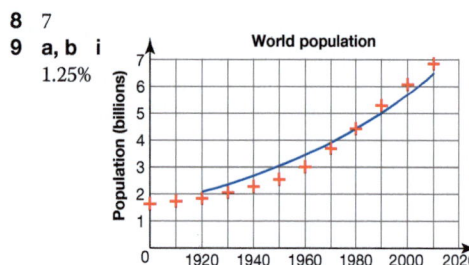

ii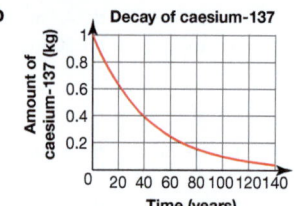

c The y-values would be multiplied by 1000.

21.4A

1 a £66.33 b £235.27 c £723.40
2 She is incorrect. The yearly interest on the half-yearly saver account is 4.04% > 4%.
3 a The number of vehicles per day when the road opens is 2400. The number of vehicles increases by 8% each year.
   b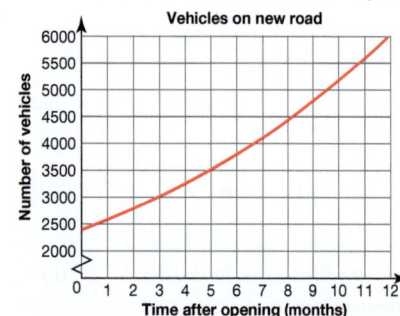
   c This implies the annual increase in vehicles continues to rise. In reality it is likely to level off at a certain value.
4 a Each year the number of trees is 70% of the previous number, × 0.7, plus the 60 new trees, + 60.
   b 208 trees
   c Smooth curve passing through (0, 250), (1, 235), (2, 225), (3, 217), (4, 212), (5, 208).
5 a 6 months   b 7.07%
6 €8934.34
7 a Liam calculated the simple interest. He should multiply by 1.045^6 to work out the compound interest.
   b Interest = $\$P(1.045^n - 1)$ c $6400
8 3.085%
9 a i 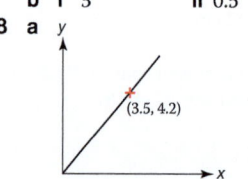   ii 0.8°C/minute
   b i Yes, the formula gives values close to Tanya's data.
      ii 20 °C is room temperature. The values are approaching this limit. 65 °C is the difference between the initial temperature of the coffee and room temperature. 0.97^t refers to the fact that the temperature decreases by 3% each minute.

*10 a

Number of years	Amount left (kg)
30 = 30 × 1	$\frac{1}{2}$
60 = 30 × 2	$\frac{1^2}{2}$
90 = 30 × 3	$\frac{1^3}{2}$
$n = 30 \times \frac{n}{30}$	$\frac{1^{\frac{n}{30}}}{2}$

Review

1 £0.21 or 21 p per 100 g
2 a i 364.5 kcal   ii 14.2 g   iii 1.2 g
   b 190 g
3 a 6700 mm   b 44 min, 11 s   c 0.45 litres
   d 25.2 km/h   e 0.067 253 m²
4 a 526.5 cm³   b 40 cm²
5 a $y = 1.2x$
   b i 9.48   ii 9.67 (3 sf)
6 a $y = 2x^2$
   b i 242   ii 5
7 a $Y = \frac{66}{X}$
   b i 3   ii 0.5
8 a 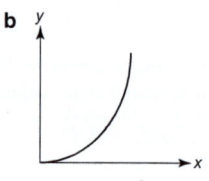 b

9 a 13 m/h (allow 12.5 – 13.5)   b 11 m/h
10 a i $1537.50   ii $1697.11
   b $V = 1500 \times 1.025^t$

Assessment 21

1 8 min 20 s
2 a 39.37 in   b $9.75   c 52 min 30 s
3 The wombat, Bolt's speed = 36.73 km/h.
4 a 7.28 g/cm³   b 23.81 cm³   c 103 g
5 a 45 tins   b Yes, 100 ÷ $\left(\frac{3}{2}\right) = 67$ days.
   c No, 30 ÷ $\left(\frac{9}{2}\right) = 7$ days.
6 a 157.5 litres   b 379 cm
7 a 69.12 kg   b $x = 24$ cm   c $y = 10$ cm
8 0.21 km²
9 a £10   b 150 000 cm³
10 a $T = 4e$ b 4.25 cm   c 18 N
11 a $A = 3.142x^2$ b Yes, 3.142 ≈ π and x = radius.
   c $A = 314.2$ d $x = 9$
12 a (Rent/Distance (km) from city centre graph, Flat 5 marked)
    b Flat 5, it lies furthest from the curve.   c 3.5 km
13 a $P = \frac{10}{\sqrt{t}}$ b 1

14 a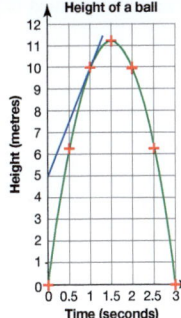

b 5 m/s

15 a $27 000, 5%   **b** $9546

Chapter 22

Check in

1 **a** -0.5 **b** $y = -0.5x + 3$
2 **a** $2x^2 - x - 15$ **b** $15x^5 - 20x^3$
3 **a** $x^3 + 2x$ **b** $x - 7$
4 **a** $t = 2.5$ **b** $x = 0, 2$

22.1S

1 **a** 13   **b** $\frac{2.5^3 - 1.5^3}{2.5 - 1.5} = 12.25$ **c** $y = 12x - 16$
2 **a** 4
   b $\frac{(2.5^3 - 2.5 \times 9) - (1.5^3 - 1.5 \times 9)}{2.5 - 1.5} = 3.25$ **c** $y = 3x - 16$
3 **a, b**

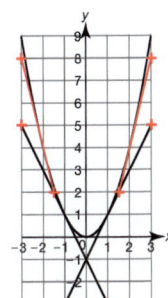

   c $-4, -2, 0, 2, 4$ **d** Gradient $= 2p$
4 **a i** 1   **ii** 0
   b $\frac{2}{3}$
   c The section with gradient 1 is twice as long as the section with gradient 0.
5 **a, d**

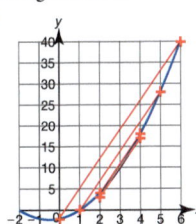

   b i 7   **ii** 7   **iii** 7
   c 7   **d** All the chords and the tangent are parallel.
6 The gradient at P is $2p$ and the equation of the tangent is $y = 2px - p^2$. This crosses the y-axis at $(0, -p^2)$, as required.
7 **a** $\frac{(4 + h)^2 - 10 - (4 - h)^2 + 10}{2h} = \frac{16h}{2h} = 8$
   b The chords which tend closer and closer to the tangent at $(4, 6)$ all have gradient precisely 8. The tangent itself therefore has gradient precisely 8.

22.1A

1 **a** -10 m/s   **b** The ball is falling 10 m each second.
2 **a** No profit is made if the price is zero or if the price is too high for potential customers.
   b At this point the profit is maximised.
   c 0.6 pounds (i.e. 60p) gives the greatest profit. At this point the profit is maximised.
3 **a** $-4, -1, -0.5$ **b** °C per minute.
   c Initially the temperature drops at approximately 4°C per minute. The temperature continues to fall at a gradually reducing rate until it reaches approximately 10°C.
4 **a**

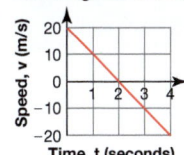

   b Acceleration   **c** -10 m/s^2
5 **a** Approximately 375 m/s   **b** Approximately 750 m/s

22.2S

1 **a** 4   **b** Students' own answers
   c 8   **d** $y' = 2x$
2 **a** 3   **b** 0
   c 12   **d** Students' own answers   **e** $y' = 3x^2$
3 e.g. For all values of c, $y = c$ is a horizontal line which has no gradient
4 **a i** 4   **ii** -2 **b i** 4   **ii** -2
   c i 12   **ii** 3   **d i** 12   **ii** 3
5 e.g. $y = x$ is a straight line with a gradient of 1
6 **a i** 3   **ii** -5 **b i** 1   **ii** -7
   c i 3   **ii** -5 **d i** 2   **ii** 26
7 **a** 28   **b** 36   **c** 57   **d** 23
8 **a** 4   **b** Students' own answers   **c** $y' = 4x^3$

22.2A

1 **a** $36x^2$ **b** $12x^5$ **c** $12x^3 + 7$ **d** $45x^2 - 8$
   e $30x^5 - 12x^3$ **f** $36x^2 + 10x$ **g** $3x^2 + 2x + 1$
   h $15x^2 + 4x - 9$
2 **a** $32x^3 + 8$ **b** $18x^5 - 8x$ **c** $18x^5 + 35x^4$
   d $120x^3 - 36x$ **e** $2x + 7$ **f** $2x + 1$
   g $2x - 8$ **h** $12x + 5$
3 **a** $14x^5$ **b** $\frac{(5x^3)}{3}$ **c** $x^2 + 0.5$ **d** $20x - 3$
   e $28x^3 - 4x$ **f** $\frac{(16x + 7)}{3}$ **g** 1   **h** 1
4 **a i** 0,   **ii** 160,   **iii** 10   **b i** 3,   **ii** 163,   **iii** 13
   c i 0,   **ii** 144,   **iii** 0   **d i** 1,   **ii** 5,   **iii** -1
   e i 0,   **ii** 84,   **iii** -24 **f i** 1,   **ii** 1,   **iii** 1
5 **a** $\frac{dP}{ds} = 900 - 4s$
   b i 80   **ii** 4   **iii** -20
   c e.g. There is a maximum between ii and iii.
6 **a** $\frac{dh}{dt} = -9.8t$
   b i -9.8 m/s   **ii** -19.6 m/s
   c e.g. The lead is falling towards the ground
7 Students' own answers, but look for comments connecting the gradient function's x-intercepts with the turning points of the function.

22.3S

1 **a i** $y = -x$ **ii** $y = 7x + 4$ **iii** $y = \frac{3}{4}x - \frac{9}{4}$
2 **a** $y = 4x - 3$ **b** $\frac{dy}{dx} = 6x^2 - 2x$
   d $y = \frac{1}{2}x - \frac{1}{4}$
3 **a** $y = -3x - 12$ **b** $\frac{dy}{dx} = 2x - 7$ **d** $y = 9x - 72$
4 $a = 1, b = 5$
5 $a = 2, b = -6$
6 $b = 4, c = 6$

22.3A

1 **a** $y = -x - 9$ **b** $y = -\frac{1}{3}x - \frac{26}{3}$
   c $y = \frac{1}{7}x + \frac{1}{7}$ **d** $y = -\frac{1}{5}x - 5$
   e $y = -\frac{1}{7}x + \frac{6}{7}$ **f** $y = \frac{1}{5}x - 6$

2 a $y = \frac{1}{2}x - 6$ b $y = -x - 6$ c $y = -\frac{1}{10}x - \frac{9}{5}$
  d $y = -x - 6$ e $y = \frac{4}{5}x + \frac{189}{40}$ f $y = -\frac{1}{10}x - \frac{51}{5}$
3 a $y = \frac{1}{2}x - \frac{5}{2}$ b $y = -\frac{1}{20}x + \frac{41}{10}$
  c $y = -0.466x - 0.641$ (3 s.f.)   d $y = \frac{1}{20}x + \frac{41}{10}$
  e $x = 0$ f $y = -\frac{1}{5}x - \frac{39}{80}$
4 a $y = 32x - 80$ b $y = -x + 5$ c $y = -\frac{1}{5}x + 1$
  d $y = 32x - 83$ e $y = -\frac{1}{5}x + 5$
5 $(-3, -1)$
6 $a = -2$ and $b = 9$
7 $a = 4$ and $b = 4$
8 $\left(-\frac{14}{3}, \frac{1}{6}\right)$

22.4S

1 a i $(2, -4)$
    ii Minimum
    iii
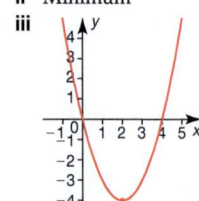

  b i $(1.5, -6.25)$
    ii Minimum
    iii
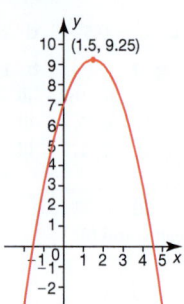

  c i $(1.5, 9.25)$
    ii Maximum
    iii
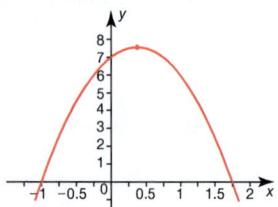

  d i $(0.375, 7.5625)$
    ii Maximum
    iii
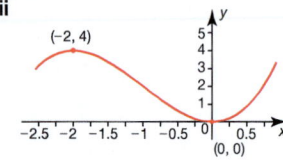

2 a i and ii Maximum at $(-2, 4)$ and minimum at $(0, 0)$
    iii

  b i and ii Maximum at $(-1, 5)$ and minimum at $(0, 4)$
    iii
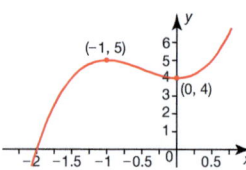

  c i and ii Point of inflection at $(0, 8)$
    iii
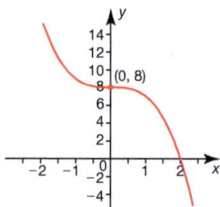

  d i and ii Maximum at $\left(-\frac{2}{3}, \frac{112}{27}\right)$ and minimum at $(0, 4)$
    iii
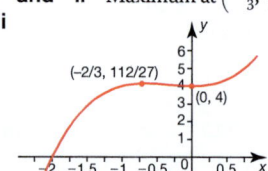

3 a $(-3, 0)$, $(0, 0)$ and $(3, 0)$
  b $(-\sqrt{3}, 6\sqrt{3})$ and $(\sqrt{3}, -6\sqrt{3})$
4 a and b
  Minimum at $(-1, -0.5)$, maximum at $(0, 6)$, minimum at $(6, -858)$
5 a $a = 4$ and $b = 5$
  b and
  c Point of inflection at $(-1, 3)$, minimum at $(1.25, -5.543)$
6 a $\frac{dy}{dx} = 3x^2 + 3$
  b The equation $3x^2 + 3 = 0$ has no solutions
7 a and b
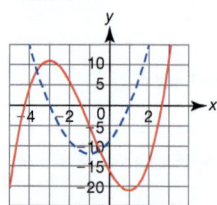

  c e.g.
  When $\frac{dy}{dx}$ is negative, y is decreasing (and vice versa). The graph of $\frac{dy}{dx}$ crosses the x-axis where y has stationary points.

22.4A

1 a $p = -48$ and $q = 9$ b 'Minimum' and correct reasoning
  c $(2, -55)$
2 a $b = -\frac{3}{4}$ and $c = \frac{17}{8}$
  b $\left(-\frac{1}{2}, \frac{19}{8}\right)$.
  c $\left(-\frac{1}{2}, \frac{19}{8}\right)$ is a maximum and $\left(\frac{1}{2}, \frac{15}{8}\right)$ is a minimum
3 a $a = -5$, $b = 8$ and $c = 4$
  b $\left(\frac{4}{3}, \frac{220}{27}\right)$
  c
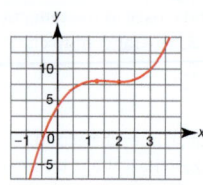

4 a $a = 4, b = 3$ and $c = 2$
  b $\left(-\frac{1}{3}, \frac{40}{27}\right)$
  c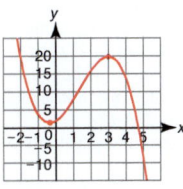

5 a $(-1.5, 3.5)$
  b Student verification
6 a $\frac{dy}{dx} = 10x - 1200$
  b Solving $10x - 1200 = 0$ will give the x-value of the minimum point, but is more efficient to realise that $x = 120$ by inspection of the completed square form.
7 a $\frac{dp}{ds} = 800 - 6s$
  b £133.33
  c £28 333.33
8 a $\frac{40}{49}$ seconds
  b $\frac{258}{49}$ metres
  c 6.93 metres
*9 $V = (30 - 2x)(20 - 2x)x = 4x^3 - 100x^2 + 600x$
  $\frac{dv}{dx} = 12x^2 - 200x + 600$
  Only one solution to $12x^2 - 200x + 600 = 0$ makes sense in the context. Therefore $x = 3.92$ cm to 2 decimal places.

Review

1 a 11 **b** 10.25
2 a 3 **b** 12
3 a 10 **b** 0
4 a 40 **b** 53
5 a $9x^2 + 3$ **b** $2x + 6$
  c $1 - 6x^2$ **d** 1
6 a $y = 9x - 7$ **b** $y = -3x - 3$
7 a $\frac{dy}{dt} = 4t + 3$ **b i** 7 **ii** 23
  c speed (m/s)
8 a i $(0, -5)$ **ii** minimum
  b i $(5, -1)$ **ii** minimum
  c i $(-1.5, 12.25)$ **ii** maximum
  d i $(-2, 28)$ **ii** maximum; $(4, -80)$, minimum
9 a $t = 1.63$s **b** 13.1m
  c 1.3 m/s

Revision 4

1 $BC = 5.28$m (3 sf)
2 7.14 km
3 161 cm
4 8.63°

5 a 17.0 **b** 17.9°
6 a 75 m **b** 36.9°
  c $BO = 180$m (3 sf), $PB = 195$m (3 sf) **d** 22.6°
7 $\overrightarrow{RQ} = -2\mathbf{b} + \mathbf{a}$, $\overrightarrow{NM} = -\mathbf{b} + \frac{1}{2}\mathbf{a}$, $\overrightarrow{NM} = \frac{1}{2}\overrightarrow{RQ}$
8 a Yes. $1^2 + 1 = 2, 2^2 + 2 = 6, 3^2 + 3 = 12, 4^2 + 4 = 20$
  b 2550, 10 100 **c** $2n + 2$ **d** 102, 202
9 a

3	4	5
6	10	15
3	6	10
9	16	25

  b i 1275 **ii** 5050
  c r^2 **d** $\frac{r(r-1)}{2}$
  e $\frac{2(2-1)}{2} = 1$ **f** $\frac{60(60-1)}{2} = 1770$
10 a Rotation through 90° anticlockwise (or 270° clockwise) about (0, 0)
  b $\begin{pmatrix} 1 & 0 \\ 0 & -1 \end{pmatrix}$
  c i $\begin{pmatrix} 0 & -1 \\ -1 & 0 \end{pmatrix}$ **ii** Reflection in $y = -x$
11 a 6 **b** Liddi 27 p, Addle 27.5 p
12 a 40 mph **b** 1 hr 26 min
  c 8 minutes **d** 10:48 am 40 miles from Ventnor
13 a 7 g/cm³ **b** 5.48 kg (3 sf)
  c 2.86 cm³ (3 sf)
14 a i $m \propto \frac{1}{d^3}$ **ii** $T \propto \sqrt{l}$ **iii** $S \propto r^2 - 3$
  b i 75 **ii** 4.84
15 1.125×10^5 Nm^{-2}
16 a $P = 1 - 0.28 - 0.35 = 0.37$
  b

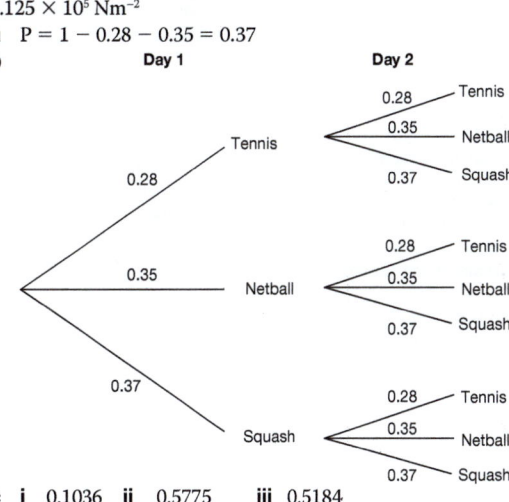

  c i 0.1036 **ii** 0.5775 **iii** 0.5184

Index

A
abundant numbers 261
acceleration 286, 292
accuracy 170–1
    implied accuracy 180
    measures and accuracy 180–3
acute angles 44
addition 8–11, 384
algebra 20–1
    algebraic fractions 34–7
    assessment 40–1
    equivalences in algebra 116–19
    expanding and factorising 30–3, 120–3
    indices 26–9
    review 39
    simplifying expressions 22–5
    summary 38
alternate segments 224
analysis of derivatives 468
angles 42–3
    acute angles 44
    alternate angles 44
    angle bisectors 228, 230
    angles and lines 44–7
    assessment 66–7
    congruence and similarity 56–9
    corresponding angles 44
    exterior angles 48
    interior angles 44, 48
    isosceles trapezium 54
    measuring lengths and angles 132–5
    obtuse angles 44
    polygon angles 60–3
    reflex angles 44
    review 65
    right angles 44
    right-angled triangles 366, 370
    subtended angles 224
    summary 64
    supplementary angles 44
    symmetry 52–5
    three-figure bearing 46
    triangles and quadrilaterals 48–51
approximation 172–5, 268
    approximate solutions 202–5
arcs 220, 222
area 446
    area scale factors 56
    formulae 484
arithmetic progressions 432
assumptions 104
averages 74–7, 324–7
axis, horizontal 336

B
bar charts 70, 78
    bar-line charts 70
    dual bar charts 73
bases 26
bearings 132
    three-figure bearing 46
BIDMAS 12
binomials 120
bisectors 224, 228, 230
box plots 328–31
business launch party 420
    forecasting 421
    future growth 421
    marketing slogan 421
    marquees 420
    number of guests 420
business plans 104–5
    business loan 105
    five year repayment formula 105
    market research 104
    projected revenue 104
    shares in the business 105
    survey 105
business start-up 236–7
    calling potential suppliers 237
    choosing tables 237
    launch party 420–1
    location 236
    number of tables 237
    price list 320
    repeat business 321
    restaurant logo 236
    restaurant maintenance 321
    restaurant recipes 320
    safety regulations 237
    stock keeping 321
    supplier meetings 320

C
calculations 2–3, 344–5
    adding and subtracting 8–11
    assessment 18–19, 360–1
    calculating with roots and indices 346–9
    calculations with fractions 92–5
    exact calculations 350–3
    multiplying and dividing 12–15
    place value and rounding 4–7
    review 17, 359
    standard form 354–7
    summary 16, 358
calculator methods 176–9
    bracket keys 176
    order of operations 176
capacity 180, 308
centre of enlargement 140
centre of rotation 140
circles 214–15, 216–19, 220–3
    assessment 234–5
    circle theorems 224–7
    circumference 216
    diameter 216
    formulae 484
    perimeter 216
    radius 216
    review 233
    summary 232
circumference 216
class intervals 78, 324
coefficients 22
collinear points 382, 384
column vectors 382
common denominator 92
common factors 30, 32, 88
compensation 8
complements 400, 402
completing the square 194
composite functions 112
compound interest 454
    formulae 484
compound units 442–5
    best buy problems 442
    rate of flow 442
    rate of pay 442
conditional probability 414–17
congruence 56–9
    congruent shapes 56, 140
constant of proportionality 450, 452
constructions 214–15
    assessment 234–5
    constructions and loci 228–31
    review 233
    summary 232
continuous data 78
converse rules 50
correlation 322–3
    scatter graphs and correlation 332–5
corresponding angles 44
corresponding sides 56
cosines 370
    cosine rule 374, 376, 378, 380
counter-examples 118
cross-sections 304
cryptography 258–9
cube numbers 432
cube roots 264
cubes 304
cuboids 58, 304
cumulative frequency graphs 328–31
curves 472–5
    gradient of a curve at a point 464, 468
    gradient of a curve between two points 464
    turning points 476–9
cyclic quadrilaterals 224
cylinders 304

D
data 68–9
    assessment 84–5, 342–3
    averages and spread 74–7, 324–7
    box plots and cumulative frequency graphs 328–31
    continuous data 78
    discrete data 78
    frequency diagrams 78–81
    grouped data 78
    representing data 70–3
    review 83, 341
    scatter graphs and correlation 332–5
    summary 82, 340
    time series 336–9
decagons 60
decay 454–7
    exponential decay 454
decimals 86–7
    assessment 102–3
    decimal places (dp) 4
    fractions, decimals and percentages 96–9
    recurring decimals 96
    review 101
    summary 100
    terminating decimals 96
deficient numbers 261
denominators 34, 268
    common denominator 92
density 180, 442
derivatives 468
derived formulae 108
diameter 216
difference of two squares 120
differentiation 462–3
    differentiating polynomials 468–71
    rates of change 464–7
    summary 480 review 481
    tangents to curves 472–5
    turning points 476–9
direct proportion 450–3
directed numbers 8
discrete data 78
distance 286
distance–time graphs 286–9
division 12–15
domain 112

E
elements 400
elevations 304
    front elevation 304
    side elevation 304
elimination 198
empty sets 402
enlargements 36, 140
equations 116, 188–9
    approximate solutions 202–5
    assessment 212–13
    completing the square 194
    equation of a straight line 278–81
    quadratic equations 194–7
    review 211
    simultaneous equations 198–201
    solving linear equations 190–3
    summary 210
equidistant points 228
equilateral triangles 48
error intervals 180
estimation 12, 172–5
    estimated mean 324
    estimates 324
Euler's identity 117
events 154, 158
    exhaustive events 162
    expected outcomes 154
    independent events 410
    mutually exclusive events 162–5
exact calculations 350–3
    exact answers 350
expanding 30–3, 120–3
exponential decay 454
exponential functions 294–7
exponential growth 454
expressions 20–1, 116
    algebraic fractions 34–7
    assessment 40–1
    expanding and factorising 30–3, 120–3
    indices 26–9
    quadratic expressions 120
    review 39
    simplifying expressions 22–5
    summary 38
exterior angles 48
extrapolation 334

F
factorising 30–3, 120–3
    canceling common factor 88
    difference of two squares 120
factors 258–9
    assessment 274–5
    common factors 30, 32, 88
    factors and multiples 260–3
    highest common factor (HCF) 260, 262
    prime factor decomposition 260
    prime factors 96, 262
    review 273
    summary 272
formulae 106–7, 108–11, 116, 484
    assessment 126–7
    derived formulae 108
    five year repayment formula 105
    review 125
    subjects 108
    summary 124
fractional indices 346
fractions 86–7
    assessment 102–3
    calculations with fractions 92–5
    common denominator 92
    common factors 88
    fractions and percentages 88–91
    fractions, decimals and percentages 96–9
    improper fractions 88
    mixed numbers 88

review 101
simplest form 88
summary 100
frequency diagrams 78–81
frequency tables, grouped 78, 324
frequency trees 408
front elevation 304
frustums 312
functions 112–15
  assessment 126–7
  composite functions 112
  review 125
  summary 124

G

gradient 278
  differentiating polynomials 468–71
  gradient function 469
  gradients and areas under graphs 290–3
  rates of change 464–7
  tangents to curves 472–5
graphs 276–7
  assessment 300–1
  cumulative frequency graphs 328–31
  equation of a straight line 278–81
  exponential and trigonometric functions 294–7
  gradients and areas under graphs 290–3
  kinematic graphs 286–9
  line graphs 336
  properties of quadratic functions 282–5
  review 299
  scatter graphs and correlation 332–5
  summary 298
  time series graphs 336
grouped data 78
grouped frequency tables 78, 324
growth 454–7
  exponential growth 454
  multipliers 454

H

hemispheres 312
heptagons 60
hexagons 60
highest common factor (HCF) 260, 262
histograms 78
horizontal axis 336
hypotenuse 366

I

identities 116
identity matrix 386
images 140
implied accuracy 180
improper fractions 88
independent events 410
indices 26–9
  calculating with indices 346–9
  fractional indices 346
  index 26
  index form 26, 264
  negative indices 346
inequalities 116, 188–9, 206–9
  assessment 212–13
  review 211
  summary 210
inputs 112
instantaneous rate of change 468
interest 248

compound interest 454
simple interest 248, 454
interior angles 44
  interior opposite angles 48
interpolation 334
interquartile range 74
intersections 260, 400, 402
inverse operations 108
inverse proportion 450–3
irrational numbers 268
isometric paper 304
isosceles trapeziums 54
isosceles triangles 48
iteration 202

K

key phrases and terms 485–6
kinematics 276–7
  formulae 484
  kinematic graphs 286–9
kites 48

L

length 132–5, 180
linear equations 190–3
linear relationships 332
linear scale factors 56
linear sequences 36, 424–7
lines 44–7
  angle bisectors 228, 230
  equation of a straight line 278–81
  gradient 278
  line graphs 336
  parallel lines 44
  perpendicular bisectors 228, 230
  perpendicular lines 44
  transversal lines 44
  y-intercept 278
liquid measurements 308
litres 308
loci (locus) 228–31
lower quartile 74
lowest common multiple (LCM) 260, 262

M

mapping diagrams 112, 114
mass 180
matrices 364–5, 386–9
  assessment 396–7
  combining transformation matrices 390–3
  identity matrix 386
  multiplication of matrices 386, 388
  order of matrix 386
  review 395
  summary 394
maximum turning points 476
mean 74, 324
  estimated mean 324
measures 170–1, 344–5
  assessment 186–7
  calculator methods 176–9
  estimation and approximation 172–5
  measures and accuracy 180–3
  review 185
  summary 184
median 74, 324
members 400
metric units 132, 136, 344–5
  converting between units 446–7
  liquid measurements 308
millilitres 308
minimum turning points 476
mirror lines 140

mixed numbers 88
modal class 324
mode 74
multiples 260–3
  lowest common multiple (LCM) 260, 262
multiplication 12–15, 384, 386, 388
multipliers 454
mutually exclusive events 162–5

N

negative indices 346
negative numbers 8
nets 304
nonagons 60
nth term 424
numerators 34

O

objects 140
obtuse angles 44
octagons 60
operations 12
  inverse operations 108
  order of operations 176
order of matrix 386
order of rotation symmetry 52
outcomes 158
  expected outcomes 154
  theoretical outcomes 154
outliers 332, 334
outputs 112

P

parallel lines 44
parallelograms 48
partitioning 8
pentagons 60
percentages 86–7, 248
  assessment 102–3
  fractions and percentages 88–91
  fractions, decimals and percentages 96–9
  percentage change 248–51
  percentage decrease 248
  percentage increase 248
  reverse percentage 250
  review 101
  summary 100
perfect numbers 261
perimeter 216
  formulae 484
periodic functions 294
perpendicular bisectors 228, 230
perpendicular lines 44
pie charts 70
place value 4–7
planes 48
plans 304
points of inflection 476
polygons 42–3
  assessment 66–7
  decagons 60
  heptagons 60
  hexagons 60
  nonagons 60
  octagons 60
  pentagons 60
  polygon angles 60–3
  regular polygons 60
  review 65
  summary 64
polynomials 468–71
position vectors 388
possibility spaces 404
powers 26, 258–9
  assessment 274–5
  powers and roots 264–7

review 273
summary 272
prime factors 96, 262
  prime factor decomposition 260
prime numbers 260
prisms 304
  cubes 304
  cuboids 304
  cylinders 304
  volume of a prism 308
probability 152–3, 398–9
  assessment 168–9, 418–19
  conditional probability 414–17
  expected outcomes 154
  mutually exclusive events 162–5
  probability experiments 154–7
  probability formula 484
  review 167, 417
  sample spaces 404–7
  summary 166, 416
  theoretical probability 158–61
  tree diagrams 408–11
  Venn diagrams 400–3
product 260
proportion 70, 238–9, 240–3, 440–1
  assessment 254–5, 460–1
  constant of proportionality 450, 452
  direct and inverse proportion 450–3
  growth and decay 454–7
  percentage change 248–51
  review 253, 459
  summary 252, 458
pyramids 304
Pythagoras' theorem 364–5, 366–9
  assessment 396–7
  formulae 484
  Pythagoras and trigonometry problems 378–81
  review 395
  summary 394

Q

quadratic equations 194–7
quadratic expressions 120
quadratic formula 194, 484
quadratic functions 282–5
  root 282
  turning points 282
  x-intercepts 282
quadratic sequences 428–31, 432
quadrilaterals 48–51
  cyclic quadrilaterals 224
  kites 48
  parallelograms 48
  rectangles 48
  rhombuses 48
  squares 48
  trapeziums 48
quartiles 74

R

radius 216
range 74, 112
  interquartile range 74
rate of flow 442
rate of pay 442
rates of change 290, 292, 464–7
  instantaneous rate of change 468
ratio 238–9, 446
  assessment 254–5
  percentage change 248–51
  ratio and scales 244–7
  review 253
  simplest form 244
  summary 252

555

rationalisation 268
reciprocals 92, 346
rectangles 48
recurring decimals 96
reflection symmetry 52
reflections 140
reflex angles 44
relative frequency 154, 156
representing data 70–3
revision 128–9, 256–7, 362–3, 482–3
    formulae 484
    key phrases and terms 485–6
rhombuses 48
right angles 44
right-angled triangles 366, 370
    hypotenuse 366
roots 258–9
    assessment 274–5
    calculating with roots 346–9
    cube roots 264
    powers and roots 264–7
    quadratic functions 282
    review 273
    square roots 264
    summary 272
    surds 268–71
rotation symmetry 52
    order of rotation symmetry 52
rotations 140
    centre of rotation 140
rounding 4–7

S

sample spaces 404–7
scalars 382
    multiplication by scalars 384
scale factors (SF) 140, 244
scalene triangles 48
scales 132
    ratio and scales 244–7
scatter graphs 332–5
sectors 220, 222
segments, alternate 224
self-similarity 130–1
sequences 422–3
    assessment 438–9
    linear sequences 424–7
    quadratic sequences 428–31, 432
    review 437
    special sequences 432–5
    summary 436
sets 400–3
    complements 400, 402

elements 400
empty sets 402
intersections 400, 402
members 400
unions 400, 402
universal sets 402
shapes 52
    congruent shapes 56, 140
    similar shapes 56, 140, 446
    three-dimensional (3D) shapes 302–3, 304–7, 308–11, 312–15, 316–19
    two-dimensional (2D) shapes 42, 130–1, 132–5, 136–9, 140–3, 144–7, 148–51
side elevation 304
significant figures (sf) 4
similarity 56–9
    similar shapes 56, 140, 446
simple interest 248, 454
simplest form 88, 244
simplifying expressions 22–5
simultaneous equations 198–201
sines 370
    sine rule 374, 376, 378, 380
solids 304
special sequences 432–5
speed 180, 286, 442
speed-time graphs 286–9
spread 74–7, 324–7
square numbers 432
square roots 264
squares 48
    unit squares 388
standard form 354–7
statistics 68–9, 322–3
substitution 198
subtended angles 224
subtraction 8–11, 384
supplementary angles 44
surds 268–71
    surd form 268
surface area 312–15
    formulae 484
symmetry 52–5
    order of rotation symmetry 52
    reflection symmetry 52
    rotation symmetry 52

T

tangents 290, 292, 370
    tangents to curves 472–5
terminating decimals 96
terms 116
    collecting like terms 22

key terms 485–6
nth term 424
position-to-term rules 424
term-to-term rules 424
tessellation 62
theoretical outcomes 154
theoretical probability 158–61
three-dimensional shapes (3D) 302–3, 304–7
    assessment 318–19
    prisms 304
    review 317
    solids 304
    summary 316
    volume and surface area 312–15
    volume of a prism 308–11
time series 336–9
    time series graphs 336
transformation matrices 390–3
transformations 140–3
translations 140
transversal lines 44
trapeziums 48
    isosceles trapezium 54
tree diagrams 408–11
trends 336
trials 158
triangles 48–51
    area of triangle 376, 378, 380
    equilateral triangles 48
    isosceles triangles 48
    right-angled triangles 366, 370
    scalene triangles 48
triangular numbers 432
trigonometry 364–5, 370–3, 374–7
    assessment 396–7
    formulae 484
    Pythagoras and trigonometry problems 378–81
    Pythagoras' theorem 366–9, 378–81
    review 395
    summary 394
    trigonometric functions 294–7
    trigonometric ratios 370
turning points 123, 282, 476–9
two-dimensional (2D) shapes 48, 130–1
    area of a 2D shape 136–9
    assessment 150–1
    measuring lengths and angles 132–5
    review 149
    summary 148
    transformations 140–3, 144–7

U

unions 400, 402
units 440–1
    assessment 460–1
    compound units 442–5
    converting between units 446–9
    review 459
    summary 458
    unit squares 388
universal sets 402
upper quartile 74

V

variables 22, 104, 108, 332
vectors 140, 364–5, 382–5
    assessment 396–7
    column vectors 382
    position vectors 388
    pre-multiplying 388
    review 395
    summary 394
Venn diagrams 400–3
    sets 400–3
volume 308–11, 446
    formulae 484
    liquid measurements 308
    volume and surface area 312–15
    volume of a prism 308
    volume scale factors 56

W

waves 294

X

x-component 382
x-intercept 282

Y

y-component 382
y-intercept 278

Z

zero 346